Handbook on Material and Energy Balance Calculations in Metallurgical Processes

REVISED EDITION

A Publication of The Minerals, Metals & Materials Society
420 Commonwealth Drive
Warrendale, Pennsylvania 15086
(412) 776-9000

Library of Congress Catalog Number 93-79857
ISBN Number 0-87339-224-8

Handbook on Material and Energy Balance Calculations in Metallurgical Processes

Revised Edition

By

H. Alan Fine
Department of Metallurgical Engineering
and Materials Science
University of Kentucky
Lexington, Kentucky

and

Gordon H. Geiger
Department of Metallurgical Engineering
University of Arizona
Tucson, Arizona

A Publication of
TMS
Minerals • Metals • Materials

Acknowledgement

The authors gratefully acknowledge Dr. Michael B. McNeill, Office of Industrial Programs of the U.S. Department of Energy, and Dr. Kenneth G. Kreider, Thermal Processes Division of the National Bureau of Standards, for their helpful comments and suggestions regarding the contents of this book. We are also grateful to the National Bureau of Standards for financial support.

Our special thanks go to Brenda Rice for typing the final edition of the manuscript. The authors sincerely thank their families for all the encouragement and patience during the long months of writing, and especially Harriet Fine for her painstaking efforts in arranging the final manuscript.

Foreword

This volume on energy and material balances inaugurates a series of textbooks to be published by The Metallurgical Society of AIME. The publication of textbooks has been undertaken since the price of such books produced commercially has risen more rapidly than inflation; further, no end is in sight.

The Metallurgical Society has also undertaken publishing textbooks in response to a survey of its members, who felt that one of the most important services it could perform would be to provide the metallurgical community with tools for continuing education. The means for The Metallurgical Society to address these needs is through conference proceedings and textbooks. Since developments have been occuring frequently, the technical committees of TMS-AIME have provided a forum for members and guests to exchange information and to stimulate new investigations. However, such direct stimuli were lost to all but the few who could attend a specific session among the many held at a meeting. The rapid publication of conference proceedings is one way of disseminating the information. The continuing success resulting from the efforts of the TMS-AIME Conference Volumes Committee has indicated that The Metallurgical Society has properly assessed this need and has been satisfying it.

By publishing new textbooks, The Metallurgical Society is attempting to satisfy the second need of its members; namely, developing texts for areas of metallurgy and materials science which are too limited to interest commercial publishers, and other material written in a style to assist with continuing education, not merely to record the state of art. Hopefully, such textbooks will enjoy a measure of longevity which the conference volumes cannot. In addition, TMS will be providing its young and future members with a valuable service.

Every profession has a patron. The one which might first come to mind is Vulcan, the blacksmith; among the Romans, he epitomized only the destructive force of fire. Hephaestus is more appropriate: he was not only the Greek god of fire -- he was endowed with many constructive attributes of energy; he was the master architect, the chariot maker, the armorer and jeweler of Mt. Olympus, and the maker of thunderbolts. He was credited with many marvels, such as fabricating golden handmaidens who could walk, talk, and think. Though lame and usually shown with as stern a countenance as Vulcan, he married Charias (grace), and Aglaia (brilliance), and later

Aphrodite. This was allegorical. It was a union, as recognized by the Greeks, of craftsmanship with grace and beauty.

It is towards such a scholarly marriage of their thoughts and of their skills that all students might well aspire. While metallurgy and material engineering may be shunned like Hephaestus by some people in the current grove of academe, these subjects are modern combinations of science with the practical arts of creating functional things and new beauty. This textbook, which deals with heat and materials, emphasizes the two aspects of Hephaestus which are usually encountered first. It is hoped, however, that current and future members, while using these two features, will also create useful and beautiful things -- the ideal which association with his name promises.

The cooperation of TMS Presidents Robert Jaffee and Robert Couch in establishing this endeavor is gratefully acknowledged. Alexander Scott and John Ballance overcame many problems of getting this book published.

James T. Waber, Chairman
TMS-AIME Book Publishing Committee
Evanston, Illinois
December, 1979

Preface

We live in a day and age when realization of the "limits to growth" and the finite extent of all of our natural resources has finally hit home. Yet our economy and our livelihoods depend on successful operation of industries that require and consume raw materials and energy. This success depends, in turn, on efficient use of the available resources, which not only allows industry to conserve materials and energy, but also allows it to compete successfully in the world markets that exist today.

The duties of the metallurgical engineer include, among many other things, development of information concerning the efficiency of metallurgical processes, either through calculation from first principles, or by experimentation. The theory of the construction of material and energy balances, from which such knowledge is derived, is not particularly complicated or difficult, but the practice, particularly in pyrometallurgical operations, can be extremely difficult and expensive.

In this handbook, we have tried to review the basic principles of physical chemistry, linear algebra, and statistics which are required to enable the practicing engineer to determine material and energy balances. We have also tried to include enough worked examples and suggestions for additional reading that a novice to this field will be able to obtain the necessary skills for making material and energy balances. Some of the mathematical techniques which can be used when a digital computer is available are also presented. The user is cautioned, however, that the old computing adage "garbage in, garbage out" is particularly true in this business, and that great attention must still be paid to setting up the proper equations and obtaining accurate data. Nevertheless, the computer is a powerful ally and gives the engineer the tool to achieve more accurate solutions than was possible just twentyfive years ago.

It is hoped that readers, particularly those who are out of practice at these kinds of calculations will ultimatley be able to perform energy balances in processes for which they are responsible, and as a result be able to improve process efficiences. A bibliography of past work on this subject is presented in an appendix to provide reference material against which results of studies can be checked. Hopefully, results reported in the future will reflect increases in efficiency.

H. Alan Fine
University of Kentucky
Lexington, Kentucky

Gordon H. Geiger
University of Arizona
Tucson, Arizona

Table of Contents

*Appendix B contains heats of formation at 298K and molecular weights of many compounds (Part 1), charts of enthalpy increments above 298K for many elements and their inorganic compounds (Part II) and Bureau of Mines Bulletin 605, Thermodynamic Properties of 65 Elements - Their Oxides, Halides, Carbides and Nitrides, by C.E. Wicks and F.E. Block (Part III).

CHAPTER 1

DIMENSIONS, UNITS, and CONVERSION FACTORS

1.0 INTRODUCTION

Most of the measurements and calculations performed in science and engineering involve the determination and/or manipulation of certain basic quantities called dimensions. Dimensions are either descriptions of non-physical concepts such as time and velocity or of physical characteristics of an object. Dimensions are specified by giving the magnitude of the dimension relative to some arbitrary standard called a unit. Therefore, the complete specification of a dimension must consist of a number and a unit.

In general, units have been developed in systems with one unit or a combination of certain basic units used to describe each dimension. The result has been the development of at least twenty-nine systems of units and a great deal of confusion for the student of engineering and even the practicing engineer. Fortunately, only three systems of units are commonly used today; and in the near future, probably only The International System of Units (Le Systeme International d' Unites, symbol SI) will be of importance.

This text will emphasize the usage of SI units. The calorie will, however, be the primary unit of energy, as the vast majority of data compilations currently available use this unit. Other CGS and American Engineering units will also be discussed, to a lesser degree, in order to demonstrate the proper usage of these units and their conversion to SI units.

1.1 THE SI SYSTEM OF UNITS

When developing a system of units, it is generally desirable to create basic units or combination of units that describe any dimension which may arise. Man's ability to create units and sets of units has led to a multitude of systems with, unfortunately, virtually no rational relationship between the various systems. To alleviate this problem, a system of units known as the metric system was developed and proposed for international usage.

In 1875, an International Bureau of Weights and Measures was established to maintain and improve the units in the metric system, the gram, the metre*, and the second. However, the proliferation of systems of units continued until 1948 when a General Conference on Weights and Measures (abbreviation CGPM from the French spelling) met to establish "a practical system of units .... suitable for adoption by all signators of the Metric Convention", which ultimately led to the adoption of the SI system of units in 1960.

---

*The spellings metre and litre as used in this text are preferred by ASTM. However, meter and liter are also widely used.

1.1.1 UNITS AND DIMENSIONS

The SI system of units consists of several basic or fundamental units from which all other units may be derived. The dimensions of interest in this text, their fundamental units, and the unit symbols are shown in Table 1.1-1.

Table 1.1-1

The Fundamental SI Units

Dimension	Name of Fundamental SI Unit	Unit Symbol
length	metre	m
mass	kilogram	kg
time	second	s
electric current	ampere	A
thermodynamic temperature	kelvin	K
amount of substance	mole	mol

As noted previously, the SI units were chosen so that all other units could be derived from these basic units. Some dimensions with derived units having special names are shown in Table 1.1-2. Additional dimensions and combinations of units which do not have special names but which are of importance in this text appear in Table 1.1-3.

Table 1.1-2

Some Auxiliary SI Units That Have Special Names

Dimension	Name of Derived SI Unit	Symbol	Definition In Terms of Base or Other Derived SI Units
frequency	hertz	Hz	1Hz = 1s^{-1}
force	newton	N	1N = 1 kg·m/s^2
pressure, stress	pascal	Pa	1Pa = 1 N/m^2
energy, work, quantity of heat	joule	J	1J = 1 N·m
power	watt	W	1W = 1 J/s
electric charge, quantity of electricity	coulomb	C	1C = 1 A·s

Table 1.1-2 (Continued)

Dimension	Name of Derived SI Unit	Symbol	Definition In Terms of Base or Other Derived SI Units
electric potential, potential difference, tension, electromotive force	volt	V	1V = 1 J C
electric capacitance	farad	F	1F = 1 C/V
electric resistance	ohm	Ω	1Ω = 1 V/A
electric conductance	siemens	S	1S = 1 Ω^{-1}

Table 1.1-3

Additional SI Units Which Do Not Have Special Names

Dimension	Units	Unit Symbols
area	square metre	m^2
volume	cubic metre	m^3
speed, velocity	metre per second	m/s
acceleration	metre per second squared	m/s^2
specific volume	cubic metre per kilogram	m^3/kg
density, mass density	kilogram per cubic metre	kg/m^3
thermal conductivity	watt per metre kelvin	W/m·K
heat transfer coefficient	watt per square metre kelvin	$W/m^2 \cdot K$
specific energy	joule per kilogram	J/kg
specific heat	joule per kilogram kelvin	J/kg·K

As can be seen in Tables 1.1-2 and 1.1-3, all of the derived units are developed by direct combination of the basic units without the need for non-unity coefficients. Consequently, the SI system is a coherent system of units. For example, the unit of force, the Newton, equals the unit of mass times the unit of acceleration. Newton's Second Law, however, states that the force, F, acting on an object is proportional to its mass, m, multiplied by its acceleration, a. Thus,

$$F \propto m \cdot a \qquad (1.1\text{-}1)$$

or

$$F = C \cdot m \cdot a \qquad (1.1\text{-}2)$$

In the SI system, the proportionality constant C in Eq. 1.1-2 must have a value of unity and dimension of $N \cdot s^2/kg \cdot m$.

In a non-coherent system of units such as the American Engineering System, the value of C in Eq. (1.1-2) is not unity. In this system, the unit of force, the pound-force (symbol lb_f), is defined as the force required to accelerate a unit of mass, the pound-mass (symbol lb_m), 32.1740 ft/s^2, the acceleration of gravity at sea level and 45° latitude[1]. Thus,

$$1\ lb_f = C \cdot 1\ lb_m \cdot 32.1740\ ft/s^2 \qquad (1.1\text{-}3)$$

and

$$C = \frac{1}{32.1740}\ \frac{lb_f \cdot s^2}{lb_m \cdot ft} \qquad (1.1\text{-}4)$$

The reciprocal of C is commonly used in many calculations and has been given a special symbol, g_c.

$$g_c = 1/C \qquad (1.1\text{-}5)$$

and

$$g_c = 32.1740\ lb_m \cdot ft/lb_f \cdot s^2 \qquad (1.1\text{-}6)$$

EXAMPLE 1.1-1: The engineering stress, σ, is defined as the ratio of the load on a sample, P, to the original cross-sectional area, A_o:

$$\sigma = P/A_o$$

a) Calculate the engineering stress in lb_f/in^2 on a 0.010 in. diameter wire which has a two pound mass hanging from it.

b) Calculate the engineering stress in N/m^2 on a 0.00025m diameter wire which has a one kilogram mass hanging from it.

c) Do a and b above for the same objects but on the moon.

Solution: Assuming that the mass of the wire is small compared to the mass of the object hung on the wire, the load on the wire will equal the force exerted by gravity on the mass being supported by the wire.

$$P = F = C \cdot m \cdot g$$

For a wire with a circular cross-section and diameter d,

$$A_o = \pi \cdot d^2/4$$

---

(1) Handbook of Chemistry and Physics, 48th edition, Chemical Rubber Co., Cleveland, Ohio, 1968.

a) If the wire is at sea level and 45° latitude, g equals 32.1740 ft/s^2.

Then, $$F = \frac{1}{32.1740} \cdot \frac{lb_f \cdot s^2}{lb_m \cdot ft} \cdot 2.00\ lb_m \cdot 32.1740\ \frac{ft}{s^2}$$

$$= 2.00\ lb_f$$

$$A_o = \pi \cdot (0.010\ in)^2/4$$
$$= 7.85 \times 10^{-5}\ in^2$$

and

$$\sigma = 2.00\ lb_f/7.85 \times 10^{-5}\ in^2$$
$$- 2.55 \times 10^4\ lb_f/in^2$$

b) In SI units, g equals 9.80621 m/s^2

Then, $$P = 1\frac{N \cdot s^2}{kg \cdot m} \cdot 1.00\ kg \cdot 9.80621\ m/s^2$$
$$= 9.81\ N$$
$$A_o = \pi \cdot (0.00025\ m)^2/4$$
$$= 4.91 \times 10^{-8}\ m^2$$

and

$$\sigma = 9.81\ N/4.91 \times 10^{-8}\ m^2$$
$$= 2.00 \times 10^8\ N/m^2$$

c) On the moon, the mean acceleration of gravity is 5.47 ft/s^2 (1.67 m/s^2).

Thus, for case a, P equals 0.34 lb_f and σ equals $4.33 \times 10^3\ lb_f/in^2$. For case b, P equals 1.67N and σ equals $0.34 \times 10^8\ N/m^2$.

It is seen in Example 1.1-1 that the unit of force in the American Engineering System of Units is defined in a manner that makes the numerical value of the measurement of force equal to the numerical value of the mass. This is only exactly true when the measurement of force is done on the earth and at sea level and 45° latitude. For most engineering calculations, a mean value for the acceleration of gravity on the earth of 32.2 ft/s^2 or 9.8 m/s^2 may be used, and pounds-mass can be assumed equal within two or three percent to pounds - force.

EXAMPLE 1.1-2: Calculate the kinetic energy of one pound-mass of oxygen that is traveling at a linear velocity of 600 ft/s.

Solution: The kinetic energy, E_k, of an object of mass m traveling at a linear velocity $\bar{v}$ equals

$$E_k = \frac{m \cdot \bar{v}^2}{2 \cdot g_c}$$

Thus, for 1.00 lb_m traveling at 600. ft/s

$$E_k = \frac{1.00_m lb \cdot (600 \frac{ft}{s})^2}{2(\frac{32.2\ lb_m \cdot ft}{lb_f \cdot s^2})}$$

$$= 5.59 \times 10^3 ft \cdot lb_f$$

EXAMPLE 1.1-3: Calculate the increase in potential energy of one kilogram of ore when it is raised to the top of a thirty metre high furnace.

Solution: The potential energy, $E_{p,1}$, of an object of mass m in a gravitational field with acceleration g and at a height z_1, above some arbitrary reference height (usually taken as the surface of the earth at approximately sea level) is

$$E_{p,1} = m \cdot Z_1 \cdot g/g_c$$

For the same object at a new height z_2,

$$E_{p,2} = m \cdot Z_2 \cdot g/g_c$$

and the change in potential energy*, ΔE_p, equals

$$\Delta E_p = E_{p,2} - E_{p,1}$$

$$= m \cdot Z_2 \cdot g/g_c - m \cdot Z_1 \cdot g/g_c$$

$$= m(Z_2 - Z_1)\ g/g_c$$

$$= m \cdot \Delta Z \cdot g/g_c$$

---

* The symbol Δ is used in this text to indicate a finite change. The symbol δ or d will be used to indicate infinitesimal, i.e., very small, changes.

Thus, for ΔZ equal 30.0m and m equals 1.00 kg

$$\Delta E_p = 1.00\ kg \cdot 30.0m \cdot 9.81\ m/s^2/(1.00\ kg \cdot m/N \cdot s^2)$$
$$= 294\ N \cdot m$$

1.1.2 PREFIXES

Another advantage of the SI System comes from the use of standard prefixes to expand the basic and derived sets of units to fit situations where larger or smaller units are desirable. The standard prefixes for SI units and their symbols are shown in Table 1.1-4.

Table 1.1-4

Standard SI Prefixes

Prefix	Symbol	Factor	
tera	T	10^{12} =	1 000 000 000 000
giga	G	10^{9}	1 000 000 000
mega	M	10^{6}	1 000 000
kilo	k	10^{3}	1 000
hecto	h	10^{2}	100
deka	da	10^{1}	10
deci	d	10^{-1}	0.1
centi	c	10^{-2}	0.01
milli	m	10^{-3}	0.001
micro	μ	10^{-6}	0.000 001
nano	n	10^{-9}	0.000 000 001
pico	p	10^{-12}	0.000 000 000 001
femto	f	10^{-15}	0.000 000 000 000 001
atto	a	10^{-18}	0.000 000 000 000 000 001

Prefixes are generally chosen so that the numerical value of a dimension lies between 0.1 and 1000. Prefixes based upon ten raised to a multiple of a power of 3 are also recommended. Combinations of prefixes such as kMJ to indicate 10^9 Joules in place of GJ should not be used.

EXAMPLE 1.1-4: Calculate the charge to a furnace when one thousand shovels full of ore are put in the furnace, and each shovel contains one kilogram of ore.

Solution: The total charge, M′, will equal the number of shovels full, n, times the mass on each shovel, m. Thus,

$$M' = n \cdot m$$

and

$$M' = 1000 \text{ shovels} \cdot \frac{1 \text{ kg}}{\text{shovel}}$$

$$= 1000 \text{ kg}$$

$$= 1000 \text{ (1000g)}$$

$$= 10^6 \text{ g}$$

$$= 1\text{Mg}^*$$

EXAMPLE 1.1-5: If waste (flue) gas is being exhausted through a 50 cm by 100 cm rectangular duct at an average velocity of 5 cm/s, calculate the volume of gas that will be exhausted in one hundred seconds. Give the answer in cubic meters.

Solution: For a gas flowing through a duct with cross-sectional area A_o and at an average linear velocity $\bar{v}$, the volumetric flowrate of gas in the duct, V, equals

$$\dot{V} = \bar{v} \cdot A_o$$

and the volume of gas, V, leaving the duct in time τ equals

$$V = \dot{V} \cdot \tau$$

$$= \bar{v} \cdot A_o \cdot \tau$$

Thus, for a rectangular duct of width w and height h

$$A_o = w \cdot h$$

and

$$V = \bar{v} \cdot w \cdot h \cdot \tau$$

*It should be noted that Mg is the symbol for magnesium, as well as a megagram, and that differentiation between these meanings can only be made by the context of its usage.

$$= \frac{5\ \text{cm}}{\text{s}} \cdot 50\ \text{cm} \cdot 100\ \text{cm} \cdot 100\ \text{s}$$

$$= 2.5 \times 10^6\ \text{cm}^3$$

$$= 2.5 \times 10^6 \cdot (1 \times 10^{-2}\text{m})^3$$

$$= 2.5 \times 10^6 \cdot (1 \times 10^{-6})\ \text{m}^3$$

$$= 2.5\ \text{m}^3$$

1.1.3 ACCEPTABLE NON-SI UNITS

While it has been strongly recommended by the CGPM that the SI system be rigidly adhered to, it was realized that several additional units which either have special meaning or well-established custom may be used. This usage should, however, be kept to a minimum.

The SI unit for volume is the cubic metre and its multiples, see Table 1.1-3. In some circumstances the litre* may be used to represent the volume of a fluid. The use of this unit should be minimized and prefixes should not be used in conjunction with this unit, though 10^{-3} litre is often referred to as a millilitre.

The SI unit for time is the second. However, for events based upon the calendar cycle, the minute, hour or day may be used. It is also realized that the customary units for velocity, kilometer/hour, may continue in use. The use of km/s is strongly recommended.

Temperature and temperature differences are expressed in terms of the thermodynamic or Kelvin temperature scale in the SI system. The unit for temperature is thus the kelvin, K. However, the Celsius temperature scale has been widely used and the expression of temperatures in degrees Celsius (symbol, °C) is also accepted in the SI system.

1.1.4 STYLE

There are many accepted practices relating to the usage of SI units, symbols and prefixes. The recommended rules of usage and style will be used in this text. For a more thorough description of the rules of style for SI units, the reader is referred to the ASTM Publication No. E 380-76, Standards for Metric Practice.

1.2 CONVERSION FACTORS

As seen in Section 1.1, it is possible to describe many dimensions using several different units. Time, for example, may

---

*See Footnote on page 2.

be expressed in the SI unit, seconds, or in the acceptable non-SI units of minutes or hours. A relationship must exist between these units to convert from one unit to another. This relationship is called a conversion factor.

Two types of conversions are possible. In the first, one unit is converted to another by simple multiplication by a conversion factor. In the second, multiplication by a conversion factor plus addition or subtraction of an additional term is necessary.

1.2.1 CONVERSION FACTOR TABLES

The relationship between the units used in different systems of units for a particular dimension are determined by treaty or from the basic definitions of the units. To assist the scientist or engineer, these conversion factors have been compiled into tables. A list of useful conversion factors is given in Table 1.2-1.

1.2.2 THE DIMENSIONAL EQUATION

As can be seen in Table 1.2-1,

$$1.000000 \, lb_m = 4.535\,924 \times 10^{-1} \, kg \qquad (1.2\text{-}1)$$

Thus, it must also be true that

$$\frac{4.535\,924 \times 10^{-1} \, kg}{1.000000 \, lb_m} = 1.000000 \qquad (1.2\text{-}2)$$

or

$$\frac{1.000000 \, lb_m}{4.535\,924 \times 10^{-1} \, kg} = \frac{2.204\,622 \, lb_m}{1.000000 \, kg} = 1.000000 \qquad (1.2\text{-}3)$$

Then, since any quantity multiplied by unity equals the original quantity

$$3.15 \, lb_m = 3.15 \, lb_m \left(\frac{4.535\,924 \times 10^{-1} kg}{1.000000 \, lb_m}\right) = 1.43 \, kg \qquad (1.2\text{-}4)$$

Manipulations of this type are easily carried out when only one conversion factor is required. When one conversion factor is not available and several factors must be used to obtain the desired result, the dimensional equation approach should be employed.

The dimensional equation contains both the number of units as well as the type of units. Conversion from the given units to new units is then accomplished by multiplication by a series of conversion factors. The dimensional equation organizes the conversion process and reduces the chance for error.

Table 1.2-1

Some Useful Conversion Factors

Dimension	Multiply the Units of	by	to Obtain the Units of
Length	inches (in)	$2.540\ 000 \times 10^{-2}$	metres (m)
	feet (ft)	$3.048\ 000 \times 10^{-1}$	metres (m)
	yards (yd)	$9.144\ 000 \times 10^{-1}$	metres (m)
	miles (mi)	$1.609\ 344 \times 10^{3}$	metres (m)
Area	square inches (in^2)	$6.451\ 600 \times 10^{-4}$	square metres (m^2)
	square feet (ft^2)	$9.290\ 304 \times 10^{-2}$	square metres (m^2)
	square yards (yd^2)	$8.361\ 274 \times 10^{-1}$	square metres (m^2)
Density	pounds per cubic inch	$2.767\ 990 \times 10^{4}$	kilograms per cubic metre (kg/m^3)
	pounds per cubic foot	$1.601\ 846 \times 10^{1}$	kilograms per cubic metre (kg/m^3)
	tons (long) per cubic yard	$1.328\ 939 \times 10^{0}$	megagrams per cubic metre (Mg/m^3)
	grams per cubic centimeter	$1.000\ 000 \times 10^{3}$	kilograms per cubic metre (kg/m^3)
Energy	British thermal unit (Btu)	$1.055\ 056 \times 10^{0}$	kilojoules (kJ)
	foot pound force ($ft \cdot lb_f$)	$1.355\ 818 \times 10^{0}$	joules (J)
	kilowatt-hour (kWh)	$3.600\ 000 \times 10^{0}$	megajoules (MJ)
	horsepower - hour (hp·h)	2.684 520	megajoules (MJ)
	calories (cal)	$4.186\ 800 \times 10^{0}$	joules (J)

(continued)

(Table 1.2-1 Cont.)

Dimension	Multiply the Units of	by	to Obtain the Units of
Force	pounds force (lb_f)	4.448 222 x 10^0	newtons (N)
	kilograms force (kg_f)	9.806 650 x 10^0	newtons (N)
	tons (short) force (ton_f)	8.896 444 x 10^3	newtons (N)
Force per Unit Area	pounds force per square inch (lb_f/in^2)	6.894 757 x 10^3	newtons per square metre (N/m^2)
Mass	ounces (oz avoirdupois)	2.834 952 x 10^1	grams (g)
	ounces (oz troy)	3.110 348 x 10^1	grams (g)
	pounds (lb_m avoirdupois)	4.535 924 x 10^{-1}	kilograms (kg)
	pounds (lb_m troy)	3.732 417 x 10^{-1}	kilograms (kg)
	hundredweight (cwt long)	5.080 235 x 10^1	kilograms (kg)
	hundredweight (cwt short)	4.535 924 x 10^1	kilograms (kg)
	long ton	1.016 047 x 10^0	megagrams (Mg)
	short ton	9.071 847 x 10^{-1}	megagrams (Mg)
	tonne	1.000 000 x 10^0	megagrams (Mg)
	metric ton	1.000 000 x 10^0	megagrams (Mg)
Mass Flow Rate	pounds per minute (lb_m/min)	7.559 873 x 10^{-3}	kilograms per second (kg/s)
	pounds per hour (lb_m/h)	1.259 979 x 10^{-4}	kilograms per second (kg/s)
Mass per Unit Area	pounds per foot squared (lb_m/ft^2)	4.882 428 x 10^0	kilograms per metre squared (kg/m^2)
Mass per Unit Length	pounds per foot (lb_m/ft)	1.488 164 x 10^0	kilograms per metre (kg/m)

Power	Btu/hr	$2.930\ 711 \times 10^{-1}$	watts (W)
	horsepower (hp)	$7.456\ 999 \times 10^{-1}$	kilowatts (kW)
Pressure Difference	pounds per square inch (psi)	$6.894\ 757 \times 10^{0}$	kilopascals (kPa)
	inches water @ 39.2°F (in. H_2O)	$2.490\ 820 \times 10^{-1}$	kilopascals (kPa)
	inches mercury @ 60°F (in. Hg)	$3.376\ 850 \times 10^{0}$	kilopascals (kPa)
	atmosphere (atm)	$1.013\ 250 \times 10^{2}$	kilopascals (kPa)
	tons (short) per square inch (ton/in^2)	$1.378\ 952 \times 10^{1}$	megapascals (MPa)
	newtons per millimetre squared (N/mm^2)	$1.000\ 000 \times 10^{6}$	kilopascals (kPa)
Temperature Interval	Fahrenheit degrees (Δ°F)	$5.555\ 556 \times 10^{-1}$	Celsius degrees (Δ°C)
	Celsius degrees (Δ°C)	$1.000\ 000 \times 10^{0}$	kelvin (ΔK)
Velocity	miles per hour (mi/h)	$1.609\ 344 \times 10^{0}$	kilometers per hour (km/h)
	feet per minute (ft/min)	$5.080\ 000 \times 10^{-3}$	metres per second (m/s)
	feet per second (ft/s)	$3.048\ 000 \times 10^{-1}$	metres per second (m/s)
Volume	cubic inch (in^3)	$1.638\ 706 \times 10^{-5}$	cubic metres (m^3)
	cubic foot (ft^3	$2.831\ 685 \times 10^{-2}$	cubic metres (m^3)
	cubic yard (yd^3)	$7.645\ 549 \times 10^{-1}$	cubic metres (m^3)
Volume (fluid)	fluid ounces (U.S.) (oz)	$2.957\ 353 \times 10^{1}$	cubic centimetres (cm^3)
	pint	$4.731\ 765 \times 10^{-1}$	cubic decimetres (dm^3)
	gallon (gal)	$3.785\ 412 \times 10^{0}$	cubic decimetres (dm^3)
	gallon (gal)	$3.785\ 412 \times 10^{-3}$	cubic metres (m^3)

(continued)

(Table 1.2-1 Cont.)

Dimension	Multiply the Units of	by	to Obtain the Units of
	litre	1.000 000 x 10^{-3}	cubic metres (m^3)
Volume Flow Rate	cubic feet per minute (ft^3/min)	4.719 474 x 10^{-4}	cubic metres per second (m^3/s)
	gallons per minute (gal/min)	6.309 020 x 10^{-5}	cubic metres per second (m^3/s)
	gallons per minute (gal/min)	6.309 020 x 10^{-2}	cubic decimetres per second (dm^3/s)
	gallons per minute (gal/min)	6.309 020 x 10^{1}	cubic centimetres per second (cm^3/s)
Time	minutes (min)	6.000 000 x 10^{1}	seconds (s)
	hours (h)	3.600 000 x 10^{3}	seconds (s)
	days	8.640 000 x 10^{4}	seconds (s)
	years	3.153 600 x 10^{7}	seconds (s)

EXAMPLE 1.2-1: The amount of energy required to heat one pound-mass of water one Fahrenheit degree* is one Btu. How many calories must be supplied to heat one gram of water one Kelvin.

Solution: The dimensional equation may be used to convert $Btu/lb_m \cdot \Delta°F$ to $cal/g \cdot \Delta K$. See Table 1.2-1 for conversion factors.

$$1.0 \frac{Btu}{lb_m \cdot \Delta°F} = 1.0 \frac{Btu}{lb_m \cdot \Delta°F}\left(\frac{1.000000 \Delta°F}{5.555556 \times 10^{-1} \Delta K}\right)\left(\frac{1.055056\ kJ}{1.000000\ Btu}\right)\left(\frac{1000\ J}{kJ}\right) \cdot$$

$$\left(\frac{1.000000\ cal}{4.186800\ J}\right) \cdot \left(\frac{1.000000\ lb_m}{4.535\ 924 \times 10^{-1}\ kg}\right)\left(\frac{1\ kg}{1000\ g}\right)$$

$$= 1.0\ cal/g \cdot \Delta K$$

Therefore, one calorie must be supplied to heat one gram of water one kelvin.

EXAMPLE 1.2-2: Determine how many kg/s of hot metal are being produced if an iron blast furnace is producing hot metal at the rate of 1000 short tons per day.

Solution:

$$\frac{1000\ ton}{day} = \frac{1000\ ton}{day}\left(\frac{9.071847 \times 10^{-1} Mg}{ton}\right)\left(\frac{1000\ kg}{Mg}\right)\left(\frac{1.0\ \ day}{8.640000 \times 10^{4}\ s}\right)$$

$$= 10.5\ kg/s$$

The dimensional equation provides a format for performing unit conversions, i.e., combining conversion factors. The choice of particular factors is arbitrary and the order of their multiplication is arbitrary. The equation does, however, permit the clear and concise representation of the conversion process. It helps eliminate errors and its use is strongly recommended, especially for persons without experience in doing these computations.

EXAMPLE 1.2-3: Repeat Example 1.2-1 using different conversion factors.

---

*The symbol that is used in this text for Fahrenheit temperature difference is Δ°F. Similarly, Δ°C, ΔK and Δ°R will be used for Celsius, Kelvin and Rankine temperature scales. The units for the temperature differences are degrees Fahrenheit or Fahrenheit degrees on the Fahrenheit scale, degrees Celsius or Celsius degrees on the Celsius scale, Kelvin on the Kelvin scale and degrees Rankine degrees on the Rankine scale.

Solution:

$$1.0\ \frac{\text{Btu}}{\text{lb}_m\cdot\Delta^\circ\text{F}} = 1.0\ \frac{\text{Btu}}{\text{lb}_m\cdot\Delta^\circ\text{F}}\left(\frac{2.326000\times10^3\ \text{J/kg}}{1.000000\ \text{Btu/lb}_m}\right)\left(\frac{1.000000\ \text{cal/g}}{4.186800\times10^3\ \text{J/kg}}\right)$$

$$\left(\frac{1.800000\Delta^\circ\text{F}}{1.000000\Delta^\circ\text{C}}\right)\cdot\left(\frac{1.000000\Delta^\circ\text{C}}{1.000000\Delta\ \text{K}}\right)$$

$$= 1.0\ \text{cal/g}\cdot\Delta\text{K}$$

1.2.3 CONVERSION EQUATIONS

Some unit conversions cannot be accomplished by use of the dimensional equation. For these conversions, the addition or subtraction of a quantity must accompany the multiplication by conversion factors. Two examples of dimensions that often require conversion equations are temperature and pressure. In these cases, the conversion equations are required because there are units with different sizes, and the zero points of the different scales have been located at different points by the inventors of the scales.

Temperature scales, such as the Fahrenheit and Celsius Scales are called relative temperature scales, because the zero points of these scales are fixed relative to an arbitrary standard, the i-e point of water. Scales that are based upon the absolute lowest temperature that is believed to be obtainable, the point at which an atom or molecule has no kinetic energy, are called absolute or thermodynamic temperature scales. Two absolute temperature scales, the Rankine and the Kelvin scales, are in common use. The latter is the preferred SI scale.

The temperature unit on the Kelvin scale, the kelvin (symbol K), is the same size as the unit of temperature on the Celsius scale or the centigrade scale, the degree Celsius or centigrade sumbol °C. Thus*,

$$1.0\ \Delta\text{K} = 1.0\ \Delta^\circ\text{C} \qquad (1.2\text{-}7)$$

Similarly,

$$1.0\ \Delta^\circ\text{R} = 1.0\Delta^\circ\text{F} \qquad (1.2\text{-}8)$$

A look at the definition of the Celsius and Fahrenheit scales shows that

$$100^\circ\text{C} - 212^\circ\text{F} \qquad (1.2\text{-}9)$$

and

$$0^\circ\text{C} = 32^\circ\text{F} \qquad (1.2.10)$$

Subtracting Eq. (1.2-10) from Eq. (1.2-9) yields

*See Footnote on Page 15 for a discussion of the notation for temperature differences. The symbols °F, °C, K and °R indicate a temperature value on the respective scales.

$$100\ \Delta°C = 180\ \Delta°F \qquad (1.2\text{-}11)$$

or

$$1.0\ \Delta°C = 1.8\ \Delta°F \qquad (1.2\text{-}12)$$

By combining Eqs. (1.2-7), (1.2-8), and (1.2-12), it is seen that

$$1\ \Delta K = 1.8\ \Delta°R \qquad (1.2\text{-}13)$$

Finally, since the zero points for the Kelvin and Celsius temperature scales are different

$$T(K) - t_o = \frac{1\Delta°R}{1\Delta°F}\,[t\,(°C) - 0°C] \qquad (1.2\text{-}14)$$

where t_o is the correction factor for shifting the zero temperature on the Celsius scale to that on the Kelvin scale. Similarly,

$$T(°R) - t_o' = \frac{1\Delta°R}{1\Delta°F}\,[t\,(°F) - 0°F] \qquad (1.2\text{-}15)$$

For conversion from Kelvin to Celsius temperatures, or visa versa t_o equals 273.15 K. For Rankine to Fahrenheit conversions, t_o' equals 459.58°R. A comparison of the Kelvin, Celsius, Fahrenheit and Rankine temperature scales is shown in Fig. 1.2-1.

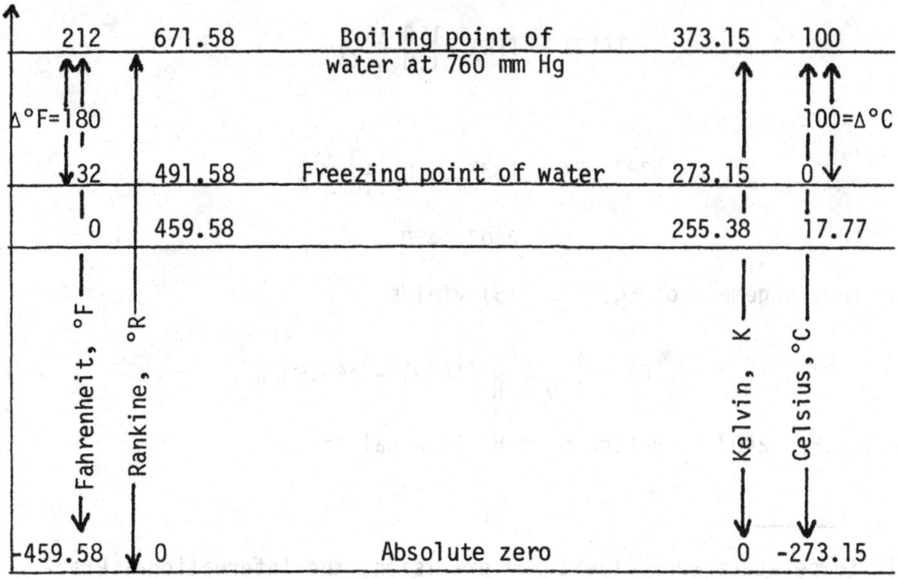

Figure 1.2-1: THE MOST COMMON TEMPERATURE SCALES
(Adapted from Reference 2, page 19)

---

2. D. M. Himmelblau, Basic Principles and Calculations in Chemical Engineering, Prentice-Hall, Englewood Cliffs, N.J. (1974).

EXAMPLE 1.2-3: The melting point of gold at 101.325 kPa pressure (1 atm) is given as 1337.58K on the International Practical Temperature scale.* Convert this value to a) °C, b) °R and c) °F.

Solution: a) Rearrangement of Eq. (1.2-14) yields

$$t(°C) = \frac{1.0\ \Delta°C}{1.0\ \Delta\ K}(T(K) - t_o)$$

Therefore, 1337.58K is equal to

$$t(°C) = \frac{1.0\ \Delta°C}{1.0\ \Delta\ K}(1337.58K - 273.15K)$$

$$= 1064.43°C$$

b) Since the Kelvin and Rankine Scales have the same zero point,

$$0\ K = 0°R$$

and

$$T(°R) - 0°R = (T(K) - 0\ K)\cdot(\frac{1.8\ \Delta°R}{1.0\ \Delta\ K})$$

or

$$T(°R) = T(K)\ (\frac{1.8\ \Delta°R}{1.0\ \Delta\ K})$$

Therefore,

$$1337.58K = 1337.58\ K(\frac{1.8\ \Delta°R}{1.0\ \Delta\ K})$$

$$= 2407.64°R$$

c) Rearrangement of Eq. (1.2-15) yields

$$t(°F) = \frac{1.0\ \Delta°F}{1.0\ \Delta°R}(T(°R) - 459.58°R)$$

Therefore, 2407.64°R (see part b) is equal to

---

*In temperature work of extreme precision, the International Practical Temperature Scale must be recognized, because many available data are based upon it. This scale, agreed upon by the CGPM in 1968 (amended edition 1975), is based on certain "defining fixed points" that permit calibration of instruments. The assignments of temperatures to these fixed points were based on the best available determinations of their thermodynamic temperatures. (Reference: Appendix SI, Standard for Metric Practice, ASTM Publication No. E 380-76.)

$$t(°F) = \frac{1.0\ \Delta°F}{1.0\ \Delta°R}(2407.64°R - 459.58°R)$$

$$= 1948.06°F$$

EXAMPLE 1.2-4: Show that the temperature reading of -40° is the same temperature on the Celsius and Fahrenheit scales.

Solution: Combining Eqs. (1.2-14) and 1.2-15) with T(°R) = 1.8T(K) yields,

$$t(°F) = \frac{1.8\ \Delta°F}{1.0\ \Delta°C}t(°C) + 32.0°F$$

Then, if t(°F) = t(°C) = X

$$X = 1.8\ x + 32°F$$

or

$$X = -40°F$$

Also,
$$-40°F = \frac{1.8\ \Delta°F}{1.0\ \Delta°C}\ t(°C) + 32°F$$

or

$$t(°C) - -40°C$$

Therefore, t(°F) and t(°C) are equivalent at -40°.

Pressure may also be measured on absolute or relative scales and with several different units. Absolute scales result in pressure readings based upon a perfect vacuum or completely evacuated reference point for zero pressure. Relative scales measure pressure in the same units, but with the zero point being the pressure of one standard atmosphere*. Several values of equivalents for the standard atmosphere are given in Table 1.2-2. The relationships between some of the most common pressure scales are shown in Fig. 1.2-2.

EXAMPLE 1.2-5: Convert a pressure measurement of one torr to a) psia, b) psig, and c) in. Hg (vac).

Solution: If pressure is defined as the force per unit area exerted by one body upon another, the pressure exerted by a column of liquid contained in a closed tube of cross-sectional area, A_o, and height, Z, on its base plate will equal

$$P = F/A_o$$

---

*The symbols psia and psig are used to indicate pounds-force per square inch on the absolute and relative guage scales, respectively.

Table 1.2-2

Equivalents for the Standard Atmosphere

1.000	atmospheres (atm)
33.91	feet of water (ft H_2O)
14.7	(14.696, more exactly) pounds per square inch absolute (psia)
29.92	(29.921, more exactly) inches of mercury (in. Hg)
760.0	millimeters of mercury (mm Hg)
1.013×10^3	Pascals (Pa)

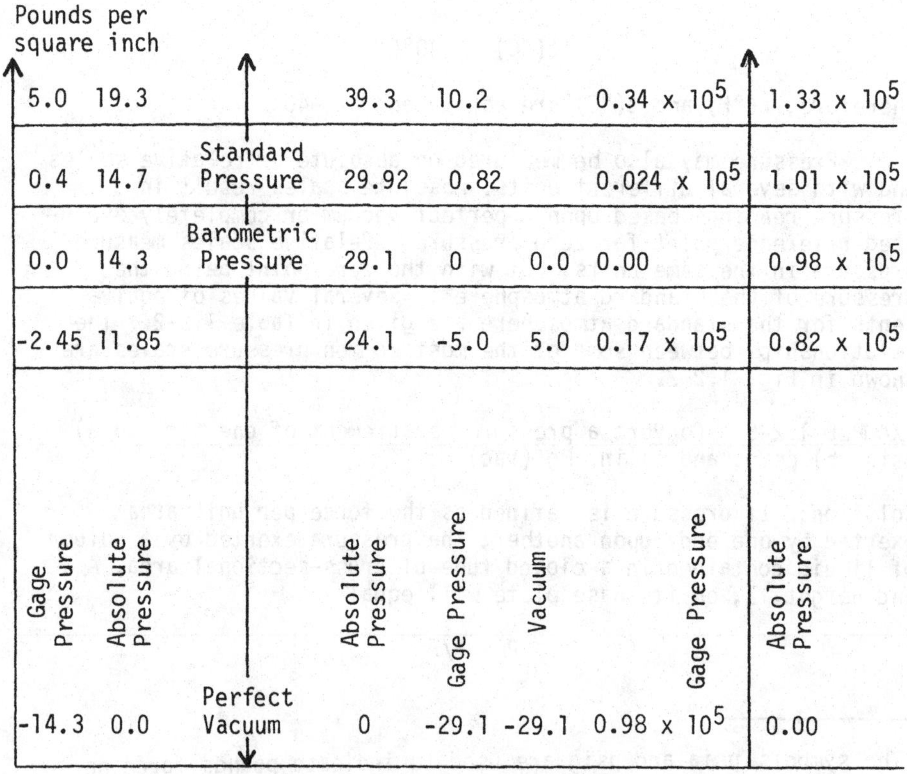

Figure 1.2-2 PRESSURE COMPARISONS WHEN BAROMETER READING IS 29.1 in. Hg. (From Reference 2, page 25)

where $$F = C \cdot M \cdot g$$

$$M = \rho \cdot V$$

and $$V = A_o \cdot Z$$

Combining the above equations yields $P = Z \cdot \rho \cdot g/g_c$ where ρ is the density of the fluid in the column and g is the acceleration of gravity. One standard atmosphere is then defined as the pressure exerted by a column of mercury 760.0 mm high at sea level, 45° lattitude and 273.15K. One torr is the pressure exerted by one millimetre of mercury at the same conditions.

a) Therefore, when the pressure is measured as one torr on an absolute scale,

$$1.0 \text{ torr} = 1.0 \text{ torr} \left(\frac{1.0 \text{ mm Hg}}{1.0 \text{ torr}}\right)\left(\frac{1.0 \text{ atm}}{760.0 \text{ mm Hg}}\right)\left(\frac{14.696 \text{ psia}}{1.0 \text{ atm}}\right)$$

$$= 1.9 \times 10^{-2} \text{ psia}$$

b) Since gauge pressure is measured with the same units but with a different zero point,

$$1.0\ \Delta\text{psia} = 1.0\ \Delta\text{psig}^*$$

and

$$p(\text{psig}) = \frac{1.0\ \Delta\text{psig}}{1.0\ \Delta\text{psia}}\ (P(\text{psia}) - 14.696 \text{ psia})$$

$$= 14.677 \text{ psig}$$

c) The vacuum pressure is generally given as the pressure difference below atmospheric pressure. By comparison to part b), the vacuum pressure in psig would be 14.677 psig (vac.). The vacuum pressure in inches of mercury, symbol in. Hg (vac), will then equal**

$$1 \text{ torr} = 14.677 \text{ psig (vac)} \left(\frac{6.894\ 757 \times 10^3 \text{ Pa}}{1.0 \quad \text{psig}}\right)\left(\frac{1.0 \text{ in. Hg (vac)}}{3.386\ 38 \times 10^3 \text{ Pa}}\right)$$

$$= 29.88 \text{ in. Hg (vac)}^{**}$$

---

*The same terminology that was used for temperature differences will be used for pressure differences. See Footnote, page 15.

** The pressure readings are for mercury at 0°C (32°F). Slightly different pressure readings will be found for mercury at a different temperature.

EXAMPLE 1.2-6: Calculate the metallostatic head (pressure) in psig at the bottom of a ladle that contains ten feet of iron at 1600°C.

Solution: The head or pressure at a given depth in a liquid is found in exactly the same manner as was described in the previous example, i.e.

$$P = Z \cdot \rho \cdot g/g_c$$

where Z is the depth of immersion and ρ the density of the liquid. Therefore, if the density of iron is $489 \cdot lb_m/ft^3$ at 1600°C, the metallostatic head will equal

$$P = (10\ ft)\ (489 \cdot lb_m/ft^3)(32.174\ ft/s^2)(1\ ft^2/144\ in^2)/(32.174\ lb_m \cdot ft/lb_f \cdot s^2)$$

$$= 34.0\ psig$$

1.3 THE MOLE UNIT

According to the definition of the mole established by the CGPM in 1960, one mole is "the amount of a substance that contains as many elementary species (i.e. atoms, molecules, etc.) as there are atoms in 0.012 kg of carbon-12". This number has been found to equal 6.02252×10^{23}, Avogadro's Number (symbol N_o). Thus, an alternative definition of the mole would be the amount of substance that contains Avogadro's Number of elementary species. In the SI system, the mole is given the symbol mol.

The masses of individual atoms are measured on a scale whose basic unit is the atomic mass unit (symbol a.m.u.). On this scale, one carbon-12 atom has a mass of 12 a.m.u. Since, by definition one mole of carbon atoms has a mass of 0.012 kg, and one mole of carbon atoms contains N_o atoms,

$$12g = 1.0\ mol\ C\ \left(\frac{N_o\ atoms\ C}{1.0\ mol\ C}\right)\left(\frac{12\ a.m.u.}{1.0\ atom\ C}\right) \qquad (1.3\text{-}1)$$

$$1.0\ g = N_o\ a.m.u. \qquad (1.3\text{-}2)$$

or

$$1\ a.m.u. = 1.66043 \times 10^{-24}\ g \qquad (1.3\text{-}3)$$

EXAMPLE 1.3-1: The mass of a uranium atom is 238.03 a.m.u. Determine the mass of one mole of uranium.

Solution: Since one mole of uranium contains N_o atoms and each atom has a mass of 238.03 a.m.u., the mass of one mole of uranium atoms, m, will equal

$$m = \frac{N_o\ atoms}{1.0\ mole} \cdot \frac{238.02\ a.m.u.}{atom} \cdot \frac{1.0\ g}{N_o\ a.m.u.}$$

$$= 238.03\ g/mol$$

Another definition of the mole can be seen from the solution to the previous example. A mole of atoms (molecules) is the amount of that substance whose mass in grams is numerically equivalent to the mass of an atom (molecule) of that substance in a.m.u. The mole unit used in the SI system was previously called the gram-mole (symbol g-mol) because it was the amount of substance whose mass in grams was equivalent to its mass per atom (molecule) in a.m.u.

Another common unit in the American Engineering System of units is the pound-mole (symbol lb-mol). This amount of substance is defined as the amount required to have a mass in pounds numerically equal to the mass per atom (molecule) in a.m.u. Similar definitions exist for the kilogram-mole (symbol kg-mol), Ton mole (symbol Ton-mol), etc.

Caution must be used when dealing with the mole unit in the SI system. It is seen from the above discussion that the mole unit is equivalent to the gram-mole even though the fundamental unit of mass in the SI system is the kilogram.

EXAMPLE 1.3-2: How many gram-moles of zinc are contained in one pound-mole of zinc? The atomic mass of zinc is 65.37 a.m.u.

Solution: According to the definition of the pound-mole,

$$1.0 \text{ lb-mol Zn} = 65.37 \text{ lb}_m \text{ Zn}$$

$$= 65.37 \text{ lb}_m \text{ Zn} \left(\frac{453.5924 \text{ g}}{1.0 \text{ lb}_m} \times \frac{1.0 \text{ g-mol Zn}}{65.37 \text{ g Zn}}\right)$$

$$= 453.5924 \text{ g-mol Zn}$$

For engineering calculations, it is assumed that there are 454 gram-moles in one pound-mole or $454 \times 6.023 \times 10^{23}$ atoms in a pound-mole. Similarly, there are 1000 gram-moles and $1000 \times 6.023 \times 10^{23}$ atoms in one kilogram-mole, etc.

1.4 DENSITY AND CONCENTRATION

As was seen in the previous section, the mole unit specifies the amount of substance in a given mass. However, it is often necessary to know how much material is contained in a given volume, i.e. its density, or how much of one material is mixed with another i.e., its concentration. The units used to describe these quantities are discussed in this section.

1.4.1 DENSITY

The density of a gas. liquid or solid is its mass per unit volume. In the SI system, the unit of density is the kg/m^3. However, the density may also be given as a pure number, i.e. a number without units. This pure number is called the specific gravity, and is equal to the ratio of the density of the material of interest to the density of a reference material.

Water is most frequently used as the reference material to specify the specific gravity of a solid or liquid. For most calculations, a value of 1.000 g/cm^3 may be used for the density of water between 0°C and 45°C. For gas phases, air is commonly used as the reference material. The density of dry air in kg/m^3 equals*

$$\rho_{air} = \frac{1.293}{1.0 + 0.00367t} \cdot \frac{H}{760} \qquad (1.4\text{-}1)$$

where t is the Celsius temperature, and H is the pressure in mm Hg. More precise values of the density of water and the density of moist air can be found in the Handbook of Chemistry and Physics**.

The specific volume may also be reported in place of the density. The specific volume is the volume per unit mass of a substance and has the units of m^3/kg in the SI system. The specific volume is obviously the reciprocal of the density.

1.4.2 CONCENTRATION

In most operations, the feed and product materials are not pure but rather mixtures or solutions of two or more materials. The concentration of the components of overall process streams are specified either by their weight or by the number of their moles in relation to the total weight or total number of moles in the overall material or process stream.

For solids and liquids, the concentration of one component in a mixture is usually given by its weight fraction or by its weight percentage. The weight fraction of component A equals

$$W_A = \frac{\text{weight of A in mixture}}{\text{total weight of mixture}} \qquad (1.4\text{-}2)$$

The weight percentage (symbol wt% or s/o) of component A equals

$$\text{wt\% A} = \text{w/o A} = 100 \cdot W_A \qquad (1.4\text{-}3)$$

Another concentration unit that is commonly used for very dilute mixtures is the Part Per Million (symbol PPM).

$$\text{PPM A} = 10^4 \cdot \text{w/o A} = 10^6 \cdot W_A \qquad (1.4\text{-}4)$$

Unless otherwise specified, concentrations for solids and

---

* The symbol ρ will be used for density and specific gravity. Differentiation between quantities will be made by specifying the units of density.

** See footnote, page 4.

liquids given in this monograph will be on a weight basis.

The molecular percentage and the mole fraction are commonly used for the specification of the analysis of a gas. The mole fraction of A equals

$$X_A = \frac{\text{moles of A in mixture}}{\text{total moles in mixture}} \qquad (1.4\text{-}5)$$

and the mole percentage of A equals

$$\%A = 100 \cdot X_A \qquad (1.4\text{-}6)$$

Conversion from weight fraction to mole fraction or vice versa can be accomplished using the following equations

$$X_i = \frac{W_i/MW_i}{(W_A/MW_A) + (W_B \cdot MW_B) + \ldots + (X_i \cdot MW_i) + \ldots} \qquad (1.4\text{-}7)$$

and

$$W_i = \frac{X_i \cdot MW_i}{(X_A \cdot MW_A) + (X_B \cdot MW_B) + \ldots + (X_i \cdot MW_i) + \ldots} \qquad (1.4\text{-}8)$$

where X_A, X_B, ..., X_i, ..., are the mole fractions of components A, B, ..., i, ..., respectively. W_A, W_B, ..., W_i, ..., are the weight fractions of components A, B, ..., i, ..., respectively. MW_A, MW_B, ..., MW_i, ..., are the molecular weights of components A, B, ..., i, ..., respectively.*

EXAMPLE 1.4-1: Calculate the crucible volume (cm^3) required to contain enough iron, chromium, and nickel powder to make ten kg of an 18% chromium, 8% nickel and 74% iron mixture. Assume that the air gaps surrounding the powder particles occupies 60% of the total volume of the crucible. The specific gravities of chromium, nickel and iron are 7.1, 8.9, and 7.7, respectively.

Solution: By combining Eqs. (1.4-1) and (1.4-2), it can be shown that the weight of A in mixture = $(\frac{w/o\ A}{100})$ (the total weight of the mixture). Thus, 1.8 kg Cr. 0.8 kg Ni and 7.4 kg Fe must be charged. The volume occupied by each is then calculated:

$$(1.8 \text{ kg Cr})\ (10^3 \text{ g/kg})(1 \text{cm}^3/7.1 \text{ g Cr}) = 254 \text{ cm}^3$$

$$(0.8 \text{ kg Ni})\ (10^3 \text{ g/kg})(1 \text{cm}^3/8.9 \text{ g Ni}) = 90 \text{ cm}^3$$

$$(7.4 \text{ kg Fe})\ (10^3 \text{ g/kg})(1 \text{cm}^3/7.7 \text{ g Fe}) = \underline{961 \text{ cm}^3}$$

Total volume occupied by metal =1305 cm^3

---

* See Chapter 2 for a complete definition of molecular weight of a substance.

$$\text{Total volume of crucible required} = \frac{1305\ \text{cm}^3\ \text{metal}}{(0.4\ \text{cm}^3\ \text{metal/cm}^3\ \text{charge})}$$

$$= 3263\ \text{cm}^3$$

EXAMPLE 1.4-2: Determine the mole percentages of an 18% chromium 8% nickel and 74% iron mixture.

Solution: For the three component system, a combination of Eqs. (1.4-2), (1.4-5) and (1.4-6) yields

$$\%A = \frac{w/oA\ /\ MW_A}{(w/oA\ /\ MW_A) + (w/oB\ /\ MW_B) + (w/oC\ /\ MW_C)}$$

and similar relationships for B and C. Using 52, 59 and 56 for the molecular weights of chromium, nickel and iron, respectively,

$$\%Cr = \frac{18/52}{18/52 + 8/59 + 74/56}$$

$$= 19\%$$

$$\%Ni = \frac{8/59}{18/52 + 8/59 + 74/56}$$

$$= 8\%$$

and by difference

$$\%Fe = 100 - 19 - 8$$

$$= 73\%$$

An alternative and sometimes useful method is to determine how many moles of each component and the total number of moles that are present. Then the combination of Eqs. (1.4-4) and (1.4-5) yields

$$\%A = 100\% \cdot \frac{\text{moles of A in mixture}}{\text{total moles in mixture}}$$

In this problem, the number of moles of chromium, N_{Cr}, equals

$$N_{Cr} = M_{Cr}/MW_{Cr}$$

$$= 1.8\ lb_m/(52\ lb_m/lb\text{-}mol)$$

$$= 0.0347\ lb\text{-}mol$$

Similarly, N_{Ni} equals 0.0136 lb-mol and N_{Fe} equals 0.1321 lb-mol. Then

$$N_T = N_{Cr} + N_{Ni} + N_{Fe}$$

$$= 0.0347 + 0.0136 + 0.1321$$

$$= 0.0184 \text{ lb-mol}$$

Finally, the molecular percentage of chromium equals

$$\%Cr = 100\% \cdot 0.0347 \text{ lb-mol}/0.1804 \text{ lb-mole}$$

$$= 19\%$$

Similar calculations can be done for nickel and iron.

Note: In this calculation, pound-moles were used. However, moles, kilogram-moles, etc. can be used as long as care is taken to be sure all units are consistent.

The analyses of gases given in this book will be given on a molecular basis, unless otherwise stated. However, as will be shown in Chapter 2, the analysis of a gas on a volume basis is virtually identical to its analysis on a mole basis, for the conditions of temperature and pressure found in most metallurgical processes. Therefore, the mole fraction of component i in a gas will also equal

$$x_i = \frac{v_i}{v_t} \tag{1.4-8}$$

where v_i is the volume of component i and v_t the total volume of the gas including component i.

Gas analyses are normally reported on a dry basis, i.e. the percentages given are for all components excluding moisture. The moisture content of the gas is then specified as an amount of moisture per unit volume or per mole of gas. The units of moisture content commonly used are grams per cubic meter or grains per cubic foot, where

$$1 \text{ grain} = 6.479\ 891 \times 10^{-5} \text{ kg} \tag{1.4-9}$$

Care must be used to determine if the moisture content is given per unit of dry or wet gas.

EXAMPLE 1.4-4: Calculate the dry analysis and the grains of moisture present per kg-mol of dry gas for a tank of gas that contains 10 kg CO, 10 kg CO_2, 0.2 kg H_2O and 78.0 kg N_2.

Solution: To obtain the dry gas analysis in molecular percent, the masses of each gas except H_2O must be converted to moles

$$N_{CO} = M_{CO}/MW_{CO}$$

$$= 10 \text{ kg CO}/(28 \text{ kg CO/kg-mol CO})$$

$$= 0.36 \text{ kg-mol CO}$$

Similarly,

$$N_{CO_2} = 0.23 \text{ kg-mol } CO_2$$

$$N_{N_2} = 2.79 \text{ kg-mol } N_2$$

Then, on a dry basis, i.e. excluding H_2O

$$N_t = N_{CO} + N_{CO_2} + N_{N_2}$$

$$= 0.36 + 0.23 + 2.79$$

$$= 3.38 \text{ kg-mol}$$

and from Eqs. (1.4-4) and (1.4-5), it can be seen that

$$\%CO = 100\% \cdot N_{CO}/N_t$$

$$= 100\% \cdot 0.36/3.38$$

$$= 10.7\%$$

$$\%CO_2 = 100\% \cdot 0.23/3.38$$

$$= 6.8\%$$

and

$$\%N_2 = 100\% \cdot 2.79/3.38$$

$$= 82.5\%$$

Finally, there are 3.38 kg-mol of gas, excluding H_2O, present, along with 0.2 kg of H_2O, or there are

$$\left(\frac{0.2 \text{ kg } H_2O}{3.38 \text{ kg-mol dry gas}}\right)\left(\frac{1 \text{ grain}}{6.48 \times 10^{-5} \text{ kg}}\right)$$

$$= 913. \text{ (grains } H_2O\text{/kg-mol dry gas)}$$

Note, there are a total of 3.39 kg-mol of gas present, including H_2O. On a wet basis, there will be

$$(0.2 \text{ kg } H_2O/3.39 \text{ kg-mol gas})(1 \text{ grain}/6.48 \times 10^{-5} \text{ kg})$$

or

$$= 910. \text{ grains } H_2O\text{/kg-mol gas}$$

present in the tank.

The moisture content may also be given as the partial pressure of a gas or as the relative humidity. The concept of partial pressure will be dealt with in detail in Chapter 2. For the present it is only necessary to assume ideal gas behavior, in which case the number of moles of H_2O, N_{H_2O}, in a gas mixture that contains N_t moles of gas including H_2O is related to the partial pressure of H_2O, P_{H_2O}, and the total pressure P_t, by the equation

$$X_i = \frac{N_{H_2O}}{N_t} \frac{P_{H_2O}}{P_t} \qquad (1.4\text{-}10)$$

The amount of moisture that can be contained as a vapor in air or in any gas is limited by the temperature and pressure of the gas. This limit is called the saturation pressure and the percentage or fraction of this maximum value actually present in the gas is called the relative humidity. Values for the saturation pressure for air are given in The Handbook of Chemistry and Physics(1).

EXAMPLE 1.4-5: On a cool winter day, the dew point of the air may reach 40°F, while on a hot summer day, the dew point may reach 80°F. Calculate the moisture content in grams per mole of dry air at these two conditions.

Solution: The dew point is the temperature at which condensation of moisture in a gas begins during cooling, i.e. the temperature at which the partial pressure of H_2O equals the saturation pressure. At 40°F, the saturation pressure is approximately 0.123 psig. Thus, if atmospheric pressure is 1 atm or 14.696 psia, the number of moles of H_2O in one mole of gas including the moisture can be found from Eq. (1.4-10).

$$\frac{N_{H_2O}}{N_t} = \frac{P_{H_2O}}{P_t}$$

or

$$N_{H_2O} = 1.0 \text{ mol}\left(\frac{0.123 \text{ psig}}{14.696 \text{ psia}}\right)\left(\frac{1.0 \text{ psia}}{1.0 \text{ psig}}\right)$$

$$= 8.37 \times 10^{-3} \text{ mol } H_2O$$

The mass of H_2O then equals

$$M_{H_2O} = 8.37 \times 10^{-3} \text{ mol } H_2O \left(\frac{18 \text{ g } H_2O}{1.0 \text{ mol } H_2O}\right)$$

$$= 0.151 \text{ g } H_2O$$

This mass of H_2O is contained in one mole of gas which includes the moisture. The mass of H_2O in one mole of dry gas must equal

$$m_{H_2O} = 0.151 \text{ g } H_2O/(1.0 - 8.37 \times 10^{-3}) \text{ mol dry gas}$$

$$= 0.152 \text{ g } H_2O/\text{mol dry gas}$$

At 80°F, the saturation pressure is approximately 0.502 psig. Using the same approach

$$m_{H_2O} = 2.84 \text{ g } H_2O/\text{mol dry gas}$$

Finally, many metallurgical processes have feed and/or product streams that consist of mixtures of solids and liquids. These mixtures are called slurries. The concentration of the solids in the slurries are given as the percentage of the total slurry weight (symbol w/o S) or as the solid percentage of the total volume of the slurry (v/o S). The relationship between the volume percent solids, weight percent solid, the specific gravity of the solid phase (symbol ρ_s) and the specific gravity of the slurry mixture (symbol ρ_m) are given below for water slurries, i.e. solids in water.

$$\text{v/o S} = \frac{100}{1 + \rho_s \left(\frac{100 - \text{w/o S}}{\text{w/o S}}\right)} \qquad (1.4\text{-}11a)$$

$$= 100 - \rho_m (100 - \text{w/o S}) \qquad (1.4\text{-}11b)$$

$$= 100 \left(\frac{\rho_m - 1}{\rho_s - 1}\right) \qquad (1.4\text{-}11c)$$

$$= \text{w/o S} \cdot \rho_m/\rho_s$$

$$\text{w/o S} = \frac{\text{v/o S} \cdot \rho_s}{1 + 0.01 \cdot \text{v/o S} \cdot (\rho_s - 1)} \qquad (1.4\text{-}12a)$$

$$= 100 \left(\frac{\rho_m - 1}{\rho_m}\right)\left(\frac{\rho_s}{\rho_s - 1}\right) \qquad (1.4\text{-}12b)$$

$$= 100 - \left(\frac{100 - \text{v/o S}}{\rho_m}\right) \qquad (1.4\text{-}12c)$$

$$= \text{v/o S} \cdot \rho_s/\rho_m \qquad (1.4\text{-}12d)$$

Any two of these four slurry variables can be determined using Fig. 1.4-2. The two known values are located on the nomograph and a straight line connecting them is drawn across the figure. Its intersections with the other variables give their values.

EXAMPLE 1.4-6: Calculate how many pounds of magnetite (symbol Fe_3O_4) must be added to 100 pounds of water to make up a slurry with specific gravity of 1.4. The specific gravity of Fe_3O_4 is 5.2.

Solution: Substituting $\rho_s = 5.2$ and $\rho_m = 1.4$ into Eq. (1.4-11b) yields

$$\text{w/o S} = 100\left(\frac{1.4 - 1.0}{1.4}\right)\left(\frac{5.2}{5.2 - 1.0}\right)$$

$$= 35 \text{ w/o } Fe_3O_4$$

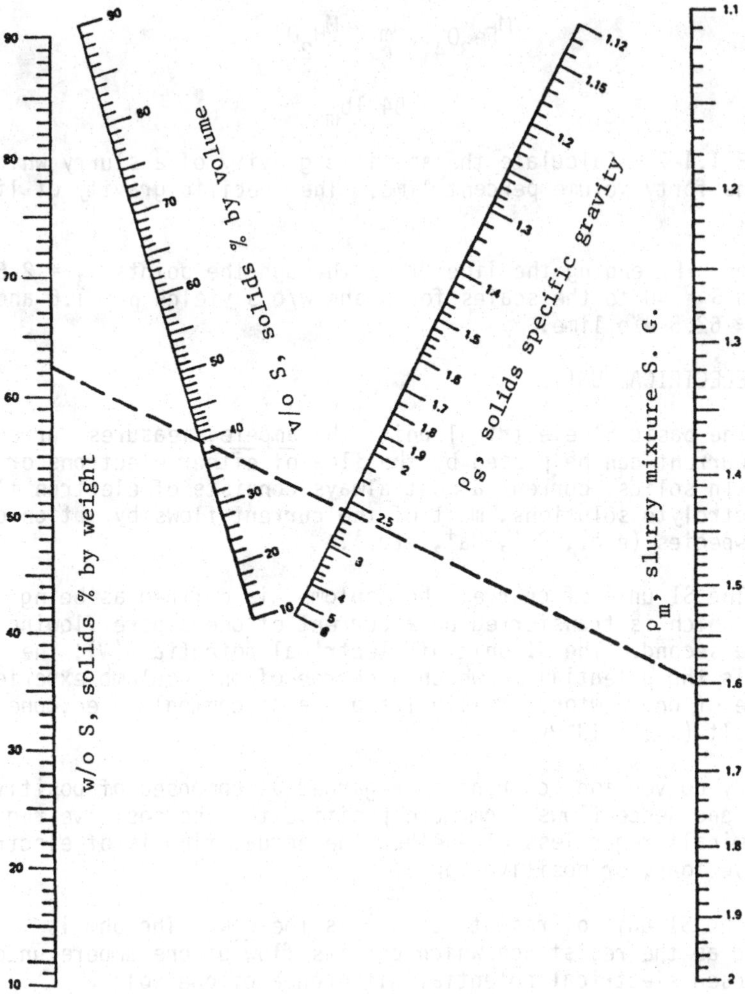

Figure 1.4-2 NOMOGRAPH FOR CALCULATION OF SLURRY PROPERTIES (Ref. F. Caplan, E/MJ, 176 No. 10 (1975), 106)

Then, if the remainder or 65% of the mass of slurry, M_m, is water,

$$\frac{M_m}{M_{H_2O}} = \frac{100\%}{65\%}$$

or

$$M_m = \frac{100\%}{65\%} \cdot 100 \text{ lb}_m$$

$$= 154 \text{ lb}_m$$

and the mass of magnetite must equal

$$M_{Fe_3O_4} = M_m - M_{H_2O}$$

$$= 54 \text{ lb}_m$$

EXAMPLE 1.4-7: Calculate the specific gravity of a slurry which contains forty volume percent lime. The specific gravity of lime is 2.6.

Solution: Extending the line drawn through the points $\rho_S = 2.6$ and v/o S = 40 to the scales for ρ and w/o S yields $\rho = 1.6$ and w/o S ≃ 62.5 w/o lime.

1.5 ELECTRICAL UNITS

The basic SI electrical unit, the ampere, measures current (i). Current can be passed by the flow of either electrons or ions. In solids, current almost always consists of electron flow; in electrolyte solutions, must of the current flows by motion of ionic species (e.g., Cu^+, Na^+, etc.).

The SI unit of charge, the Coulomb, is defined as being the charge which is transferred by a current of one ampere flowing for one second. The SI unit of electrical potential (V), the volt, is the potential in which a charge of one coulomb experiences a force of one newton. Millivolts are also commonly used; one millivolt (mv) = 10^{-3}v.

By convention, current is regarded as composed of positive charge and hence flows from more positive to less positive regions or terminals regardless of whether the actual flow is of electrons, negative ions, or positive ions.

The SI unit of resistance (R) is the ohm. The ohm is defined as the resistance which permits flow of one ampere under an imposed electrical potential difference of one volt.

Energy W is, as usual, measured in joules, and power P in watts (one watt being one joule per second).

Some basic equations of electrical flow are:

$$V = Ri \tag{1.5-1}$$

$$P = iV = i^2R \tag{1.5-2}$$

$$W = tP = i^2Rt \tag{1.5-3}$$

Resistance is connected to resistivity (specific resistance) by the relation that resistivity is resistance multiplied by the area through which the current flows and divided by the path length of the current (S = R A/l). The SI resistivity unit is the ohm-meter.

EXAMPLE 1.5-1: A 2.0 volt potential drop is applied down a circular bar of 0.1m radius which is 2m long and has a resistivity of 1.7×10^{-6} ohm-meters. What is the magnitude of the current?

Solution: The area of the bar is $\pi\ (.1)^2 = 0.0314 m^2$. Therefore, the resistance is $(1.7 \times 10^{-6}(2)/.0314 = 1.08 \times 10^{-4}$ ohms. The current is given by $i = V/R = 2/1.08 \times 10^{-4}$ amperes.

$$i = 1.85 \times 10^4 A$$

One quantity which is of importance in electrometallurgy is the <u>faraday</u>. A faraday is one mole of electrons. One faraday (F) is 96,500 coulombs. The significance of this number is that if the electricity is used to discharge ions, one faraday will discharge one gram <u>equivalent</u> of ions. A gram-equivalent of singly-charged ions in one gram-mole; an equivalent of m-charged ions is 1/m gram-mole.

EXAMPLE 1.5-2: How many gram-moles of Al^{+3} ions could be discharged in one minute by the current in Example 1.5-1, if none of the current was lost to leaks or other processes?

Solution: In one minute a current of 1.85×10^4 amperes will carry $(60)(1.85 \times 10^4) = 1.11 \times 10^5$ coulombs $= 1.11 \times 10^5/.965 \times 10^5 = 1.16$ equivalents. For Al^{+3} ions 1.16 gram-equivalents = 1.16/3 = 0.387 gram-moles.

1.6 SUMMARY

The SI system of units, which has been adopted by the majority of the countries in the world, will be emphasized in this text. This system is based upon a set of fundamental units, with additional or derived units being made up of simple combinations of the fundamental units. In all cases, the conversion factor between the derived and fundamental units are independent of the gravitational field and equal to unity.

The Conversion Factors in Table 1.2-1 are useful when converting from non-SI to SI units. When more than one conversion factor is required, the dimensional equation should be used to organize the conversion process. For the units of temperature and pressure, a conversion equation must be used to change to different size units and to shift the zero points for the different scales.

FURTHER READING

1. Y. Chin, A Dictionary for Unit Conversion, George Washington University, Washington, D.C., 1975.

2. R. A. Hopkins, Metric in a Nutshell, Polymetric Services, Inc., Tarzana, CA., 1976.

3. R. A. Hopkins, The International (SI) System and How it Works, Polymetric Services, Inc., Tarzana, CA., 1974.

4. S. H. Qasim, SI Units in Engineering and Technology, Pergamon Press, New York, 1977.

5. W. E. Glason, Lexicon of International and National Units, Elsever Publishing Co., New York, 1964.

6. Standard for Metric Practice, ASTM Publication Number E380-76, ASTM, Philadelphia, PA, 1976.

7. A. A. Klein, The World of Measurements, Simon and Schuster Publishing Co., New York, 1974.

EXERCISES

1.1 Change 400 in^3/day to cm^3/min

1.2 If a bucket holds 2 lb of NaOH (molecular wt. = 40), how many

a) lb-mol of NaOH does it contain?
b) g-mol of NaOH does it contain?

1.3 Convert the value of the gas constant R into cgs and mks (metre-kilogram-second) units. R = 0.08204 l-atom deg^{-1} $mole^{-1}$.

1.4 A crucible contains 10 lb of Cu and 15 lb of Zn. What is the weight fraction and mole fraction of each component in the crucible?

1.5 Convert 1 x 10^{-8} atm to pressure on the mm H_g (vac) scale and to torr.

1.6 Find the kinetic energy of a kilogram of water moving at 60 mph expressed as

a) joules
b) watt-sec
c) liter-atm.

1.7 The emissive power of a black body depends on the fourth power of the temperature and is given by

$$W = AT^4$$

where W = emissive power in Btu/(ft^2 h)

A = Stefan-Boltzman constant, 0.1714 x 10^{-8} Btu/(Ft^2 h °R^4)
T = temperature in °R.

What is the value of A in the units:

a) cal sec^{-1} cm^{-2} $(°C)^{-4}$

b) cal sec^{-1} cm^{-2} $(°K)^{-4}$

c) joule sec^{-1} m^{-2} $(°K)^{-4}$

1.8 Fill in the following table.

	Weight, g	Wt.%	Vol%	Mole Fraction
CO_2	12.0			
O_2	8.0			
N_2	75.0			
H_2O	5.0			

1.9 Fill in the following table.

	Wt.%	Mole Fraction	Vol.%
$CaCO_3$	22%		
$MgCO_3$	18%		
CaO	23%		
NaCl	30%		
K_2SO_4	7%		

1.10 Calculate the percentage composition by weight of a stainless steel melted in an electric furnace. The charge consisted of 2000 lb of iron, 506 lb of Cr, 225 lb of Ni, and 85 lb of Mo.

1.11 Convent 150°C to a) K, b) °F, c) °R.

1.12 The thermal conductivity of aluminum at 32°F is 117 Btu h^{-1} ft^{-1} $(°F)^{-1}$. Find the equivalent value at 0°C in terms of cal sec^{-1} cm^{-1} $(°C)^{-1}$.

1.13 Convent 340 mm Hg to:

a) newton/meter^2
b) lb/ft^2.

CHAPTER 2

STOICHIOMETRY

2.0 INTRODUCTION

Many of the material balance problems that arise in metallurgical processes can be solved with either the knowledge of the relationship between the mass of elements within a compound and the total mass of the compound, the relative masses of reactants and products involved in a chimical reaction, or the mass of compounds contained in a given volume of gas. This chapter reviews the use of stoichiometry to develop these relationships. The material that will be covered is in great part a review of material that is taught in basic chemistry. The material is, however, very important to the development of the remainder of this text, and for this reason, it is presented in some detail.

2.1 ATOMIC AND MOLECULAR MASS (WEIGHT)

As mentioned in Chapter 1, the mass of single atoms is measured in atomic mass units (symbol a.m.u.) on the Atomic Mass Scale. This scale is a relative scale with Carbon-12, i.e., the carbon atom (isotope) which contains six protons and six neutrons, being taken as the reference point. The mass of Carbon-12 is chosen to be equal to 12.0000 a.m.u. The masses of all other atoms are then measured relative to this standard.

The atomic masses, relative to the Carbon-12 standard, of some of the most common elements are given in Table 2.1-1. It should be noted that more than one isotope can exist for many atoms and the values of the atomic masses given in Table 2.1-1 reflect the average for the mixture of isotopes commonly found in nature. Thus, the atomic mass of carbon is given as 12.0115 a.m.u. and not 12.0000 a.m.u.

The terminology, atomic weight and the atomic weight unit (symbol a.w.u.) are often encountered. These terms are synonomous to the atomic mass and atomic mass unit and should not lead to any confusion.

$$1 \text{ a.w.u.} = 1 \text{ a.m.u.} \quad (2.1\text{-}1a)$$

$$= 1.6598 \times 10^{-24} \text{g} \quad (2.1\text{-}1b)$$

$$= 1.6598 \times 10^{-27} \text{kg} \quad (2.1\text{-}1c)$$

Synonyms that are frequently used to describe the amount of substance that is equal to one mole of an element are the gram-mole, the gram-atom and the gram-atomic weight.

Table 2.1-1

Table of Atomic Weights of the More Common Elements

Element	Symbol	Atomic Weight
Aluminum	Al	26.9815
Antimony	Sb	121.75
Argon	Ar	39.948
Arsenic	As	74.9216
Barium	Ba	173.34
Beryllium	Be	9.0122
Bismuth	Bi	208.980
Boron	B	10.811
Bromine	Br	79.909
Cadmium	Cd	112.40
Calcium	Ca	40.08
Carbon	C	12.01115
Chlorine	Cl	35.453
Chromium	Cr	51.996
Cobalt	Co	58.9332
Copper	Cu	63.54
Fluorine	F	18.9984
Gold	Au	196.967
Helium	He	4.0026
Hydrogen	H	1.00797
Iodine	I	126.9044
Iron	Fe	55.847
Krypton	Kr	83.80
Lead	Pb	207.19
Lithium	Li	6.939
Magnesium	Mg	24.312
Manganese	Mn	54.9380
Mercury	Hg	200.59
Molybdenum	Mo	95.94
Neon	Ne	20.183
Nickel	Ni	58.71
Nitrogen	N	14.0067
Oxygen	O	15.9994
Palladium	Pd	106.4
Phosphorus	P	30.9738
Platinum	Pt	195.09
Potassium	K	39.102
Rhodium	Rh	102.905
Selenium	Se	78.96
Silicon	Si	28.086
Silver	Ag	107.870
Sodium	Na	22.9898
Strontium	Sr	87.62
Sulfur	S	32.064
Tantalum	Ta	180.948
Tin	Sn	118.69

(continued)

Element	Symbol	Atomic Weight
Titanium	Ti	47.90
Tungsten	W	183.85
Uranium	U	238.03
Vanadium	V	50.942
Xenon	Xe	131.30
Zinc	Zn	65.37
Zirconium	Zr	91.22

All of the atomic masses can be found on the periodic table shown in Fig. 2.1-1.

Molecules are composed of atoms in characteristic numbers and arrangements. The formula that expresses the exact number of atoms of each element that make up the molecule is called the molecular formula. The mass of a molecule is found by adding the masses of all of the atoms in the molecule.

For the general compound $A_x B_y$, the molecular mass will equal x times the atomic mass of A plus y times the atomic mass of B. For a mole of the compound,

$$MW_{A_xB_y} = x \cdot MW_A + y \cdot MW_B \tag{2.1-2}$$

For one molecule,

$$m_{A_xB_y} = x \cdot m_A + y \cdot m_B \tag{2.1-3}$$

A mole of a compound consists of one mole of the molecules that make up the compound. Thus, a mole of hydrogen chloride consists of a mole of HCl molecules or a mole of hydrogen atoms combined with a mole of chlorine atoms. A mole of a substance has a mass in grams equal to its mass in a.m.u. The gram-molecular weight may be used as a synonym for the mole or gram-mole. Also, the term molecular mass or molecular weight is commonly used to specify the mass or weight of one mole of molecules (symbol MW). Differentiation between the two meanings of this term is made by the units associated with the statement, i.e. a molecular weight given in a.m.u. refers to the weight of one molecule, while if given in grams it would be the molecular weight of one mole of molecules.

A second type of formula known as the simple or empirical formula is also often used. This type of formula, however, gives a minimum amount of information about a compound as it only states the relative number of atoms in the compound. The empirical formula has subscripts that represent the set of the smallest possible integers that can show the relative number of atoms in

KEY TO CHART: Atomic Number → 50; Symbol → Sn; Atomic Weight → 118.69; Oxidation States → +2 +4; Electron Configuration → -18-18-4

Transition Elements (groups 3–12); Group 8 (Fe, Co, Ni columns)

1 H +1 −1 1.00797 1																	2 He 0 4.0026 2	K
3 Li +1 6.939 2-1	4 Be +2 9.0122 2-2											5 B +3 10.811 2-3	6 C +2 +4 −4 12.01115 2-4	7 N +1 +2 +3 +4 +5 −1 −2 −3 14.0067 2-5	8 O −2 15.9994 2-6	9 F −1 18.9984 2-7	10 Ne 0 20.183 2 8	K-L
11 Na +1 22.9898 2-8-1	12 Mg +2 24.312 2-8-2											13 Al +3 26.9815 2-8-3	14 Si +2 +4 −4 28.086 2-8-4	15 P +3 +5 −3 30.9738 2-8-5	16 S +4 +6 −2 32.064 2-8-6	17 Cl +1 +5 +7 −1 35.453 2-8-7	18 Ar 0 39.948 2-8-8	K-L-M
19 K +1 39.102 -8-8-1	20 Ca +2 40.08 -8-8-2	21 Sc +3 44.956 -8-9-2	22 Ti +2 +3 +4 47.90 -8-10-2	23 V +2 +3 +4 +5 50.942 -8-11-2	24 Cr +2 +3 +6 51.996 -8-13-1	25 Mn +2 +3 +4 +7 54.9380 -8-13-2	26 Fe +2 +3 55.847 -8-14-2	27 Co +2 +3 58.9332 -8-15-2	28 Ni +2 +3 58.71 -8-16-2	29 Cu +1 +2 63.546 -8-18-1	30 Zn +2 65.37 -8-18-2	31 Ga +3 69.72 -8-18-3	32 Ge +2 +4 72.59 -8-18-4	33 As +3 +5 −3 74.9216 -8-18-5	34 Se +4 +6 −2 78.96 -8-18-6	35 Br +1 +5 −1 79.904 -8-18-7	36 Kr 0 83.80 -8-18-8	-L-M-N
37 Rb +1 85.47 -18-8-1	38 Sr +2 87.62 -18-8-2	39 Y +3 88.905 -18-9-2	40 Zr +4 91.22 -18-10-2	41 Nb +3 +5 92.906 -18-12-1	42 Mo +6 95.94 -18-13-1	43 Tc +4 +6 +7 (97) -18-13-2	44 Ru +3 101.07 -18-15-1	45 Rh +3 102.905 -18-16-1	46 Pd +2 +4 106.4 -18-18-0	47 Ag +1 107.868 -18-18-1	48 Cd +2 112.40 -18-18-2	49 In +3 114.82 -18-18-3	50 Sn +2 +4 118.69 -18-18-4	51 Sb +3 +5 −3 121.75 -18-18-5	52 Te +4 +6 −2 127.60 -18-18-6	53 I +1 +5 +7 −1 126.9044 -18-18-7	54 Xe 0 131.30 -18-18-8	-M-N-O
55 Cs +1 132.905 -18-8-1	56 Ba +2 137.34 -18-8-2	57* La +3 138.91 -18-9-2	72 Hf +4 178.49 -32-10-2	73 Ta +5 180.948 -32-11-2	74 W +6 183.85 -32-12-2	75 Re +4 +6 +7 186.2 -32-13-2	76 Os +3 +4 190.2 -32-14-2	77 Ir +3 +4 192.2 -32-15-2	78 Pt +2 +4 195.09 -32-16-2	79 Au +1 +3 196.967 -32-18-1	80 Hg +1 +2 200.59 -32-18-2	81 Tl +1 +3 204.37 -32-18-3	82 Pb +2 +4 207.19 -32-18-4	83 Bi +3 +5 208.980 -32-18-5	84 Po +2 +4 (210) -32-18-6	85 At (209) -32-18-7	86 Rn 0 (222) -32-18-8	-N-O-P
87 Fr +1 (223) -18-8-1	88 Ra +2 (226) -18-8-2	89** Ac +3 (227) -18-9-2																-O-P-Q

*Lanthanides	58 Ce +3 +4 140.12 -19-9-2	59 Pr +3 140.907 -20-9-2	60 Nd +3 144.24 -22-8-2	61 Pm +3 (145) -23-8-2	62 Sm +2 +3 150.35 -24-8-2	63 Eu +2 +3 151.96 -25-8-2	64 Gd +3 157.25 -25-9-2	65 Tb +3 158.924 -26-9-2	66 Dy +3 162.50 -28-8-2	67 Ho +3 164.930 -29-8-2	68 Er +3 167.26 -30-8-2	69 Tm +3 168.934 -31-8-2	70 Yb +2 +3 173.04 -32-8-2	71 Lu +3 174.97 -32-9-2	-N-O-P
**Actinides	90 Th +4 232.038 -19-9-2	91 Pa +5 +4 (231) -20-9-2	92 U +3 +4 +5 +6 238.03 -21-9-2	93 Np +3 +4 +5 +6 (237) -22-9-2	94 Pu +3 +4 +5 +6 (244) -23-9-2	95 Am +3 +4 +5 +6 (243) -24-9-2	96 Cm +3 (247) -25-9-2	97 Bk +3 +4 (247) -26-9-2	98 Cf +3 (251) -28-8-2	99 Es (254) -29-8-2	100 Fm (257) -30-8-2	101 Md (256) -31-8-2	102 (254) -32-8-2	103 Lw	-O-P-Q

Figure 2.1-1 The Periodic Table. Numbers in parentheses are mass numbers of most stable isotope of that element.

the substance.

The formula weight is found in exactly the same manner as the molecular weight. The gram formula weight, the mass of one mole of the compound with the stated empirical formula, is often equated to the molecular mass. Caution must, however, be exercised in this practice.

EXAMPLE 2.1-1: The molecular formula for sphalerite is ZnS. Determine the formula weight of this compound.

Solution: The formula weight of sphalerite, m_{ZnS}, equals the atomic mass of zinc, m_{Zn}, plus the atomic mass of sulfur m_S.

$$m_{ZnS} = m_{Zn} + m_S$$

$$= 65.37 \text{ a.m.u.} + 32.064 \text{ a.m.u.}$$

$$= 97.43 \text{ a.m.u.}$$

$$= 97 \text{ a.m.u.}$$

EXAMPLE 2.1-2: Determine the mass of one mole of sphalerite.

Solution: The empirical formula for ZnS contains 1 zinc atom and 1 sulfur atom and 1 mole of ZnS contains 6.023×10^{23} molecules of ZnS or 6.023×10^{23} atoms of zinc and of sulfur. Accordingly, the mass of one mole of ZnS equals the mass of 6.023×10^{23} atoms of Zn plus the mass of 6.023×10^{23} atoms of sulfur.

$$MW_{ZnS} = 6.023 \times 10^{23}\, m_{Zn} + 6.023 \times 10^{23}\, m_S$$

$$= \left(\frac{6.023 \times 10^{23} \text{ Zn atoms}}{\text{mole ZnS}}\right)\left(\frac{65.37 \text{ a.m.u.}}{\text{Zn atom}}\right)\left(\frac{1.6598 \times 10^{-24} \text{g}}{\text{a.m.u.}}\right)$$

$$+ \left(\frac{6.023 \times 10^{23} \text{ S atoms}}{\text{mole ZnS}}\right)\left(\frac{32.064 \text{ a.m.u.}}{\text{S atom}}\right)\left(\frac{1.6598 \times 10^{-24} \text{g}}{\text{a.m.u.}}\right)$$

$$= 97.43 \text{ g/mol ZnS}$$

EXAMPLE 2.1-3: The molecular formula for benzene is C_6H_6. What are the empirical formula and gram formula weight for benzene?

Solution: The empirical formula must show the relationship between the number of C and H atoms in benzene using the smallest possible set of integers. The empirical formula is therefore CH. The gram formula weight (symbol gFw) equals:

$$gFw_{CH} = MW_C + MW_H$$

$$= 12.01115g + 1.00797g$$

$$= \underline{13.02\ g}$$

EXAMPLE 2.1-4: How many gram-atoms are contained in a mole of oxygen gas?

Solution: Oxygen gas is diatomic, i.e. each molecule has two atoms. Therefore,

$$1\ \text{mol}\ O_2 = 1\ \text{mol}\ O_2 \left(\frac{6.023 \times 10^{23}\ O_2\ \text{molecules}}{\text{mol}\ O_2}\right)\left(\frac{2.0\ \text{atoms}}{O_2\ \text{molecule}}\right).$$

$$\frac{1\ \text{g-atom O}}{6.023 \times 10^{23}\ \text{O atoms}}$$

$$= 2\ \text{g-atoms O}$$

2.2 THE CHEMICAL EQUATION

The second aspect of stoichiometry that will be considered deals with the relationship between the masses of reactants and products in chemical reactions. For the majority of processes of importance to the metallurgical engineer, atoms can neither be created nor destroyed. Thus, for any chemical reaction, the number of atoms of any element in the products must equal the number of atoms of that element in the reactants. This principle is a simple statement of the <u>law of conservation of mass</u>. The procedure followed to assure that this law is obeyed is called <u>balancing</u> the chemical equation.

Many chemical equations are balanced by inspection. It is obvious from the molecular formula for water that two hydrogen atoms must combine with one oxygen atom to form one molecule of water.

$$2H + O = H_2O \qquad (2.2\text{-}1)$$

or

$$H_2 + 1/2\ O_2 = H_2O \qquad (2.2\text{-}2)$$

or

$$2H_2 + O_2 = 2H_2O \qquad (2.2\text{-}3)$$

Eq. (2.2-1) shows that there are two atoms of hydrogen and one atom of oxygen present as reactants and products. Eq. (2.2-1) could also have been written as

$$2 \times 6.023 \times 10^{23}\ H + 6.023 \times 10^{23}\ O = 6.023 \times 10^{23}\ H_2O \qquad (2.2\text{-}4)$$

to show that two gram-atoms of hydrogen and one gram-atom of oxygen are present as reactants and products. However, the most common interpretation of Eq. (2.2-1) is that two gram-atoms of hydrogen react with one gram-atom of oxygen to form one mole of H_2O, i.e. the chemical symbols are used to indicate one mole of the substance rather than one atom or molecule. It is also common practice to write balanced chemical equations with the smallest set of integer coefficients that can be used to represent the reaction, and with the elements in their appropriate molecular formula, i.e. O_2 as opposed to O and H_2 instead of H. Thus, Eq. (2.2-3) is the preferred form for this reaction.

In order to avoid confusion and errors particularly in thermochemical calculations, the physical form of the reactants and products should also be included in parentheses after the chemical symbols in a chemical equation. For example, (g) for gas, (ℓ) for liquid, (s) for solid, or a more detailed symbol for a specific solid phase, such as (γ) for γ-iron, the face-centered cubic form of iron.

When a reaction is proceeding in one direction, an arrow ($\rightarrow$) is used to indicate the direction of the reaction. When the rate of reaction in both directions is equal, i.e. the process is at equilibrium, an equal sign (=) or double ar row ($\rightleftarrows$) is commonly used.

EXAMPLE 2.2-1: Write the chemical equation that represents the oxidation of liquid copper sulfide (Cu_2S) by gaseous oxygen to liquid copper and sulfur dioxide.

Solution: The unbalanced equation is

$$Cu_2S(\ell) + O_2(g) \rightarrow Cu(\ell) + SO_2\ (g)$$

Balancing the copper atoms yields

$$Cu_2S(\ell) + O_2(g) \rightarrow 2Cu(\ell) + SO_2\ (g)$$

At this point, sulfur and oxygen atoms are also balanced. Thus, the equation is balanced: the same number of atoms of each element appear on both sides of the equation.

The amount of information present in the balanced chemical equation may be summarized. For the equation

$$Cu_2S(\ell) + O_2(g) = 2Cu(\ell) + SO_2(g)$$

this equation shows that:

1) one mole of Cu_2S combines with one mole of oxygen to form two moles of copper and one mole of SO_2;

2) two moles of copper atoms in Cu_2S form two moles of liquid copper, one mole of sulfur atoms in Cu_2S forms one mole of sulfur dioxide which contains one mole of sulfur atoms, and one mole of oxygen gas (two gram-atoms) is required to form one mole of SO_2 which contains two gram-atoms of oxygen;

3) 159.14 g of Cu_2S combines with 32.00 g of O_2 to form 127.08 g copper and 64.06 g SO_2; and

4) 191.14 g of reactants (Cu_2S and O_2) react to form 191.14 g of products.

EXAMPLE 2.2-2: How many grams of iron can be produced from one mole of iron ore which is 100% hematite (Fe_2O_3)?

Solution: One possible reaction for this process would be

$$Fe_2O_3(s) + 3CO(g) \rightarrow 2Fe(s) + 3CO_2\ (g)$$

However, under any process in which hematite is completely reduced, one mole of Fe_2O_3 produces 2 moles of iron. Therefore, the weight of Fe is

$$2 \text{ mol Fe} = 2 \text{ mol Fe}\left(\frac{55.847 \text{ g Fe}}{\text{mol Fe}}\right)$$

$$= 111.694 \text{ g Fe}$$

$$= 112 \text{ g Fe}$$

EXAMPLE 2.2-3: How many moles of sulfur dioxide are produced when one metric ton of copper matte containing 60% Cu_2S and 40% FeS is converted to liquid copper and iron oxide?

Solution: Since the analysis of liquids is usually on a weight basis, 100 kg of matte will contain 60 kg Cu_2S and 40 kg FeS. Then,

$$1 \text{ ton matte} \equiv \left(\frac{1 \text{ ton matte}}{}\right)\left(\frac{1000 \text{ kg}}{\text{ton}}\right)\left(\frac{60 \text{ kg } Cu_2S}{100 \text{ kg matte}}\right)$$

$$\equiv 600 \text{ kg } Cu_2S$$

$$600 \text{ kg } Cu_2S \equiv \left(\frac{600 \text{ kg } Cu_2S}{}\right)\left(\frac{1 \text{ kg-mole } Cu_2S}{159 \text{ kg } Cu_2S}\right)\left(\frac{1 \text{ kg-atom S}}{\text{kg-mole } Cu_2S}\right)$$

$$\equiv 3.77 \text{ kg-atoms sulfur}$$

Similarly,

$$400 \text{ kg FeS} \equiv \frac{400 \text{ kg FeS}}{}\left(\frac{1 \text{ kg-mole FeS}}{88 \text{ kg FeS}}\right)\left(\frac{1 \text{ kg-atom S}}{\text{kg-mole FeS}}\right)$$

$$\equiv 4.55 \text{ kg-atoms sulfur}$$

Then since SO_2 contains one kg-atom of S per kg-mole of SO_2, the number of kg-moles of SO_2 will equal

$$n_{SO_2} = n_S\left(\frac{1 \text{ kg-mol } SO_2}{1 \text{ kg-atom S}}\right)$$

$$= (3.77 \text{ kg-atom S} + 4.55 \text{ kg-atom S})\left(\frac{1 \text{ kg-mol } SO_2}{1 \text{ kg-atom S}}\right)$$

$$= 8.32 \text{ kg-mol } SO_2$$

$$= 8320 \text{ mol } SO_2$$

2.3 THE IDEAL GAS LAW

Most gases involved in metallurgical processes are at sufficiently high temperature and low pressure to behave in a manner similar to ideal gases. Ideal gases obey the ideal gas law which can be represented by the equation

$$PV = nRT \qquad (2.3\text{-}1)$$

where P, V and T are the pressure, volume and absolute temperature of the gas, respectively. The number of moles of gas is n, and R is the ideal gas constant. Several useful values of the ideal gas constant are given in Table 2.3-1.

Table 2.3-1

Values of the Ideal Gas Constant

8.314	J/mol·K
1.987	cal/mol·K (Thermochemical cal)
1.986	cal/mol·K (International Table cal)
1.986	Btu/lb-mol·°R
0.08205	litre·atm/mol·K
1.545	ft·lb_f/lb-mol·°R
10.73	psi·ft^3/lb-mol·°R
0.7302	atm·ft^3/lb-mol·°R

Rearrangement of Eq. (2.3-1) yields

$$\frac{V}{n} = \frac{RT}{P} \qquad (2.3\text{-}2)$$

or at any given temperature and pressure, the volume per mole or the number of moles in a given volume of any ideal gas is the same. To date, most gas volumes have been reported at 273K and 1 atm (101.3 kPa) pressure. For these conditions, which are called the standard temperature and pressure (symbol S.T.P.),

$$1 \text{ mol gas} = 22.4 \times 10^{-3} \text{ m}^3 \quad (2.3\text{-}3a)$$

$$1 \text{ kg-mol gas} = 22.4 \text{ m}^3 \quad (2.3\text{-}3b)$$

$$1 \text{ lb-mol gas} = 359 \text{ ft}^3 \quad (2.3\text{-}3c)$$

The deviation with pressure from ideal behavior is shown in Table 2.3-2. For most engineering calculations, the error is clearly small and may be neglected. However, at high pressure and for gases with complex molecules, the error may be serious and should be accounted for.

Table 2.3-2

Corrections to Ideal-Gas Behavior at 273K

Gas	Pressure, atm			
Air	-0.06%	-0.57%	-2.27%	-3.11%
N_2	-0.05	-0.43	-1.57	-1.57
O_2	-0.09	-0.96	-4.40	-7.75
H_2	+0.06	+0.63	+3.15	+6.45
CO	-0.05	-0.45	-2.25	-2.80
CO_2	-0.33		-89.57	-79.93

To obtain actual volume, correct the volume calculated from Eq. (2.3-1) by percentages tabulated above. (Ref. R. Schuhman, Metallurgical Engineering, Addison-Wesley Pub. Co., 1952)

EXAMPLE 2.3-1: The average velocity of oxygen in a 1 inch diameter pipe is 50 ft/sec. How many moles per hour are passing through the pipe if the gas is at 273K and 100 atm (10.13 MPa)?

Solution: Assuming that oxygen is ideal at these conditions, Eq. (2.3-2) can be used to find the volume per mole of the gas. However, an alternative procedure would be to rearrange Eq. (2.3-2)

$$\frac{PV}{nT} = R = \text{constant}$$

or

$$\frac{P_1V_1}{T_1} = \frac{P_2V_2}{T_2}$$

In this example, $T_1 = T_2$, $P_1 = 1$ atm and $V_1 = 359$ ft^3 for 1 lb-mole of oxygen. For $P_2 = 100$ atm,

$$V_2 = \frac{P_1 \cdot V_1}{P_2}$$

$$= \frac{1 \text{ atm} \cdot (359 \text{ ft}^3/\text{lb-mol } O_2)}{100 \text{ atm}}$$

$$= 3.59 \text{ ft}^3/\text{lb-mol } O_2$$

Since the volumetric flow rate, $\dot{V}$, equals the linear velocity, $\bar{v}$, times the cross-sectional area, A_o,

$$\dot{V} = \bar{v} \cdot A$$

where

$$A = \pi \cdot d^2/4$$

or

$$\dot{V} = \bar{v} \cdot \pi \cdot d^2/4$$

$$= (50\frac{\text{ft}}{\text{s}}) \cdot \pi \cdot (1 \text{ in}(\frac{1 \text{ ft}}{12 \text{ in}}))^2/4$$

$$= 0.273 \frac{\text{ft}^3}{\text{s}} \cdot (\frac{3600 \text{ s}}{\text{h}}) \cdot (\frac{1 \text{ lb-mol } O_2}{3.59 \text{ ft}^3})$$

$$= 1.24 \times 10^5 \text{ lb-mol } O_2/\text{h}$$

From Table 2.3-2, it can be seen that at 100 atm the actual volume of oxygen is 7.75% lower than that predicted by the ideal gas law. Thus, the volume per lb-mole of O_2 at 100 atm is 3.31 ft^3(=0.9225 x 3.59 ft^3) and the flow rate is actually 1.35×10^5 lb-moles O_2/h.

In gas mixtures made up of ideal gases, each component of the mixture obeys the ideal gas law. Thus, for the i-th component of an ideal mixture contained in a volume V and at a temperature T,

$$P_i V = n_i \ RT \qquad (2.3\text{-}4)$$

where P_i and n_i are the partial pressure and number of moles of component i, respectively.

The partial pressure of component i is the part of the total pressure exerted by the mixture that is exerted by species i. Since ideal gases also obey Dalton's Law, the total pressure, P_t, for an n component mixture will equal

$$P_t = \sum_{i=1}^{n} P_i \tag{2.3-5}$$

Then, if

$$P_1 V = n_1 RT \tag{2.3-6a}$$

$$P_2 V = n_2 RT \tag{2.3-6b}$$

$$P_i V = n_1 RT \tag{2.3-6c}$$

$$P_n V = n_n RT \tag{2.3-6d}$$

summation of Eqs. (2.3-6a) through (2.3-6d) yields

$$V\left(\sum_{i=1}^{n} P_i\right) = \left(\sum_{i=1}^{n} n_i\right) RT \tag{2.3-7}$$

or

$$VP_t = n_t RT \tag{2.3-8}$$

where n_t is the total number of moles of gas in the mixture.

The division of Eq. (2.3-6c) by Eq. (2.3-8) yields

$$\frac{P_i}{P_t} = \frac{n_i}{n_t} \tag{2.3-9}$$

But n_i/n_t is the mole fraction of species i, X_i, in the mixture. Thus the partial pressure of i equals

$$P_i = X_i P_t \tag{2.3-10}$$

Finally, the volumes of ideal gases at constant temperature and pressure are additive, or

$$P\ V_i = n_i\ RT \tag{2.3-11}$$

Using reasoning analogous to the previous development,

$$\frac{V_i}{V_t} = \frac{n_i}{n_t} = X_i \qquad (2.3\text{-}12)$$

where V_i/V_t is the volume fraction of component i, and

$$\text{vol }\%i = 100\,\frac{V_i}{V_t} = 100\,X_i = \%\,i \qquad (2.3\text{-}13)$$

Therefore, <u>for ideal gases, volume percentage equals mole percentage</u>.

EXAMPLE 2.3-2: For most engineering calculations, air is assumed to be 21% oxygen and 79% nitrogen on a molar basis. Calculate the error associated with assuming air has the same analysis on a volume basis at a) 273K and 1 atm, and b) 273K and 100 atm.

Solution: In one hundred moles of air there will be 21 moles of oxygen and 79 moles of nitrogen. At 273K and 1 atm and if O_2 and N_2 are assumed ideal, the volume per mole of gas is given in Eq. (2.3-3a) as 22.4×10^{-3} m^3. Then using the appropriate correction factor from Table 2.3-2 yields,

$$100 \text{ mol Air} = 100 \text{ mol Air}\left(\frac{21 \text{ mol } O_2}{100 \text{ mol Air}}\right)\left(\frac{22.4 \times 10^{-3} m^3 O_2}{\text{mol } O_2}\right)(0.9991)$$

$$= 0.47 \text{ m}^3 \; O_2$$

Also,

$$100 \text{ mol Air} = 100 \text{ mol Air}\left(\frac{79 \text{ mol } N_2}{100 \text{ mol Air}}\right)\left(\frac{22.4 \times 10^{-3} \text{ m}^3 \; N_2}{\text{mol } N_2}\right)(0.9995)$$

$$= 1.77 \text{ m}^3 \; N_2$$

Then, the volume percent O_2 will equal

$$\%\,O_2 = 100\%\cdot V_{O_2}/(V_{O_2} + V_{N_2})$$

$$= \frac{100\%\cdot 0.47 \text{ m}^3}{(0.47 \text{ m}^3 + 1.77 \text{ m}^3)}$$

$$= 20.98\%\; O_2$$

$$= 21\%\; O_2 \text{ on a volume basis}$$

and

$$\% N_2 = 100 - \% O_2$$

$$= 79\% \; N_2 \text{ on a volume basis}$$

Using a similar approach for air at 100 atm, and the appropriate correction factors from Table 2.3-2;

$$V_{O_2} = 4.3 \times 10^{-3} \; m^3$$

$$V_{N_2} = 1.72 \times 10^{-2} \; m^3$$

$$\% O_2 = 20\% \text{ on a volume basis}$$

and

$$\% N_2 = 80\% \text{ on a volume basis}$$

Therefore, even at 100 atm the error associated with assuming that the analysis on a volume and molar basis for air are equivalent is small.

EXAMPLE 2.3-3: Calculate the partial pressures of CO_2 and H_2O in the gas produced by the complete combustion of methane (CH_4) with oxygen at 1000K and 1 atm.

Solution: The chemical equation for the reaction is:

$$CH_4 \text{ (g)} + 2O_2\text{(g)} \rightarrow CO_2 \text{ (g)} + 2H_2O \text{ (g)}$$

By inspection of the equation it is seen that two moles of O_2 are required to completely burn one mole of CH_4, and that the products will contain two moles of H_2O for every mole of CO_2. Thus if y moles of CH_4 are burnt, y moles CO_2 and 2y moles H_2O will form. In the product gas,

$$X_{CO_2} = \frac{n_{CO_2}}{n_{CO_2} + n_{H_2O}}$$

$$= \frac{y}{y + 2y}$$

$$= 0.33$$

$$X_{H_2O} = \frac{n_{H_2O}}{n_{H_2O} + n_{CO_2}}$$

$$= 0.67$$

$$P_{CO_2} = X_{CO_2} P_t$$

$$= 0.33 \cdot 1.0 \text{ atm}$$

$$= 0.33 \text{ atm}$$

and

$$P_{H_2O} = 0.67 \text{ atm}$$

Referring to Eq. (2.3-12), we see that the volume fraction of CO_2 is

$$\frac{V_{CO_2}}{V_t} = X_{CO_2} = 0.33$$

and

$$\frac{V_{H_2O}}{V_t} = X_{H_2O} = 0.67$$

Thus, assuming ideal gases, the partial pressure and volume fraction of each component are the same.

2.4 EXCESS AND LIMITING REACTANTS

When the path of a reaction is described by a chemical equation, the exact or stoichiometric amount of reactants that are required to produce a pre-determined quantity of product is found from the stoichiometry of the reaction, (see Example 2.3-3). In many processes, an excess of one or more reactants is supplied to "force" the complete reaction of one of the other reactants. Under these circumstances, the reactant present in the smallest stoichiometric amount is called the limiting reactant. The reactants supplied in excess of the stoichiometric amount required for complete reaction of the limiting reactant are called excess reactants.

The percent excess for any excess reactant is defined on a molar basis and equals

$$\% \text{ excess} = \frac{\text{(moles in excess)}}{\text{(moles required for complete reaction)}} \times 100\% \qquad (2.4\text{-}1)$$

The moles required for complete reaction must be re-emphasized at this point. The % excess is based upon the excess supplied over the theoretically required amount, even if the reaction does not go to completion.

By convention the composition of gases is given in terms of volume percent while the composition of solids is given in terms of weight percent. For example, when coal has the following composition:

77.6% C	4.5% O
5.3% H	2.8% S
1.5% N	8.3% Ash

5.3% H is not read as 5.3% of atomic hydrogen. It is read as "5.3% hydrogen". H, N, and O are used here as abbreviations for the words hydrogen, nitrogen and oxygen. Since coal is a solid, the analysis is given in weight percent so that 5.3%H means one has 5.3 grams of hydrogen in 100 grams of coal.

EXAMPLE 2.4-1: Calculate the theoretical air, i.e. the exact amount of air, required to completely burn 1 kg of coke. The coke analysis is 89% C, 1% H and 10% ash.

Solution: The theoretical air will contain the exact number of moles of oxygen necessary to burn carbon to carbon dioxide and hydrogen to water according to the reactions

$$C + O_2 = CO_2$$

and

$$H_2 + 1/2\ O_2 = H_2O$$

It is therefore necessary to determine the molecular composition of the coke.

$$1 \text{ kg coke} = 1 \text{ kg coke} \left(\frac{0.89 \text{ kg C}}{\text{kg coke}}\right)\left(\frac{1 \text{ kg-atom C}}{12 \text{ kg C}}\right)$$

$$= 7.4 \times 10^{-2} \text{ kg-atom C}$$

Also,

$$1 \text{ kg coke} = 1 \text{ kg coke} \left(\frac{0.01 \text{ kg H}}{\text{kg coke}}\right)\left(\frac{1 \text{ kg-mol } H_2}{2 \text{ kg H}}\right)$$

$$= 5.0 \times 10^{-3} \text{ kg-mol } H_2$$

According to the chemical reactions, the theoretical oxygen will equal

$$n_{O_2} = n_C \left(\frac{1 \text{ kg-mol } O_2}{\text{kg-atom C}}\right) + n_{H_2} \left(\frac{0.5 \text{ kg-mol } O_2}{\text{kg·mol } H_2}\right)$$

$$= 7.4 \times 10^{-2} \text{ kg-atom C} \left(\frac{1 \text{ kg-mol } O_2}{\text{kg-atom C}}\right) +$$

$$5.0 \times 10^{-3} \text{ kg-mol } H_2 \cdot \left(\frac{0.5 \text{ kg-mol } O_2}{\text{kg-atom } H_2}\right)$$

$$= 7.7 \times 10^{-2} \text{ kg-mol } O_2$$

Finally, since air is 21% oxygen, the theoretical air will equal

$$n_{air} = n_{O_2} \left(\frac{100 \text{ kg-mol air}}{21 \text{ kg-mol } O_2}\right)$$

$$= 7.7 \times 10^{-2} \text{ kg-mol } O_2 \left(\frac{100 \text{ kg-mol air}}{21 \text{ kg-mol } O_2}\right)$$

$$= 0.37 \text{ kg-mol air}$$

0.37 kg-mol or 370 mol of air are theoretically required to burn 1 kg of coke with the stated analysis.

EXAMPLE 2.4-2: If 10% excess air was supplied in the previous example, how many moles of air would be supplied per kg of coke?

Solution: Rearrangement of Eq. (2.4-1) yields

$$\text{moles in excess} = \frac{\text{\% excess}}{100} \cdot (\text{moles theoretically required})$$

10% excess air will then require the moles of excess air to be

$$\text{moles in excess} = (0.1)(\text{moles theoretically required})$$

$$= (0.1)(370 \text{ mol air/kg coke})$$

$$= 37 \text{ mol air/kg coke}$$

The total amount of air supplied must then be

$$\text{total air supplied} = (\text{moles in excess}) + (\text{moles theoretically required})$$

$$= 37 \text{ mol air/kg coke} + 370 \text{ moles air/kg coke}$$

$$= 407 \text{ mol air/kg coke}$$

EXAMPLE 2.4-3: If 407 mol of air are supplied to coke with the same analysis as in Example 2.4-1, and the carbon only burns to carbon monoxide (CO), what percentage excess air was supplied?

Solution: Since in combustion terminology the "percent excess" is based upon the theoretical requirement for complete reaction, % excess still equals 10%.

EXAMPLE 2.4-4: What is the Orsat Analysis of the gas produced by the incomplete combustion of methane with 10% excess oxygen?

Solution: The Orsat Analysis is the analysis of combustion products given by an Orsat apparatus. This analysis is given on a dry basis, i.e. all of the moisture is excluded, and may also be referred to as the flue gas analysis or gas analysis on a dry basis, (see Chapter 1).

Assuming that incomplete combustion in this example means all of the methane burns to carbon monoxide and water, it is seen from the chemical reaction

$$CH_4(g) + 1.5\ O_2(g) = CO(g) + 2H_2O(g)$$

that 1.5 moles of oxygen are stoichiometrically required for every mole of methane burned to CO and H_2O. The theoretical requirement for complete combustion is 2.0 moles of oxygen (see Example 2.3-3) and if 10% excess is supplied, 2.2 moles of oxygen must be supplied per mole of methane.

After combustion, 2.2-1.5, or 0.7 moles of oxygen will remain. One mole of carbon monoxide and two moles of water will also be present. However, the Orsat analysis will be for the oxygen and carbon monoxide only. The Orsat analysis will thus equal

$$\%\ O_2 = 100\% \cdot \frac{n_{O_2}}{n_{O_2} + n_{CO}}$$

$$= 100\% \cdot \frac{0.7}{0.7 + 1.0}$$

$$= 41\%$$

and

$$\%\ CO = 100\% - \%\ O_2$$

$$= 59\%$$

Also encountered in combustion calculations is the term stack gas. The stack gas contains all of the gases resulting from a combustion process, including moisture. (See Chapter 1 for a discussion of gas analysis on a wet basis.

Several terms are often used to describe incomplete reactions. There are no standard definitions for these terms or even standard usages. In order that a trend may be started which may lead to standard definitions, the definitions as stated in a widely used chemical engineering text on energy and material balances[1] will be used in this text.

"Conversion is the fraction of some material in the feed that is converted into products. What the basis (see Chapter 4) in the feed is and into what products the basis is being converted must be clearly specified or endless confusion results. Conversion is somewhat related to the degree of completion of a reaction, which is usually the percentage or fraction of the limiting reactant converted into products."

"Selectivity expresses the amount of a desired product as a fraction or percent of the theoretically possible amount from the feed material converted. Often the quantity defined here as selectivity is called efficiency, conversion efficiency, specificity, yield, ultimate yield or recycle yield."

"Yield, for a single reactant and product, is the weight or moles of final product divided by the weight or moles of initial reactant (P lb of product A over R lb of reactant B). This is sometimes referred to as the recovery. If more than one product and more than one reactant are involved, the reactant upon which the yield is to be based must be clearly stated."

The correct usage of the above terms is illustrated by the following example which has been adapted from the same text.

EXAMPLE 2.4-5: Antimony can be produced from its sulfide by reduction with iron according to the reaction

$$Sb_2S_3(\ell) + 3Fe(s) \rightarrow 2Sb(\ell) + 3FeS(\ell)$$

Determine a) the limiting reactant, b) the percentage of excess reactant, c) the degree or fraction of completion, d) the percent conversion, 3) the percent selectivity and f) the yield for a process in which 0.600 kg of Sb_2S_3 is mixed with 0.250 kg of iron to form 0.200 kg of Sb metal.

Solution: a) According to the above reaction, three moles or 168 grams of iron must combine with one mole or 340 grams of Sb_2S_3. Thus, if 0.600 kg of Sb_2S_3 are available to react with iron, the stoichiometric mass of iron, M_{Fe}, will equal

$$M_{Fe} = M_{Sb_2O_3}\left(\frac{168 \text{ g Fe}}{340 \text{ g } Sb_2S_3}\right)$$

---

(1) D. M. Himmelblau, Basic Principles and Calculations in Chemical Engineering, Prentice Hall Inc., Englewood Cliffs, N.J. 1974.

$$= 600 \text{ g } Sb_2S_3 \left(\frac{168 \text{ g Fe}}{340 \text{ g } Sb_2S_3}\right)$$

$$= 296 \text{ g Fe}$$

Since there are only 250 grams of iron present, there will not be enough iron for complete reaction and iron is the limiting reactant.

b) Complete reaction of the 250 grams or 4.46 moles of iron will require 4.46/3 or 1.48 moles of Sb_2S_3. However, 600 grams or 1.76 moles of Sb_2S_3 have been supplied. The % excess can then be found from Eq. (2.4-1),

$$\% \text{ excess} = \frac{(\text{moles in excess})}{(\text{moles theoretically required})} \times 100$$

$$= \frac{(1.76 - 1.48)}{1.48} \times 100$$

$$= 18.9\%$$

c) Although iron is the limiting reactant, not all of the iron reacts when only 200 grams or 1.64 moles of antimony are formed. From the stoichiometry of the reaction, it can be seen that the number of moles of iron which react equals

$$n_{Fe} = n_{Sb} \left(\frac{3 \text{ mol Fe}}{2 \text{ mol Sb}}\right)$$

$$= 1.64 \text{ mol Sb} \left(\frac{3 \text{ mol Fe}}{2 \text{ mol Sb}}\right)$$

$$= 2.46 \text{ mol Fe}$$

If the fractional degree of completion is based on the amount of iron converted to FeS,

$$\text{Fraction of completion} = \frac{\text{moles Fe converted}}{\text{moles Fe supplied}}$$

$$= \frac{2.46}{4.48}$$

$$= 0.55$$

d) The percentage conversion can be based upon the amount of Sb_2S_3 converted to Sb. As shown earlier, only 1.64 moles of Sb form, even though 3.54 moles of Sb are contained in the 1.77 moles of Sb_2S_3. Therefore,

$$\% \text{ Conversion} = \frac{\text{moles Sb produced}}{\text{moles Sb supplied}} \times 100$$

$$= \frac{1.64}{3.54} \times 100$$

$$= 46.3\%$$

e) It may be assumed that the selectivity of the conversion of Sb_2S_3 is based on the theoretical amount that can be converted, 1.49 mole:

$$\text{selectivity} = \frac{0.82}{1.49}(100) = 55\%$$

f) The yield will be stated as kilograms of Sb formed per kilogram of Sb_2S_3 that was fed to the reaction:

$$\text{yield} = \frac{0.200 \text{ kg Sb}}{0.600 \text{ kg } Sb_2S_3} = \frac{1}{3}\frac{\text{kg Sb}}{\text{kg } Sb_2S_3}$$

2.5 OXIDATION - REDUCTION REACTIONS

Many metallurgical processes contain reactions in which the valence (charge) of the atoms, groups of atoms or ions involved in the reactions changes. These chemical reactions are called oxidation-reduction or redox reactions. They must be balanced in the same manner as ordinary chemical equations. However, balancing is often more complex, as both conservation of mass and conservation of electrons or charge must be maintained. Several definitions and sets of rules for balancing redox reactions follow.

Oxidation is defined as a chemical change in which electrons are lost by an atom, group of atoms, or ions, resulting in a more positive (or less negative) valence of at least one reacting constituent. Reduction is a chemical change in which electrons are added to an atom or group of atoms, resulting in a more negative (or less positive) valence of the atom or at least one atom in the group. Oxidation and reduction always occur simultaneously and the number of electrons gained by the atom being reduced must equal the number of electrons being given up by the atoms being oxidized, i.e. electrons are conserved. The most common valences of the elements are shown on the periodic table, Figure 2.1-1.

EXAMPLE 2.5-1: Copper cementation is a process in which copper present in an aqueous solution as Cu^{2+} ions is precipitated as solid copper by replacement of the copper in the solution by iron. Write a balanced equation for this process, assuming that iron in solution has a charge of +2. Also, determine which element is being oxidized and which element is being reduced.

Solution: The reaction for copper in solution going to solid copper can be written as

$$Cu^{2+} \text{ (in aq. sol.)} \rightarrow Cu \text{ (s)}$$

This reaction is balanced for copper atoms but not for charge. Since the solid copper contains two more electrons (symbol e^-) than the copper in solution, the completely balanced equation must be

$$Cu^{2+} \text{ (in aq. sol.)} + 2e^- \rightarrow Cu \text{ (s)}$$

On the other hand, the reaction for the iron can be written as

$$Fe \text{ (s)} \rightarrow Fe^{2+} \text{ (in aq. sol.)} + 2e^-$$

Adding both equations together and cancelling the $2e^-$ which appears on both sides of the sum yields the chemical equation for the process

$$Cu^{2+} \text{ (in aq. sol.)} + Fe \text{ (s)} \rightarrow Cu \text{ (s)} + Fe^{2+} \text{ (in aq. sol.)}$$

From the partial reactions, it is seen that copper is gaining electrons, i.e. its charge is becoming less positive. Therefore, copper is being reduced. Iron is giving up electrons, i.e. its charge is becoming more positive. Therefore, iron is becoming oxidized.

To confuse matters, the element or compound that contains the element that is being reduced is often called the oxidizing agent. The element or compound that contains the element that is being oxidized is often called a reducing agent. In the previous example, iron is the reducing agent and copper the oxidizing agent.

One set of rules for balancing oxidation-reduction reactions follows. There are, however, many methods for balancing these equations and the reader is referred to any chemistry text for additional information.

To balance oxidation-reduction equations:(2)

1) Write an equation that includes those reactants and products that contain the elements undergoing a change in valence.

2) Determine the change in valence which some element in the oxidizing agent undergoes. The number of electrons gained is equal to this change times the number of atoms undergoing the change.

3) Determine the same for some element in the reducing agent.

4) Multiply each principal formula by such numbers as to make the total number of electrons lost by the reducing agent equal to

---

(2) D. Schaum, Theory and Problems of College Chemistry, Schaum Pub. Co., New York, NY, 1958.

the number of electrons gained by the oxidizing agent.

5) By inspection supply the proper coefficients for the rest of the equation.

6) Check the final equation by counting the number of atoms of each element on both sides of the equation.

EXAMPLE 2.5-2: A possible process for removing some of the sulfur found in coal involves the oxidation of sulfur present in the coal as pyrite (symbol FeS_2) with water and oxygen gas to form water-soluble $FeSO_4$ and H_2SO_4. Balance the process equation using the method outlined above.

Solution: Step 1. The unbalanced process equation is

$$FeS_2(s) + H_2O\ (\ell) + O_2(g) \rightarrow Fe^{2+}\ (\text{in aq. sol.}) + H_3O^+\ (\text{in aq. sol.}) + SO_4^{2-}\ (\text{in aq. sol.})$$

Step 2. Oxygen, in the oxidizing agent oxygen gas, goes from zero charge to -2 in the SO_4^{2-} ions. Thus, four electrons must be gained per mole of oxygen gas.

Step 3. Sulfur is oxidized from a charge of -1 to +6. Consequently, fourteen electrons must be given up by sulfur per mole of FeS_2.

Step 4. In order to equate the number of electrons given up by sulfur in FeS_2 with the number gained by oxygen, there must be 14/4 or 3.5 moles of oxygen gas reacting for each mole of FeS_2 reacting. Thus, the partially balanced process equation is

$$FeS_2(s) + H_2O\ (\ell) + 3.5\ O_2(g) \rightarrow Fe^{2+}\ (\text{in aq. sol.}) + H_3O^+\ (\text{in aq. sol.}) + SO_4^{2-}\ (\text{in aq. sol.})$$

Step 5. Balancing H atoms and S atoms yields the balanced process equation:

$$FeS_2(s) + H_2O(\ell) + 3.5\ O_2(g) \rightarrow Fe^{2+}\ (\text{in aq. sol.}) + 2H_3O^+\ (\text{in aq. sol.}) + 2SO_4^{2-}\ (\text{in aq. sol.})$$

Step 6. There are 1 Fe, 2S, 2H and 8 O atoms on both sides of the balanced equation.

An alternative approach to the balancing of redox reactions involves the gram-equivalent, as discussed in section 1.5. The gram-equivalent or gram equivalent weight of an oxidizing or reducint agent for a particular reaction is equal to the formula weight of the agent divided by the absolute value of the change in valence of the agent during the reaction:

$$\text{gram-equivalent} = \frac{\text{Formula weight}}{\text{change in valence}} \qquad (2.5\text{-}1)$$

The gram-equivalent is thus determined so that equal numbers of gram-equivalents of oxidizing and reducing agents combine exactly with one another during a balanced redox reaction.

EXAMPLE 2.5-3: How many gram-equivalents are contained in one mole of iron when ferric ions (symbol Fe^{3+}) are used as an oxidizing agent according to the reaction

$$Fe^{3+} + e^- \rightarrow Fe^{2+}$$

Solution: According to Eq. (2.5-1),

$$\text{gram-equivalent} = \frac{\text{Formula weight}}{\text{change in valence}}$$

$$= \frac{56\text{g}}{(+2) - (+3)}$$

$$= \frac{56\text{g}}{1}$$

$$= 56\text{g}$$

56 grams of iron equals one mole of iron, or one mole of iron equals one gram-equivalent of iron, <u>for this reaction</u>.

EXAMPLE 2.5-4: Balance the equation

$$K_2Cr_2O_7 + H_2S + H_2SO_4 \rightarrow Cr_2(SO_4)_3 + K_2SO_4 + S + H_2O$$

by balancing the gram-equivalents of the oxidizing agent and the reducing agent.

Solution: In $K_2Cr_2O_7$, Cr has a charge of +6. In $Cr_2(SO_4)_3$, Cr has a charge of +3. Thus, one gram-equivalent of $K_2Cr_2O_7$ will equal

$$1 \text{ gram-equivalent } K_2Cr_2O_7 = \frac{MW_{K_2Cr_2O_7}}{2(+3) - 2(+6)}$$

$$= \frac{294\text{g}}{6}$$

$$= 49\text{g}$$

Sulfur in H_2S has a charge of -2, while as S it has a charge of 0;

$$1 \text{ gram-equivalent of S} = \frac{MW_{H_2S}}{0-(-2)} = \frac{34g}{2} = 17g$$

Finally, one mole, or six gram-equivalents, of $K_2Cr_2O_7$ must combine with six gram-equivalents of H_2S, 102g or 3 moles of H_2S. Therefore, $K_2Cr_2O_7 + 3H_2S + H_2SO_4 \rightarrow Cr_2(SO_4)_3 + K_2SO_4 + S + H_2O$

To finish balancing this equation, SO_4^{2-} groups, H, S, and O atoms are balanced to yield

$$K_2Cr_2O_7 + 3H_2S + 4H_2SO_4 \rightarrow Cr_2(SO_4)_3 + K_2SO_4 + 3S + 7H_2O$$

2.6 SUMMARY

The masses of atoms are measured on the Atomic Mass Scale with Carbon-12 being taken as the reference point. Carbon-12 has been assigned a mass of 12.0000 a.m.u., and the masses of the other atoms are fixed relative to this value. The mass of a single molecule of a compound equals the sum of the masses of the atoms which make up the molecule. The formula weight is equal to the sum of the masses of the atoms in the empirical formula which represents the compound. The relationship between the masses of elements in a compound and the total mass of a compound are thus clearly defined.

The chemical equation is used to represent the relationship between the number of moles (or the masses) of the reactants and the products that participate in a chemical reaction. For a given reaction, the number of moles (or the mass) of any element in the reactants must equal the number of moles (mass) of that element in the product, i.e. mass is neither created nor destroyed. Chemical reactions that involved the change of valence of atoms in some of the reacting species are called oxidation-reduction reactions. Conservation of mass and charge (electrons) must be obeyed to have a balanced equation. The balanced chemical reaction specifies the theoretical or stoichiometric amount of products that can be obtained from a given amount of reactants.

Most gases in metallurgical processes can be considered to be ideal. For ideal gases, one mole of gas occupies $22.4 \times 10^{-3}m^3$ at 273K and 101.3 kPa (1 atm). Ideal gases also obey the ideal gas law and the partial pressures or volumes of the components in an ideal gas law and the partial pressures or volumes of the components in an ideal gas mixture can be added to get the total pressure or volume of the mixture. For ideal gases, the volume percentage exactly equals the molecular percentage.

FURTHER READING

A. General:

1. D. Himmelblau, Basic Principles and Calculations in Chemical Engineering, Prentice-Hall, Englewood Cliff, NJ, 1962.

2. E. J. Henley and H. Bieber, Chemical Engineering Calculations, McGraw-Hill Book Company, New York, 1959.

3. O. A. Hougen, K. M. Watson, and R. A. Ragatz, Chemical Process Principles, Part 1, 2nd ed., John Wiley and Sons, New York, 1956.

4. E. T. Williams, and R. C. Johnson, Stoichiometry for Chemical Engineers, McGraw-Hill Book Company, New York, 1958.

B. Combustion:

1. Combustion Handbook, North American Mfg. Co., Cleveland, Ohio, 1965.

2. Combustion Engineering, Combustion Engineering, Inc., New York, NY, 1966.

3. Steam, Babcock and Wilcox Co., New York, NY, 1955.

4. The Efficient Use of Fuel, Ministry of Technology, Her Majesty's Stationery Office, London, 1958.

5. W. K. Lewis, A. H. Radasch and H. C. Lewis, Industrial Stoichiometry, 2nd ed., McGraw-Hill, New York, 1954.

6. W. Gumz, Gas Producers and Blast Furnaces, Wiley, New York, 1958.

7. C. Davies, Calculations in Furnace Technology, Pergamon Press, London, 1970.

Exercises

2.1 A natural gas contains 60.7% CH_4, 29.03% H_2, 7.92% C_2H_6, 0.98% C_2H_4, 0.78% O_2, 0.58% CO

Find (a) Volume of air necessary to completely burn it
(b) Volume of products of combustion

both relative to one volume of gas and at the same temperature and pressure.

2.2 An annealing furnace uses a fuel oil containing 16.4% H and 83.6% C. It is proposed to fire this furnace with 25% excess air. Calculate the flue gas analysis, assuming complete combustion. Repeat the calculation assuming that 5% of the total carbon is burnt to CO only.

2.3 If pure H_2 is burned completely with 32% excess air, what is the Orsat analysis of the product gas?

2.4 A pyrite ore is reduced with hydrogen. The ore contains 10% of solid inerts (gangue). Twenty percent excess H_2 is used, and the cinder (solid residue) remaining contains 5% FeS_2 by weight.

$$FeS + 2H_2 \rightarrow Fe + 2H_2S$$

On the basis of 100 lb of ore charged, calculate the volume of furnace gases at 400°C and 1 atm. (ft^3).

2.5 A natural gas has the following composition by volume:

$$CH_4 = 94.1\%,\ C_2H_6 = 3.0\%,\ N_2 = 2.9\%$$

At a temperature of 80°F and at a pressure of 50 psia, calculate:

(a) the partial pressure of the nitrogen (psia)

(b) the volume of the nitrogen component per 100 cu. ft. of gas (ft^3)

(c) the density of the gas in kg/m^3 at the existing conditions.

You may assume that the ideal gas law is applicable.

2.6 A zinc retort is charged with 70 kg of roasted zinc concentrates containing 45% Zn, present as ZnO. Reduction takes place according to the reaction

$$ZnO + C = Zn + CO$$

One-fifth of the ZnO remains unreduced. The zinc vapor and CO pass into a condenser, from which the CO escapes and burns to CO_2 as it emerges from the mouth of the condenser. The CO enters the condenser at 300°C and 700mm Hg pressure. Calculate:

(a) The volume of CO in cubic meters entering the condenser, measured at (1) standard conditions, and (2) at actual conditions

(b) The weight of CO, in kilograms

(c) The volume of CO_2 formed when the CO burns, at 750°C and 765mm Hg.

2.7 An anthracite coal contains 89% Carbon, 3% Hydrogen, 1% Oxygen and 7% Ash. It is burned using 15 per cent more air than is theoretically needed for its perfect combustion. The ashes weigh 10 kg per 100 kg of coal burned.

Calculate:

(1) The volume of air (assumed dry and at standard conditions) used per kg of coal burned.

(2) The number of cubic feet of air per pound of coal.

(3) The percentage composition (by volume, of course) of the gas, assuming it contains no soot or unburned gas.

(4) The percentage composition of the same, if first dried and then analyzed.

(5) The number of grams of moisture carried per cubic meter of dried gas measured.

(6) The number of grains per cubic foot.

(7) The volume of the chimney gases, at standard conditions, (water assumed uncondensed) per kg of coal burned.

(8) The number of cubic feet of products per pound of coal.

(9) The volume of the products in cubic meters at 350°C and 700mm Hg pressure.

(10) The volume of the products in cubic feet at 600°F and 29 in of water pressure.

2.8 A bituminous coal contains:

Carbon	73.60 w/o	Oxygen	10.00
Hydrogen	7.30	Moisture	0.60
Nitrogen	1.70	Ash	8.05
Sulphur	0.75		100.00

It is powdered and blown into a cement kiln by a blast of air.

Calculate:

(1) The volume of dry air, at 80°F, and 29 inches barometric pressure theoretically required for the perfect combustion of one pound of the coal.

(2) The volume of the products of combustion, using no excess of air, at 550°F and 29 inches barometric pressure and their percentage composition.

2.9 A limestone is analyzed as $CaCO_3$ (93.12 wt%); $MgCO_3$ (5.38 wt%); and insoluble matter (1.50 wt%).

(a) How many kilograms of calcium oxide could be obtained from 5 tons of the limestone?

(b) How many kilograms of carbon dioxide are given off per pound of this limestone?

2.10 Pyrolusite (MnO_2) is dissolved in hydrochloric acid by the reaction

$$MnO_2 + 4HCl \rightarrow MnCl_2 + 2H_2O + Cl_2$$

The chlorine was passed into potassium iodide solution where it liberated iodine:

$$Cl_2 + 2KI \rightarrow 2KCl + I_2$$

The iodine liberate was estimated by adding sodium thiosulfate; the reaction being

$$I_2 + 2Na_2S_2O_3 \rightarrow 2NaI + Na_2S_4O_6$$

If 5.6 grams of crystallized sodium thiosulfate, $Na_2S_2O_3 \cdot 5H_2O$, were used up, how many grams of manganese were present?

2.11 How many moles of Fe_2O_3 would be formed by the action of oxygen on one kilogram of iron?

2.12 Wire silver weighing 3.48 grams was dissolved in nitric acid. What weight of silver nitrate was formed?

2.13 Iron pyrites, FeS_2, is burned in air to obtain SO_2 according to the equation:

$$4FeS + 11O_2 \rightarrow 2Fe_2O_3 + 8SO_2$$

How many liters of SO_2, measured at 300°C and 740mm of Hg, could be obtained from 100 liters of oxygen at standard conditions?

2.14 Balance the equation:

$$Cu + H_2SO_4 \rightarrow CuSO_4 + SO_2 + H_2O$$

(a) How many grams of copper sulfate would be formed for each mole of copper reacting?

(b) How many grams of Cu would be required to produce 100g of copper sulfate?

(c) How many moles of SO_2 would be found for each mole of acid reacting?

(d) How many grams of water would be formed for each mole of Cu sulfate formed?

CHAPTER 3

SAMPLING AND MEASUREMENTS

3.0 INTRODUCTION

The preparation of material and energy balances is done for a variety of reasons and at various stages in the life of a process or plant. At the design stage, various assumptions concerning efficiency of reaction, heat loss rates, etc. are made and theoretical balances computed for purposes of establishing process flow rates, temperatures, efficiencies, equipment sizes and so on. The resulting balances are exact. All equations are satisfied precisely. Sensitivity studies (how sensitive the conclusions are to the various assumptions made including assumed values of variables) may be made at the same time.

On the other hand, once a plant or process is built and operating, it is usually desirable to determine its performance for either control or accounting purposes by making material and/or energy balances on the real system. This involves measuring flow rates, chemical compositions, temperatures and the like and then determining the balances. In this situation, however the results rely on samples from heterogeneous materials, indirect measurements or calculations of flow rates, temperature measurements under extreme environmental situations, and often difficult, tedious chemical analyses, all of which have some degree of error associated with them. Therefore, the construction of actual balances often involves considerable uncertainty in the results. When it comes to trace elements, for instance, this uncertainty can be very large, rendering meaningful conclusions almost impossible to obtain.

In this Chapter the range of accuracy and sensitivity that can reasonably be expected from various measurement techniques utilized in the determination of actual metallurgical material and energy balances is examined. No attempt will be made to explain in detail how each instrument or technique works but references to such descriptions are provided.

3.1 THE IMPORTANCE OF ERRORS AND ERROR DESCRIPTION

Errors in measurements interact with each other and propagate through the development of a material or energy balance in various ways. The first problem is to define the error in a single measurement or value used.

The error of a measurement is the difference between the observed value and the true value of the dimension or quantity of interest. If the error is small compared to the magnitude of the measured quantity, the measurement is said to be accurate. For example, if the error in chemical analysis of Ni in stainless steel is ±0.05%, but the magnitude is 9.0%, the analysis is reasonably accurate. On the other hand, suppose the error in the chemical analysis of H_2 in steel is ±0.2 ppm when the magnitude is on the order of 2.0 ppm. This analysis is definitely less accurate.

The true value of a measured quantity is generally unknown. Hence, the mean, $\bar{x}$, of a series of determinations is used to describe the true value, where

$$\bar{x} = \frac{\sum_{i=1}^{i=n} x_i}{n} \quad (3.1\text{-}1)$$

The differences between observed values, x_i, and the mean of the observed values, $\bar{x}$, are referred to as residuals, d_i,

$$d_i = x_i - \bar{x} \quad (3.1\text{-}2)$$

If the residuals are small compared to the magnitude of the measured quantity, meaning that all the values are near $\bar{x}$, the measurement is said to be <u>precise</u>. (The <u>precision</u> of a measurement method or instrument is often reported by manufacturers). However, $\bar{x}$ may still not be the true value, and so <u>precise measurement is not necessarily accurate measurement</u>, as shown in Fig. 3.1-1.

Small residuals occur more often than large ones, and usually the error distribution may be adequately represented by the Gaussian or normal distribution function as in Fig. 3.1-2.

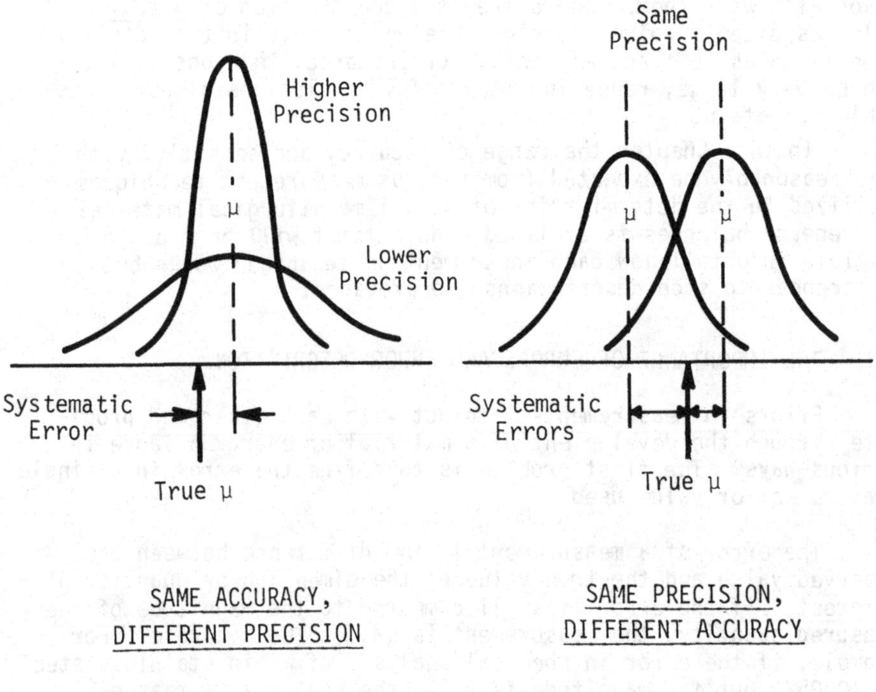

Figure 3.1-1 Comparison of precision and accuracy. μ is the mean.

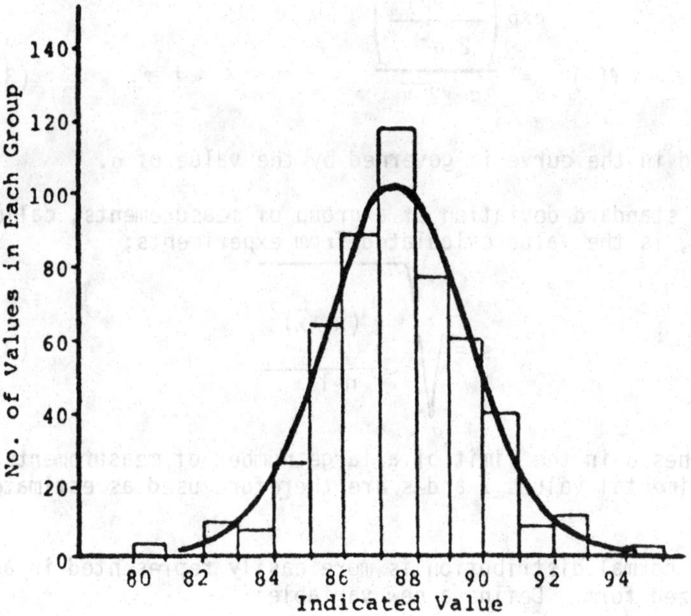

Figure 3.1-2 Typical Gaussian distribution of values.

If the true mean (or mode) is denoted by μ (which is a theoretical value approached by $\bar{x}$ when a large number of samples are taken and which is approximated by $\bar{x}$), the normal distribution is represented by

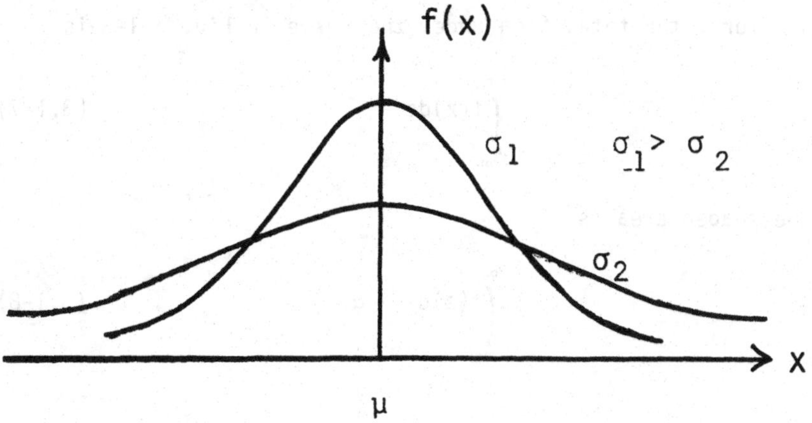

where

$$f(x) = \frac{\exp\left(\frac{-(x-\mu)^2}{2\sigma^2}\right)}{\sigma\sqrt{2\pi}}, \quad -\infty < x < +\infty \tag{3.1-3}$$

The spread in the curve is governed by the value of σ.

The standard deviation of a group of measurements, called a sample, s, is the value calculated from experiments;

$$s = \sqrt{\frac{\sum_{1}^{n} (x_i - \bar{x})^2}{n-1}} \tag{3.1-4}$$

s approaches σ in the limit of a large number of measurements. The experimental values $\bar{x}$ and s are therefore used as estimates of μ and σ.

The normal distribution is more easily represented in a standardized form. Define a new variable:

$$z = \left(\frac{x-\mu}{\sigma}\right) \simeq \left(\frac{x-\bar{x}}{s}\right) \tag{3.1-5}$$

Eq. (3.1-3) is then transformed into

$$f(z) = \frac{e^{-\left(\frac{z^2}{2}\right)}}{\sigma\sqrt{2\pi}} \tag{3.1-6}$$

With this definition, if x is a random variable distributed normally with mean μ and standard deviation σ, z is distributed normally, with its mean equal to zero and σ^2 equal to one.

In this form, the total area under the curve in Fig. 3.1-3 is

$$\int_{-\infty}^{+\infty} f(z)dz = 1 \tag{3.1-7}$$

and the shaded area is

$$\int_{z_\alpha}^{\infty} f(z)dz = \alpha \tag{3.1-8}$$

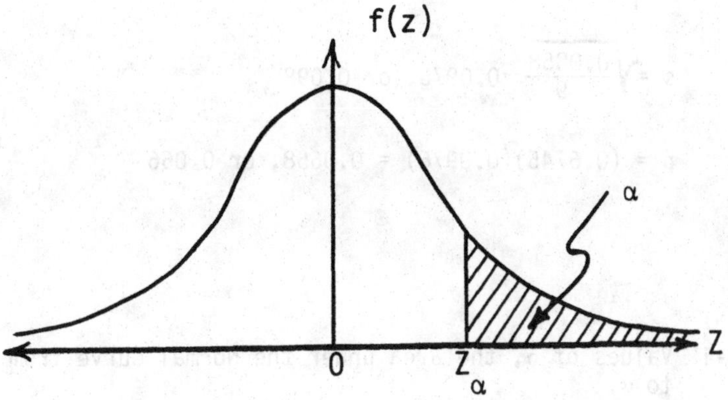

Figure 3.1-3 Normal distribution curve

where z_α is the value of z corresponding to some fraction, α, of the total area outside of the given value of z. Table 3.1-1 gives values of z_α which correspond to various values of α.

Note that when z = 0, α = 0.5, or half the area under the curve, and when z = 1 (corresponding to $(x - \mu) = \sigma$), α = 0.1587. If we want to know what fraction of the area is <u>outside</u> of both the limits $x - \sigma$ and $x + \sigma$, i.e., plus or minus <u>one standard deviation</u>, it is 2(0.1587) or 0.3174. Since the total area is 1.000, the area under the curve <u>within the limits</u> $\pm\sigma$ (or $\pm s$) is 0.6826, or, 68.3% of all x values lie within $\pm\sigma$ ($\approx\pm s$).

Sometimes reference is made to the <u>probable error</u>, p. This corresponds to the limits about $\bar{x}$ such that 50% of the population of x values are within those limits, corresponding to the situation where α = 0.25. Referring to Table 3.1-1, when α = 0.25, z lies between 0.6 and 0.7, more specifically between 0.67 and 0.68. By interpolation, z_α = 0.6745. Since

$$z = \frac{x-\bar{x}}{s} = 0.6745, \qquad (3.1\text{-}9)$$

$$p = \pm (x - \bar{x}) = 0.6745\ s \qquad (3.1\text{-}10)$$

EXAMPLE 3.1-2: Analysis of samples of an ore give the following values: 0.81, 0.72, 0.87, 0.61, 0.83, 0.94, 0.91, 0.77, 0.84, and 0.88. What are the mean, standard deviation and probable error of these data?

Solution:

$$\bar{x} = \frac{\sum_{1}^{10} x_i}{10} = \frac{8.18}{10} = 0.818 \text{ or } 0.82$$

$$\sum_{1}^{10} (x_i - 0.82)^2 = 0.0858$$

$$s = \sqrt{\frac{0.0858}{9}} = 0.0976, \text{ or } 0.098$$

$$p = (0.6745)(0.0976) = 0.0658, \text{ or } 0.066$$

Table 3.1-1 Values of α, the area under the Normal Curve from Z_α to ∞.

Z_α	.00	.01	.02	.03	.04	.05	.06	.07	.08	.09
0.0	.5000	.4960	.4920	.4880	.4840	.4801	.4761	.4721	.4681	.4641
0.1	.4602	.4562	.4522	.4483	.4443	.4404	.4364	.4325	.4286	.4247
0.2	.4207	.4168	.4129	.4090	.4052	.4013	.3974	.3936	.3897	.3859
0.3	.3821	.3783	.3745	.3707	.3669	.3632	.3594	.3557	.3520	.3483
0.4	.3446	.3409	.3372	.3336	.3300	.3264	.3228	.3192	.3156	.3121
0.5	.3085	.3050	.3015	.2981	.2946	.2912	.2877	.2843	.2810	.2776
0.6	.2743	.2709	.2676	.2643	.2611	.2578	.2546	.2514	.2483	.2451
0.7	.2420	.2389	.2358	.2327	.2296	.2266	.2236	.2206	.2177	.2148
0.8	.2119	.2090	.2061	.2033	.2005	.1977	.1949	.1922	.1894	.1867
0.9	.1841	.1814	.1788	.1762	.1736	.1711	.1685	.1660	.1635	.1611
1.0	.1587	.1562	.1539	.1515	.1492	.1469	.1446	.1423	.1401	.1379
1.1	.1357	.1335	.1314	.1292	.1271	.1251	.1230	.1210	.1190	.1170
1.2	.1151	.1131	.1112	.1093	.1075	.1056	.1038	.1020	.1003	.0985
1.3	.0968	.0951	.0934	.0918	.0901	.0885	.0869	.0853	.0838	.0823
1.4	.0808	.0793	.0778	.0764	.0749	.0735	.0721	.0708	.0694	.0681
1.5	.0668	.0655	.0643	.0630	.0618	.0606	.0594	.0582	.0571	.0559
1.6	.0548	.0537	.0526	.0516	.0505	.0495	.0485	.0475	.0465	.0455
1.7	.0446	.0436	.0427	.0418	.0409	.0401	.0392	.0384	.0375	.0367
1.8	.0359	.0351	.0344	.0336	.0329	.0322	.0314	.0307	.0301	.0294
1.9	.0287	.0281	.0274	.0268	.0262	.0256	.0250	.0244	.0239	.0233
2.0	.0228	.0222	.0217	.0212	.0207	.0202	.0197	.0192	.0188	0183
2.1	.0179	.0174	.0170	.0166	.0162	.0158	.0154	.0150	.0146	.0143
2.2	.0139	.0136	.0132	.0129	.0125	.0122	.0119	.0116	.0113	.0110
2.3	.0107	.0104	.0102	.00990	.00964	.00939	.00914	.00889	.00866	.00842
2.4	.00820	.00798	.00776	.00755	.00734	.00714	.00695	.00676	.00657	.00639
2.5	.00621	.00604	.00587	.00570	.00554	.00539	.00523	.00508	.00494	.00480
2.6	.00466	.00453	.00440	.00427	.00415	.00402	.00391	.00379	.00368	.00357
2.7	.00347	.00336	.00326	.00317	.00307	.00298	.00289	.00280	.00272	.00264
2.8	.00256	.00248	.00240	.00233	.00226	.00219	.00212	.00205	.00199	.00193
2.9	.00187	.00181	.00175	.00169	.00164	.00159	.00154	.00149	.00144	.00139

Z_α	.0	.1	.2	.3	.4	.5	.6	.7	.8	.9
3	.00135	$.0^{3}968$	$.0^{3}687$	$.0^{3}483$	$.0^{3}337$	$.0^{3}233$	$.0^{3}159$	$.0^{3}108$	$.0^{4}723$	$.0^{4}481$
4	$.0^{4}317$	$.0^{4}207$	$.0^{4}133$	$.0^{5}854$	$.0^{5}541$	$.0^{5}340$	$.0^{5}211$	$.0^{5}130$	$.0^{6}793$	$.0^{6}479$
5	$.0^{6}287$	$.0^{6}170$	$.0^{7}996$	$.0^{7}579$	$.0^{7}333$	$.0^{7}190$	$.0^{7}107$	$.0^{8}599$	$.0^{8}332$	$.0^{8}182$
6	$.0^{9}987$	$.0^{9}530$	$.0^{9}282$	$.0^{9}149$	$.0^{10}777$	$.0^{10}402$	$.0^{10}206$	$.0^{10}104$	$.0^{11}523$	$.0^{11}260$

3.2 PROPAGATION OF EXPERIMENTAL ERRORS

A final result in the determination of material and energy balances is always obtained by combining the results of different kinds of measurements, such as weights, chemical analyses, etc. The accuracy of any final result is influenced by the accuracy of the measurements of the several quantities involved. If it happens that one of the quantities involved is subject to a much greater error than the others, it will have the preponderant effect in determining the accuracy of the final result. If, however, the relative errors in the various measured quantities are of the same order of magnitude, the errors introduced by all the measured quantities must be considered. In trying to improve the accuracy of a given determination it is important to emphasize improvement of the least accurate measurement.

3.2.1 PROPAGATION OF MAXIMUM ERRORS

A simple and useful method for calculating the error in a final result is to calculate the maximum error which would be obtained if the errors in all the measured quantities had their maximum values and were in such directions that all affected the final result in the same direction. It is unlikely that the errors will combine in this way, as there is usually some compensation of errors, but it is useful to know the maximum value an error could have in an unfavorable case. When the errors are small (say a few per cent), the following methods based on differential calculus are convenient.

Addition and Subtraction. If a final result u is the sum of measured quantities x and y

$$u = x + y \tag{3.2-1}$$

$$du = dx + dy \tag{3.2-2}$$

and in terms of finite increments,

$$\Delta u = \Delta x + \Delta y \tag{3.2-3}$$

thus, if the error in measuring x is Δx and the error in measuring y is Δy, the maximum error in u is $\Delta x + \Delta y$.

EXAMPLE 3.2-1: A blend of raw materials is being made by mixing coke breeze and iron ore. The coke feeder weighs the feed rate to within $\pm 2\%$ and is reported to be feeding at 7 kg/min and the ore feeder is said to be accurate to within $\pm 3\%$ and is running at 150 kg/min. What is the mass flow rate of the resulting blend?

Solution:

$$\text{The error in coke weight} = \pm\,(0.02)(7.)$$

$$= 0.14 \text{ kg/min}$$

The error in ore weight $= \pm (0.03)(150.)$

$= \pm 4.5$ kg/min

Therefore, since the

$$\text{mass of blend} = \text{mass of ore} + \text{mass of coke}$$
$$= 150. + 7$$
$$= 157.\ \text{kg/min}$$

and

$$\Delta(\text{blend}) = \Delta(\text{ore}) + \Delta(\text{coke})$$
$$= \pm 0.14 \pm 4.5$$
$$= \pm 4.64$$

the mass flow of the blend is 157. $\pm$ 4.6 kg/min.

Since the maximum error in a case where $u = x - y$ will occur when Δy is of opposite sign, $\Delta u = \Delta x - (-\Delta y)$ also applies in this situation, so that stated in words, the rule is: The maximum error in a sum or difference is equal to the sum of the absolute values of the maximum errors in the measured quantities being added or subtracted.

Multiplication and Division. If a final result u is the product of two measured quantities x and y

$$u = xy \qquad (3.2\text{-}4)$$

$$du = xdy + ydx \qquad (3.2\text{-}5)$$

$$\frac{du}{u} = \frac{dy}{y} + \frac{dx}{x} \qquad (3.2\text{-}6)$$

If the errors in x and y are small, the error Δu in u may be calculated from Eq. (3.2-7), which also applies if $u = x/y$.

$$\frac{\Delta u}{u} = \frac{\Delta x}{x} + \frac{\Delta y}{y} \qquad (3.2\text{-}7)$$

Furthermore, in the case of calculations involving multiplication and division, it is convenient to talk about errors in terms of percentage errors, and Eq. (3.2-7) may be written as

$$\frac{\Delta u}{u} 100 = \frac{\Delta x}{x} 100 + \frac{\Delta y}{y} 100 \qquad (3.2\text{-}8)$$

Stated in words, the rule is: The maximum percentage error in a product or quotient is equal to the sum of the percentage errors in the measured quantities. This rule is accurate only when the percentage errors are small.

EXAMPLE 3.2-2: The volume flow rate of CO_2 in a gas stream is equal to the product of total gas flow times the percent CO_2 in the gas:

$$V_{CO_2} = V_{GAS} \cdot \frac{(\%CO_2)}{100}$$

If the flow of gas, V_{Gas}, is 1400 $\pm$ 70 m^3/min, and the analysis of the gas is 9.1 $\pm$ 1.8%, what is the best estimate of V_{CO_2}?

Solution: $$V_{CO_2} = (1400)(.091) = 127.4 \ m^3/min$$

But, $$\frac{\Delta(V_{CO_2})}{127.4} = \frac{70}{1400} + \frac{1.8}{9.1}$$

$$= 0.248$$

Therefore, $$\Delta(V_{CO_2}) = (0.248)(127.4) = 31.6 \ m^3/min$$

and, the value of V_{CO_2} is 127.4 $\pm$ 31.6 for an error of 24.8%.

This same result would have been obtained if the percentage errors in each measurement had been added:

$$\% \text{ error in total flow} = \frac{70}{1400} \times 100 = 5.0\%$$

$$\% \text{ error in analysis} = \frac{1.8}{9.1} \times 100 = \underline{19.8\%}$$

$$\% \text{ error in } CO_2 \text{ flow} = 24.8\%$$

3.2.2 PROPAGATION OF PROBABLE ERRORS

The probable error in a final result may be calculated if the probable errors in the various measured quantities are known. Such a calculation is a little more complicated than the calculation of the maximum error. If u is a function of the independent variables, x, y, z ..., the probable error p in u is

$$p = \pm\sqrt{\left(\frac{\partial u}{\partial x}\right)^2 \cdot p_x^2 + \left(\frac{\partial u}{\partial y}\right)^2 \cdot p_y^2 + \left(\frac{\partial u}{\partial z}\right)^2 \cdot p_z^2 + \cdots} \quad (3.2\text{-}9)$$

where p_x is the probable error in x, etc.

EXAMPLE 3.2-3: The volume of a hemispherical slag pot is given by $V = (2/3)\pi r^3$ where r is the radius of the hemisphere. If the average of several determinations of the radius is 1.21 m and the probable error in the determination of the radius is $\pm$ 0.01 m, calculate the probable error in the volume.

Solution:

$$p = \pm\sqrt{\left[\frac{\partial\left(\frac{2\pi r^3}{3}\right)}{\partial r}\right]^2 p_r^2}$$

$$= \pm\sqrt{\left(2\pi r^2\right)^2 p_r^2}$$

$$p = \pm\, 2\pi r^2 p_r = \pm\, 2\pi(1.21)^2\, 0.01 = \pm\, 0.09\text{m}^3$$

so that the calculated volume should be written 3.71 $\pm$ 0.09m^3, or 3.7 $\pm$ 0.1m^3.

Eq. (3.2-9) may be used to show that the probable error of the sum or difference of two quantities A and B, respectively, affected with probable errors $\pm\, p_a$ and $\pm\, p_b$ is

$$p = \pm\sqrt{p_a^2 + p_b^2} \qquad (3.2\text{-}10)$$

In the case of a multiplication, the probable error of the product of two quantities A and B is

$$p = \pm\sqrt{(Ap_b)^2 + (Bp_a)^2} \qquad (3.2\text{-}11)$$

The probable error of the quotient B/A of two quantities A and B is

$$p = \pm\frac{1}{A}\sqrt{p_b^2 + \frac{B^2 p_a^2}{A^2}} \qquad (3.2\text{-}12)$$

The effect of errors on large material and energy balances is not so easily determined because of the complexity of interactions of the errors. The final errors in such balances often can only be found by an analysis of the sensitivity of the results to the values used for individual parameters in the equations.

3.2.3 SIGNIFICANT FIGURES

The number of figures in any value is never exact, for the measurement by comparison with a standard unit is only as accurate as the measuring device. For example, when measuring the length of an object with a ruler which has divisions of 1mm, it is possible to measure the length to within 1mm correctly and to estimate the length to within 0.1mm. Thus, if the measurement of length was 136.1 mm, all of the numbers in the measurement would have significance. If the length was reported as 136.1352 mm, the last three digits would not have any significance. These numbers cannot be measured on this ruler with any accuracy, and consequently, must be a guess. Therefore, the digits which can be measured with certainty and the first (and first only) doubtful digit are called the significant figures of a number. The larger the number of significant figures, the more accurate is the measurement.

Zeroes are the source of some problems with respect to significant figures. Zeroes that only precede a number are not part of the significant figures. Examples would be 0.13, 0.013, etc. Zeroes that only follow a number may have significance in two cases. If they are contained in a decimal portion of a number, such as 3.70; the implication is that the number has significance to the level of the zero. In this case the first doubtful digit (and therefore the last one in the significant figure) is at the 0.01 level. Had the number been written 3.700, the implication is that there are four significant figures, that the first zero is a definite value, and that the second zero is the first doubtful one. Therefore it is clear that care must be taken not to "tack on" extra zeroes when they are not significant.

The other condition under which zeroes that follow a number may be part of the significant figure is when they precede a decimal point. Unfortunately there is often no way to tell whether they are simply setting the decimal point or whether they are significant. For example, 9100 is the same as 9100. . In the latter case the implication is that there are four significant figures, but it is still ambiguous. The ambiguity is always removed by the use of exponential notation. Writing the number 9.1×10^3 clearly indicates that there are two significant figures. Writing it as 9.100×10^3 clearly signifies four significant figures.

EXAMPLE 3.2-3: Determine the number of significant figures in a) 0.091, b) 0.910, c) 9.1, d) 910, e) 910., f) 910.005, g) 910.0.

Solution:

a) Two. The zero before the numbers is not significant.

b) Three. The zero before the decimal is not significant, but the zero following the number is.

c) Two.

d) Ambiguous. If written 9.1×10^2, the answer would be two. If 9.10×10^2 the answer would be three.

e) Three. The zero preceeds the decimal point.

f) Six. The zeroes are all significant since they either preceed a decimal point or are part of the decimal portion of the number.

g) Four. The zeroes again either precede a decimal or are a part of the decimal part of the number.

When carrying out arithmetic calculations, it is best to retain one digit beyond the least significant figure in each number and carry out the calculations using that digit in order to ensure that the least significant figure in the final answer is not altered. This figure is obtained by rounding only the final answer. The rules for rounding off are simple: starting with the digit on the far right, if greater than or equal to five, remove the digit and increase the digit to its immediate left by one. If less than five, remove the digit.

EXAMPLE 3.3-4: Round off 0.09135346 to three significant figures.

Solution: Step 1) The digit on the far right (7) is greater than 5. Increase the digit to the left by one and drop the 6. The number is now 0.0913535.

Step 2) The digit at the far right is now 5. Increase the next digit by one and drop the 5. The number is now 0.091354.

Step 3) The far right digit is now less than 5, so drop it. Now the number is 0.09135.

Step 4) Since the right-hand digit is now 5, increase the next digit by one and drop the 5. The number is thus 0.0914 and this number has three significant figures, since the zero only precedes the number.

EXAMPLE 3.3-5: Evaluate the expression

(43. + 390. - 100.1)/18. + (36. x 3.0)/3.0

Solution: Add one level to the significant numbers in the expression. (43.0 + 390.0 - 100.10)/18.0 + (36.0 x 3.00)/3.00

Evaluate step-by-step.

Step 1: 43.0 + 390.0 = 433.0

Step 2: 433.0 - 100.10 = 329.90

Step 3: 329.90/18.0 = 18.3277

Step 4: 36.0 x 3.0 = 108.0

Step 5: 108.0/3.00 = 36.00

Step 6: 18.3277 + 36.00 = 54.3277

Round final answer to least number of significant figures in original numbers, which is two. The answer is 54.

EXAMPLE 3.3-6: Average three numbers; 23.05, 23.07 and 23.07.

Solution: The average is 69.19/3 = 23.06333.

If the number is to be used in further calculations, it should be rounded to one more than the least number of significant figures in the initial set, or five figures, i.e., 23.063. If not, it should be rounded to 23.06. Note that the 3 in the denominator is really 3.0000000 ...., because one is dividing by exactly 3, the number of numbers in the original set. Thus, the least number of significant figures is in the numbers themselves.

3.3 SAMPLING METHODS AND PROCEDURES

Samples of solid, liquid, gas and two-phase process streams must be obtained in order to make real material balances. The precision of a sampling result clearly depends on the number of samples taken. Its accuracy also depends on the sampling procedure. If the phase or material being sampled is homogeneous (well-mixed), a single or very few samples may be adequate to achieve an accurate result. On the other hand, even a liquid phase such as a slag may not be well-mixed, thus, requiring more extensive sampling to improve the accuracy of their reported analyses.

3.3.1 SAMPLING GRANULAR SOLIDS

Granular solids and lump ores present the largest problems in sampling and require the use of carefully designed sampling plans or procedures. The size of sample required depends on (1) the particle size of the material and (2) on the variability of the characteristic of interest from one particle to another. Because this variability may also depend on (3) the microstructure of the material, eq., whether valuable mineral grains are uniformly distributed, finely disseminated or coarsely disseminated within a given particle, theoretical prediction of the required sample size is essentially impossible. Therefore, what has been done over the years is to develop procedures and methods based on experience for various materials.

The earliest published recommended sample weights were based on tables which utilized the relation

$$W = kD^a \tag{3.3-1}$$

where W is the sample weight required, D is the maximum (top) particle size in the sample, k is a constant depending on the material, and a is a constant, usually taken as 2, but increased to 3 when increased safety factor in the result is desired. The problem with this approach is that there is no way to predict k, which depends on the particular ore or granular solid. However, if the weight of a sample deemed adequate for a given material at one top size is known, then the sample weight of the same material, necessary at a different top size, can be estimated.

EXAMPLE 3.3-1: A copper ore is known to give a satisfactory 100 gm sample when the ore is crushed and ground to a top size of 1 mm. What size sample would be required if it was desired to sample the ore at the output from a crusher set at 12 mm opening?

Solution: Since W (= 100 gm) corresponding to a size D (= 1 mm), is known, if it is assumed that a = 2, k can be calculated.

$$k = \frac{100}{1^2} = 100 \text{ gm/mm}^2$$

Then, for the proposed top size of 12 mm,

$$W = (100)(12)^2 \text{ gm}$$

$$= 14.4 \text{ kg}$$

If a large pile, or a stream, is being sampled in order to obtain a composite sample for subsequent analysis, the probable error or the standard deviation about the mean analysis for an n cut composite sample, P_n or σ_n, is the same as the probable error for a single cut, σ or p, divided by $\sqrt{n}$. Thus, the number of individual cuts or samples needed to make a composite can be calculated, and depends on the degree of precision desired.

$$n = \left(\frac{\text{probable error based on single cut}}{\text{probable error from n-cut composite}}\right)^2 = \left(\frac{\sigma}{\sigma_n}\right)^2 \qquad (3.3\text{-}2)$$

EXAMPLE 3.3-2: If it is desired that 95% of all composite samples of coal analyzed for ash have ash values within 1% of the true mean, μ, of % ash in the coal, how many random cuts must be composited to achieve this result? Assume that the variance of the analysis of ash in coal, $\sigma^2_{\%ash}$, is known to be equal to 9.

Solution: From the attributes of a normal distribution (Section 3.1), Table 3.1-1 yields $z = \frac{x-\mu}{\sigma_n} = 1.96$ when 95% of the curve is included ($z_\alpha = 0.0250$). Since $(x-\mu)$ is desired to be $\pm$ 1%, $\sigma_n = \frac{1}{1.96} = 0.51$. Then, Eq. (3.3-2) shows that the required number is

$$n = \frac{\sigma^2}{\sigma_n^2} = \frac{9}{(0.51)^2} \simeq 35$$

Thus 35 random cuts must be composited to make the composite with the desired attribute, namely that 95% of the time this composite will have an ash analysis within 1% of the true value of the %ash.

An example of a specific sampling procedure developed for a specific material is found in ASTM Standards D492 and E105 dealing with the sampling of coal from lots of 1000 tons or less, such that 95% of the composite samples will be within 10% of the true average composition. Table 3.3-1 gives the minimum number of cuts required to make up the composite, the minimum weight of each cut, and the minimum gross or composite sample weight required in order to meet this statistical goal. In this case the samples are to be taken by "a single motion of the sampling instrument in such a way that the time for each increment is the same. The increments are collected in a random fashion such that every part of the lot or pile has a non-zero probability of selection and the entire lot is represented proportionally in the gross sample".

The sampling instrument to be used is not specified. This is because there are many such devices on the market. The choice has more to do with the physical situation and the size of sample to be taken. The only real criterion is that each cut be truly representative of the pile or stream at that point or that instant in time. Fig. 3.3-1 illustrates some of the devices available.

Once the composite sample has been collected, it must obviously be reduced to a size such that a chemical analysis can be made. This is done by making use of Eq. (3.3-1), which shows that as the particle size is reduced the weight of sample required decreases by the square of the particle size. Therefore, the entire sample is crushed and/or ground down to a smaller size and then split (i.e., itself sampled) into a smaller weight. This splitting may be accomplished by hand means (coning and quartering, shovel sampling, riffle sampling, grab sampling) or by use of mechanical splitting devices. For example, in the previous case of coal sampling, ASTM-D492 continues: "The gross sample is crushed and screened so that 95% by weight will pass through a No. 4 sieve and 100% will pass a 3/8" round hole screen. The sample is reduced to 60 lb by passing through a riffle sampler. Further crushing, screening and riffling results in a sample of not less than 1-3/4 lb sized so that 95% or more will pass a No. 8 screen".

A riffle splitter is illustrated in Fig. 3.3-2 and the procedure for coning and quartering in Fig. 3.3-3. Further procedures for sampling of specific particulate materials may be found in the ASTM standards, such as B215-60 (Metal Powder Sampling) and D197-30, (Pulverized Coal Sampling).

3.3.2 SAMPLING LIQUIDS

Ordinary liquids are sampled by various dipping procedures involving use of 1/2-gallon sampling jars. Various Federal agencies (such as the Federal Water Quality Administration) have established procedures for sampling from streams and lakes. For example, for rivers, samples are taken

(1) upstream and downstream 1/4-mile, 1/2-mile, and 3/4-mile from a given source,

(2) at both shores and in the middle, in each case, and

(3) at the surface, 1/3-, 1/2-, and 2/3- of the way to the bottom, and at the bottom, at all locations.

The composite is the chemical laboratory sample which is sampled via pipette for analysis.

Table 3.3-1

Sample Sizes Required for Sampling Coal (From ASTM Standard D492)

	Particle Size of Coal Being Sampled							
	5/8" and Under	5/8" to 1" Inclusive	1" to 2" Inclusive	Over 2" to 6" Inclusive	Over 2" to 4" Inclusive	Over 4" to 6" Inclusive	Over 6" and bottom size 5/8" or over	Over 6"
Under 8% Ash								
Min. No. of Cuts	15	15	15	15	15	15	35	35
Min. Wt. of each Cut, lbs.	2	4	6	10	10	15	30	20
Min. Gross Sample Wt., lbs.	30	60	90	150	150	225	1050	700
8 - 9.9% Ash								
Min. No. of Cuts	20	20	20	20	20	20	35	35
Min. Wt. of each Cut, lbs.	2	4	6	10	10	15	30	20
Min. Gross Sample Wt., lbs	40	80	120	200	200	300	1050	700

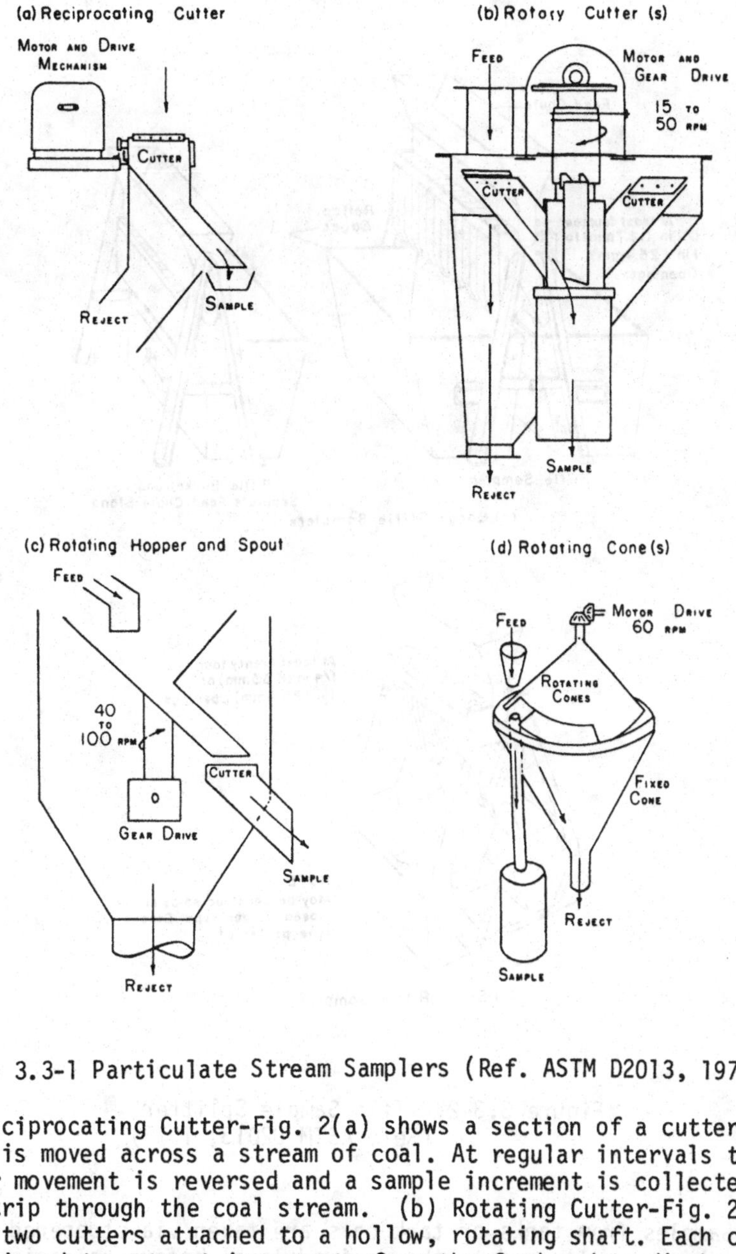

Figure 3.3-1 Particulate Stream Samplers (Ref. ASTM D2013, 1977).

(a) Reciprocating Cutter-Fig. 2(a) shows a section of a cutter which is moved across a stream of coal. At regular intervals the cutter movement is reversed and a sample increment is collected on each trip through the coal stream. (b) Rotating Cutter-Fig. 2(b) shows two cutters attached to a hollow, rotating shaft. Each cutter is designed to extract increments from the feed and to discharge these into the hollow shaft. One or more cutters may be used. (c) Rotating Hopper and Spout-Fig. 2(c) shows the totaling hopper that receives the crushed sample and discharges it through a spout over one or more stationary cutters. (d) Rotating Cone-A sampler developed by the British National Coal Board. Two slotted cones are locked together and rotated on a vertical shaft so that on each revolution the common slot operating intercepts the falling stream of coal and collects an increment.

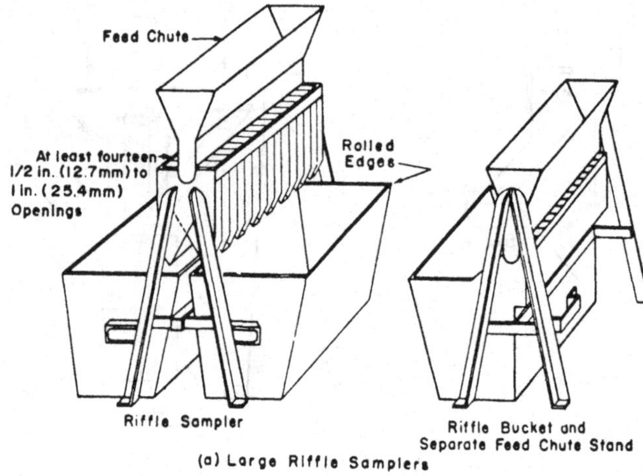

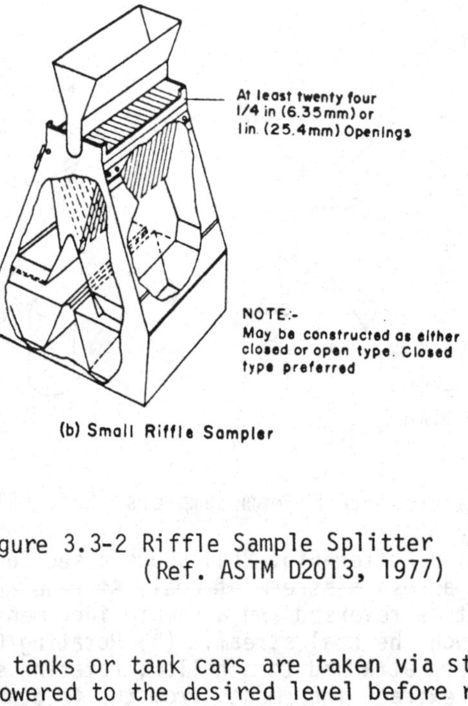

Figure 3.3-2 Riffle Sample Splitter
(Ref. ASTM D2013, 1977)

Samples from tanks or tank cars are taken via stoppered bottles which are lowered to the desired level before releasing the stopper.

Liquid metal baths are sampled in a number of different ways. Typically, sampling spoons are plunged into the bath and withdrawn with a sample of liquid metal which is then poured into a metal chill mold or other device such as one that records the thermal arrest at the liquidus during freezing, from which composition may be inferred. Other sampling devices have been developed which can be plunged deep into the metal before protective seals

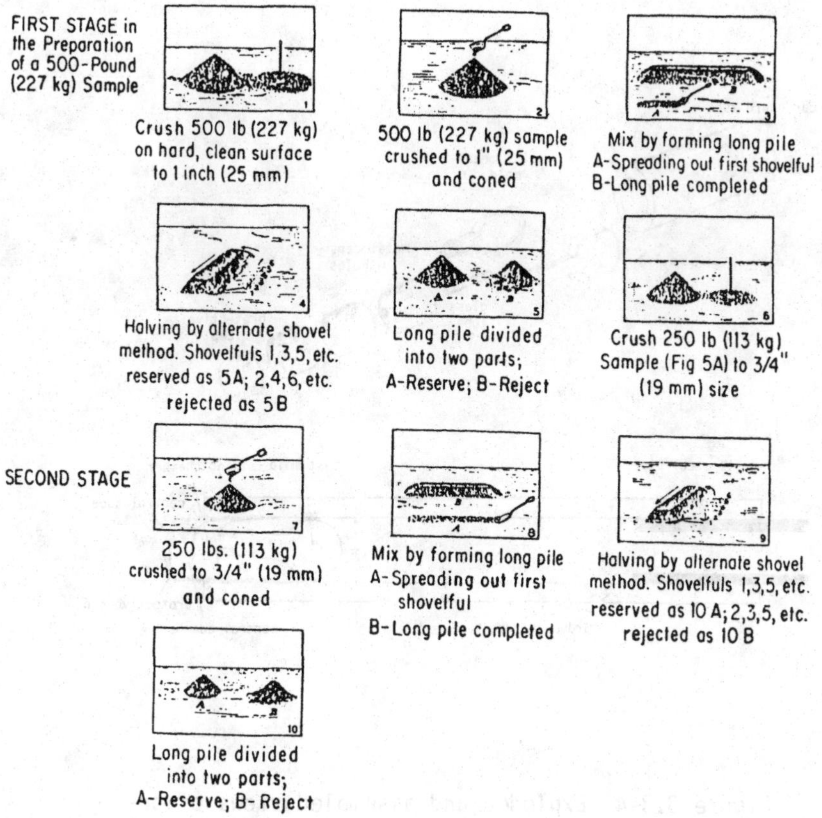
FIRST STAGE in the Preparation of a 500-Pound (227 kg) Sample
1
Crush 500 lb (227 kg) on hard, clean surface to 1 inch (25 mm)
2
500 lb (227 kg) sample crushed to 1" (25 mm) and coned
A
B
3
Mix by forming long pile
A-Spreading out first shovelful
B-Long pile completed
4
Halving by alternate shovel method. Shovelfuls 1,3,5, etc. reserved as 5A; 2,4,6, etc. rejected as 5B
A
B
5
Long pile divided into two parts;
A-Reserve; B-Reject
6
Crush 250 lb (113 kg) Sample (Fig 5A) to 3/4" (19 mm) size
SECOND STAGE
7
250 lbs. (113 kg) crushed to 3/4" (19 mm) and coned
B
A
8
Mix by forming long pile
A-Spreading out first shovelful
B-Long pile completed
9
Halving by alternate shovel method. Shovelfuls 1,3,5, etc. reserved as 10 A; 2,3,5, etc. rejected as 10 B
A
B
10
Long pile divided into two parts;
A-Reserve; B-Reject

Figure 3.3-3 Standard Method of Sampling Coke for Analysis (Ref. ASTM D346, 1977)
Necessary tools: shovel, tamper, and steel plate, broom, and rake. Use rake for raking over coke when crushing it, so that all lumps will be crushed. Sweep floor clean of all discarded coke after each time sample is halved.

or caps melt away and allow metal to be drawn into the sample mold, which is usually surrounded by a cardboard sleeve. One design of this type is illustrated in Fig. 3.3-4.

The <u>number</u> of samples from a bath of metal required to obtain <u>accurate</u> results depends on the concentration of the element in question and on the degree of homogeneity of the bath. If the bath being sampled has been subjected to vigorous boiling action, gaseous agitation, or mechanical agitation a single sample may be adequate. The minimal variability present in such cases is illustrated in Table 3.3-2, in which samples taken at the same time

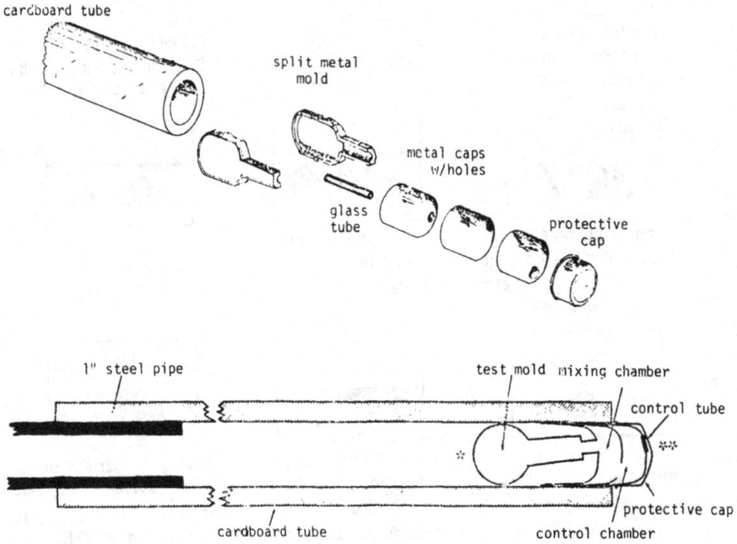

Figure 3.3-4 Exploded and assembled views of one commercial sampler for liquid metal sampling.

from different locations in electric arc furnaces were analyzed.

On the other hand, alloy additions made to "dead" baths may not properly disperse, or elements reduced from slags may be concentrated at the slag/metal interface. Samples taken near the slag/metal interface may therefore differ grossly from the mean value. This is illustrated in Table 3.3-3, for several stainless steel heats made in large electric arc furnaces.

In the case of trace elements, e.g. aluminum in steel, and dissolved gases, such as oxygen, hydrogen, and nitrogen, a statistical sampling procedure must be followed in order to obtain an accurate value as Kowal, Lewis and Wojcik* have demonstrated. Table 3.3-4 shows the variability in oxygen contents in a typical steel heat. Statistical analysis indicated that oxygen data tended to form log-normal distributions (i.e., a normal distribution was observed if the logarithim of the oxygen content was used as the variable) and therefore the geometric mean value of the data was

*R. Kowal, S. Lewis, and W. Wojcik, Proc. AIME Openhearth Conf., 1965, p. 308.

the most satisfactory estimate of the true oxygen value. The number of pin samples taken, and analytical determinations from small sections out from each sample, required to estimate the true oxygen content of the bath or batch within ± 20% of the true value, 95% of the time, is given in Table 3.3-5. Similar procedures would have to be developed for other elements and systems if quantitative estimates of accuracy are required.

The effect of the type of sampler used on the accuracy of the results is illustrated in Fig. 3.3-5, in which the variability in manganese analysis in a steel heat as a function of sampling procedure is shown.

Slag sampling has not reached the same level of sophistication as metal sampling. Slag samples are often taken by spoon and poured onto shovels or clean concrete, where the "pancake" is allowed to harden. This sample is then crushed for analysis. Care must be taken to check for entrained metallic particles, which must be removed if the "true" slag analysis is to be attained.

3.3.3 SAMPLING GASES

Samples of the process gases are often taken for either gas analysis or particulate analysis. The sampling method used depends on which analysis is required.

There are many methods of sampling gases for particulates. All depend on pulling the gas through a device which traps out the particulates and measures the total gas flow at the same time. ASTM Standard Procedure 2006 recommends using a glass cloth filter to trap the particulates, collecting them from approximately 2000 m^3 of gas drawn through the sampler over a 24 hour period. Other sample filters, such as paper, may have to be used to trap very fine fumes. For sampling hot gases, a porous alundum thimble is often used. Filters of cellulose esters are used to collect extremely fine particles such as colloids, and bacteria.

In order to get an accurate sample of the particulate content of a gas the sample must be taken under isokinetic conditions; i.e., the velocity of the dust laden stream at the entrance to the sample probe must equal the local velocity upstream from the probe. Thus, the pressure of the probe must not affect the gas velocity in the stream being sampled. The errors involved when this condition is not met are illustrated in Fig. 3.3-6.

Sampling in ducts for compositional variation should be done only after a thorough study of the flow patterns. In very large ducts, the positions corresponding to equal areas are given on the two charts in Figs. 3.3-7 and 3.3-8 and samples should be taken at each point (or every other point) in order to obtain an average sample. If flow velocities are not uniform across the duct, the total mass flow of a particular constituent must be found by multiplying the local concentrations by the local velocities and then averaging the values. After such a survey, the location of a single sample port may be established. The minimum

Table 3.3-2

Metal and Slag Composition After Oxygen Blow (Carbon Boil) in a 15-foot Diameter Electric Arc Furnace

		Metal Composition %				
Heat	Sample Location	C	Mn	Si	Ni	Cr
A	Charging Door	.014	.20	.20	13.32	2.08
	Side Door	.014	.20	.20	13.56	3.00
C	Charging Door	.04	.20	.04	12.68	2.24
	Side Door	.032	.20	.02	12.68	2.24
D	Charging Door	.023	.22	.030	11.75	4.28
	Side Door	.024	.24	.024	11.68	4.34
	Distance below the Interface					
	6"	----	---	.01	11.96	3.44
F	6"	----	---	.02	12.40	3.44
	18"	----	---	.02	12.28	3.44
	27"	----	---	.03	12.04	3.16

		Slag Composition						
Heat	Sample Location	Fe	Cr	SiO_2	CaO	MgO	Al_2O_3	MnO
A	Charging Door	20.00	26.99	6.16	14.50	6.52	1.50	6.58
	Side Door	22.00	24.71	7.10	15.17	4.52	1.83	6.71
B	Charging Door	27.60	21.12	5.44	14.84	4.95	0.71	5.95
	Side Door	27.40	21.30	5.30	14.84	5.18	2.24	6.05
C	Charging Door	21.20	25.12	7.84	15.59	6.09	2.15	4.44
	Side Door	20.40	25.14	7.86	15.93	5.40	1.95	3.12
D	Charging Door	23.66	27.94	7.52	14.11	11.64	1.99	6.13
	Side Door	23.92	28.82	7.70	14.39	3.52	1.97	6.26
E	Top of Slag	14.20	29.92	7.10	20.06	1.59	3.04	5.50
	8" below Top	14.60	30.25	6.74	19.41	1.30	2.97	5.50

Table 3.3-3

Vertical Stratification Within the Metal Bath

Heat	Depth into Metal Bath (Inches Below Slag Layer)	%Ni	%Cr
A	0 (Slag/Metal Interface)	---	34.36
	6	6.88	27.79
	12	6.88	25.62
	18	7.12	22.93
		(9.50)*	(18.08)*
B	0 (Slag/Metal Interface)	5.28	37.33
	18	15.16	11.74
	0t (Interface)	3.44	38.33
	6t	10.04	23.44
	18t	15.60	11.16
		(13.30)*	(17.24)*

* Average of bath after thorough stirring.

t Taken after five minutes of ineffective argon bubbling.

Table 3.3-4

Oxygen Analyses From Steel Samples*

Sample Location	% Oxygen Geometric Mean	% Oxygen Spread
Furnace	0.0092	0.0057 to 0.0173
Ladle	0.0068	0.0029 to 0.0135
Ingot	0.0038	0.0011 to 0.0155
Product	0.0023	0.0016 to 0.0058

* R. Kowal, S. Lewis, and W. Wojcik, Open Hearth Proceedings, AIME, 1964, p. 308.

Table 3.3-5

Sample Requirements for Oxygen Determinations in Steel Such That the Geometric Mean Value is Within 20% of the True Value 95% of the Time*

Sample Location	# of Samples	# of Analyses of Each	Total Analyses
Furnace	4	3	12
Ladle	6	3	18
Ingot	9	3	27
Sheet or Bar Product	9	1	9

* Kowal, et al., op. cit.

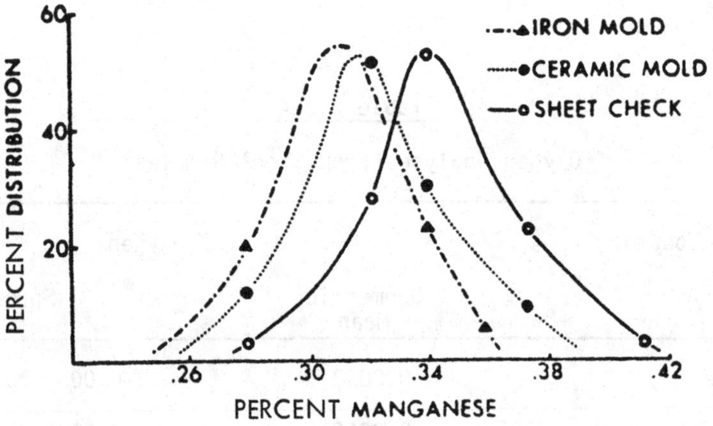

Figure 3.3-5 Manganese distribution determined by various types of samplers (liquid samples) and cut from product sheet from same heat of steel. (Ref. R. R. Strange, J. Metals, (July, 1955), p. 767.)

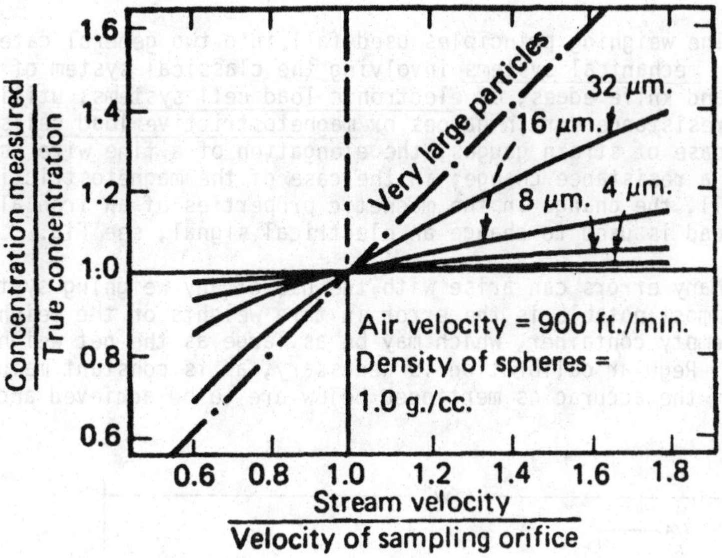

Figure 3.3-6 Errors resulting from anisokinetic sampling.

equipment requirements for a gas-sampling system are shown in Fig. 3.3-9.

EXAMPLE 3.3-3: The SO_2 flow in a 2 meter diameter flue is to be determined. How many points should be sampled and where?

Solution: The total cross-sectional area is 3.14 m^2, or 91.93 ft^2. The recommended number of test points would be 20 or more. The velocity in each equal area zone must be determined, along with the SO_2 concentration.

From Fig. 3.3-6, if the number of areas is 5, the number of points sampled on a diameter is 10, so that 20 points would be sampled if samples are taken along two diameters located at right angles to each other. The test locations would be at 4.4, 16.4, 29.0, 45.4, 68.8, 131.2, 154.6, 171.0, 183.6, and 195.6 cm from one wall, along a diameter. These values were found by multiplying the percent diameter values in the figure by the diameter.

3.4 WEIGHING

In many metallurgical processes, direct weighing of charge materials is possible. In some cases, products can also be weighed. There are a large variety of devices and instruments available for these purposes, ranging from scales with a few kilogram capacity to weighing systems with capacities of 10^5 kg or more.

The weighing principles used fall into two general categories: mechanical systems involving the classical system of levers and knife edges, or electronic load cell systems, utilizing either resistance strain gauges or magnetostrictive load cells. In the case of strain gauges, the elongation of a fine wire is read as a resistance change; in the case of the magnetostrictive load-cell, the change in the magnetic properties of an iron alloy under load is used to change an electrical signal, see Fig. 3.4-1.

Many errors can arise with the use of any weighing system, but the most notable is the error in tare weights or the weight of the empty container, which may be as large as the net weight itself. Regular calibration is necessary, as is constant maintenance, if the accuracies mentioned below are to be achieved and

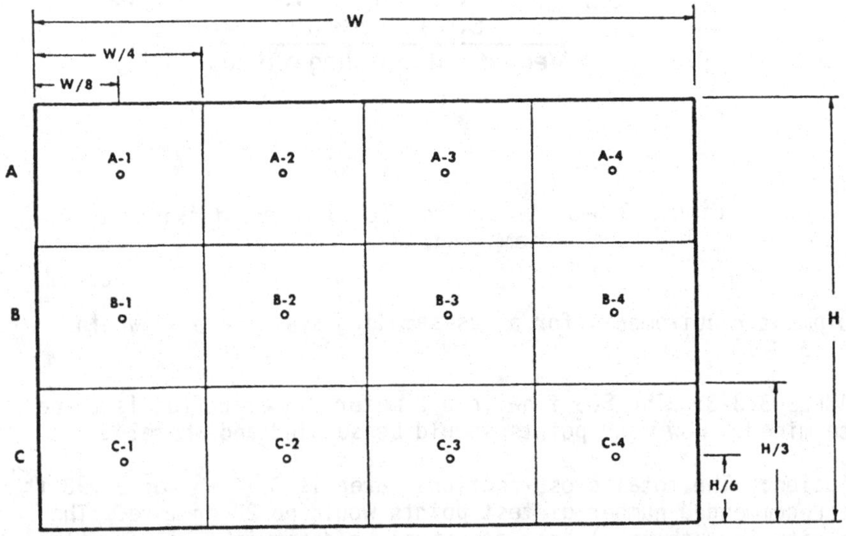

if the station chosen has fairly uniform flow, the minimum number of test points may be determined as follows:

Cross-sectional Area Square Feet	Number of Test Points
Less than 2	4
2 to 25	12
Greater than 25	20 or more

Figure 3.3-7 Typical rectangular flue with traverse layout. (Ref. Bulletin WP-50, 7th Ed., Western Precipitator Div., Joy Mfg. Co., Los Angeles).

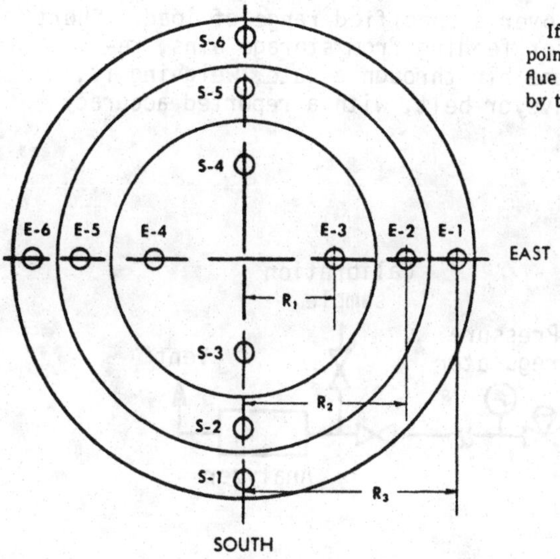

If the table below does not have enough points, the distance R from the center of the flue to any given test point, n, is calculated by the formula

$$R_n = d\sqrt{\frac{2n-1}{4N}}$$

d = Diameter of flue

N = Number of readings across a diameter

n = The n'th point from the center

R_n = Distance from center of flue to n'th test point, numbered from the center consecutively. In the diagram, n=1 for E-3, E-4, S-3, and S-4.

NOTE: Using compass points to designate stations helps to orient the test locations.

EQUAL AREA ZONES FOR VELOCITY TRAVERSE

(% of flue diameter from circumference to test point) = $\frac{d-R}{d} \times 100$

Point No.	Number of Areas 2	3	4	5	6
n = 1	6.2	4.4	3.3	2.2	2.0
2	25.0	14.7	10.5	8.2	6.7
3	75.0	29.4	19.5	14.5	11.8
4	93.8	70.6	32.1	22.7	17.7
5		85.3	67.9	34.4	25.0
6		95.6	80.5	65.6	35.4
7			89.5	77.3	64.6
8			96.7	85.5	75.0
9				91.8	82.3
10				97.8	88.2
11					93.3
12					98.0

Figure 3.3-8 Traverse positions in round flues. (Ref. Bulletin WP-50 Op cit.)

maintained. Finally, it should be pointed out that if the accuracy of a machine is given as a percentage of full capacity, for example 1%, then the error at a load less than full capacity, say 50% of capacity, will be a larger percentage of that load, in this case 2%. This is important to recognize on machines operating over a wide weighing range.

3.4.1 BELT WEIGHERS

The weight passing a given location on a moving belt is obtained by multiplying the speed of the belt times the force on one or more rollers, measured by load cells. The accuracy is usually from ± 0.25 to ± 2%, over a specified range of load. Short belt weighers are also used for feeding from storage bins, receiving their material from the bin through a gate, weighing it, and discharging it onto a conveyor belt, with a reported accuracy of ± 2%.

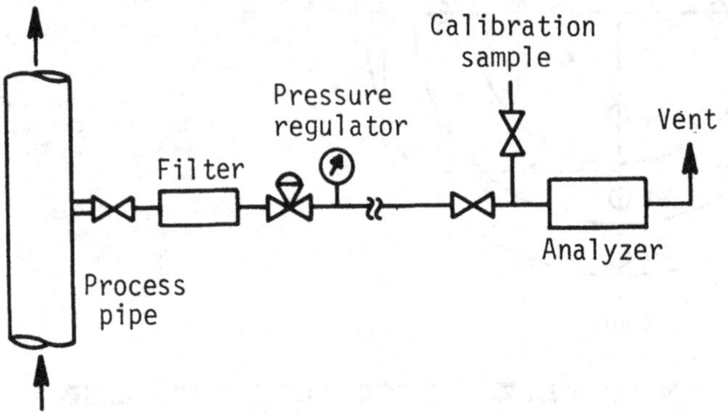

Figure 3.3-9 Minimum equipment requirements for a gas-sampling system.

3.4.2 HOPPER WEIGHERS

This general term describes the weighing of feed hoppers, charging buckets, scale cars, etc. These devices are usually weighed by load cells built into their supporting frame, with accuracies of from ± 0.1 to ± 2.0%. In general, mounting three load cells, as in Fig. 3.4-2, to support the hopper is desirable, since it is then impossible for the weight to be taken by any two cells, whereas if four load cells are used, it is possible for the load to be taken on three alone.

3.4.3 PLATFORM WEIGHERS

These scales may utilize either mechanical or load cell devices. They may be of very high capacity, such as those used to weigh ladles in steel mills, or of low capacity, such as those used to weigh a few kg of alloying additions. The accuracy of load cell platform scales (or weighbridges) generally varies between ± 0.05% and ±0.5%. Mechanical type scales are reported to be as accurate as ± 0.02%.

3.4.4 CRANE WEIGHERS

Load cells may be fitted either to the spreader beam or located on the moving crab on large bridge cranes, as in Fig. 3.4-3.

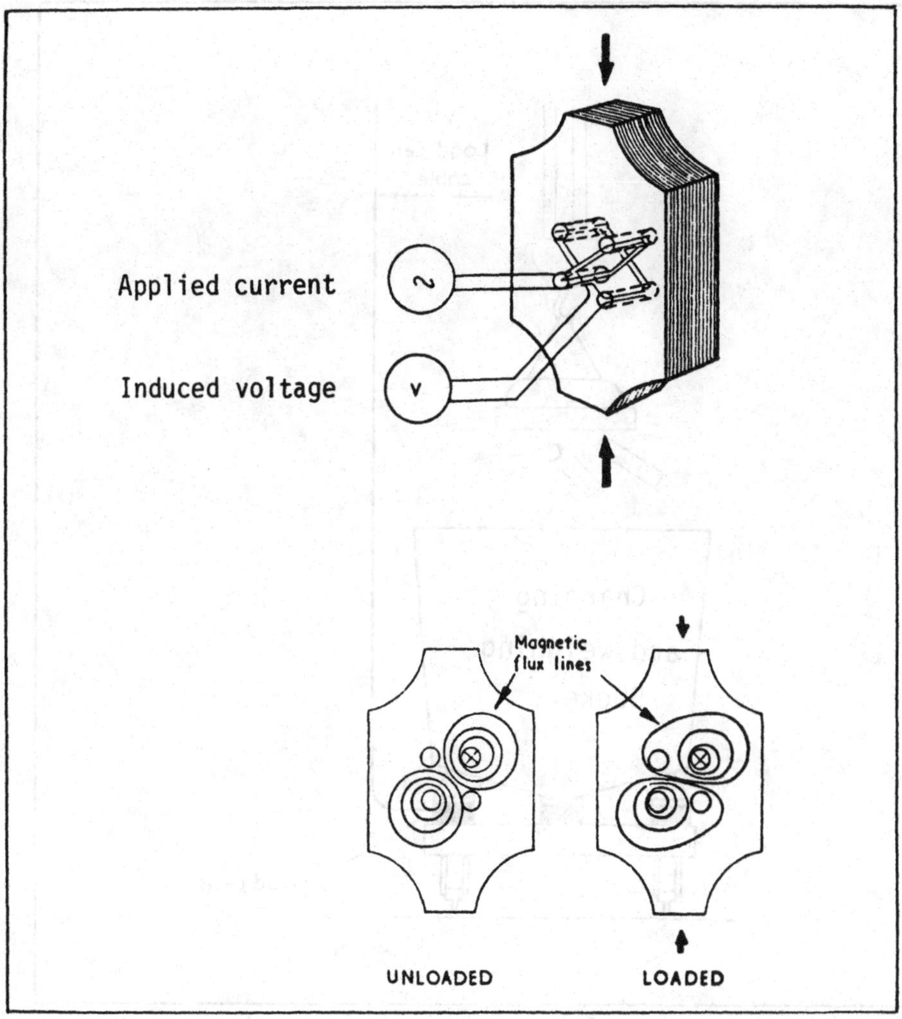

Figure 3.4-1 Magnetostrictive load cell.

After many development problems, the accuracy of the best of these systems is now about ± 0.2% of net weight or ± 0.1% of total capacity of the crane, which includes, for example, the weight of the ladle containing molten material in smelters and steel mills.

3.4.5 ROLLER CONVEYOR WEIGHERS

Slabs, ingots and other large semifinished or finished mill products may be weighed while being transported via roller conveyors. This is done by either active weigh scales, wherein a section of the roller table acts as a platform scale, or else by lifting fingers normally situated below and between the main table rollers which raise up to support the piece being weighed. The latter system can be made more accurate than the former because of the smaller ratio of tare-to-net weight. Guaranteed accuracies for both types vary from ± 0.05 to ± 0.25%.

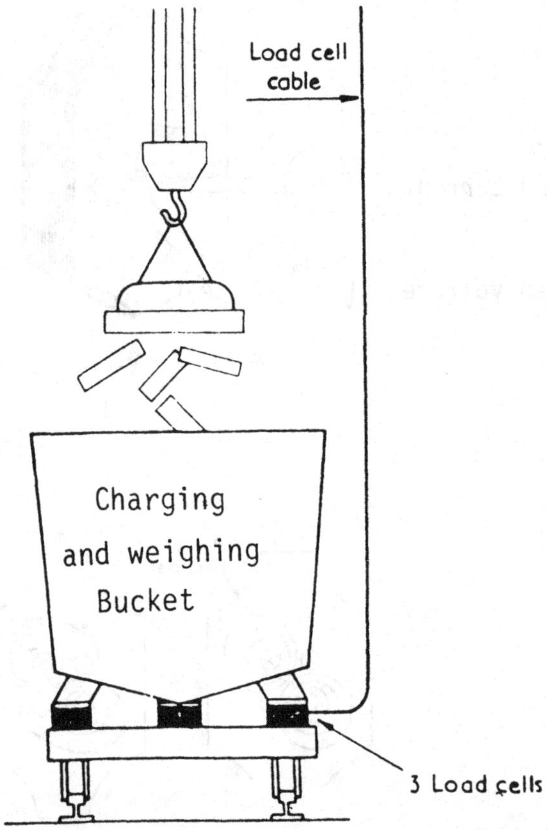

Figure 3.4-2 Weighing installation for charging electric furnace.

3.5 DENSITY MEASUREMENTS

In many processes, such as those involving molten slag and metal, mass cannot easily be measured directly, and so it is estimated by measuring or estimating the volume and then using the density to obtain the mass. This obviously requires an accurate value of the density, since the volume measurements themselves may be inaccurate.

3.5.1 LIQUID DENSITY

If the substance is a liquid at ordinary temperatures, a graduate and laboratory balance will yield the density quite accurately. However, if the material is liquid in the process, but solid at room temperature, then the use of solid sample

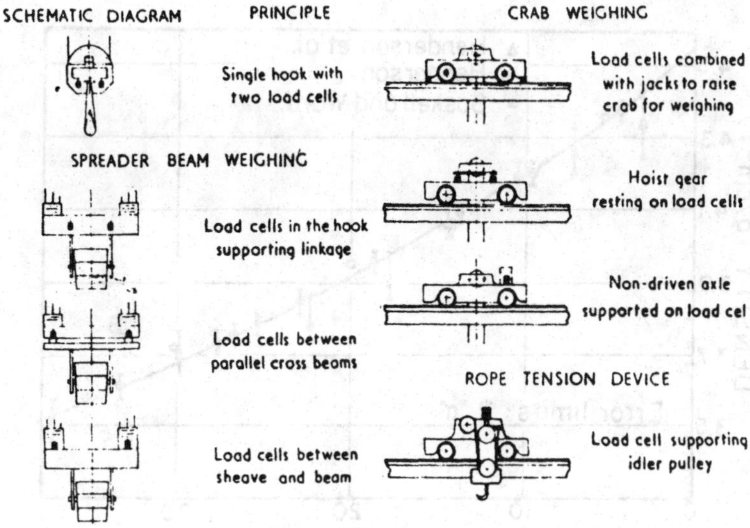

Figure 3.4-3 Typical applications of load cells in crane operations. (Ref.: N. A. Towsend and J. M. Molloy, article in Control of Composition in Steelmaking, publ. by Iron and Steel Inst. (London), 1966.)

density in combination with overall volume to estimate total weight may be very inaccurate. This is common in systems where slags are present.

In most cases, the density of the solid material is determined and an assumption of the differential shrinkage upon going from liquid to solid is made. This can have major errors associated with it. For example, gases are frequently evolved during the freezing process resulting in a "foamy" solid and an apparent density far below the true density. On the other hand, slags often have metal droplets entrained in them, resulting in potential overestimations of their true density. Finally, the composition of a slag can have a significant effect on its density, as shown in Fig. 3.5-1, and so care must be taken to use literature density data carefully.

3.5.2 BULK DENSITY

The bulk density of irregular solids, such as coke and sinter, is very hard to determine accurately. There are, however, ASTM Standard Procedures (such as D292-29 for coke) to make standardized determinations. The most important aspect of this procedure is to make sure to have a large enough sample box compared to the average particle size. In the case of coke particles

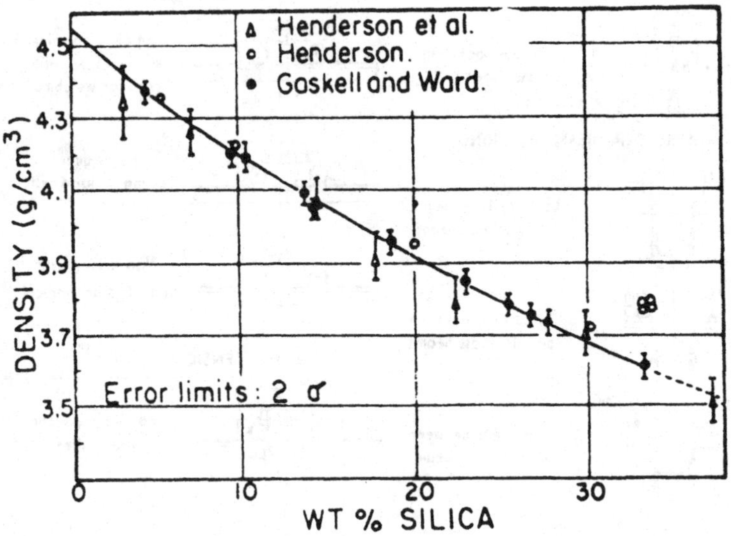

Figure 3.5-1 Density-composition relationship of iron silicates in contact with solid iron at 1410°C. (Ref.: C. Diaz, Thermodynamic Properties of Copper-Slag Systems, INCRA, New York, 1974, p. 6).

12.5 cm and smaller in diameter, the standard measuring box is 60 cm x 60 cm x 60 cm. For coke 2.5 cm or smaller, the box is 30 cm x 30 cm x 30 cm. The weighing box is filled "by means of shovel from a height of 5 centimeters above the top of the box with no attempt to spread or arrange the coke." Leveling off of excess coke is done by eye and hand. Then the box is weighed and the weight of coke per unit volume of container computed. This is the bulk density.

Bulk densities are related to the void fraction, ω, in a bed of solids, according to

$$\rho_{bulk} = (1-\omega)\rho_o \qquad (3.5\text{-}1)$$

where ρ_o is the true density of the solid phase in the bed. The void fraction depends on the method of packing, and can vary for the same material by as much as $\pm$ 0.06 depending on the weight or height of the bed, vibration, etc. Therefore, since ω for packed solids is typically on the order of 0.4, if estimated values of ω are used in conjunction with ρ_o values, the error in calculated bulk density values can easily be ± 15%.

The importance of accurate results is illustrated by the charging of an iron blast furnace, which is usually done by means of a skip hoist. The skip has a known volume. In some cases, the weight of coke contained in a skip is not determined directly, but instead the coke is filled to a consistent height in the skip and the weight is then calculated using the bulk density. This is an adequate procedure, provided the bulk density is known accurately.

EXAMPLE 3.5-1: Suppose that the volume of a skip is 20 m^3, but the control on the filling level is only to within ± 5% of this value. If the bulk density is estimated to be 1400 kg/m^3 ± 110 kg/m^3 this means that a skip load can weigh anywhere from 24,510 kg to 31,710 kg, while a value of 28,000 kg would have to be assumed for charge calculations. Obviously, any decrease in the probable error in the bulk density value will improve the accuracy of the charge calculations.

3.5.3 PULP (SLURRY) DENSITY

Pulp density is found using direct weighing of grab samples, or using nuclear density gauges. If the pulp density is known (ρ_p), and the density of the solids is known (ρ_s), then the percent solids in the stream is given by Eq. (3.5-2):

$$\text{Wt.\% solids} = \frac{\rho_s(\rho_p - 1)}{\rho_p(\rho_s - 1)} \times 100 \qquad (3.5\text{-}2)$$

The relationship between these values is given in Fig. 3.5-2. (See pages 1-29 through 1-31 for additional information.)

A nuclear density gauge operates on the principle that radiation from high-energy gamma radiation sources such as cesium-137, cobalt-60 and radium-226 provides photons whose absorbtion by a process stream depends on the mass of the stream. In most cases, a detector is placed opposite the radiation source to sense the percentage of transmitted radiation, as in Fig. 3.5-3. Nuclear density gauges can be calibrated to read in percent solids only when the density of one component, such as water, in a two-component system is fixed. An increase in pulp density will cause a predictable decrease in transmitted radiation. However, due to source decay, variability in mass absorbtion coefficients of the feed, and other problems, regular calibration is required. The final accuracy is often as good as ± 0.1% solids.

EXAMPLE 3.5-2: A sample of pulp taken from a process stream has a density of 1.35 gm/cm^3. The solids are known to have a density of 2.8 gm/cm^3. What % solids are in the pulp?

Solution: Using Fig. 3.5-2, the two values converge on the 40% solids line.

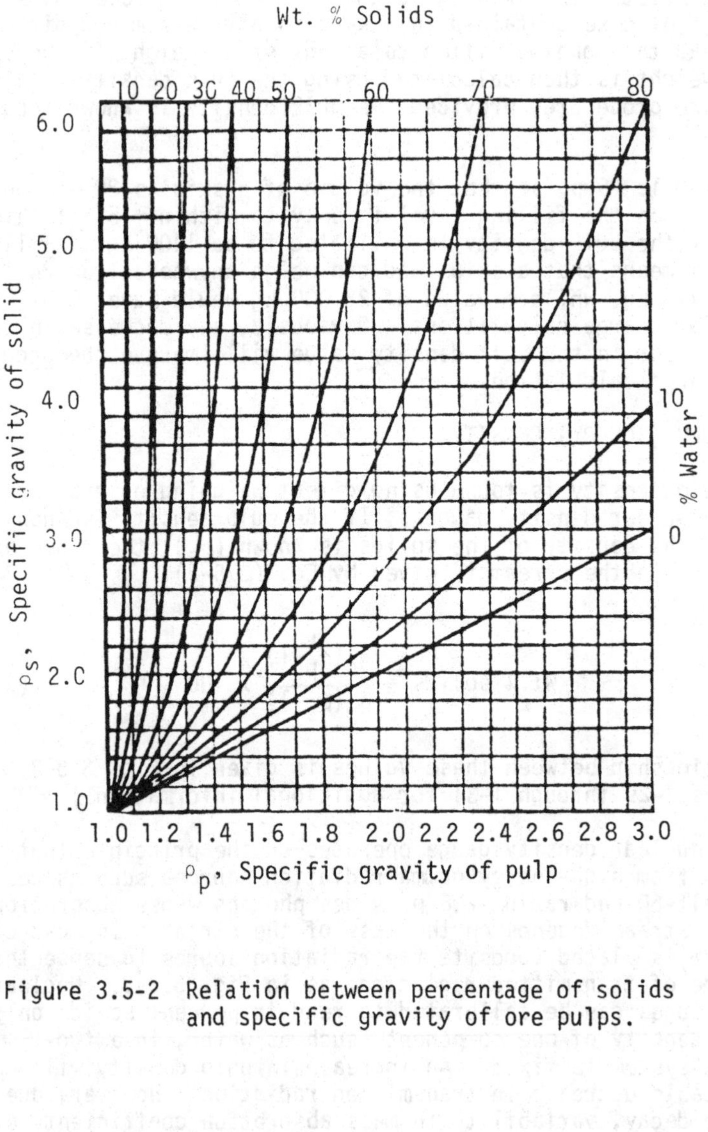

Figure 3.5-2 Relation between percentage of solids and specific gravity of ore pulps.

3.6 MEASUREMENT OF TEMPERATURE

The accurate measurement of the temperature of process input and output material streams is required in order to establish energy balances. A wide variety of temperature sensing devices are available. Some of these are given in Table 3.6-1 along with their useful temperature ranges.

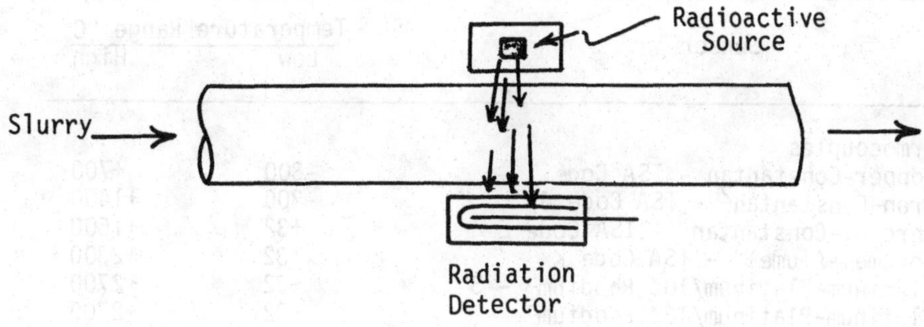

Figure 3.5-3 Schematic diagram of nuclear density gauge.

3.6.1 THERMOCOUPLES

Measurement of temperature with a thermocouple requires measurement of the emf developed by the metal couple. Common materials used in thermocouples include pure platinum, platinum-rhodium alloys, iridium, rhenium, tungsten, iron, copper, and alloys such as constantan (60% Cu, 40% Ni), chromel (10% Cr-90% Ni), and alumel (2% Al-8% Si+Mn-90% Ni). Thermocouple wire supplied by reputable manufacturers produces emf values that agree remarkably well with the published tables of emf versus temperature. Therefore, for a great deal of work it is satisfactory to make a thermocouple of this wire and use it to measure temperature without further calibration.

There is little point in using very precisely calibrated thermocouples in conjunction with automatic control equipment because, in general, the difference between the true temperature and the temperature indicated by the controller, is less than the fluctuation around the set point. It is, however, necessary to check thermocouples in permanent installations frequently to be sure that they are indicating approximately the right temperature. This is particularly true in hostile environments, where the thermocouple may be exposed to SO_2, H_2S, CO, etc. Manufacturers literature should be consulted.

For more accurate work it is occasionally necessary to have an accurate calibration. Working thermocouples are almost always calibrated by comparing them with carefully calibrated standard thermocouples. These standard couples may be either noble or base metal although, for precise work, platinum-platinum/rhodium couples are always used. Calibrated couples can be obtained from the manufacturers of pyrometric equipment, and the National Bureau of Standards will calibrate thermocouples sent to them.

Table 3.6-1

Practical Measurement Range of Temperature Sensors*

Sensor	Temperature Range °C Low	Temperature Range °C High
Thermocouples		
Copper-Constantan - ISA Code T	-300	+700
Iron-Constantan[2] - ISA Code J	-300	+1400
Chromel-Constantan[3] - ISA Code E	+32	+1600
Chromel-Alumel[4] - ISA Code K	+32	+2300
Platinum-Platinum/10% Rhodium[5] - S	+32	+2700
Platinum-Platinum/13% Rhodium[5]	+32	+2700
Iridium/Rhodium-Iridium	+1425	+3600
Tungsten-Rhenium		+3600
Resistor Bulbs		
Platinum[7]	-325	+1000
Nickel[8]	-40	+400
Copper	-330	+250
Total Radiation Pyrometers[9]		
High range	+1700	+1700
Intermediate range	+1000	+3400
Low intermediate range	+500	+1200
Low range	+100	+700
Spectracally Selective Pyrometers		
Optical or brightness pyrometer[10]	+1400	+7500
Two-color pyrometer[11]	+1400	+6500
Filled System Thermometers		
Liquid-filled (SAMA Class 1A[13])	-300	+600
Mercury-filled (Class V)	-38	+1000
Vapor-filled (Class II)	-432	+600
Gas-filled (Class III)	-450	+1400
Bimetal Thermometers[12]	-300	+1000
Pyrometric Cones	+1085	+3660
Surface-Applied Temperature-Sensitive Materials (melting and color-change types)	+100	+2500

* D. M. Considine, Chemical Engineering, January 29, 1968, p. 84.

1. The high temperature values depend on wire size and protection provided. The values given in table apply to the largest wire size normally used, encased in a suitable protecting tube.
2. Can be used intermittently up to +1575 F.
3. Can be used intermittently up to +1825 F.
4. Can be used intermittently up to +2400 F.
5. Can be used intermittently up to +2925 F.
6. Platinel is a proprietary gold-palladium alloy.
7. The figures given are those generally recommended by manufacturers. With proper precautions, platinum bulbs can be used down to -430 F.
8. With proper precautions, nickel bulbs can be used down to -240F and up to +575 F.

Since calibration is rather time consuming and requires some special equipment, it is usually cheaper and more satisfactory to have such primary calibrations made in this way.

Where high precision is required, potentiometers are used, rather than millivoltmeters. For good industrial-type potentiometers, used with thermocouples or other electrical temperature transducers, the sensitivity of the measuring system will be about $\pm$ 0.035% of fall-scale span, in terms of millivolts.

When determining heat balances in industrial environments, the temperatures of gas or air streams flowing through flues and stacks play a very important role. The measurement of the temperature of a gas is considerably different from such a measurement in liquids and solids. The principal problem is the highly practical one of causing the measuring thermocouple junction to attain the same temperature as the gas.

In measuring the temperature of flue gases, the thermocouple will usually be enclosed in a metal protection tube which is screwed into an opening in the wall of the duct. Since there is loss of heat through the duct walls, the protection tube is cooler where it enters the wall than at its tip where the thermocouple junction is placed. Thus, a flow of heat takes place through the tube and also through the thermocouple wires from the tip of the tube towards the outer duct wall. This heat must be supplied from the gas stream, and to pick up the heat the tube must be cooler than the gas. The temperature of the junction must, therefore, be lower than that of the gas. The magnitude of the resulting error will depend upon the rate of flow of the gas, the nature of the portion of the surface of the protection tube exposed to the gas, the length, diameter and materials of the tube and the thermocouple wires, and the temperature of wall to which the tube is attached. The errors are often significant.

A further source of error is radiation. In the case under consideration, a hot gas inside cooler walls, there will be a loss of heat from the tube or thermocouple junction by radiation to the walls, in addition to that by conduction through tube and wires. Experience has shown that a thermocouple located in the hot gases, where it can "see" cooler surfaces may indicate temperatures 80K too low when the actual gas temperature is 700K. The radiation error decreases as the area of the radiating surface of the junction or tube is decreased. Consequently, this area should be made as small as possible, consistent with mechanical strength.

The errors from conduction and radiation can be decreased by inducing a high velocity of flow past the sensitive element. This increases the convective transfer of heat to the element. This is the principle of the suction or aspirating pyrometer, which should be used for hot gas measurement.

Radiation errors may also be reduced by enclosing the sensitive element in a radiation shield, which may consist of two or more concentric cylinders of materials of low emissivity, which prevent the sensitive element from "seeing" any significantly cooler surface, but still permit a free flow of gas past the junction.

Finally, if the purpose of the measurement of the temperature of the gas is to work out a heat balance, it is not sufficient to measure the temperature at only one point in the gas, since the flow is invariably more or less stratified. It is also not sufficient to just measure the temperature simultaneously at a number of points in a duct, since the velocity of flow, and consequently the mass of gas represented by each measurement, is variable from point to point. The rate of flow at the various points must be measured by traversing in equal increments across two diameters at right angle to each other, as shown in Figs. 3.3-7 and 3.3-8. The points in the ducts represent locations within equal cross-sectional areas, and the average of flow and temperature measurements made at these points is then used to calculate the stream average heat flow.

Temperature measurements in molten metals are made using disposable Pt-Pt/Rh or W-Re thermocouples contained in cardboard sleeves that fit over reusuable steel lances, which contain the compensating lead wires. The accuracy of temperatures measured by this technique is thought to be ± 5°C at 1600°C, (1873K).

3.6.2 PRESSURE THERMOMETERS

The pressure or filled-system thermometer is generally a liquid- or vapor-filled bulb, A, connected by pressure tubing, B, to one of several pressure-activated devices, C, as shown in Fig. 3.6-1. The expansion and contraction of the fluid in the bulb, in response to temperature change is thus communcated as a pressure change to the helix or spiral gauge, which can then record or activate a mechanism as the application requires. The liquid-vapor system has an advantage over the other systems in that the vapor pressure is set by the bulb temperature, and consequently changes in temperature within the capillary gauge system are unimportant. Table 3.6-1 gives the ranges of usefulness of various types of pressure thermometers.

3.6.3 RADIATION PYROMETERS

Every object above absolute zero radiates electromagnetic energy, whose amount and type depend on the temperature of the

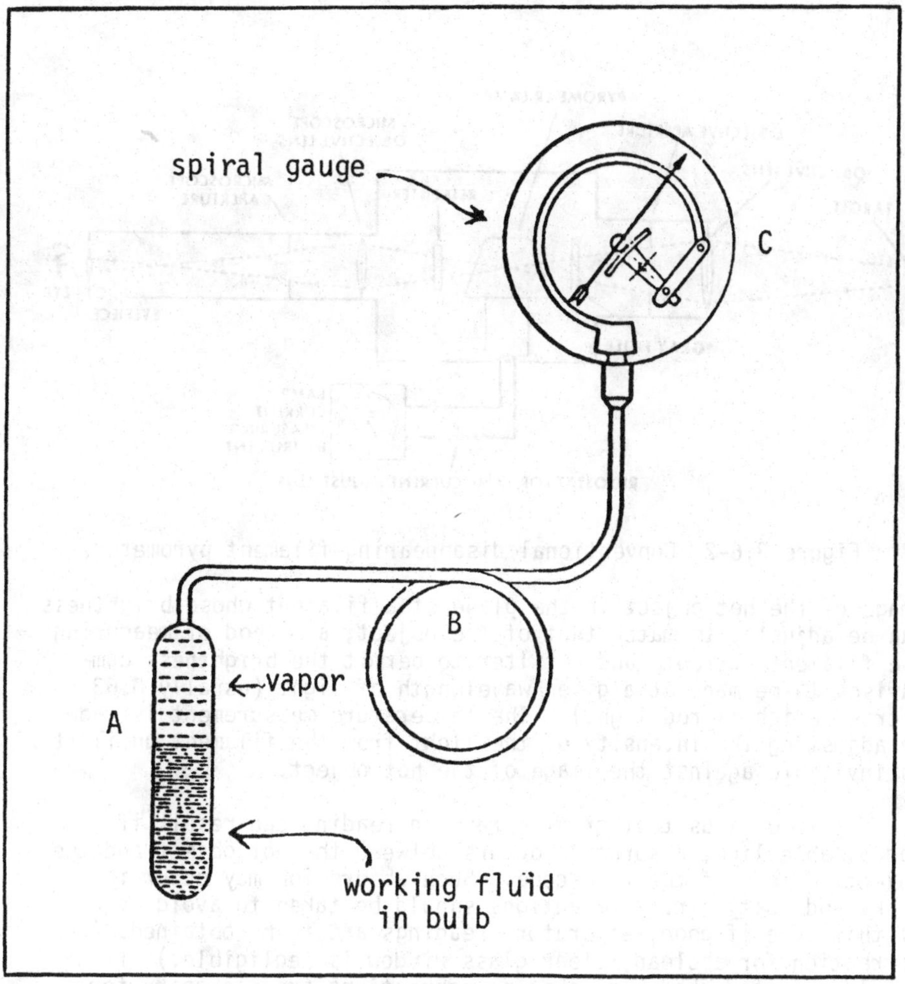

Figure 3.6-1 Principle of operation of a pressure thermometer.

object. Thus, sensing of the radiation from an object is a non-contact method for temperature measurement.

Radiation pyrometers are the only instruments that can be used to measure temperatures above those at which thermocouples can be used. Frequently they are more convenient than thermocouples for use at lower temperatures. They are of two general types: the optical pyrometer and the total radiation pyrometer.

The optical pyrometer measures the radiant emittance, N_λ, of a surface at a known wavelength. The disappearing filament optical pyrometer, Fig. 3.6-2, consists of a telescope for producing an

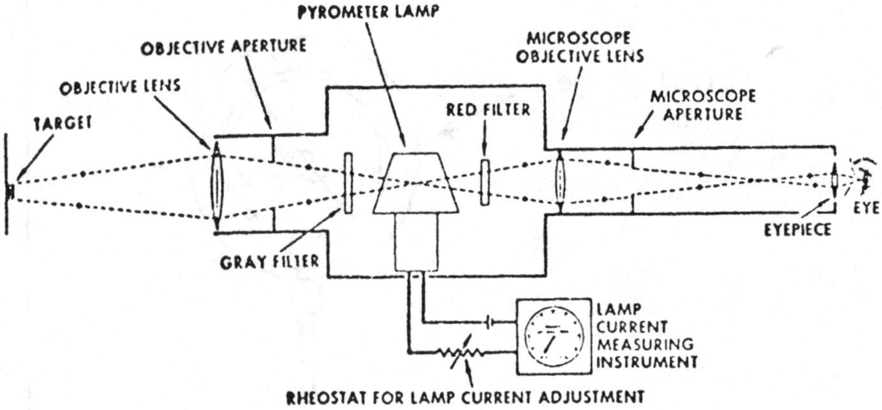

Figure 3.6-2 Conventional disappearing-filament pyrometer.

image of the hot object in the plane of a filament whose brightness can be adjusted to match that of the object; a method of measuring the filament current; and a filter to permit the brightness comparison to be made at a given wavelength of light (usually 0.63 micron, which is red light). The temperature measurement is made by adjusting the intensity of the light from the filament until it is invisible against the image of the hot object.

It is obvious that great errors in reading can result if appreciable light absorption occurs between the hot object and the objective lens of the telescope. Such absorption may occur in smoky and dusty air. Precautions should be taken to avoid errors of this type if good temperature readings are to be obtained. (The correction for a clean, clear glass window is negligible.) It should be noted, however, that no corrections are necessary for distance between the hot object and the instrument itself.

Unfortunately, the radiant emittance of most surfaces is not perfect that is, most objects are not blackbodies, and therefore the accuracy of temperature measurement by radiation pyrometers depends on the accuracy of the emissivity, ε_λ, used in Eq. (3.6-1):

$$N_\lambda = \varepsilon_\lambda C_1 \lambda^{-5} e^{-C_2/\lambda T} \tag{3.6-1}$$

where λ is the wavelength, C_1 and C_2 are constant, and T the absolute temperature. Most of the emissivity values reported in Table 3.6-2, and in the literature, are for polished metal surfaces. This is a condition usually fulfilled only by molten metals in a vacuum or in an entirely inert atmosphere. On the other hand, the emissivity of rough metal surfaces and of those with an oxide coating is considerably higher, and often approaches 1.0. Therefore, the indicated temperatures of these surfaces will be close to the actual temperature.

Table 3.6-2

Emissivity of Various Materials for Red Light ($\lambda \simeq 0.63\ \mu$)*

Material	ε	Material	ε
Silver	0.07	Cuprous oxide	0.70
Gold-solid	0.13	Iron oxide-800°C	0.98
liquid	0.22	1000°C	0.95
Platinum-solid	0.33	1200°C	0.92
liquid	0.38	Nickel oxide-800°C	0.96
Palladium-solid	0.33	1300°C	0.85
liquid	0.37	Iron-solid and liquid	0.37
Copper-solid	0.11	Nickel-solid and liquid	0.36
liquid	0.15	Iridium	0.30
Tantalum-1100°C	0.60	Rhodium	0.36
2600°C	0.48	Graphite powder	0.95
Tungsten-2000°C	0.46	Carbon	0.85
3000°C	0.43		
Nichrome-600°C	0.95		
900°C	0.90		
1200°C	0.80		

* Bureau of Standards T.P. 170.

The emissivity of an object, ε, also varies with its temperature. This is the basic limitation on the accuracy of most radiation pyrometers. If the emissivity is known, a correction to the reading may be made to obtain the true temperature, using Fig. 3.6-3.

Ratio (two-color) pyrometers are, in effect, two narrowband (brightness) pyrometers coupled electronically to compute the temperature from the ratio of the intensities at two wavelengths. Ratio pyrometers do not require emissivity compensation when the emissivity is independent of wavelength.

Optical pyrometers utilize only a short range of wavelength for observation to measure temperature. Another class of instruments, total radiation pyrometers, uses as wide a band of wavelengths as possible. Nominally these pyrometers obey the Stefan-Boltzmann law,

$$W = \varepsilon\sigma(T_1^4 - T_0^4) \tag{3.6-2}$$

where W is the total radiation, ε is the total emissivity of the radiator, T_1 is the (absolute) temperature of radiating body, T_0 is the temperature of the measuring or receiver body, and σ is a constant which equals $5.6697 \times 10^{-8}\ Wm^{-2}\ K^{-4}$. This law applies

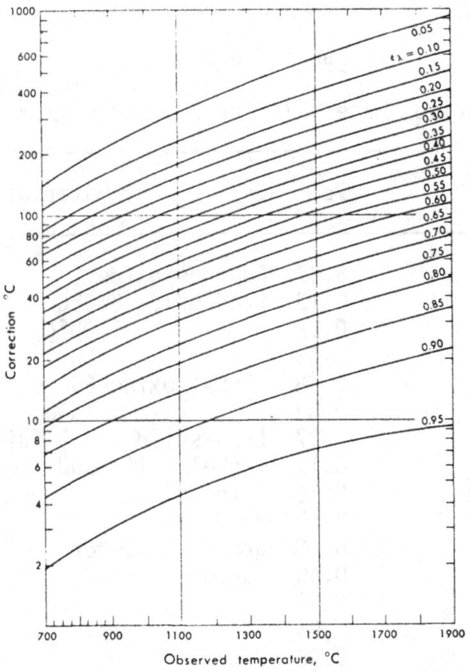

Figure 3.6-3 Corrections for Emissivity to Temperature Observed with Optical Pyrometer. The curves for emissivity values ranging from $\varepsilon = 0.18$ to $\varepsilon = 0.95$ show the additions to be made to the pyrometer reading. Computed for C_2=14,380 and $\lambda = 0.65\mu$. (Thermochemistry for Steelmaking, Vol. II, J. F. Elliott, et al., Addison-Wesley, Reading, Ma. 1963), 3-42.

for the total radiant energy of all wavelengths. In practice, total radiation pyrometers do not use all wavelengths because of absorption by the optical system. Thus they do not strictly obey the Stefan-Boltzmann law. Accordingly, total radiation pyrometers are always empirically calibrated.

Emissivity corrections must be made for total radiation pyrometers, but, as different wavelengths are involved, the corrections are different from those used for optical pyrometers. Emissivity values for use with radiation pyrometers may be found in the literature, but at best they can be considered to be only approximate. The best method to determine emissivity corrections for total radiation pyrometers is to make simultaneous readings on the surface in question with both an optical and a radiation pyrometer, and then to determine the true temperature by applying the appropriate correction from Table 3.6-2 and Fig. 3.6-3 to the optical pyrometer reading. The corrections for the optical pyrometer are much more reliable than those available for total radiation pyrometers, because the wavelength is known.

3.7 FLUID FLOW MEASUREMENT

Analysis of processes also requires information about the

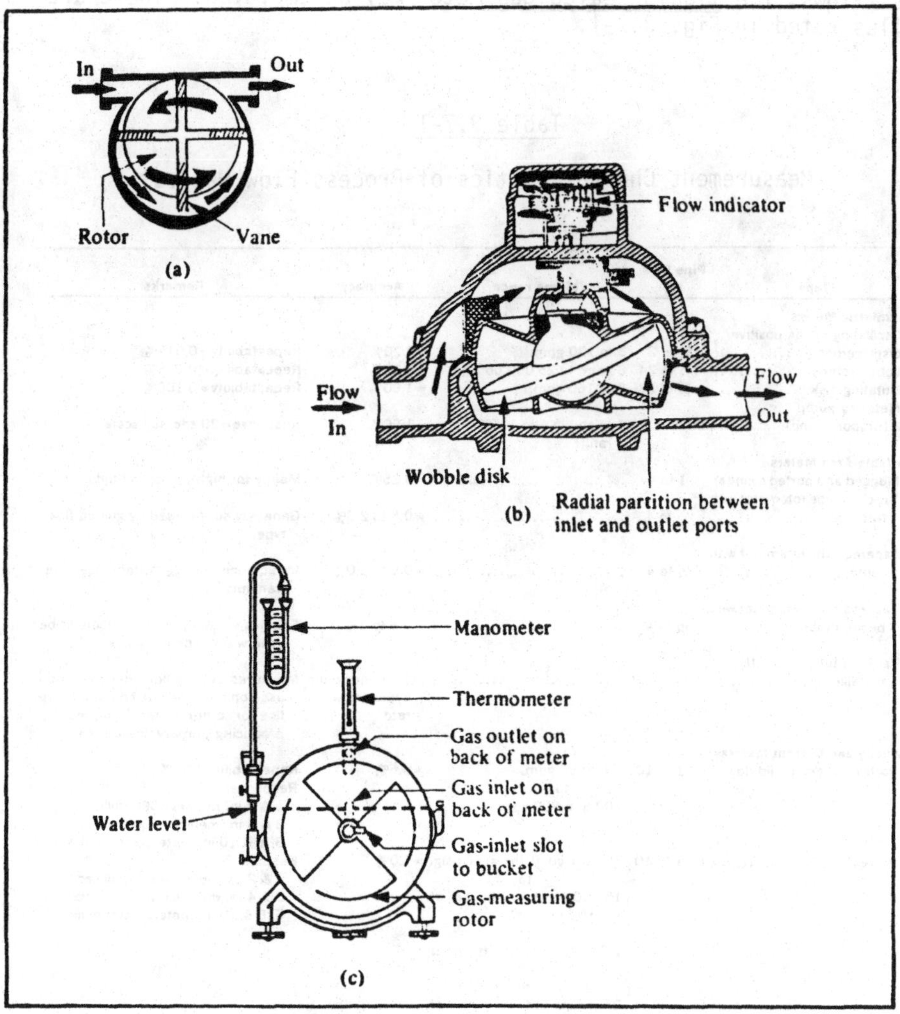

Figure 3.7-1 Various flow totalizers (a) Rotary vane meter, (b) nutating disk meter, and (c) liquid-sealed gas meter. In each case, the shaft feeds a mechanical counter.

mass-flow rates of fluids as well as solids and liquids weighed in batches. For measuring flows in closed conduits or ducts there are a wide variety of devices and techniques available. Table 3.7-1 lists some of these and their ranges and accuracies.

3.7.1 FLOW TOTALIZERS

There are many flow totalizers available which react mechanically and directly to the passage of fluid. A common type is a volumetric meter, called the rotary vane meter, which is applicable for either liquids or gases. For metering and totalizing water or other liquids, the rotating disk meter is used. For totalizing gas flows, the liquid-sealed gas meter may be employed. These are illustrated in Fig. 3.7-1.

Table 3.7-1

Measurement Characteristics of Process Flow Meters*

Type	Pipe Size In.	Volume range	Accuracy	Remarks
Volumetric Meters				
Oscillating-piston positive-displacement...............	½ to 2	3 to 120 gpm.	±0.20%	Repeatability=0.015%
Lobed-rotor.................	1½ to 24	8 gpm. to 25,000 bbl./hr.	±0.20%	Repeatability=0.015%
Nutating-disk...............	½ to 2	2 to 160 gpm.	±1.00%	Repeatability=0.100%
Metering pump (proportioning)...........	Off-line	Adjustable 4 to 100% of range	±1.00%	Response=20 sec. full scale
Variable-Area Meters				
Plugged and ported cylinder	1 to 4	....	±0.25%	Mainly for highly viscous fluids.
Tapered-tube rotameter with free bob.................	½ to 4	....	±0.5 to 2.0%	Generally superseded by guided float-type.
Tapered-tube rotameter with center guide.............	½ to 4	....	±0.5 to 2.0%	Inductance bridge often used for transmission.
Tapered-tube rotameter with beaded tube..............	up to 4	....	±0.5 to 2%	Enables viscosity-immune floats to be used without center guide.
Tapered-tube rotameter, metallic..................	....	....	2% of maximum bypass flow rate	Measures bypass flow. Hence can be used on any pipe size containing orifice or other differential-pressure producing primary flow device.
Velocity and Current Meters				
Vortex meters, liquid flow...	1 to 10	6 to 60 gpm. to 700 to 6,800 gpm.	±0.5%	Repeatability=0.1% Response: 1 & 2-in. meters=300 millisec. 3 & 4-in. meters=400 millisec. 6, 8 & 10-in. meters=500 millisec.
Vortex meters, gas..........	1 to 10	750 std. cu. ft./hr. at 20 psig. to 150,000 std. cu. ft./hr. at 1,500 psig. to 1.2 million std. cu. ft./hr. at 20 psig. to 25 million std. cu. ft./hr. at 1,500 psig.	±1.0%	Response: 1 & 2-in. meters=300 millisec. 3 & 4-in. meters=400 millisec. 6, 8 & 10-in. meters=500 millisec.
Vortex mass-flow meter, gas	1 to 10	5 to 25,000 lb./min.	±0.1% over 5:1 range ±0.3% over 25:1 range	Response=1 sec.
Velocity and Current Meters				
Turbine meter with magnetic signal pickup........	1 to 10	6 to 7,200 gpm.	±0.25%	Repeatability=±0.02%
Turbine meter (viscosity compensated)............	6 to 16	600 to 15,000 bbl./hr.	±0.15%	Reapeatability=±0.015%
Electromagnetic meter......		0.1 to 250 million gal./day	±1.0%	Responds almost instantaneously to changes in flow rate.
Current meter.............	2 to 36		±2%	Subtracts reverse flow.

*D. M. Considine, Chemical Engineering, Jan. 29, 1968, p. 84.

3.7.2 VORTEX METERS

The swirlmeter is a volume flow measuring device for use on gases. The fluid entering the swirlmeter goes through a set of stationary swirl blades which impact a swirling motion around the center axis of the meter, as in Fig. 3.7-2. A vortex forms and advances through the meter like a screw. The pitch of the vortex is determined by the swirl blade and body shape. A sensor, which can detect the passage of the vortex (via a decrease in pressure) is used to measure the flow rate: the rate of vortices passing the sensor is proportional to the fluid velocity. The gas is then deswirled at the exit. The same principle is also used in liquid vortex meters.

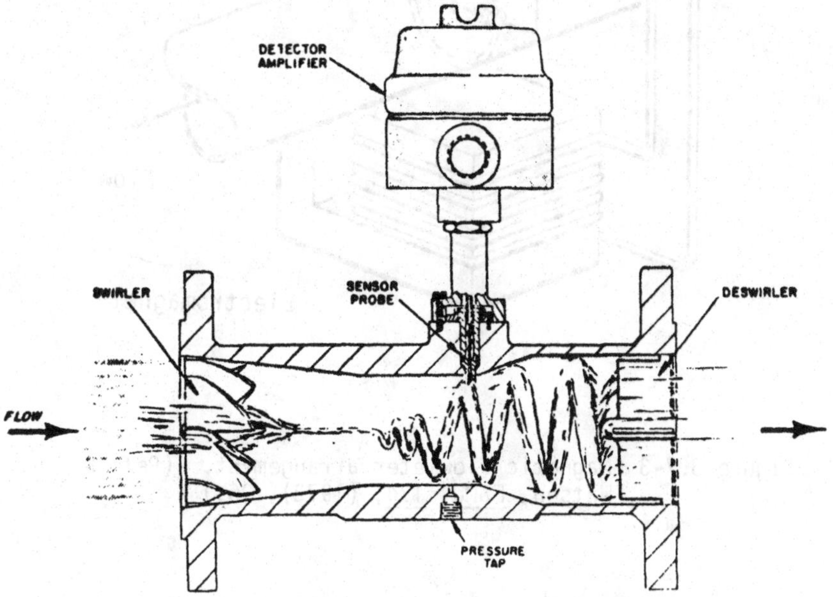

Figure 3.7-2 Cutaway view of swirlmeter.

3.7.3 MAGNETIC FLOWMETERS

Electromagnetic flowmeters have applications to the measurement of volume flow rates of liquids with conductivities of at least 20 micromhos per cm. The principle involved is Faraday's law of induction in which a conductor (the liquid) moves through a magnetic field established by electromagnets surrounding the non-conducting pipe, thereby causing a voltage to be developed across the liquid. The arrangement is shown in Fig. 3.7-3. The flow is given by the equation

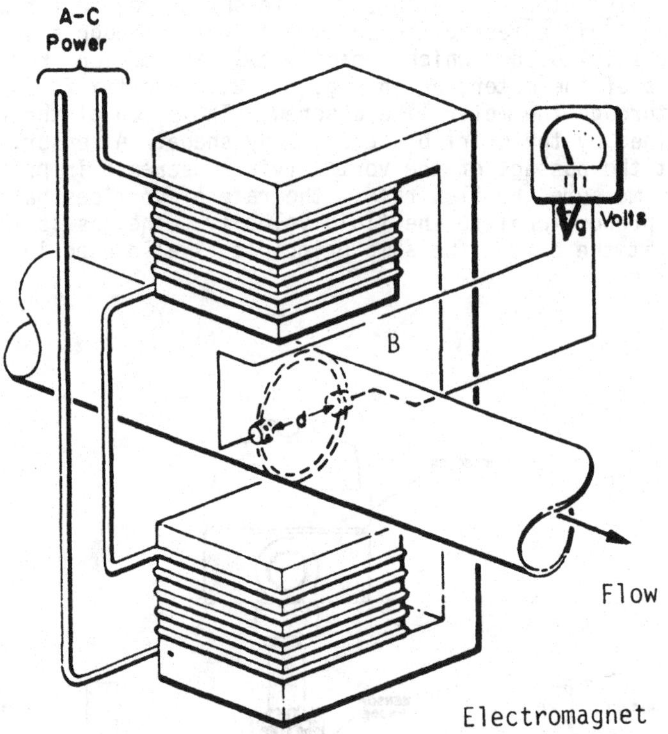

Figure 3.7-3 Magnetic flowmeter arrangement. (Ref. B. Watson, E&MJ, 174, (1973), 96.)

$$q = K\left(\frac{\pi d V_g}{4B}\right) \qquad (3.7\text{-}1)$$

where q is the volume flow rate, K is a flow coefficient, d is the flow meter diameter, V_g the voltage developed, and B the magnetic field flux density. K remains within 0.99 - 1.01 if B is kept within the proper range.

3.7.4 VARIABLE AREA METERS

These meters are based on the principle of placing a restriction in the flowing stream, creating a pressure drop and a corresponding change in flow velocity through the restricted flow area. The pressure drop stays constant and the flow area changes as the velocity changes. The most common type of area meter-called a rotameter-is illustrated in Fig. 3.7-4. In a rotameter, a fluid moves past the float and maintains it in suspension. If the float is at some equilibrium position corresponding to some mass flow rate and then the mass flow rate increases, the float rises. However, as the float rises, the tapered tube presents a larger cross-sectional area for flow, and the velocity of the fluid between the float and the tube wall decreases, so that a new equilibrium position is eventually reached.

The variety in designs of rotameters is so great that there does not exist one relationship valid for all types of rotameters to describe how the mass flow rate varies with height. However, manufacturers usually supply calibration data for their devices, each set of data being appropriate for a specific fluid.

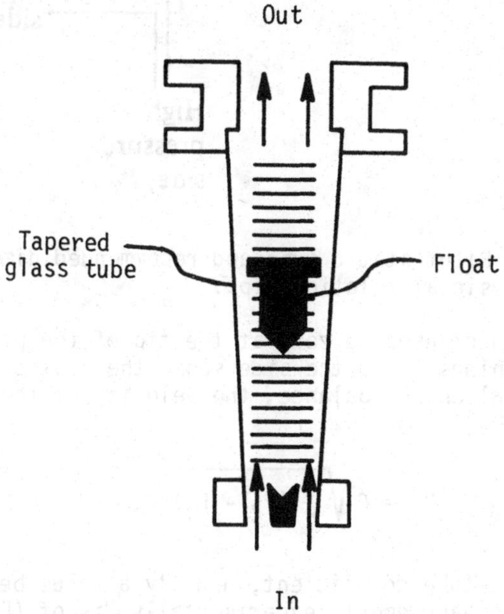

Figure 3.7-4 Schematic diagram of a rotameter.

3.7.5 VARIABLE HEAD METERS

In some circumstances, such as when large volumes of hot, dirty gases are present, direct measurements are out of the question and more indirect means of measuring the flow are required. There are two basic categories of meters for this purpose: velocity meters and head meters.

Velocity meters: A commonly encountered velocity meter is the pitot-static tube which measures local velocities. The pitot tube consists essentially of a tube with an open end facing the stream, as shown in Fig. 3.7-5. The velocity of the fluid along

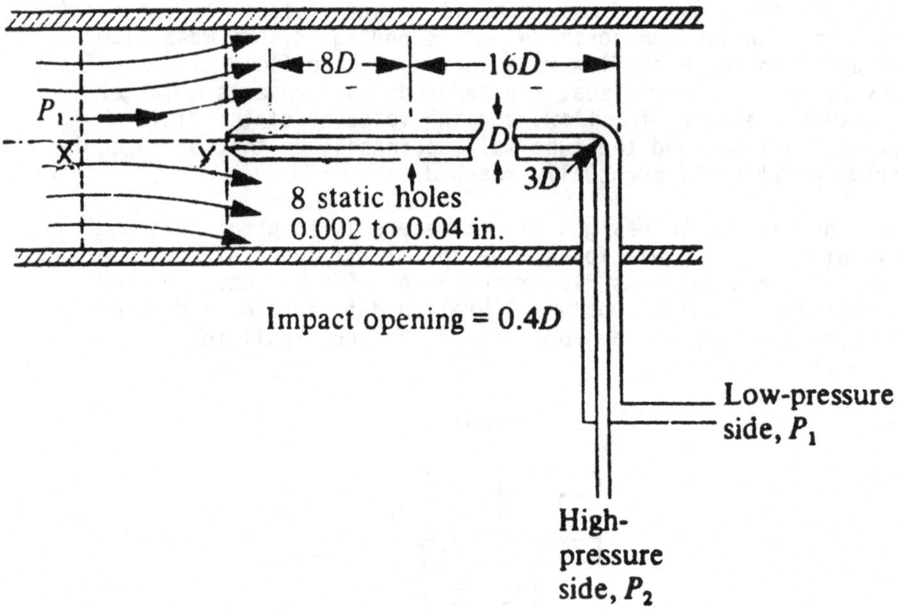

Figure 3.7-5 Pitot-static tube and recommended dimensional relationships.

the streamline x-y decreases to zero at the tip of the pitot-tube opening. Other openings along the side sense the static pressure. Applying a mechanical energy balance, the velocity of the fluid is given by

$$V_1 = C\sqrt{\frac{2}{\rho}(P_2 - P_1)} \tag{3.7-2}$$

where C is the pitot-tube coefficient, usually a value between 0.98 and 1.00. The measurement, experimentally, is of $(P_2 - P_1)$, meaning that the accuracy of V_1 depends on the accuracy of the pressure head measuring device.

Because the pitot tube measures only local velocities, a traverse of the conduit must be made in order to obtain the complete profile. To obtain the density ρ, we usually measure the temperature and pressure upstream from the probe tip. Probing is done following the same pattern as in Figs. 3.3-7 and 3.3-8. Average flow is then calculated from the results of the survey. The accuracy of velocity, and therefore mass flow rate, determinations made using pitot tube traverses is rarely better than $\pm 5\%$.

Head meters: There are essentially three types of head meters, so called because all of them place some sort of restriction in the flow line, causing a local increase in the velocity of the fluid and a corresponding decrease in the pressure head. The simplest is perhaps the orifice plate, illustrated in Fig. 3.7-6; however, the venturi meter and flow nozzle (Figs. 3.7-7 and 3.7-8) are based on the same principle, and the same group of equations apply to all of them. Theoretically, when the mechanical energy balance is solved for the average velocity, $\bar{V}_2$, at location (2) in each case,

$$\bar{V}_2 = \sqrt{\frac{2(P_1 - P_2)}{\rho\left(1 - \left(\frac{D_2}{D_1}\right)^4\right)}} \tag{3.7-3}$$

This is the theoretical velocity at the vena contracta, which disregards frictional energy losses and can never be achieved in practice. Therefore a discharge coefficient C_D which accounts for such losses and an additional geometric factor is introduced. The largest pressure drop is measured at the vena contracta, but it is more convenient to relate the velocity at the orifice plate $\bar{V}_o$ to the pressure drop, $(P_1 - P_2)$, and at the same time use the diameter of the plate opening D_o instead of the diameter of the vena contracta. Thus, a more useful form of the expression is

$$\bar{V}_o = C_D\sqrt{\frac{2\,(P_1 - P_2)}{\rho\,(1 - \beta^4)}} \tag{3.7-4}$$

in which $\beta \equiv D_o/D_i$.

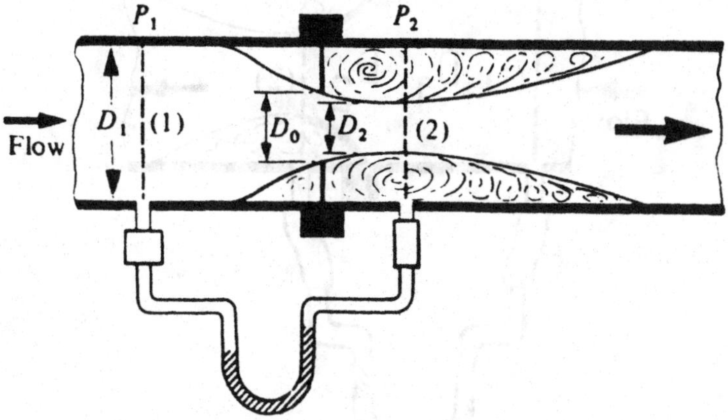

Figure 3.7-6 Orifice meter

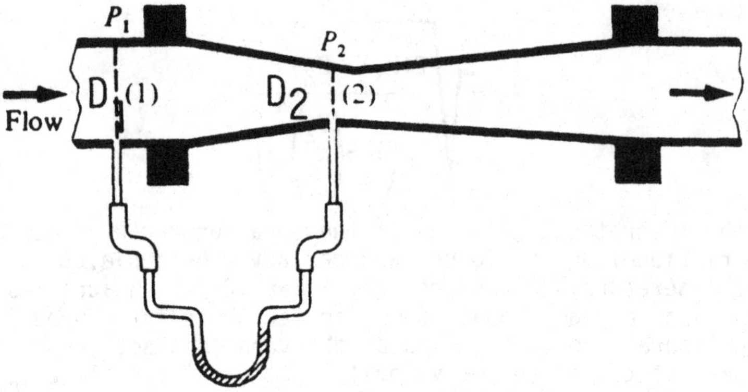

Figure 3.7-7 Venturi meter

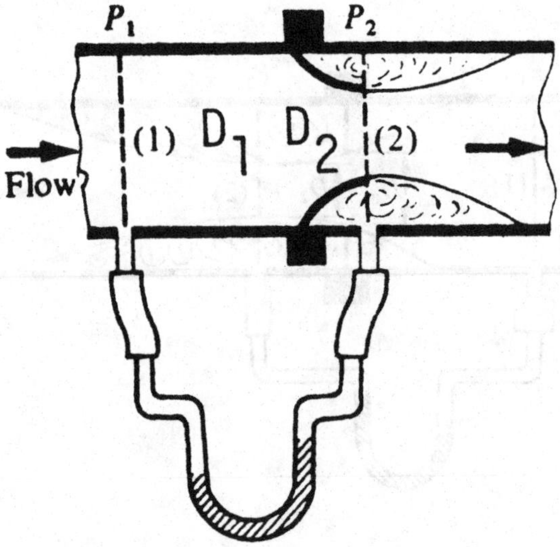

Figure 3.7-8 Flow nozzle

Orifice plates are the simplest and cheapest types of head meters, but they also cause the largest permanent pressure drop in the system, i.e. the pressure difference, $P_2 - P_1$, across the orifice is not entirely recovered.

The design of the venturi meter is such that a gradual restriction in flow area precedes the throat, which is a short, straight section; then the flow area gradually returns to the original area. A venturi tube is nearly frictionless under turbulent (high) flow conditions, so that the typical values of C_D are between 0.98 and 1.0. However, in laminar (low) flow, C_D drops rapidly and calibration data should be consulted. For venturi meters, the permanent pressure drop is much lower than for orifice plates and may be approximated as $0.1\ (P_2 - P_1)$.

Nozzles are similar to orifices in some respects, but are designed so that the discharge is preceded by a smooth contracting passage. As a result, the permanent pressure drop corresponds more closely to that of an orifice than of a venturi.

The accuracy of indicators and the size of primary devices for variable-head devices are given in Table 3.7-1.

3.8 PRESSURE MEASUREMENT

Occasionally, in systems where a gas phase is involved in the process, it is necessary to measure and record pressures. In metallurgical processes requiring material and energy balance analysis, this rarely involves pressures outside the range of 10 atm. to 10^{-6} atm. Measuring characteristics of instruments used for this purpose are presented in Tables 3.8-1 and 3.8-2.

3.8.1 POSITIVE PRESSURE

For positive pressures, a sensing device having some form of elastic element is used for pressures above 5 atm. These elements may include helixes, spirals, Bourdon elements, etc. For lower positive pressures and for compound positive/vacuum ranges, manometers and bells are commonly used. Manometers may utilize mercury or water or oil as the fluid, depending on the range to be covered. An inverted bell is illustrated in Fig. 3.8-1.

3.8.2 VACUUM

The McLeod gauge is a vacuum manometer capable of measuring intermediate vacuums.

Thermal-type vacuum gauges operate on the principle that the heat conductance of a gas changes with the density, and therefore the pressure, of the gas. Thus, a thermal-type gauge is comprised of two major elements--a heated surface and a temperature sensor. As the pressure decreases, the gas is less able to extract heat by conduction away from the heated surface, and the temperature sensor picks up this effect. Generally, one of two kinds of temperature sensors is used. In the thermocouple vacuum gauge, a thermocouple or thermopile is attached to a heater filament that is

Table 3.7-1

Mechanical Indicators and Recorders for Variable-Head Meters*

Element Type	Typical Range	Minimum Span	Accuracy	Remarks
Orifice plate	0 to 10 in. water to 0 to 400 in. water	5 to 10 in. water	± 0.25 to ± 0.5% of full scale for recorders ± 0.5 to ± 2.0% of full scale for indicators	Usually include adjustable damping Response to a step upset is from a few seconds to minutes.
Bellows type	0 to 10 in. water to 0 to 50 psi.	10 in. water	± 0.25 to ± 0.5% of full scale for recorders ± 0.5 to ± 2.0% of full scale for indicators	
Ring-balance manometer	0 to 10 in. water		about 1%	
Float and Cable	0 to 5 in. water to 0 to 36 in. water	5 in. water	± 1.0% of full scale	Relatively slow response because of mechanical nature of system.

Note: Motion-balance transmitters used with the same primary devices have approximately same characteristics as those listed above.

Table 3.7-1 (Continued)

Primary devices for variable-head meters

Flow in Conduits		Flow in Open Channels		
Type	Size	Type	Channel Width	Range of Head
	In.		Ft.	In.
Orifice plate	1/2 to 72			
Flow nozzle	3 to 24	Weir	4 to 17	2 to 18
Flow tube	3 to 48	Flume	1/4 to 50	4 to 80
Venturi tube	3 to 48			
Pitot tube	3 to 48			

Remarks: For information pertaining to constancy of coefficients, accuracy, Reynolds number range, see ASME Fluid Meters Handbook, 5th Edition, p. 188, Table 17.

Accuracy varies from 2 to 3%

* D. M. Considine, Chemical Engineering, (January 29, 1968), p. 84.

Table 3.8-1

Measurement Characteristics of Positive Pressure Sensors*

Type	Maximum Low	Range High	Minimum Span	Sensitivity	Accuracy	Response
Elastic Elements						
C bourdon	15 psi	10,000 psi**	10 psi	0.01% of span	0.05% of span	...
Helical bourdon (metallic)	100 psi	10,000 psi	100 psi	0.01% of span	0.05% of span	...
Spiral bourdon	15 psi	25,000 psi	15 psi	0.01% of span	0.05% of span	...
Spring-loaded bellows	0 psi	500 psi	5 psi	0.25% of span	0.50% of span	...
Opposed bellows	0 psi	200 psi	15 psi	0.10% of span	0.25% of span	...
Slack diaphragm	0 psi	80 in. water	1 in. water	0.25% of span	1.0 % of span	...
Metallic diaphragm	0 psi	140 psi	3 in. water	0.25% of span	1.0 % of 1.5% of span	...
Helical bourdon (fused quartz)	0 psig	500 psig	5 psi	0.00005 psi	0.010% of reading	2 min. full scale
Manometers and Bells						
Mercury manometer	0 mm	800 mm Hg	1 mm Hg	0.1 mm Hg	±0.1 mm Hg	...
Inclined draft gage	0.5 in H_2O	50 in. water	1 in. water	0.01 in. water	±0.01 in. water	...
Ledoux bell	Maximum differential is 0 to 212 in. water				±2% FS+	...
Inverted bell and balanced beam	-30 in	+2 in. water	0.1 in. water	±0.0005 in. water	±0.002 in. water	...
Electrical Sensors and Transducers						
Capacitive sensors	0 psi	5,000 psi	0.1 psi	±0.02% FS	±0.15% FS	25 millisec. for 63% step change
Strain-gage transducers						
Bellows and cantilever beam	5 psi	60 psi	30 mv	±0.1% FS	±0.25% FS	20 Hz
Catenary diaphragm	100 psi	40,000 psi	30 mv	±0.1% FS	±0.25% FS	20,000 Hz
Tubetype (flat or round)	100 psi	200,000 psi	15 to 30 mv	±0.1% FS	±0.25% FS	10,000 Hz
Beam and diaphragm	15 psi	1,000 psi	30 mv	±0.1% FS	±0.25% FS	500 Hz

* D. M. Considine, Chemical Engineering, (Jan. 29, 1968), p. 84.

** Special units can go up to 100,000 psi. or higher.

\+ Full scale.

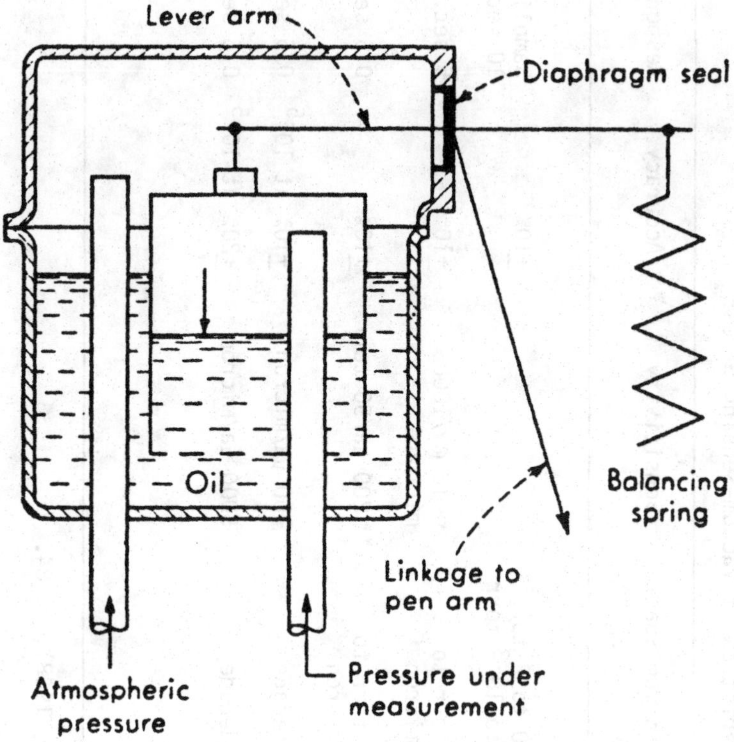

Figure 3.8-1 Inverted bell measures pressure by displacement of the bell, whose motion is transmitted through a diaphragm.

heated by a constant energy d.c. input. In the resistance-type or Pirani gauge, the same heating principle is used, but the temperature is sensed by a change in the resistance of a wire or semiconductor.

Ionization-type vacuum gauges measure the electric current that results from ionization of a gas whose pressure is being measured. The differences between various ionization gauges relate principally to the manner in which the ions are produced. To convert a gas molecule into a positive ion, an electron must be removed by supplying energy equal to the ionization potential of the given gas molecule. The required 4 to 25 electron-volts may be generated in several ways. If the energy is supplied at a constant rate, the rate of ion production will be constant. Fortunately, this is very nearly proportional to the pressure of the gas. The pressure is therefore measured by measuring currents produced.

Table 3.8-2

Measurement Characteristics of Vacuum-Sensing Gages*

Type	Maximum Range Low	Maximum Range High	Minimum Span	Sensitivity	Accuracy	Response Speed
	Torr	Torr				
McLeod	10^{-3}	150	150 to 2 torr 2 to 10^{-3} torr		±10%	Sampling: 10 sec/reading
Thermocouple	10^{-3}	1	3 torr to 1 micron	** 10 mv/1000 microns	±10%	5 sec.
Pirani	10^{-2}	50	50 torr to 1 micron	** 200 μa/50 torr	±10%	0.5 sec.
Hot-filament	2×10^{-11}	10^{-3}	1 decade	100 μa/micron	±10% at 10^{-5}	0.5 sec.
Cold cathode, Philips	2×10^{-9}	10^{-3}	1 decade	5000 μa/micron	±20% at 10^{-5}	0.5 sec.

** Full scale.

* D. M. Considine, Chemical Engineering, (Jan. 29, 1968), p. 84.

Hot-filament gauges have sufficient energy supplied to ionize the gas by means of electron bombardment. Electrons are derived from thermionic emission as in an ordinary vacuum triode. In cold-cathode gauges, a high-potential field withdraws electrons from a cold-cathode surface.

3.9 CHEMICAL ANALYSES

Once a significant sample has been obtained, an accurate chemical analysis must be made. The technique(s) used to obtain the analysis will depend on the physical condition (solid, liquid, particulate, etc.) and the nature (metallic, oxide, silicate, etc.) of the sample as well as on the level of concentration of the element in question. In turn, the technique used will govern the accuracy of the result.

Fig. 3.9-1 shows the optimum ranges for application of various analytical techniques. Table 3.9-1 indicates the usual form of the sample required in each case and some estimate of the time involved in preparation and analysis of each sample.

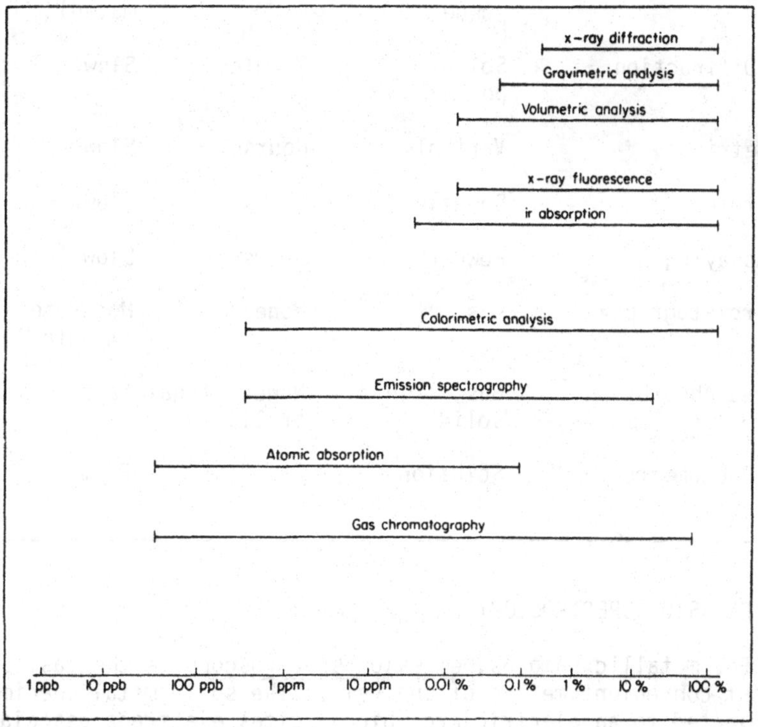

Figure 3.9-1 Ranges of application of various analytical techniques. (Ref. Undergraduate Instrumental Analysis, J. W. Robinson, Marcel Dekker, Inc., New York, N.Y. 1970)

Table 3.9-1

Sample Forms and Analytical Times for Various Analytical Techniques

Chemical Analyses	Sample	Preparation Time	Analytical Time
Emission Spectroscopy	Solid Disk (Compacted powder)	Minutes	Fast (seconds)
Atomic Absorbtion	Solution	Depends on sample-slow if solid, fast if solution.	Fast
Colorimetric Analysis	Solution	"	Moderate
X-Ray Fluorescence	Solid-powder	Minutes	Fast (if automated)
X-Ray Diffraction	Solid-powder	Minutes	Slow
Gravimetric	Variable	Hours	Slow
Volumetric	Solution	Hours	Slow
Fire Assaying	Powder	Hours	Slow (hours)
Gas Chromatography	Gas	None	Moderate (minutes)
Infrared Absorbtion	Gas, liquid, solid	None, if gas or liquid	Fast
Flame Photometry	Solution	"	Fast

3.9.1 EMISSION SPECTROSCOPY

For metallic samples, emission spectroscopy is the fastest and most convenient method of analysis. The solid metal specimen is subjected to an electric arc which excites electrons associated with the various elements present in the metal. As the electrons return to their ground states from the excited states, they emit radiation with a characteristic wavelength. The intensity of the emitted wavelengths is proportional to the concentrations of the various atoms present in the sample, subject to correction for absorption effects due to the presence of other elements in the sample (matrix effects). Using photomultiplier tubes and automatic equipment, simultaneous determination of the concentration of twelve or more elements can be made. The precision depends on the concentration. If the concentration is 1 to 10%, as little as 1% relative error can be achieved. If the concentration is on the order of 10^{-2} to 10^{-5}%, the error may be 5% of the value. It must be noted, however, that a relatively small part of the sample surface is subjected to the arc (burned) and if there is significant segregation within the sample, large errors (5-25%) could result. Brooks and Whittaker* have reported what they feel are the accuracies of analyses obtained using a modern automatic emission spectrograph (Quantovac). Their results are presented in Tables 3.9-2 and 3.9-3.

3.9.2 X-RAY TECHNIQUES

X-ray diffraction and fluorescence both require a finely ground powder sample prepared in such a way as to provide a uniform, compacted sample. In diffraction, the intensity of a beam diffracted from a particular plane of a specific crystal is measured, and by comparison to a standard intensity, the amount of that crystal phase present (not an elemental analysis) is inferred. The sensitivity limit is about 1% and the accuracy is rarely

Table 3.9-2

Accuracies of Analysis of Elements Obtainable From Sound Low-Alloy Steel Samples Using a Vacuum Emission Spectrograph

Element	Level	Accuracy
Carbon	At 1%	± 0.02%
	At 0.25%	± 0.01
Manganese	Up to 1%	± 0.02
Sulphur	Above 0.03%	± 0.004
	0.015-0.03%	± 0.002
	Less than 0.015%	± 0.001
Phosphorus	Up to 0.05%	± 0.0025
Nickel	At 1%	± 0.03
	For each 1% over add	± 0.01
Chromium	At 1%	± 0.03
	At 9%	±1% of content
Molybdenum	Up to 0.3%	± 0.015
	Above 0.3%	± 0.025
Copper	Residual	± 0.01
Tin	Residual	± 0.002

* C. H. Brooks and R. Whitaker, article in Control and Composition in Steelmaking, The Iron and Steel Institute, London, 1966, p. 103.

Table 3.9-3

Accuracies of Analysis of Stainless Steels Using the Non-Vacuum Emission Spectrograph (Quantometer)*

Element	Level	Accuracy
Manganese	Up to 2%	± 0.02%
Silicon	Up to 1%	± 0.03
Nickel	5 - 15%	± 0.15
Chromium	10 - 20%	± 0.20
Molybdenum	Up to 1.0%	± 0.02
	Up to 3.5%	± 3 of content

* From Brooks and Whitaker, op cit.

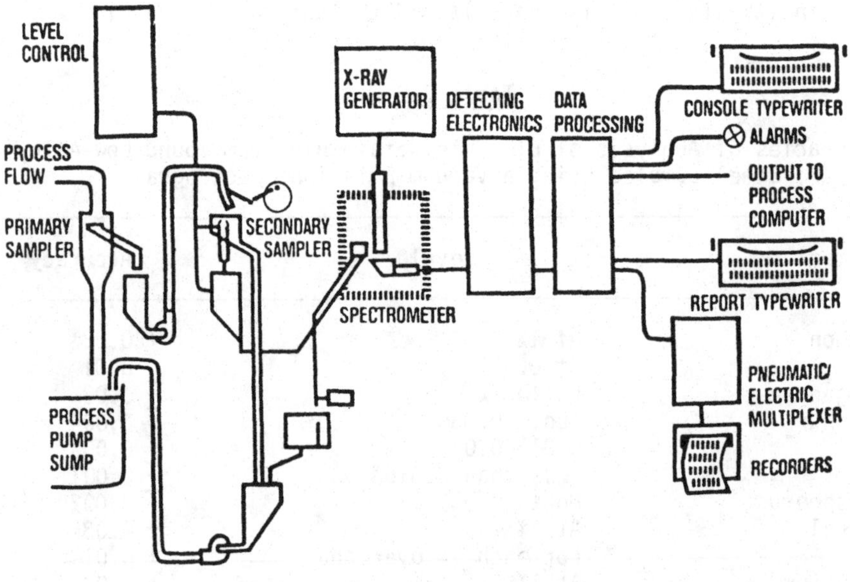

Figure 3.9-2 On Stream Analysis System utilizing x-ray fluorescence analysis on automatically collected grab samples of mineral slurry. (Ref.: Otokumpu Oy publication)

better than ± 1%. X-ray diffraction is most useful as a phase identification tool, rather than for quantitative analysis.

In x-ray fluorescence, an x-ray beam from an x-ray tube is used to excite characteristic x-radiation from the elements present in the sample. The intensity of the x-radiation from an element is then measured, and the concentration inferred, after

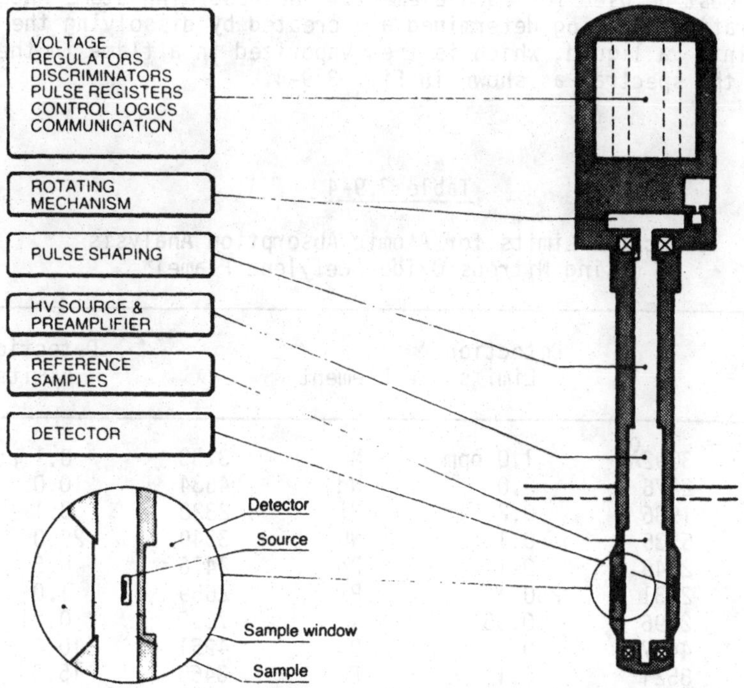

Figure 3.9-3 Radioisotope used to excite x-radiation in mineral particles contained in slurry flowing past detector. (Ref.: Otokumpu Oy publication)

inter-element correction factors have been applied. The sensitivity limit is said to be ± 0.01%, but the accuracy in complex oxide, or sulfide systems is probably not better than ± 0.5% for elements present in concentrations of 5 to 50%. This technique is usually applied to a batch sample (which may be taken by automatic sampling devices from a continuous stream of ore pulp, as in Fig. 3.9-2), but recently a technique has been developed in which both a radioisotope used to excite fluorsecent radiation from mineral particles and the radiation detector are placed directly in the process stream, as illustrated in Fig. 3.9-3.

3.9.3 ATOMIC ABSORBTION SPECTROSCOPY

In recent years, the technique of atomic absorbtion spectroscopy has found wide acceptance as an analytical tool, because it is relatively free of inter-element interactions and the resulting need for complicated corrections. Basically, the technique measures the absorbtion by metallic atoms of spectral energy characteristic of those atoms. The fraction of the spectra absorbed is directly related to the concentration of those atoms.

The spectra are generated by using hollow cathodes made of the metal whose analysis is desired, meaning that a different cathode must be used for each element. The absorbing atoms whose concentration is being determined are created by dissolving the sample into a liquid, which is then vaporized in a flame in the path of the spectra, as shown in Fig. 3.9-4.

Table 3.9-4

Detection Limits for Atomic Absorption Analysis
(Using Nitrous Oxide-Acetylene Flame)*

Element	λ	Detection Limits	Element	λ	Detection Limits
Al	3092Å	1.0 ppm	Mo	3133	0.1 ppm
Sb	2176	2.0	Nd	4634	10.0
As	1936	0.2	Ni	2320	0.1
Ba	5535	0.1	Nb	3349	20.0
Be	2349	0.1	Pd	2476	0.5
Bi	2231	250.	Pt	2659	1.0
B	2496	0.05	K	7665	0.01
Ca	4427	0.1	Pr	4951	10.0
Cs	8521	0.1	Re	3460	15.0
Cr	3579	0.1	Rh	3435	0.1
Co	2407	0.1	Rb	7800	0.1
Cu	3247	1.0	Ru	3499	0.8
Dy	4212	1.0	Sm	4297	10.0
Er	4008	2.0	Sc	3912	1.0
Eu	4594	20.0	Se	1961	1.0
Gd	3684	1.0	Si	2516	0.8
Ga	2874	2.0	Ag	3281	0.01
Ge	2652	1.0	Na	5890	0.01
Au	2428	10.0	Sr	4607	0.1
Hf	3072	2.0	Ta	4714	10.0
Ho	4104	0.1	Te	2143	0.5
In	3040	0.1	Th	3776	0.4
Fe	2483	75.	Sn	2354	0.5
La	3928	0.01	Ti	3643	1.0
Pb	2170	0.03	W	4009	1.0
Li	6707	0.001	U	3515	100.
Mg	2852	0.05	V	3184	1.0
Mn	2795	1.0	Y	3988	2.0
Hg	2537		Zn	2139	0.01
			Zr	3601	50.0

* J. W. Robinson, op cit.

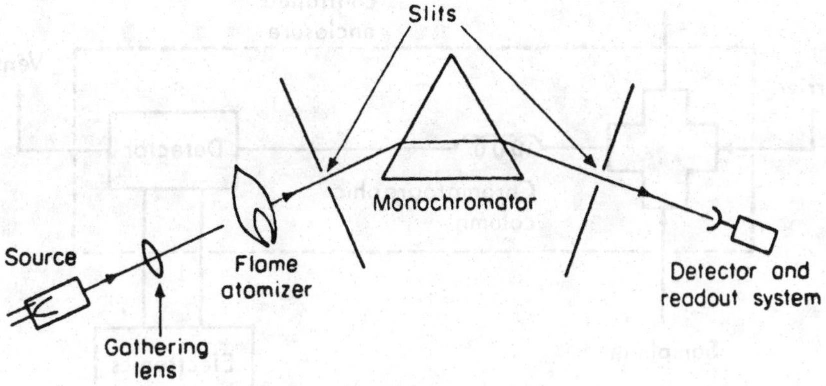

Figure 3.9-4 Schematic diagram of the equipment used for atomic absorption spectroscopy.

The elements that can be determined quantitatively by atomic absorbtion, and the lower limits of detectability are given in Table 3.9-4.

3.9.4 CHROMATOGRAPHY

The process chromatograph takes a fixed volume of sample and separates it into its molecular components by absorbing all of the molecules from the sample onto a solid absorber in a chromatographic column. Then an inert carrier gas picks up the different components as they are selectively desorbed from the column and carries them one-at-a-time through a detector.

The detector analyzes the gas discharged from the column. Usually a thermal conductivity cell is used as the detector. Fig. 3.9-5 illustrates the general layout of a chromatograph and Fig. 3.9-6 illustrates a thermal conductivity cell, which works on the principle that the heat conducted by a gas is directly affected by the gas composition. When a binary gas, the carrier plus one of the process gas constituents, passes through the sample port of the cell, the surface temperatures of the measuring filaments differ from the temperatures at the balance condition. This causes a change in resistance of the filaments and ultimately this change is recorded as a concentration of the gas constituent. Calibration for each gas constituent is required.

It should be pointed out that a chromatograph is a batch analyzer, with each sample from a process stream having to be injected separately. The accuracy depends on the level being measured and on the calibration. Trace measurements in the PPM range are possible, but typical accuracies are ± 1% of full-scale.

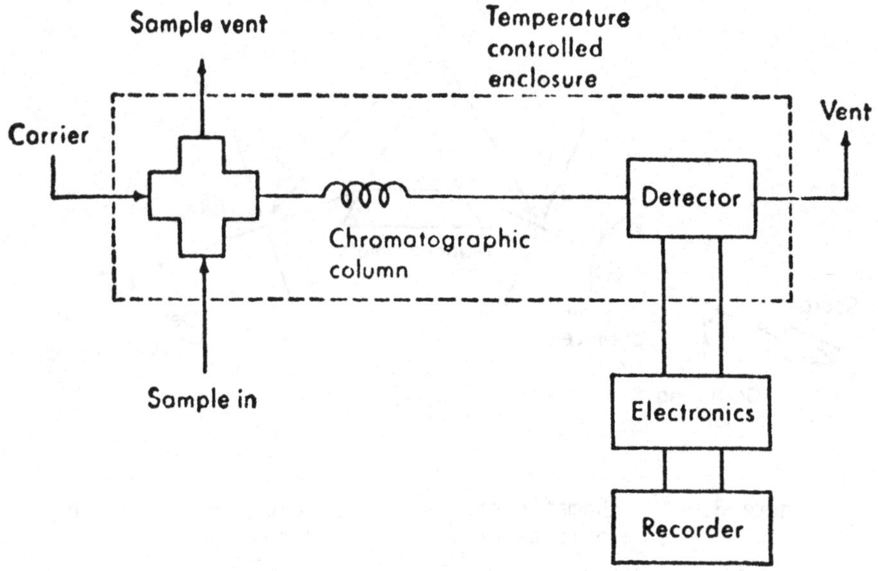

Figure 3.9-5 General layout of a gas chromatograph.

3.9.5 INFRARED ANALYZERS

Another common gas analyzer is the infrared analyzer. This works on the principle that various molecules absorb radiation when some part of the molecule naturally vibrates at the same frequency as the incident radiation. In practice, a source of irradiation provides a spectrum of frequencies from which the desired wavelength corresponding to the natural frequency of a particular molecule is chosen and passed through the sample to a detector. The detector measures the intensity of radiation after it passes through the sample, I, and, knowing the original intensity without the sample present, I_0, the ratio I/I_0 can be used to make a quantitative estimate of the concentration of the molecule absorbing that particular frequency.

By changing the frequency, the absorbtion spectra of the sample can be obtained and its complete analysis inferred. Unfortunately, in practice there are considerable difficulties with maintaining the stability of the equipment. While qualitative analysis is relatively easy, quantitative analysis is not. Infrared analyzers are used most often to monitor CO, CO_2, H_2, CH_4, etc. in heat treating and iron and steelmaking facilities.

3.9.6 GASEOUS OXYGEN ANALYZERS

The oxygen content of flue gas is an important indicator of combustion efficiency. For this reason it is often monitored on a continuous basis. The most common method makes use of the strong paramagnetic susceptibility of oxygen. In this case the process gas is passed through a thermal conductivity cell with a heated filament on either side of the stream, as shown in Fig. 3.9-7.

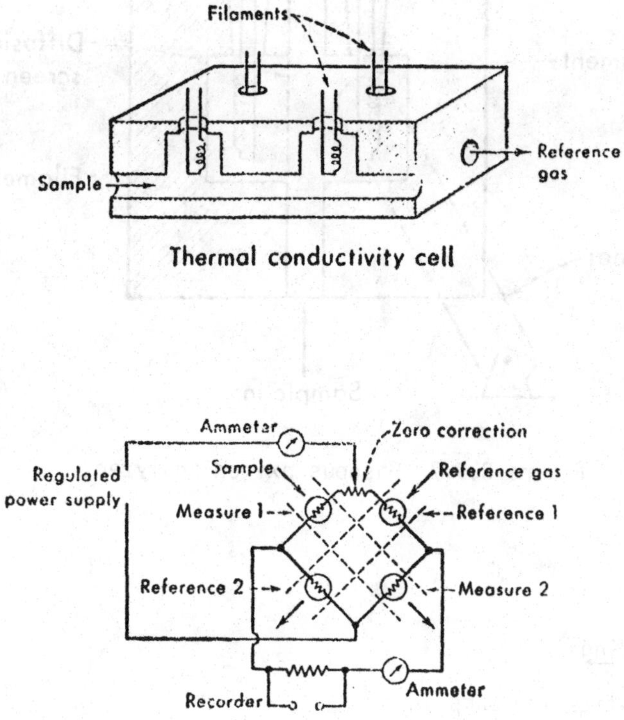

Figure 3.9-6 Thermal conductivity analyzer has four-element cells whose filaments form a Wheatstone bridge. (from Chemical Engineering, May 6, 1968, p. 165).

One filament is surrounded by a magnetic field and therefore the oxygen in the sample is drawn to that side, preferentially cooling that filament. The resulting difference in resistances of the filament is then used to infer an oxygen content of the gas.

3.10 SUMMARY

In this chapter, many of the measurement techniques and problems involved in obtaining data needed for development of material and energy balances in metallurgical plants have been discussed.

Accuracies of various measurements were presented as well as limits of applications.

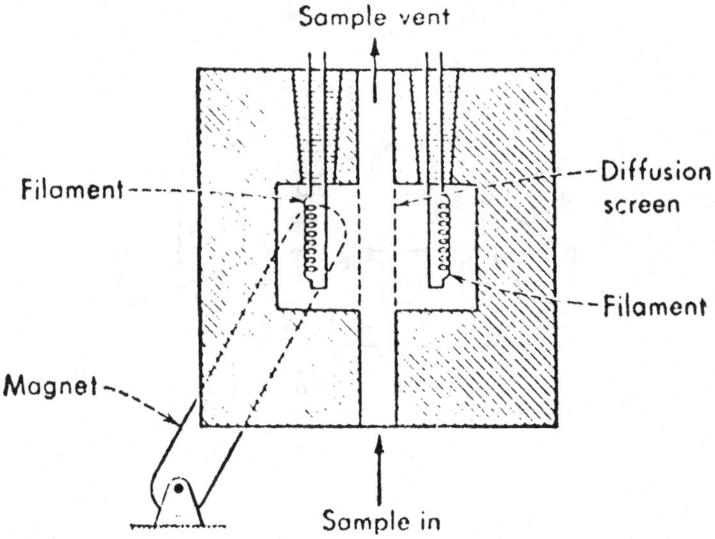

Figure 3.9-7 Gaseous oxygen analyzer.

Further Reading

A. Statistics and Sampling:

1. A. H. Bowker and G. J. Lieberman, Engineering Statistics, Prentice-Hall, 1959.

2. M. H. Belz, Statistical Methods in the Process Industries, J. Wiley, New York, N.Y., 1973.

3. J. R. Fair, B. B. Crocker, and H. R. Null, "Sampling and Analyzing Trace Quantities", Chem. Engr., Sept. 18, 1972, pp. 146-154.

4. C. A. Bicking, "The Sampling of Bulk Materials", Materials Research & Standards, ASTM, March, 1967.

5. B. H. Lloyd, "The Statistical Approach to Bulk Sampling Problems", Industrial Quality Control, vol. 9, No. 6, 1953, p. 113.

6. C. A. Bicking, "Bibliography on Sampling of Raw Materials and Products in Bulk", Tappi, vol. 47, No.5, 1964, p. 147A.

7. G. H. Jowett, "The Accuracy of Systematic Sampling from Conveyor Belts", Applied Statistics, vol. 1, No 1, 1952, p. 50.

8. A. J. Duncan, "A Case Study in the Sampling and Analysis of Bulk Material", Proceedings 32nd Session Intern. Statistical Inst., Preprint No. 29, Tokyo, 1960.

9. A. J. Duncan, "Bulk Sampling: Problems and Lines of Attack", Technometrics, vol. 4, No. 2, May 1962.

10. Sampling Plans in ASTM Specifications, ASTM Bulletin No. 223, 1957.

11. Common Rules for Methods of Sampling Bulk Materials of Mining Products, Japanese Industrial Standard M-8100-65.

12. Recommended Practice for Probability Sampling of Materials, ASTM E-105-58.

13. Standard Methods of Laboratory Sampling and Analysis of Coal and Coke, ASTM D-271-68.

14. G. E. Keller, S. J. Aresco, and J. Visman, "Sampling of Coal", Coal Preparation, AIME, New York, 1968, Chapter 2.

15. R. C. Merritt, The Extractive Metallurgy of Uranium, CSMRI, Golden, Colorado, 1971, pp. 42-44.

16. N. L. Morrow, R. S. Brief, and R. R. Bertrand, "Sampling and Analyzing Air Pollution Sources", Chem. Engr., Jan. 24, 1972, pp. 84-98.

17. A. J. Sinclair, "Sampling a Mineral Deposit for Feasibility Studies and Metallurgical Testing", Mineral Processing Plant Design, AIME, New York, 1978, pp. 115-134.

18. A. F. Taggart, Handbook of Mineral Dressing, John Wiley & Sons, New York, 1948, 3rd printing, pp. 19-01 to 19-68.

B. Error Propagation:

1. L. E. Woodman, An Application of the Theory of Measurements to Certain Engineering Problems, Bulletin, University of Mo. School of Mines and Metallurgy, vol. 15, No. 1, June 1942.

C. Instrumentation:

1. R. P. Benedict, Fundamentals of Temperature, Pressure, and Flow Measurement, 2nd Ed., Wiley-Interscience, N. Y., 1977.

2. R. G. Thompson, "Water-Pollution Instrumentation", Chem. Engr., June 21, 1976, pp. 151-154.

3. H. C. McKee, "Air-Pollution Instrumentation", Chem. Engr., June 21, 1976, pp. 155-158.

4. D. M. Considine, "Process Instrumentation", Chem. Engr., Jan. 29, 1968, pp. 84-113.

5. J. E. Lawver and W. Barbarowicz, "Automatic Controls for Flotation Plants", Froth Flotation 50th Anniversary Volume, AIME, New York, 1962, pp. 539-573.

6. M. H. Lajoy, Industrial Automatic Controls, Prentice-Hall, Inc., Englewood Cliffs, N.J., 1963, 7th printing.

D. Chemical Analyses:

1. J. W. Robinson, Undergraduate Instrumental Analysis, 2nd Ed., M. Dekker, Publisher, N. Y., 1970.

2. I. M. Kolthoff, E. B. Sandell, E. J. Meehan, S. Bruckenstein, Quantitative Chemical Analysis, 4th Ed., MacMillan, London, 1969.

3. R. S. Young, Chemical Analysis in Extractive Metallurgy, Barnes & Noble, New York, 1971.

4. J. E. Brown, "Onstream Process Analyzers", Chemical Engineering, May 6, 1968, pp. 164-176.

CHAPTER 4

MATERIAL BALANCES

4.0 INTRODUCTION

Any metallurgical process involving interactions between two or more materials may have to be subjected to an analysis involving a material balance. There are two distinctly different stages in the life of a process when balances must be made: the design stage, where efficiencies of reactions are estimated, minor losses are often ignored or not anticipated, and exact material balances are computed, and the operational stage, where for purposes of process control or performance evaluation, real balances are determined by sampling and measuring the various inputs and outputs of the process. The latter is usually very difficult, because all too often the various balances do not close on account of measuring errors or the necessity to estimate weights of various process streams. In this Chapter the principles and techniques used for calculating and developing material balances will be presented and discussed.

4.1 CONSERVATION OF MATTER (MASS)

Just as in stoichiometric calculations, the basis for all material balances is the law of conservation of matter, which states that matter cannot be created or destroyed in a given system. (Of course, this does not apply to nuclear reactions.) In the case of stoichiometric calculations, this meant that the weight of products of a reaction had to equal the weight of reactants. In the case of processes, this is not necessarily the case. It is possible to have an unsteady-state situation in which accumulation or depletion within the process may occur. In general, therefore,

$$\text{Mass Input} = \text{Mass Output} + \text{Mass Accumulation} \tag{4.1-1}$$

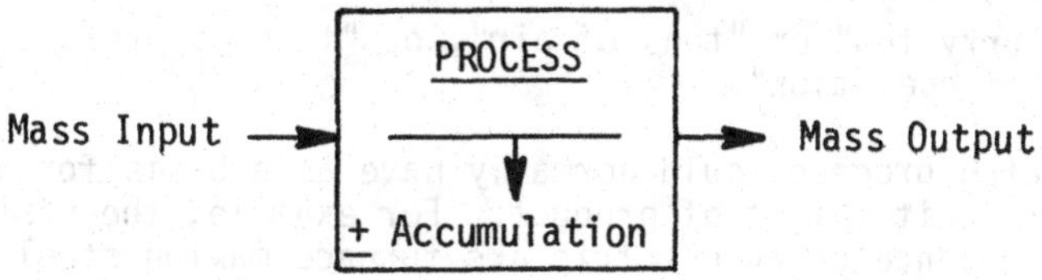

The general equation for a continuous process would be

$$\frac{\text{Mass Input}}{\text{unit time}} = \frac{\text{Mass Output}}{\text{unit time}} + \frac{\text{Mass Accumulation}}{\text{unit time}} \tag{4.1-2}$$

In a continuous process, steady-state is defined as the state of a process in which there is no change with time of any condition of the process. This includes the amount and average composition of material within the process, so that in a continuous process operating at steady-state, there can be no accumulation or depletion. Therefore,

$$\frac{\text{Mass Input}}{\text{unit time}} = \frac{\text{Mass Output}}{\text{unit time}} \qquad \text{(steady-state)} \quad (4.1\text{-}3)$$

On the other hand a batch process is never a steady-state process, but usually does not involve accumulation, and so

$$\frac{\text{Mass Input}}{\text{Batch}} = \frac{\text{Mass Output}}{\text{Batch}} \qquad \text{(Batch Process)} \quad (4.1\text{-}4)$$

It is evident that the basis eg., time, batch, etc for any material balance depends on the process. It is usually chosen for convenience. For example, the basis of a material balance for a fluidized bed roaster, into which a slurry of water and chalcopyrite ($CuFeS_2$) is continuously fed, along with enough air to roast the ore to Fe_3O_4 and $CuSO_4$, could reasonably be

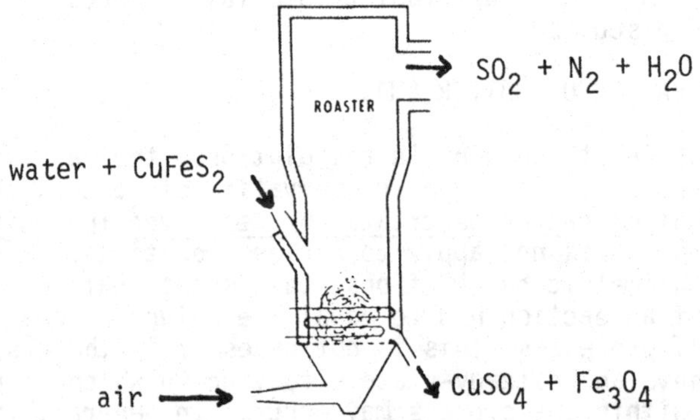

"tons of slurry fed" or "tons of air", or "tons of product solids", or "minute of operation".

A batch process would normally have as a basis for a material balance a unit weight of product. For example, the basis for a material balance on an electric arc furnace making steel would be 1 ton of steel produced.

The law of mass conservation applies not only to the total mass, but to the conservation of elements as well. That is, at steady-state or in a batch, the mass of any element put into a process must equal the mass of that element out. Since the sum of all of the elemental mass balances must equal the total mass balance, it is obvious that if there are C elements present, then there are just C independent mass balances, even though there are C+1 possible equations. However, the total balance may be utilized in place of any one elemental balance. For example, in the melting of an alloy in a laboratory furnace, calculation of the weight of charge materials needed involves making as many balances as there are elements in the alloy.

(1) A 70% Ni-30% Cu alloy being made from pure Ni, pure Cu and scrap containing 50% Ni and 50% Cu, would require two equations to determine how much of each material to use: a total weight balance and either a Cu or Ni balance, or both a Cu and a Ni balance, (but not a total balance).

(2) A 70% Ni-25% Cu-5% Sn alloy made from alloyed scrap, pure Ni, pure Cu, and pure Sn metal would require three equations: a total weight balance and two elemental balances, or three elemental balances.

In melting problems such as those above, in principle any of the elemental balances may be used. However, in other cases this may not be the case because the number of independent equations may be reduced due to additional restrictions. Consider the following example, which illustrates the role of an additional restriction, the approach to setting up balances, and the difference between a design stage balance and an operations analysis balance.

EXAMPLE 4.1-1: An ore containing Fe_3O_4 and SiO_2 is separated by a magnetic separator into two streams; one rich in magnetite (Fe_3O_4) and one depleted in Fe_3O_4. How many independent balances can be written?

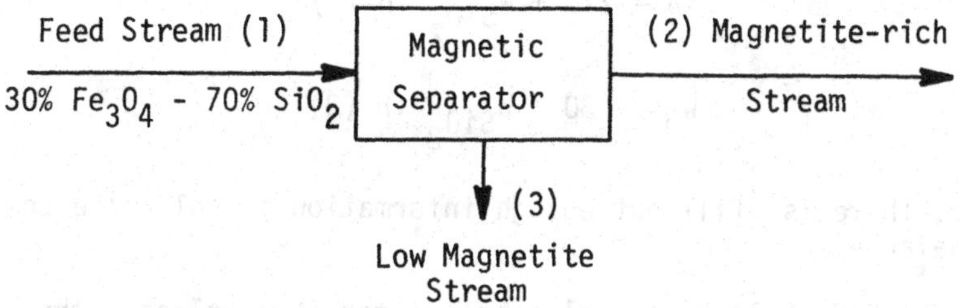

Solution: There are three elements present, Fe, Si, and O, but they are not independent, since a definite proportion of oxygen is associated with the iron in one case and the silicon in the other and no transfer of oxygen occurs between them by reaction. In other words, in this system there are only two independent elemental balances, Fe and Si, or Fe_3O_4 and SiO_2. Additionally, of course, the total balance

wt. of stream 1 = wt. of stream 2 + wt. of stream 3

can be written.

Which balances to write will depend on what other information is available. For example, if the balance is being made at the design stage, product stream analyses are not available, but a projected separation efficiency is known, eg., 90% of the input Fe_3O_4 goes into product stream (2). If the analysis of the feed stream, e.g. 30% Fe_3O_4, and the mass flow rate of that stream, e.g. 1000 Kg/h, are known, the available data may be represented as follows:

Information	Stream (1)	(2)	(3)
Mass flow rate	X	0	0
Fe_3O_4 mass flow rate	X	X	X
SiO_2 flow rate	X	0	0

Now only two independent equations can be written. For example, Equation:

Total Balance: $W_2 + W_3 = 1000$

Fe_3O_4 Balance: $[(0.9)(0.3)(1000)]_2+[(0.1)(0.3)(1000)]_3=[(0.3)(1000)]_1$

$$270 + 30 = 300$$

or $W_2 = 270 + W_{SiO_2}$ in (2)

and $W_3 = 30 + W_{SiO_2}$ in (3)

However, there is still not enough information to solve the complete balance.

The SiO_2 balance can also be written, but unless either the total mass flow of one of streams (2) or (3), or the analysis of either stream (2) or (3) are specified, the complete material balance cannot be solved. If it is assumed, i.e. assigned as a design variable, that the analysis of stream (2) is 20% SiO_2, then 270 Kg of Fe_3O_4 is 80% of (2) and W_{SiO_2} in (2) is 67.5 kg. Now the balance may be completed.

$$W_2 = 270 + 67.5 = 337.5 \text{ kg.}$$

$$W_3 = 1000 - 337.5 = 662.5 \text{ kg.}$$

The analysis of W_3 would then be (30/662.5) x 100 = 4.53% Fe_3O_4.

Suppose now, however, that such a process is running and an evaluation of its performance, i.e., its separation efficiency, is desired. In this case an analysis of each stream, but not necessarily the mass flow rate will usually be known. Recalling that only two independent balances can be written:

$$Fe_3O_4 \text{ Balance: } \left(\frac{\%Fe_3O_4}{100}\right)_1 \cdot W_1 = \left(\frac{\%Fe_3O_4}{100}\right)_2 \cdot W_2 + \left(\frac{\%Fe_3O_4}{100}\right)_3 \cdot W_3$$

$$SiO_2 \text{ Balances: } \left(\frac{\%\ SiO_2}{100}\right)_1 \cdot W_1 = \left(\frac{\%SiO_2}{100}\right)_2 \cdot W_2 + \left(\frac{\%SiO_2}{100}\right)_3 \cdot W_3$$

Now it is seen that there are two equations and three unknowns, W_1, W_2, and W_3. Therefore, at least one mass flow rate must be known before the complete balance can be solved. If, for example, W_1 is measured and is 1000 kg/h, then the balance equations become:

$$\text{Total:} \quad 1000 = 1.0\ W_2 + 1.0\ W_3$$

$$Fe_3O_4\text{: } (0.3)(1000) = (0.8)\ W_2 + (0.03)\ W_3$$

$$SiO_2\text{ : } (0.7)(1000) = (0.2)\ W_2 + (0.97)\ W_3$$

If the analysis of stream 2 is 3% Fe_3O_4 - 97% SiO_2 and that of stream 3 is 80% Fe_3O_4 - 20% SiO_2, solving the Fe_3O_4 balance for $W_2 = f(W_3)$ and substituting this into the SiO_2 balance results in

$$W_3 = 649.35 \text{ kg}$$

and

$$W_2 = 350.65 \text{ kg}$$

The separation efficiency is now found by back-calculating the Fe_3O_4 split between streams (2) and (3).

$$\text{kg } Fe_3O_4 \text{ in (2)} = (0.8)(350.65) = 280.52$$

$$\text{kg } Fe_3O_4 \text{ in (3)} = (0.03)(649.35) = 19.48$$

and the efficiency is

$$\frac{280.52}{300} = 0.935$$

or 93.5% of the incoming Fe_3O_4 reports to the product stream.

This simple example illustrates a number of points.

1. Enough information must be available to write component or total mass balances.

2. The information available depends on the situation.

3. The solution of simultaneous equations may be by different techniques.

Referring to the latter point, note that in the first case, the solution, once enough information was available, was direct, i.e., one equation was solved and its solution used directly to solve the other one. In the performance evaluation case simultaneous solution of two equations was required, since the unknowns were present in both equations. In the next section we will consider the various methods of solving equations that arise in material balance problems.

4.2 SOLUTION OF SETS OF EQUATIONS

In general, material balance problems require the solution of a set of equations such as

$$X_1 = f(X_2, X_3, \ldots\ldots X_n)$$

$$X_2 = f(X_1, X_3, \ldots\ldots X_n)$$

$$\vdots$$

$$X_n = f(X_1, X_2, \ldots\ldots X_{n-1})$$

These equations are usually linear, of the form

$$a_i X_i + a_{i+1} X_{i+1} + a_{i+2} X_{i+2} + \ldots. = b_i$$

where a_i, b_i are numerical coefficients. The linearity arises from the fact that the variables X_1, X_2, etc., are not raised to any power, nor present as a function such as log X, etc.

4.2.1 PARTITIONING OF EQUATIONS

Having written the equations, based on the information available, a test can be made to see if they can be solved one-at-a-time, or if not, whether they can be reduced to smaller subsets that may be more easily solved simultaneously. This process is called partitioning of the equations.

A simple algorithm (organized series of steps to follow) to partition the set of equations in order to determine the precedence order in which the equations are to be solved, has been developed by Lee, et al(1). Suppose that there are four equations involving unknowns to be solved:

$$F_1\ (X_1, X_3) = 0$$

$$F_2\ (X_1, X_2) = 0$$

$$F_3\ (X_3, X_4) = 0$$

$$F_4\ (X_4) = 0$$

---

(1) W. Lee, J.H. Christiensen and D.F. Rudd, A.I.Ch.E.J.,12 (1966).

We first prepare the structural array, which consists of marking in a matrix where a variable, (corresponding to a column), occurs in an equation, represented by a row. Thus, in this case, (x indicates the presence of a variable),

Equation	Variables X_1	X_2	X_3	X_4
1	x		x	
2	x	x		
3			x	x
4				x

Then apply the following algorithm:

1. Locate a column in which only one x occurs and delete that column and the corresponding equation in which the x is found.

2. Repeat 1 until no more equations can be deleted.

3. If no more equations remain, the order of equation solution for numerical value of the variables is the reverse of the order of elimination. This is the precedence order.

In this case, X_2 and Eq. 2 can be eliminated.

Equation	Variables X_1	X_2 (deleted)	X_3	X_4
1	x		x	
2 (deleted)	x	x		
3			x	x
4				x

and then X_1 and Eq. 1,

Equation	Variables X_1	X_2	X_3	X_4
1	x		x	
2				
3			x	x
4				x

and then X_3 and Eq. 3

Equation	Variables X_1	X_2	X_3	X_4
1				
2				
3			x	x
4				x

Finally X_4 and Eq. 4 are eliminated. Thus the set of four equations and four unknowns can be solved one at a time by solving, in order, Eq. 4 for X_4, Eq. 3 for X_3, Eq. 1 for X_1, and finally Eq. 2 for X_2.

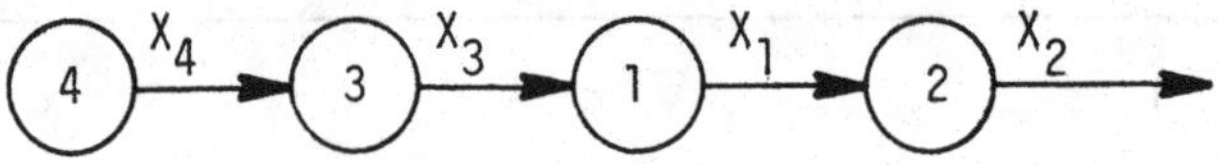

This is certainly simpler than solving all four simultaneously!

On the other hand, suppose that Eq. 4 had been $F_4(X_3, X_4)$ instead of $F_4(X_4)$. Now, when Eqs. 1 and 2 have been eliminated as before, no further elimination is possible and simultaneous solution of F_3 and F_4 for X_3 and X_4 is required. The techniques for simultaneous solution of linear equations will be discussed next, but first the reader should reflect on the usefulness of the above approach in reducing the computational effort, particularly as the problems get larger.

EXAMPLE 4.2-1: Find the sequence of equation solution needed for calculation of the proper charge to make 100 lb of alloy with composition 70% Cu - 20% Zn - 8% Sn - 2% Pb in an induction furnace. The materials available are:

Scrap brass	: 68% Cu - 32% Zn
Commercial copper	: 100% Cu
Commercial lead	: 100% Pb
Commercial tin	: 100% Sn

Solution: There are four unknowns: Weights of brass and commercial copper, lead and tin. Since there are four elements, four balance equations may be written. Assuming the four element balances are used (not the total balance), the equations would be

$$\text{Cu Balance} : 0.68\, W_{Brass} + 1.0\, W_{Com\ Cu} = (0.70)(100)$$

$$\text{Zn Balance} : 0.32\, W_{Brass} = (0.20)(100)$$

$$\text{Sn Balance} : 1.0\, W_{Com\ Sn} = (0.08)(100)$$

$$\text{Pb Balance} : 1.0\, W_{Com\ Pb} = (0.02)(100)$$

The structural array is then:

	Equation	Unknowns			
		W_{Brass}	$W_{Com\ Cu}$	$W_{Com\ Sn}$	$W_{Com\ Pb}$
1.	Cu Balance	X	X		
2.	Zn Balance	X			
3.	Sn Balance			X	
~~4.~~	~~Pb Balance~~				X

Applying the algorithm, $W_{Com\ Pb}$ and Eq. 4 can be eliminated

Equation	Unknowns		
	W_{Brass}	$W_{Com\ Cu}$	$W_{Com\ Sn}$
1	X	X	
2	X		
3			X

Then, $W_{Com\ Sn}$ and Eq. 3 are eliminated

Equation	Unknowns	
	W_{Brass}	$W_{Com\ Cu}$
1	X	X
2	X	

Finally, $W_{Com\ Cu}$ and Eq. 1 are eliminated, leaving Eq. 2 and W_{Brass}

Equation	Unknown
	W_{Brass}
2	X

Reversing the order of elimination, the solution to the overall problem is to solve Eq. 2 for W_{Brass}, followed by the others:

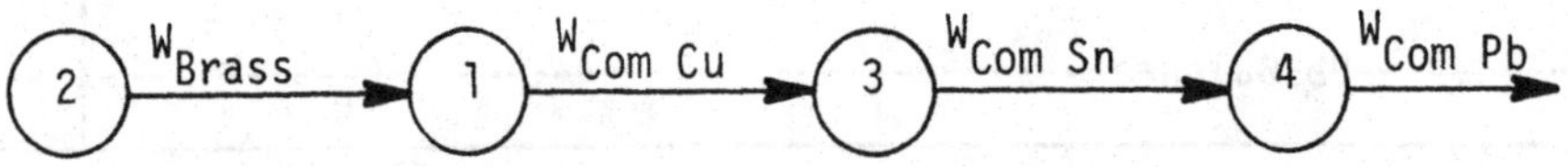

EXAMPLE 4.2-2: Suppose that in EXAMPLE 4.2-1 the four equations chosen to determine the solution to the problem had included a total balance instead of the Zn balance. What would the solution sequence be?

Solution: The equations would be:

Cu Balance : $0.68\ W_{Brass} + 1.0\ W_{Com\ Cu}$

Total Bal. : $1.0\ W_{Brass} + 1.0 W_{Com\ Cu} + 1.0 W_{Com\ Sn} + 1.0 W_{Com\ Pb} = 100.0$

Sn Balance : $1.0\ W_{Com\ Sn} = 8.0$

Pb Balance : $1.0\ W_{Com\ Pb} = 2.0$

and the structural array would be:

	Equation	Unknowns			
		W_{Brass}	$W_{Com\ Cu}$	$W_{Com\ Sn}$	$W_{Com\ Pb}$
1.	Cu Balance	x	x		
2.	Total Balance	x	x	x	x
3.	Sn Balance			x	
4.	Pb Balance				x

It is painfully clear that no unknown nor equation can be eliminated, meaning that either all four of the equations must be solved simultaneously, or one of the tearing techniques discussed in 4.2.6 must be used. This example also illustrates how an incorrect or unfortunate choice of equations to be used can result in making a relatively simple problem much more complex than necessary!

4.2.2 DETERMINANTS

Solution of simultaneous equations can be obtained through a number of means or techniques. If the equations are linear, determinants are easy to use, up to three equations and three unknowns. For instance, solution of the set

$$a_{11}\ X_1 + a_{12}\ X_2 + a_{13}\ X_3 = b_1$$

$$a_{21}\ X_1 + a_{22}\ X_2 + a_{23}\ X_3 = b_2$$

$$a_{31}\ X_1 + a_{32}\ X_2 + a_{33}\ X_3 = b_3$$

can be obtained by evaluating the determinants C_1, C_2, C_3 and D:

where:

$$D = \begin{vmatrix} a_{11} & a_{12} & a_{13} \\ a_{21} & a_{22} & a_{23} \\ a_{31} & a_{32} & 33 \end{vmatrix} = a_{11}\begin{vmatrix} a_{22} & a_{23} \\ a_{32} & a_{33} \end{vmatrix} - a_{12}\begin{vmatrix} a_{21} & a_{23} \\ a_{31} & a_{33} \end{vmatrix} + a_{13}\begin{vmatrix} a_{21} & a_{22} \\ a_{31} & a_{32} \end{vmatrix}$$

$$= a_{11}[a_{22}\cdot a_{33}-a_{32}\cdot a_{23}]-a_{12}[a_{21}\cdot a_{33}-a_{31}\cdot a_{23}]+a_{13}[a_{21}\cdot a_{32}-a_{31}\cdot a_{22}]$$

$$C_1 = \begin{vmatrix} b_1 & a_{12} & a_{13} \\ b_2 & a_{22} & a_{23} \\ b_3 & a_{32} & a_{33} \end{vmatrix}; \quad C_2 = \begin{vmatrix} a_{11} & b_1 & a_{13} \\ a_{21} & b_2 & a_{23} \\ a_{31} & b_3 & a_{33} \end{vmatrix}; \quad C_3 = \begin{vmatrix} a_{11} & a_{12} & b_1 \\ a_{21} & a_{22} & b_2 \\ a_{31} & a_{32} & b_3 \end{vmatrix}$$

Finally, after C_1, C_2, and C_3 are evaluated in the same way as D,

$$X_1 = \frac{C_1}{D}, \qquad X_2 = \frac{C_2}{D}, \text{ and } \qquad X_3 = \frac{C_3}{D}$$

EXAMPLE 4.2-3: In Example 4.1-1, the Fe_3O_4 and SiO_2 balances were solved simultaneously for W_2, the weight of stream 2 and W_3, the weight of stream 3. Solve this problem using determinants.

Solution: The equations are:

$$0.8\ W_2 + 0.03\ W_3 = 300$$

$$0.2\ W_2 + 0.97\ W_3 = 700$$

$$D = \begin{vmatrix} 0.8 & 0.03 \\ 0.2 & 0.97 \end{vmatrix} = (.8)(.97)-(.2)(.03) = 0.776-0.006 = 0.770$$

$$C_1 = \begin{vmatrix} 300 & 0.03 \\ 700 & 0.97 \end{vmatrix} = (300)(.97)-(700)(.03) = 291.0-21.0 = 270.0$$

$$C_2 = \begin{vmatrix} 0.8 & 300 \\ 0.2 & 700 \end{vmatrix} = (700)(.8)-(300)(.2) = 560.-60. = 500.0$$

$$W_2 = \frac{C_1}{D} = \frac{270}{0.77} = 350.65\ \text{kg}$$

$$W_2 = \frac{C_2}{D} = \frac{500}{0.77} = 649.35\ \text{kg}$$

These are the same values obtained in the earlier example. In principle, this technique can be used on larger sets of equations. However, the technique of solving linear equations in this way quickly becomes unmanageable as the number of equations increases.

4.2.3 GAUSSIAN ELIMINATION

For more than 3 equations and 3 unknowns, several techniques may be used. One of the most common is Gaussian elimination, illustrated here for simplicity on a three equation problem, where it is required to solve

$$X_2 + X_3 = 5$$
$$-2X_1 - X_2 + X_3 = -1$$
$$X_1 + X_2 - 2X_3 = -3$$

The notation is that of matrix algebra. For a few simple definitions relating to matrix algebra, refer to Appendix A.

Step 1: The coefficients of the unknowns are written in matrix form and the constants in each equation are written in column form (vector) and this vector is added to the coefficient matrix to form an augmented matrix. For the above equations, the augmented matrix is:

$$\begin{pmatrix} 0 & 1 & 1 & 5 \\ -2 & -1 & 1 & -1 \\ 1 & 1 & -2 & -3 \end{pmatrix}$$

$\uparrow$ X_1 vector, $\uparrow$ X_2 vector, $\uparrow$ X_3 vector, $\uparrow$ b vector

Step 2: Search down first column for largest number in absolute value (the pivot), and divide the entire row by that number. The result is:

$$\begin{pmatrix} 0 & 1 & 1 & 5 \\ 1 & 1/2 & -1/2 & 1/2 \\ 1 & 1 & -2 & -3 \end{pmatrix}$$

Step 3: Multiply the same row by a number such that when that product in each column is added to another row, a zero will appear in column 1 in the other row. Repeat until all other rows have 0 in column 1. The pivot, however, remains the same although the other rows are left as changed. In the present example, multiply row 2 by -1, then add the resulting numbers to the numbers in row 3. The result is given in the next augmented matrix.

$$\begin{pmatrix} 0 & 1 & 1 & 5 \\ 1 & 1/2 & -1/2 & 1/2 \\ 0 & 1/2 & -3/2 & -7/2 \end{pmatrix}$$

Step 4: Repeat Steps 2 and 3 for column 2.

$$\begin{pmatrix} 0 & 1 & 1 & 5 \\ 1 & 0 & -1 & -2 \\ 0 & 0 & -2 & -6 \end{pmatrix}$$

Step 5: Repeat for all columns <u>except the last.</u>

$$\begin{pmatrix} 0 & 1 & 0 & 2 \\ 1 & 0 & 0 & 1 \\ 0 & 0 & 1 & 3 \end{pmatrix}$$

Step 6: Rearrange to give an augmented <u>identity</u> matrix.

$$\begin{pmatrix} 1 & 0 & 0 & 1 \\ 0 & 1 & 0 & 2 \\ 0 & 0 & 1 & 3 \end{pmatrix}$$

Step 7: The answer is seen to be

$$X_1 + 0 + 0 = 1$$

$$0 + X_2 + 0 = 2$$

$$0 + 0 + X_3 = 3$$

Gaussian elimination is readily performed on much larger sets of linear equations by computer programs available on most scientific digital computers. The important point is that a problem that can be formulated as a set of linear equations can be solved. For examples of such problems see EXAMPLES 4.3-5 and 4.3-11.

4.2.4 MATRIX INVERSION

Another solution method involves the technique called <u>matrix inversion</u>. If we have a matrix equation*

$$\underset{(n\times n)}{A}\ \underset{(n\times 1)}{X} = \underset{(n\times 1)}{B}$$

and if

$$A^{-1}A = I$$

where A^{-1} is the inverse matrix of A and I is the identity matrix, then

---

* See Appendix A for a brief discussion on matrix algebra.

$$A^{-1}AX = A^{-1}B$$

$$IX = A^{-1}B$$

or

$$X = A^{-1}B$$

Therefore, if A has an inverse, A^{-1}, X is easily found. Solving the same set of three equations as in the previous example, we first transform $(A \mid I)$ into $(I \mid A^{-1})$. Initially,

$$\overset{(A \mid I)}{\begin{pmatrix} 0 & 1 & 1 & 1 & 0 & 0 \\ -2 & -1 & 1 & 0 & 1 & 0 \\ 1 & 1 & -2 & 0 & 0 & 1 \end{pmatrix}}$$

Use Gaussian elimination to make the transformation.

$$\begin{pmatrix} 0 & 1 & 1 & 1 & 0 & 0 \\ 1 & 1/2 & -1/2 & 0 & -1/2 & 0 \\ 1 & 1 & -2 & 0 & 0 & 1 \end{pmatrix}$$

$$\begin{pmatrix} 0 & 1 & 1 & 1 & 0 & 0 \\ 1 & 1/2 & -1/2 & 0 & -1/2 & 0 \\ 0 & 1/2 & -3/2 & 0 & 1/2 & 1 \end{pmatrix}$$

$$\begin{pmatrix} 0 & 1 & 1 & 1 & 0 & 0 \\ 1 & 0 & -1 & -1/2 & -1/2 & 0 \\ 0 & 0 & -2 & -1/2 & 1/2 & 1 \end{pmatrix}$$

$$\begin{pmatrix} 0 & 1 & 1 & 1 & 0 & 0 \\ 1 & 0 & -1 & -1/2 & -1/2 & 0 \\ 0 & 0 & 1 & 1/4 & -1/4 & -1/2 \end{pmatrix}$$

$$\begin{pmatrix} 0 & 1 & 1 & 1 & 0 & 0 \\ 1 & 0 & 0 & -1/4 & -3/4 & -1/2 \\ 0 & 0 & 1 & 1/4 & -1/4 & -1/2 \end{pmatrix}$$

$$\begin{pmatrix} 0 & 1 & 0 & 3/4 & 1/4 & 1/2 \\ 1 & 0 & 0 & -1/4 & -3/4 & -1/2 \\ 0 & 0 & 1 & 1/4 & -1/4 & -1/2 \end{pmatrix}$$

Rearrange to obtain $(I \mid A^{-1})$:

$$\begin{array}{c} (I \mid A^{-1}) \\ \left(\begin{array}{ccc|ccc} 1 & 0 & 0 & -1/4 & -3/4 & -1/2 \\ 0 & 1 & 0 & 3/4 & 1/4 & 1/2 \\ 0 & 0 & 1 & 1/4 & -1/4 & -1/2 \end{array}\right) \end{array}$$

As a check to see if A^{-1} found above is correct, test if $A^{-1}A$ is I.

$$AA^{-1} = \begin{pmatrix} 0 & 1 & 1 \\ -2 & -1 & 1 \\ 1 & 1 & -2 \end{pmatrix} \begin{pmatrix} -1 & -3 & -2 \\ 3 & 1 & 2 \\ 1 & -1 & -2 \end{pmatrix} 1/4$$

$$= \begin{pmatrix} 4 & 0 & 0 \\ 0 & 4 & 0 \\ 0 & 0 & 4 \end{pmatrix} 1/4 = \begin{pmatrix} 1 & 0 & 0 \\ 0 & 1 & 0 \\ 0 & 0 & 1 \end{pmatrix} = I$$

Then, obtain $X = A^{-1}B$.

$$\begin{pmatrix} X_1 \\ X_2 \\ X_3 \end{pmatrix} = 1/4 \begin{pmatrix} -1 & -3 & -2 \\ 3 & 1 & 2 \\ 1 & -1 & -2 \end{pmatrix} \begin{pmatrix} 5 \\ -1 \\ -3 \end{pmatrix} = 1/4 \begin{pmatrix} 4 \\ 8 \\ 12 \end{pmatrix} = \begin{pmatrix} 1 \\ 2 \\ 3 \end{pmatrix}$$

and again observe that $X_1 = 1$, $X_2 = 2$, and $X_3 = 3$.

4.2.5 MATRIX CONDITIONING

While both matrix inversion and Gaussion elimination (and other variations of these basic techniques) will obtain solutions to large, complex, sets of simultaneous equations, they are not without pitfalls. Solutions, in come cases, may be totally unrealistic, and yet mathematically correct. Consider the solution of the set,

$$1.00\ x + 1.00\ y = 1$$

$$1.00\ x + 1.01\ y = 2$$

The solution is $x = -99$ and $y = 100$. Now, suppose that a 2% error in one of the coefficients, resulting, for example, from a measuring error, changes the equations to

$$1.00\ x + 1.00\ y = 1$$

$$1.00\ x + 0.99\ y = 2$$

The solution is now $x = 101$ and $y = -100$. This difference in solutions is very large and in the case of material balance equations would render the solution totally useless. Yet it has nothing to do with the mathematical solution technique. It is the result of the equation set and the resulting coefficient matrix being <u>ill-conditioned</u>, i.e., the solution of the set is super-sensitive to small changes in the values of the coefficients. Thus, care must be exercised to make sure that, if a large matrix of material balance equations is utilized in the solution of a material balance problem, it is not ill-conditioned.

Various tests for whether or not a matrix is ill-conditioned have been devised by mathematicians. The simplest test is to calculate the determinant of the coefficient matrix, det. A. If the determinant is very small, the matrix is ill-conditioned. This is easily seen from the preceding example, where

$$A = \begin{vmatrix} 1.00 & 1.00 \\ 1.00 & 1.01 \end{vmatrix}$$

$$\text{and det } A = -0.01$$

Unfortunately, when the coefficient matrix becomes larger than 3 x 3, evaluation of det A becomes very tedious.

The true test is to compute the von Neumann P-condition number, P(A). Unfortunately, this is very difficult to compute and so some more easily evaluated condition numbers have been established that bound P(A). One such number is Turing's M-condition number, M(A):

$$M(A) = n \max|a_{ij}| \max|\alpha_{ij}|$$

where A is an nxn matrix, a_{ij} are the elements in A, and α_{ij} are the elements in the inverse matrix, A^{-1}. This obviously requires that A^{-1} be found by a method such as in the preceding section. Once M(A) is found, the bounds on P(A) are given by

$$\frac{M(A)}{n} \le P(A) \le nM(A)$$

Another limiting value is given by Turing's N-condition number, N(A):

$$N(A) = \frac{||A|| \cdot ||A^{-1}||}{n}$$

where $||A|| = (\sum_{ij} a_{ij}^2)^{1/2}$ and $||A^{-1}|| = (\sum_{ij} \alpha_{ij}^2)^{1/2}$

again requiring that A be inverted. In this case,

$$N(A) \leq P(A) \leq nN(A)$$

If the P(A) value of a matrix is on the order of 10^{-3} or 10^{-4}, it is probable that solutions of the set of linear equations associated with the matrix are meaningless, irrespective of the number of significant digits. Usually, the metallurgist making material or energy balances will be utilizing generalized computer programs for solving sets of simultaneous equations and not be involved in the writing of these programs in the first place. However, it is important to understand the concepts presented above and the limitations on their utilization. The sensitivity of solutions of sets of linear equations to coefficient variations should be tested and the results interpreted with the aid of common sense.

4.2.6 TEARING EQUATIONS

Finally, if the equations are non-linear or if the setting up of matrix solutions is not desirable, and yet partitioning of the equations has detected a set of equations that must be solved simultaneously, there is another technique that may be used to ease solution effort. It is called tearing.

The basic idea is that a set of equations to be solved simultaneously may be solved in sequence, one-at-a-time, if the value of one of the unknowns is assumed initially, and then recalculated as a last result in the sequence. This calculated value is then compared to the assumed value and if they are the same, a solution has clearly been reached: if not, the assumed value is adjusted and the procedure repeated, until agreement is reached. Rudd and Watson[2] present an algorithm to find the most efficient place to tear the set of equations.

Step 1: Apply the partitioning algorithm in 4.2.1 to the structural array of the equation set. If it does not delete all equations, go to Step 2.

Step 2: Define k, equal to the minimum value of $p(X_i)-1$, where $p(X_i)$ is the number of equations in which variable X_i appears.

Step 3: Identify sets of k equations which have the property that, when that set is deleted, an array remains with at least one variable appearing in only one equation.

(2) D. Rudd and C. C. Watson, Strategy of Process Engineering, Wiley, N.Y., N.Y. 1968.

Step 4: Delete the equation set identified in Step 3.

Step 5: Reapply the partitioning algorithm to the remaining array.

Step 6: If no precedence order is obtained in Step 5, try a different set of k equations.

Step 7: If the deletion of all possible sets of k equations does not result in an array that can be precedence ordered, increase k by one and return to Step 3.

By way of illustration, suppose the set of equations is

$$F_1 (X_1, X_3) = 0$$

$$F_2 (X_2, X_3) = 0$$

$$F_3 (X_1, X_3, X_4) = 0$$

$$F_4 (X_1, X_5) = 0$$

$$F_5 (X_1, X_4, X_5) = 0$$

Step 1 eliminates Eq. 2, leaving:

Equations	Variables				
	X_1	X_2	X_3	X_4	X_5
1	x		x		
3	x		x	x	
4	x				x
5	x			x	x

Step 2 shows that $p(X_1) = 4$, $p(X_3) = p(X_4) = p(X_5) = 2$ and so $k = (2-1) = 1$. By inspection, it is seen that if <u>any</u> of the remaining equations is deleted, at least one variable will appear in only one remaining equation. Try deleting Eq. 1. The result is in the next table. Now eliminate X_3 and Eq. 3.

Equation	Variables				
	X_1	X_2	X_3	X_4	X_5
3	x		x	x	
4	x				x
5	x			x	x

followed by X_4 and Eq. 5.

Equation	Variables				
	X_1	X_2	X_3	X_4	X_5
4	x				x
5	x			x	x

X_1 and X_5 remain. The solution is then found by choosing one of these remaining variables as the design variable, to which we assign a value. In this case, a value is assigned to X_5 (perhaps a good idea of what its value should be is known). Then the steps are retraced backwards to a solution.

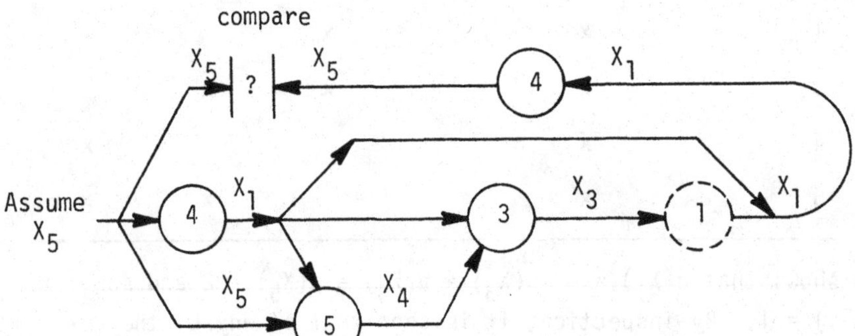

If X_1 instead of X_5 had been assigned a value, the iteration would have been thus:

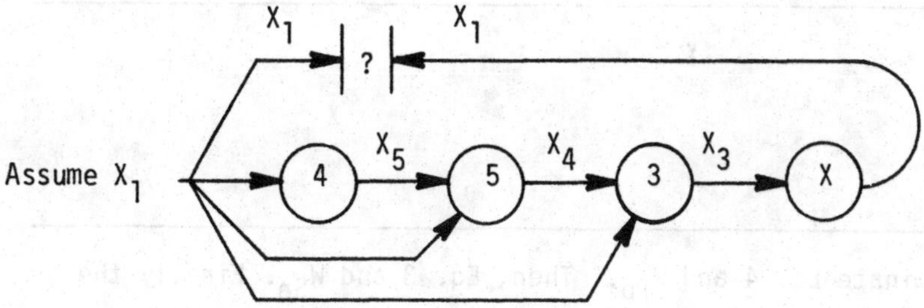

In both cases, once the comparison between estimated and computed values of X_1 or X_5 is satisfactory, the value of X_3 from the last iteration can be used in Eq. 2 (the first elimination using the algorithm in 4.2.1) to calculate X_2, and the problem is solved.

EXAMPLE 4.2-4: In EXAMPLE 4.2-2 the choice of three element balances plus the total balance gave a set of equations that could not be solved in a sequence of one-at-a-time solutions. The structural array was

Equation	Unknowns			
	W_{Brass}	$W_{Com\ Cu}$	$W_{Com\ Sn}$	$W_{Com\ Pb}$
1. Cu Balance	x	x		
2. Total Balance	x	x	x	x
3. Sn Balance			x	
4. Pb Balance				x

Utilize the concept of tearing to help solve this set of equations without having to use a matrix method such as Gaussian elimination.

Solution: Applying the algorithm, $p(W_{Brass}) = 2$, $p(W_{Com\ Cu}) = 2$, $p(W_{Com\ Sn}) = 2$, and $p(W_{Com\ Pb}) = 2$.

Therefore, $\min\ [p(x_i)-1] = 2-1 = 1$; $k = 1$. Now, eliminate one equation to produce an array that can be precedence ordered. It turns out that elimination of any one of the four equations will do the trick. We will eliminate Eq. 2. The array is then

Equation	Unknowns			
	W_{Brass}	$W_{Com\ Cu}$	$W_{Com\ Sn}$	$W_{Com\ Pb}$
1	X	X		
3			X	
4				X

Next, eliminate Eq. 4 and W_{Pb}. Then, Eq. 3 and W_{Sn}. Finally the array is

Equation	Unknowns	
	W_{Brass}	$W_{Com\ Cu}$
1	X	X

The final solution thus requires that a value of one of the remaining variables be assumed. The sequence then is:

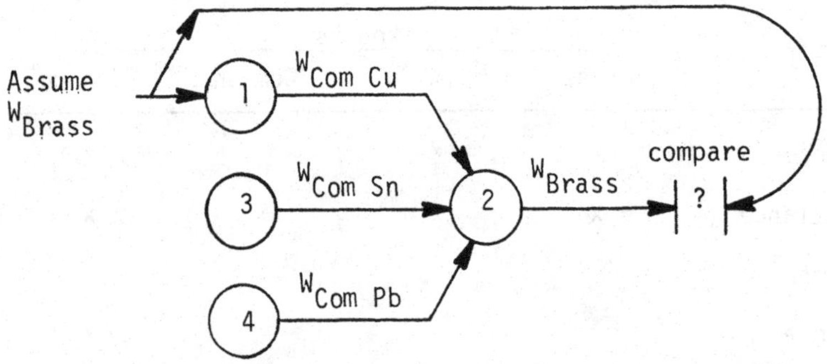

The assumed value of W_{Brass} is compared to the calculated one from Eq. 2 and adjusted, if necessary, until the comparison is close enough.

4.3 MATERIAL BALANCES

Analysis of material balance problems must proceed in an orderly fashion. The first step is to <u>define the system</u>. This might be, for example,

a) a furnace into which scrap and alloying additions are charged and from which a heat of alloy and some slag are removed,

b) a sinter plant into which you continuously charge coke breeze and ore and remove screened sinter,

c) an entire plant into which ore, scrap, fuel, electrical energy, etc., flow and finished products leave, as shown in Fig. 4.3-1.

Draw a schematic flow diagram of the system showing all inward and outward directed material flows.

Select a useful basis for the calculation. In the case of Fig. 4.3-1 the basis is 1,000,000 tons of cast steel slabs.

Place all known data on the diagram, including whatever is known about any chemical reactions which occur within the system.

The next step involves writing the various balances and any restrictive equations that apply to the system. This is the most critical step in the entire sequence, because great care must be taken to make sure that only independent relationships are included. The systematic approach to this problem is presented in the remainder of this section.

4.3.1 DEGREES OF FREEDOM

During any design exercise, the designer has, as was seen earlier, some variables that may be specified. These design variables are equal in number to the degrees of freedom, F, which is the difference between the number of variables, V, and the number of restrictions or formal relationships between variables, R, that exist in the process. In other words, since V is the total possible number of unknowns, a solution to the problem will only be found if V-R degrees of freedom are specified, so that the number of unknowns and number of equations are the same.

$$F = V - R \qquad (4.3\text{-}1)$$

All items of information required to define any process stream entering or leaving a process unit will be called variables. This means in general, temperature T, pressure P, mass flow W, and complete chemical or mineralogical analysis. For a C component mixture, C-1 percentages are required to define the complete analysis. Therefore the maximum number of variables (unknowns) per stream is C + 2, e.g. the C-1 percentages plus temperature, pressure and mass flow. Therefore, the maximum total number of variables for a system having n stream equals

$$V = \sum_{i=1}^{n} (C+2) \qquad (4.3\text{-}2)$$

Remember that the stream need not be continuous: the analysis applies to batch as well as continuous processes.

Restrictions include any quantitative relationship that can

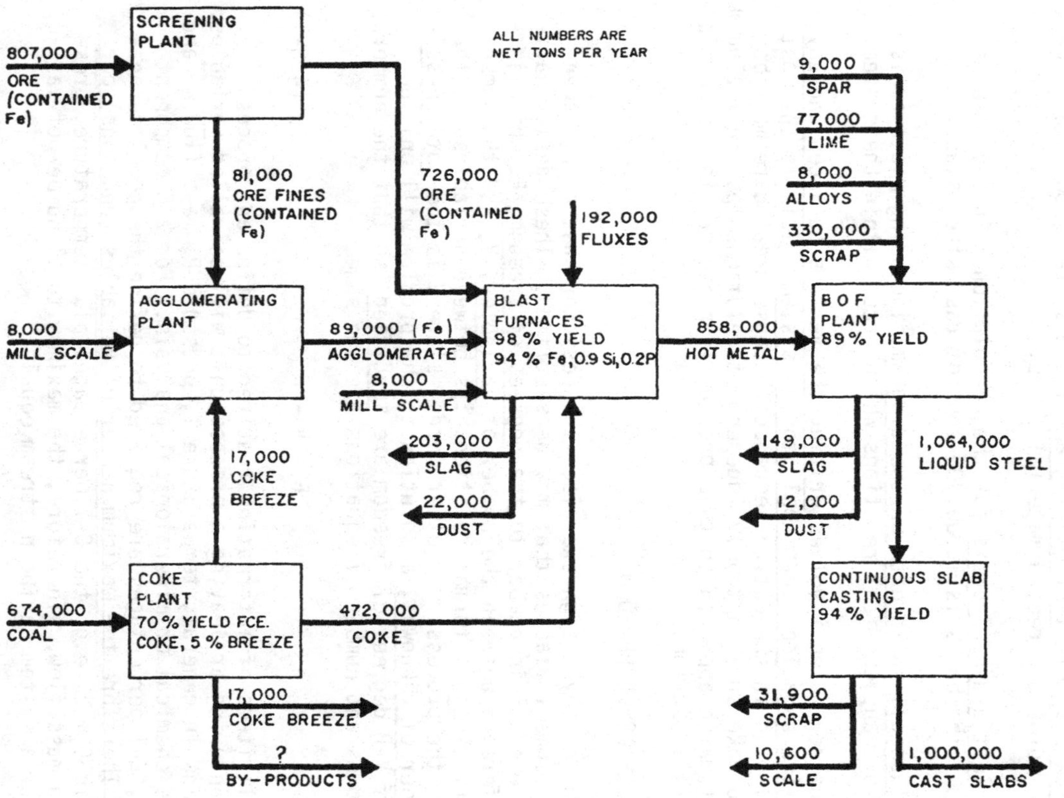

Figure 4.3-1 Block diagram of the primary production area of an integrated steel mill

or must be imposed on the process. The most obvious quantitative relationship that can be written down and that must be satisfied are C material balances for a C-component system. If the process is purely physical, involving no chemical reactions, then the number of components, C, is equal to the number of distinct chemical species that exist in the system. If the process involves chemically reactive species, then the number of components is the number of species S, minus the minimum number of independent relationships relating the species to each other, R. A simple rule for determining. this number is as follows:

1. Write chemical equations for formation, from their constituent atoms, of all species regarded as being present in significant amount in the system. (This obviously requires that the chemistry of the system be well known.)
2. Combine these equations in such a way as to eliminate from them any free atoms which are not actually present.
3. The resulting number of equations is the minimum number of independent chemical equations necessary to represent the stoichiometry of the system, R. Then, C equals S minus R.

Other restrictions include the following:

(1) If a species is not present in a given stream, then its concentration is implicitly specified to be zero and the C+2 variables in that stream are reduced to C+1. In other words, a restriction of 1 is placed on the system, or R=1 for each missing component.

(2) If two or more species are present in a fixed ratio, either from thermodynamic equilibrium considerations or because they are present in a compound that does not participate in any reactions, that is a restriction on the system. R=1 for each fixed ratio.

(3) If only a material balance is being made, temperature and pressure data on each stream are not needed and so the number of variables should be reduced by 2 (number of streams). R=2x(no. of streams) for material balance problems.

The schematic 4-component process illustrated below, for which only a material balance is being made, would have

$$V = \sum_{1}^{5} (C+2)$$

$$= 5\ (4+2)$$

$$= 30$$

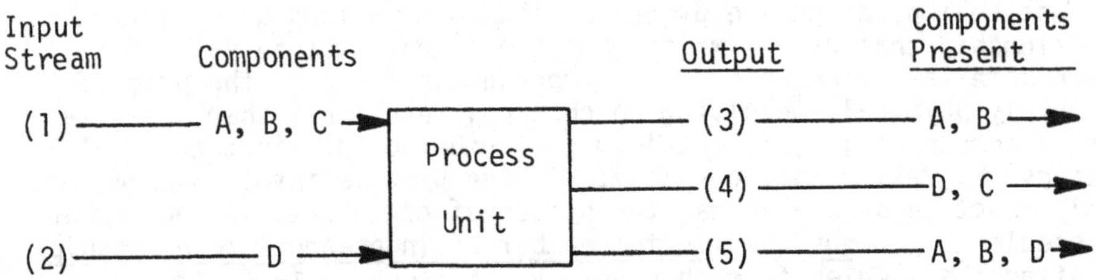

This total can be reduced by considering the restrictions. First, there are 4 components, so 4 material balances may be written, $R_1 = 4$. There are missing components; 1 in stream (1), 3 in stream (2), 2 in stream (3), 2 in stream (4), and 1 in stream (5) for a total of 9. Therefore $R_2 = 9$. Since we are making only a material balance, neither T nor P is needed for any stream, so we can further reduce the variables by $R_3 = 2 \times 5$ streams = 10. The total number of restrictions is 23. Thus, the degrees of freedom = 30 - 23 = 7. This means that if we know 5 chemical analyses and 2 mass flows or 5 mass flows and 2 analyses, or any combination of at least 1 mass flow and chemical analysis information totalling 7 pieces of information, in principle the material balance can be completely determined. Whether the solution can be obtained by solving one equation at a time or will require simultaneous solution of 4 equations, depends on which pieces of information are available and can be determined using the techniques presented in 4.2. The 7 required pieces of information may come from sampling the process, or estimating values, or by arbitrarily assuming certain values. The significance of the 7 degrees of freedom is that much information must be known before a solution can be obtained.

EXAMPLE 4.3-1: In the roasting of chalcocite, Cu_2S, the reaction might be

$$3Cu_2S + 6O_2 = 2Cu_2O + SO_2 + 2CuSO_4$$

There are 5 species present. There are only 3 elements. Determine the number of components C for use in material balance calculations.

Solution: Determine the number of independent relations, R, between species, S.

Step 1: Form the species from atoms:

$$2Cu + S \rightarrow Cu_2S$$

$$2Cu + O \rightarrow Cu_2O$$

$$S + 2O \rightarrow SO_2$$

$$Cu + S + 4O \rightarrow CuSO_4$$

$$O \rightarrow 1/2\ O_2$$

Step 2: Since O, S, and Cu are presumed not to occur in the system, eliminate first O:

$$2Cu + S \rightarrow Cu_2S$$
$$2Cu + 1/2O_2 \rightarrow Cu_2O$$
$$S + O_2 \rightarrow SO_2$$
$$Cu + S + 2O_2 \rightarrow CuSO_4$$

then S:

$$2Cu + 1/2O_2 \rightarrow Cu_2O$$
$$Cu_2S + O_2 \rightarrow 2Cu + SO_2$$
$$Cu_2S + 2O_2 \rightarrow CuSO_4 + Cu$$

and then Cu:

$$Cu_2S + \frac{3}{2} O_2 \rightarrow Cu_2O + SO_2$$
$$Cu_2S + \frac{9}{4} O_2 \rightarrow CuSO_4 + \frac{1}{2} Cu_2O$$

The result is two independent equations relating the species, so $R = 2$, and so $C = 5 - 2 = 3$. In this case, C is the same as the number of elements.

EXAMPLE 4.3-2: In the calcining of $CaCO_3$, according to the reaction

$$CaCO_3 \rightarrow CaO + CO_2$$

there are 3 species. Find C.

Solution: Step 1: Applying the algorithm, form the species from atoms:

$$Ca + C + 3O \rightarrow CaCO_3$$
$$Ca + O \rightarrow CaO$$
$$C + 2O \rightarrow CO_2$$

Step 2: Since Ca, C, and O are not present, eliminate C:

$$Ca + CO_2 + O \rightarrow CaCO_3$$
$$Ca + O \rightarrow CaO$$

and then Ca and O:

$$CaO + CO_2 \rightarrow CaCO_3$$

Thus, $R = 1$, and $C = 3-1 = 2$, even though there are 3 elements present.

However, if all CaO and CO_2 comes from the decomposition of $CaCO_3$, there is only 1 independent component balance because of the additional stoichiometric relationship between the species.

EXAMPLE 4.3-3: In the production of metal M from its oxide, MO, a furnace is fed with carbon and oxygen to produce a reducing atmosphere. For thermodynamic reasons, the ratio of CO to CO_2 in the output gas is fixed at a constant value, K. In order to derive the material balances for this process, how many design variables must the process designer specify?

Solution: The flow diagram is as follows:

$$K = \frac{CO}{CO_2}$$

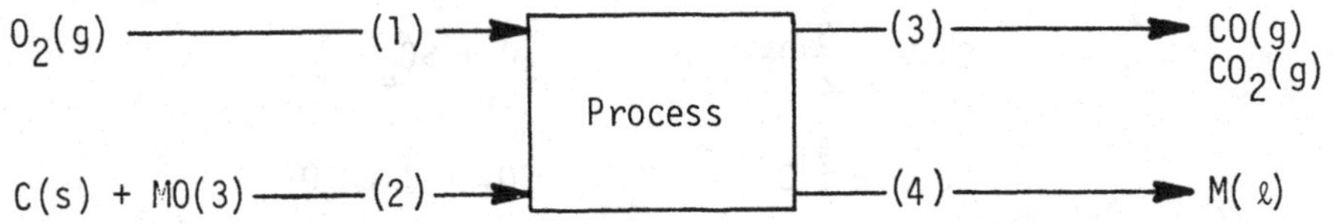

The total number of possible unknowns, $V = \sum_{1}^{4} (5) = 20$. The restrictions or formal relationships that can be written between variables will be found and added up.

Missing components:

Stream 1: C and M	$R_1 = 2$
Stream 2: none	
Stream 3: M	$R_2 = 1$
Stream 4: C and O	$R_3 = 2$
No T or P data needed:	$R_4 = 8$
Fixed Ratio:	
$W_{CO} = K\, W_{CO_2}$	$R_5 = 1$
Independent Component Balances:	
C Balance	$R_6 = 1$
O Balance	$R_7 = 1$
M Balance	$R_8 = 1$
$\Sigma R =$	17

The degrees of freedom are:

$$F = V - R$$
$$= 20 - 17$$
$$F = 3$$

This means that, to solve the set of equations for all weights of materials in and out of the process, <u>three</u> of the unknowns must be <u>specified</u> by the designer. The unknowns include weights of each component in each stream, total stream weights, or the ratio of C to O_2 in stream 1. For example, if the weight of MO fed and M removed and the weight of CO in the output gas is given, the remaining weights can be obtained directly. The reader should satisfy himself that this is true.

EXAMPLE 4.3-4: Fluxed sinter is being made on a sinter strand by adding dolomitic limestone (75% $CaCO_3$ - 25% $MgCO_3$) to the sinter mix at a rate of 12000 kg/h. If the initial analysis of the mix is 2.5% CaO and the final product analyzes 12.5% CaO, what is the production rate of the plant? Assume no losses except for CO_2 from the dolomite.

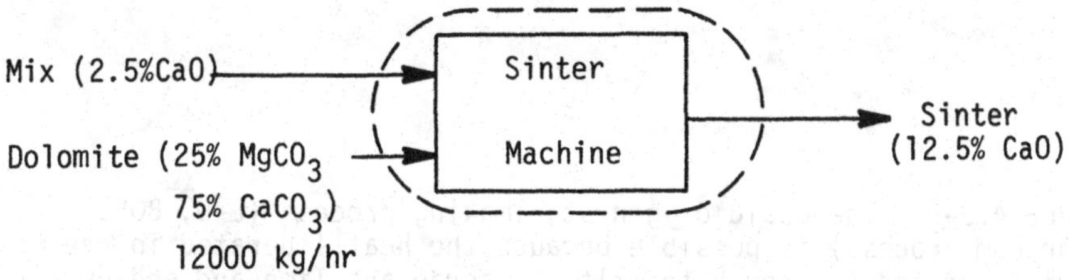

Basis: 1 hour of production.

Solution: Since all of the CO_2 will be removed from the dolomite, initially we will recalculate the feed rate to put it on the basis of MgO and CaO.

	$MgCO_3$	→	$MgO + CO_2$
Mol. Wt.:	84.33		40.32
Mass Flow Rate, kg/h:	3000.	→	1435.
	$CaCO_3$	→	$CaO + CO_2$
Mol. Wt.:	100.09		56.08
Mass Flow Rate, kg/h:	9000.	→	5040.

In this problem, there are three components; CaO, MgO and "other solids" and thus, V equals fifteen. There are also fourteen restrictions on the system; a CaO balance, an MgO balance, the analysis for CaO in the mix, the analyses for CaO and MgO in the dolomite, no "other solids in the dolomite, the dolomite flow rate, the analysis for CaO in the sinter, and six restrictions since only a mass balance is being made. There is therefore one degree of freedom. Thus, to completely solve the problem one more variable, such as the concentration of MgO in the mix or sinter must be fixed. However, if only a partial solution, e.g. the flow rates of the mix and sinter, are required, two equations with two unknowns can be written and solved without specifying the additional quantity.

Total Mass Balance:

$$W_{Mix} + 1435 + 5040 = W_{product}$$

CaO Balance:

$$(0.025)W_{Mix} + 5040 = (0.125)W_{product}$$

Solving these simultaneously,

$$W_{product} = 49{,}905 \text{ kg/h}$$

and

$$W_{Mix} = 43{,}430 \text{ kg/h}$$

EXAMPLE 4.3-5: The basic oxygen steelmaking process (L-D, BOF, BOP or OSM process) is possible because the heat liberated in the refining process is enough to melt the scrap and lime and end up with steel at a satisfactory temperature for "teeming" into ingots.

Information available for the production of <u>one metric ton of steel</u> is shown on the following diagram. Infiltrate air is included because gas analysis of the waste gas is not possible except in the hood, where the air has already infiltrated. The gas analysis reported is the average during the blow, the value which is necessary in order to set up the equations.

Calculate the complete over-all material balance for the process; that is

a) The weight of all input materials per ton of steel, and

b) The weight of all output materials per ton of steel.

Basis: 1000 kg of steel.

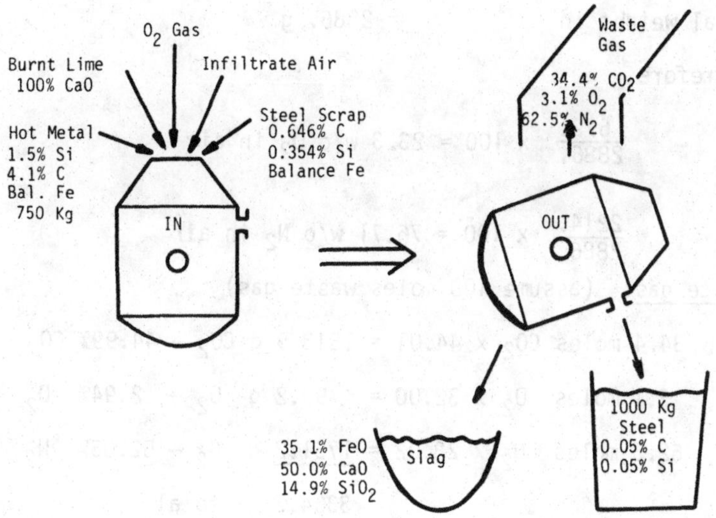

Solution: There are a total of 6 element balances (Fe, C, Si, O_2, N_2, Ca) that can be written, plus the over-all balance. Therefore, the number of unknowns cannot exceed 6, the number of independent relationships between variables. The variables are:

weight of hot metal	, W_{HM}	=	750 kg.
weight of steel scrap	, W_{SC}	=	?
weight of oxygen gas	, W_{O_2}	=	?
weight of burnt lime	, W_{LM}	=	?
weight of infiltrate air	, W_{AIR}	=	?
weight of waste gas	, W_{GAS}	=	?
weight of steel	, W_{STEEL}	=	1000 kg.
weight of slag	, W_{SLAG}	=	?

There are 6 unknowns; therefore, any 6 equations of the 7 possible are required.

First, convert the gas analysis from volume % to weight %:

<u>infiltrate air</u>: (assume 100 moles air)

$$\left(\frac{21 \text{ moles } O_2}{100 \text{ moles air}}\right)\cdot\left(\frac{32.00 \text{ g } O_2}{\text{mole } O_2}\right) = 672. \text{ g } O_2$$

$$\left(\frac{79 \text{ moles } N_2}{100 \text{ moles air}}\right)\cdot\left(\frac{28.02 \text{ g } N_2}{\text{mole } N_2}\right) = \underline{2214. \text{ g } N_2}$$

Total weight in = 2886. g

Therefore,

$$\frac{672.}{2886.} \times 100 = 23.3 \text{ w/o } O_2 \text{ in air}$$

$$\frac{2214.}{2886.} \times 100 = 76.71 \text{ w/o } N_2 \text{ in air}$$

waste gas: (assume 100 moles waste gas)

34.4 moles CO_2 x 44.01 = 1513.9 g CO_2 → 44.99% CO_2

3.1 moles O_2 x 32.00 = 99.2 g O_2 → 2.94% O_2

62.5 moles N_2 x 28.02 = 1751.2 g N_2 → 52.05% N_2

3364.3 g total

OVERALL BALANCE:

$$W_{HM} + W_{SC} + W_{O_2} + W_{LM} + W_{AIR} = W_{GAS} + W_{SLAG} + W_{STEEL}$$

↑ 750 kg (under W_{HM}); ↑ 1000 kg (under W_{STEEL})

or $$W_{SC} + W_{O_2} + W_{LM} + W_{AIR} - W_{GAS} - W_{SLAG} = 250 \text{ kg}$$

IRON BALANCE:

$$0.944\, W_{HM} + 0.99\, W_{SC} + 0\cdot W_{O_2} + 0\cdot W_{LM} + 0\cdot W_{AIR} = 0\cdot W_{GAS} + (.351)(.777) W_{SLAG} + 0.999 W_{STEEL}$$

$$708 \text{ kg} + 0.99\, W_{SC} + 0\cdot W_{O_2} + 0\cdot W_{LM} + 0\cdot W_{AIR} = 0\cdot W_{GAS} + 0.273\, W_{SLAG} + 999 \text{ kg}$$

$$0.99\, W_{SC} + 0\cdot W_{O_2} + 0\cdot W_{LM} + 0\cdot W_{AIR} - 0\cdot W_{GAS} - 0.273\, W_{SLAG} = 291 \text{ kg}$$

SILICON BALANCE:

$$0.015 W_{HM} + 0.00354 W_{SC} + 0\cdot W_{O_2} + 0\cdot W_{LM} + 0\cdot W_{AIR} = 0\cdot W_{GAS} + (.149)(.467) W_{SLAG} + .0005 W_{STEEL}$$

↑ 11.25 kg (under W_{HM}); ↑ .5 kg (under W_{STEEL})

$$0.00354\ W_{SC} + 0 \cdot W_{O_2} + 0 \cdot W_{LM} + 0 \cdot W_{AIR} - 0 \cdot W_{GAS} - .0696\ W_{SLAG} = -10.75 \text{ kg}$$

CARBON BALANCE:

$$0.041 W_{HM} + 0.00646 W_{SC} + 0 \cdot W_{O_2} + 0 \cdot W_{LM} + 0 \cdot W_{AIR} = (.4499)(.273) W_{GAS} + 0 \cdot W_{SLAG} + .0005 W_{STEEL}$$

$$0.00646 W_{SC} + 0 \cdot W_{O_2} + 0 \cdot W_{LM} + 0 \cdot W_{AIR} - .1228 W_{GAS} - 0 \cdot W_{SLAG} = -30.25 \text{ kg}$$

LIME (CaO) BALANCE:

$$1.00\ W_{LM} = 0.50\ W_{SLAG}$$

$$0 \cdot W_{SC} + 0 \cdot W_{O_2} + 1.00\ W_{LM} + 0 \cdot W_{AIR} - 0 \cdot W_{GAS} - 0.50\ W_{SLAG} = 0$$

OXYGEN BALANCE:

$$1.00 W_{O_2} + 0.233\ W_{AIR} = (.4499)(.727) W_{GAS} + (0.0294) W_{GAS} + (.149)(.533) W_{SLAG} + (.351)(.223) W_{SLAG}$$

$$0 \cdot W_{SC} + 1.00 W_{O_2} + 0 \cdot W_{LM} + 0.233 W_{AIR} - 0.3565 W_{GAS} - 0.1577 W_{SLAG} = 0$$

NITROGEN BALANCE:

$$0.7671\ W_{AIR} = 0.5205 W_{GAS}$$

$$0 \cdot W_{SC} + 0 \cdot W_{O_2} + 0 \cdot W_{LM} + .7671\ W_{AIR} - 0.5205 W_{GAS} - 0 \cdot W_{SLAG} = 0$$

Now write the augmented matrix for solution of simultaneous equations. (Pick any six of the following seven for solution to determine the six unknowns.)

BALANCES	UNKNOWN PARAMETERS						CONSTANTS
	W_{SCRAP}	W_{O_2}	W_{LIME}	W_{AIR}	W_{GAS}	W_{SLAG}	
OVERALL	1.0	1.0	1.0	1.0	-1.0	-1.0	250.0
IRON	0.99	0.	0.	0.	0.	-.273	291.0
SILICON	.00354	0.	0.	0.	0.	-.0696	-10.75
CARBON	.00646	0.	0.	0.	-.1228	0.	-30.25
LIME(CaO)	0.	0.	1.0	0.	0.	-.50	0.
OXYGEN	0.	1.0	0.	.233	-.3565	-.1577	0.
NITROGEN	0.	0.	0.	.7671	-.5205	0.	0.

The final answer is, using the top six equations, and rounding off to the nearest kg,

hot metal (input)	=	750 kg
steel (output)	=	1000 kg
steel scrap (input)	=	341.kg
oxygen (input)	=	80.kg
lime (input)	=	86.kg
infiltrate air (input)	=	179.kg
waste gas (output)	=	264.kg
slag (output)	=	172 kg

A check shows that the total input = total output = 1436 kg.

4.3.2 PROBLEMS HAVING DIRECT SOLUTIONS

Problems of this type are usually found in non-reacting systems with few components. The problem solution involves setting up of direct algebraic solutions.

EXAMPLE 4.3-6: In the sinter plant shown schematically in Fig. 4.3-2 the sinter "mix" is moistened while being pelletized on a flying disk before feeding to the sinter strand.

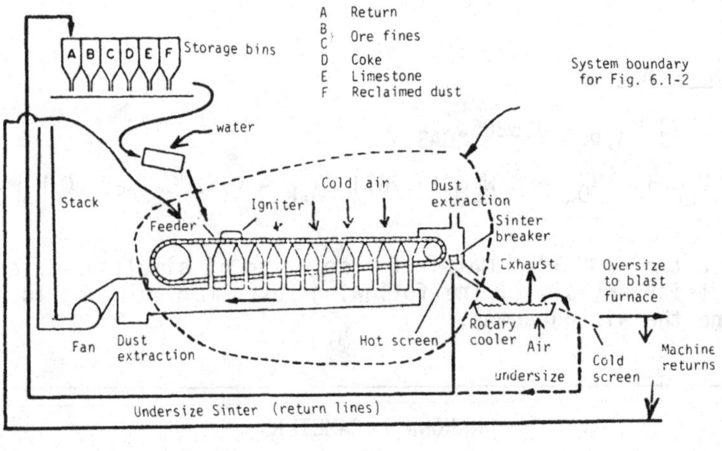

Figure 4.3-2 General arrangement of a sinter plant. (Ref. Aglomeration of Iron Ores, J. Ball, J. Dartnell, J. Davison, A. Grieve, R. Wild, Elsevier, NY, 1973, 40)

If the mix has 1% H_2O initially, and 9% H_2O is desired after pellitizing, how many kg of water are required per metric ton of mix (dry basis)?

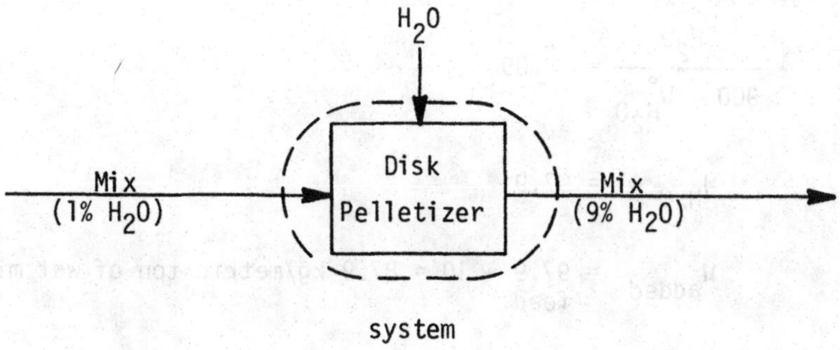

Basis: 1 metric ton of dry mix

Solution: Let $W^{\circ}_{H_2O}$ = total weight of H_2O in the output stream.

Then,

$$\frac{W^{\circ}_{H_2O}}{1000 + W^{\circ}_{H_2O}} = 0.09$$

$$W^{\circ}_{H_2O} = 90 + 0.09\, W^{\circ}_{H_2O}$$

$$W^{\circ}_{H_2O} = 98.9 \text{ kg}$$

Now let W'_{H_2O} = weight of H_2O in the mix feed stream.

$$\frac{W'_{H_2O}}{1000 + W'_{H_2O}} = 0.01$$

$$W'_{H_2O} = 10.1 \text{ kg}$$

Therefore, by water balance,

$$W_{added} = W^{\circ}_{H_2O} - W'_{H_2O}$$

$$= 98.9 - 10.1$$

$$= 88.8 \text{ kg/metric ton of solids.}$$

EXAMPLE 4.3-7: If the preceding problem had asked for an answer in kg H_2O per metric ton of wet feed, what would the solution be?

Solution:

$$W'_{H_2O} = (.01)\ (1000) = 10 \text{ kg}$$

$$\frac{W^{\circ}_{H_2O}}{900 + W^{\circ}_{H_2O}} = .09$$

$$W^{\circ}_{H_2O} = 97.9$$

$$W_{added} = 97.9 - 10 = 87.9 \text{ kg/metric ton of wet mix feed.}$$

4.3.3 TIE COMPONENTS

Another approach to solution of simpler problems involves the use of tie components. A tie component is a component (element or compound) that travels through the system without engaging in any reaction or change and without having any of that component added or withdrawn during its passage through the process.

The simplest example is that of N_2 in combustion processes, when the N_2 enters with air. Likewise, since Ni is not oxidized in the presence of Fe during the refining of steel, Ni becomes a good tie element during ferrous melting process analysis.

EXAMPLE 4.3-8: A furnace calcining $CaCO_3$ is burning coal with air and the airflow is controlled and metered so that it is known that $100 m^3/min$ (STP) of air enters the furnace. If the flue-gas analysis (STP) is 55% N_2, what is the volume (STP) of the gases?

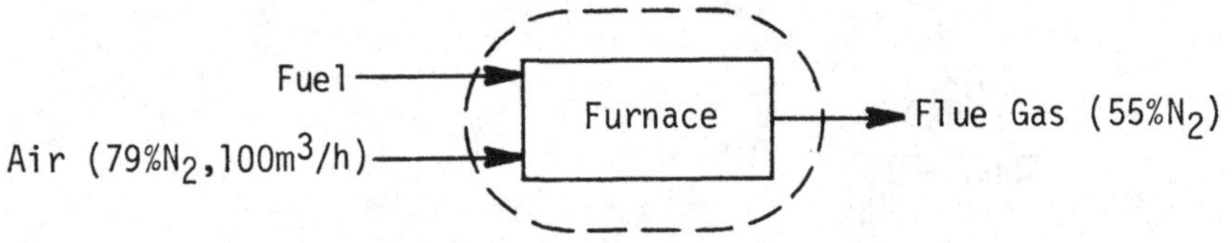

Basis: $100 \ m^3$ air/min

Solution: N_2 in = N_2 out = $(0.79 m^3 \ N_2/m^3 \text{ air})(100 \ m^3 \text{ air/min})$

$$= 79 \ m^3 N_2/\text{min}.$$

The 79 m^3/min of N_2 corresponds to 55% of the waste gases, so the

total volume of waste gases $= \frac{79}{.55} = 143.6 \ m^3/\text{min}.$

EXAMPLE 4.3-9: One of the more difficult problems faced by a melter making a heat of alloy in a large furnace is that of calculating alloy additions to meet a desired composition. The difficulty usually arises because of lack of direct knowledge of the total weight of liquid metal in the furnace at any given moment. Suppose, for example, that an electric furnace contains a charge of plain carbon steel scrap (containing no Ni) and 2000 kg of electrolytic Ni cathodes. When melted, the bath analyzes 8.3% Ni. After blowing the bath with oxygen to remove C, Si, and inevitably some Fe, Mn, and Cr, but not Ni, the bath analyses 8.8% Ni and 16.3% Cr. If the desired Cr level is 17.5%, how much ferrochrome (FeCr, containing 75% Cr) should now be added?

Solution: Using Ni as a tie element, make a Ni balance:

$$\underline{\text{Ni in}} = \underline{\text{Ni out}}$$

$$2000 = (0.088)(W_{bath,\ after\ O_2\ injection})$$

Solving, W_{bath} = 2000/0.088 = 22727. kg.

Now make a Cr balance using the value of W_{bath} calculated above:

$$\underline{\text{Cr in}} = \underline{\text{Cr out}}$$

$$(0.163)(22727) + 0.75\ W_{FeCr} = 0.175\ (22727 + W_{FeCr})$$

Solving for W_{FeCr},

$$W_{FeCr} = 474.\ \text{kg.}$$

Since the bath weight after this FeCr addition will now be

$$22727 + 474 = 23201\ \text{kg}$$

the final % Ni will be reduced to

$$\%\ \text{Ni} = (2000/23201) \times 100 = 8.62\%$$

It is also possible to calculate how much was oxidized from the bath into the slag or gases during the oxidation stage by means of Ni as a tie element. The initial bath weight after melting would have been

$$W_{bath},\ \text{melted} = 2000/.083 = 24096.\ \text{kg}$$

Since the weight of the bath after oxidation was 22727 kg, the weight of C, Si, Mn, Cr, Fe, etc. lost is the difference:

$$W_{lost} = 24096 - 22727 = 1369\ \text{kg}$$

It is important to test the sensitivity of tie element calculations. In most cases, an element that is present in large proportions in the streams in which it occurs should be used as a tie element. In Example 4.3-8, if the flue gases really contained 53% instead of 55% N_2, the calculated value of flue gas volume would be 149.0 m^3/min, an error of 3.7%. In Example 4.3-9, an analytical error of 0.2% in the Ni content of the bath after oxidation would translate into a calculated error of 11 extra kg in the required weight of ferrochrome. This in turn would cause the resulting analysis to be 17.52% Cr, not 17.50%, an insignificant error.

On the other hand, suppose that a melting furnace contains aluminum with no copper in it. If 5.0 kg of Cu is added to the bath, and the analysis after mixing is reported to be 0.75% Cu, the bath weight would be calculated to be 666.7 kg. If, however, the true analysis is 0.65%, (a 13.3% error in analysis) the true weight would be 769.2 kg, a large 13.3% error in absolute value. This would cause serious problems if further alloying additions were required.

4.3.4 RECYCLE CALCULATIONS

Recycle calculations throw a new wrinkle into problem solution. The important aspect of this problem is to place the system boundary correctly.

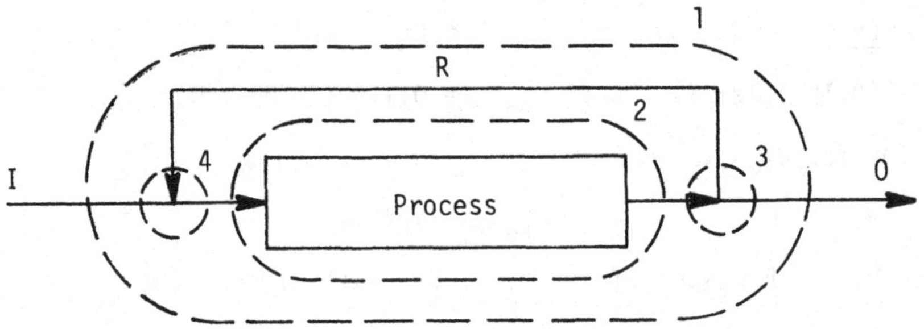

In the above schematic, several system boundaries are shown. System 1 ignores all of the internal aspects of the process, including the recycle, and just gives the net input, I, and output, 0. In the case of System 2, the process is looked at as in previous sections, but the entering and leaving streams both include the recycle stream and will be larger than streams I and 0. Systems 3 and 4 involve recycle streams and may be regarded as mixing (separation) process units. The reader should satisfy himself that only three of the four resulting balances are independent. Any three will give the fourth.

It is possible to draw another system boundary, but this is not a very intelligent choice, since R both leaves and enters, and therefore cancels itself and the net is the same as system 1 in the previous figure.

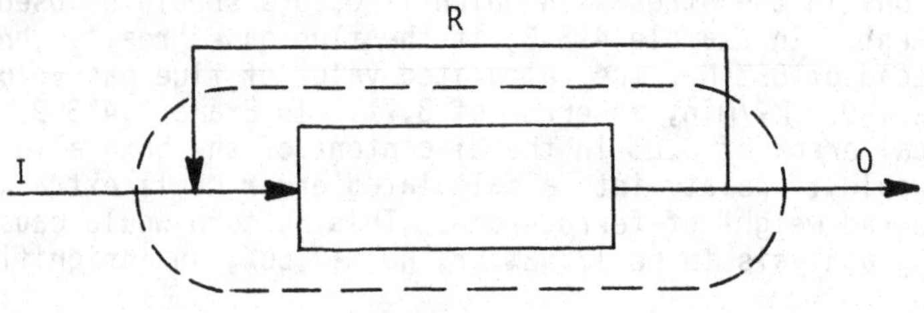

EXAMPLE 4.3-10: Consider a powdered metal plant with a screening system on the product stream. If 100 tonnes of fresh feed material are put into the plant everyday, all of which is converted into product P, and a batch screen test on P shows that 10% of the total is not passing a 48-Mesh screen, while 20% of what passes the 48-Mesh screen also passes a 300-Mesh screen, calculate the net production, plant yield and actual amount produced, P.

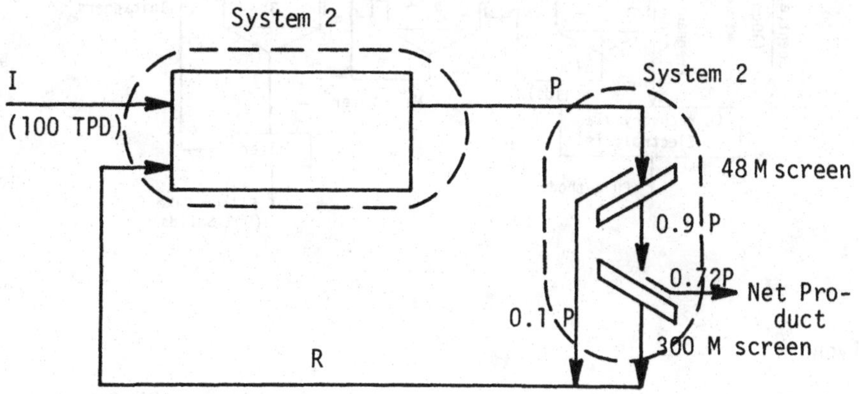

Solution: If the plant is to be in steady state operation, the net product must also be 100 TPD.

P or R are not known, so two equations are needed. Make two system boundaries, 1, around the process unit, and 2, around the screening system. Write mass balances around each:

(1) $100 + R = P$

(2) $P = 0.1\ P + 0.2(0.9P) + 100$

Solving these two equations simultaneously,

$P = 138.9$

$R = 38.9$

and the yield = 100/139 = 72%

EXAMPLE 4.3-11: In the separation of liquid from solids in a leach-wash counter-current decantation (C.C.D.) circuit for recovery of Cu from $CuSO_4$ in ore, the ore is leached with water to remove the $CuSO_4$, and then counter-current washed in three thickeners. The Cu is then recovered by electrolysis from the pregnant $CuSO_4$ solution.

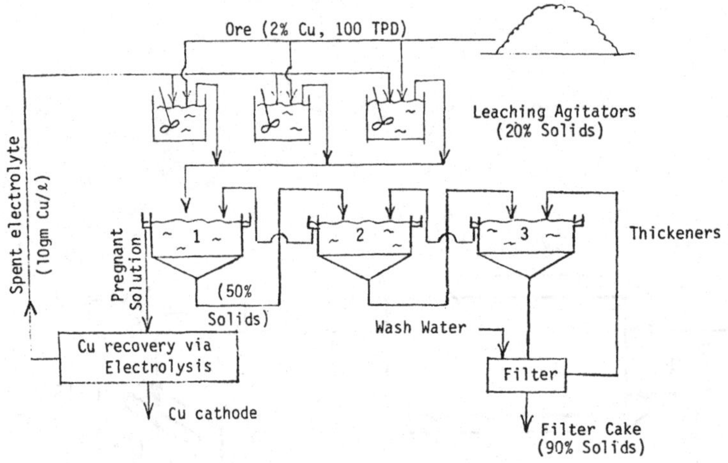

Given:

1. 100 metric tonnes of ore per day containing 2% Cu as $CuSO_4$ is the feed.
2. Agitators maintained at 20% solids by weight. All $CuSO_4$ dissolves into water.
3. The thickeners operate with clear overflow and 50% solids by weight in the underflow from each.
4. The filter cake is 90% solids by weight.
5. The spent electrolyte contains 10 gm Cu/l.

<u>Calculate</u>

a) Water balance around circuit.

b) Weight of copper lost per day in filter cake.

c) The concentration of Cu in the pregnant solution.

Solution: Break the total system into units and place all known or directly computed data on the diagram.

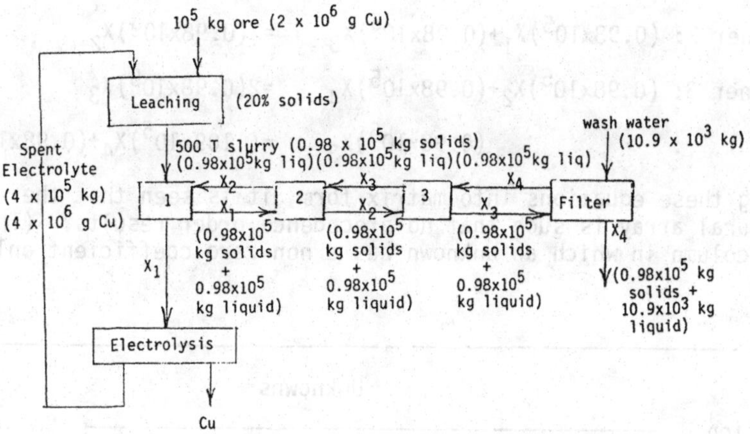

Step 1: Since the leach agitators are kept at 20% solids, 400 tonnes of spent electrolyte ($4x10^5$kg) must enter with 100 tonnes of ore, and 98 tonnes of solids leave in the slurry.

Step 2: A <u>water balance</u> shows that the only place water enters or leaves the system is at the filter, so make a water balance around the filter:

$$\underline{\text{In}} \qquad\qquad\qquad \underline{\text{Out}}$$

$$W_{\text{wash water}} + 0.98\text{x}10^5 = 0.98\text{x}10^5 + W_{\text{water in filter cake}}$$

or the wash water must balance the water lost in the cake; 10% of the total filter cake weight.

$$W_{\text{water in filter cake}} = 0.1(0.98\text{x}10^5 + W_{\text{water in filter cake}})$$

$$0.9W_{\text{water in filter cake}} = 9.8 \text{ x } 10^3$$

Therefore, $W_{\text{wash water}} = 10.9 \text{ x } 10^3$ kg/day

Step 3: Place the solid and liquid flow in and out of each thickener on the diagram. The unknowns are the compositions of streams X_1, X_2, X_3 and X_4.

Step 4: Since the system is at steady state, it is possible to make Cu balances around each thickener in the CCD circuit. Working in units of grams Cu/kg(=grams Cu/l).

	In	=	Out
Thickener 1:	$(4x10^5)(10)+2x10^6+(.98x10^5)X_2$	=	$(0.98x10^5)X_1+(4x10^5)X_1$
Thickener 2:	$(0.98x10^5)X_1+(0.98x10^5)X_3$	=	$2(0.98x10^5)X_2$
Thickener 3:	$(0.98x10^5)X_2+(0.98x10^5)X_4$	=	$2(0.98x10^5)X_3$
Filter :	$(0.98x10^5)X_3$	=	$(.109x10^5)X_4+(0.98x10^5)X_4$

Putting these equations into matrix form, it is seen that the structural array is such that no precedence order results. (There is no column in which an unknown has a non-zero coefficient only once.)

Equation	Unknowns				Constant
	X_1(gCu/kg)	X_2	X_3	X_4	
Cu Balance around Thickener 1	$4.98x10^5$	$-0.98x10^5$	0	0	6×10^6
Cu Balance around Thickener 2	$0.98x10^5$	$-1.96x10^5$	$0.98x10^5$	0	0
Cu Balance around Thickener 3	0	$0.98x10^5$	$-1.96x10^5$	$0.98x10^5$	0
Cu Balance around Filter	0	0	$-0.98x10^5$	$1.089x10^5$	0

Therefore, a computer is used to solve the set of equations simultaneously, with the result that

X_1 = 14.69 grams Cu/l (Concentration in pregnant solution)

X_2 = 13.47 "

X_3 = 12.24 "

X_4 = 11.02 "

Thus, the copper lost per day, in the filter cake, is:

$(10.9x10^3$ kg solution/day)(11.02 g Cu/kg solution)=$1.2x10^5$ g Cu/day

This amounts to 6% of the Cu into the process.

4.4 REPRESENTATION OF MATERIAL BALANCES

Representation of material balances can be by the type of diagrams or tables used so far, or it can be by a different type of diagram, in which the magnitude of a stream is represented by the width of its arrow. In this way, the relative amounts of materials entering and leaving a process may be easily visualized. Such a diagram is shown below for the sinter plant depicted in Fig. 4.3-2.

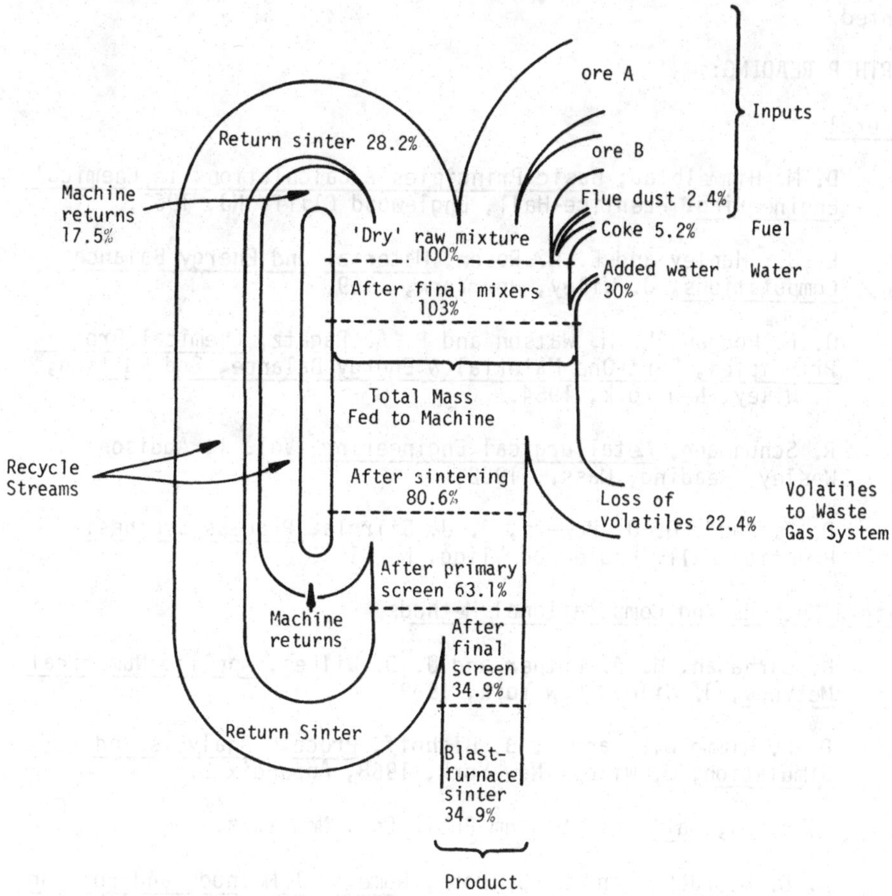

Figure 4.4-1 Material flow diagram for sinter plant depicted in Fig. 4.3-2. (Ref. Ball et al, op cit, p. 96)

The "machine returns" are larger sized pieces of sinter that are purposely recycled to provide an unreactive layer of solids to protect the sinter machine pallets from the high temperatures generated in the bed. The "return sinter" is undersize material (usually < 1 cm) that cannot be used in the blast furnace. The startling factor that clearly can be seen from such a diagram as

Fig. 4.4-1 is that in order to produce 1000 tonnes per day of sinter, 2865 tonnes per day of dry solids must be processed by the machine.

4.5 SUMMARY

Material balances on metallurgical processes generally result in a set of simultaneous linear equations and the number of independent equations in this set fixes the maximum number of unknowns that may be determined for any problem. The procedures for setting up material balances, determining the number of independent equations and obtaining solutions to the sets of equations are presented.

FURTHER READING:

General

1. D. M. Himmelblau; Basic Principles & Calculations in Chemical Engineering, Prentice-Hall, Englewood Cliff, NJ, 1963.

2. E. J. Hanley and E. M. Rosen, Material and Energy Balance Computations, J. Wiley, New York, 1969.

3. O. F. Hougon, K. M. Watson and R. A. Ragatz, Chemical Process Principles, Part-One Material & Energy Balance, 2nd Edition, J. Wiley, New York, 1954.

4. R. Schuhmann, Metallurgical Engineering, Vol. 1, Addison-Wesley, Reading, Mass., 1952.

5. D. F. Rudd, G. J. Powers, J. J. Siirola, Process Synthesis, Prentice-Hall, Englewood Vligg, NJ, 1973.

Matrix Methods and Computational Methods

1. B. Carnahan, H. A. Luther and J. D. Wilkes, Applied Numerical Methods, J. Wiley, New York, 1969.

2. D. M. Himmelblau and K. B. Bishoff, Process Analysis and Simulation, J. Wiley, New York, 1968, Appendix 2.

3. F. Ayres, Matrices, Schanm Publ. Co., New York.

4. D. D. McCracken and W. S. Dorn, Numerical Methods and Fortran Programming, J. Wiley, New York, 1964.

5. J. T. Golden, Fortran IV: Programming and Computing, Prentice-Hall, Englewood Cliffs, NJ, 1965.

6. R. W. Hornback, Numerical Methods, Quantum Publishers, Inc., New York, 1975.

EXERCISES

4.1 FeS_2 is roasted to Fe_3O_4 and SO_2 in a flash roaster.

FeS_2 occurs in a pyrite ore, which in this case analyses 48%S. If the air supply is controlled at 10 lb air/lb ore, calculate the expected exit gas analysis.

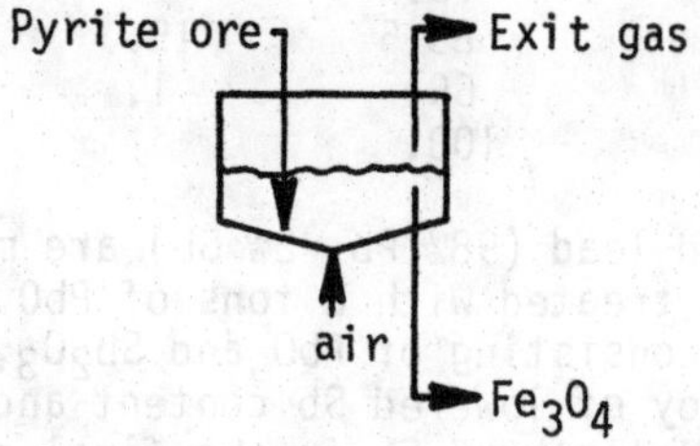

4.2 Show how the following sets of equations can be partitioned into smaller subsets and indicate the order of solution of the subsets.

(a) $f_1(x_1,x_2,x_3) = 0$

$f_2(x_2) = 0$

$f_3(x_2,x_3) = 0$

(b) $f_1(x_1,x_2,x_4) = 0$

$f_2(x_1,x_2,x_3) = 0$

$f_3(x_4) = 0$

$f_4(x_3,x_4) = 0$

(c) $f_1(x_1,x_3,x_5) = 0$

$f_2(x_2,x_6) = 0$

$f_3(x_5,x_6,x_7) = 0$

$f_4(x_1) = 0$

$f_4(x_2,x_4) = 0$

$f_6(x_2,x_6) = 0$

$f_7(x_3,x_5,x_7) = 0$

4.3 Suggest how the following sets of equations may be "torn":

(a) $f_1(x_1,x_3) = 0$

$f_2(x_2,x_3) = 0$

$f_3(x_1,x_2) = 0$

(b) $f_1(x_1,x_3,x_4) = 0$

$f_2(x_1,x_2,x_3,x_4) = 0$

$f_3(x_1,x_2,x_3) = 0$

$f_4(x_1,x_2,x_4) = 0$

4.4 You are working in a laboratory and are requested to melt 100 lb of stainless steel, AISI Type 304, with an aim analysis of 17.5% Cr, 8.5%Ni, and 0.5%Mn. Assuming no losses during melting, compute the charge that you would use, for the following available materials:

%:	Cr	Ni	Fe	Mn
alloy scrap	68	20	10	2
ferrochromium	75		25	
electro Ni		100		
electro Fe			100	

4.5 A Cu - 5% Sn - 10% Bi - 5% Zn alloy is to be melted. If there is a 10% loss of Zn from the charge during melting, and the following alloys are available, how many lb of each alloy would you charge to make 100 lb of alloy?

	% Cu	% Sn	% Zn	% Bi
Alloy A	70.		30.	
Alloy B	83.5	15.		1.5
Alloy C	69.	1.		30.
Pure Cu	100.			

4.6 100 tons of hard lead (98% Pb, 2% Sb) are melted in a steel kettle and then treated with 2 tons of PbO. The products are (1) a slag consisting of PbO and Sb_2O_3, analyzing 20% Sb, (2) a Pb-Sb alloy of lowered Sb content and negligible oxygen content. Calculate the % Sb in the final alloy. The reaction involved is:

$$3PbO + 2\ Sb(in\ Pb) = Sb_2O_3 + 3\ Pb$$

4.7 The plant in the flow sheet below employs H_2 to reduce 2000 kg/h of Fe_2O_3 according to

$$Fe_2O_3 + 3H_2 \rightarrow 3H_2O + 2Fe$$

The hydrogen in the recycle is mixed with the H_2 in the fresh feed before entering the reactor. The purge stream, P, is bled off to prevent CO_2 buildup in excess of 2.5% at the inlet. The ratio of R to fresh feed gas is 4:1. Calculate the amount and composition of the purge stream.

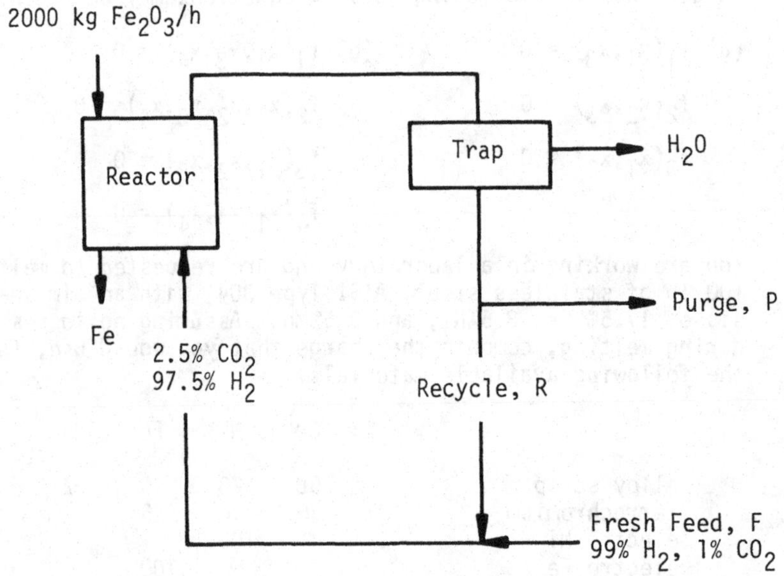

4.8 In a new process for copper melting, the old smelting furnace is bypassed and the concentrated ore is fed directly to the converter in a continuous stream where it is melted and blown with air to form blister copper in one operation. This concentrate must, however, be dried, since moisture in contact with liquid metal can easily explode. Assuming that the feed rate is 30 tons of concentrate/h, on a wet basis with 7% moisture, and that the moisture after drying (before charging to the converter) is 1/2%, calculate:

a) the dry feed rate,

b) the weight of moisture removed/h

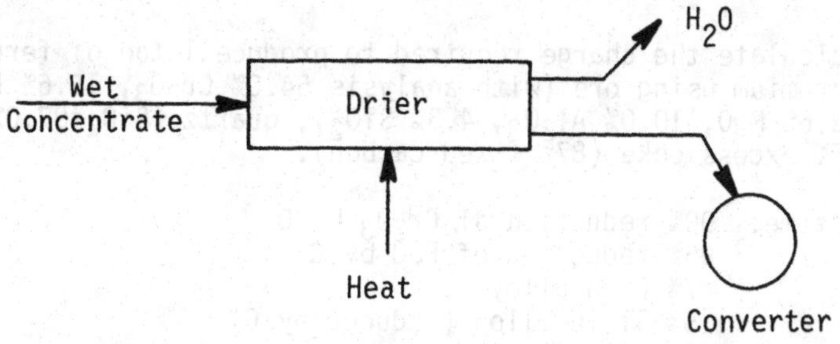

4.9 Suggest how the following sets of equations may be "torn":

$$f_1(x_1,x_3,x_5) = 0$$
$$f_2(x_2,x_6) = 0$$
$$f_3(x_5,x_6,x_7) = 0$$
$$f_4(x_1) = 0$$
$$f_4(x_2,x_4) = 0$$
$$f_6(x_2,x_6) = 0$$
$$f_7(x_3,x_5,x_7) = 0$$

4.10 The following materials are to be used in an iron blast furnace:

	Ore A (%)	Ore B (%)	Limestone (%)	Coke (%)
Fe_2O_3	78	70	4	1
SiO_2	11	16	4	5
Al_2O_3	2	12	2	2

(cont'd)

	Ore A (%)	Ore B (%)	Limestone (%)	Coke (%)
CaO	5	2	46	2
MgO	4		4	
CO_2			40	
C				90

1 ton of coke is used per ton of pig iron made.
Pig iron will contain 95% Fe, 4% C, and 1% Si.
The slag produced is to have a ratio of $\Sigma CaO:Al_2O_3:SiO_2$=55:15;30 where ΣCaO is the sum of all CaO plus 56/40 times the MgO present.
Determine the weight of each material (in lb) to be charged, per 2000 lb of pig iron made.

4.11 Calculate the charge required to produce 1 ton of ferrochromium using ore (with analysis 54.0% Cr_2O_3, 15.6% FeO, 13.6% MgO, 10.0% Al_2O_3, 4.5% SiO_2), quartz (SiO_2=97.0%), and 25% excess coke (87% fixed carbon).

Assume: 90% reduction of Cr_2O_3 by C
95% reduction of FeO by C
7% C in alloy
1.5% Si in alloy (reduced by C)
30% SiO_2 in slag

4.12 A reduction process requires a feed of C, O_2 and MO and produces liquid M and a gas phase containing CO and CO_2. For thermodynamic reasons, the ratio of CO to CO_2 in the product gas is desired to be K. (i.e., $K = \frac{W_{CO}}{W_{CO_2}}$). If the weight of MO fed to the system, and therefore M out of the system is given, set up the equations that must be solved to find the weight of C(W_C), weight of O_2(W_{O_2}), W_{CO} and W_{CO_2} in the input and output streams. Precedence order the set of equations, or, in the event they cannot be precedence ordered, find a tearing sequence that will result in solution of the set.

CHAPTER 5

THERMOCHEMISTRY

5.0 INTRODUCTION

The primary tool for the solution of energy balance problems is the first law of thermodynamics. In this chapter, several mathematical forms of the first law which are particularly useful for energy balance calculations are developed. The types and sources of data required for the complete solution of these problems are also discussed. Since the vast majority of data available in these compilations are in calories, this unit and not the Joule will be emphasized in this and the following chapters.

5.1 THE SYSTEM

Energy balances are usually prepared for a collection of objects or region of space that is of particular interest when analyzing a process. This portion of the universe which is identified for a study is called the system. The remainder of the universe is called the surroundings; and the real or imaginary wall that separates the system from the surroundings is called the boundary. For any given process, many system boundaries can be identified. The selection of the boundary often has an influence on the difficulty of the subsequent calculations.

Rules for choosing the easiest system cannot be enumerated. The choice must be made based upon the experience gained from the solution of similar problems. However, the novice to this field should not be discouraged. Even if the "wrong" system boundary is chosen, the desired results can usually be obtained through additional calculations.

5.1.1 TYPES OF THERMODYNAMIC SYSTEMS

Thermodynamic systems may be characterized as open, closed or isolated; simple or complex; steady-state or unsteady-state; homogeneous or heterogeneous.

When both mass and energy can pass through the boundary of a system, the system is an open system. A system is a closed system when only energy can pass through its boundary. An isolated system can exchange neither mass nor energy with its surroundings.

EXAMPLE 5.1-1: Is liquid nitrogen contained in a Dewar Flask an isolated system?

Solution: A Dewar Flask is a double-walled jar which has an evacuated space between the walls. This type of enclosure may be considered a perfect thermal insulator for most calculations. Thus, when stoppered, energy and mass may not enter the volume inside of the flask. Accordingly, the volume inside the flask is an isolated system. The liquid is, however, not necessarily an isolated system, since, if space exists above it in the flask, mass may be transferred to or from the vapor phase above the liquid. The liquid may therefore be an open system.

EXAMPLE 5.1-2: If an energy balance is being made on a refining furnace which contains liquid metal, liquid slag, and a gas phase, what closed system can be chosen for the analysis?

Solution: Alternative selection of the system is possible dependind on the reactions to be studied: 1) If chemical reactions occur within the liquid metal to form a slag product, the system then is slag and metal. 2) If the refractory also reacts with either slag or metal, it may have to be included as part of the system. 3) If the gas atmosphere over the melt affects the results, it also may become part of the system.

Properties of thermodynamic systems are the definable or measurable characteristics of the system. Thermodynamic properties are a function of the current characteristics or state of the system, and for this reason are sometimes called state properties or variables of state. Properties of systems which are independent of the size of the system, such as temperature and pressure, are called intensive properties. Properties such as the mass or volume which are dependent on the size of the system are called extensive properties.

Solid, liquid, or gas are terms that describe recognizable states of matter. Unfortunately, this type of classification is inadequate for describing the state of a thermodynamic system. Experience has taught us that an adequate specification of the state of a system can only be achieved by specifying a limited number of state properties of the system.

Denbigh[1] shows that for a simple system, the specification of only two intensive properties is adequate to describe all of the other intensive properties, such as specific heat, density, surface tension, etc. Thus:

$$I_i \ (i = 3,4,\ldots n) = f(I_1, I_2) \qquad (5.1\text{-}1)$$

where I refers to intensive properties. While if extensive properties, E_1, E_2, ... E_p are to be defined, at least one extensive property such as mass, m, as well as two intensive properties must be specified.

$$E_i \ (i = 2,3,\ldots.p) = f(I_1, I_2, E_1) \qquad (5.1\text{-}2)$$

Eqs. (5.1-1) and (5.1-2) are generalized equations of state. Specific functional relationships of this type can be used to denote the interrelations between the properties of a system. The well known gas law, Eq. (2.3-1), is an example of an equation of state for gases. While such equations of state are readily assignable to gases and some liquids, their formulation for solids, especially metals, is not readily accomplished, since complete

(1) K. Denbigh, Principles of Chemical Equilibrium, Cambridge Press, England, 1961.

specification of the condition of the solid requires inclusion of the past history of the material.

For an unsteady-state system, the properties of the system vary with time. Thus, a system which is undergoing a temperature change is an unsteady-state system. If, however, the same system were brought back to its original temperature and other thermodynamic properties, or if the properties of the system are constant with time, the system could be considered to be a steady-state system, since the state of the initial system and the final system are the same.

EXAMPLE 5.1-3: When can a batch-type refining furnace be considered a steady-state system?

Solution: A batch-type refining furnace meets the definition of a steady-state open system when a sufficient number of successive heats are considered to provide the average quality that is essential for steady-state analysis. A single heat in a refining furnace, obviously does not represent steady-state conditions.

Finally, systems are also broken down further into homogeneous systems and heterogeneous systems. Metallurgical terms can conveniently be used to describe these two. Liquid metal (or a solid solution) when considered alone is a homogeneous system, since it has the same properties throughout. Two solid solutions or two liquids or any combination of two or more homogeneous phases which result in abrupt changes in properties at phase boundaries and which could, theoretically, be mechanically separated, constitute a heterogeneous system.

5.2 THE FIRST LAW OF THERMODYNAMICS

The first law of thermodynamics is concerned with energy changes within a system. Energy changes in a system can occur in a variety of ways: The addition or removal of heat from a system represents one way. Putting heat into a steam engine and getting out work is another. When a liquid freezes, heat is evolved, whereas when the liquid is vaporized heat is required. A blast furnace is heated not only by the combustion of the fuel but also by the sensible heat acquired from the preheated air and from the raw materials. Therefore energy changes can occur with or without an accompanying change in the mass of the system. The evolution of heat resulting from a combustion process represents an exothermic process while other reactions are known to require heat during the (endothermic) process. The oxidation of carbon is exothermic while the reduction of CO_2 by C is endothermic. When a gas is compressed, as in a tire pump, work is done on the system. A battery does electrical work when it is used. A magnet does work when it lifts an iron bar. Surface energy causes grain junctions in metals to acquire specific configurations which tend to minimuze this energy. Mechanical working of metals contributes energy either in the form of heat, which may be lost gradually, or in the form of a distorted or cold worked structure which may be retained indefinitely. The question is whether or not all these and a myriad of other changes are subject to some universal law. The answer is "yes" and the law is the First Law of Thermodynamics.

Since a device for directly measuring energy is not available, energy changes can only be inferred from the changes in properties of the system by using an equation similar to Eq. (5.1-1) or Eq. (5.1-2). The first law also provides a means for calculating energy changes and developing relationships between the system's energy changes and changes in the properties of the system.

5.2.1 GENERAL CASE

The simplest statement of the First Law of Thermodynamics is that energy is conserved. For any law of conservation,

$$\text{Input} = \text{Output} + \text{Accumulation} \tag{5.2-1}$$

Thus, for the conservation of energy,

$$\begin{array}{c}\text{Input}\\ \text{of Energy}\\ \text{to the}\\ \text{System}\end{array} = \begin{array}{c}\text{Output}\\ \text{of Energy}\\ \text{From the}\\ \text{System}\end{array} + \begin{array}{c}\text{Accumulation}\\ \text{of Energy}\\ \text{Within the}\\ \text{System}\end{array} \tag{5.2-2}$$

The First Law applies to the conservation of all forms of energy: kinetic energy, the energy associated with the motion of a system; potential energy, the energy associated with the position of a system in an electric, gravitational or magnetic field; internal energy, the energy which is stored within a system due to the relative motion and positions of the atoms within the system; energy in transit, such as heat, the energy flow across the boundary of the system due to a temperature difference between the system and surroundings, and work, the energy flow due to any other potential difference such as pressure. The energy created or destroyed by nuclear processes also can be covered by the first law; however, for our purposes the mass-energy equivalence question need not be considered.

The total energy of a system (symbol E') is an extensive property of the system. The total energy of a system of mass m is given by

$$E' = U' + E_k' + E_p' \tag{5.2-3}$$

where U' is the internal energy of the system. The kinetic energy of the system (symbol E_k') is

$$E_k' = \frac{m\bar{v}^2}{2g_c} \tag{5.2-4}$$

where $\bar{v}$ is the average velocity of the system. The potential energy equals

$$E_p' = \frac{m\,g\,Z}{g_c} \tag{5.2-5}$$

where Z is the relative position of the system.

Combining Eqs. (5.2-3) through (5.2-5) yields

$$E' = U' + \frac{m\bar{V}^2}{2g_c} + \frac{m\,g\,Z}{g_c} \qquad (5.2\text{-}6)$$

The <u>specific energy</u> (symbol e), the <u>energy per unit mass</u> of the system equals

$$e = u + \frac{\bar{V}^2}{2g_c} + \frac{gZ}{g_c} \qquad (5.2\text{-}7)$$

where u is the specific internal energy, the internal energy per unit mass.

To formulate a mathematical representation of the First Law, it is now simply necessary to apply the law of conservation of energy to an arbitrary system, such as the one shown in Fig. 5.2-1, for which energy flows can be summarized as follows:

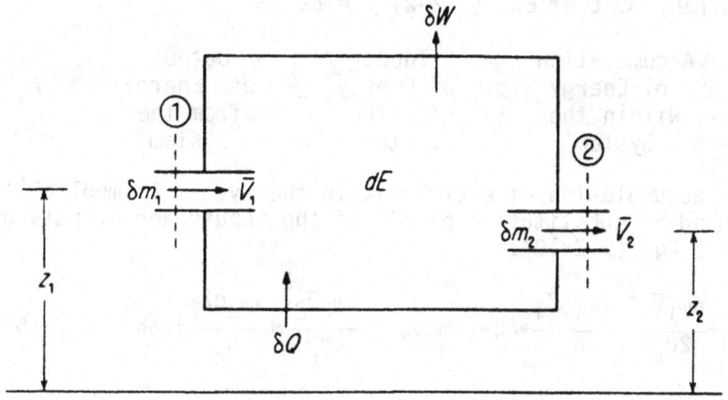

Figure 5.2-1 Energy and mass flows in an arbitrary, open system. (From Ref. 2)

Energy inputs to the system:

1) The energy carried in by mass δm_1 entering the system at 1,

$$\delta m_1\, e_1 = \delta m_1\, u_1 + \frac{\delta m_1 \bar{v}_1^2}{2g_c} + \frac{\delta m_1\, g\, Z_1}{g_c} \qquad (5.2\text{-}8)$$

---

2. M. Mark, <u>Thermodynamics: An Auto-Instructional Text</u>, Prentice-Hall, Inc., New Jersey, 1967.

2) The work done on the system to move δm_1 across 1 and into the system, and

3) The heat added to the system, δq.

Energy outputs from the system:

1) The energy carried out by δm_2 at 2,

$$\delta m_2 e_2 = \delta m_2 u_2 + \frac{\delta m_2 \bar{v}_2^2}{2g_c} + \frac{\delta m_2 \, g \, Z_2}{g_c} \tag{5.2-9}$$

2) The work done by the system to push δm_2 across 2 and out of the system, and

3) Other work done by the system, δw

The conventions that are used in this text for the signs for work and heat are: heat is positive when <u>added to</u> or absorbed by the system and work is positive when <u>done by</u> the system. Thus a positive δq is an input, while a positive δw is an output from the system.

Rearrangement of Eq. (5.2-2) yields

Accumulation of Energy Within the System	=	Input of Energy to the System	-	Output of Energy from the System	(5.2-10)

Thus, the accumulation of energy within the system (symbol dE')* can be found by substitution of all of the inputs and outputs of energy into Eq. (5.2-10).

$$dE' = \delta m_1 u_1 + \frac{\delta m_1 \bar{v}_1^2}{2g_c} + \frac{\delta m_1 g Z_1}{g_c} + \delta q - \left(\delta m_2 u_2 + \frac{\delta m_2 \bar{v}_2^2}{2g_c} + \frac{\delta m_2 g Z_2}{g_c}\right) - \delta W \tag{5.2-11}$$

where δW, the net work done by the system, equals

$$\delta W = \delta w + \text{Flow work to push out } \delta m_2 - \text{Flow work to add } \delta m_1 \tag{5.2-12}$$

If there is a pressure P_1 at the entrance to the system at 1 and the entrance has an area A, a force <u>infinitesimally greater</u> than $P_1 A$ must be exerted to move material into the system. Calculation of the flow work, δw_{f1}, to move a unit mass of the fluid at

---

*For intensive and extensive properties of the system, "d" will be used to indicate a small change in the value of the property and Δ to indicate a large change. Energy in transit, heat and work, are not properties of the system, and their increments are designated by δ to differentiate them from changes in the properties of the system.

1 into the system can be done by assuming that for all practical purposes the force which must be exerted is equal to P_1A. The work then equals

$$\delta w_{f_1} = \text{Force} \cdot \text{distance} \tag{5.2-13}$$

where the distance for this calculation is chosen so that the volume per unit mass of the fluid at 1, v_1, equals

$$v_1 = \text{distance} \cdot A \tag{5.2-14}$$

or

$$\text{distance} = v_1/A \tag{5.2-15}$$

then

$$\delta w_{f_1} = P_1 A \cdot (v_1/A) \tag{5.2-16}$$

or

$$\delta w_{f_1} = P_1 v_1 \tag{5.2-17}$$

Similarly, the flow work to move a unit mass out of the system at 2 equals

$$\delta w_{f_2} = P_2 v_2 \tag{5.2-18}$$

For δm_1 and δm_2 moving into and out of the system

$$\delta W = \delta w + \delta m_2\ \delta w_{f_2} - \delta m_1\ \delta w_{f_1} \tag{5.2-19}$$

Substitution of Eqs. (5.2-17), (5.2-18), and (5.2-19) into Eq. (5.2-11) and rearrangement of the terms yields

$$dE' = \delta q - \delta w + \delta m_1 \left(u_1 + P_1 v_1 + \frac{\bar{V}_1^2}{2g_c} + \frac{gZ_1}{g_c}\right) - \delta m_2 \left(u_2 + P_2 v_2 + \frac{\bar{V}_2^2}{2g_c} + \frac{gZ_2}{g_c}\right) \tag{5.2-20}$$

For a system with a combination of n inlets and outlets, a general statement of the First Law is

$$dE' = \delta q - \delta w + \sum_{i=1}^{n} \delta m_i \left(u_i + P_i v_i + \frac{\bar{V}_i^2}{2g_c} + \frac{gZ_i}{g_c}\right) \tag{5.2-21}$$

where δm_i is the mass in stream i and is positive for an inlet stream and negative for an outlet.

The rate of change of energy in the system can be found by dividing both sides of Eq. (5.2-21) by the time increment $\Delta\tau$,

$$\frac{dE'}{\Delta\tau} = \frac{\delta q}{\Delta\tau} - \frac{\delta w}{\Delta\tau} + \sum_{i=1}^{n} \frac{\delta m_i}{\Delta\tau} \left(u_i + P_i v_i + \frac{\bar{v}_i^2}{} + \frac{gZ_i}{}\right) \qquad (5.2\text{-}22)$$

Then letting $\Delta\tau$ approach zero,

$$\frac{dE'}{d\tau} = \dot{q} - \dot{w} + \sum_{i=1}^{n} \dot{m}_i \left(u_i + P_i v_i + \frac{\bar{v}_i^2}{2g_c} + \frac{gZ_i}{g_c}\right) \qquad (5.2\text{-}23)$$

where $dE'/d\tau$ is the instantaneous rate of change of the energy in the system. The heat flux to the system is $\dot{q}$. The power output from the system is $\dot{w}$, and $\dot{m}_i$ is the mass flow rate in stream i.

Usage of the First Law for solution of energy balance problems is demonstrated in the remainder of this section, in Section 5.3, and in Chapter 6.

5.2.2 SPECIAL CASES

A. *Isolated Systems*

For an isolated system, the First Law states that the total energy of the system remains constant. The law does not restrict the flow of energy from one part of the system to another nor does it restrict the conversion of one form of energy to another. When viewed within the limitations imposed by an isolated system, it is apparent why the First Law is referred to as the law of conservation of energy.

A mathematical expression of the First Law for an isolated system is obtained when the restrictions on an isolated system; $\delta q = 0$, $\delta w = 0$, and $\delta m_i = 0$ are substituted into Eq. (5.2-21). Thus, for an isolated system

$$dE' = 0 \qquad (5.2\text{-}24)$$

Where there are n different energy changes occurring <u>within</u> the isolated system, the First Law is

$$\sum_{i=1}^{n} dE_i' = 0 \qquad (5.2\text{-}25)$$

EXAMPLE 5.2-1: Many liquids can be cooled below their normal solidification temperature, i.e. supercooled. However, once solidification begins, the liquid will completely solidify only if it is free to exhaust all of the energy given off during solidification to the surroundings. Determine what fraction of one mole of liquid copper which has been supercooled to 1307K and placed into an isolated system will solidify. The normal melting point (freezing point) of copper is 1357 K. The energy requirement for heating liquid copper from 1307 K to 1357 K is 375 cal/mole. The energy release for liquid copper at 1307 K going to solid copper at 1357 K is 2745 cal/mole.

Solution: For the isolated system, the total energy given off by the y moles of copper which solidifies and heats up must equal the energy gained by the 1-y moles of copper which only heats up, since

$$\Delta E' = 0$$

The change in energy of the liquid copper will equal

$$(1\text{-}y) \text{ moles} \cdot 375 \text{ cal/mole} = (1\text{-}y)\ 375 \text{ cal}$$

For the copper which freezes and then heats up, the change of energy equals

$$y \text{ moles} \cdot (-2745 \text{ cal/mole}) = -2745y \text{ cal}$$

Therefore,

$$375(1\text{-}y) + y(-2745) = 0$$

or $$y = 0.12 \text{ moles Cu solidified}$$

Thus, in the isolated system, 0.12 moles of copper will freeze and the remaining 0.88 moles of liquid plus 0.12 moles of solid will be heated to the normal melting point.

B. *Closed Systems*

For closed systems, transfer of mass across the system's boundary is not permitted. Thus, Eq. (5.2-21) simplifies to

$$dE' = \delta q - \delta w \tag{5.2-26}$$

The implications of the first law for closed systems can be demonstrated by considering Joule's experiments which converted mechanical energy into heat.(3) In this experiment, a container of water is brought to constant temperature by exposing it to a constant temperature bath, e.g. a water-ice mixture, See Fig. 5.2-2. The water in the container is now stirred by a paddle actuated by a falling weight in such a manner that the work output of the falling weight can be measured exactly. The system is then allowed to come to equilibrium so that the water in the container returns to its original temperature of 273K. The amount of heat added to the system by this process is measured by the amount of ice which melts, allowing for the normal melting which would have occurred during the test interval. When this test is carried out with different work inputs and with sufficient precision, it is found that the amount of heat extracted by the constant-temperature bath is equal

---

(3) J. P. Joule, Philisophical Magazine, 31 (1847), 173; 35 (1849), 533.

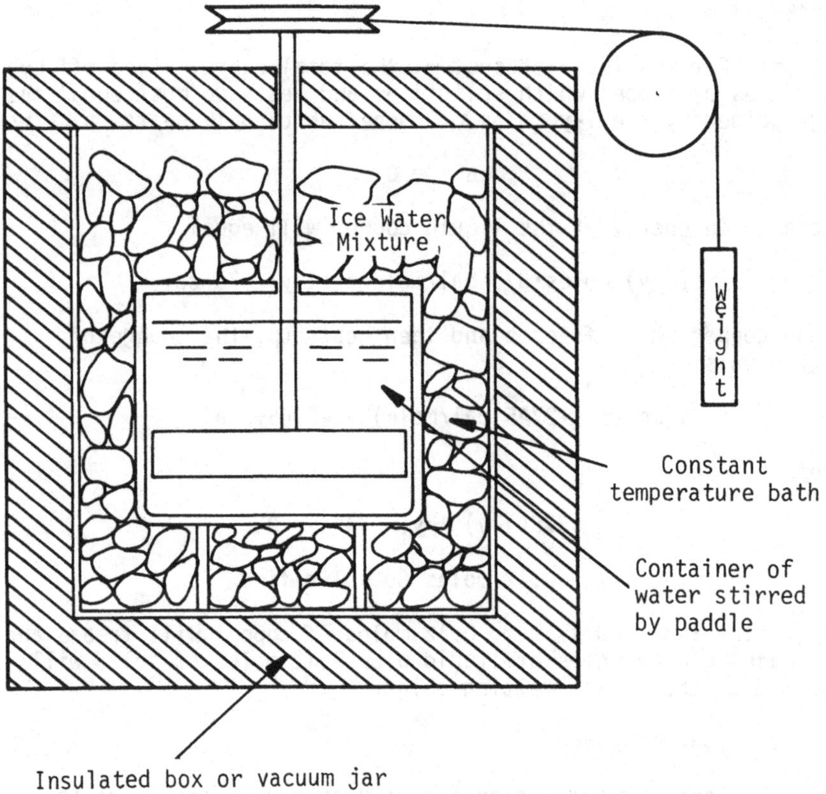

Figure 5.2-2 Schematic illustration of Joule's experiment converting mechanical energy into heat.

to the work input. If the water in the container is considered to be the system under consideration, we note that the process of heating the water by the rotation of the paddle and the return of the water to its original temperature by transferring the heat to the ice-water mixture <u>does not change the state of the system</u> since it has the same state properties of temperature, pressure, and volume after the test as it had before. Therefore, since the energy of the system is a state property of the system, the energy content of the system must not change, i.e. $dE' = 0$, and Eq. (5.2-26) yields

$$-\delta q = -\delta W \tag{5.2-27}$$

Thus, the work energy input can be equated directly to the heat lost to the ice-water mixture.

It should not take too much imagination to visualize that this experiment could be modified to include several forms of input (work) energy. In all cases, the relationship between work input and heat lost by the system (shown in Eq. 5.2-27) must be obeyed. It should also be possible to combine several forms of work input to effect the same results. As a consequence, we can generalize Eq. (5.2-27) to include the energy added to the system by electrical work, δw_e; by magnetic work, δw_m, etc.

$$-\delta q = (\delta w_e + \delta w_m + \ldots.) \qquad (5.2\text{-}28)$$

If, the container of water is kept out of the constant temperature bath and insulated from its surroundings during an experiment, it will heat up from the action of the paddle wheel. Since work was done <u>on</u> the system <u>adiabatically</u>, that is, no heat was transmitted to or from the surroundings,

$$dE' = 0 - (-\delta w) = \delta w \qquad (5.2\text{-}29)$$

The increase in energy represented by dE' is associated with a temperature rise of the water. Since it represents an energy increase that cannot be accounted for by any of the recognized forms of energy such as kinetic, potential, electrical, etc., this energy increase must be an increase in <u>internal energy</u> of the system. Therefore, <u>under the conditions of constant kinetic and potential energy</u>,

$$dE' = dU' \qquad (5.2\text{-}30)$$

and

$$dU' = \delta q - \delta w \qquad (5.2\text{-}31)$$

The internal energy of a system is fixed at a given value if a sufficient number of its state variables, such as temperature, pressure and specific volume, are fixed. Changing the state of the system to another set of values of the state variable will result in a change in internal energy to another value. Had the temperature rise of the water in the previous experiment been noted, and suitable additional experiments performed, it could have been established that the internal energy <u>increase</u> of a system is directly related to its temperature <u>increase</u>. As a matter of fact, from kinetic theory it can be shown that, for a perfect gas:

$$\Delta U' = 3/2\ nR\Delta T \qquad (5.2\text{-}32)$$

where R is the ideal gas constant, n is the number of mols of gas in the system, and the absolute temperature change of the system is ΔT. For more complex substances such as real gases or solids or liquids, these simple relations do not suffice, and other variables, such as pressure, also influence internal energy. It is, therefore, not always easy to develop a precise definition of internal energy in terms of state variables. However, the First Law provides the means to develop the relationship between internal energy changes and the changes in the state properties of the materials which make up the system.

Work comprises the various energy terms that are listed in Eq. (5.2-28). Some texts imply that work is equivalent to the work of expansion of a gas. This is a specialized case and is liable to be confusing when other forms of work such as electrical work, are considered. The work term can be separated into two components: δw_{pv} work of expansion done by the system, and δw^* useful work other than the work of expansion. Following this convention,

$$dU' = \delta q - \delta w$$

$$= \delta q - \delta w_{pv} - \delta w^* \quad (5.2\text{-}33)$$

The work of expansion of a system equals

$$\delta w_{pv} = PdV' \quad (5.2\text{-}34)$$

thus,

$$dU' = \delta q - PdV' - \delta w^* \quad (5.2\text{-}35)$$

If only work of expansion is considered, Eq. (5.2-35) can be rearranged as follows:

$$dU' + PdV' = \delta q \quad (5.2\text{-}36)$$

At constant pressure,

$$dU' + PdV' = dU' + d(PV') = d(U' + PV') \quad (5.2\text{-}37)$$

or

$$d(U' + PV') = \delta q \quad (5.2\text{-}38)$$

The combination of U' + PV' occurs so frequently that it has been found convenient to designate it by a special symbol*, H',

$$H' = U' + PV' \equiv \text{enthalpy} \quad (5.2\text{-}39)$$

---

* The symbol H' will be used to indicate the enthalpy of the total system, an extensive property. The enthalpy per unit mass, an intensive property, will be designated by h, while H will be used to designate the enthalpy per mole, also an intensive property. If the enthalpy is that corresponding to a specific element or compound i, the subscript i is used, H_i, and if it is for the element or compound in its standard state, the superscript o is used, e.g., H_i^o. The standard state of an element or compound is taken to be the pure substance in its most stable form at 101 kPa (1 atm) and the temperature of interest.

Combining Eq. (5.2-38) and Eq. (5.2-39),

$$dH' = \delta q \tag{5.2-40}$$

Thus, under the specific condition of pressure held constant and the only work being work of expansion, the enthalpy change is equal to the heat absorbed. This accounts for the fact that ΔH has been called the "heat content". Obviously, this is only a specialized case, where other forms of work are not considered or are absent. Therefore, it is better to use the term "enthalpy change" for ΔH rather than "heat content". Enthalpy is a thermodynamic state property since it is defined by thermodynamic state properties, Eq. (5.2-39). Heat exchanged does not represent a thermodynamic state property, since it can vary depending on how the change of state occurred.

EXAMPLE 5.2-2: Determine the quantity of heat which is given off to the surroundings from a closed system at a constant pressure of 101 kPa (1 atm) and at 1000 K when one mole of magnetite (Fe_3O_4) is reduced by carbon to form iron and carbon monoxide.

Solution: For a closed system at constant pressure and doing no work other than the work of expansion

$$dH' = \delta q$$

or

$$\delta H' = q$$

The enthalpy change required for the reaction

$$Fe_3O_4(s) + 4C(s) \rightarrow 3Fe(s) + 4CO(g)$$

at 1000 K is + 153,800 cal/mole of Fe_3O_4 reduced. In order for the system to have an increase in enthalpy, heat must be gained from the surroundings, i.e., this is an endothermic reaction. Consequently, this reaction must receive 153,800 cal from the surroundings per mole of magnetite reduced.

EXAMPLE 5.2-3: How much heat must be removed at constant pressure from a mole of water at 273 K to produce a mole of ice at 273 K?

Solution: For a closed system at constant pressure and doing no work other than the work of expansion

$$dH' = \delta q$$

or

$$\Delta H' = q$$

For water going to ice at 273 K

$$\Delta H = q = -1436 \text{ cal/mole}$$

or 1436 calories must be given off to the surroundings in order to make one mole of ice from one mole of water, when both are at 273K and 101 kPa.

C. *Open Systems at Constant Pressure*

For most metallurgical processes, the kinetic and potential energy of the inlet and outlet streams are small relative to the other terms in the energy equation and are usually ignored. Thus, for a metallurgical system that is at constant pressure, combination of Eqs. (5.2-21), (5.2-30) and (4.2-37) yields

$$dH' = \delta q - \delta w^* - \sum_{i=1}^{n} \delta m_i h_i \qquad (5.2\text{-}41)$$

This final equation is one of the most useful forms of the first law for metallurgical engineers.

As mentioned earlier, the First Law can only be used to evaluate energy changes. For convenience, therefore, in handling expressions like Eq. (5.2-41) we shall regard the enthalpy of the pure elements in their most stable form at 298K and 101kPa (1 atm) as zero. With this reference point, values of all internal energies and all other enthalpies can be determined for use in Eq. (5.2-41).

EXAMPLE 5.2-4: A continuous process for the reduction of magnetite by carbon to make iron and carbon monoxide operates at 1000K and consumes 50 moles per hour of magnetite. If the enthalpies at 1000 K per mole of magnetite, iron, carbon and carbon monoxide are -232,450, +5830, +2810 and -21,230 cal, respectively, what is the heat requirement of the process?

Solution: If it is assumed that the process is at constant pressure and operating at steady-state, and that no work other than the work of expansion is being done,

$$dH' = 0 \text{ and } \delta w^* = 0$$

If kinetic and potential energy changes are also ignored, the energy balance can be found using Eq. (5.2-41):

$$0 = \delta q + \sum_{i=1}^{n} \delta m_i h_i$$

or

$$-\delta q = \sum_{i=1}^{n} \delta m_i h_i$$

When enthalpy data, H_i, is given per mole instead of per unit mass, the enthalpy per unit mass can be found since

$$h_i = H_i/MW_i$$

However,

$$\delta q = \sum_{i=1}^{n} \delta m_i \left(\frac{H_i}{MW_i}\right)$$

$$= \sum_{i=1}^{n} \frac{\delta m_i}{MW_i} \cdot H_i$$

$$= \sum_{i=1}^{n} \delta n_i H_i$$

where δn_i is the number of moles of i in the stream. For flowing systems,

$$-\delta \dot{q} = \sum_{i=1}^{n} \delta \dot{n}_i H_i$$

and

$$-\dot{q} = \sum_{i=1}^{n} \dot{n}_i H_i$$

where $\dot{n}_i$ is the flow rate (moles/unit time). In this problem, 50 moles per hour of magnetite enter the process. From the stoichiometry of the reaction (see Example 5.2-2), 200 moles per hour of carbon must be supplied and 150 moles of iron and 200 moles of carbon monoxide will be produced. Therefore,

$$-\dot{q} = (50 \text{ mol } Fe_3O_4/\text{hour}) \cdot H_{Fe_3O_4} + (200 \text{ mol C/hour}) \cdot H_C$$

$$-(150 \text{ mol Fe/Hour}) \cdot H_{Fe} - (200 \text{ mol CO/hour}) \cdot H_{Co}$$

$$= (50) \cdot (-232{,}450) + (200) \cdot (2810) - (150) \cdot (5830)$$

$$-(200) \cdot (-21{,}230)$$

$$= -7.69 \times 10^6 \text{ cal/hour}$$

or

$$\dot{q} = 7.69 \times 10^6 \text{ cal/hour}$$

This process is endothermic, as shown in Example 5.2-2, and for the given flow rate requires 7.69×10^6 cal/hour or $7.69 \times 10^6/50$ = 153,800 cal per mole of magnetite reduced.

It should not be surprising that the same energy requirement is calculated for this open system as for the closed system in Example 5.2-2. As mentioned in Section 5.1, there are many ways of breaking down any process into systems.

5.3 ENTHALPY

Because most metallurgical systems are at nearly constant pressure and do not involve the production of useful work, most energy balance calculations reduce to balancing the enthalpy changes occurring within the system. Methods for calculating individual enthalpy changes are discussed in this Section. Sources of data for calculating enthalpy changes are discussed in Section 5.4. Chapter 6 describes system enthalpy balances.

5.3.1 ENTHALPY INCREMENTS

The enthalpy per mole, an intensive variable, can be represented by an equation similar in form to Eq. (5.1-1);

$$H = f(I_1, I_2) \qquad (5.3\text{-}1)$$

where I_1 and I_2 are any two intensive properties of the system. I_1 and I_2 are usually chosen to be the temperature and pressure; hence,

$$H = f(T, P) \qquad (5.3\text{-}2)$$

Since enthalpy is a thermodynamic property, its differential equals*

$$dH = \left(\frac{\partial H}{\partial T}\right)_P dT + \left(\frac{\partial H}{\partial P}\right)_T dP \qquad (5.3\text{-}3)$$

For a system at constant pressure,

$$dP = 0 \qquad (5.3\text{-}4)$$

and Eq. (5.3-3) becomes

$$dH = \left(\frac{\partial H}{\partial T}\right)_P dT \qquad (5.3\text{-}5)$$

The differential enthalpy change at constant pressure is thus proportional to the differential temperature change of the system:

$$dH \propto dT \qquad (5.3\text{-}6)$$

---

* The differentials of thermodynamic or state properties are called exact differentials and have special properties because the changes are independant of path. For more details see p. 709-714 in G. B. Thomas, Calculus and Analytical Geometry, Addison-Wesley Publishing Co., Reading, Mass., 1962.

or

$$dH = C_p \, dT \tag{5.3-7}$$

where C_p is the proportionality constant and is called the (molar) specific heat.

The specific heat is equal to the amount of heat required to raise the temperature of one mole of a substance one temperature unit. The specific heat is an intensive property. Other names that are used synonomously are heat capacity and molar heat capacity. The heat capacity of a system is, C_p', an extensive property, and its use should be restricted to describing an entire system. The relationship between individual heat capacities and system heat capacity is:

$$C_p' = nC_p \tag{5.3-8}$$

or, for a system with several components,

$$C_p' = \sum_{\text{all } i} n_i C_{p_i} \tag{5.3-9}$$

where n_i is the number of moles of i in the system, and C_{p_i} has units of energy per mole i per unit temperature.

The values of C_p or C_p' are for systems at constant pressure only. For systems at constant volume

$$dU' = C_v' \, dT \tag{5.3-10}$$

or

$$dU = C_v \, dT \tag{5.3-11}$$

where C_v is the specific heat at constant volume and the heat capacity of the system at constant volume equals C_v'.

For ideal gases,

$$C_p = C_v + R \tag{5.3-12}$$

For other substances

$$C_p = C_v + VT\alpha^2/\beta \tag{5.3-13}$$

where V is the molar volume, α is the coefficient of thermal expansion of the material,

$$\alpha = \frac{1}{V}\left(\frac{\partial V}{\partial T}\right)_T \tag{5.3-14}$$

and β is the isothermal compressibility,

$$\beta = \frac{1}{V}\left(\frac{\partial V}{\partial P}\right)_T \tag{5.3-15}$$

EXAMPLE 5.3-1: Calculate the percentage difference between C_p and C_v for iron at room temperature. The properties of iron are:

$\alpha = 3.51 \times 10^{-6}/K$, $\beta = 0.52 \times 10^{-6}/atm$, $V = 0.00711$ ℓ/mole and $C_p = 5.98$ cal/mol·K.

Solution: Rearrangement of Eq. (5.3-13) yields

$$\frac{C_p - C_v}{C_p} \times 100\% = \frac{\alpha^2 TV/\beta}{C_p} \times 100\%$$

Substitution of the data, T = 298 K and conversion factor from ℓ-atm to calories yields

$$\frac{C_p - C_v}{C_p} \times 100\% = 2.04\%$$

Thus, C_p is approximately two percent greater than C_v for iron at room temperature.

The enthalpy increment, is the amount of heat which can be stored in an object by raising its temperature from T_1 to T_2. The enthalpy increment, $[H(T_2,P) - H(T_1,P)]$, is found by integrating Eq. (5.3-7). The result of this integration is

$$[H(T_2,P) - H(T_1,P)] = \int_{T_1}^{T_2} C_p \, dT \tag{5.3-16}$$

Specific heat data is normally presented in polynomial form, such as $C_p = a + bT$, or $C_p = a + bT + cT^{-2}$, making the integration in Eq. (5.3-16) quite simple.

EXAMPLE 5.3-2: Calculate the enthalpy of copper at 1000K and the enthalpy increment for heating copper from 298K to 1000K. The specific heat of copper at 101 kPa (1 atm) can be represented by the equation

$$C_p = 5.41 + 1.50 \times 10^{-3} T \text{ (cal/mol·K)}$$

Solution: The enthalpy increment is found using Eq. (5.3-16). For copper,

$$[H(1000\text{ K, 1 atm}) - H(298\text{ K, 1 atm})] = \int_{298}^{1000} (5.41 + 1.50 \times 10^{-3}T)\, dT$$

$$= 5.41T \Big|_{298}^{1000} + \frac{1.50 \times 10^{-3}}{2} T^2 \Big|_{298}^{1000}$$

$$= 5.41(1000-298) + 0.75 \times 10^{-3}(1000^2 - 298^2)$$

$$= 4480 \text{ cal/mol}$$

The enthalpy of copper at 1000 K can be found, since the enthalpy of the pure element at 298 K and 1 atm equals zero by convention. Therefore, for copper

$$H_{Cu}\ (1000\text{ K, 1 atm}) = 4480 \text{ cal/mol}$$

In certain instances mean specific heats are used to facilitate calculations. The mean specific heat is defined as

$$\bar{C}_p = \frac{\int_{T_1}^{T_2} C_p\, dT}{\int_{T_1}^{T_2} dT} = \frac{\int_{T_1}^{T_2} C_p\, dT}{T_2 - T_1} \qquad (5.3\text{-}17)$$

This value, as well as the values of specific heats should only be used within the temperature ranges for which it has been determined to be valid.

When the mean specific heat is used,

$$[H(T_2,P) - H(T_1,P)] = \bar{C}_p \cdot (T_2 - T_1) \qquad (5.3\text{-}18)$$

EXAMPLE 5.3-3: Determine the mean specific heat for copper between 298 K and 1000 K.

Solution: Rearrangement of Eq. (5.3-18) yields

$$\bar{C}_p = \frac{H(T_2,P) - H(T_1,P)}{T_2 - T_1}$$

Using the results from the previous example,

$$\bar{C}_p = \frac{4480\ \text{cal/mol}}{1000\text{K}-298\ \text{K}}$$

$$= 6.38\ \text{cal/mol·K}$$

EXAMPLE 5.3-4: Calculate the enthalpy increments for copper between 298 K and 500 K, 700 K and 900 K using the true and mean specific heats. Estimate the maximum error.

Solution: The enthalpy increments are found with the true and mean specific heats using Eq. (5.3-16) and Eq. (5.3-18), respectively. The results are tabulated below.

Temperature	Enthalpy Increment (H_T-H_{298}) Calculated from Eq. (5.3-16)	Enthalpy Increment (H_T-H_{298}) Calculated from Eq. (5.3-18)
500 K	1215 cal/mole	1290 cal/mole
700 K	2480 cal/mole	2560 cal/mole
900 K	3800 cal/mole	3480 cal/mole

The maximum error occurs at approximately 650 K where the value calculated from the mean specific heat data is about 90 cal/mole too high (approximately 4%).

Synonyms that are used for the enthalpy increment are heat content and sensible heat. Alternative symbols such as $\Delta H_{sensible}$, $\Delta H(T_1 \rightarrow T_2)$, and $H_{T_2} - H_{T_1}$ are used to indicate the enthalpy increment. Enthalpy increments are tabulated for elements and compounds. Examples of such tabulations are discussed in Section 5.4.

EXAMPLE 5.3-5: Determine the energy requirement for heating one mole of copper from 500 K to 900 K.

Solution: Since

$$[H_{900} - H_{500}] = \int_{500}^{900} C_p\ dT$$

$$\text{and} \quad \int_{500}^{900} C_p dT = \int_{298}^{900} C_p dT - \int_{298}^{500} C_p dT$$

$$[H_{900} - H_{500}] = [H_{900} - H_{298}] - [H_{500} - H_{298}]$$

Using the results from the previous example

$$[H_{900} - H_{500}] = 3800 \text{ cal/mol} - 2480 \text{ cal/mol}$$

$$= 1320 \text{ cal/mol}$$

5.3.2 LATENT HEAT

When a first-order phase transformation occurs within a system, energy must be added to or subtracted from the system at the temperature of the transformation. The temperature of the system does not change during this process, but the enthalpy of the system does change by the amount of heat required to produce the transformation. This enthalpy quantity is called the latent heat of transformation (symbol ΔH_t). Particular examples of latent heats of transformation given separate symbols are the latent heat of fusion (melting), ΔH_m, and the latent heat of vaporization, ΔH_v. Melting and vaporization are always first-order phase transformations. Some solid-solid phase transformations (e.g. ferrite-austenite) are first order, but others (e.g., order-disorder transformations) are second-order and have no latent heat.

The energy added to a system to produce a phase change increases the enthalpy of the system at the transformation temperature of the system. Thus, when a phase transformation occurs at T_t while heating an object from T_1 past T_t to T_2, the change in enthalpy of the object will equal the enthalpy increments plus the latent heat of transformation:

$$\Delta H_{(T_1 \rightarrow T_2)} = \int_{T_1}^{T_t} C_{p\,(\text{phase }1)}\,dT + \Delta H_t + \int_{T_t}^{T_2} C_{p(\text{phase }2)}\,dT \qquad (5.3\text{-}19)$$

The integration in this case must be broken into two parts because, in general, the specific heats of phase 1 and phase 2 are different. Again, care must be taken to be certain that only specific heat data appropriate to the temperature range for each phase is used. Also, it is a good practice to use subscripts in the equation to identify the data to be used, i.e. $C_{p\,(\text{phase }1)}$, etc.

EXAMPLE 5.3-6: Calculate and plot a graph of enthalpy increment for heating copper from 500 K to 1500 K. Copper melts at 1357 K with a heat of fusion of 3120 cal/mole. For liquid copper,

$$C_p = 7.50 \text{ cal/mole}$$

Solution: The enthalpy increment can be found using Eq. (5.3-19).

$$\Delta H_{(500K \rightarrow 1500K)} = \int_{500}^{1357} C_{p_{Cu(s)}}\,dT + \Delta H_m + \int_{1357}^{1500} C_{p_{Cu(\ell)}}\,dT$$

$$= \int_{500}^{1357} (5.41 + 1.5 \times 10^{-3} T)\, dt + 3120 + \int_{1357}^{1500} 7.50\, dT$$

$$= 5.41\, T \Big|_{500}^{1357} + 0.75 \times 10^{-3} T^{-2} \Big|_{500}^{1357} + 3120 + 7.50\, T \Big|_{1357}^{1500}$$

$$= 5.41\,(1357 - 500) + 0.75 \times 10^{-3}\,(1357^2 - 500^2) + 3120 + 7.50\,(1500 - 1357)$$

$$= 10{,}020 \text{ cal/mole}$$

The enthalpy increments are shown in Fig. 5.3-1.

EXAMPLE 5.3-7: Determine how much heat is given off during a constant pressure process in which one mole of zinc vapor at 1181 K is converted into one mole of solid zinc at 298 K. The required data are summarized below.

Boiling Point_{Zn} = 1181 K; Melting Point_{Zn} = 693 K

$\Delta H_{g \to \ell}$ = -27,560 cal/mole; $H_{\ell \to s}$ = -1,765 cal/mole

$C_{p(liq.)}$=7.5 cal/mole·K; $C_{p(s)}$=5.35+2.40 x 10^{-3}T cal/mole·K

Solution: For a system at constant pressure and doing no work other than the work of expension, the heat given off from the system will equal the change in enthalpy of the system.

$$dH = \delta q$$

For a system which undergoes n transformations, Eq. (5.3-19) can be generalized to yield

$$\Delta H_{(T_1 \to T_2)} = \int_{T_1}^{T_{t_1}} C_{p(\text{phase } 1)}\, dT + \Delta H_{t_1} + \int_{T_{t_1}}^{T_{t_2}} C_{p(\text{phase } 2)}\, dT + \Delta H_{t_2}$$

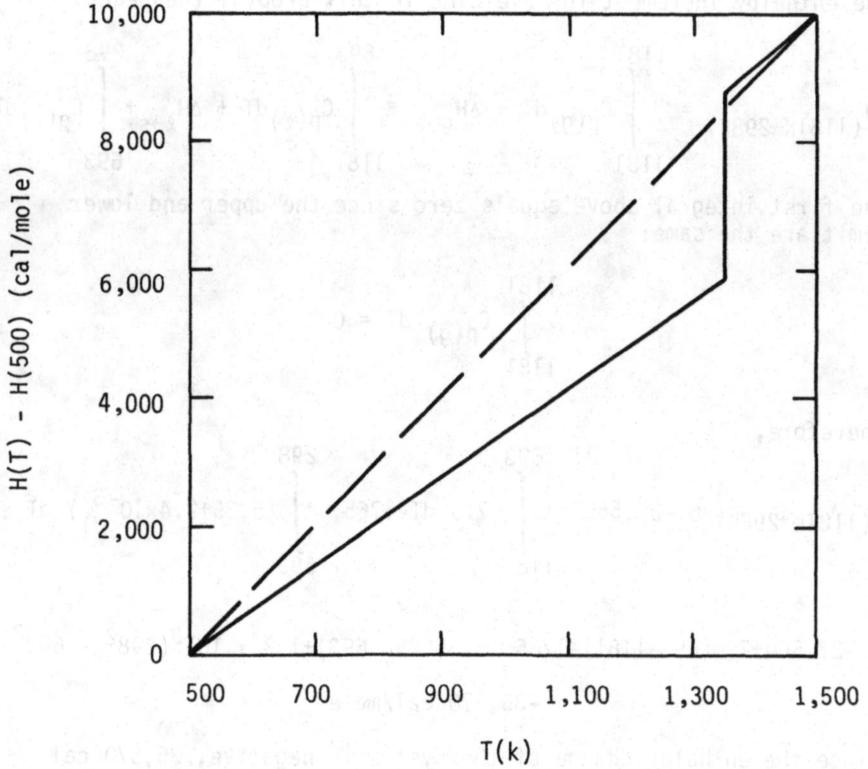

Figure 5.3-1 Enthalpy increments for copper.

$$+ \; . \; . \; . \; . \; . + \int_{T_{t_{n-1}}}^{T_{t_n}} C_{p(\text{phase } n-1)} dT + \Delta H_{t_n} + \int_{T_{t_n}}^{T_2} C_{p(\text{phase } n)} dT$$

For only two transformations,

$$\Delta H_{(T_1 \rightarrow T_2)} = \int_{T_1}^{T_{t_1}} C_{p(\text{phase } 1)} dT + \Delta H_{t_1} + \int_{T_{t_1}}^{T_{t_2}} C_{p(\text{phase } 2)} dT$$

$$+ \Delta H_{t_2} + \int_{T_{t_2}}^{T_2} C_{p(\text{phase } 3)} dT$$

The enthalpy increment for the zinc in this problem then equals

$$\Delta H_{(1181K\to 298K)} = \int_{1181}^{1181} C_{p(g)}dT + \Delta H_{g\to\ell} + \int_{1181}^{693} C_{p(\ell)}dT + \Delta H_{\ell\to s} + \int_{693}^{298} C_{p(s)}dT$$

The first integral above equals zero since the upper and lower limit are the same:

$$\int_{1181}^{1181} C_{p(g)}\ dT = 0$$

Therefore,

$$H_{(1181K\to 298K)} = 0\ -27{,}560\ + \int_{1181}^{693} 7.5\ dT - 1765\ + \int_{693}^{298} (5.35+2.4\times10^{-3}T)\ dT$$

$$= -27{,}560+7.5(693-1181)-1765+5.35(298-693)+1.2\times10^{-3}(298^2-693^2)$$

$$= -35{,}570\ \text{cal/mole}$$

Since the enthalpy change of the system is negative, 35,570 cal per mole of zinc is being given off to the surroundings.

5.3.3 HEAT OF FORMATION AND REACTION

The heat of formation of a compound (symbol ΔH_f) is the heat evolved from or added to a system when one mole of the compound is formed at the temperature of interest and at 101 kPa (1 atm) from the pure elements in their most stable forms at 101 kPa and the temperature in question. The heat of formation is the heat generated for a particular type of reaction, i.e. pure elements going to one compound.

The heat of reaction (symbol $\Delta H_R'$) is the heat generated or absorbed by a system due to any reaction within the system. The heat of reaction is usually given per mole of a particular product formed or reactant consumed. However, care must be employed to be sure that the appropriate units are used and the extent of reaction is known.

Since enthalpy is a state property, the enthalpy change for a reaction equals the enthalpy of the final state minus the enthalpy of the initial state of the system. For the reaction

$$aA + bB \to cC + dD \qquad (5.3\text{-}20)$$

the final state is c moles of C and d moles of D, the initial state is a moles of A and b moles of B. The heat of reaction equals

$$\Delta H_R' = cH_C + dH_D - (aH_A + bH_B) \qquad (5.3\text{-}21)$$

For the same reaction, the heat of reaction per mole of C equals

$$\Delta H_R = H_C + \frac{d}{c} H_D - (\frac{a}{c} H_A + \frac{b}{c} H_D) \qquad (5.3\text{-}22)$$

For the special case of the formation reaction,

$$\Delta H_R = \Delta H_f = H_{(product)} - \sum_{elements} n_i H_i \qquad (5.3\text{-}23)$$

EXAMPLE 5.3-8: Pure oxygen and pure iron are placed in a container at 1000 K and 101 kPa (1 atm) and allowed to react to form pure hematite (Fe_2O_3) at 1000 K and 101 kPa. What is the heat of formation of hematite if 192,600 calories are given off from the container for every mole of hematite produced.

Solution: If it is assumed that the oxygen, iron and hematite are in their most stable forms at 1000 K and 101 kPa, the enthalpy change of the system must equal the heat of formation of the hematite. Also since the system is at constant pressure

$$dH = \delta q$$

or

$$\Delta H = q$$

since no work other than the work of expansion is being done. Then, since the enthalpy change equals the heat of formation,

$$\Delta H_f = q$$

or

$$\Delta H_{f_{Fe_2O_3}} = -102{,}600 \text{ cal/mole } Fe_2O_3$$

EXAMPLE 5.3-9: The heat of formation of uranium dioxide (UO_2) is -259,200 cal/mole at 298 K. What is the enthalpy of UO_2 at 298 K?

Solution: According to Eq. (5.3-23),

$$\Delta H_f = H_{(product)} - \sum_{elements} H_{(element)}$$

or, for the reaction

$$U(s) + O_2(g) \rightarrow UO_2(s)$$

$$\Delta H_f = H_{UO_2} - (H_U + H_{O_2})$$

At 298 K, the enthalpies of pure U and pure O_2 are taken by convention to be equal to zero. Thus,

$$\Delta H_f = H_{UO_2} = -259{,}200 \text{ cal/mole}$$

The enthalpy of a compound equals the heat of formation of the compound only at 298 K, because only at this temperature are the enthalpies of the elements equal to zero.

EXAMPLE 5.3-10: The enthalpies of hydrogen (H_2), carbon monoxide (CO), methane (CH_4) and water (H_2O) at 500 K are 1406, -24,998, -15,910 and -56,143, respectively. What is the heat of reaction at 500 K for

$$CO(g) + 3H_2(g) \rightarrow CH_4(g) + H_2O(g)$$

Solution: For this reaction, the enthalpy of the products equals $H_{CH_4} + H_{H_2O}$, the enthalpy of the reactants equals $H_{CO} + 3H_{H_2}$, and

$$\Delta H_R = [H_{CH_4} + H_{H_2O}] - [(H_{CO} + 3H_{H_2})]$$

$$= [(-15{,}910)+(-56{,}143)] - [-24{,}998 + 3(1406)]$$

$$= -51{,}273 \text{ cal}$$

This heat of reaction is per mole of CO consumed. The heat of reaction per mole of hydrogen consumed equals -51,273 ÷ 3 or -17,091 cal/mole H_2.

When all of the heats of formation of the compounds involved in a reaction are known at the temperature of interest, the heat of reaction may also be found by application of Hess' Law,

$$\Delta H_R' = \sum_{\text{products}} \Delta H_{f(\text{products})} - \sum_{\text{reactants}} \Delta H_{f(\text{reactants})} \quad (5.3\text{-}24)$$

EXAMPLE 5.3-11: The heats of formation at 400 K for carbon monoxide (CO) and water (H_2O) are -26,300 and -58,050 cal/mole respectively. Determine the heat of reaction for the reaction

$$H_2O(g) + C(s) \rightarrow CO(g) + H_2(g)$$

Solution: According to Eq. (5.3-24), the heat of reaction in this problem must equal

$$\Delta H_R = \Delta H_{f,CO} + \Delta H_{f,H_2} - \Delta H_{f,H_2O} - \Delta H_{f,C}$$

The heat of formation is the heat gained or evolved when one mole of a pure compound in its most stable form at the temperature in question is formed at 101 kPa (1 atm), from the pure elements in their most stable forms at 101 kPa and the same temperature. According to this definition, the heat of formation of an element at any temperature must equal zero, since the product and reactant are the same, i.e. the pure element in its most stable form at the temperature in question and 101 kPa. Thus, at 400 K,

$$\Delta H_{f,C} = \Delta H_{f,H_2} = 0$$

or

$$\Delta H_{R,400\ K} = \Delta H_{f,CO,400\ K} - \Delta H_{f,H_2O,400\ K}$$

$$= -26{,}300 - (-58{,}050)$$

$$= 31{,}750\ \text{cal}$$

EXAMPLE 5.3-12: The enthalpies of carbon (C), carbon monoxide (CO), hydrogen (H_2), and water (H_2O) at 400 K are +250, -25,705, +707, and -56,972, respectively. Determine the heat of reaction for

$$H_2O(g) + C(s) \rightarrow CO(g) + H_2(g)$$

Solution: In Example 5.3-10, it was shown that

$$\Delta H_R = \Sigma H_{(products)} - \Sigma H_{(reactants)}$$

Therefore,

$$\Delta H_R = H_{CO} + H_{H_2} - H_{H_2O} - H_C$$

$$= (-25{,}705) + 707 - (-56{,}972) - (250)$$

$$= 31{,}720\ \text{cal}$$

The error of approximately 0.1% between this result and the result in Example 5.3-11 is caused by small errors in the data and is insignificant. The two methods for calculating the heat of reaction should yield identical results since the heat of formation of CO equals

$$\Delta H_{f,CO} = H_{CO} - H_C - 1/2H_{O_2}$$

and the heat of formation of H_2O equals

$$\Delta H_{f,H_2O} = H_{H_2O} - H_{H_2} - 1/2H_{O_2}$$

Consequently,

$$\Delta H_R = \Delta H_{f,CO} - \Delta H_{f,H_2O}$$

$$= (H_{CO} - H_C - 1/2H_{O_2}) - (H_{H_2O} - H_{H_2} - 1/2H_{O_2})$$

$$= H_{CO} + H_{H_2} - H_{H_2O} - H_C$$

5.3.4 HEATS OF MIXING

To this point, only pure substances have been discussed. However, mixtures of substances, e.g. solutions, are of primary importance in the metallurgical industries.

When a mole of species i at temperature T is added to a large amount of a solution at temperature T, heat may be absorbed from or evolved to the surroundings. This quantity of heat, which is the difference between the heat content of one mole of the substance i in the solution (symbol $\bar{H}_i$), and one mole of pure i, H_i°, is called the <u>heat of mixing</u> (symbol $\Delta\bar{H}_i^m$)*:

$$\Delta\bar{H}_i^m = \bar{H}_i - H_i^\circ \tag{5.3-25}$$

Then for an r-component solution which contains n_1 moles of 1, n_2 moles of 2, etc., the <u>total enthalpy content of the solution</u> is the sum of the enthalpies of each component:

$$H' = \sum_{i=1}^{r} n_i\bar{H}_i \tag{5.3-26}$$

or

$$H' = \underset{\text{(Enthalpy of Mixing)}}{\sum_{i=1}^{r} n_i\Delta\bar{H}_i^m} + \underset{\text{(Enthalpy of unmixed Constituents)}}{\sum_{i=1}^{r} n_iH_i^\circ} \tag{5.3-27}$$

For ideal gases, the heats of mixing are zero. The heats of mixing for components of solid and liquid solutions may be positive zero, or negative. However, in many systems, these terms are small and can be ignored in engineering calculations.

EXAMPLE 5.3-13: Determine the enthalpy of one mole of a 20% manganese -80% iron alloy at 1873 K. (Analysis is on an atomic basis.) The heats of mixing for manganese and iron at this composition and

---

* Heats of mixing, $\Delta\bar{H}_i^m$, are defined relative to the enthalpy of substances in their most stable form, i.e., their standard state, at the temperature of interest. Since they may be solid at that temperature, but dissolve into a liquid, care must be taken to make sure that the correct standard state is utilized.

temperature are $\Delta\bar{H}^{m}_{Mn}$ = 640 and $\Delta\bar{H}^{m}_{fe}$ = 40 cal/mole, and H°_{Fe} = 18550 cal/mole and H°_{Mn} = 19380 cal/mole.

Solution: In general, the enthalpy of n_t moles of a solution can be found using Eq. (5.3-27).

$$H' = \sum_{i=1}^{m} n_i \Delta\bar{H}^{M}_{i} + \sum_{i=1}^{m} n_i H^{\circ}_{i}$$

The enthalpy per mole of solution will equal

$$H = \frac{H'}{n_t} = \sum_{i=1}^{m} \frac{n_i}{n_t} \Delta\bar{H}^{m}_{i} + \sum_{i=1}^{m} \frac{n_i}{n_t} H^{\circ}_{i}$$

$$= \sum_{i=1}^{m} X_i \Delta\bar{H}^{m}_{i} + \sum_{i=1}^{m} X_i H^{\circ}_{i}$$

Thus, for the solution of interest in this problem,

$$H = X_{Mn} \Delta\bar{H}^{m}_{Mn} + X_{Fe} \Delta\bar{H}^{m}_{Fe} + X_{Mn} H^{\circ}_{Mn} + X_{Fe} H^{\circ}_{Fe}$$

$$= 0.2(640) + 0.8(40) + 0.2(19380) + 0.8(18550)$$

$$= 18{,}880 \text{ cal/mole of solution}$$

In this case, the heat of mixing changes the total enthalpy of the solution by less than 1%.

EXAMPLE 5.3-15: One notable exception to the generality concerning negligible heats of mixing, stated above, is that of silicon in iron. The heat of mixing of silicon in a 20% silicon - 80% iron mixture, $\Delta\bar{H}^{m}_{Si}$, is -24,100 cal/mole at 1853 K. The enthalpy of pure liquid silicon at this temperature is 20,900 cal/mole. Determine if heat has to be supplied to a system in which one mole of silicon at 298 K is added to a large bath of 20% Si - 80% Fe solution at 1853 K in order to maintain the resulting bath at 1853 K.

Solution: The total enthalpy change for the system is the change in enthalpy required to heat the silicon from 298 K to 1853 K (which includes melting it at 1683 K) plus the enthalpy change due to mixing.

$$\Delta H_{(process)} = \Delta H_{Si}\ (298K \rightarrow 1853K) + \Delta\bar{H}^{m}_{Si}$$

$$= (H^{\circ}_{Si,1853} - H^{\circ}_{Si,298}) + \Delta\bar{H}^{m}_{Si}$$

$$= 20{,}900 + (-24{,}100)$$

$$= -3200 \text{ cal/mole Si}$$

Since this process is clearly exothermic, heat does not have to be supplied. The mixing process liberates enough heat to heat up the silicon and still give off 3200 calories to the surroundings. This heat of mixing cannot be ignored in energy balance calculations.

The heat capacity of a solution is also affected by the mixing process. As a first approximation, the heat capacity of a solution can be estimated from Kopp's Law

$$C_p' = \sum_{i=1}^{m} n_i C_{p_i} \tag{5.3-28a}$$

or

$$C_p = \sum_{i=1}^{m} X_i \, C_{p_i} \tag{5.3-28b}$$

where C_{p_i} is the specific heat of pure component i. However, this approximation is only completely true for mechanical mixtures, i.e. substances which do not combine with one another upon mixing, and extreme caution should be used when using Eq. (5.3-28) to estimate data for solutions.

EXAMPLE 5.3-15: Air is assumed to be 21% oxygen and 79% nitrogen in most engineering calculations. Estimate the heat capacity of five moles of air at 298 K using Kopp's law, given that the specific heats of oxygen and nitrogen are 7.16 and 6.95 cal/mol·K, respectively.

Solution: Five moles of air will contain 5 x 0.21 = 1.05 moles of oxygen and 3.95 moles of nitrogen. Using Eq. (5.3-28a), it can be seen that the heat capacity of five moles of air equals

$$C_p' = 1.05\, C_{p(O_2)} + 3.95\, C_{p(N_2)}$$

$$= 1.05\,(7.16) + 3.95\,(6.95)$$

$$= 35.0 \text{ cal/K}$$

The experimentally determined specific heat of air is 6.94 cal/mol·K at 298 K. Based upon this value and Eq. (5.3-9)

$$C_p' = 5(6.94)$$

$$= 34.7 \text{ cal/K}$$

The good agreement (approximately 1% error) between the two calculations results from the fact that oxygen and nitrogen behave almost ideally and mixtures of ideal gases are, in essence, mechanical mixtures, i.e., there is no interaction between the various components. Much larger errors can be expected for non-ideal solutions of gases or condensed phases.

5.4 SOURCES OF ENTHALPY DATA

Energy balance calculations can only be successfully completed when all of the data required in the calculation are available or can be reasonably estimate. The two main forms for enthalpy data are either data compilations or thermodynamic relationships from which enthalpy data can be calculated. Use of these data sources is discussed in this section.

5.4.1 DATA COMPILATIONS

Enthalpy data are available in many different compilations of thermodynamic data*. It is impossible to list all of the compilations; however, some of the most useful are described below.

A. *National Bureau of Standards Publications*

NBS Technical Notes 270-3 through 270-7 entitled Selected Values of Chemical Thermodynamic Properties are published by the National Bureau of Standards.** These compilations contain heats of formation at 0 K and 298 K, enthalpy increments from 0 K to 298 K and specific heat data at 298 K for a number of the more common elements and some of their inorganic compounds. Additional compilations in this series may soon be available.

EXAMPLE 5.4-1: Determine the mean specific heat for sulfur dioxide between 0 K and 298 K and the specific heat of sulfur dioxide at 298 K from the data given in Fig. 5.4-1.

Solution: The data in Fig. 5.4-1 shows that the specific heat of sulfur dioxide (symbol SO_2) at 298 K equals 9.53 cal/mol·K and that the enthalpy increment between 0 K and 298 K equals 2.521 kcal/mol. The mean specific heat is found using Eq. (5.3-18):

$$\bar{C}_p = \frac{(2.521\ \text{kcal/mol})\cdot(1000\ \text{cal/kcal})}{(298\text{K} - 0\ \text{K})}$$

$$= 8.46\ \text{cal/mol}\cdot\text{K}$$

NBS Circular 500, Series II (1952), Selected Values of Chemical Thermodynamic Data, contains heats and temperatures of trans-

---

* Data given in tables is usually for the pure substance at 101 kPa (1 atm) and the most stable form at the temperature in question. Data for these conditions are often designated by a superscript 0 and called the standard enthalpy change. For example, ΔH_R^0 is the symbol for the standard heat of reaction, which is the heat of reaction when the reactants and products are all pure substances at 101 kPa and in their most stable forms at the temperature of the reaction.

** Available from the Superintendant of Documents, U.S. Government Printing Office, Washington, D.C., 20402, as publications.

Table 14(3)

Enthalpy and Gibbs Energy of Formation; Entropy and Heat Capacity

SULFUR

Substance			ΔHf°_0	ΔHf°	ΔGf°	$H^\circ_{298} - H^\circ_0$	S°	C_p°
			0°K	298.15°K (25°C)				
Formula and Description	State	Formula Weight	kcal/mol				cal/deg mol	
S_7^+	g			248.				
S_8	g	256.512	25.35	24.45	11.87	7.531	102.98	37.39
S_8^+	g		247.	247.				
SO	g	48.0634	1.5	1.496	-4.741	2.087	53.02	7.21
SO_2	liq	64.0628		-76.6				
	g		-70.336	-70.944	-71.748	2.521	59.30	9.53
undissoc.; std. state, m = 1	aq			-77.194	-71.871		38.7	
in 100 H_2O	aq			-78.054				
150 H_2O	aq			-78.226				
200 H_2O	aq			-78.355				
250 H_2O	aq			-78.458				
300 H_2O	aq			-78.547				
400 H_2O	aq			-78.691				
500 H_2O	aq			-78.811				
750 H_2O	aq			-79.036				

Figure 5.4-1 Example of data compiled in NBS Technical Notes 270-3 to 270-7.

formations, fusion and vaporization of the elements and some of their compounds.

NBS Monograph 68 (1963) is a Compilation of the Melting Points of Metal Oxides.

B. *JANAF Thermochemical Tables*

This compilation represents an on-going effort by the National Standard Data Reference System to provide an up-to-date compilation of thermodynamic information. The JANAF (Joint Army-Navy-Air Force) Thermochemical Tables and 1974, 1975 and 1978 supplements are printed in book form.* This compilation contains enthalpy increments and heats of formation for many of the elements and their compounds at all temperatures where data are available.

EXAMPLE 5.4-2: Determine the enthalpy of one mole of CuO at 1000 K from the data given in Fig. 5.4-2.

Solution: As shown in Example 5.3-9, the heat of formation of a compound at 298 K equals the enthalpy of that compound at 298 K. The data in Fig. 5.4-2 shows that at 298 K, the heat of formation of CuO equals

$$\Delta H_f^\circ = -37.250 \text{ kcal/mol}$$

since

$$H_{298} = \Delta H_f^\circ$$

and

$$H_{1000} - H_{298} = 8.847 \text{ kcal/mol}$$

The enthalpy of CuO at 1000 K therefore equals

$$H_{1000} = 8.847 - 37.25$$

$$= -28.40 \text{ kcal/mol}$$

C. *U. S. Bureau of Mines Bulletins*

U. S. Bureau of Mines Bulletin 584 entitled, Contributions to the Data on Theoretical Metallurgy by K. K. Kelley, was published in 1960 by the Bureau of Mines, U. S. Department of the Interior, Washington, D. C. This volume contains high-temperature heat content and heat capacity data for the elements and their inorganic compounds.

EXAMPLE 5.4-3: Determine the enthalpy of SO_2 at 950 K.

---

* Available from the American Chemical Society, 1155 16th Street, N.W., Washington, D.C. 20036.

Copper Monoxide (CuO)

(Crystal) GFW = 79.5394

T, °K	Cp° (gibbs/mol)	S° (gibbs/mol)	−(G°−H°298)/T (gibbs/mol)	H°−H°298 (kcal/mol)	ΔHf° (kcal/mol)	ΔGf° (kcal/mol)	Log Kp
0	.000	.000	INFINITE	-1.695	-36.713	-36.713	INFINITE
100	3.945	2.335	17.732	-1.540	-37.063	-34.988	76.466
200	8.321	6.460	11.058	-.920	-37.272	-32.810	35.853
298	10.066	10.183	10.183	.000	-37.250	-30.622	22.446
300	10.094	10.245	10.183	.019	-37.249	-30.581	22.278
400	11.244	13.323	10.595	1.091	-37.129	-28.374	15.503
500	11.960	15.914	11.406	2.254	-36.948	-26.204	11.454
600	12.500	18.145	12.348	3.478	-36.734	-24.076	8.770
700	12.980	20.108	13.319	4.752	-36.494	-21.984	6.864
800	13.430	21.871	14.280	6.073	-36.229	-19.930	5.445
900	13.870	23.478	15.214	7.438	-35.938	-17.909	4.349
1000	14.310	24.962	16.115	8.847	-35.621	-15.923	3.480
1100	14.750	26.347	16.983	10.300	-35.273	-13.970	2.776
1200	15.190	27.649	17.818	11.797	-34.896	-12.050	2.195
1300	15.630	28.882	18.622	13.338	-34.488	-10.161	1.708
1400	16.070	30.057	19.397	14.923	-37.233	-8.206	1.281
1500	16.510	31.180	20.146	16.552	-36.789	-6.147	.896
1600	16.950	32.260	20.869	18.225	-36.305	-4.121	.563
1700	17.390	33.301	21.570	19.942	-35.779	-2.125	.273
1800	17.830	34.307	22.250	21.703	-35.212	-.160	.019
1900	18.270	35.283	22.910	23.508	-34.605	1.770	-.204
2000	18.710	36.231	23.553	25.357	-33.956	3.668	-.401

June 30, 1966

$\Delta Hf^\circ_0 = -36.71$ kcal/mol

$S^\circ_{298.15} = 10.18 \pm 0.1$ gibbs/mol $\qquad \Delta Hf^\circ_{298.15} = -37.25 \pm 0.5$ kcal/mol

Td = 1395°K

Heat of Formation.

The more consistent data relative to the heat of formation are reviewed below. The selected value of -37.25 kcal/mol was derived from $\Delta Hf^\circ_{298} = -40.7$ kcal/mol for $Cu_2O(c)$ and $\Delta Hr^\circ_{298} = 33.80$ kcal/mol for $2CuO(c) = Cu_2O(c) + 0.5\ O_2$. Four different sets of oxygen dissociation pressure measurements yield heats of reaction in close agreement with each other and with the aqueous calorimetry of Thomsen. Thomsen's data may be reduced to the heat of reduction of CuO(c) with H_2 by combining heats of reaction for $CuO + H_2SO_4$, $Fe + H_2SO_4$, and $Fe + CuSO_4(aq)$. Direct calorimetric measurement of the heat of reduction by von Wartenberg and Werth yields $\Delta Hf^\circ_{298} = -38.04 \pm 0.2$ when corrected for incomplete condensation of water. Direct measurement is difficult and this value may correspond to incomplete reduction. The value of -33.02 reported by Wöhler and Jochum, Z. physik. Chem. 167A, 169 (1933), is unreasonable. Equilibrium constants derived from Balesdent and Chiche yield -37.6 kcal/mol by third law analysis, but a serious entropy error is present. This is no doubt due to uncertainties in the activity of Cu in Cu-Au alloys. Other data have been reviewed by Randall, Nielsen and West, Ind. Eng. Chem. 23, 388 (1931).

Author	Method	Reaction**	Temp.	No. of Points	ΔHr°_{298} (kcal/mol) 2nd Law	3rd Law	Drift (eu)	ΔHr°_{298}* (kcal/mol)
1. Thomsen (1883)	Aqueous Calorimetry	A	291	—	-31.15		—	-37.16
2. Wartenberg et al. (1932)	Calorimetry	A	295	—	-30.27±0.2		—	-38.04
3. Balesdent (1955)	Equilibrium Data	B	1011-1156	32	68.2 ±0.7	75.18	6.4±0.7	-37.59
4. Assayag (1955)	Kp	C	1047-1342	11	33.28±0.15	33.81	0.4±0.1	-37.25
5. Becker (1927)	Kp	C	1193-1293	6	33.84±0.27	33.80	0.0±0.2	-37.25
6. Roberts et al. (1921)	Kp	C	1189-1358	20	33.96±0.07	33.76	-0.1±0.1	-37.23
7. Foote et al. (1908)	Kp	C	1223-1323	7	33.19±0.17	33.74	0.4±0.1	-37.22
8. Combined 4,5,6,7	Kp	C	1047-1358	45	33.70±0.10	33.77	0.04±0.09	-37.24

*Based on third law where possible.

**Reaction A: $CuO(c) + H_2 = Cu(c) + H_2O(l)$ B: $2CuO(c) = 2Cu(c) + O_2$

C: $2CuO(c) = Cu_2O(c) + 0.5\ O_2$

1. J. Thomsen, "Thermochemische Untersuchungen," vol. I, III, Barth, Leipzig, 1883.
2. H. von Wartenberg and H. Werth, Z. elektrochem. 38, 51 (1932).
3. D. Balesdent, Compt. rend. 240, 760 (1955); 1884 (1955); P. Chiche, Ann. chim. 7, 361 (1952).
4. P. Assayag, Ann. chim. (Paris) 12, 637 (1955).
5. F. Becker, Dissertation, Darmstadt, 1927; cf. Wöhler and Jochum, Z. physik. Chem. 167A, 169 (1933).
6. H. S. Roberts and F. H. Smyth, J. Am. Chem. Soc. 43, 1061 (1921); 42, 2582 (1920).
7. H. W. Foote and E. K. Smith, J. Am. Chem. Soc. 30, 1344 (1908).

Heat Capacity and Entropy.

Low temperature values are based on data (15-297°K) from J. Hu and H. L. Johnston, J. Am. Chem. Soc. 75, 2471 (1953). Earlier data from R. W. Millar (71-302°K), J. Am. Chem. Soc. 51, 215 (1929), are in satisfactory agreement, while those of K. Clusius and P. Harteck (30-200°K), Z. physik. Chem. 134, 243 (1928), are higher by several percent. A small anomaly in the heat capacity is observed in the region 210-230°K. Magnetic measurements of M. O'Keefe and F. S. Stone, Phys. Chem. Solids 23, 261 (1962), and neutron diffraction studies of B. N. Brockhouse, Phys. Rev. 94, 781 (1954), suggest that this is a Néel point associated with antiferromagnetism. The entropy was obtained from the heat capacities based on $S^\circ_{15} = 0.016$ eu.

Figure 5.4-2 Example of data compiled in JANAF Thermochemical Tables.

Solution: The enthalpy increment for SO_2 can be found by interpolation between the values at 900 K and 1000 K or by use of the equation given in Fig. 5.4-3. Linear interpolation yields

$$[H_{950}-H_{298}] = H_{900} + (\frac{950-900}{1000-900})[(H_{1000}-H_{298}) - (H_{900}-H_{298})]$$

$$= 6940 + \frac{50}{100}[(8230) - (6940)]$$

$$= 7590 \text{ cal/mol}$$

Use of the equation yields

$$[H_{950}-H_{298}] = 11.04(950) + 0.94 \times 10^{-3}(950)^2 + 1.84 \times 10^5(950)^{-1} - 3992$$

$$= 7540 \text{ cal/mol}$$

This small difference is well within the stated precision of the data and can be ignored.

Finally, to calculate the enthalpy of SO_2 at 950 K, the enthalpy at 298 K must be found. This value is not given in this compilation but can be found in NBS Technical Note No. 270-3; see Fig. 5.4-1. For $SO_2(g)$,

$$\Delta H_{f_{298}} = H_{298} = -70{,}944$$

$$[H_{950} - H_{298}] = 7540$$

$$H_{950} = 7540 - 70944$$

$$= -63400 \text{ cal/mol}$$

U. S. B. M. Bulletin 605, Thermodynamic Properties of 65 Elements-Their Oxides, Halides, Carbides and Nitrides, by C. E. Wicks and F. E. Block, was published by the U. S. Bureau of Mines in 1963.* This compilation contains heats of formation, enthalpy increments and specific heat data for many elements and their compounds with oxygen, carbon, nitrogen and the halides at 298K and above.

EXAMPLE 5.4-4: Determine the heat of formation of magnetite (symbol Fe_3O_4). The table shows

$$\Delta H_f = -260{,}800 \text{ cal/mol}$$

(When using any data compilation be sure to check on the

---

* Reproduced in Appendix B.

DIOXIDE

References: Avdeeva (*27*) (298°–1,400°); *Cross* (*128*) (298°–1,800°); *Evans and Wagman* (*173*) (298°–1,500°); *Gordon* (*218*) (298°–2,800°); and *Justi and Lüder* (*326*) (273°–3,273°).

TABLE 762.—*Heat content and entropy of $SO_2(g)$*

[Base, ideal gas at 295.15° K.; mol. wt. 64.07]

T, ° K.	$H_T-H_{298.15}$, cal./mole	$S_T-S_{298.15}$, cal./deg. mole	T, ° K.	$H_T-H_{298.15}$, cal./mole	$S_T-S_{298.15}$, cal./deg. mole
400	1,015	2.92	1,300	12,215	17.24
500	2,090	5.32	1,400	13,565	18.24
600	3,240	7.41	1,500	14,925	19.18
700	4,440	9.26	1,600	16,290	20.06
800	5,675	10.91	1,700	17,660	20.89
900	6,940	12.40	1,800	19,035	21.68
1,000	8,230	13.76	1,900	20,415	22.42
1,100	9,545	15.01	2,000	21,800	23.13
1,200	10,875	16.17			

$SO_2(g)$:

$$H_T-H_{298.15}=11.04T+0.94\times10^{-3}T^2+1.84\times10^5T^{-1}$$
$$-3,992 \text{ (0.8 percent; 298°–2,000° K.)};$$
$$C_p=11.04+1.88\times10^{-3}T-1.84\times10^5T^{-2}.$$

TRIOXIDE

References: Chernobaev (*96*) (273°–1,473°); *Evans and Wagman* (*173*) (298°–1,500°); and *Stockmayer, Kavanagh, and Mickley* (*695*) (298°–1,200°).

TABLE 763.—*Heat content and entropy of $SO_3(g)$*

[Base, ideal gas at 298.15° K.; mol. wt. 80.07]

T, ° K.	$H_T-H_{298.15}$, cal./mole	$S_T-S_{298.15}$, cal./deg. mole	T, ° K.	$H_T-H_{298.15}$, cal./mole	$S_T-S_{298.15}$, cal./deg. mole
400	1,330	3.82	1,000	11,860	19.52
500	2,820	7.13	1,100	13,860	21.43
600	4,450	10.11	1,200	15,900	23.20
700	6,190	12.79	1,300	17,980	24.86
800	8,010	15.23	1,400	20,090	26.42
900	9,900	17.46	1,500	22,230	27.88

$SO_3(g)$:

$$H_T-H_{298.15}=13.90T+3.05\times10^{-3}T^2+3.22\times10^5T^{-1}$$
$$-5,495 \text{ (0.7 percent; 298°–1,500° K.)};$$
$$C_p=13.90+6.10\times10^{-3}T-3.22\times10^5T^{-2}.$$

CHLORIDES

References: *Luft and Todhunter* (*447*) (S_2Cl_2, 298°–1,000°); McDowell and Moelwyn-Hughes (*472*) (SCl_2, 298°–1,000°); and *Stamreich, Forneris, and Sone* (*677*) (molecular constant data for SCl_2).

TABLE 764.—*Heat content and entropy of $S_2Cl_2(g)$*

[Base, ideal gas at 298.15° K.; mol. wt., 135.05]

T, ° K.	$H_T-H_{298.15}$, cal./mole	$S_T-S_{298.15}$, cal./deg. mole	T, ° K.	$H_T-H_{298.15}$, cal./mole	$S_T-S_{298.15}$, cal./deg. mole
400	1,840	5.30	800	9,590	18.66
500	3,730	9.52	900	11,580	21.03
600	5,660	13.03	1,000	13,570	23.12
700	7,610	16.04			

$S_2Cl_2(g)$:

$$H_T-H_{298.15}=19.43T+0.41\times10^{-3}T^2+1.87\times10^5T^{-1}$$
$$-6,457 \text{ (0.2 percent; 298°–1,000° K.)};$$
$$C_p=19.43+0.82\times10^{-3}T-1.87\times10^5T^{-2}.$$

TABLE 765.—*Heat content and entropy of $SCl_2(g)$*

[Base, ideal gas at 298.15° K.; mol. wt., 102.98]

T, ° K.	$H_T-H_{298.15}$, cal./mole	$S_T-S_{298.15}$, cal./deg. mole	T, ° K.	$H_T-H_{298.15}$, cal./mole	$S_T-S_{298.15}$, cal./deg. mole
400	1,275	3.67	1,000	9,340	15.93
500	2,580	6.58	1,200	12,090	18.44
600	3,905	9.00	1,400	14,845	20.56
700	5,250	11.07	1,600	17,610	22.41
800	6,610	12.89	1,800	20,380	24.04
900	7,975	14.50	2,000	23,155	25.50

$SCl_2(g)$:

$$H_T-H_{298.15}=13.68T+0.07\times10^{-3}T^2+1.39\times10^5T^{-1}$$
$$-4,551 \text{ (0.1 percent; 298°–2,000° K.)};$$
$$C_p=13.68+0.14\times10^{-3}T-1.39\times10^5T^{-2}.$$

HEXAFLUORIDE

References: Gaunt (*195*) (298°–500°); *Lagemann and Jones* (*409*) (molecular constant data); Meyer and Buell (*474*) (298°–5,000°); and *Yost* (*791*) (molecular constant data).

TABLE 766.—*Heat content and entropy of $SF_6(g)$*

[Base, ideal gas at 298.15° K.; mol. wt., 146.07]

T, ° K.	$H_T-H_{298.15}$, cal./mole	$S_T-S_{298.15}$, cal./deg. mole	T, ° K.	$H_T-H_{298.15}$, cal./mole	$S_T-S_{298.15}$, cal./deg. mole
400	2,595	7.46	1,000	22,420	37.20
500	5,515	13.96	1,200	29,600	43.47
600	8,675	19.72	1,400	36,890	49.36
700	11,980	24.81	1,600	44,250	54.27
800	15,390	29.36	1,800	51,640	58.63
900	18,870	33.46	2,000	59,070	62.54

$SF_6(g)$:

$$H_T-H_{298.15}=31.89T+2.10\times10^{-3}T^2+9.01\times10^5T^{-1}$$
$$-12,717 \text{ (0.8 percent; 298°–2,000° K.)};$$
$$C_p=31.89+4.20\times10^{-3}T-9.01\times10^5T^{-2}.$$

Figure 5.4-3 Example of data compiled in Bureau of Mines Bulletin 584.

T, ° K.	H_T-H_{298}	S_T	ΔH°_T	ΔF°_T
298	....	13.74	−63,800	−58,670
400	1,210	17.22	−63,700	−57,000
500	2,440	19.97	−63,400	−55,300
600	3,700	22.07	−63,250	−53,550
700	4,980	24.24	−63,150	−51,100
800	6,280	25.97	−63,150	−50,500
900	7,590	27.52	−63,200	−48,950
1,000	8,920	28.92	−63,400	−47,350
1,100	10,280	30.21	−63,800	−45,700
1,200	11,670	31.42	−63,900	−43,950
1,300	13,080	32.55	−63,900	−42,350
1,400	14,520	33.62	−63,750	−40,750
1,500	15,980	34.62	−63,700	−39,050
1,600	17,460	35.58	−63,550	−37,400
1,700	26,510	41.06	−56,050	−36,000
1,800	28,140	42.00	−55,900	−34,750

Triiron Tetraoxide, Fe_3O_4 (*c*)

$\Delta H^\circ_{298} = -267{,}800$ calories per mole (*24*)
$S_{298} = 35.0$ *e.u.* (*83*)
$T.P. = 900°$ K. (*24*)
$\Delta H_T = 0$ calories per mole
$M.P. = 1{,}870°$ K. (*30*)
$\Delta H_M = 33{,}000$ calories per mole

Zone I (α) (298°–900° K.)

$$C_p = 21.88 + 48.20\times10^{-3}T \quad (27)$$
$$H_T - H_{298} = -8{,}640 + 21.88T + 24.10\times10^{-3}T^2$$

Zone II (β) (900°–1,800° K.)

$$C_p = 48.0 \quad (27)$$
$$H_T - H_{298} = -12{,}650 + 48.00T$$

Formation: $3Fe + 2O_2 \longrightarrow Fe_3O_4$

Zone I (298°–900° K.)

$$\Delta C_p = -2.55 + 24.90\times24.90\times10^{-3}T - 0.49\times10^5T^{-2}$$
$$\Delta H_T = -268{,}300 - 2.55T + 12.45\times10^{-3}T^2 + 0.49\times10^5T^{-1}$$
$$\Delta F_T = -268{,}300 + 2.55T\ln T - 12.45\times10^{-3}T^2 + 0.24\times10^5T^{-1} + 73.07T$$

Zone II (900°–1,033° K.)

$$\Delta C_p = 23.57 - 23.30\times10^{-3}T - 0.49\times10^5T^{-2}$$
$$\Delta H_T = -272{,}760 + 23.57T - 11.65\times10^{-3}T^2 + 0.49\times10^5T^{-1}$$
$$\Delta F_T = -272{,}760 - 23.57T\ln T + 11.65\times10^{-3}T^2 + 0.24\times10^5T^{-1} + 234.0T$$

Zone III (1,033°–1,179° K.)

$$\Delta C_p = 2.48 - 2.00\times10^{-3}T + 0.80\times10^5T^{-2}$$
$$\Delta H_T = -262{,}950 + 2.48T - 1.00\times10^{-3}T^2 - 0.80\times10^5T^{-1}$$
$$\Delta F_T = -262{,}950 - 2.48T\ln T + 1.00\times10^{-3}T^2 - 0.40\times10^5T^{-1} + 89.38T$$

Zone IV (1,179°–1,674° K.)

$$\Delta C_p = 19.13 - 11.00\times10^{-3}T + 0.80\times10^5T^{-2}$$
$$\Delta H_T = -277{,}000 + 19.13T - 5.50\times10^{-3}T^2 - 0.80\times10^5T^{-1}$$
$$\Delta F_T = -277{,}000 - 19.13T\ln T + 5.50\times10^{-3}T^2 - 0.40\times10^5T^{-1} + 162.62T$$

Zone V (1,674°–1,800° K.)

$$\Delta C_p = 2.78 - 2.00\times10^{-3}T + 0.80\times10^5T^{-2}$$
$$\Delta H_T = -262{,}500 + 2.78T - 1.00\times10^{-3}T^2 - 0.80\times10^5T^{-1}$$
$$\Delta F_T = -262{,}500 - 2.78T\ln T + 1.00\times10^{-3}T^2 - 0.40\times10^5T^{-1} + 91.0T$$

T, ° K.	H_T-H_{298}	S_T	ΔH°_T	ΔF°_T
298	....	35.0	−267,800	−243,300
400	3,990	46.48	−267,200	−234,900
500	8,320	56.12	−266,300	−226,900
600	13,060	64.75	−265,300	−219,100
700	18,340	72.88	−264,000	−211,500
800	24,260	80.77	−262,300	−204,200
900	30,550	88.18	−260,500	−197,000
1,000	35,350	93.24	−260,800	−189,900
1,100	40,350	97.81	−261,500	−182,600
1,200	44,950	101.99	−262,000	−175,500
1,300	49,750	105.83	−261,500	−168,300
1,400	54,550	109.39	−261,000	−161,100
1,500	59,350	112.70	−260,900	−154,000
1,600	64,150	115.80	−260,500	−146,800
1,700	68,950	118.71	−260,800	−139,800
1,800	73,750	121.45	−260,800	−133,000

Diiron Trioxide, Fe_2O_3 (*c*)

$\Delta H^\circ_{298} = -196{,}800$ calories per mole (*112*)
$S_{298} = 21.5$ *e.u.* (*112*)
$T.P. = 950°$ K. (*24*)
$\Delta H_T = 160$ calories per mole
$T.P. = 1{,}050°$ K. (*24*)
$\Delta H_T = 0$ calories per mole
Decomposes = 1,730° K. (*24*)

Zone I (α) (298°–950° K.)

$$C_p = 23.49 + 18.60\times10^{-3}T - 3.55\times10^5T^{-2} \quad (84)$$
$$H_T - H_{296} = -9{,}020 + 23.49T + 9.30\times10^{-3}T^2 + 3.55\times10^5T^{-1}$$

Zone II (β) (950°–1,050° K.)

$$C_p = 36.0 \quad (84)$$
$$H_T - H_{298} = -11{,}980 + 36.0T$$

Zone III (γ) (1,050°–1,730° K.)

$$C_p = 31.71 + 1.76\times10^{-3}T \quad (84)$$
$$H_T - H_{298} = -8{,}450 + 31.71T + 0.88\times10^{-3}T^2$$

Formation: $2Fe + 3/2O_2 \longrightarrow Fe_2O_3$

Zone I (298°–950° K.)

$$\Delta C_p = 6.01 + 2.90\times10^{-3}T - 3.81\times10^5T^{-2}$$
$$\Delta H_T = -200{,}000 + 6.01T + 1.45\times10^{-3}T^2 + 3.81\times10^5T^{-1}$$
$$\Delta F_T = -200{,}000 - 6.01T\ln T - 1.45\times10^{-3}T^2 + 1.90\times10^5T^{-1} + 108.4T$$

Zone II (950°–1,033° K.)

$$\Delta C_p = 18.52 - 15.7\times10^{-3}T - 0.26\times10^5T^{-2}$$
$$\Delta H_T = -203{,}300 + 18.52T - 7.85\times10^{-3}T^2 + 0.26\times10^5T^{-1}$$
$$\Delta F_T = -203{,}300 - 18.52T\ln T + 7.85\times10^{-3}T^2 + 0.13\times10^5T^{-1} + 189.0T$$

Zone III (1,050°–1,179° K.)

$$\Delta C_p = 0.17 + 0.26\times10^{-3}T + 0.60\times10^5T^{-2}$$
$$\Delta H_T = -193{,}100 + 0.17T + 0.13\times10^{-3}T^2 - 0.60\times10^5T^{-1}$$
$$\Delta F_T = -193{,}100 - 0.17T\ln T - 0.13\times10^{-3}T^2 - 0.30\times10^5T^{-1} + 60.07T$$

Zone IV (1,179°–1,674° K.)

$$\Delta C_p = 11.27 - 5.74\times10^{-3}T + 0.60\times10^5T^{-2}$$
$$\Delta H_T = -202{,}600 + 11.27T - 2.87\times10^{-3}T^2 - 0.60\times10^5T^{-1}$$
$$\Delta F_T = -202{,}600 - 11.27T\ln T + 2.87\times10^{-3}T^2 - 0.30\times10^5T^{-1} + 142.29T$$

Figure 5.4-4 Example of data compiled in Bureau of Mines Bulletin 605.

notation that is being used by the authors. In this compilation, ΔH°_T is used to designate the heat of formation at temperature T. In the JANAF Thermochemical Tables, ΔH°_f was used.)

Several older, but often useful, Bureau of Mines Bulletins are: High Temperature Specific Heat Equations for Inorganic Substances, Bulletin 371, (1935); Heats of Fusion for Inorganic Substances, Bulletin 393, (1936); The Thermodynamic Properties of Sulfur and Its Inorganic Compounds, Bulletin 406, (1937); The Thermodynamic Properties of Metal Carbides and Nitrides, Bulletin 407, (1937); High Temperature Heat Content, Heat Capacity and Entropy Data for Inorganic Compounds, Bulletin 476, (1949); Heats and Free Energies of Formation of Inorganic Oxides, Bulletin 542, (1954); and Reprint of Bulletins 383, 384, 393 and 406, Bulletin 601, (1962).

D. *Thermochemistry for Steelmaking*

Thermochemistry for Steelmaking, Vol. I and II, by John F. Elliott, M. Gleiser and K. Ramakrishna* contains compilations of data useful for steelmaking and many other high-temperature metallurgical systems. Of particular interest in this compilation are the specific heat, temperature and heat of transformation, and heat content data for the elements (Volume I), heat content and heat of formation data for selected inorganic compounds (Volume I) and the heat of mixing data for metallic solutions (Volume II).

The Making, Shaping and Treating of Steel, 9th Edition, 1971,** contains many thermodynamic data in Chapter 13, Section 6, including heat capacity equations, heats of transformation, heats of solution in iron, and heats of reactions. BOF Steelmaking, Vol. 2-Fundamentals, (Chapter 4)*** contains essentially the same data.

EXAMPLE 5.4-5: Determine the enthalpy increment for iron between 1000 K and 2000 K.

Solution: Since enthalpy is a state property,

$$(H_{2000} - H_{1000}) = (H_{2000} - H_{298}) - (H_{1000} - H_{298})$$

Using the data given in Fig. 5.4-5,

$$[H_{2000} - H_{1000}] = 19{,}714 - 7{,}135$$

$$= 12{,}580 \text{ cal/mol}$$

* Published in 1960 by Addison-Wesley Publishing Co., Reading, Mass.

** Published by U.S. Steel Corp., Pittsburgh, PA.

***Published in 1975 by American Institute of Mining, Metallurgical and Petroleum Engineers, 345 E. 47th St., New York, NY.

Fe, gfw 55.85 gm

Ideal monatomic gas.

$(H^\circ_{298.15} - H^\circ_0) = 1638$ cal/gfw.

T, °K	C°_P, cal/°/gfw	$H^\circ_T - H^\circ_{298.15}$, cal/gfw	S°_T, cal/°/gfw	$\frac{-(F^\circ_T - H^\circ_{298.15})}{T}$
298	6.14	0	43.12	43.12
300	6.14	11	43.16	43.12
400	6.10	625	44.93	43.37
500	5.95	1228	46.27	43.82
600	5.79	1815	47.34	44.32
700	5.64	2386	48.23	44.82
800	5.53	2945	48.97	45.29
900	5.44	3493	49.62	45.74
1000	5.38	4034	50.19	46.15
1100	5.33	4569	50.70	46.54
1200	5.30	5101	51.16	46.91
1300	5.29	5631	51.58	47.25
1400	5.30	6160	51.98	47.58
1500	5.31	6690	52.34	47.88
1600	5.34	7223	52.69	48.17
1700	5.38	7760	53.01	48.45
1800	5.43	8301	53.32	48.71
1900	5.49	8846	53.62	48.96
2000	5.55	9398	53.90	49.20
2100	5.61	9956	54.17	49.43
2200	5.68	10521	54.43	49.65
2300	5.75	11093	54.69	49.86
2400	5.82	11672	54.93	50.07
2500	5.90	12258	55.17	50.27
2600	5.97	12851	55.41	50.46
2700	6.04	13452	55.63	50.65
2800	6.12	14060	55.85	50.83
2900	6.19	14675	56.07	51.01
3000	6.27	15298	56.28	51.18

Figure 5.4-5 Example of data compiled in Thermochemistry for Steelmaking.

E. *International Copper Research Association Publications*

The International Copper Research Association in New York, N.Y., publishes the INCRA Series on the Metallurgy of Copper, a series of Monographs that contain data of importance in the metallurgy of copper. Monographs which have already been published are Selected Thermodynamic Values and Phase Diagrams for Copper and Some of its Binary Alloys (Monograph I), Thermodynamic Properties of Copper and its Inorganic Compounds (Monograph II), The Thermodynamic Properties of Copper - Slag Systems (Monograph III), and The Thermodynamic Properties of Aqueous Inorganic Copper Solutions (Monograph IV). Monographs which will be published in the future will contain thermodynamic data for ternary and higher-order alloys.

F. *American Society for Metals*

An excellent collection of data, presented in tabular and graphical form is found in two books published by the ASM*,

Selected Values of the Thermodynamic Properties of the Elements, P. Desai, D. Hawkins, M. Gleiser, K. K. Kelley and D. D. Wagman, (1973).

Selected Values of the Thermodynamic Properties of Binary Alloys, R. Hultgren, P. Desai, D. Hawkins, M. Gleiser, and K. K. Kelley, (1973).

EXAMPLE 5.4-6: What is the enthalpy associated with the process of dissolving Si in liquid iron at 1873 K and at a composition of 1 wt % Si?

Solution: The data in Fig. 5.4-6 are used to obtain the answer to this. The reaction is

$$Si(\ell, \text{pure}, 1873K) \rightarrow \underline{Si}(1 \text{ w/o Fe alloy}, 1873K)$$

for which the heat effect is $\Delta\bar{H}^M_{Si}$, or, in the notation of Fig. 5.4-6, $\Delta\bar{H}_{Si}$. Using (Eq. 1.4-6), the mole fraction of silicon in the solution equals

$$X_{Si} = \frac{1/28.08}{1/28.08 + 99/55.85} = .02$$

then, linear interpolation between -31,400 for X_{Si} = 0 and -29,897 for X_{Si} = 0.1 yields $\bar{H}_{Si}$ = -31,100 cal/mol Si.

G. *Thermodynamics Texts*

Most thermodynamics text books contain data in Appendices. Of particular note is Metallurgical Thermodynamics by O. Kubaschewski, E. L. Evans and C. B. Alcock**. This text contains

* American Society for Metals, Metals Park, Ohio 44073.

** Published by Pergamon Press, New York, N. Y. (1967).

Partial Molar Quantities for Liquid Alloys at 1873°K

Fe Component $Fe_{(l)} = Fe(\text{in alloy})_{(l)}$

x_{Fe}	a_{Fe}	γ_{Fe}	$\Delta\bar{G}_{Fe}$	$\Delta\bar{G}^{xs}_{Fe}$	$\Delta\bar{H}_{Fe}$	$\Delta\bar{S}_{Fe}$	$\Delta\bar{S}^{xs}_{Fe}$
1.0	1.000	1.000	0	0	0	0.000	0.000
0.9	0.860	0.955	- 562	- 170	- 99	0.247	0.038
0.8	0.621	0.776	- 1777	- 946	- 786	0.529	0.085
0.7	0.333	0.476	- 4090	- 2762	- 3096	0.530	-0.178
0.6	0.128	0.214	- 7644	- 5743	- 7038	0.324	-0.691
0.5	0.0511	0.102	-11070	- 8490	-11456	-0.206	-1.584
	(±.019)	(±.038)	(±1750)	(±1750)	(±1500)	(±1.23)	(±1.23)
0.4	0.0249	0.0622	-13745	-10335	-15549	-0.963	-2.784
0.3	0.0128	0.0428	-16211	-11730	-19065	-1.524	-3.916
0.2	0.00614	0.0307	-18959	-12969	-21751	-1.491	-4.689
0.1	0.00222	0.0222	-22740	-14170	-23727	-0.527	-5.103
0.0	0.000	0.0162	- ∞	-15349	-25037	∞	-5.172

Si Component $Si_{(l)} = Si(\text{in alloy})_{(l)}$

x_{Si}	a_{Si}	γ_{Si}	$\Delta\bar{G}_{Si}$	$\Delta\bar{G}^{xs}_{Si}$	$\Delta\bar{H}_{Si}$	$\Delta\bar{S}_{Si}$	$\Delta\bar{S}^{xs}_{Si}$
0.0	0.000	0.00132	- ∞	-24682	-31400	∞	-3.587
0.1	0.00030	0.00297	-30228	-21658	-29897	0.177	-4.399
0.2	0.00190	0.00950	-23322	-17332	-26214	-1.544	-4.742
0.3	0.0122	0.0406	-16407	-11926	-19399	-1.597	-3.990
0.4	0.0713	0.178	- 9827	- 6417	-12096	-1.211	-3.032
0.5	0.223	0.446	- 5580	- 3000	- 6642	-0.567	-1.944
	(±.053)	(±.105)	(±1000)	(±1000)	(±2000)	(±1.19)	(±1.19)
0.6	0.406	0.677	- 3354	- 1453	- 3266	0.047	-0.968
0.7	0.581	0.830	- 2022	- 694	- 1346	0.361	-0.348
0.8	0.742	0.928	- 1109	- 278	- 435	0.360	-0.084
0.9	0.885	0.983	- 455	- 63	- 76	0.202	-0.007
1.0	1.000	1.000	0	0	0	0.000	0.000

Figure 5.4-6 Example of data available in Selected Values of the Thermodynamic Properties of Binary Alloys, Hultgren et al.

$$C_p = a + bT + cT^{-2} \text{ (cal} \cdot \text{deg}^{-1} \cdot \text{mole}^{-1}\text{)}$$

Substance	C_p in cal · deg⁻¹ · mole⁻¹			Remarks	Temp.-Range °K	Ref.
	a	b · 10^3	c · 10^{-5}			
⟨Ag⟩	5·09	2·04	0·36		298–m.p.	124
{Ag}	7·30	—	—		m.p.–1600	124
⟨AgCl⟩	14·88	1·00	−2·70		298–m.p.	124
{AgCl}	16·0	—	—		m.p.–900	124
⟨AgBr⟩	7·93	15·40	—		298–m.p.	124
{AgBr}	14·9	—	—		m.p.–900	124
⟨AgI⟩α	5·82	24·10	—		298–423	124
⟨AgI⟩β	13·5	—	—		423-600	124
⟨Ag_2O⟩	14·18	9·75	−1·0		298–500	138
⟨Ag_2S⟩α	10·13	26·40	—		298–452	124
⟨Ag_2S⟩β	21·64	—	—		452–850	124
⟨Ag_2SO_4⟩	*23·1*	*27·9*	—		298–m.p.	124
⟨Ag_2Se⟩α	15·35	15·58	—		298–406	124
⟨Ag_2Se⟩β	20·4	—	—		406–500	124
⟨Ag_3Sb⟩	19·53	16·0	—		298–700	124
⟨Ag_2CO_3⟩	18·97	25·85	—		298–450	138
⟨Ag-Al⟩	additive				273–773	247
⟨Ag-Mg⟩	additive				298–773	247
⟨Ag-Au⟩	additive				298–m.p.	151
⟨Al⟩	4·94	2·96	—		298–m.p.	124
{Al}	7·00	—	—		m.p.–1273	124
(AlF)	8·9	—	−1·45		298–2000	77
⟨AlF_3⟩α	17·27	10·96	−2·30		298–727	314
⟨AlF_3⟩β	20·93	3·0	—		727–1400	314
(AlCl)	9·0	—	−0·68		298–2000	77
⟨$AlCl_3$⟩	13·25	28·00	—		273–m.p.	124
{$AlCl_3$}	31·2	—	—		m.p.–500	124
($AlCl_3$)	*19·8*	—	*−2·64*		298–1800	77, 255
⟨$AlBr_3$⟩	18·74	18·66	—		298–m.p.	124
{$AlBr_3$}	29·5	—	—		m.p.–500	124
⟨AlI_3⟩	16·88	22·66	—		298–m.p.	124
{AlI_3}	29·0	—	—		m.p.–500	124
⟨Al_2O_3⟩	25·48	4·25	−6·82		298–1800	124, 480, 129
⟨$Al_2(SO_4)_3$⟩	87·55	14·96	−26·68		298–1100	124
⟨AlN⟩	5·47	7·80	—		298–900	124
⟨Al_4C_3⟩	24·08	31·6	—		298–600	124
⟨Al_2SiO_5⟩	40·09	5·86	−10·13	sillimanite	298–1600	124
⟨Al_2SiO_5⟩	46·24	—	−12·53	andalusite	298–1600	124
⟨Al_2SiO_5⟩	45·32	2·34	−16·00	kyanite	298–1700	124
⟨Al_2TiO_5⟩	43·63	5·30	−11·21		298–1800	280

Figure 5.4-7 Example of data compiled in Metallurgical Thermodynamics, by Kubaschewski, Evans, and Alcock.

extensive tables of heats of transformation, heats of formation, specific heats and heats of mixing for metallic solutions at 298 K and above.

EXAMPLE 5.4-7: What is the specific heat of alumina (symbol Al_2O_3) at 1000 K?

Solution: The specific heat for alumina is given in Fig. 5.4-7. In this compilation, all specific heats have the form

$$C_p = a + bT + cT^{-2} \text{ (cal/mol·K)}$$

For alumina,

$$a = 25.48$$

$$b \times 10^3 = 4.25, \text{ or } b = 4.25 \times 10^{-3}$$

$$c \times 10^{-5} = -6.82, \text{ or } c = -6.82 \times 10^5$$

Thus,

$$C_{p,Al_2O_3} = 25.48 + 4.25 \times 10^{-3}T - 6.82 \times 10^5 T^{-2}$$

and at 1000 K

$$C_p = 29.05 \text{ cal/mol K}$$

Another book which is particularly useful for finding enthalpy data for metallic solutions, i.e. heats of mixing, is Thermodynamic Data of Alloys by O. Kubaschewski and J. A. Catterall, Pergamon Press, London (1956).

H. *Additional Data Sources*

Many engineering handbooks contain thermodynamic data relating to common gases, liquids and some solids. Of particular use are the following:

Perry's Chemical Engineer's Handbook, Fourth Edition (1963), (published by McGraw-Hill, N.Y., N.Y.) contains a great deal of thermochemical data on elements, organic compounds, and inorganic compounds, including heats of formation, transformation, fusion and vaporization, heat capacity data, enthalpy increments, heats of combustion, and heats of solution of inorganic and organic compounds in water.

Free Energy of Formation of Binary Compounds - An Atlas of Charts for High Temperature Chemical Calculations, by Thomas B. Reed, (available from M.I.T. Press, Cambridge, Mass.) Contains heat of formation, heat of vaporization and heat of ionization data for some hydrides, nitrides, oxides, sulfides, selenides, tellurides, and halides of the elements as functions of temperature.

Handbook of Thermochemical Data for Compounds and Aqueous Species by H. E. Barner and R. V. Scheuerman (Published by Wiley-Interscience, Somerset, N.J.)
A comprehensive compilation of thermodynamic data as a function of temperature for ions and neutral complexes in aqueous solution. Tables present values of heat of formation, free energy of formation, enthalpy and entropy for ions over the temperature range of 25°C to 300°C; similar data are presented for neutral complexes in aqueous solution over the temperature range of 25°C to 200°C. Data on a large number of compounds and minerals over the temperature range of 25°C to 300°C are also given.

International Atomic Energy Agency (Vienna, Austria) Publications which include The Chemical Thermodynamics of Actinide Elements and Compounds and a series of articles containing reviews of thermodynamic data on individual elements and their compounds by O. Kubachevski.

Thermochemical Properties of Inorganic Substances by I. Barin and O. Knacke (Published by Springer-Verlay) which contains specific heat and enthalpy data for many compounds at elevated temperatures. Great care must, however, be exercised when using this book because the tabulated enthalpy data is not in a standard format. (see Page X of the introduction to the book for more details.)

Thermodynamic Properties of Minerals and Related Substances at 298.15K and 1 Bar (10^5 Pascals) Pressure and at Higher Temperatures by R. A. Robie, B. Hemingway and J. R. Bisher (U.S. Geological Survey Bulletin 1452, 1978) available from the Superintendant of Documents, U.S. Government Printing Office, Washington, D.C. 20402, contains a summary of the thermodynamic data for minerals including specific heats and enthalpy increments at temperatures up to 1800 K.

Thermodynamic Data for Inorganic Sulfides, Sellenides and Tellurides by K. C. Mills (Published by Halsted Press, New York, 1974).

Rare Earth Information Center (Energy and Materials Resources Research Institute, Iowa State University, Ames, Iowa 50011) Publications including Thermochemistry of the Rare Earths (#IS-RIC-6) by K. A. Gschneider, Jr. et al. and Thermochemistry of the Rare Earth Carbides, Nitrides and Sulfides for Steelmaking (#IS-RIC-5) by K. A. Gschneider, Jr. and N. Kipperhan as well as many other papers by K. A. Gschneider et al. on the thermodynamics of the rare earth elements and their compounds.

The CODATA Tables (available from the Office of CODATA, 51 Boulevard De Montmorency, 75016 Paris, France) are a critical review of the heats of formation at 298K and heat content between OK and 298K of selected compounds. Bulletin 28 (April, 1978) summarizes this work and gives reference to the sources of the data.

Thermochemical Constants of Compounds (in Russian, Senior Editor V. P. Glushko and available from VINITI, Oktyaberskaya Prospect 403, Moscow, U.S.S.R.) is a compilation similar to NBS Circular 500. It contains heats of formation at 298K, specific heat, heat content and heats of transformation data for many elements and their compounds. The data presented in this compilation is not as well reviewed as in the NBS circular however, references to the sources of the data are given.

There are many articles in the technical literature which contain specific heat and enthalpy data. Several reviews of data by L. Hepler et al which are particularly useful are on the platinum group metals and their compounds (Chemical Reviews 68 (1968), 229.); aqueous ions (J. Phys. Chem. 72 (1968), 2902.); Manganese and its compounds (Chemical Reviews 68 (1968), 737.); Silver and its compounds (Englehard Ind. Tech. Bulletin IX (1969), 117.); gold and its compounds (Englehard Ind. Tech. Bulletin X (1969), 5.); compounds and aqueous ions of Niobium and Tantalum (Chemical Reviews 71 (1971), 127.); Scandium (Thermochim. Acta 15 (1976), 89.); Lanthanum (Thermochim. Acta 16 (1976), 95.) and Chromium, Molybdenum and Tungsten (Chemical Reviews 76 (1976), 283.). Many additional sources for data can be found in Chemical Abstracts and in the papers listed in Appendixes B and C.

5.4.2 DATA FROM THERMODYNAMIC RELATIONSHIPS

Many thermodynamic relationships can be used to find enthalpy data when other thermodynamic data such as free energy data, are available. The uses of some of these relationships are demonstrated below.

A. *Kirchoff's Square*

Because enthalpy is a state property, the change in enthalpy for any cyclical process, i.e. a process in which the final state is the same as the initial state, must equal zero. Thus, for a cyclical process involving m steps

$$\sum_{j=1}^{m} \Delta H_j = 0 \qquad (5.4\text{-}1)$$

where ΔH_j is the enthalpy change for the jth step in the cyclical process. This relationship is often called Kirchoff's Law.

The simplest cyclical process that can be imagined is the heating of an object from T_1 to T_2 and cooling back to T_1. The methods for calculation of the enthalpy increment $\Delta H_{(T_1 \rightarrow T_2)}$ were discussed in Section 5.3. Using Eq. (5.4-1), $\Delta H_{(T_2 \rightarrow T_1)}$ can be found since

$$\Delta H_{(T_1 \rightarrow T_2)} + \Delta H_{(T_2 \rightarrow T_1)} = 0 \qquad (5.4\text{-}2)$$

Therefore,

$$\Delta H_{(T_2 \rightarrow T_1)} = -\Delta H_{(T_1 \rightarrow T_2)} \qquad (5.4\text{-}3)$$

Using similar reasoning, it can be seen that the latent heat for the transformation from phase 2 to phase 1 equals the negative of the latent heat for phase 1 going to phase 2.

EXAMPLE 5.4-7: Determine how much heat is evolved when one mole of iron is frozen at 1812 K.

Solution: The enthalpy increment for iron melted from solid δ-Fe at 1812 K is the heat of fusion, ΔH_m = 3670 cal/mol. Since the process of melting followed by the process of freezing amounts to a complete cycle, $\Delta H_m + \Delta H_{freezing} = 0$ and therefore $\Delta H_{freezing} = -\Delta H_m$ = -3670 cal/mol.

When using complicated cyclical processes to determine enthalpy data, a schematic diagram of the process should be made. This diagram, which is often called a Kirchoff Square, should indicate all of the phases which are present at each state in the process and a brief description of the state, i.e. temperature, pressure, and composition. Eq. (5.4-1) can then be applied in a logical manner to the complicated process and the enthalpy change for a single step within the overall process determined.

EXAMPLE 5.4-9: Determine the heat of vaporization of water at 298K. The heat of vaporization at the normal boiling point, i.e. 373 K, equals 9770 cal/mol and the specific heats of liquid water and steam are 18.04 and 7.30 + 2.46 x 10^{-3}T cal/mol·K, respectively.

Solution: A Kirchoff Square for a cyclical process that will allow the calculation of ΔH_v at 298 K is shown in Fig. 5.4-8. A cyclical process would then be

$$H_2O\ (\ell\ @\ 298K) \xrightarrow{\text{I}} H_2O\ (\ell\ @\ 373K) \xrightarrow{\text{II}} H_2O\ (g\ @\ 373K)$$

$$\xrightarrow{\text{III}} H_2O\ (g\ @\ 298K) \xrightarrow{\text{IV}} H_2O\ (\ell\ @\ 298K)$$

with the enthalpy changes being equal to

$$\Delta H_1 = \int_{298}^{373} c_p(\ell)dT \qquad \text{for Step I}$$

$$\Delta H_{v,373} \qquad \text{for Step II}$$

Similar results to thos obtained using the kirchoff Square method can be derived for calculation of the heat of reaction at T_2 from the heat of reaction at T_1 by combination of Eq. (5.4-1) and Eq. (5.3-18)

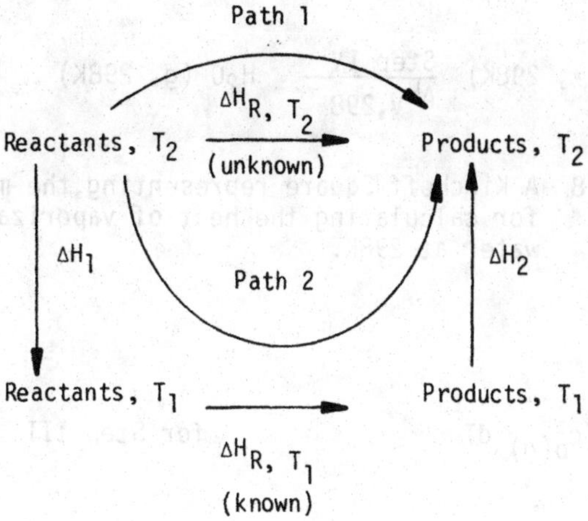

Since (Path 1) ≡ (Path 2)

$$\Delta H_{R,T_2} = \Delta H_{R,T_1} + \Delta H_1 + \Delta H_2 \qquad (5.4\text{-}4)$$

where

$$\Delta H_1 = \sum_{\text{Reactants}} n_i \; (H_{T_1} - H_{T_2})_i = \sum_{\text{Reactants}} n_i \; (H_{T_2} - H_{T_1})_i \qquad (5.4\text{-}5a)$$

and

$$\Delta H_2 = \sum_{\text{Products}} n_i \; (H_{T_2} - H_{T_1})_i \qquad (5.4\text{-}5b)$$

EXAMPLE 5.4-10: Do Example 5.4-9 using Eq. (5.4-4).

Solution: A close examination of Example 5.4-9 shows that since ΔH_1 and ΔH_2 only involve integrating heat capacity data, only minor modifications need be made to perform the calculations in a manner similar to Eq. (5.4-4). Since

$$\Delta H_{v,298} = \Delta H_{v,373} + \int_{298}^{373} C_p\,(\ell)\,dT + \int_{373}^{298} C_p\,(g)\,dT,$$

and

$$H_2O\ (\ell,\ 373K)\ \xrightarrow[\text{Step II}]{\Delta H_{v,373}}\ H_2O\ (g,\ 373K)$$

Step I ΔH_1 Step III ΔH_2

$$H_2O\ (\ell,\ 298K)\ \xrightarrow[\Delta H_{v,298}]{\text{Step IV}}\ H_2O\ (g,\ 298K)$$

Figure 5.4-8 A Kirchoff Square representing the method for calculating the heat of vaporization of water at 298K.

$$\Delta H_2 = \int_{373}^{298} c_{p(g)}\, dT \qquad \text{for Step III}$$

and

$$-\Delta H_{v,298} \qquad \text{for Step IV}$$

Substitution of these enthalpy changes into Eq. (5.4-1) yields

$$\Delta H_1 + \Delta H_{v,373} + \Delta H_2 + (-\Delta H_{v,298}) = 0$$

or

$$\Delta H_{v,298} = \Delta H_1 + \Delta H_{v,373} + \Delta H_2$$

$$= \int_{298}^{373} 18.04TdT + 9770 + \int_{373}^{298} (730 + 2.46 \times 10^{-3}T)dT$$

$$= 1354 + 9770 - 608$$

$$= 10{,}510 \text{ cal/mol}$$

Another way to look at the Kirchoff Square in this case would be to consider that the unknown enthalpy change is for the reaction path from $H_2O(\ell, 298) \rightarrow H_2O(g, 298)$ and that this state change can be achieved either directly, or by going through steps I → II → III. Either route will give the same enthalpy effect, so they must equal each other:

$$\Delta H_{v,298} = \Delta H_1 + \Delta H_{v,373} + \Delta H_2$$

This problem illustrates an important piece of data that arises often in conjunction with the calculation of enthalpy effects associated with any reaction in which water is a reactant or product. If the product of the reaction is $H_2O(g)$, it is more convenient to use the ΔH_f of $H_2O(g)$ at 298 K, i.e, to add the heat of vaporization at 298 K to the heat of formation. Then the enthalpy increment for $H_2O(g)$ from 298 K can be found without having to add in the ΔH_v at 373 K. Data for this calculation are given in Figure 5.4-9.

[Base, ideal gas at 298.15° K.; mol. wt., 18.016]

T,° K.	$H_T-H_{298.15}$, cal./mole	$S_T-S_{298.15}$, cal./deg. mole	T,° K.	$H_T-H_{298.15}$, cal./mole	$S_T-S_{298.15}$, cal./deg. mole
400	825	2.38	2,000	17,370	18.13
500	1,655	4.23	2,100	18,600	18.73
600	2,510	5.79	2,200	19,845	19.31
700	3,390	7.14	2,300	21,100	19.87
800	4,300	8.36	2,400	22,370	20.41
900	5,240	9.47	2,500	23,650	20.93
1,000	6,210	10.49	2,750	26,895	22.17
1,100	7,210	11.44	3,000	30,200	23.32
1,200	8,240	12.34	3,250	33,545	24.38
1,300	9,295	13.18	3,500	36,930	25.38
1,400	10,385	13.99	3,750	40,350	26.33
1,500	11,495	14.76	4,000	43,805	27.22
1,600	12,630	15.49	4,250	47,275	28.06
1,700	13,785	16.19	4,500	50,770	28.86
1,800	14,965	16.86	4,750	54,290	29.62
1,900	16,160	17.51	5,000	57,825	30.34

$H_2O(g)$:

$$H_T-H_{298.15}=7.30T+1.23\times10^{-3}T^2-2,286$$

(0.7 percent; 298°–2,750° K.);

$$C_p=7.30+2.46\times10^{-3}T.$$

Figure 5.4-9 Heat content and entropy of $H_2O(g)$ (From: Bureau of Mines Bulletin 584, by K. K. Kelley, p. 80).

$$\int_{373}^{298} C_{p(\ell)}\, dT = - \int_{298}^{373} C_{p\,(g)}\, dT$$

then

$$\Delta H_{v,298} = \Delta H_{v,373} + \int_{373}^{298} C_{p\,(g)}\, dT - \int_{373}^{298} C_{p\,(\ell)}\, dT$$

$$= \Delta H_{v,373} + \int_{373}^{298} (C_{p\,(g)} - C_{p\,(\ell)})\, dT$$

$$= \Delta H_{v,373} + \int_{373}^{298} \Delta C_p\, dT$$

where

$$\Delta C_p = C_{p\,(g)} - C_{p\,(\ell)}$$

$$= \sum_{\text{products}} C_{p,i} - \sum_{\text{reactants}} C_{p,i}$$

For the vaporization reaction

$$H_2O\ (\ell) \rightarrow H_2O\ (g)$$

Using the data in the previous example,

$$\Delta C_p = 7.30 + 2.46 \times 10^3 - 18.04$$

and

$$\Delta H_{v,298} = 10{,}310 \text{ cal/mol } H_2O$$

The heat of formation of $H_2O(g)$ at 298K is found to be the sum of the heat of formation of $H_2O(\ell)$ at 298 (-68,320 cal/mol) and the heat of vaporization at 298, (+10,510), or

$$\Delta H_{f,H_2O(g),298} = -57{,}800 \text{ cal/mol.}$$

EXAMPLE 5.4-11: Determine the heat of reaction at 500 K for the reaction

$$H_2O\ (g) + C\ (s) \rightarrow H_2(g) + CO\ (g)$$

Data:	$\Delta H^\circ_{f,\ 298}$	$\bar{C}_p$
$H_2O(\ell)$	-68,320 cal/mol	18.03 cal/mol-K
C (s)	0	2.82 cal/mol-K
$H_2(g)$	0	6.96 cal/mol-K
CO (g)	-26,400 cal/mol	7.02 cal/mol-K
H_2O (g)	-57,800 cal/mol	8.99 cal/mol-K

Solution: Using the heat of formation of $H_2O(\ell)$, the heat of reaction at 298K can be calculated,

$$\Delta H_{R,298} = \sum_{\text{products}} H^\circ_f - \sum_{\text{reactants}} H^\circ_f$$

$$= \Delta H^\circ_{f,CO} + \Delta H^\circ_{f,H_2} - (\Delta H^\circ_{f,H_2O(1)} - \Delta H^\circ_{f,C})$$

$$= -26,400 - (-68,320)$$

$$= 41,920 \text{ cal/mol}$$

and since the enthalpy increments between 298 and 500 K can be calculated;

$$\Delta C_p = \sum_{\text{products}} C_p - \sum_{\text{reactants}} C_p$$

and

$$= C_{p_{H_2(2)}} + C_{p,CO(g)} - (C_{p,H_2O(g)} + C_{p,c(s)})$$

$$= -6.86 \text{ for } 298 \text{ K} < T \leq 373 \text{ K}$$

$$= 2.17 \text{ for } 373 \text{ K} < T \leq 500 \text{ K}$$

then the Kirchoff's Square shown in Fig. 5.4-10(a) can be drawn and the heat of reaction at 500K can be found:

$$\Delta H_{R,500} = \Delta H_{R,298} + \int_{298}^{500} \Delta C_p \, dT + (-\Delta H_{v,H_2O})$$

$$= 41,920 + \int_{298}^{373} -6.86 \, dT + \int_{373}^{500} 2.17 \, dT - \Delta H_{v,373}$$

$$= 41,920 - 514 + 275 - 9770$$

$$\Delta H_{R,\ 500} = 31{,}910 \text{ cal.}$$

Another approach to the same problem utilizes the results of Example 5.4-10, the data in Fig. 5.4-9, and the Kirchoff Square in Fig. 5.4-10(b).

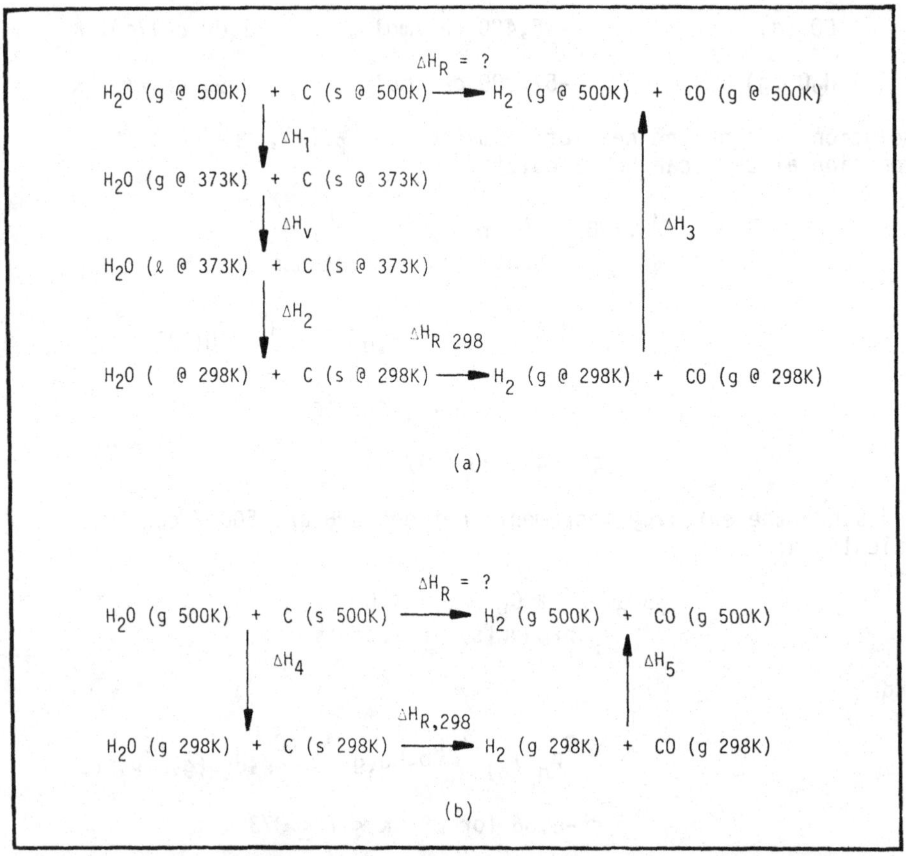

Figure 5.4-10 Kirchoff Squares for determination of the heat of reaction at 500 K from the data given in Example 5.4-11.

The heat of reaction at 298 K is calculated using $H_2O(g)$ instead of $H_2O(\ell)$:

$$\Delta H_{R,298} = \Delta H^\circ_{f,CO} + \Delta H^\circ_{f,H_2} - (\Delta H^\circ_{f,H_2O(g)} + \Delta H_{f,C})$$

$$= -26{,}400 - (-57{,}800)$$

$$= 31{,}400 \text{ cal}$$

Instead of integrating ΔC_p data, in this case use enthalpy increments.

$$\Delta H_4 = \sum_{reactants} (H_{298} - H_{500}) = \sum_{reactants} -(H_{500} - H_{298})$$

$$= -(H_{500} - H_{298})_{H_2O} - (H_{500} - H_{298})_C$$

$$= -1655 - 570$$

$$= -2225 \text{ cal.}$$

$$\Delta H_5 = \sum_{products} (H_{500} - H_{298})$$

$$= (H_{500} - H_{298})_{H_2} + (H_{500} - H_{298})_{CO}$$

$$= 1406 + 1418$$

$$= 2824 \text{ cal.}$$

Therefore, $\Delta H_{R,500} = \Delta H_4 + \Delta H_{R,298} + \Delta H_5$

$$= -2225 + 31{,}400 + 2824$$

$$= 32{,}000 \text{ cal.}$$

The difference between the two answers is caused by the use of $\bar{C}_p$ data combined with round-off errors.

B. *Maxwell Relationships*

Because the differentials of thermodynamic properties are exact differentials, certain relationships, known as the Maxwell Relationships, exist between the partial derivatives of the thermodynamic properties. These Maxwell Relationships can be used to determine the effect of pressure on enthalpy.

All data compilations give the enthalpy data at 101 kPa (1 atm) and the temperature in question. Enthalpy is not tabulated as a function of pressure for gases because most of the gases found in metallurgical processes are at temperatures and pressures at which they can be assumed to be ideal. (c.f., Section 2.3) The enthalpy of ideal gases is independent of pressure. The enthalpy of liquids and solids is dependent on pressure. However, this dependence is small and in most cases can be ignored.

When the effect of pressure must be included, the enthalpy change associated with an increase of pressure from P_1 to P_2 can be found, using the following relationship:

$$H(T, P_2) - H(T,P_1) = \int_{P_1}^{P_2} V(1 - \alpha T)\, dP \qquad (5.4\text{-}6)$$

where α, the coefficient of thermal expansion, is defined by Eq. (5.3-14), and V is the molar volume. Generally, α and V, are not functions of pressure, and Eq. (4.4-6) becomes

$$H(T, P_2) - H(T, P_1) = V(1-\alpha T)(P_2 - P_1) \quad (5.4\text{-}7)$$

When pressure and temperature have an important effect on the enthalpy, the effects are additive:

$$H(T_2,P_2) - H(T_2,P_1) = \underbrace{H(T_2,P_1) - H(T_1,P_1)}_{\text{The effect of temperature at constant } P_1} + \underbrace{H(T_2,P_2) - H(T_2,P_1)}_{\text{The effect of pressure at constant } T_2} \quad (5.4\text{-}8)$$

or

$$H(T_2,P_2) - H(T_1,P_1) = \underbrace{H(T_2,P_2) - H(T_1,P_2)}_{\text{The effect of temperature at constant } P_2} + \underbrace{H(T_1,P_2) - H(T_1,P_1)}_{\text{The effect of pressure at constant } T_1} \quad (5.4\text{-}9)$$

Eq. (5.4-8) and Eq. (5.4-9) yield the same result because enthalpy is a state property and the change must be independent of the path by which the calculation is made. However, one subtle difference does exist between these two equations; α and V must be known at T_2 and C_p at P_1 for the calculations as outlined in Eq. (5.4-8), while α and V at T_1 and C_p at P_2 are required in Eq. (5.4-9).

EXAMPLE 5.4-12: Determine the increase in enthalpy of one mole of iron when the pressure is raised from one to one hundred atmospheres at 298 K. The values of α and V for iron at 298 K are 0:3 x $10^{-4}K^{-1}$ and 7.1 cm^3/mol, respectively.

Solution: If it is assumed that α and V are independent of pressure

$$H(298K, 100\ atm) - H(298K, 1\ atm) = V(1 - \alpha T)(100-1)$$

$$= (7.1\ cm^3/mol)\cdot(1 - 0.3 \times 10^{-4}\ K^{-1}\cdot 298K)\cdot(99\ atm)$$

$$= 697\ cm^3\cdot atm/mol$$

$$= 16.9\ cal/mol$$

EXAMPLE 5.4-13: Determine how much the iron in the previous example would have to be cooled to lower its enthalpy to the original value. The heat capacity for iron at 298 K and 1 atm is 6.0 cal/mol K. It may be assumed that this value is constant over a small temperature range and can be used at 100 atm.

Solution: The required enthalpy change must equal

$$\Delta H = C_p\ \Delta T$$

or

$$\Delta T = \Delta H/C_p$$

$$= -16.9\ (\text{cal/mol})/6.0\ (\text{cal/mol K})$$

$$= -2.8\ \text{K}$$

C. *Gibbs-Helmholtz Equations*

It will simply be stated that if data are given for the standard Gibbs Free Energy* change of a reaction (symbol $\Delta G°$ or $\Delta F°$) the enthalpy change for the reaction can be found from the Gibbs-Helmholtz equation:

$$\Delta H^0 = \left.\frac{\partial(\Delta G^0/T)}{\partial(1/T)}\right|_p = -T^2\ \frac{\partial(\Delta G^0/T)}{\partial T} \qquad (5.4\text{-}10)$$

EXAMPLE 5.4-14: Determine the heat of reaction at 800 K for

$$2Cu(s) + 1/2\ O_2(g) \rightarrow Cu_2O\ (s)$$

given that the standard free energy change of the reaction is

$$\Delta G° = -40{,}500 - 1.70T\ \ell n\ T + 29.5\ T$$

Solution: The heat of reaction can be found using Eq. (5.4-10).

$$\Delta H_R = -T^2\ \frac{\partial(\Delta G°/T)}{\partial T}$$

$$= -T^2\ \frac{\partial(-40{,}500/T - 1.70\ell nT + 29.5\ T)}{\partial\ T}$$

$$= -T^2\ (40{,}500/T^2 - 1.70/T)$$

$$= -40{,}500 + 1.70\ T$$

At 800 K,

$$\Delta H_R = -39{,}140\ \text{cal.}$$

D. *Clausius - Clapeyron Equation*

The heat of sublimation or heat of vaporization can be found from vapor pressure data using the Clausius-Clapeyron Equation. When the logarithm of the vapor pressure, ℓnP, is known as a function of temperature,

$$\frac{\partial\ \ell n\ P}{\partial T} = \frac{\Delta H_v}{RT^2} \qquad (5.4\text{-}11)$$

---

* See a standard thermodynamics text for the definition of Gibbs Free Energy.

If the vapor pressure is given over the solid, the heat of sublimation, i.e. the solid to vapor transformation, is found. The heat of vaporization is found from vapor pressure data over the liquid.

EXAMPLE 5.4-15: Determine the heat of vaporization of zinc at 1000 K. The vapor pressure over the liquid zinc is represented by

$$\log P_{Zn}\ (\text{mmHg}_1) = -6620/T - 1.255 \log T + 12.34$$

Solution: The heat of vaporization is found using Eq. (5.4-11):

$$\Delta H_v = RT^2 \frac{\partial \ell nP}{\partial T} = 2.303\ RT^2 \frac{\partial \log P}{\partial T}$$

$$= 2.303\ RT^2 \frac{\partial(-6620/T - 1.255 \log T + 12.34)}{\partial T}$$

$$= 2.303\ RT^2 (+6620/T^2 - 0.545/T)$$

$$= +30{,}300 - 2.49T$$

At 1000 K

$$\Delta H_v = +30{,}300 - 2.49\ (1000)$$

$$= 27{,}810 \text{ cal/mol}$$

Figure 5.4-11 shows vapor pressure data for pure metals plotted as log P versus temperature. The slopes of these curves are remarkably similar, i.e., $\partial \log P/\partial T$ is similar for many metals, which leads to the conclusion that ΔH_{sub} and ΔH_v are similar for most metals at the same temperature.

E. *Gibbs-Duhem Equations*

Relationships do exist between both the heats of mixing of individual components in a solution and between the heats of mixing of individual components and the total heat of mixing of the solution. It is beyond the scope of this text to develop these relationships or to describe the use of these relationships in detail. The reader is referred to any good thermodynamics text for these. It should, however, be pointed out that when either one of the individual (partial) heats of mixing or the total (integral) heat of mixing for a solution are known as a function of composition other properties of the solution can be calculated from the following relationships:

$$d\,\bar{H}_B^{\,m} = -\frac{X_A}{X_B}\,d\,\bar{H}_A^{\,m} \qquad (5.4\text{-}12)$$

$$\Delta H^m = X_A \bar{H}_A^{\,m} + X_B \bar{H}_B^{\,m} \qquad (5.4\text{-}13)$$

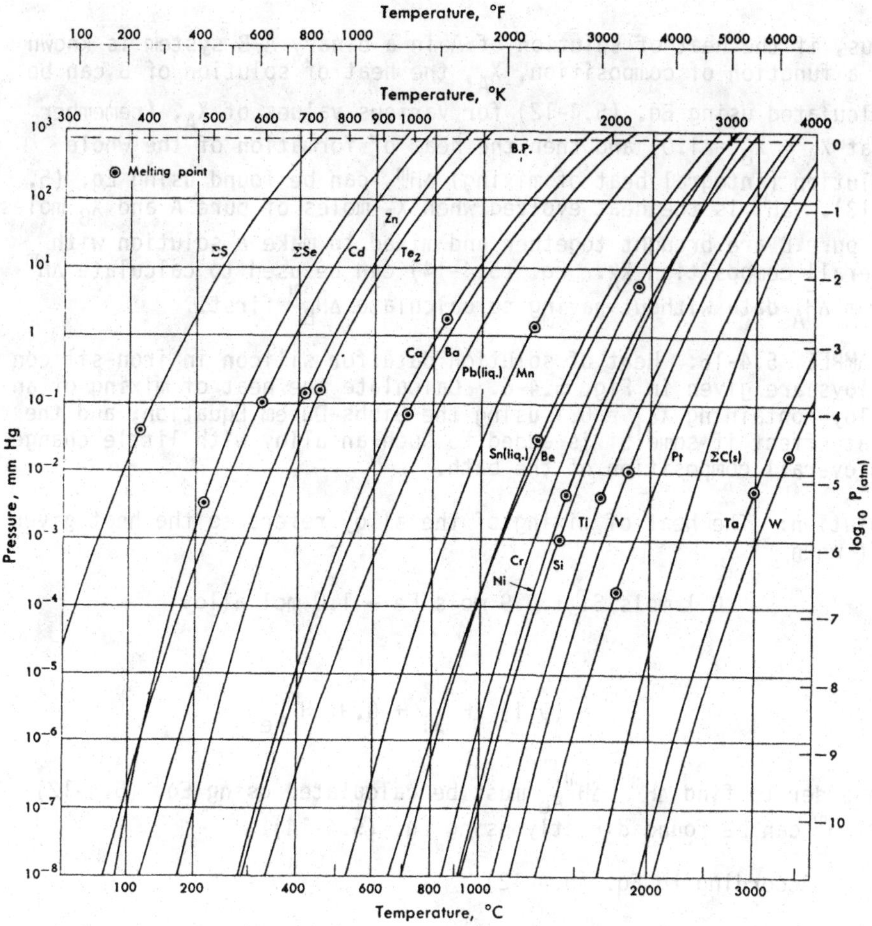

Figure 5.4-11 Vapor pressures of pure metals (From: Thermochemistry for Steelmaking, Vol. I by J. F. Elliott and M. Gleiser, Addison-Wesley Pub. Co., Reading, MA (1960) p. 270).

and

$$\Delta H^m = X_B \int_0^{X_A} \frac{\bar{H}_A^m}{X_B^2} dX_A \qquad (5.4\text{-}14)$$

Thus, if the heat of solution of A in a binary A-B system is known as a function of composition, X_A, the heat of solution of B can be calculated using Eq. (5.4-12) for various values of X_A, (remember that $X_A + X_B = 1.0$) and then the heat of formation of the whole solution (integral heat of mixing) ΔH^m, can be found using Eq. (5.4-13). ΔH^m is the heat evolved when X_A moles of pure A and X_B moles of pure B are brought together and mixed to make a solution with overall composition X_A. Eq. (5.4-14) can be used to calculate ΔH^m from $\Delta \bar{H}_A$ data without having to calculate $\Delta \bar{H}_B^m$ first.

EXAMPLE 5.4-16: Heat of solution data for silicon in iron-silicon alloys are given in Fig. 5.4-6. Calculate the heat of mixing of an alloy containing $X_{Si} = 0.1$ using the Gibbs-Duhem Equation, and the heat effect if some Si is added to such an alloy with little change in overall composition of the bath.

Solution: The heat of mixing of the alloy refers to the heat given off when

$$0.1 \text{ mols Si} + 0.9 \text{ mols Fe} \rightarrow 1.0 \text{ mol alloy}$$

or

$$\Delta H^m = (0.1)\, \Delta H^m_{Si} + 0.9\, \Delta H^m_{Fe}$$

In order to find ΔH^m, $\Delta \bar{H}^m_{Fe}$ must be calculated using Eq. (5.4-12) or ΔH^m can be found directly using Eq. (5.4-14).

According to Eq. (5.4-12),

$$d(\Delta \bar{H}^m_{Fe}) = - \frac{X_{Si}}{X_{Fe}}\, d\, (\Delta \bar{H}^m_{Si})$$

or

$$\int_{\Delta \bar{H}^m_{Fe} \,@\, X_{Fe} = 1}^{\Delta \bar{H}^m_{Fe} \,@\, X_{Fe}} d(\Delta \bar{H}^m_{Fe}) = - \int_{\Delta \bar{H}^m_{Si} \,@\, X_{Fe} = 1}^{\Delta \bar{H}^m_{Si} \,@\, X_{Fe}} \frac{X_{Si}}{X_{Fe}}\, d\, (\Delta \bar{H}^m_{Si})$$

The integration on the left yields

$$(\Delta\bar{H}^m_{Fe} \ @\ X_{Fe}) - (\Delta\bar{H}^m_{Fe} \ @\ X_{Fe} = 1) = \Delta\bar{H}^m_{Fe}$$

The integration on the right can be done graphically by plotting X_{Si}/X_{Fe} versus $\Delta\bar{H}^m_{Si}$ and determining the area under the curve, as shown below.

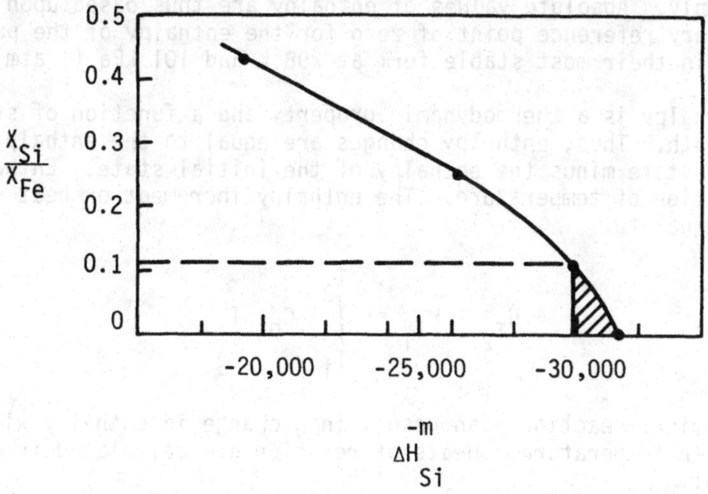

$$\Delta\bar{H}^m_{Fe} = - \text{area under the curve}$$

$$\sim -90 \text{ cal/mol}$$

and

$$\Delta H^m = 0.1\ (-29{,}900) + 0.9\ (-90)$$

$$= -\ 3070 \text{ cal/mol of solution.}$$

The value of -90 cal/mol for $\Delta\bar{H}^m_{Fe}$ is in good agreement with the published value of -99 cal/mol shown in Fig. 5.4-6. The error results from inaccuracies in graphical integration.

5.5 SUMMARY

The statement energy is conserved is the basis of the first law of thermodynamics. Application of this postulate to a general system yields the relationship

$$dE' = \delta_q - \delta w + \sum_{i=1}^{n} \delta m_i \left(u_i + P_i v_i + \frac{\bar{V}_i^2}{2g_c} + \frac{gZ_i}{g_c}\right)$$

For most metallurgical systems kinetic and potential energy changes can be ignored and pressure may be assumed to be constant. For these conditions

$$dH' = \delta q - \delta w^* + \sum_{i=1}^{m} \delta n_i H_i$$

where dH' is the enthalpy change of the system and H_i is the enthalpy per mole of material in stream i. Enthalpy is defined as $H = U + PV$.

The First Law provides a means for determination of enthalpy changes only. Absolute values of enthalpy are thus based upon an arbitrary reference point of zero for the enthalpy of the pure elements in their most stable form at 298 K and 101 kPa (1 atm).

Enthalpy is a thermodynamic property and a function of state and not path. Thus, enthalpy changes are equal to the enthalpy of the final state minus the enthalpy of the initial state. Enthalpy is a function of temperature. The enthalpy increment or heat content is equal to

$$H_{T_2} - H_{T_1} = \int_{T_1}^{T_2} C_p \, dT$$

Chemical reactions can result in a change in enthalpy without a change in temperature. Heats of reaction are calculated from Hess' Law:

$$\Delta H_R = \Sigma \Delta H_{f(\text{products})} - \Sigma \Delta H_{f(\text{reactants})}$$

or from the enthalpies of the products and reactants

$$\Delta H_R = \Sigma H_{(\text{products})} - \Sigma H_{(\text{reactants})}$$

Enthalpy changes also result from mixing processes and pressure changes.

Finally, enthalpy data are available from many thermodynamic data compilations. Enthalpy data can also be calculated from other thermodynamic data using many useful thermodynamic relationships.

FURTHER READING:

1. D. R. Gaskell, Intro. to Metallurgical Thermodynamics, McGraw-Hill, New York, 1973.

2. R. H. Parker, An Introduction to Chemical Metallurgy, Pergamon Press, London, 1967.

3. C. Bodsworth and A. S. Appleton, Problems in Applied Thermodynamics, Longmans, London, 1965.

4. O. Kubachawski, E. L. Evans, and C. B. Alcock, Metallurgical Thermochemistry, 45h Edition, Pergamon Press, London, 1967.

5. L. Darken and R. W. Gurry, Physical Chemistry of Metals, McGraw-Hill, New York, 1953.

6. O. F. Hougen, K. M. Watson and R. A. Ragatz, Chemical Process Principles, Part Two-Thermodynamics, 2nd Edition, V. Wiley, New York, 1954.

7. I. Klotz, Introduction to Chemical Thermodynamics, W. A. Benjamin, New York, 1965.

EXERCISES

5.1 Discuss some of the factors that make it difficult to arrive at an equation of state for a solid metal.

5.2 Identify the type of system, that is, whether open, closed, isolated, etc. Discuss your selection.

a. The thermite reaction as applied to weld a steel rail joint.
b. A flashlight that is switched on.
c. The last stages of the refining period in an open-hearth furnace.
d. The zone-refining process.
e. The thermite reaction carried on in a bomb calorimeter.
f. A guided missile.
g. A fuel element in an atomic reactor.
h. A thermos bottle of hot soup.
i. A thermocouple measuring the temperature of a furnace.
j. Eighty pounds of copper and twenty pounds of zinc melted in a gas-fired crucible furnace.
k. Steel heat-treated in an argon atmosphere.
l. Steel heat-treated in an endothermic gas atmosphere.
m. The reaction $CO_{(g)} + H_2O_{(g)} + CO_{2(g)}$
n. A burst balloon.
o. The spot welding process.
p. Consumable electrode process.

5.3 Show that surface tension in dynes per cm. is numerically equivalent to surface energy in ergs per square centimeter.

5.4 Make a list of intensive properties of matter.

5.5 Make a list of extensive properties of matter.

5.6 Give two examples of simple systems other than gases.

5.7 Give two examples of complex systems.

5.8 What are the solid state forms of titanium?

5.9 Given the temperature and pressure of water, list as many as possible of its other intensive properties which are thereby established.

5.10 Which of the following systems are homogeneous and which are heterogeneous.

a. A mixture of CO, CO_2, H_2O, and H_2.
b. A lead-antimony alloy of 50% lead at room temperature.
c. Liquid steel.
d. Liquid steel exposed to: 1) a gas, 2) a slag.
e. Solid steel.

5.11 Distinguish between E and U as these functions are described in this chapter.

5.12 Design a system similar to Joule's experiment in which magnetic energy is converted to thermal energy.

5.13 Distinguish between exact and inexact differentials.

5.14 Based on the definitions provided in this chapter can we define H by the equation, H = E + PV? Why or why not?

5.15 Indicate whether ΔU or ΔH or both are measured by the heat evolved or required by the following reactions when carried out in a calorimeter at constant pressure?

a. 3C (graphite) + Fe_2O_3 (s) $\rightarrow$ 3CO (g) + 2Fe (s)
b. Cu (l) $\rightarrow$ Cu (s)
c. HCl + NaOH $\rightarrow$ NsCl + H_2O all in a dilute aqueous solution.
d. Ten grams of H_2O at 90°C added to 90 grams of H_2O also at 90°C.

5.16 In vaporizing water at 1 atmosphere and 100°C the heat of vaporization per mole is about 9730 cal. The volume of one mol of vapor under these conditions is 29730 cc, while the volume of the liquid is negligible. What is the difference in internal energy per mol between the vapor and liquid?

5.17 $\Delta H = q$ when pressure is held constant. What other conditions must be met for this equality to hold true.

5.18 Show that for the isothermal expansion of an ideal gas $\delta w = RTd\ln V$.

5.19 Determine ΔH as a function of temperature for the reaction

$$C\ (graphite) + CO_2(g) = 2\ CO(g)$$

5.20 Determine ΔH as a function of temperature for the freezing of water.

5.21 Calculate ΔH_{298} for the reaction

$$Fe_2O_3(s) + 3CO(g) = 2Fe(s) + 3CO_2(g)$$

5.22 If the $\Delta H^\circ_{f,298K}$ value for $Fe_2O_3(s)$ is -197,000 cal/mol, calculate the ΔH°_f at 0K.

5.23 Calculate ΔH_f for water at 120°C when the $\Delta H^\circ_{f,298K}$ = -68,320 cal/mole.

5.24 $\Delta H^\circ_{f,298K}$ for CaO is -151,900 cal/mole and $\Delta H^\circ_{f,298K}$ for SiO_2 is -205,400 cal/mole. What is the $\Delta H_{R,298K}$ for forming $CaSiO_3$ from a) the elements and b) the compounds?

5.25 From the first law of thermodynamics, show that δq in $du = \delta q - \delta w$ is exact for a process operating at constant pressure where only mechanical work is done.

5.26 How much heat (calories) is liberated or required when ZnO is reduced by C to form Zn and CO at 1300K?

5.27 (a) How many calories are required to heat 1 mole of $NiCl_2$ to 1300K?
(b) How many calories are required to heat 1 mole of $NiCl_2$ to 1304K?
(c) How many calories are evolved or adsorbed when one lb-mole of Ni reacts with 1 lb-mole of Cl_2 to form $NiCl_2$ at 1300K?

5.28 What is the internal energy change when 1 mole of CO_2 reacts with 1 mole of C (gr) to form CO (g)

(a) in a constant volume reaction vessel and
(b) in a constant pressure situation? Assume the gases are ideal and the temperature is 298K.

5.29 How much would a piece of iron containing 0.5% C by weight heat up, or cool down, if all of the carbon wan initially present as Fe_3C and then the Fe_3C adiabatically decomposed to Fe plus C (graphite) at 800K?

5.30 Carbon deposits as soot in the presence of iron oxide at about 800K. How much heat is liberated or absorbed by the reaction

$$2CO \rightarrow CO_2 + C$$

per lb of C deposited?

5.31 (a) Calculate the heat of reaction of

$$CuCl_2 + Zn \rightarrow Cu + ZnCl_2$$

at 1200K. Assume 1 mole of $CuCl_2$ and 1 mole Zn are reacted.

(b) How much heat is required to raise the reactants in

(a) above from 298K to 1200K?

5.32 Calculate the work done when 1 mole of water is vaporized at 100°C and at 1 atm pressure. In carrying out the calculation:

(a) use the observed increase in volume of 30.19 l/mole vaporized;
(b) use the volume increase which would occur if the water vapor were a perfect gas.

Compare the results with the latent heat of vaporization (9706 cal/mole) and consider, in terms of molecular forces, why the heat is so much larger than the work.

5.33 Calculate the enthalpy increment of one mole of Cr between 298K and 1200K.

CHAPTER 6

ENERGY BALANCES

6.0 INTRODUCTION

In the previous chapter, the basis for process energy balance calculations was presented, namely, the Law of Conservation of Energy. The energy effects of various reactions, transformations, and changes of state were presented. In this chapter all of these effects are brought together as metallurgical process plants and operations are analyzed.

As in the case of material balances, the purpose of the energy balance must be considered before the problem can be properly formulated. The purpose may be to compare several complete processes which produce the same product, to see which requires less total energy from the universe. Or, it may be to compare alternative modes of operation of an existing process, or it may be to develop information needed for control of a process. Each purpose requires differing details, and will require different data. Consider for example, the basic oxygen steelmaking process. If the process is being compared with the electric furnace steelmaking process for purposes of total energy consumption comparison, the energy consumption of the process would have to include the electrical energy required to tilt the vessel, the electrical energy to hoist the oxygen lance, the energy required to produce the oxygen from air, the energy to run the waste gas recovery system, etc.

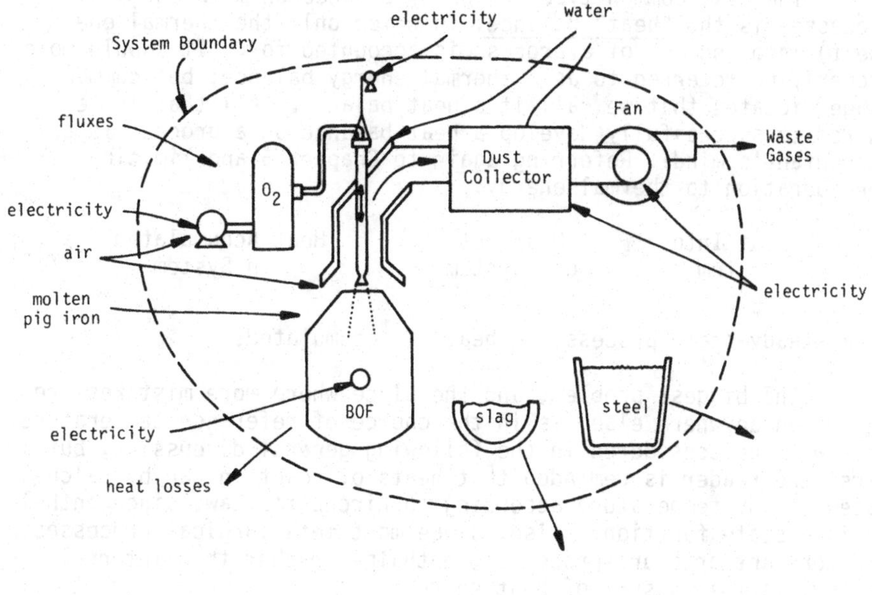

On the other hand, if an energy balance was being developed to be used to control the process, many of these inputs would not be required. In other words, we would draw the system boundary differently:

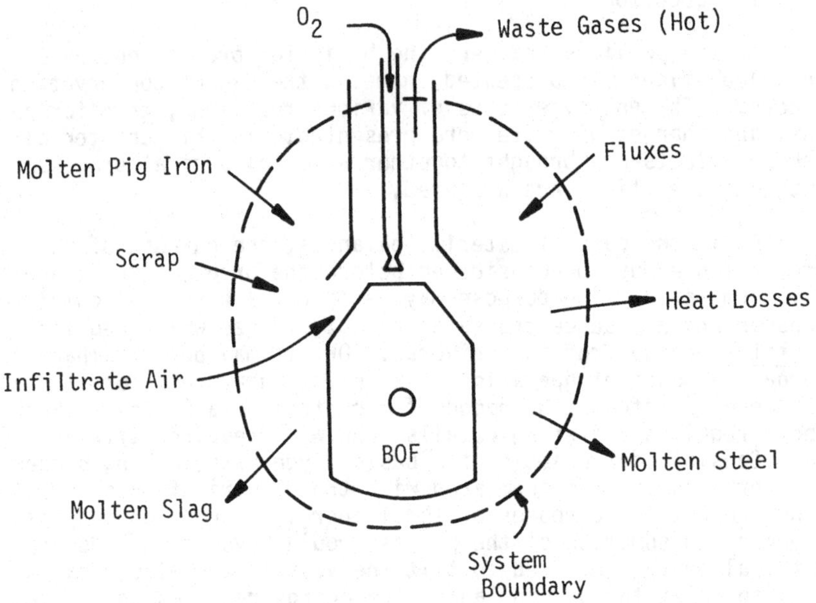

6.1 THE HEAT BALANCE

The most common type of energy balance on metallurgical processes is the "heat" balance, in which only the thermal energy (heat) into and out of a process is accounted for. It should more properly be referred to as a thermal energy balance, but common usage dictates that we call it a heat balance. All that is required to successfully develop a heat balance on a process is an accountant's mind! Referring again to Chapter 5 and limiting consideration to thermal energy:

$$\text{Heat Into System} = \text{Heat out of System} + \text{Heat Accumulated in System} \qquad (6.1\text{-}1)$$

In a steady-state process, no heat is accumulated.

The biggest problem, and the place where more mistakes are made than anywhere else, is in the choice of reference temperature. This will be considered in the following general discussion, but first the reader is reminded that heats of reaction can be calculated at any temperature according to Kirchoff's Law, since enthalpy is a state function. Also, since most metallurgical processes are constant pressure processes, enthalpy, rather than internal energy, is the measure of heat energy.

Consider a process in which solid materials, fuels and preheated air enter a process, react, and release process gases and products at elevated temperatures, and loose energy by radiation and convection to the surroundings:

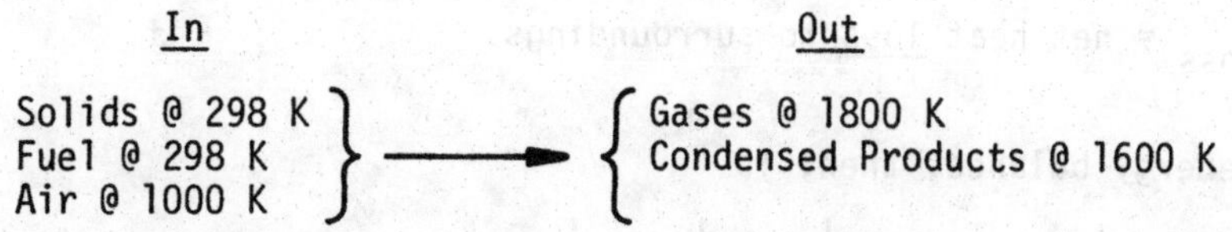
In
Out
Solids @ 298 K
Fuel @ 298 K
Air @ 1000 K
Gases @ 1800 K
Condensed Products @ 1600 K

The heat balance for such a process can be represented as follows, if the reference temperature is taken as 298K:

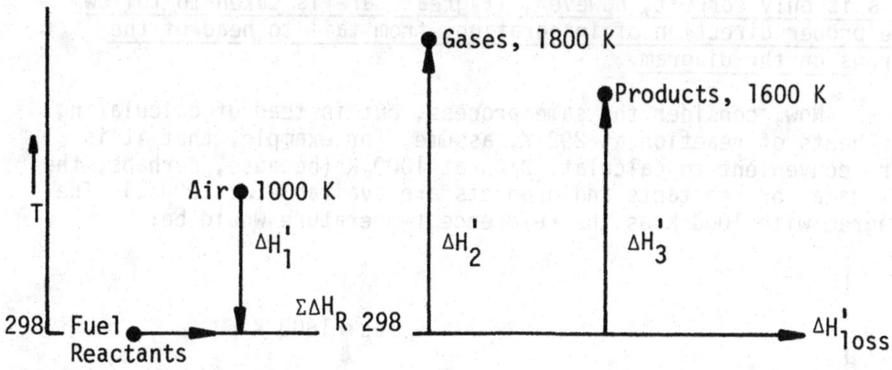

The enthalpy terms would be as follows, with the sign of the term being as indicated.

Term		Sign of Term
$\Delta H_1'$	$= \int_{1000}^{298} C_{p,air}' dT$	-
$\Delta H_2'$	$= \int_{298}^{1800} C_{p,gases}' dT$	+
$\Delta H_3'$	$= \int_{298}^{1600} C_{p,products}' dT$	+

		Sign of Term
$\Sigma\Delta H_{R,298}$ =	Sum of the Heats of Reaction at 298K to form the output products and gases.	+ if endothermic - if exothermic
$\Delta H'_{loss}$ =	net heat loss to surroundings.	+

The energy balance, then, is

$$+\Delta H'_1 + \Sigma\Delta H_R + \Delta H'_2 + \Delta H'_3 + \Delta H'_{loss} = 0 \qquad (6.1\text{-}2)$$

This is only correct, however, if great care is taken to follow the proper direction of integration, from tail to head of the arrows on the diagram.

Now, consider the same process, but instead of calculating the heats of reaction at 298 K, assume, for example, that it is more convenient to calculate $\Sigma\Delta H_R$ at 1000 K (because, perhaps, the ΔH_f data for reactants and products are available at 1000K.) The diagram with 1000 K as the reference temperature would be:

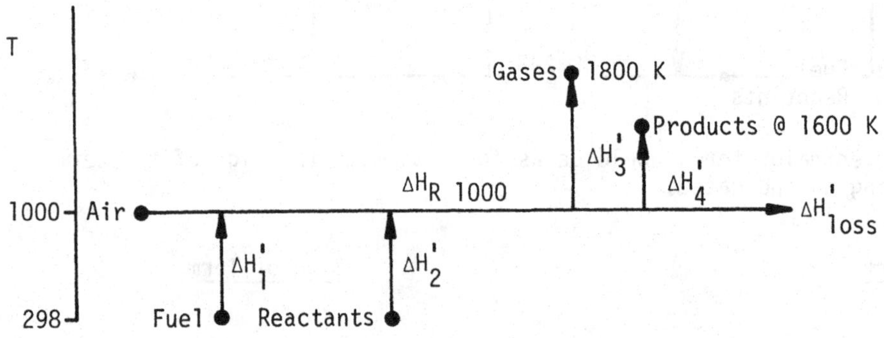

The terms in this case would be

Term	Sign of Term
$\Delta H'_1 = \int_{298}^{1000} C'_{p,Fuel}\,dT$	+
$\Delta H'_2 = \int_{298}^{1000} C'_{p,reactants}\,dT$	+
$\Delta H'_3 = \int_{1000}^{1800} C'_{p,gases}\,dT$	+

Term	Sign of Term
$\Delta H_4' = \int_{1000}^{1600} C_{p,products}' dT$	+
$\Sigma\Delta H_{R,1000} = \underset{products}{\Sigma\Delta H_{f,1000}} - \underset{reactants}{\Sigma\Delta H_{f,1000}}$	+ or -
$\Delta H_{loss}'$	+

Again, as long as the integration direction follows the arrows, the enthalpy balance is:

$$\Delta H_1' + \Delta H_2' + \Delta H_3' + \Delta H_4' + \Sigma\Delta H_{R,1000} + \Delta H_{loss}' = 0 \qquad (6.1\text{-}3)$$

This time, however, note that there is no term involving the sensible heat contained in the incoming air and that none of the terms (except possibly $\Sigma\Delta H_{R,1000}$) is a negative term.

The argument can be carried further. The result would always be the same. <u>To set up the balance properly, use the following algorithm</u>:

Step 1: Decide at which temperature heats of reaction will be calculated. This is the reference temperature.

Step 2: Draw a temperature scale.

Step 3: Place the reactants at their initial temperatures on the scale and draw an arrow from that temperature to the reference temperature.

Step 4: Repeat Step 3 for all products leaving the system.

Step 5: Calculate all sensible heat terms by integrating heat capacity data in the direction of the arrow to the reference temperature.

Step 6: Calculate the heats of reactions, solution, etc., at the reference temperature.

Step 7: Add all of the terms obtained in previous steps.

Step 8: The heat loss term is the result. A positive value means that there is a loss from the system to the surroundings.

EXAMPLE 6.1-1: Show how the thermal energy balance can be obtained from the First Law of Thermodynamics.

Solution: The most general form of the First Law is given in

Eq. (5.2-2). For a system at constant pressure, this formulation becomes

$$d(E' + PV') = \delta q - \delta w^* + \sum_{i=1}^{n} \delta m_i \left(h_i + \frac{\bar{V}_i^2}{2g_c} + \frac{g}{g_c} Z_i\right)$$

Since the thermal energy or heat balance considers changes in enthalpy only, the changes in kinetic and potential energy of the system can be assumed to be equal to zero and

$$d(E' + PV') = d(U' + PV') = dH' = \delta q + \Sigma\delta m_i\, h_i - \delta w^*$$

If no electrical work is put into the system or produced by the system,

$$dH' = \delta q + \Sigma m_i h_i$$

For a system at steady-state there is no change in the state function:

$$dH' = 0$$

or, therefore, the heat put into or lost from the system,

$$\delta q = -\Sigma\delta m_i h_i$$

or

$$\delta q = \left(\begin{matrix}\text{enthalpy of outputs}\\ \text{from system}\end{matrix}\right) - \left(\begin{matrix}\text{enthalpy of inputs}\\ \text{to the system}\end{matrix}\right)$$

On a molar basis, $\delta q = -\Sigma\delta n_i H_i$

To show that this approach is equivalent to the heat balance given in Eq. (6.1-2), consider the example with a reference temperature of 1000 K. For the process,

$$\Delta H_1' + \Delta H_2' + \Delta H_3' + \Delta H_4' + \Sigma\Delta H_{R,1000} = \Delta H_{loss}'$$

However,

$$\Delta H_1' = \int_{T_{in}=298}^{T_{Ref}=1000} C'_{p,fuel}\, dT = [H'_{fuel} @(T_{Ref}=1000) - H'_{fuel} @(T_{in}=298)]$$

$$\Delta H_2' = \int_{T_{in}=298}^{T_{Ref}=1000} C_{p,reactants}\,dT = [H'_{reactants}@(T_{Ref}=1000) - H'_{reactants}@(T_{in}=298)]$$

$$\Delta H_3' = \int_{T_{Ref}=1000}^{T_{out}=1800} C'_{p,gas}\,dT = [H'_{gas}@(T_{out}=1800) - H'_{gas}@(T_{Ref}=1000)]$$

$$\Delta H_4' = \int_{T_{Ref}=1000}^{T_{out}=1600} C'_{p,prod.}\,dT = [H'_{prod.}@(T_{out}=1600) - H'_{prod.}@(T_{Ref}=1000)]$$

and

$$\Sigma\Delta H_{R,1000} = \sum_{products} \Delta H_F @ (T_{Ref}=1000) - \sum_{reactants} \Delta H_F @ (T_{Ref}=1000)$$

$$= \sum_{products} H @ (T_{Ref}=1000) - \sum_{reactants} H (T_{Ref}=1000)$$

$$= H'_{products} @ (T_{Ref}=1000) + H'_{gas} @ (T_{Ref}=1000)$$

$$-[H'_{fuel}@(T_{Ref}=1000) + H'_{reactants}@(T_{Ref}=1000) + H'_{air}(T_{Ref}=1000)]$$

Finally, combination of these equations yields

$$\Delta H_1' + \Delta H_2' + \Delta H_3' + \Delta H_4' + \Sigma\Delta H_{R,1000}$$

$$= [H'_{fuel}@(T_{Ref}) - H'_{fuel}@(T_{in})]$$

$$+ [H'_{reactants}@(T_{Ref}) - H'_{reactants}@(T_{in})]$$

$$+ [H'_{gas}@(T_{out}) - H'_{gas}@(T_{Ref})]$$

$$+ [H'_{products}@(T_{out}) - H'_{products}@(T_{Ref})]$$

$$+ [H'_{prod}@(T_{Ref}) + H'_{gas}@(T_{Ref}) - H'_{fuel}@(T_{Ref})$$

$$- H'_{reactants}@(T_{Ref}) - H'_{air}(T_{Ref})]$$

$$= H'_{products} @ (T_{out}) + H'_{gas} @ (T_{out})$$

$$- [H'_{fuel} @ (T_{in}) + H'_{reactants} @ (T_{in}) + H'_{air} @ (T_{Ref})]$$

Since the inlet temperature of the air was the reference temperature

$$\Delta H'_1 + \Delta H'_2 + \Delta H'_3 + \Delta H'_4 + \Sigma\Delta H_{R,1000}$$

$$= [H'_{products} @ (T_{out}) + H'_{gases} @ (T_{out})]$$

$$-[H'_{fuel} @(T_{in}) + H'_{reactants} @(T_{in}) + H'_{air} @(T_{in})]$$

= (enthalpy of outputs from system)

- (enthalpy of inputs to system)

Finally, the negative of the heat loss from the system must correspond to the heat gained by the system

$$- \Delta H'_{loss} = \delta q$$

or

δq = enthalpy of products leaving system - enthalpy of inputs to system

which is equivalent to the statement of the First Law.

The previous example demonstrates that the First Law and the heat balance are equivalent. It also shows that the choice of the reference temperature is completely arbitrary since all terms based upon this value cancel out during the calculation. The heat loss term will be the same, regardless of what reference temperature is chosen, if the problem is set up properly.

EXAMPLE 6.1-2: A furnace for reheating billets, prior to their being rolled, uses fuel oil and air to produce hot gases which pass over the billets, transferring heat to both the billets and the furnace walls. Some of the heat in the walls is re-radiated to the billets and some travels through the walls and is lost to the surrounding. The gases finally exit from the furnace at a reduced temperature.

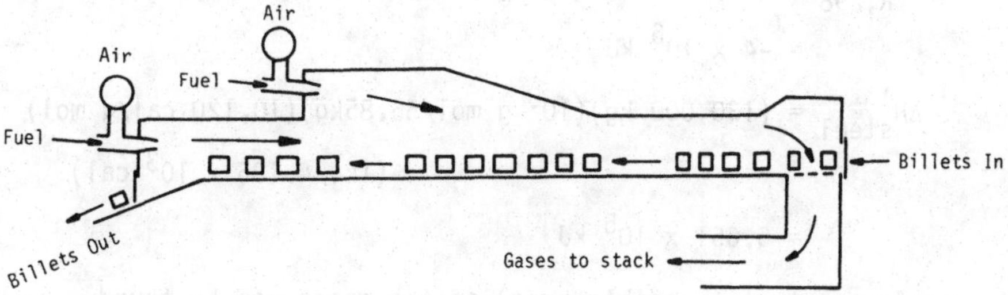

The net heating value* of the fuel oil is 40,000 kJ/kg, 25% excess air is supplied, and the waste gases analyze 4.6% O_2. The billets enter the furnace at 298 K and leave at 1400 K at a rate of 130,000 kg/hr. The waste gases leave at 1500 K. The oil is supplied at a rate of 100 kg/hr, and analyzes 85% C, 14% H and 1% S. Compute the heat losses and the thermal efficiency. For purposes of the example, use a mean heat capacity of flue gases = 1.05 kJ/kg·K. (=0.27 Btu/lb-°F).

Solution: The heat balance reference temperature is clearly 298 K, since that is the temperature at which the heat of combustion of the fuel is available.

The diagram thus becomes:

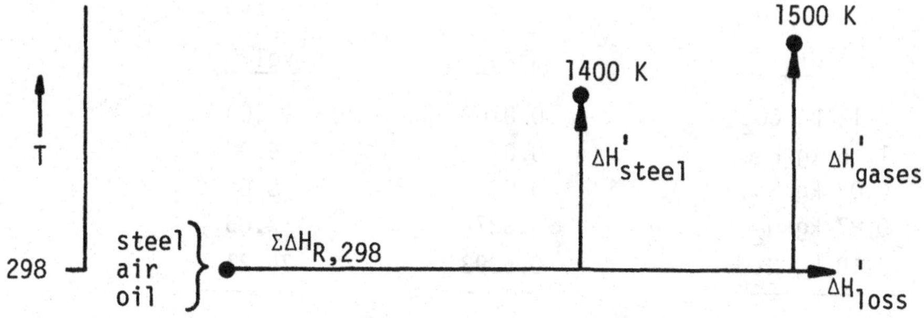

---

* The net heating value is the thermochemical energy liberated by complete combustion of the fuel to $H_2O(g)$ and CO_2 at 298 K. The gross heating value of a fuel refers to the energy liberated by complete combustion to $H_2O(\ell)$ and CO_2 at 298 K. The difference is obviously the heat of vaporization of water at 298 K and the heat of vaporization of water. For fuels made up or pure elements, or a mixture of compounds, the calorific power can be calculated from heats of combustion of the individual compounds, but for fuels such as coal or oil, which are indefinite in composition, experimentally determined values must be used.

The enthalpy terms are:

$$\Sigma\Delta H_{R,298} = \Delta H \text{ Combustion} = -(100 \text{ kg})(40{,}000 \text{ kJ/kg})$$

$$= -4 \times 10^6 \text{ kJ}$$

$$\Delta H'_{steel} = (130{,}000 \text{ kg})(10^3 \text{ g mol}/55.85\text{kg})(10{,}120 \text{ cal/g mol})(1\text{kJ}/4.186 \times 10^3 \text{ cal})$$

$$= 5.651 \times 10^5 \text{ kJ}$$

The enthalpy content (sensible heat) of the gases can be found, after combustion calculations are made in order to find the amount of infiltrate air (unmeasurable otherwise). For 1 kg oil:

	products	required O_2
(0.85 kg C)·(44/12) →	3.12 kg CO_2	2.34 kg
(0.14 kg H_2)·(18/2) →	1.26 kg H_2O	1.12
(0.01 kg S)·(64/32) →	0.02 kg SO_2	0.01
1.00 kg oil	4.40 kg	3.47 kg O_2

The required air would bring 11.62 kg N_2 with it. The 25% excess air would bring 0.87 kg O_2 and 3.48 kg N_2 with it, so that the final products of combustion would contain:

wt.	moles	vol %
3.12 kg CO_2	0.0709	10.03
1.26 kg H_2O	0.07	9.90
0.02 kg SO_2	0.0003	0.0004
0.87 kg O_2	0.0272	3.85
15.10 kg N_2	0.5393	76.21
20.37 kg gases	0.7077	100

Since the oxygen analyzer indicates 4.6% O_2, there is clearly infiltrate air present. Further calculations show that this analysis, on a volume basis, will occur if 1 additional kg of air per mole of oil enters the system. Thus, the total amount of waste gases must amount to 21.37 kg gases/kg oil, and

$$\Delta H'_{flue\ gases} = (2137 \text{ kg gas})(1.05 \text{ kJ/kg·K})(1500 \text{ K}-298 \text{ K})$$

$$= 2{,}697{,}100 \text{ kJ/hr.}$$

The balance then is:

$$-4{,}000{,}000 + 565{,}100 + 2{,}697{,}100 = -\Delta H'_{loss}$$

$$\Delta H'_{loss} = 737{,}800 \text{ kJ/hr.}$$

The thermal efficiency of the process can be measured or described in several ways. Since the requirement of the process is to heat steel from 298K to 1400K, the efficiency has to be measured against this requirement. However, the base of this efficiency can vary.

For instance, in order that the steel reach 1400K, the energy required to raise it to that temperature must be available at temperatures above 1400K. The gases cannot transfer energy to the metal unless they are hotter than the metal. In a counter-current system such as in this Example, the gases could exit from the furnace at any temperature above 298K. Therefore, the potential available energy is 4.0 x 10^6 kJ, and the efficiency is

$$\text{Efficiency} = \frac{\text{energy absorbed by product}}{\text{energy supplied}} \times 100$$

$$= \frac{565{,}100}{4{,}000{,}000} \times 100$$

$$= 14.1\%$$

On the other hand, had this been a batch type furnace, with a cold charge going into the furnace and staying there until the entire charge reaches 1400K, the energy available to be transferred to the metal would have been less than 4.0 x 10^6 kJ, because as the stock heated up, the waste gases could not transfer energy at temperatures below the stock temperature. This would cause the numerator in the efficiency equation to become smaller, increasing the apparent efficiency.

Finally, the efficiency of the furnace could be measured by the effectiveness of the insulation, in which case the efficiency would be

$$\text{Efficiency} = \frac{\text{Energy In - Heat Loss}}{\text{Energy In}} \times 100$$

$$= \frac{4.0 \times 10^6 - 0.74 \times 10^6}{4.0 \times 10^6}$$

$$= 81.5\%$$

6.1.1 SANKEY DIAGRAMS

An excellent technique for visualizing the distribution of energy in a process, or for that matter in an entire plant, is the Sankey diagram. All incoming energy sources and all outgoing energy streams are placed on the diagram; the width of each stream being proportional to the amount of energy contained in it. For example, the Sankey diagram for the furnace in Example 6.1-2 would be as follows:

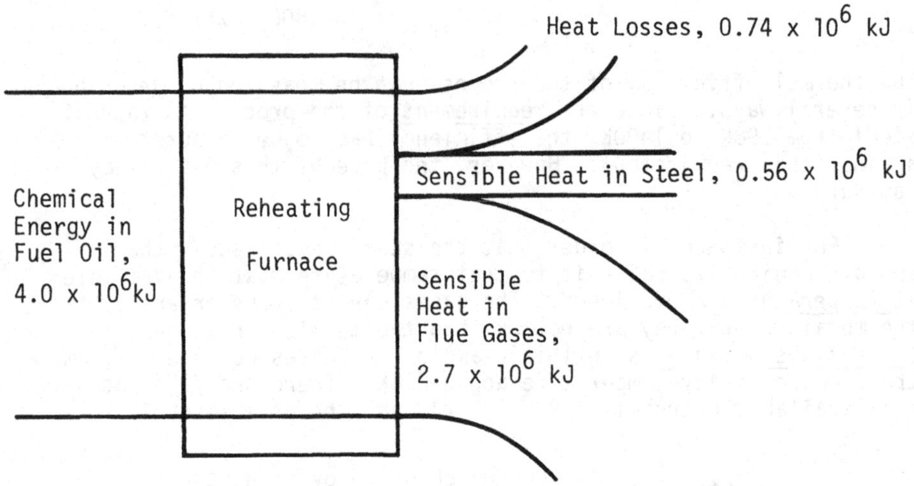

It is immediately obvious that, in this case, a large amount of the energy put into the furnace is lost from this furnace as sensible heat in the flue gases, and that the efficiency of heating of the steel is low. Thus, it is easy to visualize where efforts ought to be made to increase the energy efficiency of the operation. Redesign of the furnace, for example, to decrease heat losses or improve heat transfer so that less fuel is required would be one way, but even more obvious would be to utilize the heat in the flue gases. This might be accomplished by using the gases to produce steam in a boiler and then using the steam to produce electricity via a turbine. The use of waste-heat boilers is common in many large process plant complexes.

Another way to utilize the energy in the flue gas would be to transfer its energy to the incoming combustion air, thus recirculating some of the energy and reducing the requirement for fuel oil. This is done by means of a recuperator, in which energy is "recouped", the French word for regained. Recuperators take many forms, such as shell and tube heat exchangers, crossflow or counter-current in arrangement, and they may be continuous or intermittent in operation. As far as energy efficiency is concerned, it is only the efficiency of heat exchange that is of interest here.

EXAMPLE 6.1-3: Taking the situation in Example 6.1-2, examine what happens if the heat recuperator shown in Fig. 6.1-1 was put on the flue gas stream and it was assumed that the recuperator was 50% efficient at heat exchange.

Solution: Since the use of preheated air will decrease the requirement for chemical heat, the fuel requirements must be recalculated. Also, since less fuel is required, less combustion air will be required and fewer kg of waste gases will be available. The heat balance is somewhat different now:

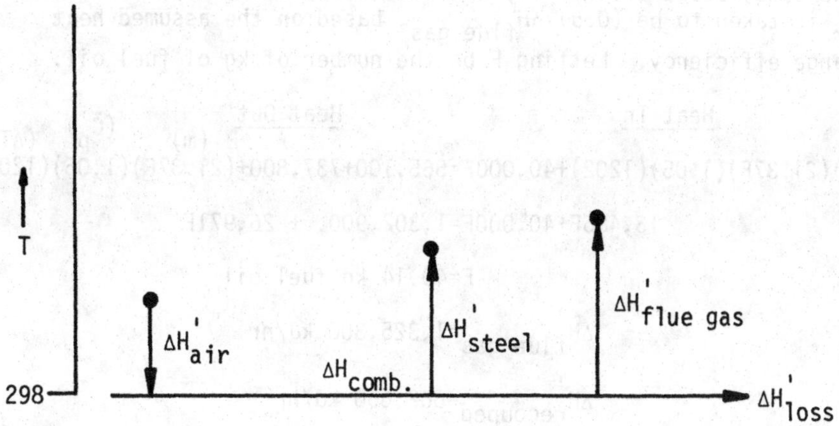

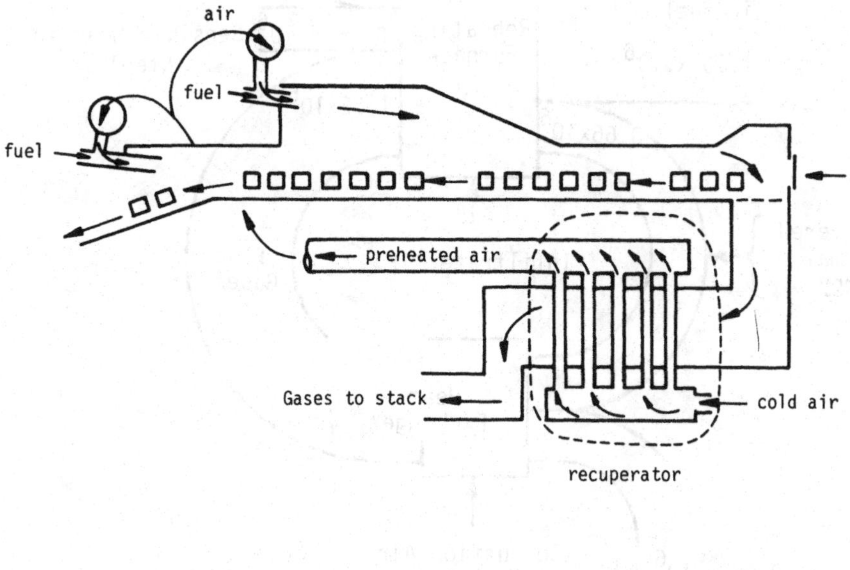

Fig. 6.1-1 Billet heating furnace with recuperator.

$\Delta H'_{steel}$ is the same, and for the moment, assume that $\Delta H'_{loss}$ is the same, because virtually the same temperatures are involved. $\Delta H'_{air}$ is taken to be (0.5) $\Delta H'_{flue\ gas}$, based on the assumed heat exchange efficiency. Letting F be the number of kg of fuel oil.

$$\underline{\text{Heat In}} \quad = \quad \underline{\text{Heat Out}}$$

$$(0.5)(21.37F)(1.05)(1202)+40{,}000F=565{,}100+737{,}800+\underset{(m)}{(21.37F)}\underset{(\bar{C}_p)}{(1.05)}\underset{(\Delta T)}{(1202)}$$

$$13{,}485F+40{,}000F=1{,}302{,}900. + 26{,}971F$$

$$F=49.14 \text{ kg fuel oil}$$

$$\Delta H'_{flue\ gas}=1{,}325{,}300 \text{ kJ/hr}$$

$$\Delta H'_{recouped}=662{,}650 \text{ kJ/hr}$$

$$\Delta H_{combustion}=1{,}965{,}500 \text{ kJ/hr.}$$

The Sankey diagram for this situation would look as follows:

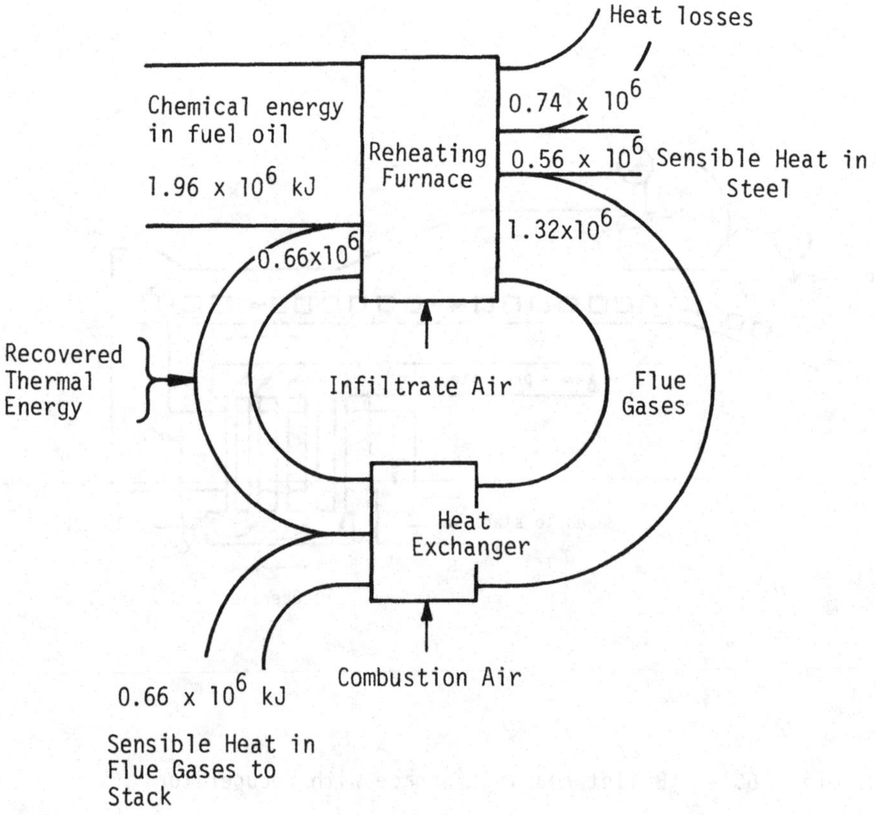

Now, total fuel consumption has been drastically decreased, from 100 kg oil/hr to 49.1 and the efficiency is up to 21%. Heat recovery of this type should be practiced as often as possible, and usually is, consistent with local economic conditions.

The Sankey diagram for processes involving chemical reactions other than combustion of fuel also helps to highlight the potential areas for fuel savings. Such a diagram, for the sinter plant illustrated in Fig. 4.3-2 is given in Fig. 6.1-2. Notice first that the heat losses, as a percentage of the energy in, are small. This is typical of most gas-solid packed bed systems, such as sinter plants and shaft furnaces. Notice also that the energy to evaporate moisture is not insignificant. Hydrated ores, such as $Fe_2O_3 \cdot H_2O$ require extra energy to process. The only potential energy savings can be made by recouping heat from the waste gases or the hot sinter itself. These amount to about 50% of the overall thermal output of the process. Some sinter plants do have sinter coolers in which the heat recovered from the hot sinter by air is transferred to the sinter on the machine via huge ducts that empty just above the sinter bed. Some fuel savings are effected in this manner.

One might be tempted to think that if these sensible heat streams could be removed as outputs, the fuel required would be decreased. They do look like "losses". However, this would be a mistake, because they result from the fact that within the process, certain temperatures must be attained, and in order to reach those temperatures, the fuel is necessary in the first place. Furthermore, none of the streams leaving the system is at a temperature anywhere near the maximum temperatures reached in the process itself, and may not be useful, except in neighboring processes or plants. This is a result of where the system boundary has been drawn. In other words, the "quality" of the energy being lost is too low for the energy to be useful. Thus, the use of Sankey diagrams from the standpoint of understanding a process can be misleading. More on the subject of energy quality is presented in Section 6.4.

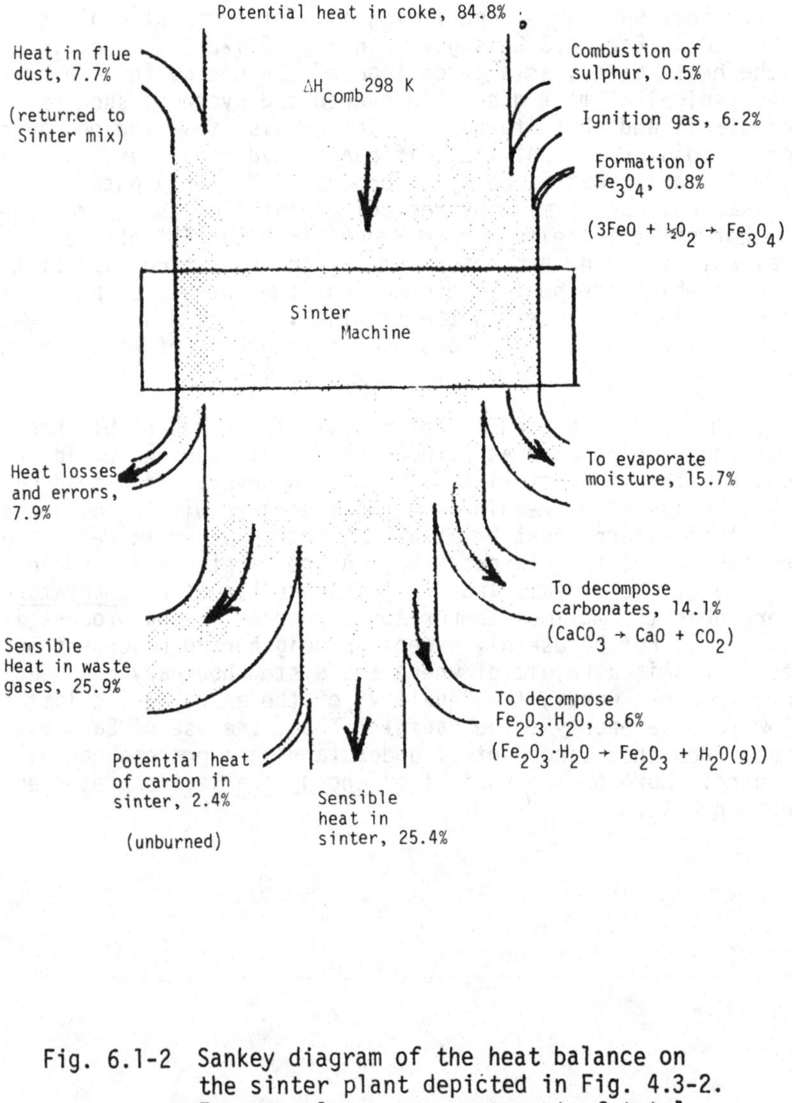

Fig. 6.1-2 Sankey diagram of the heat balance on the sinter plant depicted in Fig. 4.3-2. Energy values are as percent of total energy input and output. (Adapted from Agglomeration of Iron Ores, D. F. Ball, et al., American Elsevier, New York, 1973.)

EXAMPLE 6.1-4: In Example 4.3-7, the material balance for a "heat" of basic oxygen steel was developed. Using the results of that calculation prepare a heat balance for the same "heat". Additional pertinent data are on the figure below:

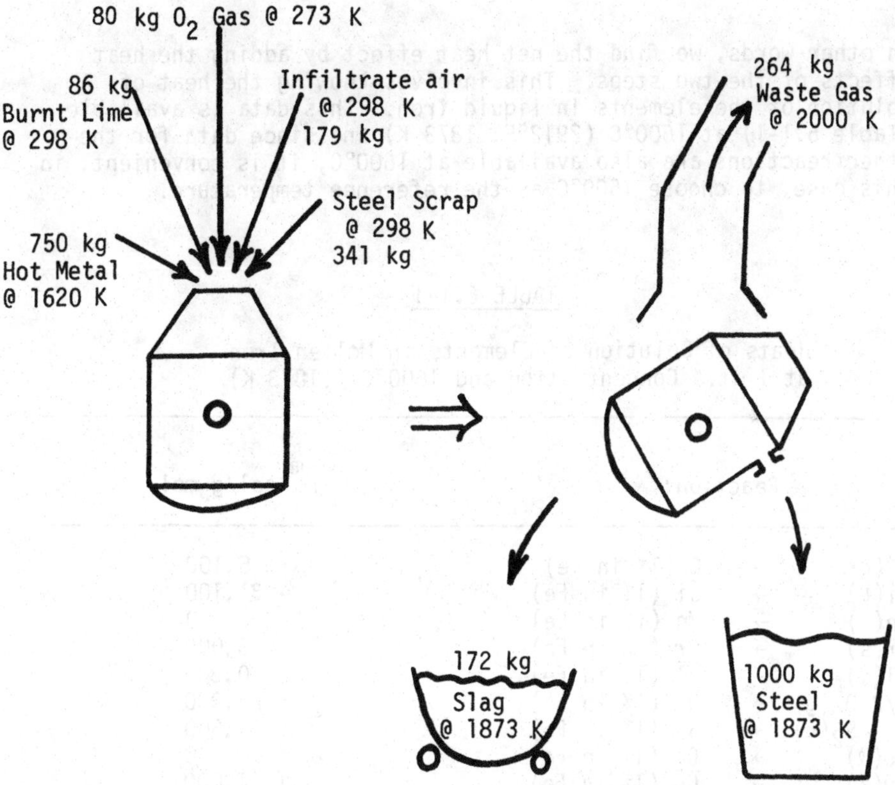

Solution: The choice of a reference temperature requires that which reactions are taking place and what the heats of those reactions are be known. In this case the reactions are

$$\underline{C} + O_2(g) \rightarrow CO_2 \qquad (a)$$

$$\underline{Si} + O_2(g) \rightarrow SiO_2(\ell) \qquad (b)$$

$$Fe(\ell) + 1/2\ O_2(g) \rightarrow FeO(\ell) \qquad (c)$$

$$1.65\ CaO(S) + SiO_2\ (\ell) \rightarrow 1.65\ CaO\cdot SiO_2(\ell) \qquad (d)$$

where the underline, eg., $\underline{Si}$, indicates that the element is dissolved in molten iron. From Chapter 5, we saw that since Eqs. (a), (b) and (c) involve two steps, eg.,

$$\underline{Si} \rightarrow Si(\ell) \qquad -\Delta\bar{H}^M_{Si} = +\ 31{,}100\ \text{cal/mol}$$

$$Si(\ell) + O_2(g) \rightarrow SiO_2(\ell) \qquad \Delta H_{f,SiO_2} = -\ 217{,}700\ \text{cal/mol}$$

the sum of the two steps gives the net reaction and its enthalpy:

$$\underline{Si} + O_2(g) \rightarrow SiO_2(\ell) \qquad \Delta H_{R,(b)} = -\ 186{,}600\ \text{cal/mol}\ SiO_2$$

In other words, we find the net heat effect by adding the heat effects of the two steps. This involves knowing the heat of solution of the elements in liquid iron. This data is available (Table 6.1-1) at 1600°C (2912°F, 1873 K) and since data for the other reactions are also available at 1600°C, it is convenient, in this case, to choose 1600°C as the reference temperature.

TABLE 6.1-1

Heats of Solution of Elements in Molten Iron at 1 wt.% Concentration and 1600°C (1873 K)

Reaction:			$\bar{H}_i^m$ cal/g mol
C (gr)	→	$\underline{C}$ (1% in Fe)	+ 5,100
$Si(\ell)$	→	$\underline{Si}$ (1% in Fe)	- 31,100
$Mn(\ell)$	→	$\underline{Mn}$ (1% in Fe)	0
Cr(s)	→	$\underline{Cr}$ (1% in Fe)	+ 5,000
$Al(\ell)$	→	$\underline{Al}$ (1% in Fe)	- 10,300
$1/2\ O_2(g)$	→	$\underline{O}$ (1% in Fe)	- 28,000
$1/2\ S_2(g)$	→	$\underline{S}$ (1% in Fe)	- 31,500
$Co(\ell)$	→	$\underline{Co}$ (1% in Fe)	0
$Cu(\ell)$	→	$\underline{Cu}$ (1% in Fe)	+ 8,000
$Ni(\ell)$	→	$\underline{Ni}$ (1% in Fe)	- 5,000

At 1600°C, the heat effects of reactions (a), (b), and (c) are:

	Reaction	$\Delta H_{R,\ 1873}$
(a)	$\underline{C} + O_2(g) \rightarrow CO_2(g)$	- 99,940 cal/mol
(b)	$\underline{Si} + O_2(g) \rightarrow SiO_2(\ell)$	-186,000 "
(c)	$Fe + 1/2\ O_2(g) \rightarrow FeO(\ell)$	- 55,780 "

In the case of reaction (d), this reaction is the closest that we can come to estimating the heat of mixing of the slag from CaO, SiO_2, and FeO. The FeO - SiO_2 liquid system is nearly ideal from a thermodynamic standpoint, so the heat of mixing of those two constituents is small and will be ignored, but it is known (however inaccurately) that there is a heat evolution when CaO(s) and $SiO_2(\ell)$ are mixed to form a liquid. The heat effect in this case will be approximated by:

(d) $1.65CaO(s) + SiO_2(\ell) \rightarrow 1.65CaO \cdot SiO_2(\ell)$, $\Delta H_R = -20{,}000$ cal/mol

Now, draw the heat balance diagram:

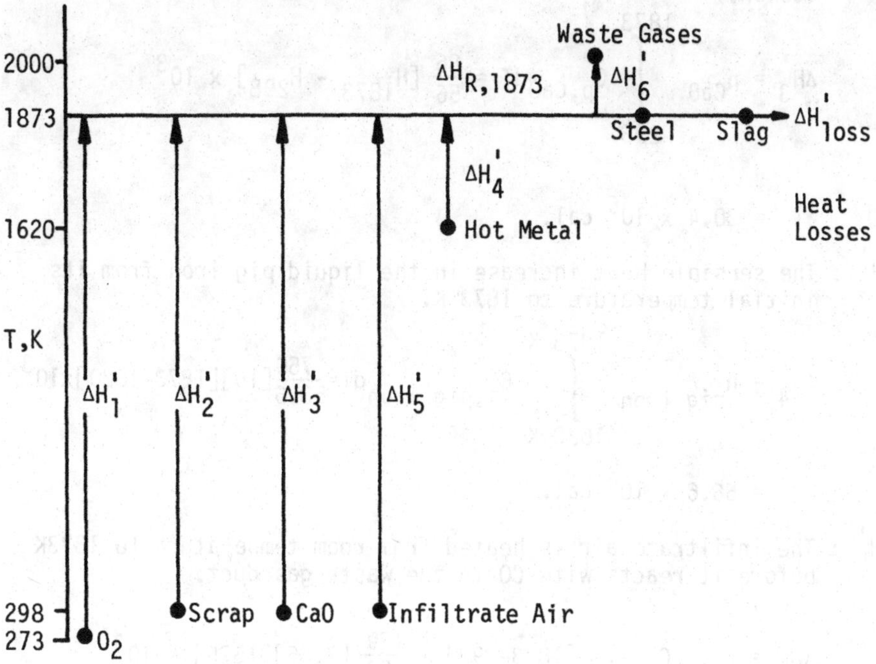

The heat balance is then:

$$\Delta H_1' + \Delta H_2' + \Delta H_3' + \Delta H_4' + \Delta H_5' + \Delta H_{R,1873} = -\Delta H_{loss}'$$

$\Delta H_1'$: This is the heat to bring the oxygen to 1873 K.

$$\Delta H_1 = n_{O_2} \int_{273}^{1873} C_{p,O_2} dT = \frac{86}{32} [H_{1873} - H_{273}]_{O_2} \times 10^3$$

$$= 35.0 \times 10^6 \text{ cal.}$$

$\Delta H_2'$: This is the heat needed to heat and melt scrap steel and bring it to 1873 K.

$$\Delta H_2' = n_{scrap} [H_{1873} - H_{298}]_{Fe} = \frac{341}{55.85} [18550] \times 10^3$$

$= 113.1 \times 10^6$ cal.

H'_3 : This is the heat required to bring the lime to 1873 K. Since this lime is going to react with $SiO_2(\ell)$ to form the slag, and since the heat of formation, equation (d), is written in terms of CaO(s), there is no need to add the heat of fusion.*

$$\Delta H'_3 = n_{CaO} \int_{298}^{1873} C_{p,CaO} dT = \frac{86}{56} [H_{1873} - H_{298}] \times 10^3$$

$= 30.4 \times 10^6$ cal.

$\Delta H'_4$: The sensible heat increase in the liquid pig iron from its initial temperature to 1873 K.

$$\Delta H'_4 = n_{\text{pig iron}} \int_{1620\text{ K}}^{1873\text{ K}} C_{p,\text{pig iron}} dT = \frac{750}{53.6} [16][1873-1620] \times 10^3$$

$= 56.6 \times 10^6$ cal.

$\Delta H'_5$: The infiltrate air is heated from room temperature to 1873K before it reacts with CO in the waste gas duct.

$$\Delta H_5 = n_{air} \bar{C}_{p,air} [1873-298] = \frac{179}{29} [7.85][1575] \times 10^3$$

$= 76. \times 10^6$ cal.

$\Sigma\Delta H_{R,1873}$: Now that all of the reactants are at the reference temperature, compute the amounts of reactants, the heats of reaction and add:

Based on the waste gas analysis,

$$\frac{(0.4499 \text{ kg } CO_2)(264 \text{ kg gas})(10^3 \text{ mol } CO_2)}{(1.0 \text{ kg gas}) \qquad (44 \text{ kg } CO_2)} = 2699 \text{ mol } CO_2 \text{ are formed.}$$

Based on the slag analysis,

$$\frac{(172 \text{ kg slag})(0.35 \text{ kg FeO})(10^3 \text{ mol FeO})}{(1.00 \text{ kg slag})(72 \text{ kg FeO})} = 836 \text{ mol FeO,}$$

and

* See footnote on page 210.

$$\frac{(172 \text{ kg slag})(0.149 \text{ kg SiO}_2)(10^3 \text{ mol SiO}_2)}{(1.0 \text{ kg slag})(60 \text{ kg SiO}_2)} = 427 \text{ mol SiO}_2$$

are formed.

Heats of Reaction at 1873 K:

$$\text{(a) } (2699 \text{ mol C}) \frac{-99{,}940 \text{ cal}}{\text{mol CO}_2 \text{ formed}} = -269.7 \times 10^6 \text{ cal}$$

$$\text{(b) } (427 \text{ mol SiO}_2) \frac{-186{,}000 \text{ cal}}{\text{mol SiO}_2} = -79.4 \times 10^6 \text{ cal}$$

$$\text{(c) } (836 \text{ mol FeO}) \frac{-55{,}780 \text{ cal}}{\text{mol FeO}} = -46.6 \times 10^6 \text{ cal}$$

$$\text{(d) } (427 \text{ mol SiO}_2) \frac{-20{,}000 \text{ cal}}{\text{mol 1.65 CaO SiO}_2} = -8.54 \times 10^6 \text{ cal}$$

$$\Sigma\Delta H_{R,\,1873} = -404.2 \times 10^6 \text{ cal}$$

$\Delta H_6'$: This represents the enthalpy contained in the waste gases above 1873 K.

$$\Delta H_6' = n_{O_2}[H_{2000}-H_{1873}]_{O_2} + n_{CO_2}[H_{2000}-H_{1873}]_{CO_2} + n_{N_2}[H_{2000}-H_{1873}]_{N_2}$$

$$= 10.7 \times 10^6 \text{ cal.}$$

Finally,

$$-\Delta H'_{loss} \times 10^{-6} = 35.0 + 113.1 + 30.4 + 56.6 + 76.3 + 10.7 - 404.2$$

$$= +322.1 - 404.2.$$

$$= -82.1$$

or

$$\Delta H'_{loss} = 82.1 \times 10^6 \text{ cal or } 20.3\% \text{ of heat input.}$$

This amount of heat is lost via convection and/or radiation to the surroundings, including the water-cooled hood, up to the point where the 2000K gas temperature was measured, or is absorbed by the refractory lining of the furnace.

A Sankey diagram for the BOF process with a reference temperature of 1873K would look as follows:

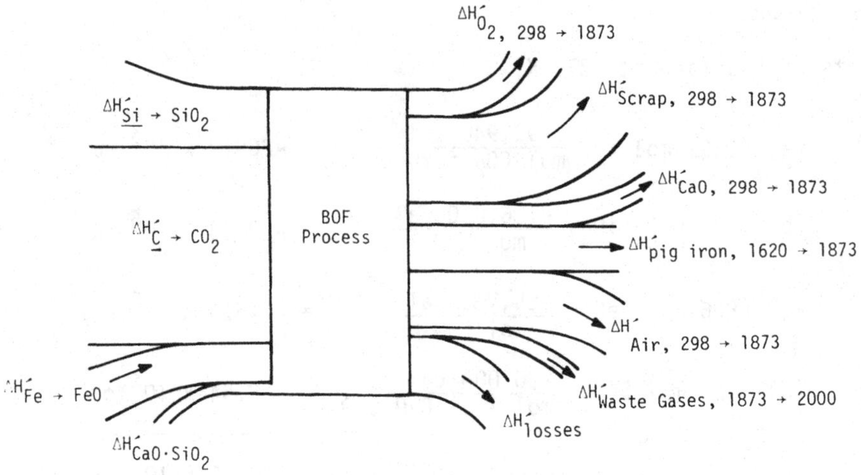

On the other hand, we could have chosen 298 K as the reference temperature, in which case the heat balance diagram would have been:

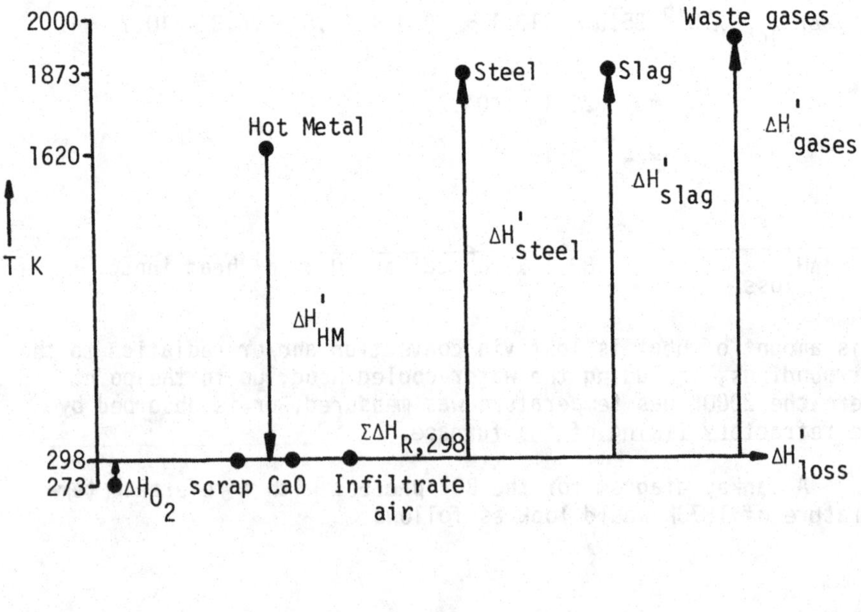

and the Sankey diagram would be:

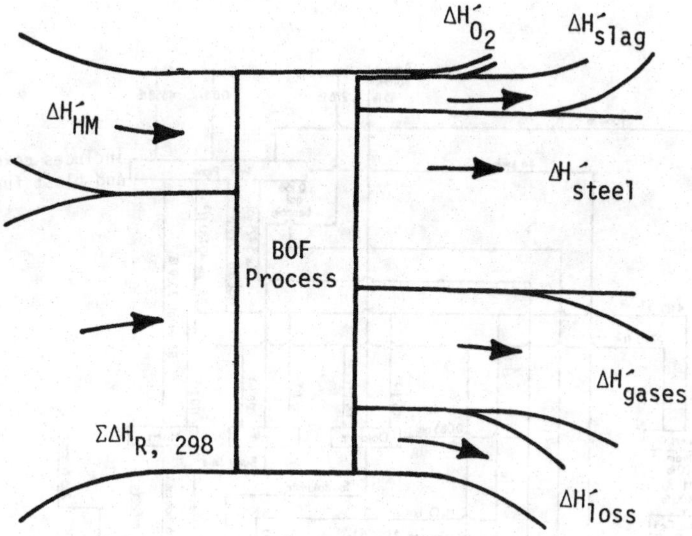

Figures 6.1-3 and 6.1-4 illustrate the material flows and a Sankey diagram of the energy balance for an entire integrated steel plant, incorporating the preceding diagram for the BOF process.

Finally, another way to graphically present a heat balance is by means of a bar graph. This is more useful for a batch process than for a continuous one. For the process in Example 6.1-4, such a diagram would be as follows, for the reference temperature of 1873 C:

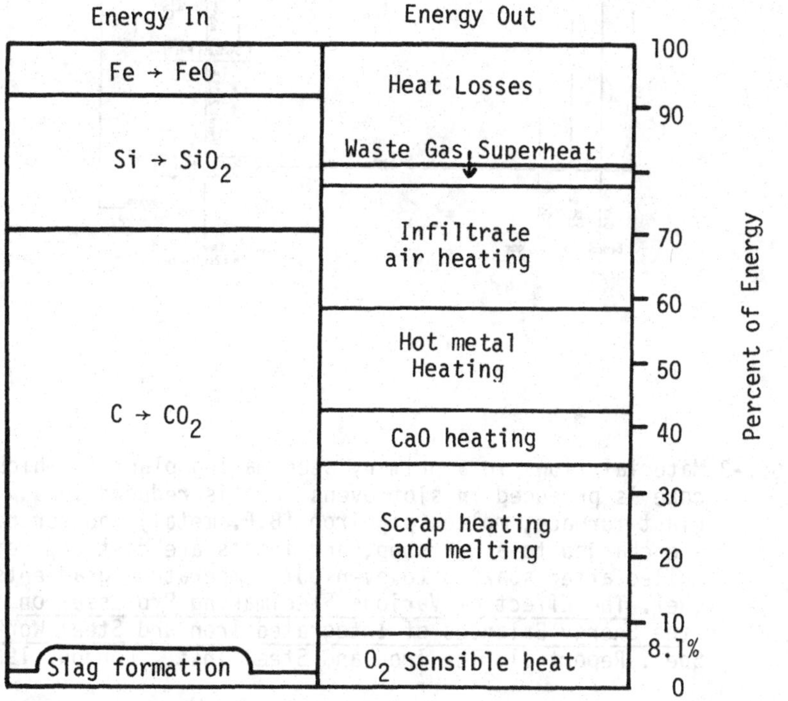

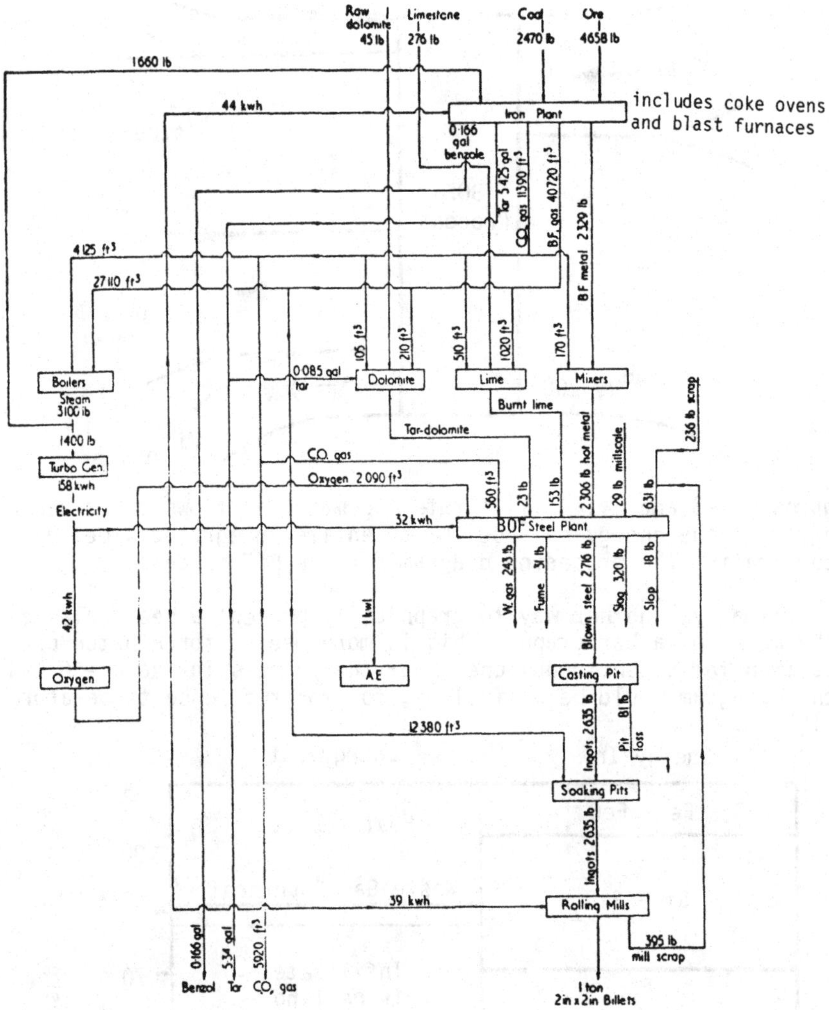

Fig. 6.1-3 Material flows in a primary steelmaking plant in which coke is produced in slot ovens, ore is reduced in a blast furnace, molten pig iron (B.F. metal) and scrap are charged to a BOF shop, and ingots are cast and hot-rolled after soaking to even out temperature gradients. (Ref. The Effect of Various Steelmaking Processes on the Energy Balances of Integrated Iron and Steel Works, Spec. Report 71, The Iron and Steel Inst., London, 1962)

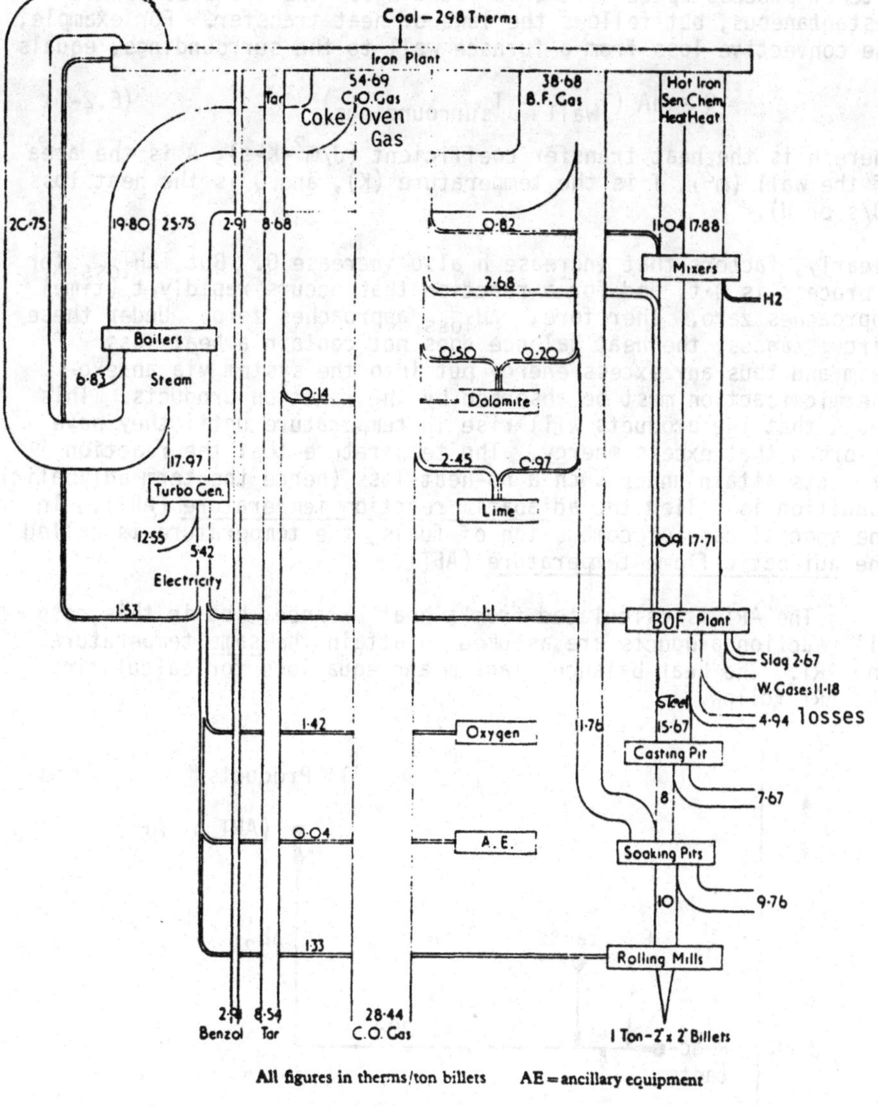

Fig. 6.1-4 Sankey diagram for plant shown in Fig. 6.1-3, with a reference temperature of 298 K joules (ref. Spec. Report 71, op cit.) 1 Therm - 100,000 BTU = 1.055 x 10^8 joules.

6.2 ADIABATIC REACTION TEMPERATURES

In the previous section a heat balance was set up for a process which included a term for heat lost. Heat loss is time dependent, i.e., heat is lost by being transferred from the reaction site or process space to the surroundings. The transfer is not instantaneous, but follows the laws of heat transfer. For example, the convective loss from a furnace wall to the surroundings, equals

$$Q = hA\,(T_{wall} - T_{surroundings}) \tag{6.2-1}$$

where h is the heat transfer coefficient (J/m^2-K-s), A is the area of the wall (m^2), T is the temperature (K), and Q is the heat loss (J/s or W).

Clearly, factors that increase h also increase Q. But, ΔH_{loss} for a process is Q·t, and for a reaction that occurs rapidly, t (time) approaches zero. Therefore, ΔH_{loss} approaches zero. Under these circumstances, the heat balance does not contain a heat loss term and thus any excess energy put into the system via an exothermic reaction must be absorbed by the reaction products. This means that the products will rise in temperature until they have absorbed that excess energy. The temperature that the reaction products attain under such a no-heat-loss (hence the term adiabatic) condition is called the <u>adiabatic reaction temperature</u> (ART). In the special case of combustion of fuels, the temperature is called the <u>adiabatic flame temperature</u> (AFT).

The ART is calculated from a heat balance, but in this case <u>all</u> reaction products are assumed to attain the <u>same</u> temperature, the ART. The heat balance diagram and equations for calculating the ART follow.

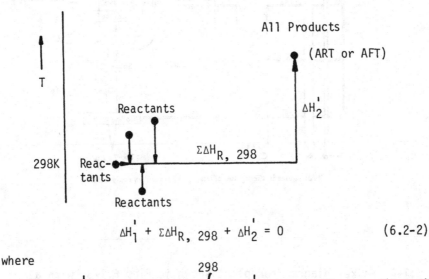

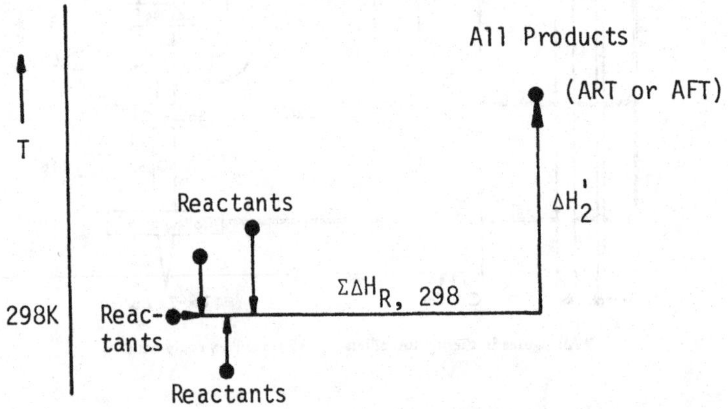

$$\Delta H_1' + \Sigma\Delta H_{R,\,298} + \Delta H_2' = 0 \tag{6.2-2}$$

where

$$\Delta H_1' = \sum_{\substack{\text{all}\\\text{reactants}}} \int_{\text{initial T}}^{298} C_{p,\text{reactant}}'\,dT \tag{6.2-3}$$

and

$$\Delta H_2' = \sum_{\text{all products}} \int_{298}^{ART} C'_{p,\text{product}}\, dT \qquad (6.2\text{-}4)$$

and

$$C_p' = nC_p$$

The ART or AFT are important in several instances. In the thermite welding os steel sections, the ART of a mixture of steel shot, Fe_2O_3 and aluminum powder governs the extent of heating and the size and location of the heat-affected zone in the neighboring steel. The reaction, $2Al + Fe_2O_3 \rightarrow 2Fe + Al_2O_3$, once started, occurs so rapidly that it is justifiable to calculate the ART.

In the production of many ferro-alloys, the reduction of metal oxides to metal is carried out in batches by reaction with a reactive metal such as silicon or calcium. The ART of such a reaction governs the safety of the process, as well as the purity, since the final temperature achieved after the very rapid reaction may exceed the maximum safe operating temperature of the containing vessel.

In the operation of shaft furnaces with air or air-fuel injection, the reaction temperature, or AFT, at the injection tuyeres may govern the behavior of the overall process. The effect of changes in the composition of the injected air/fuel mixtures must be calculated in order to control the AFT.

EXAMPLE 6.3-1: What is the AFT for the combustion of a gas containing 96% CH_4, 0.8% CO_2, 3.2% N_2, when a) theoretical air is used, and b) when air enriched to 30% oxygen is used? Assume that both the gas and air enter the combustion reaction chamber at 298 K.

Solution: As a basis for the calculation, use 100 mols of gas. Since the analysis of a gas is on a volume basis, this means that it is also on a mol basis. Therefore, the gas contains 96 mols of CH_4, 0.8 mols CO_2, and 3.2 mols N_2.

The theoretical air required would have to provide enough oxygen to burn all of the CH_4 to CO_2 and H_2O.

$$CH_4(g) + 2\,O_2(g) \rightarrow CO_2(g) + 2H_2O(g)$$

$$\frac{96 \text{ mols } CH_4}{100 \text{ mols gas}} \left| \frac{2 \text{ mols } O_2}{\text{mol } CH_4} \right. = \frac{192 \text{ mols } O_2 \text{ required}}{100 \text{ mols gas}}$$

Since air contains 79 mols N_2 per 21 moles of O_2, the nitrogen that will enter the system along with the 192 mols O_2 is

$$\left(\frac{192 \text{ mols } O_2}{100 \text{ mols gas}}\right)\left(\frac{79 \text{ mols } N_2}{21 \text{ mols } O_2}\right) = 722 \text{ mols } N_2$$

Therefore, the products of the reaction will include 722 mols N_2 from the air plus 3.2 mols N_2 from the gas.

Since the gas and air both enter at 298 K, 298 K becomes a convenient reference temperature. Neither gas nor air brings or requires any sensible heat.

The reaction

$$CH_4 \text{ (g)} + 2\ O_2 \text{ (g)} \rightarrow CO_2 \text{ (g)} + 2\ H_2O \text{ (g)}$$

$$96 \text{ mols} + 192 \text{ mols} \rightarrow 96 \text{ mols} + 192 \text{ mols}$$

liberates heat. $\Sigma\Delta H'_{R,\ 298}$ is found from

$$\Sigma\Delta H'_{R,\ 298} = 96\Delta H^\circ_{f,CO_2} + 192\ \Delta H^\circ_{f,H_2O(g)} - [96\Delta H^\circ_{f,CH_4} + 192\Delta H^\circ_{f,O_2}]$$

$$= 96\ (-94.05) + 192\ (-57.80) - 96\ (-17.89) - 192\ (0)$$

$$= -18{,}408.\ \text{kcal}$$

This heat must be absorbed by the reaction products as sensible heat. The heat capacity of the reaction products in the system is:

$$C'_p = n_{CO_2} \cdot C_{p,CO_2} + n_{H_2O} \cdot C_{p,H_2O} + n_{N_2} \cdot C_{p,N_2}$$

$$= 96.8(10.57 + 2.10 \times 10^{-3}T - 2.06 \times 10^5 T^{-2}) + 192(7.30 + 2.46 \times 10^{-3}T)$$

$$+ 725.2\ (6.83 + 0.90 \times 10^{-3}T - 0.12 \times 10^5 T^{-2})$$

$$= 7377.9 + 1.301T - 286.4 \times 10^5 T^{-2}$$

Now, Eq. (6.2-2) becomes

$$\Sigma\Delta H'_{R,298} = \int_{298}^{T_{AFT}} (7378 + 1.30T - 2.86 \times 10^7 T^{-2})dT$$

$$= (7378T + 0.650T^2 + 2.86 \times 10^7 T^{-1})\Big|_{298}^{T_{AFT}}$$

or,

$$18{,}408{,}900 = 7378T_{AFT} + 0.650T_{AFT}^2 + 2.86 \times 10^7\ T_{AFT}^{-1} - 2.352 \times 10^6$$

Since this is a non-linear equation, the value of the AFT can be determined by either successive approximations or graphically. In the graphical procedure, a graph of enthalpy (calories) vs. temperature is made, $\Delta H_R'$ is entered as a horizontal line, and the right side of the above equation is plotted as a function of temperature. The two lines intersect at an AFT of 2330 K, as shown in Fig. 6.2-1.

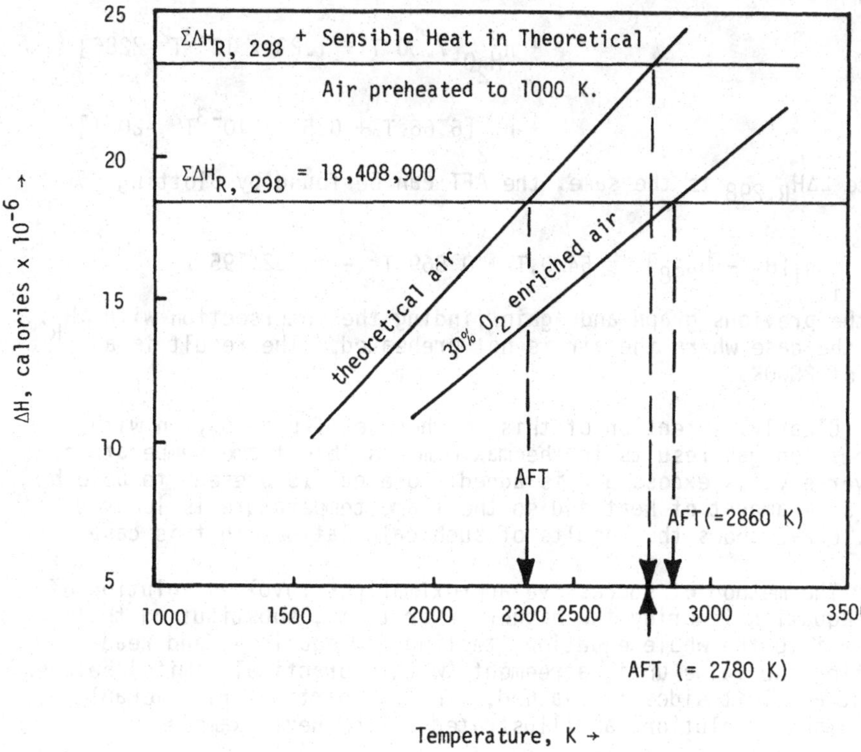

Fig. 6.2-1 Graphical determination of ART.

If the air had been preheated, the sensible heat brought into the system above the reference temperature would be added to the heat of reaction, to be absorbed by the reaction products, resulting in a higher AFT of 2780K, as shown in Fig. 6.2-1.

Similarly, if the air has been enriched with gaseous oxygen, resulting in less N_2 carried through the system to absorb reaction heat, the AFT would be higher, because fewer moles of nitrogen will be in the products. In the present example, enrichment of the air to 30% O_2 would mean that only

$$\left(\frac{192 \text{ moles } O_2}{100 \text{ mols gas}}\right)\left(\frac{70 \text{ moles } N_2}{30 \text{ moles } O_2}\right) = \frac{448 \text{ mols } N_2}{100 \text{ mols gas}}$$

would enter in the air. This time, instead of integrating heat capacity data, use the already integrated functions, $[H_T - H_{298}]$:

$$\int_{298}^{T_{AFT}} n_i C_{p_i} dT = \sum_i n_i [H_T - H_{298}] = n_{CO_2}[-3926 + 10.55\ T + 1.08 \times 10^{-3} T^2]$$

$$+ n_{H_2O}[7.30\ T + 1.23 \times 10^{-3} T^2 - 2286]$$

$$+ n_{N_2}[6.66\ T + 0.51 \times 10^{-3} T^2 - 2031]$$

Since $\Sigma\Delta H_{R,298}$ is the same, the AFT can be found by plotting

$$\sum_i n_i [H_T - H_{298}]_i = 5420.T + 0.569\ T^2 - 1{,}732{,}195$$

on the previous graph and again finding the intersection with $\Delta H_R'$. for the case where the air is not preheated. The result is an AFT of 2860K.

Clearly, extension of this to the use of pure oxygen with combustion gas results in the maximum possible flame temperature. Conversely, as excess air is added, more gas is present to absorb the same amount of heat and so the flame temperature is lowered. Fig. 6.2-2 shows the results of such calculations in this case.

The method of successive approximations involves solution of the equation ignoring the higher order terms, substituting that value into the whole equation, testing for equality, and readjusting the value until agreement (within practical limits) between right and left sides is reached. This is particularly amenable to computer solution, as illustrated in the next example.

EXAMPLE 6.2-2: Some processes operate only if the temperature at certain locations in the furnace is maintained at a particular temperature. One such process is iron ore reduction in the iron blast furnace, whose typical temperature profile is shown in Fig. 6.2-3. The maximum possible temperature is reached at the tuyere, where a blast of air at some temperature is forced into the "bosh" zone of the furnace. This blast may be augmented by the addition or injection of O_2, CH_4, fuel oil, powdered coal or other supplementary fuel. The blast will contain some moisture either from atmospheric humidity or due to steam injection. The

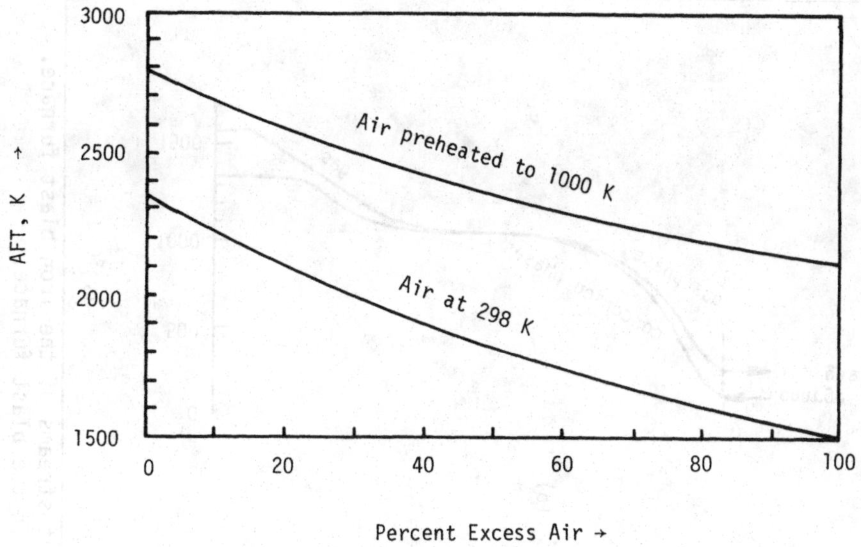

Fig. 6.2-2 Adiabatic flame temperatures for combustion of natural gas with air.

blast reacts with coke which continually enters the tuyere zone from above at a temperature approximating the gas and solid temperatures in the zone through which the gaseous reaction products move. The temperature of the flame produced by reaction at elevated temperatures between the coke, the supplementary fuels, and the oxygen, controls the rate of heating and melting of the product of reduction in the shaft of the furnace, and thus indirectly controls the overall production rate. Of course the AFT is an ideal temperature; it is the highest temperature possible under the stated conditions, but practical observations have shown its usefulness. Develop a computer program to calculate the AFT for various blast conditions.

Solution:

Reactions:

$$C + 1/2\ O_2 = CO \qquad (a)$$

$$H_2O(g) + C = CO + H_2 \qquad (b)$$

$$CH_4(g) + 1/2\ O_2 = CO + 2H_2 \qquad (c)$$

(An excess of C prevents any CO_2 from forming.)

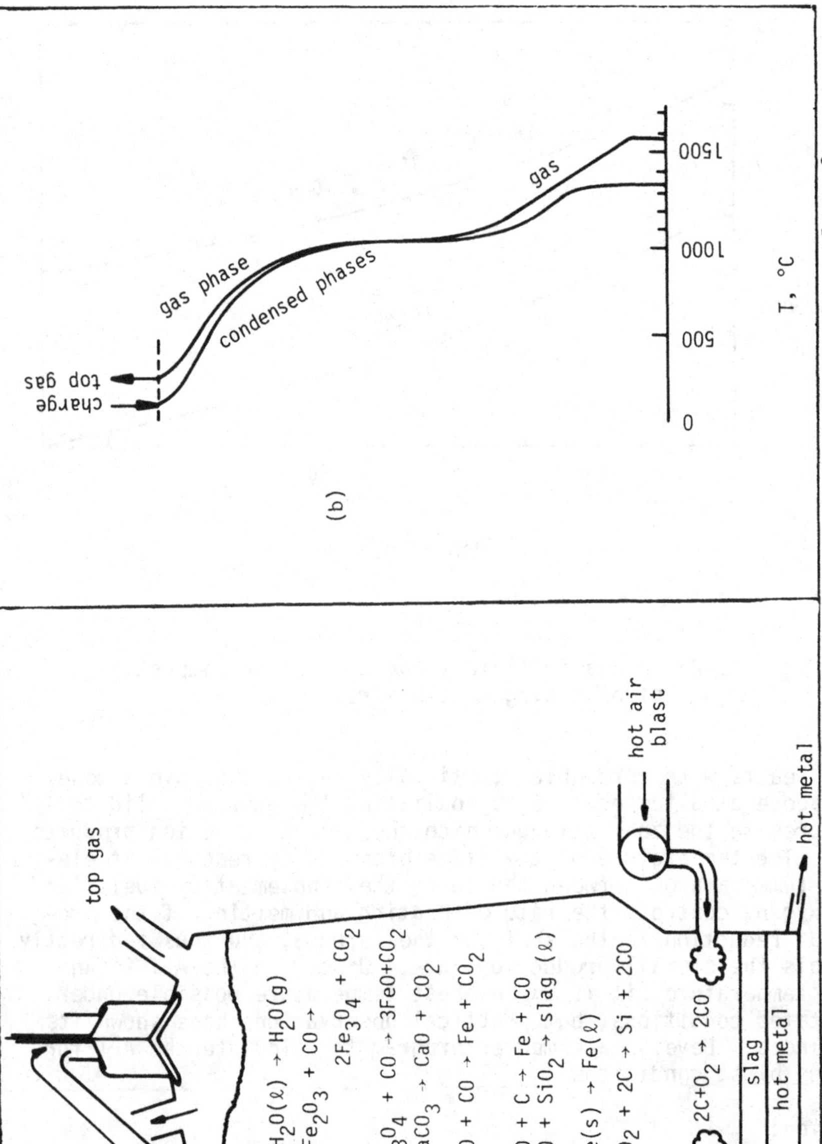

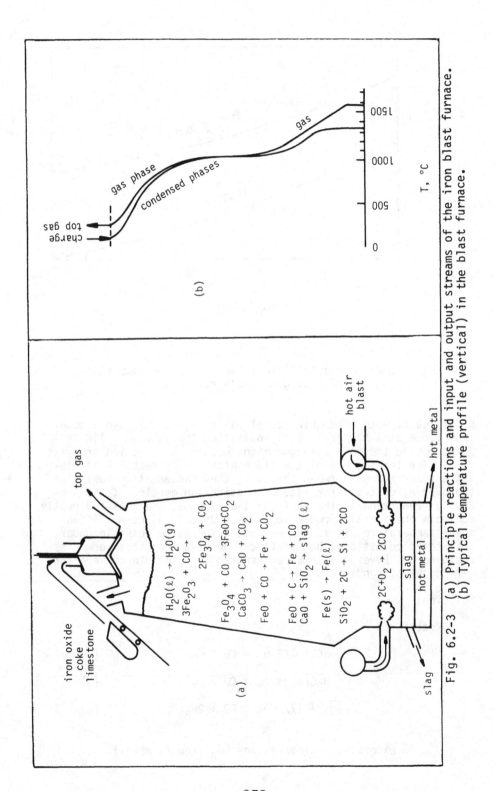

Fig. 6.2-3 (a) Principle reactions and input and output streams of the iron blast furnace.
(b) Typical temperature profile (vertical) in the blast furnace.

Known or assumed conditions. (May be read in as input to a computer program.)

$$\dot{n}_{O_2}, \dot{n}_{N_2}, \dot{n}_{H_2O}, \text{ at } T_{blast}$$

$$\dot{n}_{gas}, \dot{n}_{oil}, \text{ or } \dot{n}_{coal} \text{ at 298 K.}$$

$$\text{Coke at } T_{coke}$$

Reference Temperature:

The simplest reference temperature is 298 K since ΔH_R for reactions (a), (b) and (c) are easily found. The diagram for the adiabatic flame temperature calculation would then be:

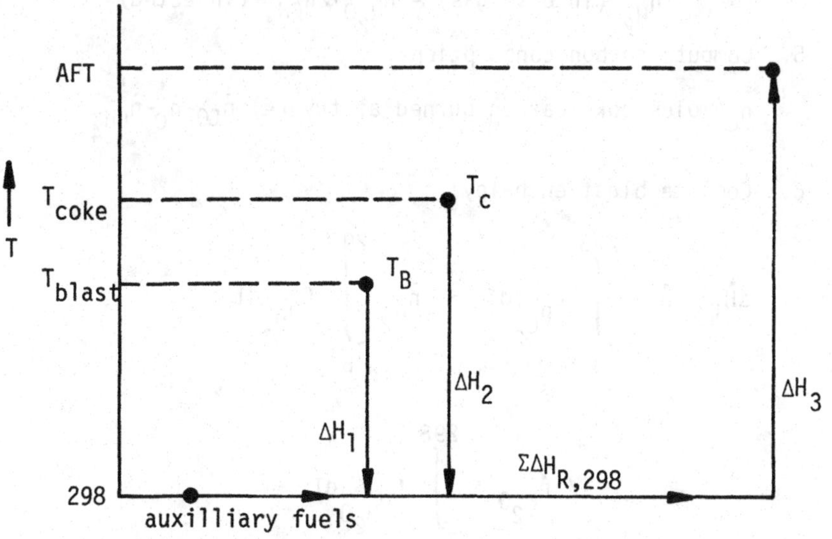

or for any instant of time,

$$\Delta\dot{H}_1 + \Delta\dot{H}_2 + \Sigma\Delta\dot{H}_{R,298} + \Delta\dot{H}_3 = 0$$

The procedure used to determine the AFT is outlined below:

1. Read in moles O_2, N_2, H_2O in the blast per minute, $\dot{n}_{O_2}$, $\dot{n}_{N_2}$, $\dot{n}_{H_2O}$.

2. Read in blast temperature, T_B, and assumed coke

temperature, T_C.

3. Read in moles O_2, CH_4, C injected at room temperature per minute, $\dot{n}'_{O_2}$, $\dot{n}'_{CH_4}$, $\dot{n}'_C$.

4. Compute bosh gas (reaction product) composition (assuming excess C)
O_2 balance:

$$\dot{n}_{CO} \text{ (in bosh gas)} = 2\dot{n}_{O_2} + \dot{n}_{H_2O} + 2\dot{n}'_{O_2}$$

N_2 balance:

$$\dot{n}_{N_2} \text{ (in bosh gas)} = \dot{n}_{N_2}$$

H_2 balance:

$$\dot{n}_{H_2} \text{ (in bosh gas)} = \dot{n}_{H_2O} + 2\dot{n}_{CH_4} \text{ (injected)}$$

5. Compute carbon consumption

$$\dot{n}_C \text{(moles coke carbon burned at tuyere)} = \dot{n}_{CO} - \dot{n}'_C - \dot{n}'_{CH_4}$$

6. Compute blast enthalpy:

$$\Delta\dot{H}_1 = \dot{n}'_{O_2} \int_{T_B}^{298} C_{p_{O_2}} dT + \dot{n}_{N_2} \int_{T_B}^{298} C_{p_{N_2}} dT$$

$$+ \dot{n}_{H_2O} \cdot \int_{T_B}^{298} C_{p_{H_2O}} dT$$

7. Compute coke enthalpy:

$$\Delta\dot{H}_2 = \dot{n}_C \int_{T_{coke}}^{298} Cp_C \, dT$$

8. Compute heats of reaction: $\Delta\dot{H}_{R,\,298} =$

$$(\dot{n}_C + \dot{n}'_C - \dot{n}_{H_2O}) \cdot \Delta H_{R(a),298} + \dot{n}_{H_2O} \cdot \Delta H_{R(b),298} + \dot{n}'_{CH_4} \cdot \Delta H_{R(c),298}$$

9. Compute the enthalpy of the bosh gas:

$$= \Delta\dot{H}_1 + \Delta\dot{H}_2 + \Sigma\Delta\dot{H}_{R,\ 298}$$

10. Compute the enthalpy function for the bosh gas from heat capacity equations:

$$\Delta\dot{H}_3 = \int_{298}^{AFT} [\dot{n}_{CO}(a+bT+cT^{-2})_{CO} + \dot{n}_{H_2}\cdot(a+bT+cT^{-2})_{H_2} + \dot{n}_{N_2}\cdot(a+bT+cT^{-2})_{N_2}]dT$$

$$= (\dot{n}_{CO}a_{CO}+\dot{n}_{H_2}a_{H_2}+\dot{n}_{N_2}a_{N_2})(AFT - 298) + (\dot{n}_{CO}b_{CO}+\dot{n}_{H_2}a_{H_2}+\dot{n}_{N_2}b_{N_2})^{1/2}$$

$$\cdot(AFT^2 - 298^2) - (\dot{n}_{CO}c_{CO} + \dot{n}_{H_2}c_{H_2} + \dot{n}_{N_2}c_{N_2})\left(\frac{1}{AFT} - \frac{1}{298}\right)$$

$$= A(AFT - 298) + B(AFT^2 - 298^2) - C\left(\frac{1}{AFT} - \frac{1}{298}\right)$$

$$= A(AFT) + B(AFT)^2 - C/AFT + D$$

11. $\Delta\dot{H}_3$ is then equated to a $-(\Delta\dot{H}_1 + \Delta\dot{H}_2 + \Sigma\Delta\dot{H}_{R\ 298})$.

Since the resulting equation is not a linear or quadratic equation, the solution is found most easily by an iterative procedure:

(a) Let the B and C terms = 0.

(b) Solve for $AFT_1 = (\Delta\dot{H}_3 - D)/A$.

(c) Calculate $B(AFT_1)^2$ and C/AFT_1.

(d) Solve for $AFT = (\Delta\dot{H}_3 - D - B(AFT_1)^2 - C/AFT_1)/A$.

(e) Compare AFT and AFT_1. If they are within some predetermined limit, such as 1 K, then AFT has been found. If not, set AFT = AFT_1 and return to step (d).

The flow chart for this calculation is given in Fig. 6.2-4 and Fig. 6.2-5 gives some typical results for a blast furnace situation.

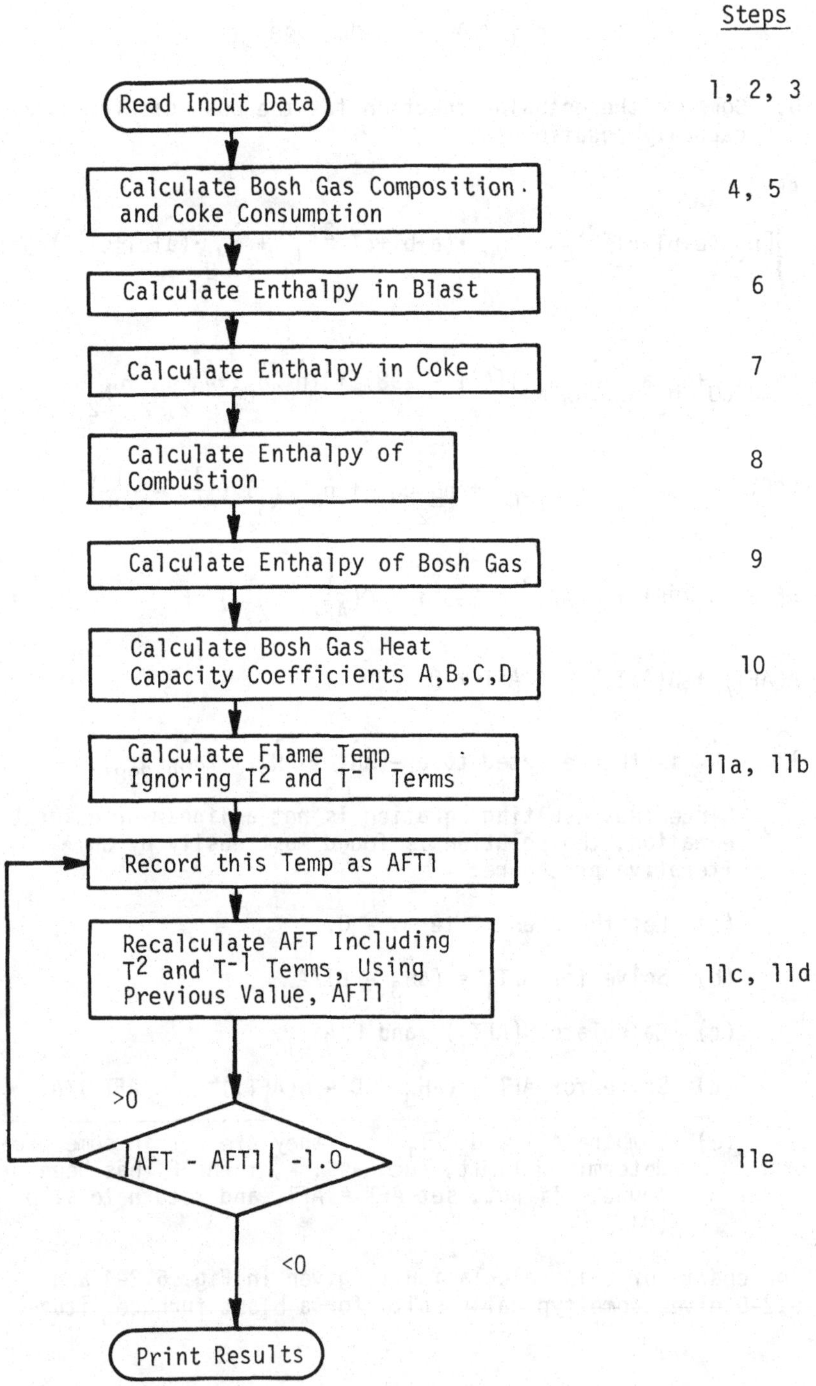

Fig. 6.2-4 Flow diagram for computer program to calculate AFT for shaft furnace burning coke.

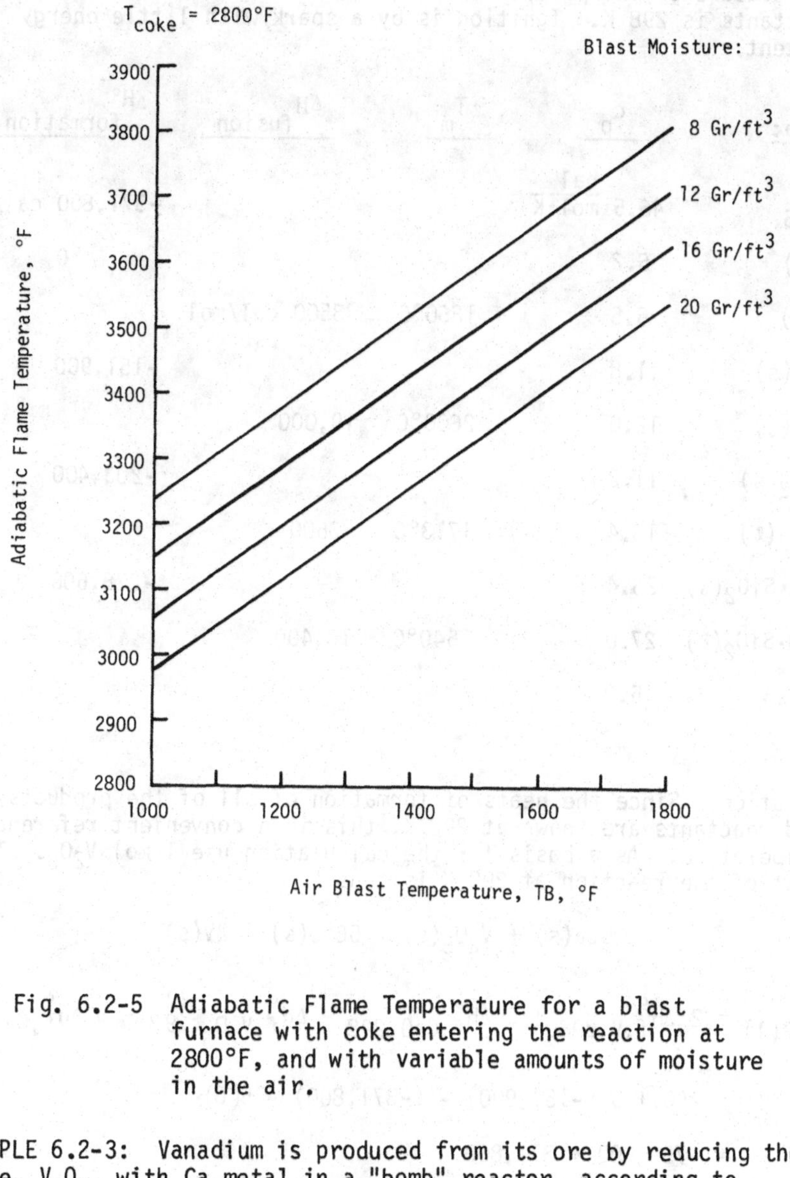

Fig. 6.2-5 Adiabatic Flame Temperature for a blast furnace with coke entering the reaction at 2800°F, and with variable amounts of moisture in the air.

EXAMPLE 6.2-3: Vanadium is produced from its ore by reducing the oxide, V_2O_5, with Ca metal in a "bomb" reactor, according to

$$5Ca + V_2O_5 \rightarrow 5CaO + 2V \quad (1)$$

In order to absorb some of the excess heat evolved in this reaction, SiO_2 is added to the stoichiometric reactant mixture. This SiO_2 combines with CaO to form a silicate:

$$CaO + SiO_2 = CaO \cdot SiO_2 \qquad (2)$$

If 5 moles of SiO_2 are added to the original stoichiometric mix of Ca and V_2O_5, per mole of V_2O_5, calculate the adiabatic reaction temperature of the products. The initial temperature of all reactants is 298 K. Ignition is by a spark with little energy content.

Data:	c_p	T_m	ΔH_{fusion}	$\Delta H^\circ_{formation,298}$
V_2O_5	46.5 $\frac{cal}{mol \cdot K}$			-371,800 cal/mol
V(s)	5.2			0
V(ℓ)	6.5	1860°C	3500 cal/mol	
CaO(s)	11.8			-151,900
CaO(ℓ)	12.0	2600°C	19,000	
SiO_2(s)	11.2			-203,400
SiO_2(ℓ)	13.4	1713°C	3600	
CaO·SiO_2(s)	26.4			-378,600
CaO·SiO_2(ℓ)	27.0	1540°C	13,400	
Ca(s)	5.3			0

Solution: Since the heats of formation of all of the products and reactants are known at 298 K, this is a convenient reference temperature. As a basis for the calculation use 1 mol V_2O_5. The heat of the reaction at 298 K is thus

$$5Ca(s) + V_2O_5(s) = 5CaO(s) + 2V(s)$$

$$\Delta H_{R(1)} = 2\,\Delta H^\circ_{f,V,298} + 5\,\Delta H^\circ_{f,CaO,298} - \Delta H^\circ_{f,V_2O_5,298} - 5\,\Delta H^\circ_{f,Ca,298}$$

$$= 2(0) + 5\,(-151,900) - (-371,800) - 5(0)$$

$$= -759,500 + 371,800$$

$$= -387,700 \text{ cal}$$

In addition, the 5 moles of SiO_2 added to the mixture form a slag with the CaO in the reaction production. The heat effect of this reaction is approximated by the heat of formation of CaO·SiO_2 from CaO and SiO_2:

$$\Delta H_{R,(2)} = 5\Delta H^{o}_{f,CaO\cdot SiO_2,298} - 5\Delta H^{o}_{f,CaO} - 5\Delta H^{o}_{f,SiO_2}$$

$$= 5\ (-378,600) - 5\ (-151,900) - 5\ (-203,400)$$

$$= -116,500 \text{ cal}$$

$$\Sigma\Delta H_{R,298} = -387,700 - 116,500 = -504,200 \text{ cal}$$

The reaction products at 298 K, are thus

2 mols V(s)

5 mols $CaO\cdot SiO_2(s)$

and they have to absorb a total of 504,200 calories.

The temperature reached by the products (ART) is found by raising the temperature of all of the products, and calculating the system enthalpy increment, until a system enthalpy increment equal to 504,200 cal. is found:

Product	n	x	$\bar{C}_p$	x	ΔT	=	$[H_{T_2} - H_{298}]$	$\Sigma\ [H_T - H_{298}]$
Heat to T = 1540°C:								
V(s)	2	x	5.2	x	1515	=	15,756	
$CaO\cdot SiO_2(s)$	5	x	26.4	x	1515	=	199,980	
							215,736	215,736
Melt the $CaO\cdot SiO_2$ at 1540°C:								
$CaO\cdot SiO_2(\ell)$	5	x	13,400			=	67,000	282,736
Heat to 1860°C:								
V(s)	2	x	5.2	x	320	=	3,327	
$CaO\cdot SiO_2(\ell)$	5	x	27.0	x	320	=	43,200	
							46,528	329,264
Melt the V at 1860°C:								
$V(\ell)$	2	x	3500			=	7,000	336,264
Heat to 2400°C:								
$V(\ell)$	2	x	6.5	x	540	=	7,020	
$CaO\cdot SiO_2(\ell)$	5	x	27.0	x	540	=	72,900	
							79,920	416,184

(continued)

Product	n		$\bar{C}_p$		ΔT	=	$[H_{T_2} - H_{298}]$	$\Sigma[H_T - H_{298}]$
Heat to 2900°C:								
$V(\ell)$	2	x	6.5	x	500	=	6,500	
$CaO{\cdot}SiO_2(\ell)$	5	x	27.0	x	500	=	67,500	
							74,000	490,184
Heat to 3000°C:								
$V(\ell)$	2	x	6.5	x	100	=	1,300	
$CaO{\cdot}SiO_2(\ell)$	5	x	27.00	x	100	=	13,500	
							14,800	504,984

The ART is therefore close to 3000°C, and easily can be found by graphing the last two data points.

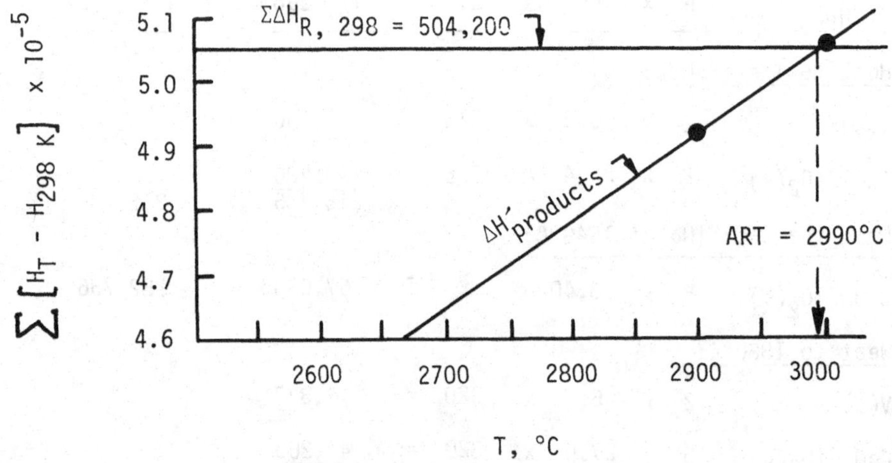

This example, besides illustrating how to find an adiabatic reaction temperature, also illustrates that such metallothermic reactions can result in very high temperatures! If they are rapid, very little heat loss can occur and these theoretical temperatures may actually be approached. In order that such a temperature as the one above is not reached, which would be very dangerous, additional inert material must be added which will absorb heat but not participate in the reaction. In this case, adding more lime (CaO) and sand to the mixture would provide additional slag, which would absorb much more heat than it would liberate when it is formed, resulting in a lower adiabatic reaction temperature.

6.3 ELECTRICAL ENERGY

Many metallurgical processes utilize electricity as an energy source. Examples include resistance heated furnaces, induction furnaces, arc melting furnaces, arc welding processes, electroslag remelting furnaces, electron beam furnaces, and electrolytic cells. In these cases the electrical energy utilized in the process must be included in the energy balance. Referring to the statement of the First Law given by Eq. 5.2-41,

$$dH' = \delta q - \delta w^* - \Sigma \delta m_i h_i \qquad (5.2\text{-}41)$$

The term δw^* is the non-PV work done by the system. If electrical work is done on the system, δw^* is negative and the equation becomes, for a steady-state process,

$$\delta m_i h_i = \delta q - (-\delta w^*) \qquad (6.3\text{-}1)$$

Now the electrical work done on the system must be added to the heat transferred to the system as an energy input in an energy or heat balance.

6.3.1 ELECTRIC POWER FUNDAMENTALS

Electrical power, P, is the work done by a steady current of one ampere flowing for one second under the driving force of one volt.

$$P = I \cdot V \qquad (6.3\text{-}2)$$

The product of volt·amperes is called the watt. Power is also the rate of doing work or absorbing energy, so that the energy absorbed by a process is the product of the power times the interval over which it is applied:

$$\text{Energy} = \text{Power} \times \text{time} \qquad (6.3\text{-}3)$$

The watt-hour or kilowatt-hour (abbreviated kwh) is the usual measure of electrical energy.

Power calculated using Eq. (6.3-2) is the total power provided by electricity. In a direct current system it is also the power which is converted to thermal energy, since the power that goes to thermal energy is given by

$$P = I^2 R \qquad (6.3\text{-}4)$$

where R is the pure resistance to current flow of the current path. The amount of heat created at a particular location by a current I flowing through a circuit will, therefore, depend on the resistance of the conducting path at that location.

If the electricity used is alternating current, the voltage and current fluctuate f times per second; f = 60 cps (60 Hertz) in most U.S. supplies. The current is then given by

$$I = I_{max} \cos (2\pi ft) \qquad (6.3\text{-}5)$$

and the voltage varies according to

$$V = V_{max} \cos (2\pi ft) \qquad (6.3\text{-}6)$$

The effective current (r-m-s current) in ac systems is then found by averaging the current, and is

$$\bar{I} = 0.707\ I_{max} \qquad (6.3\text{-}7)$$

and the effective voltage is

$$\bar{V} = 0.707\ V_{max} \qquad (6.3\text{-}8)$$

$\bar{I}$ and $\bar{V}$ are the values measured and indicated by ammeters and voltmeters connected to ac circuits.

In terms of effective current and voltage, the total power into an ac system is

$$P = \bar{V}\cdot\bar{I} \qquad (6.3\text{-}9)$$

and the power converted to heat in an ac system is

$$P = I^2R \qquad (6.3\text{-}10)$$

where R is the resistance of the current path to current flow.

In addition to pure resistance which impedes current flow and causes heat generation according to Eq. (6.3-10), another impedance to current flow in ac electrical circuits is inductance, X. Inductance arises when an ac current passing through a conductor sets up an alternating magnetic field around the conductor. The changing field, in turn, induces a voltage in any adjacent portions of the conducting path. This induced voltage, however, is opposite in sign to the primary voltage and the net result is a decrease in the effective primary voltage, equivalent to an increase in resistance to current flow, but without the effect of causing additional I^2R heating. Since the voltage initially provided to the system has thus been effectively decreased, the watts that are available to be converted into thermal energy via I^2R heating are also decreased. Therefore, in order to evaluate the thermal efficiency of an ac electrical system, the relationship between the initial power put into the system and the watts available to produce heat must be known. The relationship is

$$\text{(Available Thermal Watts) in an ac system} = \bar{V}\cdot\bar{I}\cdot\text{(Power Factor)} \qquad (6.3\text{-}11)$$

where the Power Factor is a measure of the amount of inductive impedance in the electrical system. The power factor is expressed as cos θ, where θ is the angle between a vector representing the total impedance, Z, in the circuit, and the resistive component of impedance, Z_R. The total impedance, Z, equals

$$Z = \sqrt{Z_R^2 + Z_I^2} \qquad (6.3\text{-}12)$$

where Z_I equals X, the inductive impedance, and Z_R equals R, the resistive impedance. Z_I and Z_R are vectors at right angles to each other on an impedance diagram.

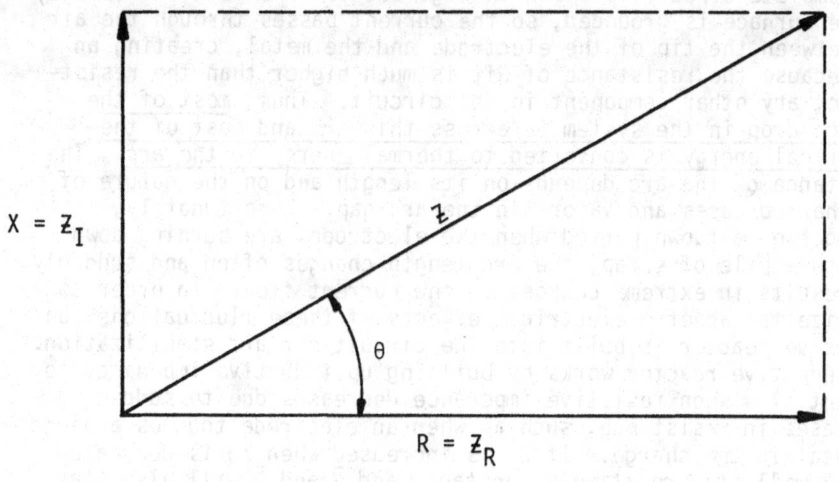

From this diagram, it can be seen that cos θ is Z_R/Z, or the ratio of pure resistance to total impedance. Since thermal energy is only produced by resistive impedance,

$$\text{Available Thermal Watts} = \bar{V}\cdot\bar{I}\cdot\cos\theta \qquad (6.3\text{-}13)$$

When the resistive component is zero, θ = 90°, cos θ = 0, and no I^2R heating occurs. When no inductive impedance is present in the system, θ = 0°, cos θ = 1, the power factor is 1.0 and all of the power into the system, $\bar{V}\cdot\bar{I}$, is available for production of I^2R watts. However, virtually all systems have some inductance. Thus, the person making a heat balance must know the power factor for the system.

In most applications, the principal exception being electrolytic cells for electrowinning or electrorefining, electrical work is put into the system by converting the electrical energy into thermal energy. To avoid converting electrical power, P = V·I, into thermal energy before it reaches the customer, power companies transmit power at elevated voltages and reduced currents, i.e., they minimize I^2R.

As used in metallurgical systems, however, the electrical energy is usually at a relatively low voltage and high current, resulting in high conversion to thermal energy (high I^2R). The high voltage received from the transmission lines must therefore be converted to low voltage by means of a step-down transformer at the plant. The step-down results in some losses, both thermal and inductive.

After step-down, the electrical energy reaches the process. This is usually the point at which the metallurgist starts to make an energy balance.

6.3.2 ELECTRIC ARC FURNACES

Electric arc furnaces utilize 3-phase ac power which enters the furnace from the secondary circuit via carbon electrodes. The complete circuit is shown in Fig. 6.3-1. The metallic charge, in the furnace is grounded, so the current passes through the air gap between the tip of the electrode and the metal, creating an arc because the resistance of air is much higher than the resistance of any other component in the circuit. Thus, most of the voltage drop in the system is across this arc and most of the electrical energy is converted to thermal energy in the arc. The resistance of the arc depends on its length and on the nature of the charged gases and vapors in the arc gap. Unfortunately, during the meltdown period when the electrodes are burning down through a pile of scrap, the arc length changes often and suddenly, and results in extreme changes in the current flow. In order to minimize the adverse electrical effects of these fluctuations, an inductive reactor is built into the circuit for arc stabilization. The inductive reactor works by building up inductive impedance to current flow when resistive impedance decreases due to sudden decreases in resistance, such as when an electrode touches a piece of metal in the charge. If Z_I is increased when Z_R is decreased then Z will stay relatively constant, and $\bar{V}$ and $\bar{I}$ will also stay at steady values. This makes life much easier for power companies. However, this increases the inductive reactance of the circuit itself, reducing the amount of initial power than can be converted to thermal energy for melting.

The voltage applied to the secondary circuit is controlled by the choice of tap on the transformer and is fixed at the level chosen. The current is controlled by varying the resistance of the circuit. This is done by moving the electrodes up or down to vary the arc length until the current giving the optimum power input is reached. If the arc length is longer than the optimum length, the arc voltage will increase but the current will fall in greater percentage, so the arc wattage is less. Conversely, if the arc is shorter than optimum, the current rises, but the arc voltage drops even more, and again the arc wattage is less.

Power and energy consumption are measured in several ways. The kilowatt-hour meter, which gives the total power ($\bar{V}\cdot\bar{I}$) into the system, is connected ahead of the step-down transformer using primary voltage and current transformers. It is usually a polyphase meter, totaling all of the energy put into the transformer via all three phases, and it is the basis for calculating energy costs.

The secondary voltage, between each phase and ground, is usually measured as close to the electrodes as possible, as this is the current flowing in each electrode circuit. There is also a wattmeter connected to each secondary phase which gives a closer measure of the actual instantaneous power delivered to each

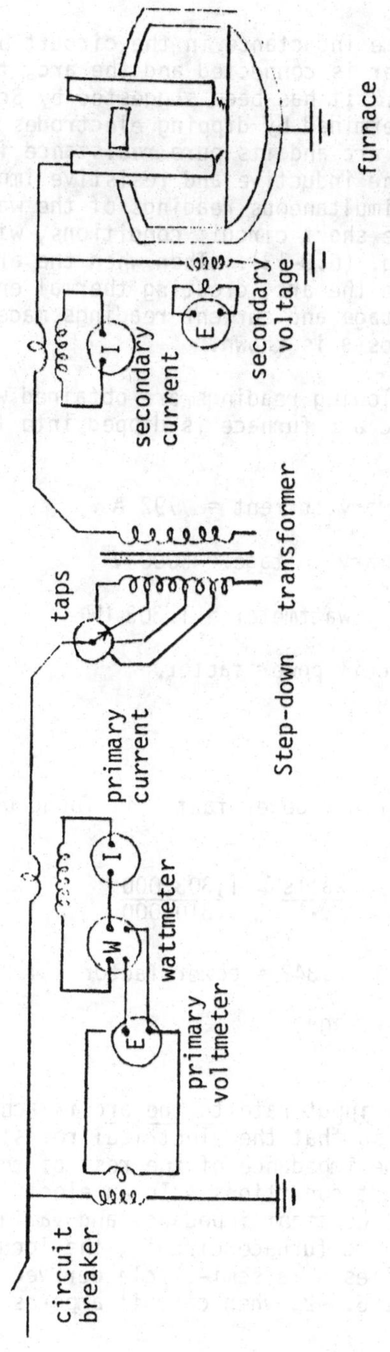

Fig. 6.3-1 Electrical circuit for electric arc furnace. Only one phase of a three-phase circuit is shown.

electrode, in kilowatts. If a poly-phase wattmeter is used, its reading may be divided by three to obtain the average power being delivered to each electrode.

Since there is some inductance in the circuit between the point where the wattmeter is connected and the arc, the power factor is less than 1.0. It has been suggested by Schwabe* that the power factor be determined by dipping electrodes into the melt, thereby eliminating the arc and its pure resistance from the circuit and leaving only the inductive and resistive impedance in the rest of the circuit. Simultaneous readings of the watts, voltage and amperage under these short circuit conditions, will allow calculation of cos θ via Eq. (6.3-13). Then when the arc is present, the watts being put into the arc, creating thermal energy, can be calculated from the voltage and current readings made when the arc is present, since cos θ is known.

EXAMPLE 6.3-1: The following readings are obtained when the electrode in an electric arc furnace is dipped into the liquid metal bath:

Primary current = 792 A

Primary voltage = 5000 V

Primary wattmeter = 1.303 MW

Calculate the short circuit power factor.

Solution: The short circuit power factor is found using Eq. (6.3-13):

$$\cos\theta = \frac{\text{watts}}{V \cdot I} = \frac{1{,}303{,}000}{3{,}810{,}000}$$

$$\cos\theta = 0.342 = \text{power factor } (34.2\%)$$

(Note that $\theta = 70°$)

The maximum power input rate <u>to the arc</u> is achieved when the arc length is adjusted so that the electrical resistance of the arc, Z_R, is equal to the impedance of the rest of the circuit, Z_I, found under short circuit conditions. In an electrical circuit of constant input voltage, constant impedance and variable resistance, such as in an electric arc furnace circuit, the locus of all possible current vectors lies on a semi-circle derived as described below and shown in Fig. 6.3-2, when circuit amperes is plotted vs. volts.

---

* W. E. Schwabe, Iron and Steel Engineer, June, 1954, p. 87.

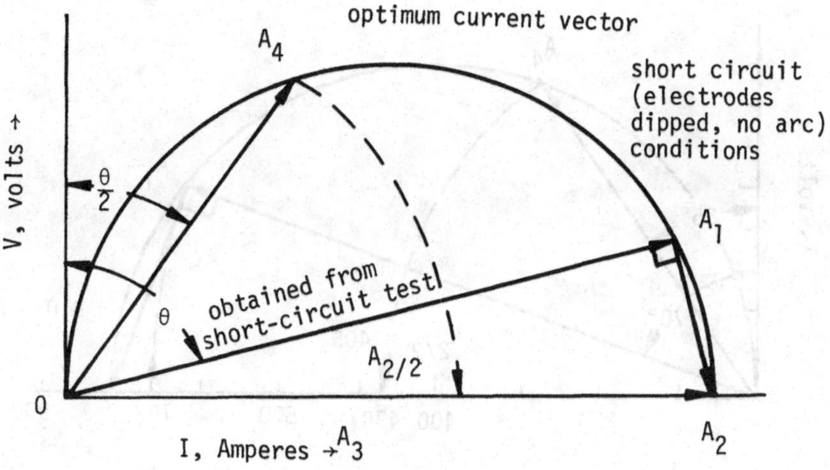

Figure 6.3-2 Vector diagram to determine optimum current.

θ, obtained from short-circuit conditions, is used to locate the vector OA_1, of length equal to the current under short-circuit conditions, A_1. A line perpendicular to OA_1 is drawn from A_1 intersecting the current axis at A_2. Then the semi-circle of radius $A_2/2$ is drawn. It is the locus of all possible current vectors for that circuit. The current at maximum arc power is found when the vector OA_4 is drawn at θ/2 angle. This value of θ/2 becomes the angle whose cosine is the power factor at optimum operation. In many cases, the optimum angle is 38° and the optimum power factor turns out to be 0.79, or the ratio of available watts for arc heating to total electrical watts, is 0.79.

EXAMPLE 6.3-2: For the electric arc furnace in Example 6.3-1, find the optimum power factor, the current at optimum conditions, and the amount of electrical energy converted to thermal energy in the arc, if the furnace is operated at optimum conditions.

Solution: The optimum power factor for this system is at 1/2 of the short circuit angle, θ/2, or 35°, for which the power factor cos θ, is 0.82. A_1 = 792 Amps. Constructing the vector diagram, A_2 = 810 amps and the optimum current vector, OA_4, drawn at an angle of 35°, results in a vector of length corresponding to an optimum primary current of 470 Amps.

The thermal power produced by the arc at optimum conditions (tap still at 5000 volts) would then be

$$\text{Available Thermal Watts} = (5000)(470)(0.82)$$

$$= 1.927 \text{ megawatts}$$

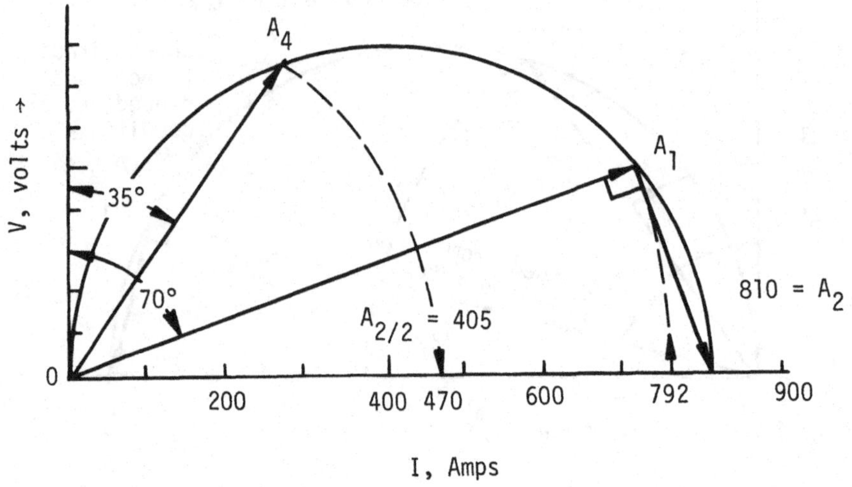

or, if applied over a period of one hour,

$$\delta w^* = 1927 \text{ kwh}$$

During meltdown, most of the heat generated in the arc is absorbed by the metal (scrap) charge, because the electrodes "bore" through the loose scrap and the radiation from the arc "sees" only metal charge which absorbs all the radiation. As the amount of metal melted increases and a molten pool is formed, more and more of the arc radiation misses the metal charge and strikes the furnace wall instead. Although this wall reradiates some energy, some is absorbed and lost by conduction through the wall. The amount lost increases in proportion to the increase in temperature of the melt. When a flat bath has been achieved, with any unmelted material under the surface of the liquid metal, the arc radiation to the walls and roof becomes intense, because of reflection off the surface of the bath as well as direct radiation. Therefore, the efficiency of heat transfer from arc to metal bath decreases as meltdown proceeds.

Fig. 6.3-3 illustrates this concept, along with the idea that there is a maximum bath temperature, T_∞, that can be achieved after an infinite time, when the energy in via the arc just balances the heat losses through the furnace walls. The thermal efficiency decreases as heat losses increase. Since heat losses are time dependent, this provides an incentive to put electrical energy into the furnace as fast as possible in order to bring the metal charge to the tapping temperature rapidly. This is the incentive for the development of ultra-high power (UHP) electric arc furnaces. Fig. 6.3-4 compares the time-temperature profiles

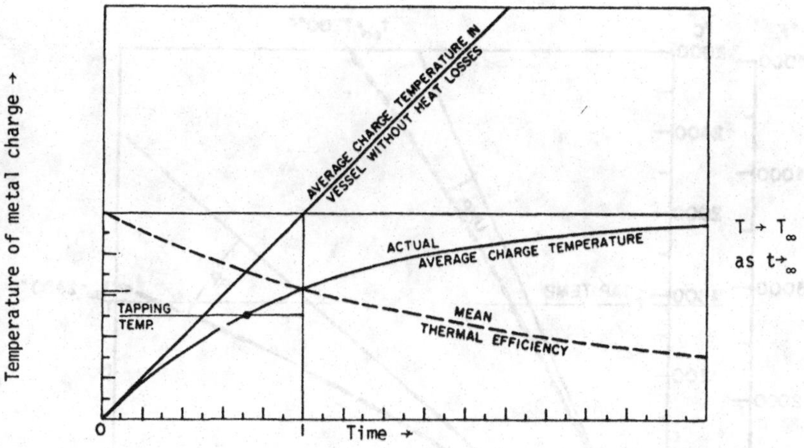

Figure 6.3-3: Time-temperature profile in an electric arc furnace. Temperature of charge approaches a limiting value, T_∞, depending on the rate of power input and the heat losses from the furnace.

of regular power (RP) and UHP furnaces, and Fig. 6.3-5 shows the range of power input levels in industrial use.

For a typical regular-power arc furnace, melting steel, the energy balance, based on power supplied to the primary side of the furnace transformer, would be

$$W^* + \Delta H_{\substack{\text{exothermic}\\\text{reactions}}} = \Delta H_{\substack{\text{endothermic}\\\text{reactions}}} + \Delta H_{\text{loss}} + \sum_{\text{phases}} (H_{T,\text{product}} - H_{298})$$

Energy In	MJ/metric ton	Energy Out	MJ/metric ton
Electrical Energy	1,984	Melting & Superheating Metal	1,381
ΔH(carbon oxidation)	119		
ΔH(silicon oxidation)	48	Melting Slag	120
ΔH(iron oxidation)	52	Electrical Losses (18%)	358
ΔH(electrodes burning)	10	Heating Infiltrate Air and Process Gases	245
Total	2,213		
		Thermal Losses (10%)	199
		Total	2,213

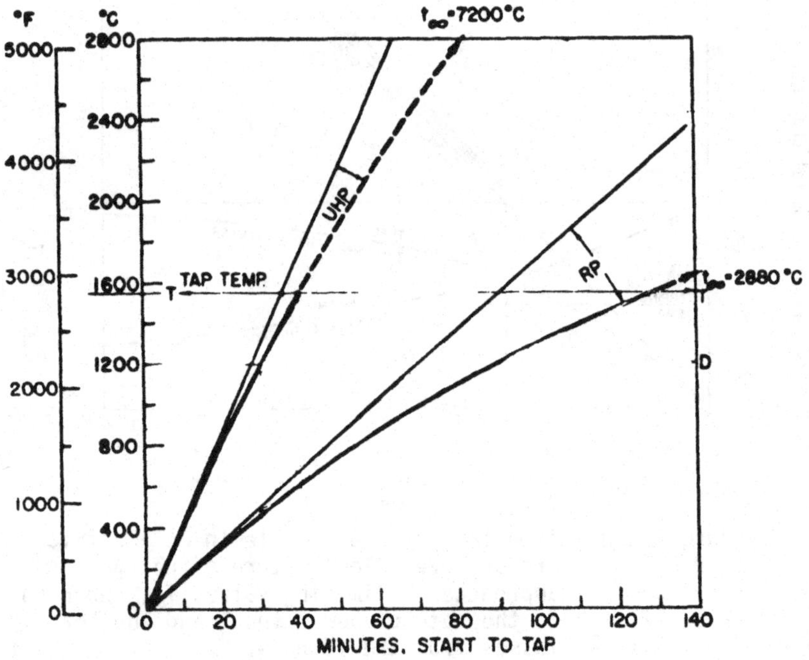

Figure 6.3-4 Comparison between regular power (RP) and ultra-high power (UHP) electric arc furnace time-temperature profiles.

<u>Energy efficiency</u> may be expressed in a number of ways; (1) based upon energy contained only in the desired product; (2) based on energy contained in product and unavoidable associated phases such as slag and process gases; (3) based on all thermal consumption (little of which is avoidable due to the nature of heat flow): each of the above may be made relative to either (a) the electrical energy in or (b) the total energy in. In the present case, therefore, the energy efficiency could be said to be any of the following:

Case			Efficiency
(1)/(a):	(1,381)/(1,984)	=	69%
(1)/(b):	(1,381)/(2,213)	=	62%
(2)/(a):	(1,746)/(1,984)	=	88%
(2)/(b):	(1,746)/(2,213)	=	79%
(3)/(a):	(1,945)/(1,984)	=	98%
(3)/(b):	(1,945)/(2,213)	=	88%

Clearly, great care must be taken in defining energy efficiency!

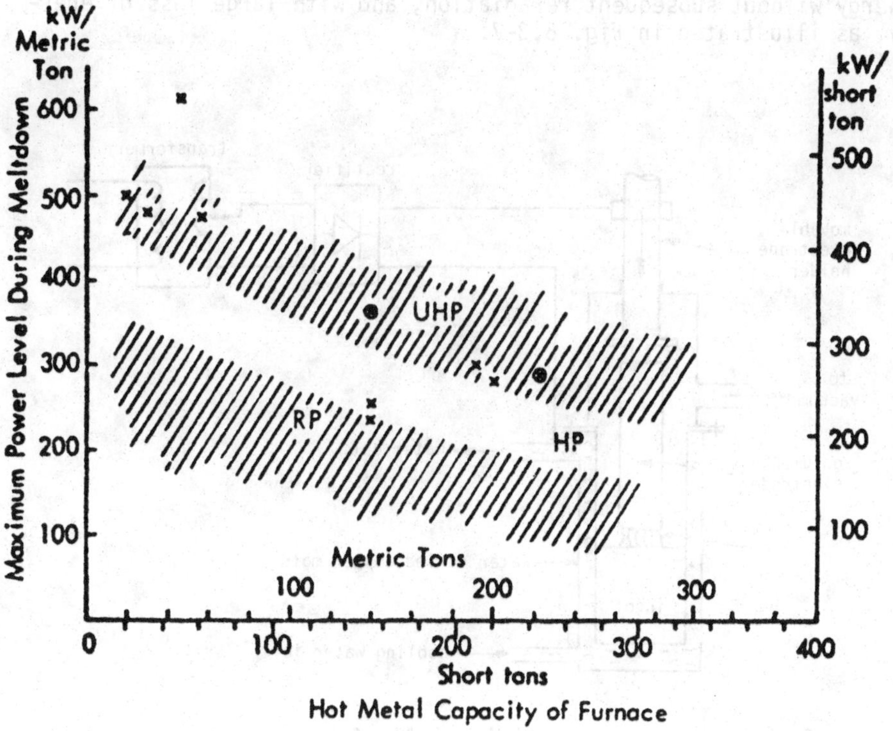

Figure 6.3-5 Range of power input levels found in regular and ultra-high power electric arc furnaces. (Ref. Schwage, J. Metals)

6.3.3 VACUUM ARC REMELTING FURNACES

The power into vacuum arc remelting (VAR) furnaces, Fig. 6.3-6, is measured and calculated in a manner similar to that for an electric arc furnace operating in air, except that usually it is direct current (dc). In this case the ac high voltage power into a plant must be rectified (using silicon rectifiers) to produce direct current, after having been dropped in voltage through a transformer. Therefore, there are still electrical losses associated with the transforming and rectification process.

However, energy supplied to the arc is more easily determined, since the power factor relating current, voltage and kilowatts into the arc is 1.0 for dc, <u>if the electrical parameters are measured at the furnace.</u>

The distribution of electrical energy in the system is different from that in the air melt arc furnace, however, because in this case the I^2R heating of the arc results in considerable radiation directly to the water-cooled mold, which absorbs this energy without subsequent reradiation, and with large loss of energy, as illustrated in Fig. 6.3-7.

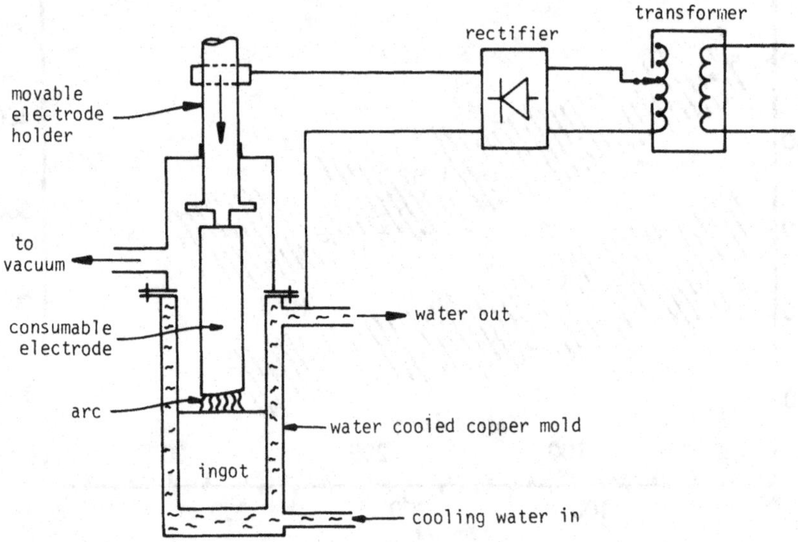

Figure 6.3-6 Vacuum arc remelting (VAR) furnace.

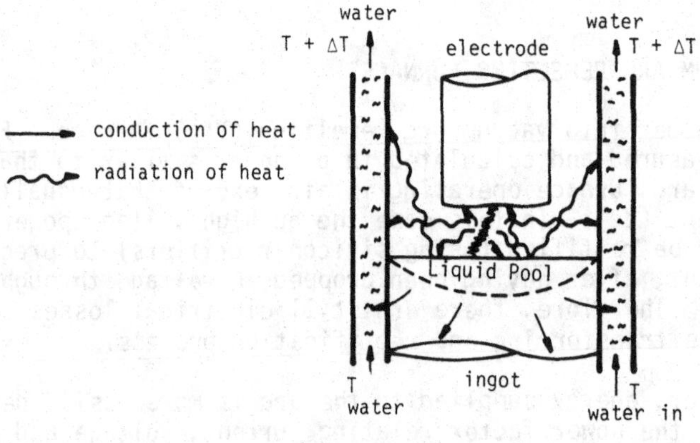

Figure 6.3-7 Distribution of energy from VAR arc. All energy is removed, ultimately, by cooling water.

6.3.4 ELECTROSLAG REMELTING FURNACES

In electroslag melting (ESR), either ac or dc power may be used. In this case the heating is caused by the resistance of the slag, and no arc is involved. However, since the slag is in contact with the walls and is at the highest temperature in the system, some of the energy put into the slag, I^2R_{slag}, is lost by

radiation from the slag surface up the annulus between the electrode and the water-cooled mold, some to the water-cooled mold wall above the slag, and the rest by conduction through the slag or the ingot to the cooling water, as illustrated by Fig. 6.3-8. About 50% of the heat energy produced by the I^2R heating of the slag is lost by radial conduction through the slag to the wall.

In this case, the <u>over-all</u> heat balance <u>will not reflect how the process operates at all</u>. The only input will be electrical energy, and the only outputs will be radiant losses to the surroundings, an increase in the heat content of the cooling water, and a slight amount of heat still in the ingot when it is removed!

Again, the heating of the slag is caused by passing a very large current, 10^3-10^4 A, through it, and using a slag compositon that will result in a high resistance. The I^2R heating for this process is usually in the neighborhood of 4,750,000 kJ/metric ton, while the enthalpy needed to melt the electrode is approximately 1,580,000 kJ/metric ton, so about 67% of the energy put into the furnace as I^2R heat goes <u>directly</u> to losses by radiation or conduction. However, the process would not work any other way, and since the refining it performs is necessary, its low energy efficiency is tolerated.

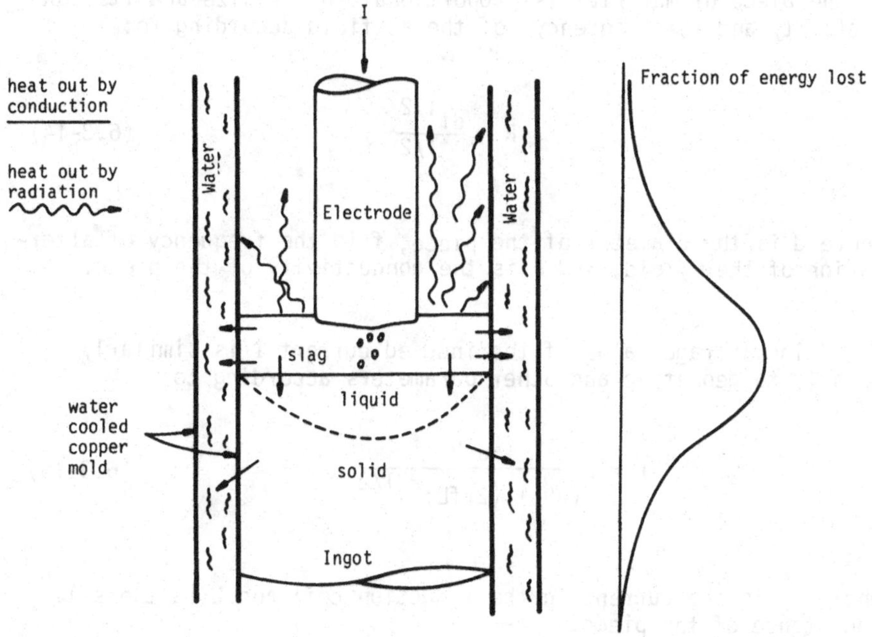

Figure 6.3-8 Schematic representation of an electroslag furnace.

Notice that the heat of fusion (ΔH_m) of the electrode is completely recovered as heat of solidification ($-\Delta H_m$) in the ingot, and the heat of fusion of new slag continuously added from above is recovered as heat of solidification of the slag along the mold walls, resulting, as mentioned above, in a typical overall heat balance of:

Energy In	(MJ/metric ton)	Energy Out	(MJ/metric ton)
Electrical Energy	4,750	$\Delta H'_{cooling\ water}$	3,950
		$\Delta H'_{radiation\ losses}$	600
		$\Delta H'_{ingot}$	200
		Total	4,750

6.3.5 INDUCTION MELTING AND HEATING FURNACES

Induction melting is based on the process of passing electrical current through a coil which causes a magnetic field to be formed. This field, in turn, is made to pass through charge material or molten metal and in turn induces a current to flow in this material, as illustrated in Fig. 6.3-9.

Since a current flowing in a conductor tends to concentrate near the surface, the first part of a piece of solid material to be heated in the magnetic field will be the surface, with subsequent heat conduction toward the center. The effective resistance of the piece of material is proportional to its size and its conductivity and the frequency of the ac field according to:

$$R \propto \frac{df^{1/2}}{\sigma^{1/2}} \qquad (6.3\text{-}14)$$

where d is the diameter of the piece, f is the frequency of alternation of the field, and σ is the conductivity of the piece.

The average value of the induced current I is similarly related to geometric and other parameters according to:

$$I \propto \frac{I_0\, d^2\, f}{(R^2 + (2\pi fL)^2)^{1/2}} \qquad (6.3\text{-}15)$$

where I_0 is the current in the induction coil and L is the self-inductance of the piece.

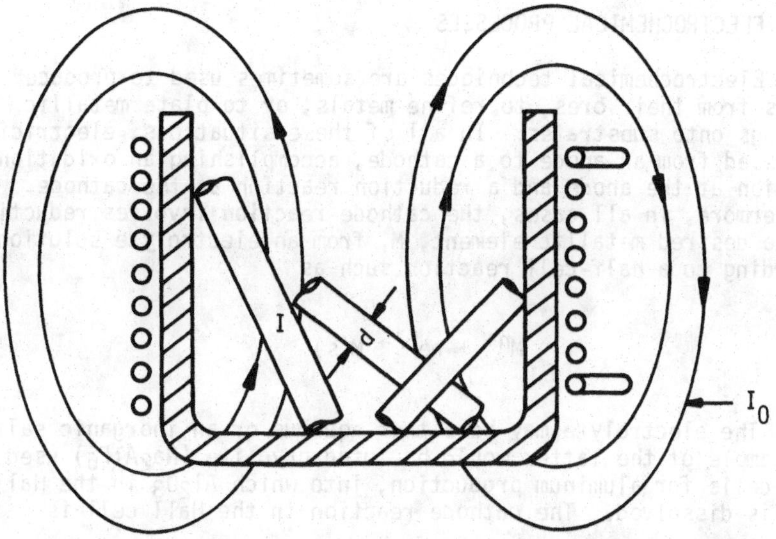

Figure 6.3-9: Schematic representation of an induction furnace.

Thus, the inductive transfer of electrical energy from an induction coil to a charge of metal pieces in a crucible is dependent on a great many factors. For example, a high conductivity metal, such as copper, is hard to induction heat because it doesn't offer much resistance. The product of the diameter times the frequency is also important, and is the reason why small diameter particles require high frequency for melting. On the other hand, once a charge is molten, the only energy needed is that necessary to replace heat losses, which are relatively small. Therefore, the energy efficiency of an induction melting process is, in many respects, the opposite of that of the electric arc furnace, being low when the charge is cold and improving with time. A typical energy balance on an induction furnace melting a cold steel charge to produce a heat of molten steel which is then tapped from the furnace would be:

Energy In	(MJ/metric ton)	Energy Out	(MJ/metric ton)
Electrical Energy	2,000	Melting and Superheating Metal	1,381
		Electrical Losses due to Inductive Reactance, and Coupling Inefficiency	370
		Electrical Transmission Losses from Supply to Furnace	94
		Heat Losses	155
			2,000

6.3.6 ELECTROCHEMICAL PROCESSES

Electrochemical techniques are sometimes used to produce metals from their ores, to refine metals, or to plate metallic coatings onto substrates. In all of these situations, electricity is passed from an anode to a cathode, accomplishing an oxidation reaction at the anode and a reduction reaction at the cathode. Furthermore, in all cases, the cathode reaction involves reduction of the desired metallic element, M, from an electrolyte solution, according to a half-cell reaction such as

$$M^{n+} + ne^- \rightarrow M(s)$$

The electrolyte may be either aqueous or an inorganic salt. An example of the latter would be fused cryolite (Na_2AlF_6) used in Hall cells for aluminum production, into which Al_2O_3 in the Hall cell is dissolved. The cathode reaction in the Hall cell is

$$Al^{+3} + 3e^- \rightarrow Al(l)$$

The anode reactions vary, depending on the process. In the case of electro-refining, the anode material is the impure source of M and the anode reaction is

$$M \rightarrow M^{n+} + ne^-$$

In the case of electrowinning, the metal M enters the cell in the electrolyte as M^{n+} ions, so that in order for a current to flow from the anode to the cathode and there accomplish reduction of M^{n+} from solution, the anode reaction will involve the electrolyte itself. In aqueous systems, the anode reaction is

$$H_2O = 2H^+ + 1/2\ O_2(g) + 2e^-$$

and oxygen gas is evolved on the surface of an insoluble electrode such as lead which serves essentially as a current carrier.

In the case of aluminum, the oxygen ions in the cryolite participate in the anode reaction at the surface of a graphite electrode;

$$C + 2(O)^{2-} \rightarrow CO_2\ (g) + 4e^-$$

Thus, the total reaction in an aluminum reduction cell (Fig. 6.3-10) is the sum of the two half-cell reactions, properly balanced:

$$4\ [\ Al^{+3} + 3e^- \rightarrow Al(l)\ ]$$

$$3\ [\ C + 2\ O^{2-} \rightarrow CO_2(g) + 4e^-\ ]$$

$$4\ Al^{+3} + 3\ C + 6\ O^{2-} \rightarrow 3CO_2(g) + 4\ Al\ (l)$$

In an electrowinning or electroplating cell, (Fig. 6.3-11), the sum of the reactions might be,

$$H_2O = 2H^+ + 1/2\ O_2(g) + 2e^-$$

$$M^{+2} + 2e^- = M(s)$$

$$M^{+2} + H_2O = 2H^+ + 1/2\ O_2(g) + M(s)$$

In an electrorefining cell, the reactions are

$$M(s,impure) \rightarrow M^{n+} + ne^-$$

$$M^{n+} + ne^- \rightarrow M(s,pure)$$

$$M(s,impure) \rightarrow M(s,pure)$$

It is beyond the scope of this handbook to go into a detailed discussion of electrochemistry, but for purposes of making energy balances on electrometallurgical processes, it is important for the reader to know that there is a voltage drop, E_{cell}, that is required by thermodynamics in order to carry out the desired reactions under perfect conditions. Associated with each of these reactions is a change in the thermodynamic function called the Gibbs Free Energy, ΔG, of the system*. ΔG is related to the voltage drop across the cell when the reaction is taking place very slowly (i.e., reversibly) by

$$\Delta G_{cell} = -n\ F\ E_{cell}$$

where F is Faraday's constant, defined in Section 1.5. E_{cell} can therefore be determined from thermodynamics and a knowledge of the reactions taking place and represents the voltage that must be

---

* Consult any standard thermodynamics textbook for the definition of the Gibbs Free Energy.

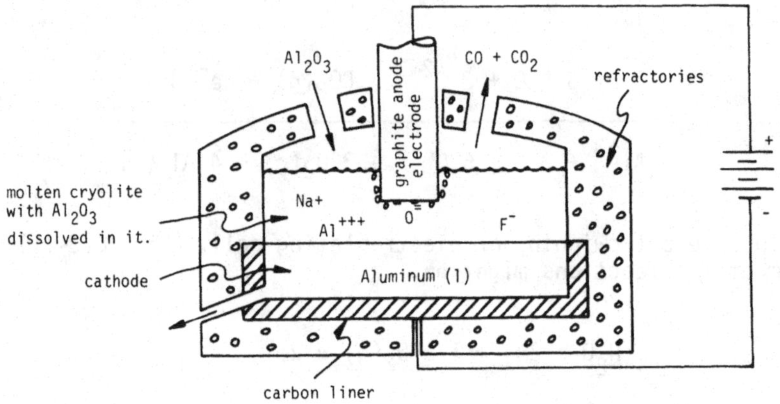

Figure 6.3-10 Schematic diagram of Hall cell for alumina reduction to aluminum.

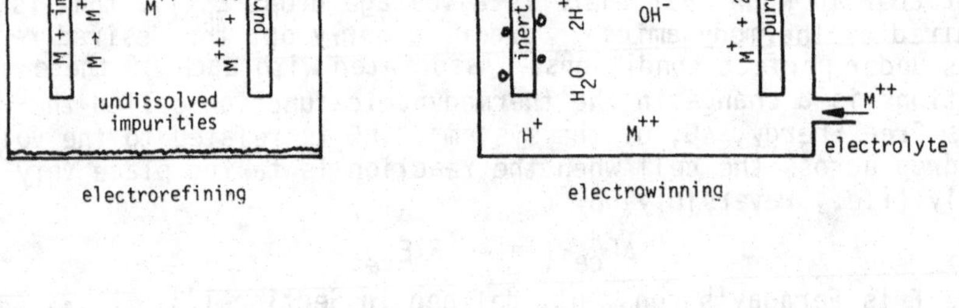

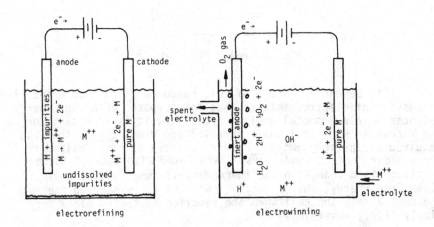

Figure 6.3-11 Schematic diagram of electrorefining and electrowinning cells.

applied to the cell to cause the reactions to take place under ideal, reversible, conditions.

A number of other factors, however, enter into the cell voltage requirement. As the current density is increased, since the electrolyte is an ionic conductor, there is a diffusive motion of cations toward the cathode and anions toward the anode. If the rate of reaction is faster or slower than the ion transport rate there will be a build up or depletion of ions in the electrolyte near the electrode surface, relative to the bulk composition, with the result that there is a change in the over-all cell voltage by an amount V_c, the concentration overpotential, as shown in Fig.6.3-12.

This concentration polarization increases as current density increases, and in the extreme, limits the current density that can be applied to the cell.

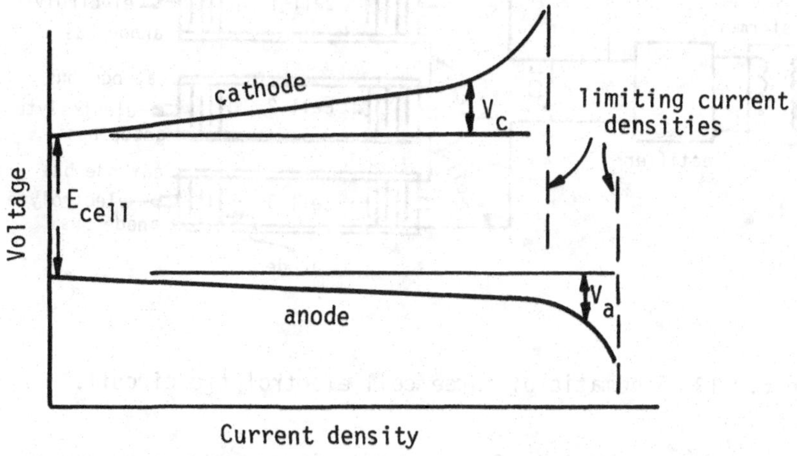

Fig. 6.3-12 Current density voltage relationship

Other overpotentials will be involved, such as those associated with activation, crystallization, oxide formation on an anode, or gas evolution. The sum of concentration and all other overpotentials is the cathode or anode overpotential, $V_{cathode}$ and V_{anode}, as shown on Fig. 6.3-12.

Finally, there is the ohmic resistance to current flow through the cell itself (electrodes plus electrolyte) that causes an IR_{cell} voltage drop. R_{cell} depends on the resistance of the electrolyte and can be decreased by various additions to the electrolyte and/or control of pH.

The total voltage required for a cell is the sum of the various contributions above:

$$V_{cell} = E_{cell} + V_{cathode} + V_{anode} + IR_{cell}$$

In the commercial operation of an electrowinning, electro-refining, or electrodeposition plant, the AC electrical energy is provided by the utility at high voltage. It must be transformed to the lower voltage needed and rectified to DC for use in electrolysis. From the rectifier the high current flows to the electrolytic cells. This is illustrated schematically in Fig. 6.3-13.

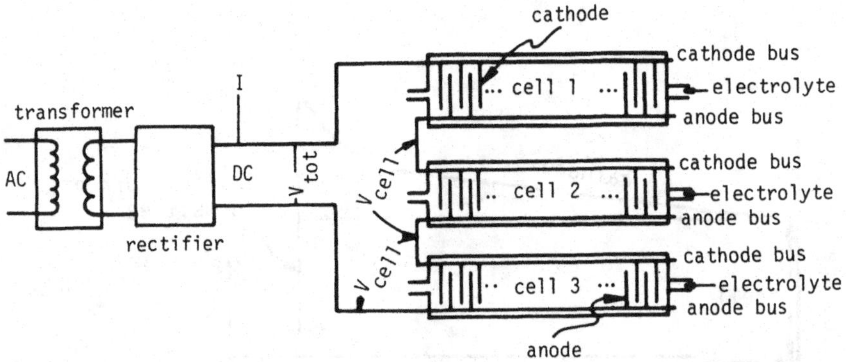

Figure 6.3-13 Schematic of three cell electrolytic circuit.

From the standpoint of energy efficiency and energy consumption, there are the usual induction and resistance losses from the utility to the rectifier. Beyond the rectifier, only resistance losses are involved, since there are no inductive losses in DC circuits. In order to minimize these losses, which obviously could be large since the currents can run into the tens of thousands of amperes, the bus bars are made of the lowest resistivity (ρ) material available, copper, and the cross-sectional area, A, is made as large as practical, since $R_{bus} = \frac{\rho L}{A}$.

The power used by a single cell is IV_{cell}: for a series of n cells, $V_{total} = \sum_{1}^{n} V_{cell}$ and the power is IV_{total}.

As mentioned in Section 1.5, 1 faraday is equivalent to one mole of electrons or 96,500 coulombs of electricity. The current efficiency, θ_C, of an electrochemical process is defined as the ratio of the amount of metal actually deposited to the amount that could be theoretically deposited by the current supplied to the cell:

$$\Theta_C = \frac{\left(\frac{\text{gms}\cdot\text{metal}}{\text{MW}}\right)\ n}{\frac{I\cdot t}{96{,}500}} = \frac{\left(\frac{\text{gms}\cdot\text{metal}}{\text{cell}}\right)\cdot n\cdot 96{,}500}{I\cdot t\cdot \text{MW}}$$

where t is in seconds.

The electrical energy consumed is

$$(\text{watt-hrs}) = I\cdot V_{cell}\cdot t'$$

where t' is in hours.

The specific energy consumption per gm-mole of metal produced is

$$(\text{kwh/gm-mole}) = \frac{I\cdot V_{cell}\cdot t'\cdot n\cdot 96{,}500}{I\cdot t\cdot \Theta_C\cdot 1000}$$

or

$$\frac{\text{kwh}}{\text{gm-mole}} = \frac{V_{cell}\cdot n\cdot 0.02681}{\Theta_C}$$

In terms of lb-moles,

$$\text{Electrical Energy, (kwh/lb-mole deposited)} = \frac{12.16\ n\ V_{cell}}{\theta_c}$$

Some representative cases are as follows:

Process	E,volts	n	θ_c	kwh/lb-mole	kwh/lb
Zn electrowinning	~3.5	2	~0.90	94.58	1.45
Cu electrowinning	~2.5	2	~0.85	71.53	1.12
Cu electrorefining	~0.3	2	~0.95	7.68	0.12
Al electroreduction	~4.5	3	~0.85	193.0	7.1

EXAMPLE 6.3-3: The Hall process for electrolytic production of aluminum is based on the reaction:

$$2\ Al_2O_3 + 3\ C \rightarrow 4\ Al + 3\ CO_2$$

However, some CO is produced as a result of the parasitic back-reaction

$$2\ Al + 3\ CO_2 \rightarrow Al_2O_3 + 3\ CO$$

If all other side reactions can be neglected, and the gas escaping from the cell is 88% CO_2 and 12% CO, what is the current efficiency?

Solution: The reaction can be written

$$2\ Al_2O_3 + 3C \rightarrow (4-2x)\ Al + (3-3x)\ CO_2 + Al_2O_3 + 3x\ CO$$

The gas emitted by this reaction will have a composition ratio

$$\frac{CO}{CO_2} = \frac{3x}{3-3x} = \frac{x}{1-x}$$

In this case

$$\frac{x}{1-x} = \frac{12}{88} = 0.136$$

$$x = 0.119$$

The current efficiency can be expressed as the ratio of Al really produced to the theoretically possible Al

$$\Theta_c = \frac{(4-2x)}{4} = 1 - \frac{x}{2}$$

$$= 1-0.0595$$

$$\Theta_c = 0.94, \text{ or } \underline{94\%\ \text{efficiency}}.$$

EXAMPLE 6.3-4

A copper refinery has 200 tanks (cells) in series, each containing 20 cathodes and 22 anodes. The current is 10,000 amperes. If V_{cell} = 0.3 volts and the daily metal production is 115,000 lbs., what is the energy consumption per pound of copper produced and the energy efficiency? What is the power requirement?

Solution: The total voltage is (0.3)(200) = 60 volts. Therefore, the power required is

$$P = \frac{(60\ V)\ (10{,}000\ A)}{1000} = 600\ kw$$

Theoretically, 10,000 amps flowing for 24 hours should deposit

$$\frac{(10{,}000)(24)(3{,}600)}{(2)\ 96{,}500} = 4476.\ \text{gm. moles Cu}$$

or

$$\frac{(4476)(63.5)}{454} = 626.\ \text{lbs. Cu in a cell.}$$

On the average, the amount of copper deposited per cell in 24 hours in this refinery is:

$$\frac{115{,}000}{200} = 575\ \text{lbs. Cu per cell.}$$

Thus,

$$\Theta_c = \frac{575}{626} = 0.92$$

Knowing Θ_c, the energy consumption is

$$(\text{kwh/lb-mole}) = \frac{(12.16)(2)(0.3)}{0.92}$$

$$= 7.93$$

The energy consumption per lb is

$$\frac{7.93}{63.5} = 0.12\ \text{kwh/lb. Cu}$$

or, in S.I. units,

$$\frac{0.12\ \text{kwh}}{\text{lb Cu}}\quad \frac{3.6\ \text{MJ}}{\text{kwh}}\quad \frac{1.0\text{lb}}{0.454\ \text{kg}} = 0.99\ \text{MJ/kg Cu.}$$

6.4 STAGED HEAT BALANCES

As was pointed out in earlier sections of this chapter, overall heat or energy balances do not necessarily give any hint as to how a process works or what thermal conditions must really be met. In this section, the use of staged heat balances to analyze the behavior of various components or portions of a process is examined. In so doing, the concept of energy quality will be presented.

6.4.1 CRITICAL TEMPERATURES AND ENERGY QUALITY REQUIREMENTS

Some metallurgical processes operate well only if portions of the charge are maintained at or above some critical temperature. This is because phase changes, such as melting, or changes in physical properties of constituents, such as slag viscosities, may not occur unless the temperature is above some critical temperature, such as a melting point. In other cases, while it may be thermodynamically possible for a chemical reaction to occur at lower temperatures, the rate of the reaction may not be fast enough, below some critical temperature, for economical operation of a process requiring that reaction: the volume of the process vessel might have to be immense, in order to give a long residence time if the reaction is slow, in order to reach the desired daily production. An illustration of this is given in Fig. 6.4-1, where the time for 80% completion of the reaction

$$2Fe_3O_4 + 1/2\ O_2 \rightarrow 3Fe_2O_3$$

in air is plotted as a function of initial temperature of the reactants. This reaction liberates heat and the heat is included in the heat balance of iron ore pellet plants. It may be reasonably assumed that for practical purposes, 1400°F is the critical temperature for this reaction, i.e., the reaction is too slow below this temperature for any significant reaction to occur in a plant operating at commercial throughput rates.

The iron blast furnace, depicted in Fig. 6.2-3, is essentially a continuous process. The preheated air blast reacts with coke at the tuyeres to form CO and H_2 which reacts in turn with iron oxides as it travels upwards in the furnace, reducing them to metal. The temperature at the tuyeres, as calculated in Example 6.2-2, is above that needed to melt the iron, so the iron melts

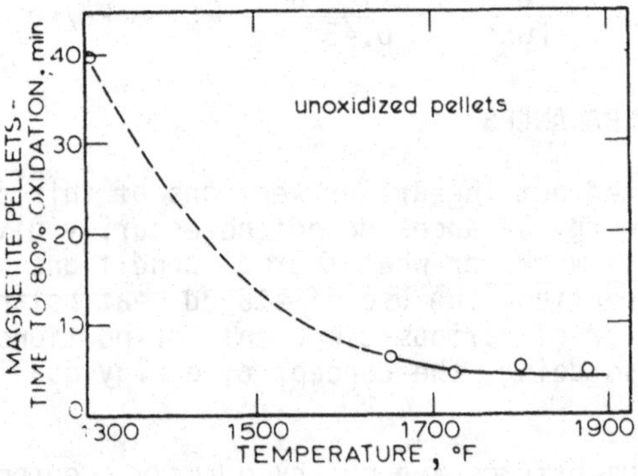

Fig. 6.4-1: Time to 80% completion of the reaction $2Fe_3O_4 + 1/2\ O_2 \rightarrow 3Fe_2O_3$ for 5/8" diameter pellets in air. Temperature is initial temperature of solid. (Ref., P. O. Pape, R. D. Frans, G. H. Geiger, Ironmaking and Steelmaking, 3 (1976), p. 138.)

as does the slag. Both melting processes, however, require energy but, more importantly, energy available at a temperature necessary to do the job.

The amount of energy needed to melt one mol of iron is 3670 calories. If one mol of solid iron at 1537°C is brought in contact with an infinite supply of gas at 1536°C (containing an infinite amount of thermal energy), it will not melt, because none of the energy will transfer to it. Thermal energy only transfers from a higher to a lower temperature. If it is put in contact with gas at 1538°C, the gas will give energy to it until the gas temperature is 1537°C. For a gas with a specific heat of 15 cal/mol-°C), 245 mols of gas, initially at 1538°C, are required to melt one mol of iron. If the gas was presented to the iron at 1547°C, only 1/10th as much gas would be required, and so on, until very little gas, if initially hot enough, is needed to give up enough thermal energy, $n_{gas}\,[H_{T_{gas}} - H_{1537}]_{gas}$, to equal 3670 calories. The point is that the availability of 3670 calories, at a temperature below which it cannot perform the required job is of no use.

Therefore, not only does the quantity of energy required have to be considered, but also the "quality" of that energy. Usually, the quality is described as the amount of energy that can be released by a substance at temperatures above some critical level, 1537°C in the above sample.

Many processes requiring high temperatures to carry out desired phase changes or chemical reactions have products or wastes which contain considerable total energy, but of low quality, meaning that it is available at a lower temperature and can only be used to carry out functions for which only low quality energy is required.

Care must always be exercised in analyzing a process to be sure that the critical temperatures required are achieved within the process, and that the energy requirements at these temperatures are met, i.e., adequate energy quality is present. To do so often requires that the process be broken into stages, usually separated by temperature requirements. The most well-known example is that of the iron blast furnace, whose productivity has increased by over 100% during the period 1950 - 1977, largely because of improved understanding of its operation which came from utilizing such techniques as staged heat balances.

Referring to Fig. 6.4-2, an overall heat balance on the blast furnace can be constructed, with a reference temperature of 298K, considering only the incoming and outgoing streams. The diagram for such a balance would be:

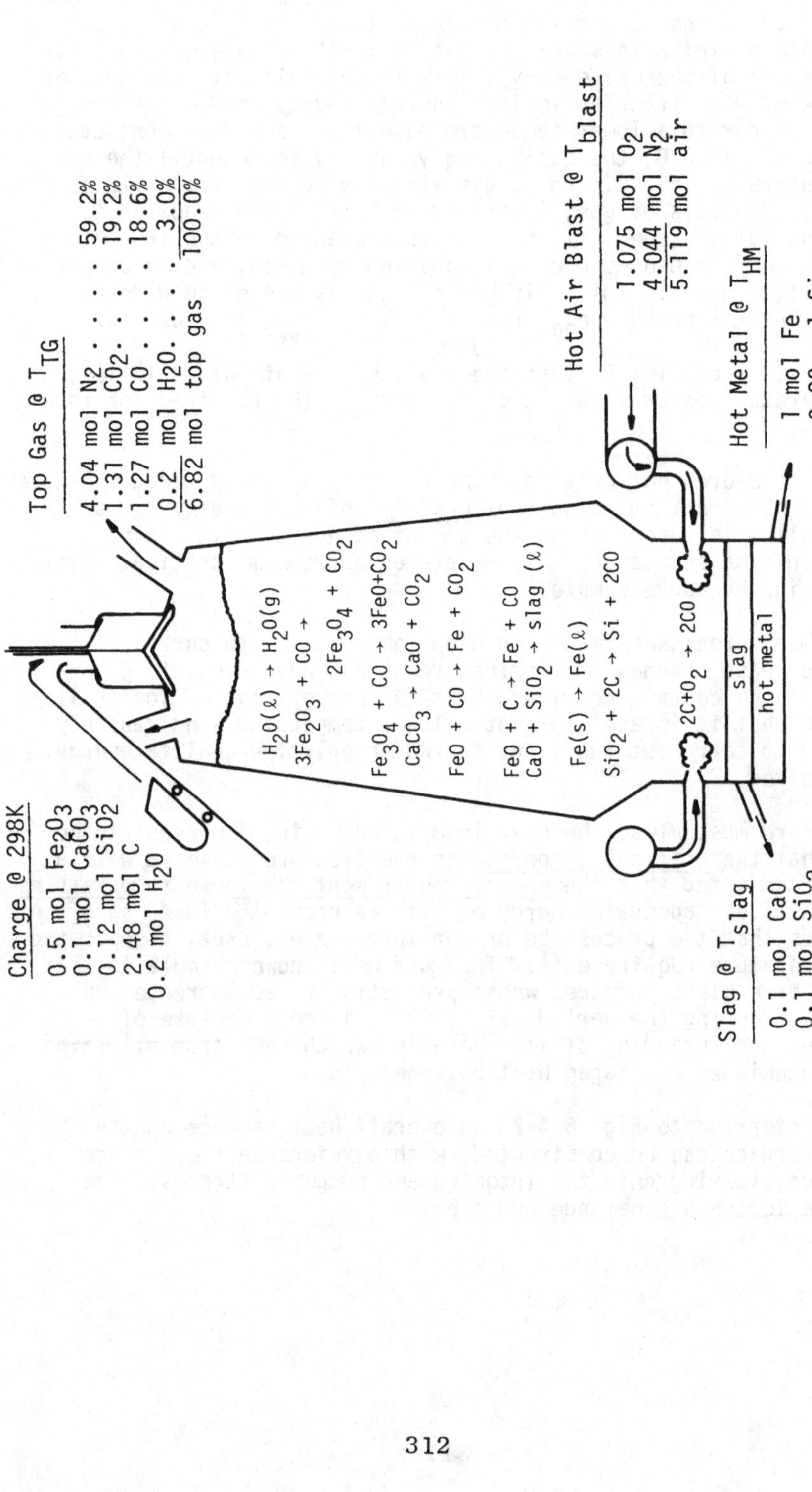

Fig. 6.4-2 Analyses, conditions, and amounts of materials entering and leaving a blast furnace.

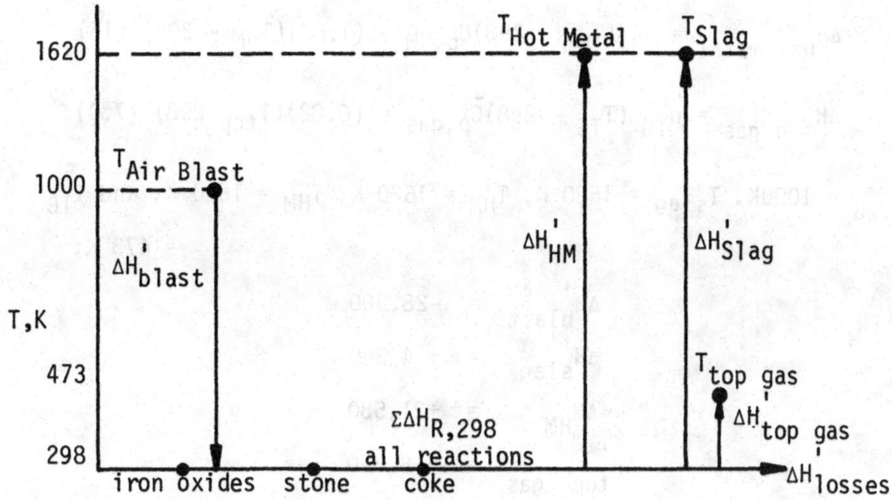

From an overall viewpoint, the oversimplified model depicted in Fig. 6.4-2, identifies the major reactions. The heats of reaction can then be written so that the product stream analyses reflect the proportions of each reaction that occur relative to 1 mol Fe:

mol reactant / mol iron	Reaction		$\Delta H^\circ_{R,298}$/mol reactant	$\Delta H_{R,298}$
1.27	$C + 1/2\ O_2$	$\rightarrow CO$	- 26,400	- 33,530
1.21	$C + O_2$	$\rightarrow CO_2$	- 94,050	-113,800
0.2	$H_2O(\ell)$	$\rightarrow H_2O(g)$	+ 10,510	+ 1,260
0.10	$CaCO_3$	$\rightarrow CaO + CO_2$	+288,450	+ 28,845
0.50	Fe_2O_3	$\rightarrow 2Fe + 3/2\ O_2$	+196,800	+ 98,400
0.02	SiO_2	$\rightarrow Si + O_2$	+209,500	+ 4,190
			$\Sigma\Delta H_{R,298}$	-14,630

The sensible heats contained in the air blast, slag, hot metal, and top gas are estimated to be:

$$\Delta H'_{blast} = -n_{air}\,(T_{blast} - 298)\bar{C}_{p,air} = -(5.11)(298-T_B)(7.8)$$

$$\Delta H'_{slag} = n_{slag}(T_{slag}-298)\bar{C}_{p,slag} = (0.2)(T_{slag}-298)(16.5)$$

$$\Delta H'_{hot\ metal} = n_{HM}(T_{HM} - 298)\bar{C}_{p,HM} = (1.02)(T_{HM} - 298) \quad (16)$$

$$\Delta H'_{top\ gas} = n_{TG}\,(T_{TG} - 298)\bar{C}_{p,gas} = (6.82)(T_{top}-298) \quad (7.0)$$

If T_B = 1000K, T_{slag} = 1620 K, T_{HM} = 1620 K, T_{HM} = 1620 K, and T_{TG} = 473 K;

$$\Delta H'_{blast} = -28{,}000$$
$$\Delta H'_{slag} = +\ 4{,}360$$
$$\Delta H'_{HM} = +21{,}580$$
$$\Delta H'_{top\ \ gas} = +\ 8{,}350$$

Then, the <u>overall heat balance</u> is

$$+\Delta H'_{blast} + \Sigma\Delta H_{R,298} + \Delta H'_{HM} + \Delta H'_{slag} + \Delta H'_{top\ gas} + \Delta H_{loss} = 0$$

or

$$\Delta H'_{HM} + \Delta H'_{slag} + \Delta H'_{top\ gas} + \Delta H'_{blast} + \Sigma\Delta H_{R,298} = -\Delta H'_{loss}$$

or

$$21{,}580 + 4{,}360 + 8{,}350 + (-28{,}000) + (-14{,}630) = \Delta H'_{loss}$$
$$8{,}340 \text{ cal/mol Fe} = \Delta H'_{loss}$$

The heat loss is 4.8% of the energy into the process, which in this case would be 175,330 cal, the sum of all exothermic reactions plus the sensible heat in the air blast.

Notice that nowhere in the preceeding calculation was there any mention of having to provide any temperature above that at which hot metal and slag emerge from the furnace. But, if some phase (gas, in this case) was not present <u>above</u> that temperature to <u>transfer heat</u> to the iron and slag, the iron and slag would never reach the necessary temperatures. Therefore, thermal energy must be provided with the proper <u>quality</u>. The overall balance does not even hint at this!

To see these needs more clearly, staged heat balances have been developed by many engineers and researchers in this field. Each differs slightly from the others based on the assumption of temperatures at which various reactions are assumed to occur, releasing or absorbing heat. For the pruposes of this monograph, a simplified schematic approach will be taken.

Based on a wide variety of measurements and observations made on operating furnaces, and laboratory simulations of the furnace shaft, the general picture that has emerged is that three temperature regions within the furnace should be considered: with certain reactions occurring within each region.

Region	Temperature	n(moles)	Reactions	$\Delta H'$ (cal/mole Fe)
	298 K	0.2	$(H_2O(\ell) \rightarrow H_2O(g))$	+ 1,260
		0.1	$(CaCO_3 \rightarrow CaO + CO_2)$	+28,850
		0.5	$(Fe_2O_3 + CO \rightarrow 2FeO + CO_2)$	+ 925
Low Temperature				
		0.71	$(FeO + CO \rightarrow Fe + CO_2)$	- 2,300
		+	Sensible heat absorbed by solids to reach 1273K	+24,510
		+	Heat losses (10% of total heat loss)	835
			Net	+54,080
	1273 K			
		0.29	$(FeO + C \rightarrow Fe + CO)$	+11,470
		1.0	$(Fe(s) \rightarrow Fe(\ell))$	+ 3,700
Intermediate		0.1	$(CaO + SiO_2 \rightarrow slag\ (\ell))$	- 2,000
		+	Sensible heat absorbed by condensed phases to reach 1425K from 1273K	+ 3,720
		+	Heat losses (20% of total heat loss)	+ 1,670
			Net	+18,560
	1425 K			
		0.02	$(SiO_2 + 2C \rightarrow Si + 2CO)$:	+ 3,060
High Temperature			Superheating hot metal:	+ 3,180
			Superheating of slag:	+ 620
			Heat losses (70% of total heat loss):	+ 5,840
	1620 K			
			Net	+12,700
			Total	+85,340

The only energy input is via coke reacting with the air blast. The constituents in the blast react with coke, as in Example 6.2-2, releasing heat, raising the reaction products to the AFT and putting a certain number of moles, n_g, of gases into the furnace. This gas stream rises, giving up heat to the descending liquids and, above them, the solids.

In the construction of staged heat balances, in this case, what is done is to calculate the net heat requirements of the solids within each zone, including the heat losses from the furnace in that zone and see what energy is therefore required to be given up by the hotter gases as they move countercurrently through the zone. In effect, we make a heat balance on each zone.

In the present case, in the Low Temperature zone, the heat balance would look as follows:

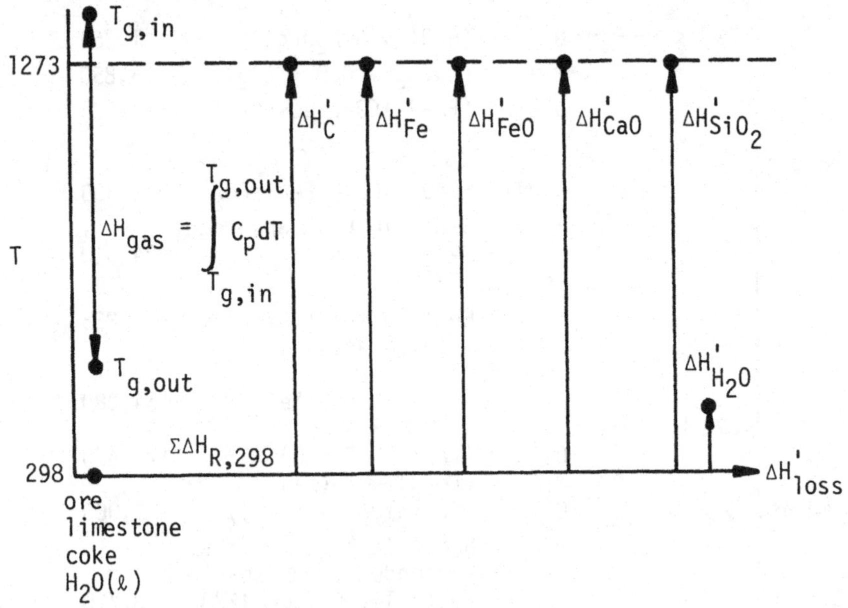

and, as indicated in the preceeding table, the sum of all of the heat requirements is 54,080 cal/mol Fe, which will have to come from the gas.

In the next stage, the thermal requirements to finish reducing the FeO, this time using C instead of CO, and to heat the Fe, the remaining coke, and slag, to the melting point of a carbon-saturated iron, and melt both the iron and slag, must be met. They amount to 18,560 cal/mol Fe. Finally, in the High Temperature zone, energy to superheat the slag and metal, provide for heat losses to cooling water and the hearth, and reduce some SiO_2 to Si dissolved in the iron must be provided, all by gases at temperatures above the temperatures of the condensed phases.

Fig. 6.4-3 shows the enthalpy needs of the condensed phases graphed as a function of temperature. In temperature regions where there are large heat requirements, such as in the Intermediate region, the slope decreases, indicating that considerable energy is needed <u>in those temperature regions</u>.

The total needed is 85,340 cal/mol Fe, if the metal and slag only reach 1620K and only 0.02 mol SiO_2 is reduced. If the metal and slag are to be hotter and more SiO_2 is to be reduced, more

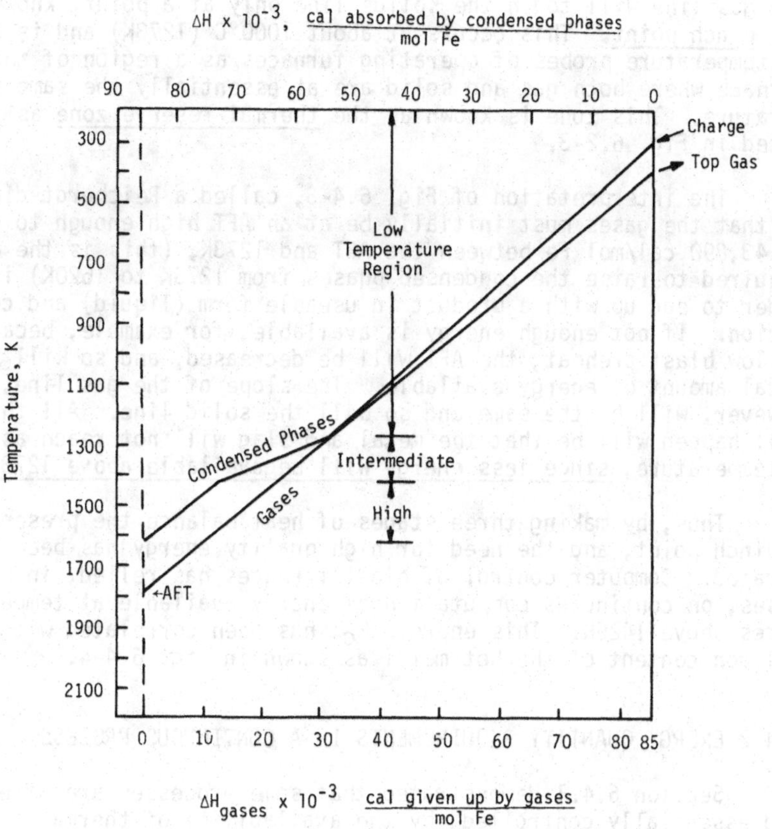

Fig. 6.4-3: Staged heat balance for blast furnace with enthalpy requirements of condensed phases plotted right to left on upper scale, and enthalpy yielded by the gases as they cool from the AFT plotted on the lower scale.

total energy will be needed <u>and more high quality energy</u> will be needed.

On an overall basis, enough energy has to be supplied by the blast through combustion and sensible heat in the preheated air to equal this total required. In addition, enough energy has to be available to bring the condensed phases up to the necessary intermediate temperatures. This is tested by plotting, also on Fig. 6.4-3 the change in sensible heat of the gases as they decrease from the AFT. Since the heat capacity and volume of the gases doesn't change significantly as the gas cools from the AFT, the result is essentially a straight line with slope equal to $C_p \cdot n_g$. Since the gases are the heat source for the condensed phases, <u>this line can never be at a lower temperature than the condensed phase</u>

line. Because of the shape of the condensed phase enthalpy plot, the gas line will touch the solids line only at a point, known as the pinch point. This occurs at about 1000°C (1273K) and is found in temperature probes of operating furnaces as a region of the furnace where both gas and solid are at essentially the same temperature. This zone is known as the thermal reserve zone as indicated in Fig. 6.2-3.

The interpretation of Fig. 6.4-3, called a Reichardt diagram, is that the gases must initially be at an AFT high enough to give up 43,890 cal/mol Fe between the AFT and 1273K, (this is the energy required to raise the condensed phases from 1273K to 1620K) in order to end up with a product in useable form (liquid) and composition. If not enough energy is available, for example, because of low blast preheat, the AFT will be decreased, and so will the total amount of energy available. The slope of the gas line, however, will be the same and so will the solid line. All that will happen will be that the metal and slag will not reach as high a temperature, since less energy will be available above 1273K.

Thus, by making three stages of heat balance the presence of a pinch point, and the need for high quality energy has been demonstrated. Computer control of blast furnaces has relied, in some cases, on continuous computation of energy available at temperatures above 1425K. This energy, E_c, has been correlated with the silicon content of the hot metal as shown in Fig. 6.4-4.

6.4.2 ENERGY QUANTITY REQUIREMENTS IN A CONTINUOUS PROCESS

Section 6.4.1 demonstrated that some processes are affected, and essentially controlled, by the availability of thermal energy at or above some critical temperature. Other processes, however, may have a different problem: too much energy is available at elevated temperatures, resulting in an unavoidable waste of energy or overheating of the material in the process upstream from that point, unless extraordinary measures are taken. An example of this follows:

In iron ore pelletizing, lower grade iron ore containing magnetite (Fe_3O_4) is upgraded by various mineral processing techniques to produce a concentrate consisting mostly of the iron mineral plus from 2 to 6% SiO_2. This concentrate is very finely divided and must be "put back together" in order to be handled and subsequently fed into a blast furnace or other reduction device. The process of putting it together is called agglomeration, and usually is done by adding clay and water to the concentrate, then forming pellets of this wet mixture, drying the pellets, and finally heating them to 1620K (2450°F) in order to cause sintering of the fine particles, thus imparting strength to the resulting

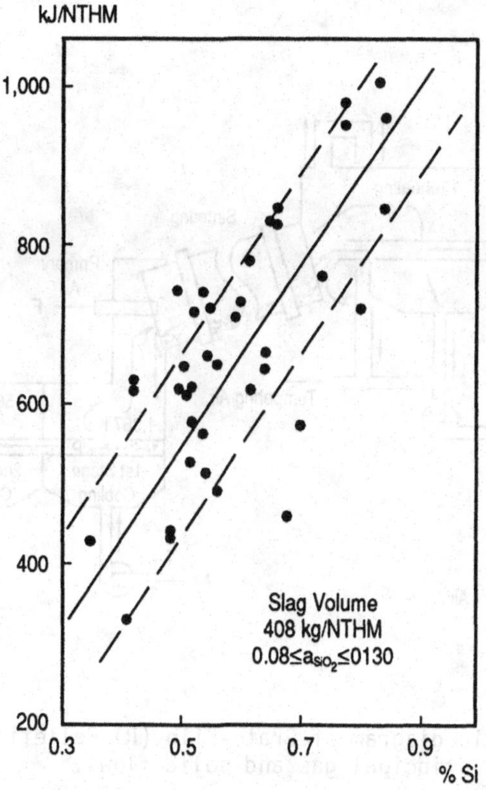

Fig. 6.4-4: E_C vs % Si (adapted from J. M. Vanlangan, A. Poos, R. Vidal, J. Metals, Dec., 1965, p. 9).

pellet. These later operations are referred to as <u>pelletizing</u>, and are performed in a variety of plants utilizing different schemes for contacting hot gases with the solids.

One such scheme is shown in Fig. 6.4-5 in which the wet pellets are placed on a moving grate, dried with gases and heated to 533 K (260°C) in the drying zone, then preheated to 1273 K (1000°C) by gases that enter the preheat chamber at 1328 K (1055°C). The pellets drop off the grate into a rotary kiln, which is heated by fuel burned at the opposite end. The pellets reach a temperature of 1620 K (1347°C) in the kiln and are then discharged into a cooler. In the cooler, cold air is forced up through the bed of pellets, recovering some of the sensible heat in them and carrying this heat back into the kiln as combustion air. Some of the heat is not recovered in the first stage cooler and is put into the atmosphere by the secondary cooling air.

In this process, if Fe_2O_3 were the original oxide, the only heat input would be the heat of combustion of the fuel. In the case where Fe_3O_4 is

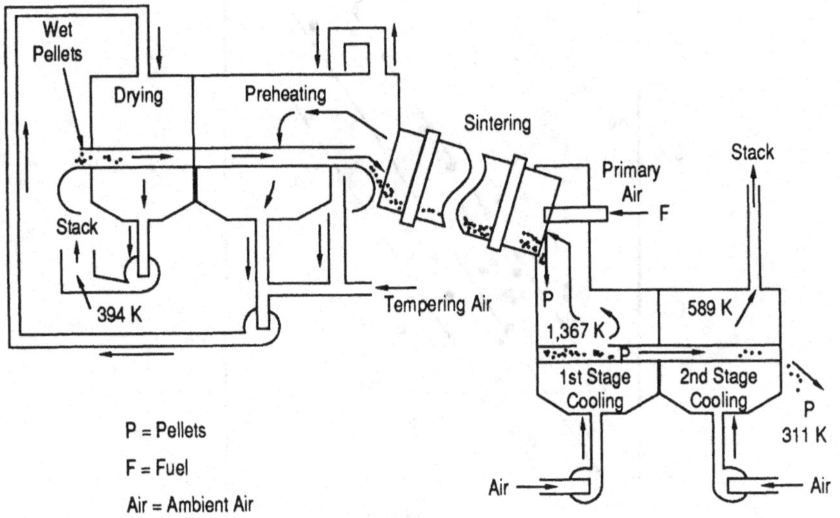

Fig. 6.4-5: Schematic diagram of Grate-Kiln (R) Pelletizing Plant showing principal gas and solid flows.*

present, however, the O_2 in the process gases will react with it, at temperatures where the reaction kinetics are significant (see Fig. 6.4-1), forming Fe_2O_3 according to the reaction

$$2Fe_3O_4 + 1/2\ O_2 \rightarrow 3Fe_2O_3$$

which gives off 114718 J per mole of Fe_3O_4 reacted. This becomes another source of heat, but it is released in the preheating zone, a different location than the fuel-based heat, where the critical reaction temperature is first reached by the solids.

A Sankey diagram, Fig. 6.4-6, illustrates the flow of energy in this process, in the case where magnetite is being processed.

Focus on the preheat zone and consider what would happen if the side stream (A) of heat was not taken out of stream (B) and sent to the drying zone. That portion of the diagram would be as shown in Fig. 6.4-6A on page 322.

* Grate-Kiln is a registered tradename of Allis-Chalmers Mfg. Co., Milwaukee, Wisc.

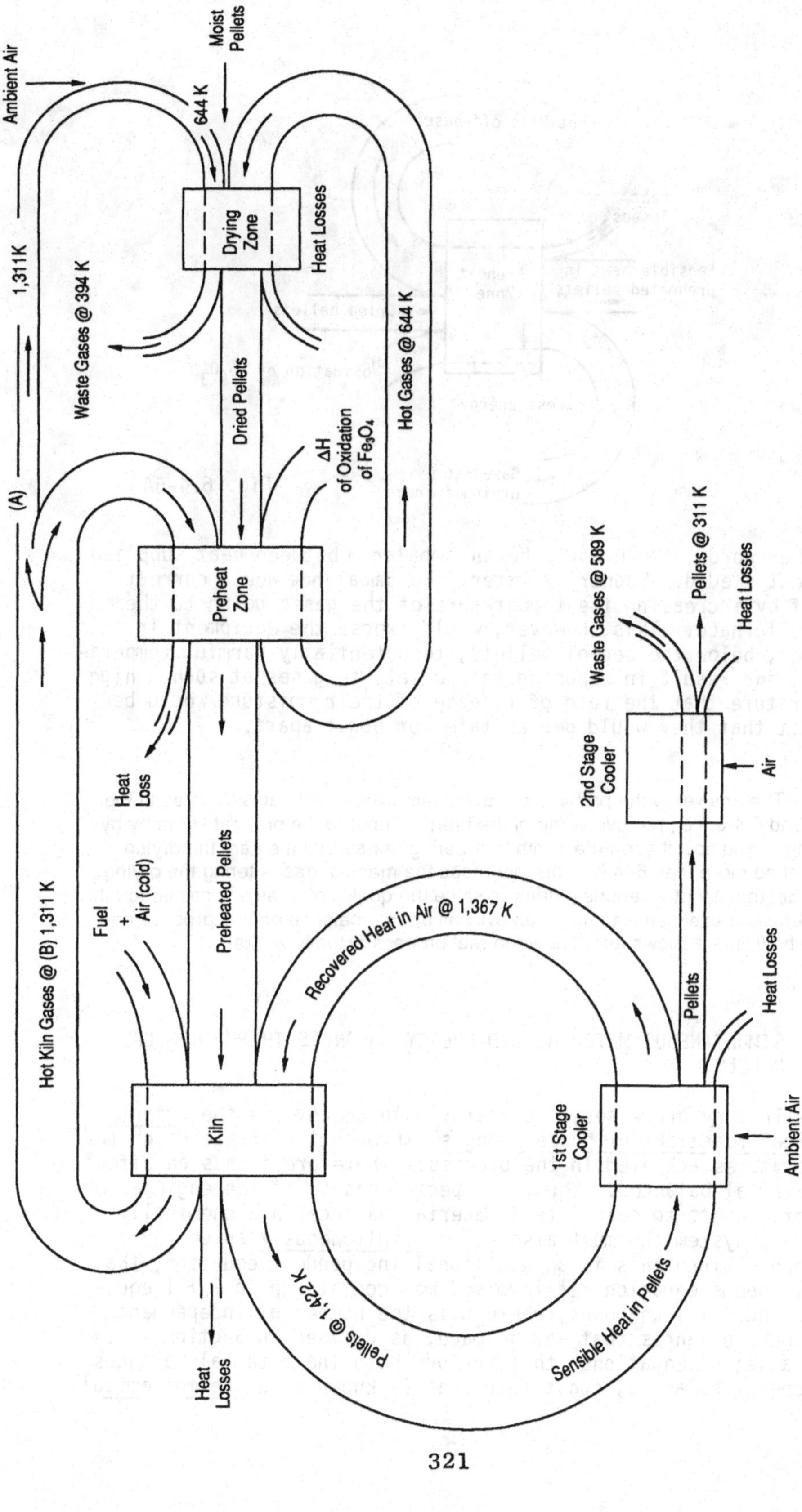

Fig. 6.4-6 Schematic Sankey diagram for Grate-Kiln R Iron ore Pelletizing Plant shown in Fig. 6.4-5.

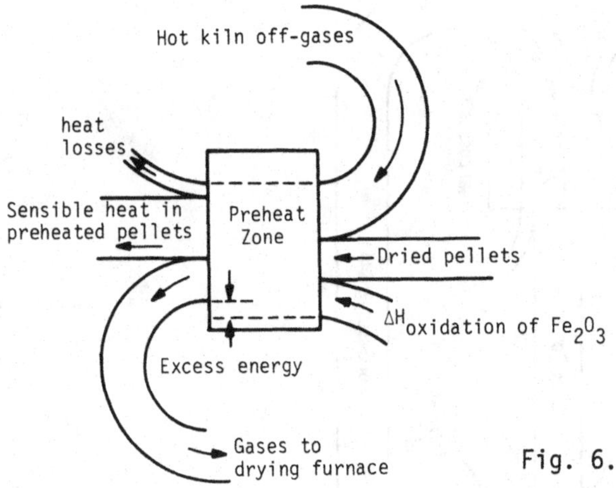

Fig. 6.4-6A

In other words, there would be an imbalance between heat supplied and heat needed. Sooner or later, the imbalance would correct itself by increasing the temperature of the gases going to the drying furnace. This, however, would expose the equipment in that chamber, below the bed of pellets, to potentially harmful temperatures, and result in exposing wet pellets to gases at such a high temperature that the rate of release of their moisture would be so fast that they would decrepitate, or burst apart.

The answer to the problem of having this excess of heat is to do as in Figs. 6.4-5 and 6.4-6, i.e., remove some of the thermal input to the preheat stage by bypassing it, and cool it separately with "tempering" air so that it enters the drying stage at no more than 644 K. This increases the mass of gas entering the drying zone, but that is not a serious detriment since the quality of energy is still adequate. An interesting exercise is to make an over-all energy balance on this process and notice how little it shows about how the real process actually works!

6.5 SIMULTANEOUS MATERIAL AND ENERGY BALANCES-THERMOCHEMICAL MODELS

In many processes, the energy balance governs the extent of chemical reactions or phase changes, through its influence on the temperatures achieved in the process. Therefore it has an effect on material balances. Thus, in special cases, if the engineer or operator wants to calculate a material balance in a chemically reactive system, he must also solve simultaneously an energy balance. Since this is an additional independent equation, the simultaneous equation set involved may contain up to C + 1 equations and C + 1 unknowns, where C is the number of independent component balances that may be made, as defined in Section 4.3. Such a set of equations, that include both the material balances and energy balances, constitute what is known as a thermochemical model of a process.

In Example Problems 4.3-7 and 6.1-3 a basic oxygen furnace was analyzed. In the first example, certain chemical analyses and/or weights were assumed to be known, and a set of six material balance equations were solved for the remaining unknown weights of materials or analyses.

In the latter case, the results obtained in the first example were used to determine the heat loss from the system by means of a heat balance. In effect, a heat balance was solved which simultaneously satisfied the material balance already determined.

In the following example, rather than following the pattern of the previous BOF examples, i.e., analysis of a given process, the use of simultaneous heat and energy balances for prediction of the process' performance is illustrated. Data such as the heat loss term needed to close the heat balance is presumed to have been developed by experimental measurements and analysis.

The development of such a model is usually undertaken because it is desired to be able to predict, either for purposes of design or process control, the effects of changes in raw material chemistry or temperature, desired product chemistry or temperature, or other process variables, on material requirements or other operating variables that may be manipulated to maintain the process at desired conditions. Most often a model is programmed onto a digital computer so that variables may be manipulated and the effects of variations determined. In some cases access to the computer is by remote terminals used by operators of a process, with instantaneous results used by them to make control decisions.

As in the case of material balances alone, when constructing a thermochemical model all of the relationships between variables that are to be taken into account as well as the component and energy balance must be written down, and the entire system analyzed to see how many degrees of freedom are involved. Certain variables, those whose values are to be found, and not numbering more than C + 1, are left to be determined. All of the other variables (called design variables, and equal to N - (C + 1)) are then assigned values.

The methods of Section 4.3 may be used to help reduce the system of equations to the point where the fewest number of simultaneous equations have to be solved.

EXAMPLE 6.5-1: The basic oxygen furnace first described in Example 4.3-5, page 164, has to utilize hot metal with variable silicon content and temperature. Develop a thermochemical model to be used to predict the proper charge and oxygen requirements to make one ton of steel with 0.05% Si, at 1600°C, if the temperature and silicon content of the hot metal are available, but variable.

Solution: Identify those other analyses or conditions that are known, a priori, and not considered to be variable:

1. Analysis of scrap -
2. Analysis of lime -
3. Analysis of oxygen -
4. Probable slag analysis -
5. Slag temperature - T_{SL}
6. Heat loss - ΔH_{loss}
7. Steel composition -
8. Steel temperature - T_{steel}
9. Steel weight - W_{steel}
10. Composition of air -
11. Weight of waste gas - W_{gas} (See discussion below)

Identify the unknowns to be determined:

1. Weight of hot metal - W_{HM}
2. Weight of scrap - W_{SC}
3. Weight of lime - W_{CaO}
4. Weight of oxygen - W_{O_2}
5. Weight of CO_2 - W_{CO_2}
6. Weight of slag - W_{slag}
7. Weight of infiltrate air - W_{air}

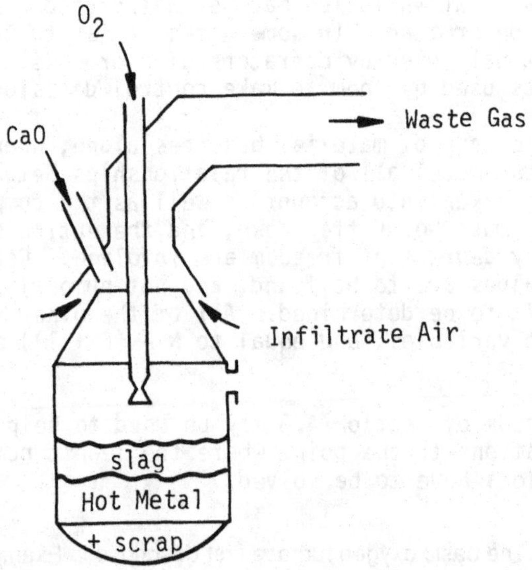

(Since most exhaust system fans run at a constant speed, the mass flow of gas pulled by the fan is relatively constant. Therefore, it is assumed that W_{gas} is constant at 264 kg/t of steel produced, based on the previously determined material balance, and that the heat loss determined in Example 6.1-4, page 263, can be utilized. This was based on a system in which CO from the decarburization reaction was burned to CO_2 above the mouth of the vessel by infiltrate air. Since the weight of CO_2 will vary, depending on the ratio of hot metal to scrap, the amount of infiltrate air, W_{air}, will also vary. Analysis of the material balance in Example 4.3-5, page 164, shows that the oxygen supplied (W_{oxygen}) to remove the carbon exceeds that needed to form CO, after FeO and SiO_2, by 22.25%.)

The carbon content of the hot metal, F_C, is a function of temperature, T_{HM}, and silicon content, F_{Si}, according to

$$F_C = 2.5 \times 10^{-5}\ T_{HM} - 0.5\ F_{Si} + 8. \times 10^{-3}$$

based on the Fe-C-Si phase diagram. There are seven unknowns and therefore seven equations are needed. Since the energy balance must be incorporated, this means that six material balances will be required. They are as follows:

<u>CARBON BALANCE</u>:

$$F_C \cdot W_{HM} + 0.00646\cdot W_{SC} = 0.0005\cdot \overset{1000}{\cancel{W_{Steel}}} + 0.272\ W_{CO_2}$$

$$F_C \cdot W_{HM} + 0.00646\cdot W_{SC} - 0.272\cdot W_{CO_2} = 0.5 \qquad (1)$$

<u>SILICON BALANCE</u>:

$$F_{Si} \cdot W_{HM} + 0.00354 W_{SC} = (28/60)(0.149) W_{SL} + 0.0005\ \overset{1000}{\cancel{W_{Steel}}}$$

$$F_{Si} \cdot W_{HM} + 0.00354 W_{WC} - 0.06965\cdot W_{SL} = 0.5 \qquad (2)$$

<u>IRON BALANCE</u>:

$$(1.0-F_C-F_{Si})W_{HM} + 0.99\cdot W_{SC} = (55.85/71.85)(0.351)W_{SL} + 0.99\cdot \overset{1000}{\cancel{W_{Steel}}}$$

$$(1.0-F_C-F_{Si})W_{HM} + 0.99\cdot W_{SC} - 0.2728\cdot W_{SL} = 999. \qquad (3)$$

<u>OXYGEN BALANCE</u>:

$$1.0\cdot W_{OX} + 0.233\cdot W_{Air} = \underbrace{(0.233\cdot W_{air} - 0.2827\ W_{CO_2})}_{\text{Wt. of } O_2} \underbrace{+0.727\cdot W_{CO_2}}_{\text{Wt. of } O_2 \text{ in } CO_2 \text{ in waste gas}}$$

$$+ (32/60)(F_{SiO_2})\cdot W_{SL} + (16/71.85)(F_{FeO})\cdot W_{SL}$$

$$1.0 \cdot W_{OX} - 0.44\ W_{CO_2} - 0.1575 \cdot W_{SL} = 0 \qquad (4)$$

LIME BALANCE:

$$1.0 \cdot W_{CaO} - 0.5\ W_{SL} = 0 \qquad (5)$$

TOTAL MATERIAL BALANCE:

$$W_{HM} + W_{CaO} + W_{Oxy} + W_{Air} + W_{SC} = W_{SL} + \overset{1000}{W_{Steel}} + \overset{264}{W_{Waste\ Gas}}$$

$$1.0\ W_{HM} + 1.0\ W_{CaO} + 1.0\ W_{Oxy} + 1.0\ W_{Air} + 1.0\ W_{SC} - 1.0\ W_{SL} = 1264 \qquad (6)$$

ENERGY BALANCE:

As shown in Example 6.1-4, the heat input relative to a base temperature of 1600°C (1873°K) is from the oxidation of carbon to CO_2, Si to SiO_2, Fe to FeO and the formation of slag:

For carbon oxidation:

$$\frac{W_{CO_2}\ \text{kg } CO_2}{} \cdot \frac{\text{kg-mol } CO_2}{44\ \text{kg } CO_2} \cdot \frac{10^3\ \text{g-mol}}{\text{kg-mol}} \cdot \frac{-418{,}148.96\ \text{J}}{\text{g-mol}} = -9.5034 \times 10^6 W_{CO_2}$$

For silicon oxidation:

$$\frac{0.149\ \text{kg } SiO_2}{\text{kg slag}} \cdot \frac{W_{SL}, (\text{kg slag})}{} \cdot \frac{10^3\ \text{g}}{\text{kg}} \cdot \frac{\text{g-mol } SiO_2}{60.08\ \text{g } SiO_2} \cdot \frac{-778{,}224.0\ \text{J}}{\text{g-mol } SiO_2}$$

$$= -1.930 \times 10^6 W_{SL}$$

For iron oxidation:

$$\frac{0.351\ \text{kg FeO}}{\text{kg slag}} \cdot \frac{W_{SL}, \text{kg slag}}{} \cdot \frac{10^3\ \text{g}}{\text{kg}} \cdot \frac{\text{g-mol FeO}}{71.85\ \text{g FeO}} \cdot \frac{-233{,}383.52\ \text{J}}{\text{g-mol FeO}} = -1.140 \times 10^6 W_{SL}$$

For slag formation:

$$\frac{0.149\ \text{kg } SiO_2}{\text{kg slag}} \cdot \frac{W_{SL}, \text{kg slag}}{} \cdot \frac{10^3\ \text{g}}{\text{kg}} \cdot \frac{\text{g-mol } SiO_2}{60.08\ \text{g } SiO_2} \cdot \frac{\frac{-83{,}680.0\ \text{J}}{\text{g-mol}}}{1.65\ CaO \cdot SiO_2} = -2.0753 \times 10^5\ W_{SL}$$

Sensible heat terms are as follows:

Hot Metal:

$$\Delta H'_{HM} = \frac{W_{HM}}{53.6}[66.944\,J/mol\cdot K][1873 - T_{HM}]\times 10^3 = \left[2.339\times 10^6 - 1248.9T_{HM}\right]\times W_{HM}$$

$$\Delta H'_{HM} = \left[2.339\times 10^6 - 1248.9\,T_{HM}, J\right]\times W_{HM}$$

Lime:

$$\Delta H'_{CaO} = \frac{W_{CaO}}{M.W.CaO}\overbrace{\left[H_{1873} - H_{298}\right]_{CaO}}^{82{,}801.36}$$

$$\Delta H'_{CaO} = 1.476\times 10^6\, W_{CaO}, J.$$

Oxygen:

$$\Delta H'_{O_2} = \frac{W_{Oxy}}{32}\overbrace{\left[H_{1873} - H_{273}\right]_{O_2}}^{54{,}601.2}$$

$$\Delta H'_{O_2} = 1.706\times 10^6\, W_{Oxy}, J.$$

Scrap:

$$\Delta H'_{SC} = \frac{W_{SC}}{55.85}\overbrace{\left[H_{1873} - H_{298}\right]_{SC}}^{77{,}613.2}$$

$$\Delta H'_{SC} = 1.3897\times 10^6\, W_{SC}, J.$$

Air:

$$\Delta H'_{Air} = \frac{W_{Air}}{29.}\overbrace{\left[H_{1873} - H_{298}\right]_{Air}}^{51714.2}$$

$$\Delta H'_{Air} = 1.7836\times 10^6\, W_{Air}, J.$$

Waste Gas:

O_2 in Waste Gas:

$$\Delta H_{O_2} = 3.515\times 10^4\, W_{Air} - 5.4861\times 10^4\, W_{CO_2}$$

N_2 in Waste Gas:

$$\Delta H_{N_2} = 4.328\times 10^7 - 3.819\times 10^4\, W_{Air} - 1.043\times 10^5\, W_{CO_2}$$

CO_2 in Waste Gas:

$$\Delta H_{CO_2} = 1.773\times 10^5\, W_{CO_2}$$

$$\Delta H'_{waste\ gas} = 4.328\times10^7 - 3{,}040.0\ W_{Air} + 18{,}186.27\ W_{CO_2}$$

$$\Delta H'_{loss} = 3.435\times10^8$$

Thus, since input = output,

$$9.5034\times10^6\ W_{CO_2} + 3.2775\times10^6\ W_{SL} = \left[2.339\times10^6 - 1248.9\ T_{HM}\right] \cdot W_{HM} + 1.476\times10^6\ W_{CaO} + 1.706\times10^6\ W_{Oxy} + 1.7836\times10^6\ W_{Air} + 1.3897\times10^6\ W_{SC} + 4.328\times10^7 - 3040.0\ W_{Air} + 18{,}186.27\ W_{CO_2} + 3.435\times10^8$$

OR:

$$-\left[2.339\times10^6 - 1248.9\ T_{HM}\right] W_{HM} - 1.476\times10^6\ W_{CaO} - 1.706\times10^6\ W_{Oxy} - 1.7806\times10^6\ W_{Air}$$

$$-1.3897\times10^6\ W_{SC} + 9.4852\times10^6\ W_{CO_2} + 3.2775\times10^6\ W_{SL} = 3.8678\times10^8 \quad (7)$$

The resulting 7 x 7 matrix is as follows:

unknown / balance	W_{HM}	W_{CaO}	W_{O_2}	W_{Air}	W_{SC}	W_{CO_2}	W_{SL}	Constant
carbon	F_C	0	0	0	0.00646	-0.272	0	0.5
silicon	F_{Si}	0	0	0	0.00354	0	–0.06965	0.5
iron	$1.0-F_C-F_{Si}$	0	0	0	0.99	0	–0.2728	999.
oxygen	0	0	1.0	0	0	-0.44	–0.1575	0
lime	0	1.0	0	0	0	0	–0.5	0
total	1.0	1.0	1.0	1.0	1.0	0	–1.0	1264.
energy	$-[2.339\times10^6 - 1248.9\ T_{HM}]$	-1.476×10^6	-1.706×10^6	-1.7806×10^6	-1.3897×10^6	$+9.4852\times10^6$	$+3.2775\times10^6$	$+3.8678\times10^8$

Solution of the seven simultaneous equations with various values of F_C, F_{Si} and T_{HM} allows study of the effects of variations of F_{Si} and T_{HM} on the material requirements for making steel via the BOF. The values of the variables corresponding to a set of F_C, F_{Si}, and T_{HM} values are shown in the final table, along with the computed value of the %Scrap in the metallic charge.

F_{Si}	T_{HM}	F_C	W_{HM}	W_{CaO}	W_{Oxy}	W_{Scrap}	W_{Slag}	%Scrap
.015	1620K	.041	744.2	85.4	79.0	346.5	170.7	31.7
.010	1620K	.043	763.3	59.1	74.1	311.6	118.2	29.0
.005	1620K	.046	772.2	31.4	69.5	286.1	62.8	27.0
.010	1670K	.044	735.0	57.8	73.2	338.6	115.6	31.5
.010	1720K	.046	699.7	56.1	72.8	372.8	112.2	34.7

The effect of increasing the silicon content of the hot metal, at the same temperature, on the ability to consume scrap is clearly shown in Fig. 6.5-1(a) and the effect of increases in hot metal temperature are shown in Fig. 6.5-1(b).

Changes in assumed slag chemistry, scrap chemistry, etc. can be studied in a similar manner, thus providing the process analyst with a powerful tool.

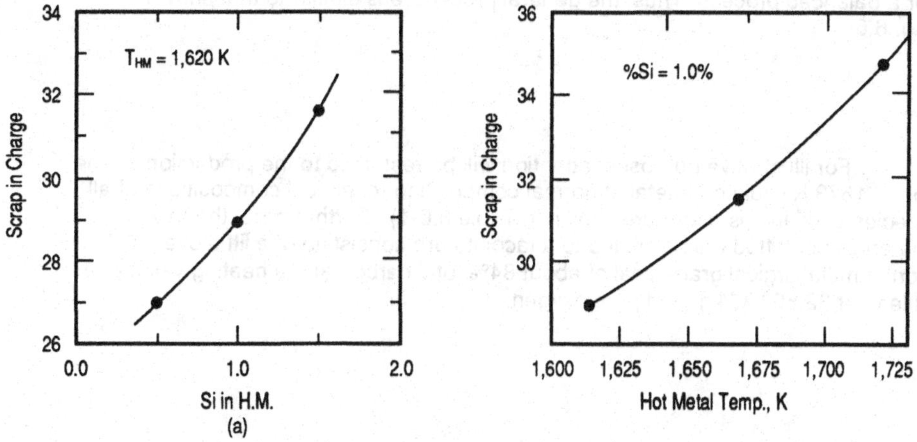

Fig. 6.5-1: Effect of % Si and hot metal temperature on % Scrap in metallic charge to BOF.

6.6 PROCESS ANALYSIS

In this final section two different approaches to the analysis of energy requirements for processes are presented. They are the work of other authors, acknowledged in each section, with only minor changes to fit the format of this monograph.

In the first approach, Dr. Robins demonstrates a step-by-step approach to the analysis of a process, starting with a very simplistic viewpoint, making it ever more complex, as allowances for inefficiences, realistic material ratios, etc., are added to the basic, theoretically perfect process, and in the end seeing the effect of these more realistic assumptions on the energy and material balances.

In the second section, Professor Kellogg presents his approach to the calculation of total process energy requirements, in which he takes a global viewpoint and includes the energy required, not only for the process itself, but also required to produce the electricity and the reagents used.

6.6.1 THEORETICAL ENERGY REQUIREMENTS FOR IRONMAKING*

Theoretically, as shown in Fig. 6.6-1, the reduction of one ton of iron from hematite requires 166.3 kg of carbon; and an additional 150.7 kg of carbon and 268.3 kg of oxygen are required to supply the heat required for reduction and melting. The purpose of this section is to show how the amounts of carbon and oxygen required to produce one ton of liquid iron change as we move, step by step, from the purely ideal process represented by Fig. 6.6-1 toward a more realistic process. The general procedure will be to approach each process as one approaches a reaction--i.e., the energy required will be determined as the difference between the heats of formation of the products and the heats of formation of the reactants. The net energy requirement so determined will then be satisfied by burning additional coal with oxygen, so as to arrive at the total coal requirement for a balanced process. Thus, the general procedure is similar to that shown in Fig. 6.6-1.

For illustrative purposes, attention will be restricted to the production of one ton of 1673 K liquid hot metal of normal composition (chemical compositions of all species used in this paper are shown in Table 6.6-1). Furthermore, the raw materials permitted will be limited to a taconite ore consisting of a little over 30% iron, a metallurgical grade coal of about 84% total carbon and a heating value of a little over 32,564 kJ/kg, and pure oxygen.

---

*Written by N. A. Robins, Vice President for Research, Inland Steel Company, East Chicago, Indiana. Published in Iron and Steelmaker, Vol. 2 (1976), p. 39. Reprinted with permission of the author and the Iron and Steel Society of AIME.

$$0.5\,Fe_2O_3\,(s, 298\,K) + 0.75C\,(s, 298\,K) \rightarrow$$

$$1.0\,Fe\,(L, 1672\,K)^* + 0.75\,CO_2\,(g, 298\,K) \rightarrow$$

$$\Delta H = 3305.3\ kJ/kg$$

$$C\,(s, 298\,K) + O_2\,(g, 298\,K) \rightarrow CO_2\,(g, 298\,K)$$

$$\Delta H = 32{,}814.7\ kJ/kg\ C$$

$$C\ FOR\ REDUCTION = \frac{1000}{55.85} \times (0.75 \times 12.01) = 161.3\ kg/t$$

$$C\ FOR\ HEAT = (1000 \times 3305.3) + 32{,}814.7 = 100.7\ kg/t$$

$$TOTAL\ C\ REQUIRED = 262.0\ kg/t$$

$$O_2\ FOR\ HEAT = \frac{100.7}{12.01} \times 32 = 268.3\ kg/t$$

* ALTHOUGH PURE IRON IS NOT LIQUID AT 1672 K, THIS STATE IS ASSUMED FOR COMPATIBILITY WITH LATER CALCULATIONS USING HOT METAL; THE HEAT CONTENT IS ESTIMATED FROM HIGHER-TEMPERATURE DATA.

Fig. 6.6-1: Theoretical carbon and oxygen requirements for the production of liquid iron from hematite.

In order to insure that the minimum amount of energy is determined in each case, the assumptions shown in Table 6.6-2 are used, except when otherwise noted. The first assumption is "100% chemical efficiency", which requires that all useful elements be in their lowest-energy chemical state. The second assumption is "100% thermal efficiency", which requires that all of the sensible heat be recovered and that there be no heat losses. The third assumption is "100% mechanical efficiency", which requires that there be no frictional losses and that all potential and kinetic energy be recovered.

For the first example, let us suppose that the reduction takes place in one step, as shown at the top of Fig. 6.6-2. The reactants are ore and coal at ambient temperature, and the products are liquid hot metal at 1673 K and oxides and gas at ambient temperature. By the assumption of 100% chemical efficiency, all of iron in the ore—both as magnetite and hematite—will end up in the hot metal. Thus, in order to produce one ton of hot metal containing 937.2 kg of iron, 2964.58 kg of taconite ore, also containing 937.2 kg of iron, are required. Then, the coal required to reduce the ore (including the Si, Mn, and P required in the hot metal) and to supply carbon to the hot metal is calculated as 196.0 kg. Finally, the quantities of gas and oxides produced are determined by material balance. The gas is composed of CO_2, H_2O, SO_3, and N_2—nitrogen, it is assumed, not being oxidizable. All of the oxygen in the gas comes from the ore and the coal. The oxides come from the gangue in the ore and the ash in the coal. It should be noted that the amount of coal required is only that required for chemical reaction and hot metal saturation; the coal required for heat will be discussed later.

TABLE 6.6-1

Chemical Composition of Various Species

Ore	Coal	Coke from Total C
28.58% Fe_3O_4	84.0% Total C	93.96% C
15.63% Fe_2O_3	64.9% Fixed C	3.58% SiO_2
47.54% SiO_2	5.0% H	2.46% AL_2O_3
1.87% MnO_2	3.7% O	
0.52% P_2O_5	1.4% N	Coke from Fixed C
1.62% Al_2O_3	0.5% S	
2.02% CaO	3.2% SiO_2	92.32% C
2.22% MgO	2.2% Al_2O_3	4.55% SiO_2
		3.13% Al_2O_3
Pellets	**Slag**	**Hot Metal**
91.50% Fe_2O_3	53.4% $CaO \cdot SiO_2$	93.72% Fe
6.47% SiO_2	28.1% $2CaO \cdot SiO_2$	4.60% C
1.15% MnO_2	13.4% $2CaO \cdot Al_2O_3 \cdot SiO_2$	0.80% Si
0.13% P_2O_5	3.0% $2MgO \cdot SiO_2$	0.80% Mn
0.25% Al_2O_3	2.0% MnO_2	0.80% P
0.25% CaO		
0.25% MgO	$\frac{CaO + MgO}{SiO_2 + Al_2O_3} = 1.1$	

TABLE 6.6-2

Assumptions for Calculating Process Energy Requirements

Assumption	Consequence
1. 100% Chemical Efficiency	A. All useful elements recovered in the products B. All side products in lowest-energy chemical state
2. 100% Thermal Efficiency	A. All sensible heat recovered B. No heat losses
3. 100% Mechanical Efficiency	A. No frictional losses B. All potential and kinetic energy recovered

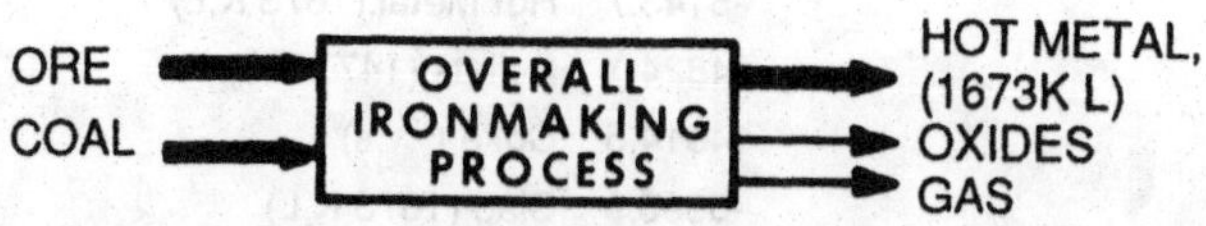

FIG.6.6-2

MATERIAL	WEIGHT (kg)	ENERGY (kJ)
INPUTS		
Ore	2964.6	-30,011,238.7
Coal	196	-246,450.4
Total	3160.6	-30,257,689.1
OUTPUTS		
Hot metal	1000	1,354,200
Oxides	1632.9	-23,355,830
Gas	527.8	-5,080,211.6
Total	3160.7	-27,081,841.6
NET		3,175,847.5

	NET ENERGY	Coal	O_2	Material Processed
Process	3,175,847.5	196	0	3160.7
Requirement	-3,175,847.5	95.6	249.2	344.9
Total	0	291.6	249.2	3505.6

Energy Equivalent: 9,174,052.0 kJ

Fig. 6.6-2: Energy Balance for ideal one-step ironmaking process.

As indicated earlier, the new energy requirement for this process is determined as the difference between the heats of formation of the products and those of the reactants. The appropriate heats of formation are indicated in the "Energy" column in the

Table 6.6-3

	Heat of Formation at 298 k(kJ/kg)		Heat Contents(kJ/kg)
Fe_2O_3	-5143.7	Hot Metal,(1673 K,L)	1354.2
Fe_3O_4	-4824.1	Pellets,(1477 K,s)	1089.5
SiO_2	- 14614.5	Coke (,s)	1644.7
MnO_2	-5993.9	Slag (1673 K,L)	1592.1
P_2O_5	-10371.9	Fe (1673 K,L)	1239.1
CaO	-11332.5		
$CaCO_3$	-12109.9		
MgO	-14932		Heat Value(kJ/kg)
Al_2O_3	-16441.8	Coal	33,176.7
Ore	-10123.2		
Pellets	-5841.3		
Coal	-1257.4		
Coke	-731.8		
CO_2	-8947.4		
H_20	-13431.9		
SO_3	-4935.8		
$CaO{\cdot}SiO_2$	-765.9		
$2CaO{\cdot}SiO_2$	-724.3	From the Oxides	
$2CaO{\cdot}Al_2O_3.SiO_2$	-380.3	at 298K	
$2MgO{\cdot}SiO_2$	-449.8		

table in the middle of Fig. 6.6-2 (heats of formation and sensible heats of all species used in this paper are shown in Table 6.6-3). It should be noted that the entry for coal is its heat of formation—not its heat of combustion, which would be over 5,275,000.0 kJ. Also, the entry for hot metal is mostly due to its sensible heat fusion. It can be seen that the net energy requirement for this one-step ironmaking process is 2,650,408.4 kJ/t hot metal.

The short table at the bottom of Fig. 6.6-2 summarizes these calculations on the line labeled "Process." In order to balance the process thermally, there is a requirement of 2,404,450.5 kJ, which can be met by burning 95.6 kg of coal with 249.2 kg of O_2. Thus, in order to satisfy the chemical and energy requirements of this ideal ironmaking process, a total of 291.7 kg of coal are required. This represents a total heat input of about 9,652,226.6 kJ/t of hot metal. One might compare this with the actual energy requirement for the production of one ton of hot metal in a blast furnace which, if one considers all of the energy required to provide the various inputs to the blast furnace process, runs about 26,747,134.0 kg/t. Thus, if the heat requirement derived in this exercise is considered to be ideal for these raw materials, one might say that the thermal efficiency of the blast furnace process is about 36%.

For the next example, let us take cognizance of the fact that we do not want to put a taconite ore and a metallurgical grade coal directly into the reduction process. Rather, we will put the ore through a pelletizing process and the coal through a coking process, and then feed the pellets and the coke to the reduction process. However, as before, the conditions of 100% chemical, thermal, and mechanical efficiency will be assumed for each process.

In the pelletizing process, shown in Fig. 6.6-3, pellets of the analysis shown in Table 6.6-1 are produced. As will be seen shortly, the amount required in the reduction process to produce one ton of hot metal is 1464.4 kg. Based on the assumption of 100% chemical efficiency, all of the iron in the ore will end up in the pellets; so 2964.6 kg of ore will be required, and the 847.3 kg of magnetite in the ore will have to be oxidized to hematite, which requires 29.2 kg of oxygen. All of the material that does not end up in the pellets ends up in the gangue. As before, the heats of formation of the products and reactants are shown in the "Energy" column in Fig. 6.6-3. Taking the difference indicates that 420,288.6 kJ are generated in this process. This is the equivalent to the heat generated by the oxidation of the magnetite.

The coking process is shown in Fig. 6.6-4. The amount of coke that will be required in the reduction process is 215.7 kg. Because of the assumption of 100% chemical efficiency, all of the carbon in the coal must end up in the coke. Therefore, 241.3 kg of coal are required. The coke consists only of carbon and ash, and the rest of the elements in the coal are oxidized to their lowest chemical state with 88.5 kg of oxygen to produce 113.6 kg of gas, mostly H_2O. Since the heat of formation of the gases is so low, there is energy production of over 1 million kJ in this process.

Finally, looking at the reduction process shown in Fig. 6.6-5, the 1464.4 kg of pellets and the 215.7 kg of coke are required by material balance and stoichiometry to make one ton of hot metal. In the process, 105.8 kg of oxides are generated from the gangue in the pellets and the ash in the coke; and 574.2 kg of gas, all carbon dioxide, are generated from the carbon in the coke in the course of reducing the pellets. The net result for the reduction process is a requirement of 3,412,873.3 kJ.

Fig. 6.6-6 is a summary of the three processes combined. Together the three processes require 1,676,822.1 kJ, which can be met by burning 50.5 kg of coal with 131.5 kg of oxygen. The result is that, in the three-step ironmaking process, a total of 291.8 kg of coal and 249.2 kg of oxygen are required, which is precisely the same as that required for the one-step ironmaking process. The only difference between the two cases is that, in the one-step process, only 3500.0 kg of material are processed per ton of hot metal; whereas in the combined three-step process, over 5000 kg of material must be processed for each ton of hot metal.

Fig. 6.6-3

MATERIAL	WEIGHT (kg)	Energy (kJ)
INPUTS		
Ore	2964.6	–30,011,238.7
O_2	29.2	0
Total	2993.8	–30,011,238.7
OUTPUTS		
Pellets	1464.4	–8,553,999.7
Gangue	1529.5	–21,877,527.6
Total	2993.9	–30,431,527.3
NET		–420,288.6

Fig. 6.6-3: Energy balance for ideal pelletizing process.

Fig.6.6-4

MATERIAL	WEIGHT (kg)	ENERGY (kJ)
INPUTS		
Coal	241.3	–303,410.6
O_2	88.5	0
Total	329.8	–303,410.6
OUTPUTS		
Coke	215.7	–157,849.3
Gas	114.1	–1,461,323.9
Total	329.8	–1,619,173.2
NET		–1,315,762.6

Fig. 6.6-4: Energy balance for ideal coking process.

In the next example, we take one more step back from ideality by recognizing that, in the pelletizing process, only the magnetite component of the ore can be used to make pellets and, in the coking process, only the fixed-carbon component of the coal can be used to make coke. Thus, for these two processes, we are relaxing the assumption of 100% chemical efficiency by no longer requiring that all of the iron or carbon in the reactants enter the products. However, we will still retain this assumption for the reduction process. This three-step process is referred to as "sub-ideal".

The sub-ideal pelletizing process is shown in Fig. 6.6-7. It can be seen that a little over 4500.0 kg of ore are required to make the same 1464.4 kg of pellets, because only the magnetite component of the ore is being recovered in the pellets. Since all of the iron in the pellets is coming from the magnetite, there is more magnetite to oxidize than in the previous case. Thus, slightly more oxygen is used—44.7 kg versus 29.2 kg in the previous case. The net result is a slight increase in energy available from the process because of more oxidation taking place.

In the sub-ideal coking process, shown in Fig. 6.6-8, the coal being used has 65% fixed carbon and 84% total carbon. Thus, when only the fixed carbon is allowed to end up in the coke, considerably more coal must be used to make the same amount of coke. It should be noted that 219.5 kg of coke are being produced in this case as opposed to 215.7 kg of coke in the previous case. The difference is in the ash content of the coke, which in these calculations is determined completely by the ratio of silica and alumina to carbon in the coal. Since less of the carbon in the coal is ending up in the coke in this case, the ash content must of necessity be higher. However, the effect of this small difference on the material and energy balances is negligible.

The significant change in the coking process is the large increase in the amount of gas produced, which is mostly carbon dioxide from the carbon in the coal that is now not going into the coke. 366.2 kg of gas are produced, as opposed to 114.0 kg produced previously; and this results in a net energy available in the coking process of over 3 million kJ, which is almost three times what it was in the previous case.

Finally, looking at the sub-ideal reduction process, shown in Fig. 6.6-9, the only difference between this case and the previous case is the approximately four more kilograms of ash in the coke, which show up as four more kilograms of oxide in the products; however, since these oxides do not change energy state, there is no change in the net energy required for the reduction process.

FIG. 6.6-5

MATERIAL	WEIGHT (kg)	ENERGY (kJ)
INPUTS		
Pellets	1464.4	–8,553,999.7
Coke	215.7	–157,849.3
Total	1680.1	–8,711,849
OUTPUTS		
Hot Metal	1000	1,354,200
Oxides	105.8	–1,515,865.3
Gas	574.2	–5,137,310.4
Total	1680	–5,298,975.7
NET		3,412,873.3

Fig. 6.6-5: Energy balance for ideal reduction process using pellets and coke.

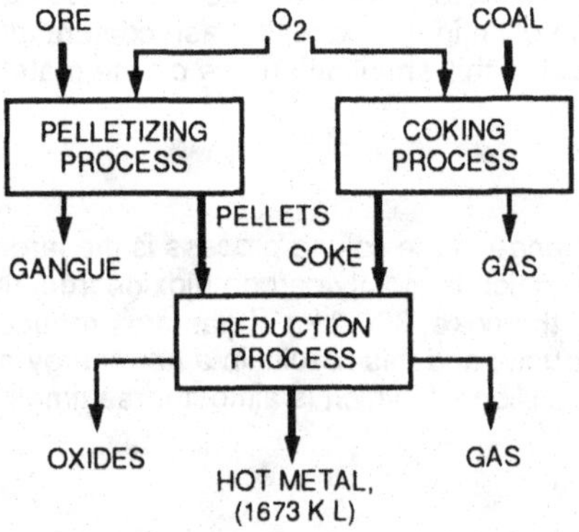

	NET ENERGY	COAL	O_2	MATERIAL PROCESSED
Pelletizing	–420,288.6	0	29.2	2993.9
Coking	–1,315,762.6	241.3	88.5	329.8
Reducting	3,412,873.3	0	0	1680
Requirement	–1,676,822.1	50.5	131.5	182
Total	0	291.8	249.2	5185.7
Totals from One Step Example		291.6	249.2	3505.6

Fig. 6.6-6: Combined energy balance for three-step ideal iron-making process.

MATERIAL	WEIGHT (kg)	ENERGY (kJ)
INPUTS		
Ore	4531.7	–45,875,305.4
O_2	44.7	0
Total	4576.4	–45,875,305.4
OUTPUTS		
Pellets	1464.4	–8,553,999.7
Gangue	3112	–37,962,788.8
Total	4576.4	–46,516,788.5
NET		–641,483.1

Fig. 6.6-7: Energy balance for sub-ideal pelletizing process.

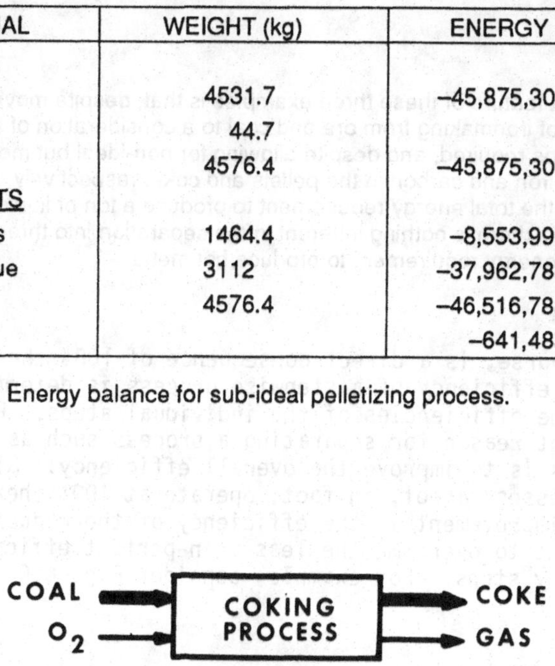

MATERIAL	WEIGHT (kg)	ENERGY (kJ)
INPUTS		
Coal	312.3	–392,686
O_2	273.5	0
Total	585.8	–392,686
OUTPUTS		
Coke	219.5	–258,750
Gas	366.2	–3,847,052.5
Total	585.7	–4,105,802.5
NET		–3,713,116.5

Fig. 6.6-8: Energy balance for sub-ideal coking process.

Fig. 6.6-10 gives the summary for this example. It can be seen that, because of the amount of coal required in the coking process to satisfy the material balance, there is more than enough energy available in total to satisfy this three-step ironmaking process. The extra energy is equivalent to 27.2 kg of coal which, if credited to the process, means that the total requirement of the three-step process is 285.0 kg of coal. This is about 7 kg less than in the previous cases, which is due to the fact that some of the required energy is now being supplied by the additional oxidation of magnetite.

The significant feature of these three examples is that, despite moving from a simple overall view of ironmaking from ore and coal to a consideration of the three individual process steps required, and despite allowing for non-ideal but more realistic recoveries of iron and carbon in the pellets and coke, respectively, there is virtually no change in the total energy requirement to produce a ton of liquid hot metal at 1673 K. Thus, there is nothing inherent in the separation into three process steps that affects the energy requirement to produce hot metal.

This, of course, is a direct consequence of 100% thermal efficiency. The efficiency of a stepwise process is determined by the product of the efficiencies of the individual steps. Presumably, an important reason for separating a process such as ironmaking into steps is to improve the overall efficiency. Since the preliminary processes cannot, in fact, operate at 100% thermal efficiency, the improvement in the efficiency of the reduction step must be sufficient to overcome the less-than-perfect efficiencies of the preliminary steps. For example, consider Fig. 6.6-11. If

MATERIAL	WEIGHT (kg)	ENERGY (kJ)
INPUTS		
Pellets	1464.4	−8,553,999.7
Coke	219.5	−258,750
Total	1683.9	−8,812,749.7
OUTPUTS		
Hot Metal	1000	1,354,200
Oxides	109.7	−1,574,941.6
Gas	574.2	−5,137,310.4
Total	1683.9	−5,358,052
NET		3,454,697.7

Fig. 6.6-9: Energy balance for sub-ideal reduction process using pellets and coke.

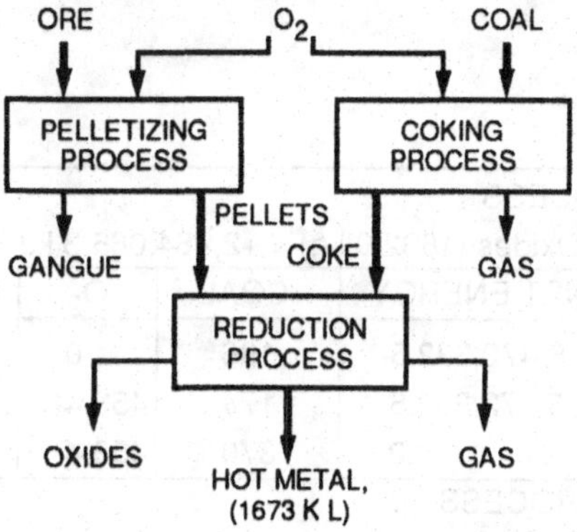

	NET ENERGY	COAL	O_2	MATERIAL PROCESSED
Pelletizing	−641,483.1	0	44.7	4576.4
Coking	−3,713,116.5	312.3	273.5	585.8
Reduction	3,454,697.7	0	0	1683.9
Requirement	−899,901.9	−27.2	−70.9	0
Total	0	285.1	247.3	6846.1
Totals from One-Step Example		291.6	249.2	3505.6

Fig. 6.6-10: Combined energy balance for three step sub-ideal ironmaking process.

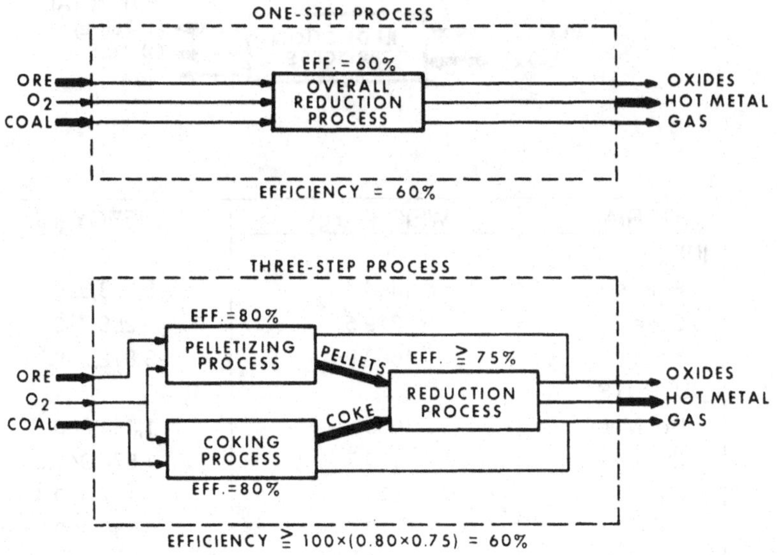

Fig. 6.6-11: Schematic representation of process efficiencies.

ONE-STEP PROCESS Heat Content of Oxides (1632.9 kg) = +2,464,065.3 kJ				
	NET ENERGY	COAL	O_2	MATERIAL PROCESSED
Process	5,472,332.5	196	0	3160.7
Requirement	–5,472,332.5	174	453.4	627.5
Total	0	370	453.4	3788.2
THREE-STEP PROCESS Heat Content of Pellets (1464.4 kg = +1,512,125.2 kJ Heat Content of Coke (219.5 kg = +342,261.9 kJ Heat Content of Oxides (109.7 kg = +165,564.4 kJ				
	NET ENERGY	COAL	O_2	MATERIAL PROCESSED
Pelletizing	902,226.6	0	44.7	4576.4
Coking	–3,177,248.7	312.3	273.5	585.8
Reduction	3,438,822.8	0	0	1683.9
Requirement	–1,163,800.7	37	96.4	133.4
Total	0	349.3	414.6	6979.5

Fig. 6.6-12: Combined energy balances for ironmaking processes with solids heat loss.

the overall reduction process had an efficiency of 60%, and if the pelletizing and coking processes each had an efficiency of 80%, then the process of producing hot metal from pellets and coke would have to have an efficiency greater than 75% in order to make the three-step process more efficient than the one-step process.

As an example of what happens to the energy requirements when inefficiencies are incorporated into the idealized calculations already given, the assumption of 100% thermal efficiency can be weakened by saying that the heat content in the solids produced in each process step cannot be recovered. The three-step process can then be compared with the overall process, as shown in Fig. 6.6-12. Considering the loss of heat from coke at 1367 K, pellets at 1477 K, and oxides* from the reduction process at 1673 K, the energy requirement of the three-step process goes up to 349.3 kg of coal and the energy requirement for the overall process goes up to 370.1 kg of coal. Thus, considering only this one inefficiency in the system, it is apparent that the three-step process is more efficient than the single overall process.

As a final example of this approach, let us make the reduction part of the three-step process slightly more realistic by including the fact that the slag must have a basicity of 1.1 in order for the process to work. As shown in Fig. 6.6-13, this is accomplished by providing 179.6 kg of limestone, which is considered to be pure calcium carbonate, as one of the reactants in the reduction process. On the output side, allowance is now made for the formation of complex oxides in the slag, which provides some energy to compensate for the energy required to calcine the limestone. The net result is an increase of 327,821 kJ in the total heat required for reduction (compare with Fig. 6.6-12). As shown in Fig. 6.6-14, this leads to an increase in the coal requirement to 359.7 kg of coal.

Fig. 6.6-15 shows a summary of the coal and oxygen requirements for each of the examples considered. This table demonstrates more clearly how the assumptions of 100% efficiency result in nearly constant requirements for coal and oxygen, despite the splitting of the overall process into individual steps. The table also shows, of course, how the amount of material processed must increase substantially when the process is split. In actual practice, this necessity for material processing is a negative factor in the economics of splitting up the process, and presumably must be overcome by efficiencies elsewhere.

*It is assumed that the oxides at 1673 K have the same heat content per kilogram as that of a typical liquid blast furnace slag at 1673 K, (see Table 6.6-3).

MATERIAL	WEIGHT (kg)	ENERGY (kJ)
INPUTS		
Pellets	1464.4	−8,553,999.7
Coke	219.5	−258,750
Limestone	179.6	−10,874,698.6
OUTPUTS		
Hot Metal	1000	1,354,200
Slag	210.3	−2,391,534.4
Gas	653.2	−5,539,351.9
Total	1863.5	−6,576,686.3
NET		4,298,012.3

Fig. 6.6-13: Energy balance for sub-ideal reduction process using pellets and coke, with realistic slag and solids heat loss.

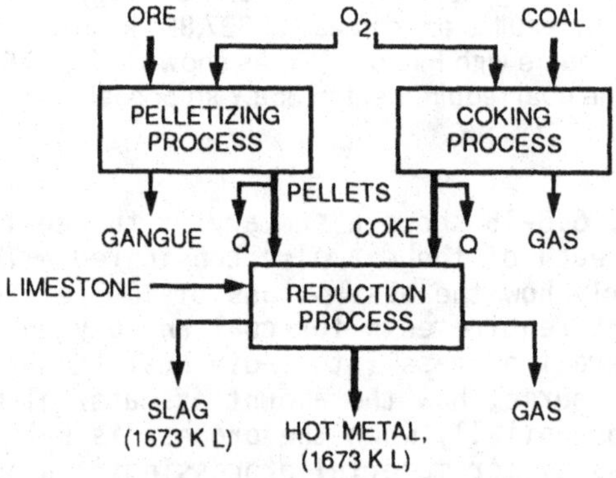

	NET ENERGY	COAL	O_2	MATERIAL PROCESSED
Pelletizing	902,226.6	0	44.7	4576.4
Coking	−3,177,248.7	312.3	273.5	585.8
Reduction	4,298,012.3	0	0	1863.5
Requirement	1,491,622.6	47.4	123.6	171
Total	0	359.7	441.8	7196.7

Fig. 6.6-14: Combined energy balance for three-step sub-ideal ironmaking process, with realistic slag and solids heat loss.

In conclusion, it should be stated that the simplistic nature of the analysis given here is recognized, but it should also be obvious that the approach can be extended to whatever complexity is desired. One particularly useful extension is to include steelmaking in the process and then to compare the pellet-coke-hot metal-BOF process to the pellet-reduced pellet-electric furnace process. Possibilities for complicating the individual process steps also come readily to mind. In any case, it is felt that the approach can be useful in separating some of the wheat from the chaff in many discussions of energy requirements for iron and steelmaking.

ONE-STEP PROCESS			
	COAL (kg)	O_2 (kg)	MATERIAL PROCESSED
Ideal	291.6	249.2	3505.6
Ideal + Heat Loss	370	453.4	3788.2
THREE-STEP PROCESS			
Ideal	291.8	249.2	5185.7
Sub-Ideal	285.1	247.3	6846.1
Sub-Ideal + Heat Loss	349.3	414.6	6979.5
Sub-Ideal + Heat Loss + Slag	359.7	441.8	7196.7

Fig. 6.6-15: Summary of coal, oxygen, and material processing requirements for examples considered.

6.6.2 ENERGY EFFICIENCY IN THE AGE OF SCARCITY*

A measure of process efficiency that has special relevance for future process planning may be called "energy efficiency". It should be a measure not only of the fuel and electric power directly used by the process, but also of the energy used to produce the purchased reagents, fluxes and supplies consumed by the process. Two related measures of this kind would seem to have particular value for description of metallurgical processes and products. The first, Process Fuel Equivalent (PFE), would measure only the energy resources consumed by the process in question. The second, Material Fuel Equivalent (MFE), would measure the total of all energy resources used to produce the material (product of the process) from ultimate raw material (ore in the ground). Both PFE and MFE are inverse measures of efficiency--low values of these quantities correspond to high efficiencies. The defining equations are:

$$PFE(\text{kJ/unit of product}) = F + E + S - B \qquad (6.6\text{-}1)$$

*Written by H. H. Kellogg, Stanley-Thompson Professor of Metallurgy of Columbia University, and originally published in Journal of Metals, June, 1974. Reprinted here with permission of the author.

$$MFE(kJ/unit\ of\ product) = PFE + R \tag{6.6-2}$$

where F = direct fuel consumption of the process, kJ/unit of product.
E = fuel equivalent of electric energy used by the process, kJ/unit of product.
S = sum of the total fuel resources, kJ/unit of product, used to produce the reagents, fluxes, and other major supplies consumed by the process.
B = sum of the useful surplus heat and the fuel equivalent of saleable byproducts of the process, kJ/unit of product.
R = fuel equivalent of the raw material feed to the process, kJ/unit of product.

FUEL EQUIVALENT OF ELECTRIC ENERGY

Before proceeding to examples of PFE and MFE calculation, the somewhat debatable conversion of electric energy to equivalent fuel energy must be decided. Steam- and diesel-powered electric generating plants come in a variety of efficiencies—generally ranging from a low of 25% to a high of 40% overall efficiency. With a theoretical conversion of 3599.7 kJ/kWh, these efficiencies correspond to a range of 14,400.8 to 8999.1 kJ of fuel per kWh of electrical energy generated. For nuclear plants, the value usually quoted is 11,256.9 kJ/kWh. The U.S. Bureau of Mines quotes a national average heat rate for fossil-fueled steam-electric plants of 11,071.2 kJ/kWh for the years 1970 and 1971. All calculations in this paper are based on this latter value (rounded to 11,077.5 kJ/kWh). Arguments for a higher value (lower efficiency) can logically be based on the fact that transmission losses (ranging from near zero to perhaps 15%) are not included in the value chosen. A lower value (higher efficiency) might be justified by recognition that a small fraction of our electric power is generated from the potential energy of water at much higher efficiency.

PROCESSING OF COPPER ORE TO METAL

To illustrate the meaning and general usefulness of PFE and MFE we will first consider the major steps in production of copper, from ore in the ground to wire-bars. In this hypothetical case we assume a sulfide copper ore of 0.7%·Cu, mined by an openpit method, with a waste-rock to ore ratio of 2.5/1., concentrated by flotation (80% recovery), smelted and refined by conventional methods (98% recovery of copper in smelting and refining). For simplicity we will assume that there are no valuable byproducts (Mo, Au, Ag, etc.). The data will be presented without details to better illustrate the overall significance of PFE and MFE.

Mining data from several sources[2,3] indicate that the energy equivalent for open-pit mining (including the energy equivalent of explosives, truck tires, drill bits and other supplies) amounts to about 46,516.8 kJ/t of total material mined. For a waste rock to ore ratio of 2.5, this becomes $3.5 \times 46{,}516.8 = 162{,}808.6$ kJ/t of ore mined. For ore of 0.7% Cu, we find:

162,808.6/.007 = 23.3 x 10^6 kJ/t of copper in ore mined.

PFE = 23,260.0 kJ/kg of copper in ore mined.

Since the raw material for mining is ore in the ground, it 'possesses a zero value of fuel equivalent. Therefore, R in Equation 6.6-2 is zero, and it follows that:

MFE = PFE = 23,260.0 kJ/kg of copper in mined ore.

Beneficiation data on flotation beneficiation gathered by the U.S. Bureau of Mines[4] can be used to show that the equivalent fuel used for simple copper ore beneficiation is about 267,471.3 kJ/t of ore treated. This includes electric energy for crushing, grinding, flotation, dewatering and material transport, as well as the equivalent fuel value of steel consumed for mill liners, balls and rods, and equivalent fuel value of flotation reagents used. It does not include the equivalent fuel value of the ore treated (the energy for mining). If we convert this value to the basis of kJ/kg of copper in the concentrate, assuming 80% recovery we find:

$$\begin{aligned} PFE &= \frac{267{,}471.3}{1000 \times .007 \times .8} \\ &= 47{,}762.7 \text{ kJ / kg copper in concentrate} \end{aligned}$$

47,762.7 kJ/kg copper recovered is the contribution of the beneficiation process, considered by itself (PFE). But the raw material to the beneficiation process is mined ore, which itself required energy to produce. Therefore, R (Eq. 6.6-2) is finite, and since it required 0.625 kg of copper in mined ore to produce 0.5 kg of copper in concentrate (80% recovery) we find:

$$\begin{aligned} R &= \text{MFE for mined ore} \times 1.25 \\ R &= 23{,}260.0 \times 1.25 = 29{,}075.0 \text{ kJ/kg Cu} \\ MFE &= PFE + R = 47{,}762.7 \text{ kJ/kg} + 29{,}075.0 \text{ kJ/kg} \\ &= 76{,}837.7 \text{ kJ/kg copper in concentrate} \end{aligned}$$

Smelting and refining data from the 1967 Census of Manufacturing Industries[5] suggests that the PFE value for conversion of concentrate to wire bar is about 46,520 kJ/kg of wire-bar copper. Although the author has some doubt about the accuracy of this figure, it will be accepted here for want of more reliable data.

PFE = 46,520 kJ/kg of wire bar

The raw material for this stage of processing is copper concentrate (MFE = 76,837.7 kJ/kg of Cu) and it requires 0.45 kg/.98 = 0.46 kg of copper in concentrate to make 0.45 kg of wire bar. Hence

R = 76,837.7 × 1.02 = 78,374.5 kJ/kg Cu

and

$$MFE = 46,520 + 78,374.5 = 124,894.5 \text{ kJ/kg of wire bar}$$

The calculated values of PFE and MFE are summarized in Table 6.6-4. Inspection of these figures shows that PFE indicates the energy resources used by a particular processing step (it is the proper criterion for comparing rival processes), whereas MFE accumulates the energy resources consumed by each processing step, with proper allowance for incomplete recovery, to yield the total energy resources consumed in producing the particular product from ultimate raw material (ore in the ground).

TABLE 6.6-4

Fuel Equivalent for Recovery of Copper

PROCESS	kJ/kg Cu	PRODUCT	kJ/kg Cu
Mining	23,260	Mined Ore	23,260
Beneficiation	47,776	Concentrate	76,851
Smelting	46,520	Wire Bars	124,906

GENERAL FEATURES OF PFE

The justification for inclusion of the fuel equivalent of major reagents, fluxes and supplies used by the process (the quantity S in eq. 6.6-1) in calculation of PFE can be made clear by simple examples. Consider a smelting process which uses 0.000379 m^3 of residual fuel oil to smelt 0.45 kg of metal when air is used for fuel combustion. It is determined that the fuel consumption can be reduced to 0.000265 m^3 of oil if the combustion air is enriched with 0.084 m^3 of 99% oxygen per kilogram of metal smelted. If direct fuel consumption, F, is alone considered, the use of oxygen enrichment shows an obvious energy advantage. But it requires energy (electricity or process steam) to produce the oxygen, so the question must be asked: is the saving in oil more or less than the extra fuel required to produce oxygen? Calculation of PFE for both alternatives would answer this question.

As a second example, consider the choice between two alternate routes for recovery of copper from dilute dump leach solutions: the first by solvent extraction, and electrowinning and the second by cementation with iron, followed by smelting and refining of cement copper. The former route consumes LIX reagent and kerosene, both of which require energy for their production, and electric energy for electrowinning and general process energy (pumping, agitation, etc.). The latter route consumes iron that requires energy for its production, plus fuel, fluxes, and electric power for smelting and electrorefining. Calculation of PFE, with proper account of the fuel equivalent of supplies (LIX, kerosene, iron, etc.) can answer which of these very different processes makes greater demands on our scarce energy resources.

The formulation of Eq. 6.6-1 for PFE considers only the fuel equivalent of consumable supplies for the process. It can be argued that a complete energy analysis should also include the energy equivalent of capital equipment amortized over the expected life of the plant. The validity of this view cannot be questioned, but, in the interest of simplicity and usefulness, the more restricted definition of PFE should prove more useful. The fuel consumed in transport of materials between mine and plant has also been omitted from PFE and MFE for purposes of expediency.

TABLE 6.6-5

MFE For Reagents, Fluxes and Supplies

TABLE 6.6-5	
ITEM	MFE (KJ/Kg)
Limestone	116
Oxygen	3722
Sulfuric Acid	3256
Lime	5815
Coke Breeze	23,260
Coke, Lump	34,890
Iron for Cementation	23,260
Explosives	34,890
Steel Mill	43,031
Soda Ash	44,194
Flotation	46,520
Ammonia	58,150
Chlorine	33,727
Sodium Hydroxide	33,727

The byproduct factor, B, in Eq. 6.6-1 is included in order to give proper credit to processes which produce useful byproducts-useable surplus heat, fuel, electric power, or byproduct materials which can be sold or used outside of the process in question. Recovered heat or other byproducts used within the process should not be included in B. Some processes produce very significant amounts of byproducts (production of coke in slot-ovens produces coke, breeze, coke-oven gas, tar, light oil and other byproducts; the Imperial Smelting Process produces lead as well as zinc), and, in such cases, the value of B can exert a strong influence on PFE for the main product.

Calculation of either S or B (Eq. 6.6-1) requires knowledge of the MFE for each item and the amount of each item used in the process. To aid those who may wish to make such calculations,

Table 6.6-5 lists some of the author's rough estimates of MFE values for supplies commonly used in metallurgical processing.

TABLE 6.6-6

PFE and MFE for Iron Production in the Blast Furnace*
(average performance of U.S. industry in 1970)

TABLE 6.6-6			
	Amount/t pig	MFE kJ/unit	kJ/t pig
Fuels			
Residual Oil	0.0064 m^3	41,805,357.4 kJ/m^3	277,587.6
Tar and Pitch	0.00193 m^3	41,805,357.4 kJ/m^3	80,684.3
Natural Gas	15.1 m^3	37,250 kJ/m^3	562,475
Coke Oven Gas	2.8 m^3	20,487.5 kJ/m^3	57,365
Blast Furnace	541.2 m^3	3538.75 kJ/m^3	1,915,171.5
Coke	0.7169 t	34,890.0 kJ/Kg	25,012,641
Total Fuel			27,905,924.4
Flux and Reag			
Oxygen	4.6 m^3	4917 kJ/m^3	22,618.2
Limestone	0.1713 t	116291.9 kJ/t	19,920.8
Total Flux			42,539
Byproducts			
Blast Furnace	1537.6 m^3	3538.75 kJ/m^3	5,441,182
Coke Breeze	0.0188 t	23,260.0 kJ/Kg	437,288
Total Byprod.			5,878,470
Raw Materials			
Iron ore	0.4588 t	116,291.9 kJ/t	53,354.7
Manganese ore	0.0101 t	116,291.9 kJ/t	1174.5
Agglomerates	1.3086 t	1,395,502.6 kJ/t	1,826,154.7
Scrap	0.0388 t	–	–
Cinder, Scale	0.0651 t	–	–
Total Raw Mat.			1,880,683.9

PFE = F + S − B = 19,785,901.7 kJ/t (19,785.9 kJ/kg).
MFE = PFE + R = 21,491,903.7 kJ/t (21,491.9 kJ/kg).

* Based on data from Minerals Yearbook, 1970 (6).

PRODUCTION OF IRON IN THE BLAST FURNACE

In Table 6.6-6 the iron blast furnace has been used as an example of detailed calculation of PFE and MFE, including the effects of the terms S, B, and R. The data employed do not apply to any single furnace. Rather they represent composite data for the average performance of all U.S. blast furnaces in 1970, as reported by the U.S. Bureau of Mines.[6]

The wide variety of fuels listed in Table 6.6-6 reflects both the composite nature of the data employed and the fact that the blast furnace process uses fuel for a variety of purposes—injection through the tuyeres, preheat of blast, blast compression, and to generate electrical power for plant operations. Coke remains the predominant fuel and reducing agent. The value of MFE for coke used in Table III (34,890.0 kJ/kg), represents not the heating value of coke (about 30,470.6 kJ/kg) but rather the fuel equivalent required to make coke from coal, with proper allowance for the byproducts of slot-oven coking.

No item of purchased electric power is listed for the blast furnace operation by the Bureau of Mines review,[6] so that E (Eq 6.6-1) has been assumed to be zero in Table 6.6-6. This assumption will only be correct provided that all electric power actually used is generated from the fuels listed in the table.

Blast furnace gas and coke breeze are recovered as byproducts of the blast furnace process. The excess of recovered gas over that used for blast preheat is used in other processes (steelmaking, fabrication, etc.), and represents an energy credit to the blast furnace process, as shown in Table 6.6-6. Coke breeze also represents a small credit to the process.

The pig iron produced by the process comes from the variety of raw materials listed in Table 6.6-6. Scrap, cinder and scale are shown with zero values of MFE, because they are recycled products, but for ore and agglomerates estimates of the fuel equivalent for mining, beneficiation and agglomeration have been used to derive finite values of MFE. The increasing use of high grade pellets and sinter (agglomerates) as feed to the blast furnace has helped to increase furnace capacity and to reduce coke rate, but, as Table 6.6-6 shows, the energy saving on coke rate is partly compensated by increased energy consumption for preparation of agglomerates.

The PFE for the blast furnace process (~19,771.0 kJ/kg of pig) is low for a metallurgical process, indicating a relatively efficient operation. MFE for pig iron is only marginally larger than PFE because the raw materials are derived from relatively high-grade ores that require only modest energy inputs <u>per unit of contained iron</u>. This contrasts with the case of copper, where the low grade of ore causes a wide spread between PFE and MFE for the smelting and refining step.

CONCLUSIONS

Over the past several years the author has used calculations of PFE and MFE to compare rival copper smelting process[7] and to compare the energy used for production of various primary and secondary metals.[8] Some conclusions of interest from these studies are discussed below.

For smelting of copper sulfide concentrate, the flash smelting process[9,10] uses less than half the equivalent fuel of the reverberatory. When commercial oxygen is used in flash smelting[11] the equivalent fuel consumption is still lower. The new continuous copper smelting processes (Noranda[12] and Mitsubishi[13]) also use significantly less equivalent fuel than the reverberatory. When these energy advantages are coupled with the advantages that these new processes enjoy with respect to SO_2 recovery (continuous production of 10-12% SO_2 gas) the combination offers a powerful challenge to the conventional reverberatory.

Reliable operating data for new hydrometallurgical flowsheets for recovery of copper from sulfide concentrates have not yet been published. Reasonable estimates of energy consumption for the major unit operations of several such processes, however, show values of PFE which are about 2 times larger than for conventional smelting and refining. Thus the popular view that hydrometallurgy uses less energy than pyrometallurgy may be in serious error, particularly for those flowsheets employing the energy-intensive step of electrowinning.

Most of our current metal extraction processes were conceived at a time when energy resources were plentiful and cheap, and scant attention was paid to energy conservation in these process designs. Accordingly we possess excellent opportunities for energy conservation through redesign of processes. Although much can be done to improve the energy efficiency of existing processes through recovery of waste heat, more significant contributions will come from radical new designs, conceived from the start with a view toward energy conservation and pollution abatement. As evidence of a trend toward redesign we have not only the new copper smelting processes noted above, but also the recent announcement from Alcoa[14] of a new chloride electrolysis for aluminum manufacture which uses 30% less energy than the very energy-intensive Hall process.

The author hopes that PFE and MFE, or some similar measures of energy efficiency, will become widely used yardsticks for the comparison of new and old metallurgical processes and products. Purveyors of new processes owe it to their clients, the profession and the public at large to provide ready comparison of their processes to others, both with respect to cost and equivalent fuel consumption. PFE provides a quantitative basis for the latter comparison. MFE for fabricated metals provides the materials engineer with the data necessary to gauge the effect of material substitution on energy consumption.

REFERENCES TO SECTION 6.6.2

1. Minerals Yearbook, 1971, U.S. Dept. of Interior, U.S. Gov. Print. Office, 1973, pg. 22, Vol. I.

2. Census of the Mineral Industries, 1963, U.S. Bureau of Census.

3. "An Economic Appraisal of the Supply of Copper from Primary Domestic Sources", Inf. Cir. 8598, 1973, U.S. Bureau of Mines.

4. Minerals Yearbook, 1970, U.S. Dept. of Interior, U.S. Gov. Print. Office, 1972, pp. 80-104, Vol. I.

5. Census of the Manufacturing Industries, 1967, U.S. Bureau of Census.

6. Reference 4, pp. 593-610, Vol. I.

7. "Prospects for the Pyrometallurgy of Copper", by H. H. Kellogg, Proceedings of the Latin American Congress on Mining and Extractive Metallurgy, Santiago, Chile, Aug. 1973, forthcoming.

8. "Energy Considerations in Metal Production, Selection and Utilization", by H. H. Kellogg and J. Tien, paper given at AIME-ASM meeting, Chicago, Oct., 1973.

9. "New Developments in Outokumpu Flash Smelting", by S. U. Harkki and J. T. Juusela, paper presented at AIME meeting, Dallas, Feb., 1974.

10. T. Fujii, M. Ando, Y. Fujiwara; paper No. T IV d 5, joint meeting MMIJ-AIME, Tokyo, May, 1972.

11. R. R. Saddington, W. Curlook, P. Queneau; in Pyrometallurgical Processes in Non-ferrous Metallurgy, ed. J. N. Anderson, P. E. Queneau, Gordon and Breach, N.Y., 1967, pp. 262-67.

12. N. J. Themelis, et al; Journal of Metals, Vol. 24 (4), pp. 25-32, 1972.

13. T. Suzuki, T. Nagano; paper No. T IV e 4, joint meeting MMIJ-AIME, Tokyo, May 1972.

14. Aluminum Co. of America, Pitts., Pa., news release dated Jan. 11, 1973.

Further Reading

Appendix C contains a Bibliography of Material and Energy Balances on Metallurgical Cement, and Lime Processes. The reader is referred to that document for many further sources of detailed information on energy balances.

For general references, refer to the general texts listed in Chapter 4.

Practice Problems

6.1 The process flow diagram for a straight-grate iron ore pelletizing plant is shown below. Sketch a Sankey diagram for this process.

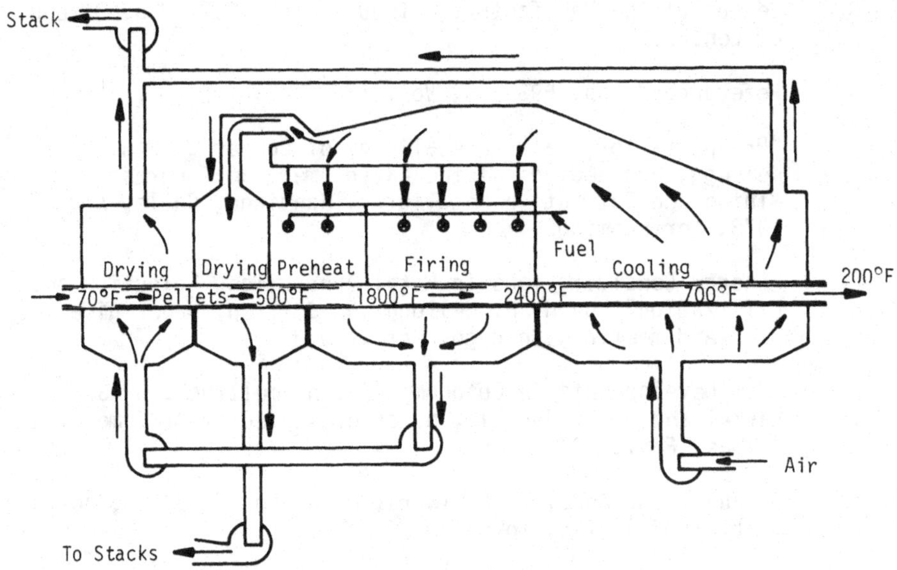

6.2 Write a Fortran Program to find the AFT for the problem described in Example 6.2-2.

6.3 Find the ART for a mixture of 30% steel shot, 70% stoichiometric Al + Fe_2O_3. How much steel shot can be mixed with the stoichiometric Al + Fe_2O_3 mixture to achieve an ART of 1800 C?

6.4 At what temperature will the condensed phases leave the (idealized) process in Fig. 6.4-3 if the AFT is reduced to 1650 K?

6.5 Make an overall energy balance on the process shown in Fig. 6.4-5. Assume that the H_2O content of the wet pellets is 9%, and that the fuel used is oil with a net heating value of 40,000 kJ/kg, that 80% of the theoretical air for oil combustion is supplied as primary air, that the 1st & 2nd stage cooling fans run at the same air flow rate, and that 1.1 kg of process gases are provided per 1.0 kg Fe_2O_3.

6.6 You are assigned to cool off a heat of steel which is liquid, but too hot. It weighs 90 tons and is at 3100°F. You want to add cold (77°F) steel to it to lower its temperature to 2900°F. Assuming no heat loss from the furnace, how many lbs. would you add? Assume all thermal properties are those of iron.

6.7 Calculate how much coke breeze is needed to heat a sinter bed to 2400°F. Assume the material is 90% Fe_2O_3, 6% $CaCO_3$ and 4% SiO_2.

Assume the reactions $CaCO_3 \rightarrow CaO + CO_2$

$$\text{and } CaO + SiO_2 \rightarrow CaO \cdot SiO_2$$

$$C + O_2 \rightarrow CO_2$$

go to completion. Assume the incoming and outgoing air are at room temperature. Assume an adiabatic system. Use as a basis for your calculations 1 short ton of sinter mix containing 10% H_2O.

6.8 Producer gas of the following composition:

Carbon monoxide	28%	Methane	2%
Carbon dioxide	4%	Water Vapor	1%
Hydrogen	4%	Nitrogen	61%

is burned with 10 per cent more air than theoretically required, both air and gas being preheated to 1000°C.

Calculate the theoretical maximum temperature of the flame.

6.9 In iron foundry cupola 14 metric tons of pig iron are melted in one hour, using 1.5 tons of coke (90 per cent carbon). The gases passing away contain by volume CO 13 per cent., CO_2 13 per cent., nitrogen 74 per cent., and leave the cupola at 500°C. One ton of slag is formed and tapped. Assume typical cupola slag and pig iron analyses.

Calculate:

(1) The net melting efficiency of the cupola.
(2) The proportion of the calorific power of the coke lost.

(a) By the sensible heat of the hot gases escaping.
(b) By the imperfect combustion of the coke.
(c) By radiation from bottom and walls of the cupola.

6.10 Leach liquor is partially stripped of Zn ions by electrolysis. The initial analysis is 80.6 grams per litre Zn and the after-electrolysis analysis is 25.0 grams per litre. The current efficiency is 64.8%, and the voltage is 2.55V. What is the electrical energy consumed per kg of Zn deposited?

APPENDIX A

Basic Properties of Matrices

by

Daniel T. Hanson

A matrix is defined in terms of a set of elements. The mathematician asserts that the elements may be from a collection of objects called a field. However, this involves an additional step of abstraction that seems neither necessary nor useful in this introductory series. The elements of concern here are from the set of objects called real numbers or from the set called complex numbers.

If the matrices have elements that are all real numbers, they are said to be defined over the real field. If the elements are either real or complex, the matrices are said to be defined over the complex field.

Thus, the reader will be working with examples of more-general kinds of objects. While many of the rules carry over without change to general fields, there are instances where they do not. This should be borne in mind by the reader who may wish to extend the ideas presented in this series.

The term "element" (as used here) is a real or a complex number. This may be a constant or the value of a function. An array of elements that are (or, can be) arranged in rows and columns is a matrix. If there are m rows and n columns, it is called an m by n matrix. Typical examples are:

$$\begin{bmatrix} a_{11} & a_{12} \cdots a_{1n} \\ a_{21} & a_{22} \cdots a_{2n} \\ \cdot & \\ \cdot & \\ a_{m1} & a_{m2} \cdots a_{mn} \end{bmatrix}, \begin{bmatrix} 1 & 3 & 2 \\ 2 & 4 & 1 \end{bmatrix}, \begin{bmatrix} x_1 \\ x_2 \\ \cdot \\ \cdot \\ x_m \end{bmatrix}, [x_1, x_2, \cdots, x_m] \quad (1)$$

Thus, the arrays given by Eq. (1) are $m \times n$, 2×3, $m \times 1$, and $1 \times m$ matrices, respectively. The symbol a_{ij} denotes the number in the ith row and j^{th} column of the matrix. In the second matrix of Eq. (1), $a_{12} = 3$. Hence, a matrix is specified once the number of rows m, the number of columns n, and a real or complex number is specified at the intersection of each row and each column. The following symbols are used to denote such a matrix:

$$A = [a_{ij}] = \begin{bmatrix} a_{11} & a_{12} \cdots a_{1n} \\ a_{21} & a_{22} \cdots a_{2n} \\ \cdot & \cdot \quad \cdot \\ \cdot & \cdot \quad \cdot \\ a_{m1} & a_{m2} \cdots a_{mn} \end{bmatrix} \quad (2)$$

Note that brackets are used to ensure that a matrix will not be confused with a determinant. A determinant is a single number; a matrix, an array of numbers.

It will prove particularly convenient to list the following properties for matrices:

- Equality—Two matrices are equal if, and only if, they have the same number of rows and columns, and corresponding elements are equal.
- Addition—Two matrices can be added if, and only if, they have the same number of rows and columns. Their sum is the matrix formed by adding corresponding elements of the original two matrices. Addition is commutative, that is: $A + B = B + A$. This is established by:

$$[a_{ij}] + [b_{ij}] = [a_{ij} + b_{ij}] = [b_{ij} + a_{ij}] = [b_{ij}] + [a_{ij}]$$

The first equality is from the definition of addition. The second is from commutativity of addition for real or complex numbers (the element is unchanged by the order of addition).

The third equality is from the definition of addition for matrices:

$$(A + B) + C = A + (B + C) = A + B + C$$

Hence addition of matrices is associative. This is established from the definition of addition, and the known properties of real or complex numbers:

$$([a_{ij}] + [b_{ij}]) + [c_{ij}] = [a_{ij} + b_{ij}] + [c_{ij}] = [a_{ij} + b_{ij} + c_{ij}] = [a_{ij}] + [b_{ij} + c_{ij}] = [a_{ij}] + ([b_{ij}] + [c_{ij}]) = [a_{ij}] + [b_{ij}] + [c_{ij}]$$

In short, the order of addition is immaterial.

The addition of three matrices, by whatever order the addition of corresponding elements is accomplished, is denoted by:

$$A + B + C$$

• The Zero matrix, **0**, is defined as that matrix which has zero as each element.

• Multiplication by a scalar—The scalar multiple of a matrix is that matrix in which each element of the original matrix is multiplied by the scalar. If k represents the scalar, then:

$$k[a_{ij}] = [ka_{ij}] = [a_{ij}]k$$

Note that multiplication by a scalar is commutative. The left-most and right-most sides of this expression mean the same thing, and either is defined by the central term.

• Negative (additive inverse) of a matrix: $-\mathbf{A}$ means $(-1)\mathbf{A}$.

Multiplication of Matrices

Two matrices, **A** and **B**, are said to be conformable in the given order if **A** has the same number of columns as **B** has rows.

Multiplication is defined only for conformable matrices, and only in the order in which they are conformable.

If **A** has m rows and p columns, and **B** has p rows and n columns, then the product **AB** is a matrix that has m rows and n columns. This product is defined by:

$$\underset{m \times p}{[a_{ij}]}\,\underset{p \times n}{[b_{ij}]} = \underset{m \times n}{\left[\sum_{k=1}^{n} a_{ik} b_{kj}\right]} \tag{3}$$

In Eq. (3), the dimensions are shown below each matrix. In expanded notation, the product is given by:

$$\begin{bmatrix} a_{11}a_{12} \cdots a_{1p} \\ a_{21}a_{22} \cdots a_{2p} \\ \vdots \\ a_{i1}\ a_{i2} \cdots a_{ip} \\ \vdots \\ a_{m1}\ a_{m2} \cdots a_{mp} \end{bmatrix} \begin{bmatrix} b_{11}\ b_{12} \cdots b_{1j} \cdots b_{1n} \\ b_{21}\ b_{22} \cdots b_{2j} \cdots b_{2n} \\ \vdots \\ b_{p1}\ b_{p2} \cdots b_{pj} \cdots b_{pn} \end{bmatrix} = \begin{bmatrix} & j^{\text{th}} \text{ column} \\ & \downarrow \\ & \vdots \\ i^{\text{th}} \text{ row} \longrightarrow & \cdots \sum_{k=1}^{n} a_{ik} b_{kj} \cdots \end{bmatrix} \tag{4}$$

Thus, the element in the ith row and j^{th} column of the product is formed by what is called the "dot product" of the ith row-vector of **A** with the jth column-vector of **B** (multiply corresponding elements of each, and sum).

By reference to Eq. (4), it can be seen that problems may arise in cases where a row vector of **A** does not have the same number of elements as a column vector of **B**. Multiplication of corresponding elements of such vectors is not defined because there is not a one-to-one correspondence of elements.

The above ideas may be made clear by a pair of problems. Problem 1: Is the product **AB** equal to the product **BA**? (In other words, is multiplication commutative for these matrices?)

$$A = [a_{11}\quad a_{12}\quad a_{13}], \quad B = \begin{bmatrix} b_{11} \\ b_{21} \\ b_{31} \end{bmatrix}$$

The product **AB** is a single number (a 1×1 matrix):

$$AB = a_{11}b_{11} + a_{12}b_{21} + a_{13}b_{31}$$

On the other hand, the product **BA** is a 3×3 matrix:

$$BA = \begin{bmatrix} b_{11}a_{11} & b_{11}a_{12} & b_{11}a_{13} \\ b_{21}a_{11} & b_{21}a_{12} & b_{21}a_{13} \\ b_{31}a_{11} & b_{31}a_{12} & b_{31}a_{13} \end{bmatrix}$$

Problem 2: Is the product **AB** equal to the product **BA** for the following matrices?

$$A = \begin{bmatrix} a_{11} & a_{12} & a_{13} \\ a_{21} & a_{22} & a_{23} \end{bmatrix} \quad B = \begin{bmatrix} b_{11} \\ b_{21} \\ b_{31} \end{bmatrix}$$

Note that **A** and **B** are conformable in that order. They are not conformable in the order **B**, **A**. Thus, while the product **AB** is defined, the product **BA** is not.

Generally, multiplication is not commutative. This must always be kept in mind when multiplying matrices (or matrix-valued functions). There are a few important cases in which commutativity does hold. Some of these will be discussed.

The following properties of matrix multiplication follow from the definition of the product, and the properties of real or complex numbers.

Multiplication is distributive over addition, that is:

$$(A + B)C = AC + BC$$
$$D(A + B) = DA + DB$$

This first equality is shown by:

$$([a_{ij}] + [b_{ij}])[c_{ij}] = [a_{ij} + b_{ij}]\,[c_{ij}]$$
$$= [\sum_k \{(a_{ik} + b_{ik})c_{kj}\}] = [\sum_k a_{ik}c_{kj} + \sum_k b_{ik}c_{kj}]$$
$$= [\sum_k a_{ik}c_{kj}] + [\sum_k b_{ik}c_{kj}] \equiv AC + BC$$

The second equality is similarly demonstrated.

Multiplication is associative, that is:

$$(AB)C = A(BC) = ABC$$

If ordering of the matrices is preserved, the sequence in which the multiplications are accomplished is immaterial. This is shown as follows:

$$(\underset{m \times p}{[a_{ij}]}\quad \underset{p \times q}{[b_{ij}]})\quad \underset{q \times n}{[c_{ij}]} = \underset{m \times q}{\left[\sum_{k=1}^{p} a_{ik}b_{kj}\right]}\ \underset{q \times n}{[c_{ij}]}$$

$$= \left[\sum_{l=1}^{q}\sum_{k=1}^{p} (a_{ik}b_{kl})c_{lj}\right] = \left[\sum_{k=1}^{p} a_{ik}\left(\sum_{l=1}^{q} b_{kl}c_{lj}\right)\right]$$

[To avoid confusion, an index l is used in the second set of sums.] The last equality follows since the order of summation of the numbers is immaterial (note that this is an $m \times n$ matrix). The right-hand side is just $A(BC)$; we started with $(AB)C$. The common result is, therefore, just as well denoted by ABC.

Notation and Common Matrices

A matrix is called a square matrix if it has the same number of rows as it has columns. If a matrix is say, 4×4, it is called a 4th-order matrix. If $n \times n$, it is an nth-order matrix. The term "order" is used only for square matrices.

For an nth-order (square) matrix $[a_{ij}]$, the elements of the set $\{a_{ii}, i = 1,...,n\}$ are said to be on the main diagonal of the matrix. The main diagonal runs from upper left to lower right, and is formed by elements at the intersection of each ith row with the ith column. The term "diagonal element" refers only to an element on the main diagonal.

A square matrix with each diagonal element not zero, and zeros elsewhere, is called a diagonal matrix. Thus, the following is a fourth-order diagonal matrix:

$$\begin{bmatrix} a_{11} & 0 & 0 & 0 \\ 0 & a_{22} & 0 & 0 \\ 0 & 0 & a_{33} & 0 \\ 0 & 0 & 0 & a_{44} \end{bmatrix}$$

Eq. (4) will be helpful in verifying that:

1. Premultiplication of a matrix $[b_{ij}]$ by a conformable diagonal matrix $[a_{ii}]$ gives a matrix in which each element in the ith row of $[b_{ij}]$ is multiplied by a_{ii}, or:

$$[a_{ii}][b_{ij}] = [a_{ii}b_{ij}]$$

2. Postmultiplication of a matrix $[b_{ij}]$ by a conformable diagonal matrix $[a_{jj}]$ gives a matrix in which each element in the jth column of $[b_{ij}]$ is multiplied by a_{jj}. That is:

$$[b_{ij}][a_{jj}] = [a_{jj}b_{ij}]$$

The diagonal matrix with $a_{ii} = 1$ for each i, is called the identity matrix and is denoted by I. The terms "idem" matrix and "unit" matrix are also common, but will not be used here. From the relations just reviewed, it follows that $I[b_{ij}] = [b_{ij}]$, and $[b_{ij}]\ I = [b_{ij}]$ for any matrix $[b_{ij}]$—provided that I is the corresponding conformable identity matrix. Note that if B is a square matrix, then I commutes with B under multiplication.

A triangular matrix is a matrix in which all elements above or below the main diagonal are zeros. Such a matrix is called a lower triangular matrix, or an upper triangular matrix, respectively. The following are examples of a fourth-order lower triangular matrix and a fourth-order upper triangular matrix:

$$\begin{bmatrix} a_{11} & 0 & 0 & 0 \\ a_{21} & a_{22} & 0 & 0 \\ a_{31} & a_{32} & a_{33} & 0 \\ a_{41} & a_{42} & a_{43} & a_{44} \end{bmatrix} \quad \text{and} \quad \begin{bmatrix} a_{11} & a_{12} & a_{13} & a_{14} \\ 0 & a_{22} & a_{23} & a_{24} \\ 0 & 0 & a_{33} & a_{34} \\ 0 & 0 & 0 & a_{44} \end{bmatrix}$$

The determinant of either a triangular or a diagonal matrix is the product of the elements on the main diagonal.

The inverse of a matrix (if there is one) is defined only for square matrices. For the moment, the inverse of a square matrix A is defined as that matrix which when postmultiplied or premultiplied by A produces the identity matrix. That is, given an nth-order matrix A, a second n^{th}-order matrix A^{-1} is the inverse of A if:

$$A^{-1}A = I = AA^{-1}$$

(Existence and uniqueness of the inverse follow from a more basic definition that will be given later. For the moment, it is merely stated that if a matrix has an inverse, then the inverse is unique.)

The transpose of a matrix is defined for matrices of any dimensions, and is formed by interchanging its rows and columns. The transpose is denoted by a superscript T, or otherwise. For example, given a matrix:

$$A = \underset{m \times n}{\begin{bmatrix} a_{11} & a_{12} & a_{13} \cdots a_{1n} \\ a_{21} & a_{22} & a_{23} \cdots a_{2n} \\ \cdot & \cdot & \cdot \\ \cdot & \cdot & \cdot \\ a_{m1} & a_{m2} & a_{m3} \cdots a_{mn} \end{bmatrix}} = [a_{ij}]$$

Then its transpose, and the element in the jth column and i^{th} row of A^T are:

$$A^T = \underset{n \times m}{\begin{bmatrix} a_{11} & a_{21} \cdots a_{m1} \\ a_{12} & a_{22} \cdots a_{m2} \\ a_{13} & a_{23} \cdots a_{m3} \\ \cdot & \cdot & \cdot \\ \cdot & \cdot & \cdot \\ a_{1n} & a_{2n} & a_{mn} \end{bmatrix}} = [a_{ji}]$$

A symmetric matrix is one for which $A = A^T$ or, equivalently $a_{ij} = a_{ji}$ for each j and i. A symmetric matrix must be square; otherwise the equality does not make sense.

For example, the following represent fourth-order symmetric matrices:

$$\begin{bmatrix} 1 & 5 & 4 & 2 \\ 5 & 2 & 1 & 6 \\ 4 & 1 & 3 & 1 \\ 2 & 6 & 1 & 4 \end{bmatrix}, \begin{bmatrix} a_{11} & a_{21} & a_{31} & a_{41} \\ a_{21} & a_{22} & a_{32} & a_{42} \\ a_{31} & a_{32} & a_{33} & a_{43} \\ a_{41} & a_{42} & a_{43} & a_{44} \end{bmatrix}, \begin{bmatrix} a_{11} & a_{12} & a_{13} & a_{14} \\ a_{12} & a_{22} & a_{23} & a_{24} \\ a_{13} & a_{23} & a_{33} & a_{34} \\ a_{14} & a_{24} & a_{34} & a_{44} \end{bmatrix}$$

A vector is an ordered n-tuple of numbers, and may be represented as:

$$x = \begin{bmatrix} x_1 \\ x_2 \\ \cdot \\ \cdot \\ \cdot \\ x_n \end{bmatrix} \quad \text{or } x^T = [x_1, \ x_2, \ \cdots, \ x_n]$$

The two representations above are the same vector except for the way in which matrices operate on (or are operated on by) the vector. Thus, the vector x

is called a column vector (an $n \times 1$ matrix—sometimes called a column matrix). On the other hand, the transpose of x, x^T, is called a row vector (a $1 \times n$ matrix—often called a row matrix). The transpose symbol will be used throughout this series to denote a row vector. Otherwise, the vector is to be considered a column vector.*

Row and Column Operations

For a given matrix **A**, it has been demonstrated that $IA = AI = A$ for the conformable identity matrices, that is, pre- or postmultiplication by **I** does not change the matrix.

On the other hand, consider the matrix formed by interchanging two rows, say the kth and l^{th} rows, of **I**. Call the resulting matrix $I_{k,l}$. Premultiplication of a given matrix **A** by the conformable $I_{k,l}$ matrix interchanges the kth and l^{th} rows of **A**. For example, $I_{2,4}$ as a premultiplier interchanges the second and fourth rows of a matrix as illustrated:

$$\begin{bmatrix} 1&0&0&0&0\\ 0&0&0&1&0\\ 0&0&1&0&0\\ 0&1&0&0&0\\ 0&0&0&0&1 \end{bmatrix} \begin{bmatrix} a_{11}&a_{12}&a_{13}&a_{14}\\ a_{21}&a_{22}&a_{23}&a_{24}\\ a_{31}&a_{32}&a_{33}&a_{34}\\ a_{41}&a_{42}&a_{43}&a_{44}\\ a_{51}&a_{52}&a_{53}&a_{54} \end{bmatrix} = \begin{bmatrix} a_{11}&a_{12}&a_{13}&a_{14}\\ a_{41}&a_{42}&a_{43}&a_{44}\\ a_{31}&a_{32}&a_{33}&a_{34}\\ a_{21}&a_{22}&a_{23}&a_{24}\\ a_{51}&a_{52}&a_{53}&a_{54} \end{bmatrix}$$

Similarly, postmultiplication of **A** by $I_{k,l}$ interchanges the kth and l^{th} columns of **A**.

Suppose that an element in the second row and fourth column of **I** is replaced by a scalar β. As a premultiplier, this matrix changes the second row of **A** by adding to it (element by element), β times the corresponding element of the fourth row:

$$\begin{bmatrix} 1&0&0&0&0\\ 0&1&0&\beta&0\\ 0&0&1&0&0\\ 0&0&0&1&0\\ 0&0&0&0&1 \end{bmatrix} \begin{bmatrix} a_{11}&a_{12}&a_{13}&a_{14}\\ a_{21}&a_{22}&a_{23}&a_{24}\\ a_{31}&a_{32}&a_{33}&a_{34}\\ a_{41}&a_{42}&a_{43}&a_{44}\\ a_{51}&a_{52}&a_{53}&a_{54} \end{bmatrix}$$

$$= \begin{bmatrix} a_{11}&a_{12}&a_{13}&a_{14}\\ a_{21}+\beta a_{41}&a_{22}+\beta a_{42}&a_{23}+\beta a_{43}&a_{24}+\beta a_{44}\\ a_{31}&a_{32}&a_{33}&a_{34}\\ a_{41}&a_{42}&a_{43}&a_{44}\\ a_{51}&a_{52}&a_{53}&a_{54} \end{bmatrix}$$

In general, let $I(k, l, \beta)$ denote the identity matrix with the element in the kth row and l^{th} column replaced by a scalar β. Given **A**, the conformable premultiplier $I(k, l, \beta)$ multiplies elements of the lth row by β and adds them to corresponding elements of the kth row, providing $k \neq l$. If $k = l$, a diagonal element of **I** has been replaced by β (in the kth row and k^{th} column). Thus, $I(k, k, \beta)$ as a premultiplier, multiplies elements of the kth row by β.

As a conformable postmultiplier, $I(k, l, \beta)$, multiplies the kth column of **A** by β, and adds it to the lth column, provided $k \neq l$. If $k = l$, then $I(k, k, \beta)$ as a postmultiplier, multiplies elements of the kth column of **A** by β.

Gauss-Jordan Reduction

Given a matrix **A**, the premultiplier $I\,[1, 1, (1/a_{11})]$ multiplies each element in the first row of **A** by $1/a_{11}$ (unless $a_{11} = 0$). If the resulting matrix is then premultiplied by $I(2, 1, -a_{21})$, the second row is changed by adding to each of its elements $-a_{21}$ times the corresponding element of the new first row. That is:

$$I(2,1,-a_{21})I\left(1,1,\frac{1}{a_{11}}\right)A = \begin{bmatrix} 1 & a_{12}^{(1)} & a_{13}^{(1)}\cdots a_{1n}^{(1)}\\ 0 & a_{22}^{(1)} & a_{23}^{(1)}\cdots a_{2n}^{(1)}\\ a_{31} & a_{32} & a_{33} \quad \cdots a_{3n}\\ \vdots & & \\ a_{m1} & a_{m2} & a_{m3} \quad \cdots a_{mn} \end{bmatrix} \quad (5)$$

If the result of Eq. (5) is now premultiplied by the product of the sequence of matrices:

$$I(m,1,-a_{m1})I(m-1,1,-a_{m-1,1})\cdots I(4,1,-a_{41})I(3,1,-a_{31})$$

the result is:

$$\begin{bmatrix} 1 & a_{12}^{(1)} & a_{13}^{(1)}\cdots a_{1n}^{(1)}\\ 0 & a_{22}^{(1)} & a_{23}^{(1)}\cdots a_{2n}^{(1)}\\ 0 & a_{32}^{(1)} & a_{33}^{(1)}\cdots a_{3n}^{(1)}\\ \vdots & & \\ 0 & a_{m2}^{(1)} & a_{m3}^{(1)}\cdots a_{mn}^{(1)} \end{bmatrix} \quad (6)$$

Eq. (6) is the first step of either Gaussian or of Gauss-Jordan elimination. These procedures were presented in previous articles (*Chem. Eng.*, Apr. 6, 1970, p. 119; May 18, 1970, p. 153) in terms of a system of simultaneous linear algebraic equations (SLAE's).

If in the matrix of Eq. (6), the number $a_{22}^{(1)}$ is not zero, then the following sequence [as a premultiplier of Eq. (6)]:

$$I(m,2,-a_{m2}^{(1)})\cdots$$

$$I(3,2,-a_{32}^{(1)})I(1,2,-a_{12}^{(1)})I\left(2,2,\frac{1}{a_{22}^{(1)}}\right) \quad (7)$$

gives:

$$\begin{bmatrix} 1 & 0 & a_{13}^{(2)}\cdots a_{1n}^{(2)}\\ 0 & 1 & a_{23}^{(2)}\cdots a_{2n}^{(2)}\\ 0 & 0 & a_{33}^{(2)}\cdots a_{3n}^{(2)}\\ \vdots & \vdots & \vdots\\ 0 & 0 & a_{m3}^{(2)}\cdots a_{mn}^{(2)} \end{bmatrix} \quad (8)$$

In general, once the $(l-1)^{st}$ matrix is formed in which each of the first $(l-1)$ columns has a one

* Note that lower-case bold-face symbols denote vectors. Capital letters in bold face denote matrices that may have more than one column and more than one row.

as the diagonal element and zeros elsewhere], the next sequence of premultipliers is:

$$\left\{\prod_{\substack{k=1\\k\neq l}}^{m} I(k,l,-a_{kl}^{(l-1)})\right\} I\left(l,l,\frac{1}{a_{ll}^{(l-1)}}\right) \tag{9}$$

provided that:

$$a_{ll}^{(l-1)} \neq 0$$

Note that Gaussian elimination is described by replacing the product symbol in Eq. (9) with:

$$\prod_{k=l+1}^{m}$$

By reference to the set in Eq. (5), (7) or (9), it can be seen that difficulties arise if the $(l-1)^{st}$ calculated array has the diagonal element $a_{ll}^{(l-1)}$ equal to zero. In this case, count elements down the lth row until we find an element that is not zero (or else, find all zeros). Note that to count subsequent elements in the row is to count by columns. If a nonzero element is found, then that column is interchanged with the lth column. This interchange can be accomplished by the appropriate postmultiplier. The computations then proceed as previously outlined. However, if the SLAE's $Ax = c$, are being solved by elimination, two of the variables in vector x should be renumbered if two columns of A are interchanged.

In the previous paragraph, suppose that $a_{ll}^{(l-1)} = 0$, and as we count down the lth row we find nothing but zeros. This whole row of zeros can be interchanged with a nonzero row below it (if there is one). On the other hand, if all elements in the lth row as well as all elements in the row below it are zeros, the process is finished.

In general, given an $m \times n$ matrix A, the Gauss-Jordan reduction will produce a matrix of the form:

$$\begin{bmatrix} 1 & 0 & 0 \cdots 0 & a_{1,r+1}^{(r)} & a_{1,r+2}^{(r)} & \cdots a_{1,n}^{(r)} \\ 0 & 1 & 0 \cdots 0 & a_{2,r+1}^{(r)} & a_{2,r+2}^{(r)} & \cdots a_{2,n}^{(r)} \\ \vdots & \vdots & \vdots & \vdots & \vdots & \vdots \\ 0 & 0 & 0 \cdots 1 & a_{r,r+1}^{(r)} & a_{r,r+2}^{(r)} & \cdots a_{r,n}^{(r)} \\ 0 & 0 & 0 \quad 0\ 0 & 0 & 0 \\ \vdots & \vdots & \vdots & \vdots & \vdots \\ 0 & 0 & 0 \quad 0\ 0 & 0 & 0 \end{bmatrix} \tag{10}$$

($m - r$ = rows of zeros)

Note that the number r in Eq. (10) is determined by this procedure. In his own problems, the reader can count the number of rows of zeros. The numbers m and n are known since A is given.

The integer r turns out to be a rather important little devil. It is normally calculated as a byproduct of an elimination process. The idea of rank will be considered in the next article.

A Collection of Miscellaneous Relations

The number of representations of a given mathematical object are perhaps as numerous as the number of authors.

For example, suppose that we were considering the following set of m sums:

$$\begin{matrix} a_{11}\alpha_1 + a_{12}\alpha_2 + \cdots + a_{1n}\alpha_n \\ a_{21}\alpha_1 + a_{22}\alpha_2 + \cdots + a_{2n}\alpha_n \\ \vdots \\ a_{m1}\alpha_1 + a_{m2}\alpha_2 + \cdots + a_{mn}\alpha_n \end{matrix} \tag{11}$$

where α_1 through α_n are arbitrary (perhaps unknown) constants, and a_{ij} (some number) is the coefficient of α_j in the ith sum of Eq. (11). On the other hand, α_j is the coefficient of a_{ij} for any i from 1 through m in Eq. (11).

Certainly, Eq. (11) can be written as follows, because the definition of multiplication assures that Eq. (12) is the same as Eq. (11):

$$\begin{bmatrix} a_{11} \cdots a_{1n} \\ \vdots \\ a_{m1} \cdots a_{mn} \end{bmatrix} \begin{bmatrix} \alpha_1 \\ \vdots \\ \alpha_m \end{bmatrix} \tag{12}$$

Of course, Eq. (12) can be written as:

$$A\alpha \tag{13}$$

On the other hand, all of the coefficients of α_i in Eq. (12) can be arranged in a column and multiplied by α_1, the coefficients of α_2 can be arranged in a column and multiplied by α_2, etc., and the results added, to give the representation:

$$\alpha_1\begin{bmatrix} a_{11} \\ a_{21} \\ \vdots \\ a_{m1} \end{bmatrix} + \alpha_2\begin{bmatrix} a_{12} \\ a_{22} \\ \vdots \\ a_{m2} \end{bmatrix} + \cdots + \alpha_n\begin{bmatrix} a_{1n} \\ a_{2n} \\ \vdots \\ a_{mn} \end{bmatrix} \tag{14}$$

The definition of multiplication by a scalar, and of addition of vectors or matrices, assures that Eq. (14) means the same thing as Eq. (11), (12), and (13).

The bracketed columns (column vectors) in Eq. (14) are just the columns of matrix A [the left-most matrix of Eq. (12)]. Since matrix A undergoes no particular change in meaning in respect to the way it is written, we could just as well write A as:

$$\begin{bmatrix} a_{11} \\ a_{21} \\ \vdots \\ a_{m1} \end{bmatrix}, \begin{bmatrix} a_{12} \\ a_{22} \\ \vdots \\ a_{m2} \end{bmatrix}, \cdots, \begin{bmatrix} a_{1n} \\ a_{2n} \\ \vdots \\ a_{mn} \end{bmatrix} \tag{15}$$

Matrix A can be considered as a collection of vectors. The elements of a vector are often called the components of the vector. Thus, a_{23} is the second component in the third vector, and a_{ij} is the ith component of the jth vector in Eq. (15). [A collection of such vectors will later be called a vector subspace.]

If we agree to denote the first column of A, or the first vector of Eq. (15), by a_1, the second by a_2, ..., the jth by a_j, ... etc., then Eq. (15) might be written:

$$a_1, a_2, \cdots, a_j, \cdots a_n \tag{16}$$

Eq. (16) also represents the matrix A. [If the reader

likes brackets when discussing matrices, Eq. (16) will not be changed if enclosed in brackets.]

Having agreed to denote the jth column by a_j, Eq. (14) can be written as:

$$\alpha_1 a_1 + \alpha_2 a_2 + \cdots + \alpha_n a_n \qquad (17)$$

The representation of Eq. (17) has precisely the same meaning as Eq. (11), (12), (13) and (14). Eq. (17) can also be written as either:

$$\sum_{j=1}^{n} \alpha_j a_j \qquad (18)$$

or:

$$\sum_{j=1}^{n} \alpha_j a_{ij} \quad i = 1, 2, \cdots, m \qquad (19)$$

Eq. (11)-(14), (17)-(19) represent the same object. The particular expression that is used is a matter of individual choice. However, one expression may be more useful in a particular problem than another. What object do these represent? Each of them represents a vector of m components; the vector is described when each a_{ij} (hence, m and n) and each a_j are specified. The terminology that is used in conjunction with Eq. (11) through (19) is abundant.

An author wishes to denote some object by the symbol $\square$. The obvious choice for such an object would be a box. However, the author wishes to choose an object for which multiplication is well defined. Therefore, let $\square_j$ denote the jth rabbit in a hutch of n rabbits ($\square_j$ = 1 rabbit, the jth one). At some later time, the author may find that the number of rabbits in his hutch has increased by:

$$\alpha_1 \square_1 + \alpha_2 \square_2 + \cdots + \alpha_n \square_n = \sum_{j=1}^{n} \alpha_j \square_j \qquad (20)$$

where the α's are constants (some may be zeros depending on gender or other factors).

Any objects, each multiplied by a constant and then summed, are called a linear combination of them, provided that scalar multiplication followed by addition is defined. Thus, Eq. (20) might be called a linear combination of rabbits. If α_1 through α_n are constants, then Eq. (14), (17) and (18) are a linear combination of vectors.

In order to define linear dependence, suppose that:

$$\sum_{j=1}^{n} \alpha_j a_j = 0 \qquad (21)$$

If there are constants α_1 through α_n, not all zeros, for which Eq. (21) holds, the vectors a_1 through a_n are said to be linearly dependent. Otherwise, the vectors are said to be linearly independent.

Note that if there is at least one nonzero constant for which Eq. (11) holds, then at least one of the linearly dependent vectors can be expressed as a linear combination of others in the set. Also note that the question of whether or not a set of vectors is a linearly dependent set is to ask if there is a solution of:

$$A\alpha = 0 \qquad (22)$$

Eq. (22) always has the solution $\alpha = 0$. This is called the trivial solution. The question of nontrivial solutions ($\alpha \neq 0$) will be discussed in subsequent articles. ■

Problems to Test Your Knowledge

Here is an opportunity to check your skills on the material covered in this installment. The completed answers will be published in the next CE Refresher, July 27, 1970.

Problem 1: Postmultiply the following matrix A by the matrix I (2, 4, β)

$$\begin{bmatrix} a_{11} & a_{12} & a_{13} & a_{14} \\ a_{21} & a_{22} & a_{23} & a_{24} \\ a_{31} & a_{32} & a_{33} & a_{34} \\ a_{41} & a_{42} & a_{43} & a_{44} \end{bmatrix} \begin{bmatrix} 1 & 0 & 0 & 0 \\ 0 & 1 & 0 & \beta \\ 0 & 0 & 1 & 0 \\ 0 & 0 & 0 & 1 \end{bmatrix} = ?$$

Problem 2: What is the difference between the final matrices obtained when premultiplication or postmultiplication is performed on matrix A by a conformable matrix I (k, 1, β)?

Here are the answers to the problems given in Part 2 of this series, May 18, 1970.

Problem 1:

$$\begin{bmatrix} 2 & 4 & -4 \\ 3 & 8 & -2 \\ 2 & 1 & -2 \end{bmatrix} \begin{bmatrix} -7/16 & 1/8 & 3/4 \\ 1/16 & 1/8 & -1/4 \\ -13/32 & 3/16 & 1/8 \end{bmatrix} = \begin{bmatrix} A & B & C \\ D & E & F \\ G & H & J \end{bmatrix}$$

For convenience (in referring to the element in the product matrix), each element is identified by the letters A, B, C, etc.

$$\begin{aligned}
A &= 2(-7/16) + 4(1/16) + (-4)(-13/32) = 1 \\
B &= 2(1/8) + 4(1/8) + (-4)(3/16) = 0 \\
C &= 2(3/4) + 4(-1/4) + (-4)(1/8) = 0 \\
D &= 3(-7/16) + 8(1/16) + (-2)(-13/32) = 0 \\
E &= 3(1/8) + 8(1/8) + (-2)(3/16) = 1 \\
F &= 3(3/4) + 8(-1/4) + (-2)(1/8) = 0 \\
G &= 2(-7/16) + 1(1/16) + (-2)(-13/32) = 0 \\
H &= 2(1/8) + 1(1/8) + (-2)(3/16) = 0 \\
J &= 2(3/4) + 1(-1/4) + (-2)(1/8) = 1
\end{aligned}$$

Problem 2:

$$\begin{bmatrix} -7/16 & 1/8 & 3/4 \\ 1/16 & 1/8 & -1/4 \\ -13/32 & 3/16 & 1/8 \end{bmatrix} \begin{bmatrix} 2 \\ 1 \\ 13 \end{bmatrix} = \begin{bmatrix} 9 \\ -3 \\ 1 \end{bmatrix}$$

$$\begin{aligned}
-7/16(2) + 1/8(1) + 3/4(13) &= 9 \\
1/16(2) + 1/8(1) + (-1/4)(13) &= -3 \\
-13/32(2) + 3/16(1) + 1/8(13) &= 1
\end{aligned}$$

Problem 3:

The augmented matrix for the set of simultaneous equations is:

$$\left[\begin{array}{ccc|c} 4 & -3 & 2 & -7 \\ 6 & 2 & -3 & 33 \\ 2 & -4 & -1 & -3 \end{array}\right]$$

The reduced matrix is:

$$\left[\begin{array}{ccc|c} 1 & -3/4 & 1/2 & -7/4 \\ 0 & 1 & -12/13 & 87/13 \\ 0 & 0 & 1 & -4 \end{array}\right]$$

Hence; $x_1 = 5/2$, $x_2 = 3$, and $x_3 = -4$.

Appendix B

ENTHALPY DATA

Appendix B contains heats of formation at 298K and molecular weights of many compounds (Part 1), charts of enthalpy increments above 298K for many elements and their inorganic compounds (Part II) and Bureau of Mines Bulletin 605, Thermodynamic Properties of 65 Elements - Their Oxides, Halides, Carbides and Nitrides, by C. E. Wicks and F. E. Block (Part III). While it is realized that the data given in this appendix are in general more than twenty years old, it is believed that the accuracy of these data is sufficient for most industrial problems. If more precise information is warranted, the reference given in Chapter 5 should be consulted.

Parts I and II of this appendix are reproduced from Principles of Extractive Metallurgy by T. Rosenqvist with the permission of the author and McGraw-Hill Publishing Company. Part III is reproduced with the permission of the U. S. Bureau of Mines.

Part I - Heats of formation at 298K and molecular weights of many compounds*

Substance	Mole weight	$-\Delta H^\circ_{298}$ kcal/mole	Accuracy ±kcal	
Ag(s)	107.9	0		
AgCl(s)	143.3	30.3	0.2	
Ag_2O(s)	231.7	7.3	0.1	
Ag_2S(s)	247.8	7.6	0.2	
Al(s)	27.0	0		
AlF(g)	46.0	61.0	2.0	
AlF_3(s)	84.0	356.0	1.0	
AlCl(g)	62.4	11.6	0.8	
$AlCl_3$(s)	133.3	168.6	0.5	
Al_2O_3(s)	102.0	400.0	1.5	
AlN(s)	41.0	76.5	1.0	
Al_4C_3(s)	144.0	51.5	2.0	
Andalusite	162.1	1.3†	0.5	†Al_2SiO_5 from oxides.
Kyanite	162.1	1.9†	0.5	
Sillimanite	162.1	0.6†	0.5	
Mullite	426.0	−7.0‡	0.5	‡$Al_6Si_2O_{13}$ from oxides.
As(s)	74.9	0		
As_2O_3(s)	197.8	156.6	1.0	
As_2O_5(s)	229.8	218.5	1.5	
As_2S_3(s)	246.0	30.0	3.0	
As_2S_5(s)	310.1	35.0	3.0	
B(s)	10.8	0		
BN(s)	24.8	60.5	0.8	
Ba(s)	137.3	0		
$BaCl_2$(s)	208.2	205.4	0.6	
BaO(s)	153.3	139.0	2.0	
BaO_2(s)	169.3	152.5	3.0	
$BaSO_4$(s)	233.4	350.2	5.0	
$BaCO_3$(s)	197.3	290.0	7.5	
C(graphite)	12.0	0		
C(diamond)	12.0	−0.454	0.03	
C(coke etc.)	12.0	−3.0	1.5	
CH_4(g)	16.0	17.89	0.1	
CCl_4(l)	153.8	33.3	0.5	
CCl_4(g)	153.8	25.5	0.4	
$COCl_2$(g)	98.9	53.3	1.5	
CO(g)	28.0	26.40	0.03	
CO_2(g)	44.0	94.05	0.01	
CS_2(l)	76.1	−21.0	1.0	
CS_2(g)	76.1	−27.7	1.0	
COS(g)	60.1	33.9	1.0	
Ca(s)	40.1	0		
CaF_2(s)	78.1	292.0	3.5	
$CaCl_2$(s)	111.0	191.4	1.0	
CaO(s)	56.1	151.6	0.4	
CaS(s)	72.1	110.0	2.5	
$CaSO_4$(s)	136.1	342.4	3.5	

Substance	Mole weight	$-\Delta H^\circ_{298}$ kcal/mole	Accuracy ±kcal	
Ca_3P_2(s)	182.2	120.0	6.0	
CaC_2(s)	64.1	14.1	2.0	
$CaCO_3$(s)	100.1	288.4	0.7	
CaSi(s)	68.2	36.0	2.0	
$CaSi_2$(s)	96.3	36.0	3.0	
Ca_2Si(s)	108.3	50.0	3.0	
$CaSiO_3$(s)	116.2	21.5*	0.3	
Ca_2SiO_4(s)	172.3	30.2*	1.5	*From oxides
Ca_3SiO_5(s)	228.3	27.0*	1.5	
$CaAl_2$(s)	94.0	54.0	3.0	
$CaAl_2O_4$(s)	158.0	3.7*	0.4	*From oxides.
$Ca_3Al_2O_6$(s)	270.2	1.6*	0.4	
Cd(s)	112.4	0		
$CdCl_2$(s)	183.3	93.0	0.5	
CdO(s)	128.4	61.1	0.7	
CdS(s)	144.5	34.5	0.5	
$CdSO_4$(s)	208.5	221.4	1.0	
Ce(s)	140.1	0		
CeO_2(s)	172.1	260.2	2.5	
Co(s)	58.9	0		
$CoCl_2$(s)	129.8	77.8	4.0	
CoO(s)	74.9	57.1	0.5	
CoS(s)	91.0	21.1	1.0	
Co_3S_4(s)	305.0	75.0	3.0	
CoS_2(s)	123.1	33.5	4.0	
$CoSO_4$(s)	155.0	207.5	6.0	
Cr(s)	52.0	0		
$CrCl_2$(s)	122.9	97.0	3.5	
$CrCl_3$(s)	158.4	132.0	5.0	
Cr_2O_3(s)	152.0	270.0	2.5	
CrO_3(s)	100.0	138.5	2.5	
Cr_4C(s)	220.0	16.4	1.5	
Cr_7C_3(s)	400.0	42.5	2.5	
Cr_3C_2(s)	180.0	21.0	2.0	
Cu(s)	63.5	0		
CuCl(s)	99.0	32.2	0.7	
$CuCl_2$(s)	134.4	49.2	2.5	
Cu_2O(s)	143.1	40.0	0.7	
CuO(s)	79.5	37.1	0.8	
Cu_2S(s)	159.1	19.6	0.4	
CuS(s)	95.6	12.1	0.5	
$CuSO_4$(s)	159.6	184.0	2.5	
Fe	55.8	0		
$FeCl_2$(s)	126.8	81.8	0.2	
$FeCl_3$(s)	162.2	95.7	0.2	
FeO(s)	71.9	63.2§	0.3	§$Fe_{0.95}O$
Fe_3O_4(s)	231.6	266.9	1.0	

*Reproduced from <u>Principles of Extractive Metallurgy</u> by T. Rosenqvist.

Substance	Mole weight	$-\Delta H^\circ_{298}$ kcal/mole	Accuracy ±kcal	
Fe_2O_3(s)	159.7	196.3	0.8	
FeS(s)	87.9	22.8	0.3	
FeS_2(s)	120.0	42.4	1.5	
$FeSO_4$(s)	151.9	220.5	6.0	
Fe_4N(s)	237.4	2.6	2.0	
Fe_2N(s)	125.7	0.9	2.0	
Fe_3P(s)	198.5	39.0	2.0	
Fe_3C(s)	179.6	−5.4	1.5	
$FeCO_3$(s)	115.9	178.7	3.0	
FeSi(s)	83.9	19.2	1.5	
Fe_2SiO_4(s)	203.8	2.6‖	2.5	‖From 0.1 Fe + .2 $Fe_{0.95}O$ + SiO_2.
$FeCr_2O_4$(s)	223.9	1.3*	2.5	*From oxides.
H_2(g)	2.016	0		
HF(g)	20.0	64.8	0.5	
HCl(g)	36.5	22.0	0.1	
H_2O(g)	18.0	57.80	0.01	
H_2O(l)	18.0	68.32	0.01	
H_2S(g)	34.1	4.9	0.1	
Hg(l)	200.6	0		
HgCl(s)	236.0	31.5	0.3	
$HgCl_2$(s)	271.5	55.0	1.5	
HgO(s, red)	216.6	21.7	0.2	
HgS(s, red)	232.7	13.9	1.5	
K(s)	39.1	0		
KF(s)	58.1	134.5	1.0	
KCl(s)	74.6	104.2	0.2	
K_2O(s)	94.2	86.4	2.0	
KOH(s)	56.1	101.8	0.5	
K_2SO_4(s)	174.3	342.6	1.0	
K_2CO_3(s)	138.2	188.55	1.5	
K_2SiO_3(s)	154.3	62.5*	7.0	*From oxides.
Mg(s)	24.3	0		
$MgCl_2$(s)	95.2	153.4	0.2	
MgO(s)	40.3	143.7	0.2	
MgS(s)	56.4	83.0	2.0	
$MgSO_4$(s)	120.4	305.5	5.0	
Mg_2C_3(s)	84.7	−19.0	8.0	
MgC_2(s)	48.3	−21.0	5.0	
$MgCO_3$(s)	84.3	262.0	3.0	
Mg_2Si(s)	76.7	19.0	1.0	
Mg_2SiO_4(s)	140.7	15.1*	1.0	*From oxides.
$MgSiO_3$(s)	100.4	8.7*	0.7	
Mn(s)	54.9	0		
$MnCl_2$(s)	125.8	115.2	0.5	
MnO(s)	70.9	92.0	0.5	
Mn_3O_4(s)	228.8	331.4	1.0	
Mn_2O_3(s)	157.9	228.7	1.2	
MnO_2(s)	86.9	124.3	0.5	
Mn_2O_7(s)	221.9	174.1	2.5	
MnS(s)	87.0	49.0	0.5	
MnS_2(s)	119.1	49.5	2.5	
$MnSO_4$(s)	151.0	254.2	1.0	
Mn_3C(s)	176.8	3.6	3.0	
$MnCO_3$(s)	115.0	213.9	1.2	

Substance	Mole weight	$-\Delta H^\circ_{298}$ kcal/mole	Accuracy ±kcal	
$MnSiO_3$(s)	131.0	5.9*	0.4	*From oxides.
Mn_2SiO_4(s)	202.0	11.8*	0.7	
Mo(s)	95.9	0		
MoO_2(s)	127.9	140.0	0.9	
MoO_3(s)	143.9	178.2	0.5	
Mo_2S_3(s)	288.1	92.5	4.0	
MoS_2(s)	160.1	60.4	3.0	
Mo_2N(s)	205.9	16.6	0.5	
Mo_2C(s)	203.9	−4.2	5.0	
N_2(g)	28.0	0		
NH_3(g)	17.0	11.0	0.2	
N_2O(g)	44.0	−19.6	0.4	
NO(g)	30.0	−21.6	0.4	
NO_2(g)	46.0	−8.0	0.2	
Na(s)	23.0	0		
NaF(s)	42.0	137.3	1.0	
NaCl(s)	58.5	98.6	0.2	
Na_2O(s)	62.0	100.7	1.2	
NaOH(s)	40.0	102.3	1.2	
Na_2S(s)	78.1	92.4	2.0	
Na_2SO_4(s)	142.1	333.5	2.0	
Na_2CO_3(s)	106.0	271.6	2.5	
Na_2SiO_3(s)	122.1	55.5*	3.5	*From oxides.
$Na_2Si_2O_5$(s, α)	182.2	60.5*	3.5	
Nb(s)	92.9	0		
NbO(s)	108.9	98.5	2.0	
NbO_2(s)	124.9	190.4	1.5	
Nb_2O_5(s)	265.8	455.0	2.0	
NbC(s)	104.9	33.7	1.5	
Ni(s)	58.7	0		
NiCl(g)	129.6	−55.0	9.0	
$NiCl_2$(s)	129.6	73.0	0.5	
NiO(s)	74.7	57.5	0.5	
Ni_3S_2(s)	240.3	47.5	2.5	
NiS(s)	90.8	22.2	1.4	
NiS_2(s)	122.8	34.0	4.0	
$NiSO_4$(s)	154.8	212.5	5.0	
Ni_3C(s)	188.1	−9.0	1.5	
$NiCO_3$(s)	118.7	162.7	3.0	
$Ni(CO)_4$(g)	170.8	36.4¶	1.0	¶From Ni + 4CO.
P(s, white)	31.0	0		
P(s, red)	31.0	4.4	0.4	
PCl_3(g)	137.3	66.4	1.0	
PCl_3(l)	137.3	74.3	0.8	
PCl_5(g)	208.2	87.3	3.0	
PCl_5(s)	208.2	106.0	3.0	
P_2O_5(s)	141.9	356.6	2.5	
Pb(s)	207.2	0		
$PbCl_2$(s)	278.1	85.8	0.5	
PbO(s, red)	223.2	52.4	0.2	
Pb_3O_4(s)	685.6	175.6	4.0	
PbO_2(s)	239.2	66.1	1.0	
PbS(s)	239.3	22.5	0.5	
$PbSO_4$(s)	303.3	219.5	0.8	
$PbCO_3$(s)	267.2	167.3	2.5	
Pb_2SiO_4(s)	506.5	7.0*	3.5	*From oxides.
$PbSiO_3$(s)	283.3	2.5*	2.0	

Substance	Mole weight	$-\Delta H^\circ_{298}$ kcal/mole	Accuracy ±kcal	
S(s,rh.)	32.1	0		
S(s,monocl.)	32.1	−0.07	0.01	
S(g)	32.1	−56.8	1.5	
S_2(g)	64.1	−31.0	1.0	
SCl_4(l)	173.9	13.6	3.0	
SO_2(g)	64.1	70.95	0.1	
SO_3(g)	80.1	94.4	0.3	
Sb(s)	121.8	0		
$SbCl_3$(s)	228.1	91.4	0.5	
$SbCl_5$(l)	299.0	104.8	3.0	
Sb_2O_3(s)	291.5	169.4	1.0	
Sb_2S_3(s, black)	339.7	40.5	5.0	
$Sb_2(SO_4)_3$(s)	531.7	575.3	8.0	
Si(s)	28.1	0		
SiF_4(g)	104.1	385.0	3.0	
$SiCl_4$(l)	169.9	164.0	1.5	
SiO(g)	44.1	23.2	2.5	
SiO_2(s)**	60.1	217.0	1.0	**α-quartz.
SiO_2(s)††	60.1	216.1	1.0	††β-cristobalite.
SiS_2(s)	92.2	49.0	6.0	
SiS(g)	60.2	−28.0	10.0	
SiC(s)	40.1	15.0	1.0	
Sn(s, white)	118.7	0		
Sn(s, gray)	118.7	0.50	0.05	
$SnCl_2$(s)	189.6	83.6	1.5	
$SnCl_4$(l)	260.5	130.3	1.5	

Substance	Mole weight	$-\Delta H^\circ_{298}$ kcal/mole	Accuracy ±kcal	
SnO_2(s)	150.7	138.7	0.2	
SnS(s)	150.8	25.1	1.2	
SnS_2(s)	182.8	40.0	4.0	
Ti(s)	47.9	0		
$TiCl_2$(s)	118.8	122.8	3.0	
$TiCl_4$(g)	189.7	181.7	0.8	
$TiCl_4$(l)	189.7	191.6	0.6	
TiO(s)	63.9	123.9	0.8	
Ti_2O_3(s)	143.8	362.9	0.8	
Ti_3O_5(s)	223.7	586.9	1.5	
TiO_2(s)	79.9	225.5	1.0	
TiC(s)	59.9	43.9	1.5	
V(s)	50.9	0		
V_2O_3(s)	149.9	293.0	7.0	
V_2O_5(s)	181.9	372.3	4.5	
Zn(s)	65.4	0		
$ZnCl_2$(s)	136.3	99.5	0.3	
ZnO(s)	81.4	83.2	0.3	
ZnS(s)	97.4	48.2	2.0	
$ZnSO_4$(s)	161.4	233.9	2.0	
$ZnCO_3$(s)	125.4	194.2	0.3	
Zn_2SiO_4(s)	222.8	7.0*	1.5	*From oxides.
Zr(s)	91.2	0		
ZrO_2(s)	123.2	259.5	1.5	
ZrN(s)	105.2	87.3	0.5	
ZrC(s)	103.2	44.1	1.6	

Part II - Enthalpy Increments Above 298K*

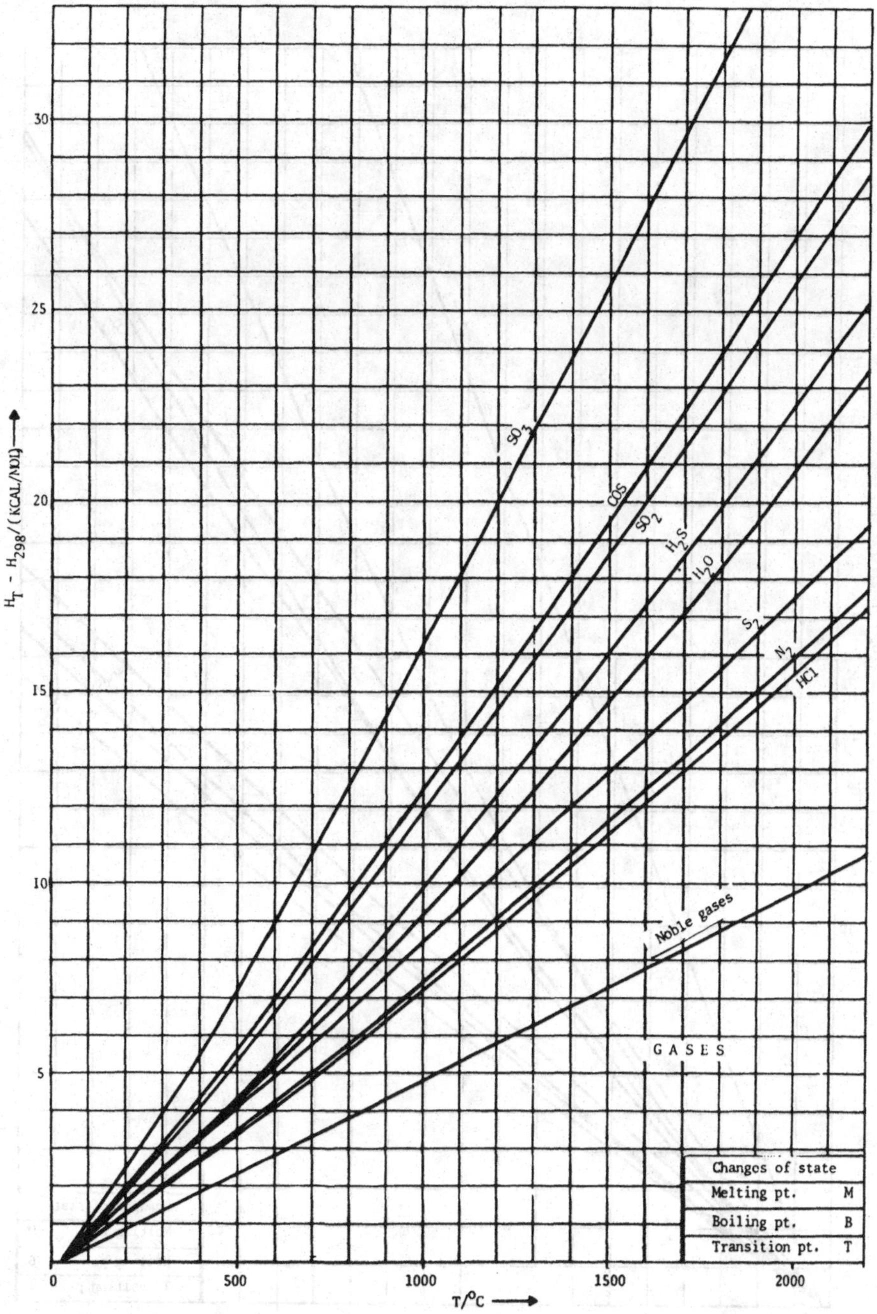

*The sources for the data in this part are as follows: Kelley, K.K.: Bureau of Mines Bulletin, 584, 1960. Wicks, C.E. and Block, F.E.: Bureau of Mines Bulletin, 605, 1963. JANAF Thermochemical Tables, 1965-68. Reproduced from Principles of Extractive Metallurgy by T. Rosenqvist.

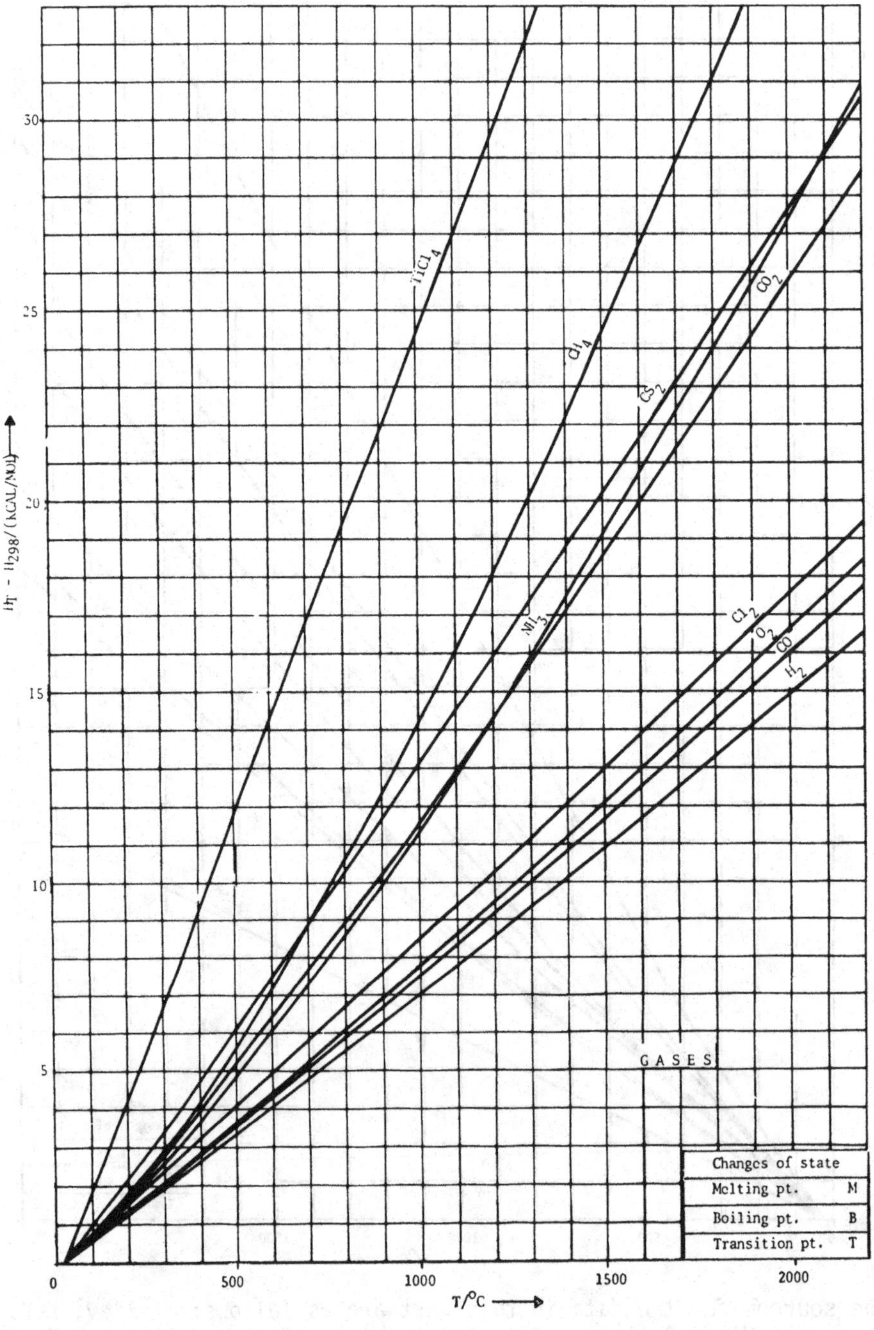
$H_T - H_{298}$/(KCAL/MOL)
30
25
20
15
10
5
0
500
1000
1500
2000
T/°C
$TiCl_4$
CCl_4
CS_2
CO_2
NH_3
Cl_2
O_2
CO
H_2
GASES
Changes of state
Melting pt. M
Boiling pt. B
Transition pt. T

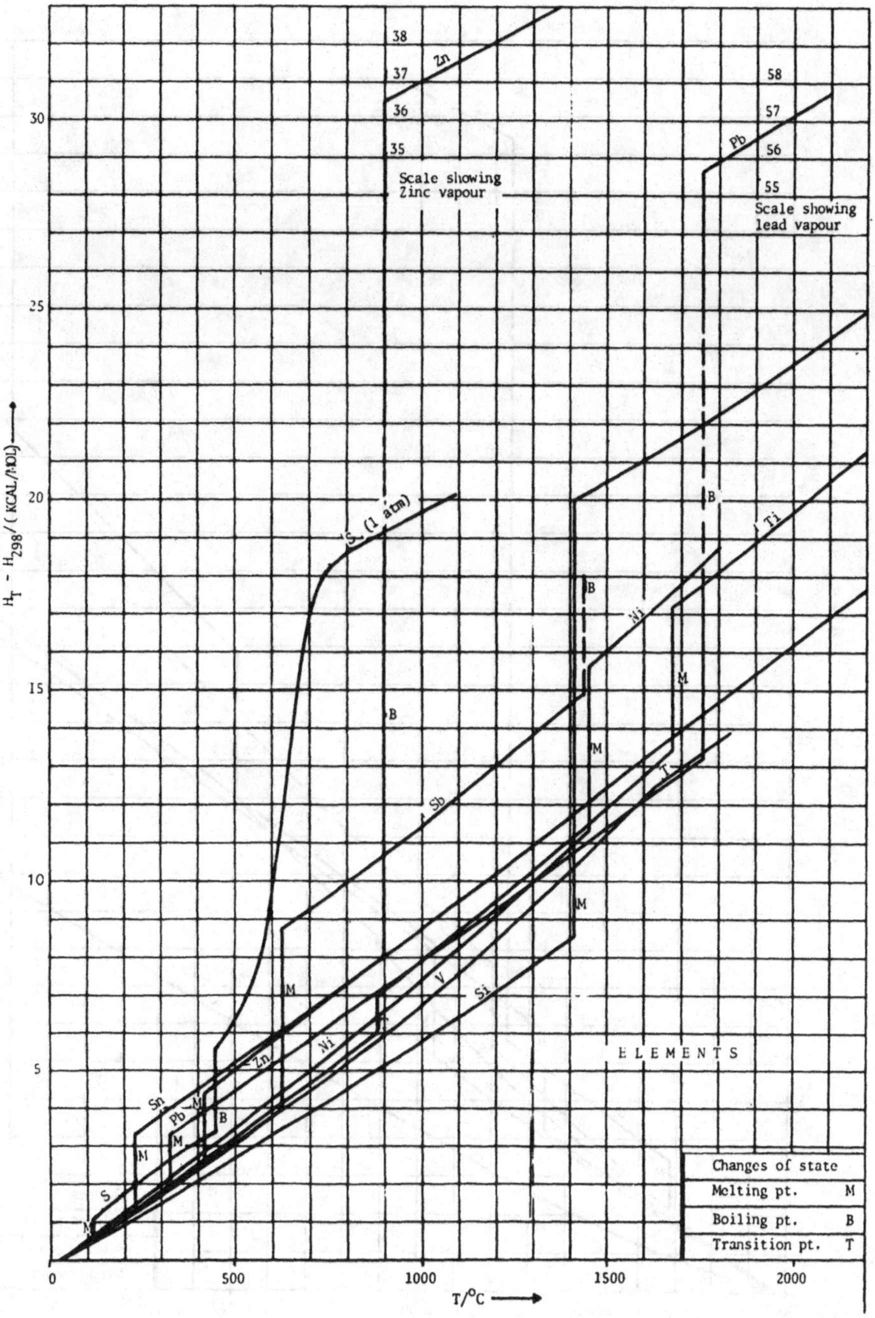

38
37
36
35
Zn
Scale showing Zinc vapour
58
57
56
55
Pb
Scale showing lead vapour
30
25
20
15
10
5
$H_T - H_{298}$/(KCAL/MOL)
S (1 atm)
B
M
Sb
Ni
Ti
T
V
Si
Zn
Sn
Pb
S
ELEMENTS
Changes of state
Melting pt. M
Boiling pt. B
Transition pt. T
0
500
1000
1500
2000
T/°C

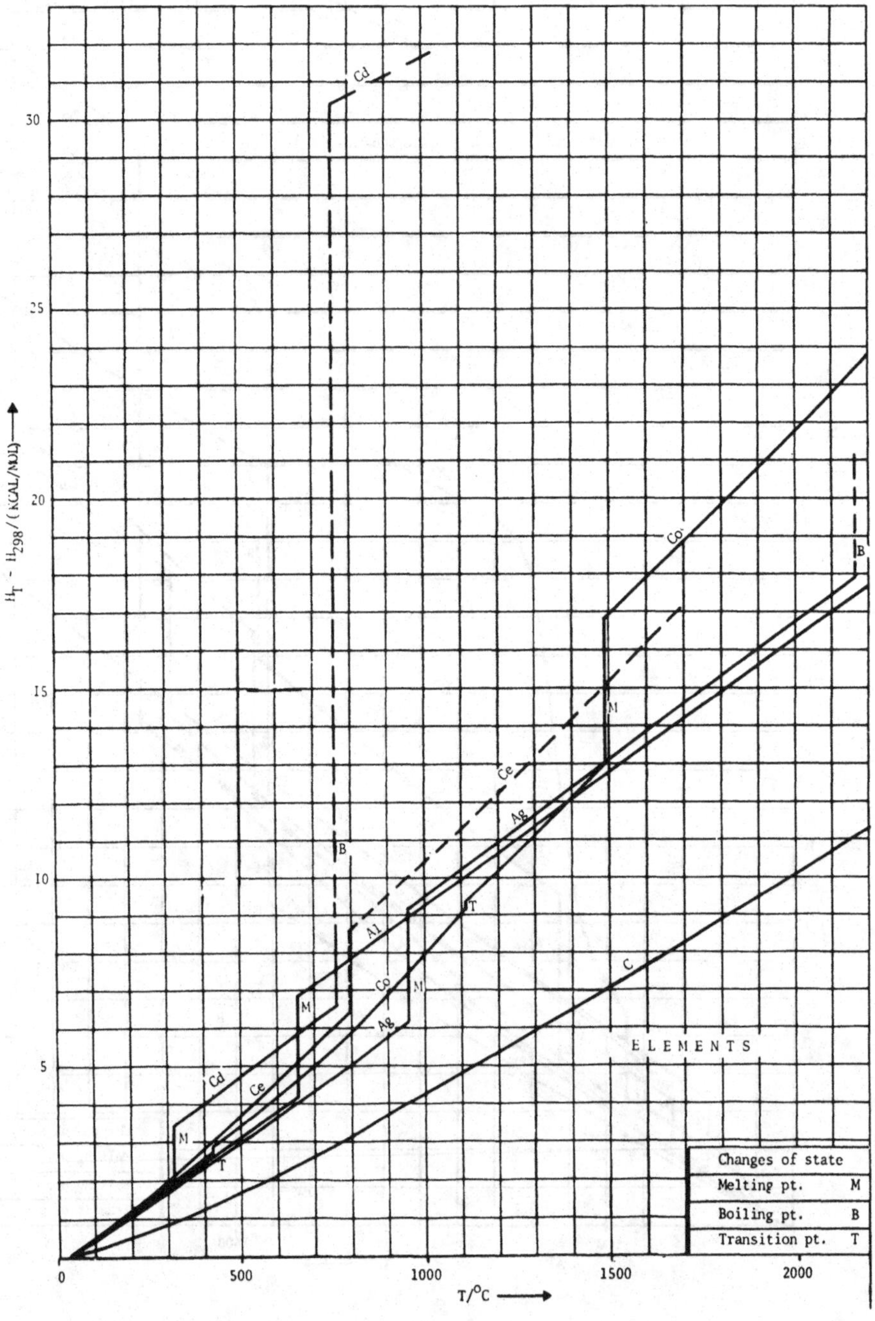

$H_T - H_{298}$/(KCAL/MOL)
T/°C
ELEMENTS
Changes of state
Melting pt. M
Boiling pt. B
Transition pt. T
Cd
Ce
Co
Ag
Al
C

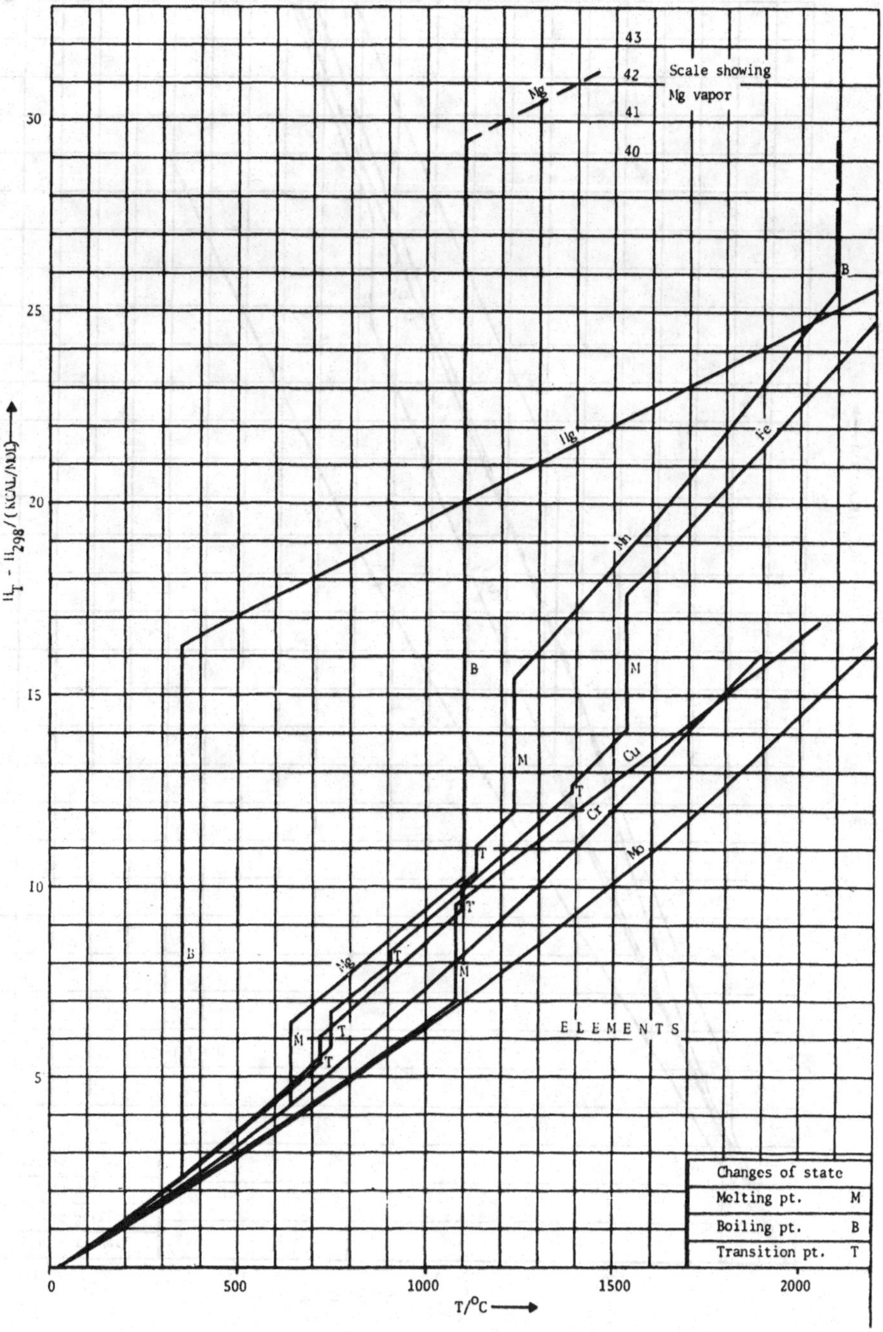

Scale showing
Mg vapor
43
42
41
40
Mg
Hg
Mn
Fe
Cu
Cr
Mo
M
B
T
ELEMENTS
Changes of state
Melting pt. M
Boiling pt. B
Transition pt. T
$H_T - H_{298}$/(KCAL/MOL)
T/°C
0
500
1000
1500
2000
5
10
15
20
25
30

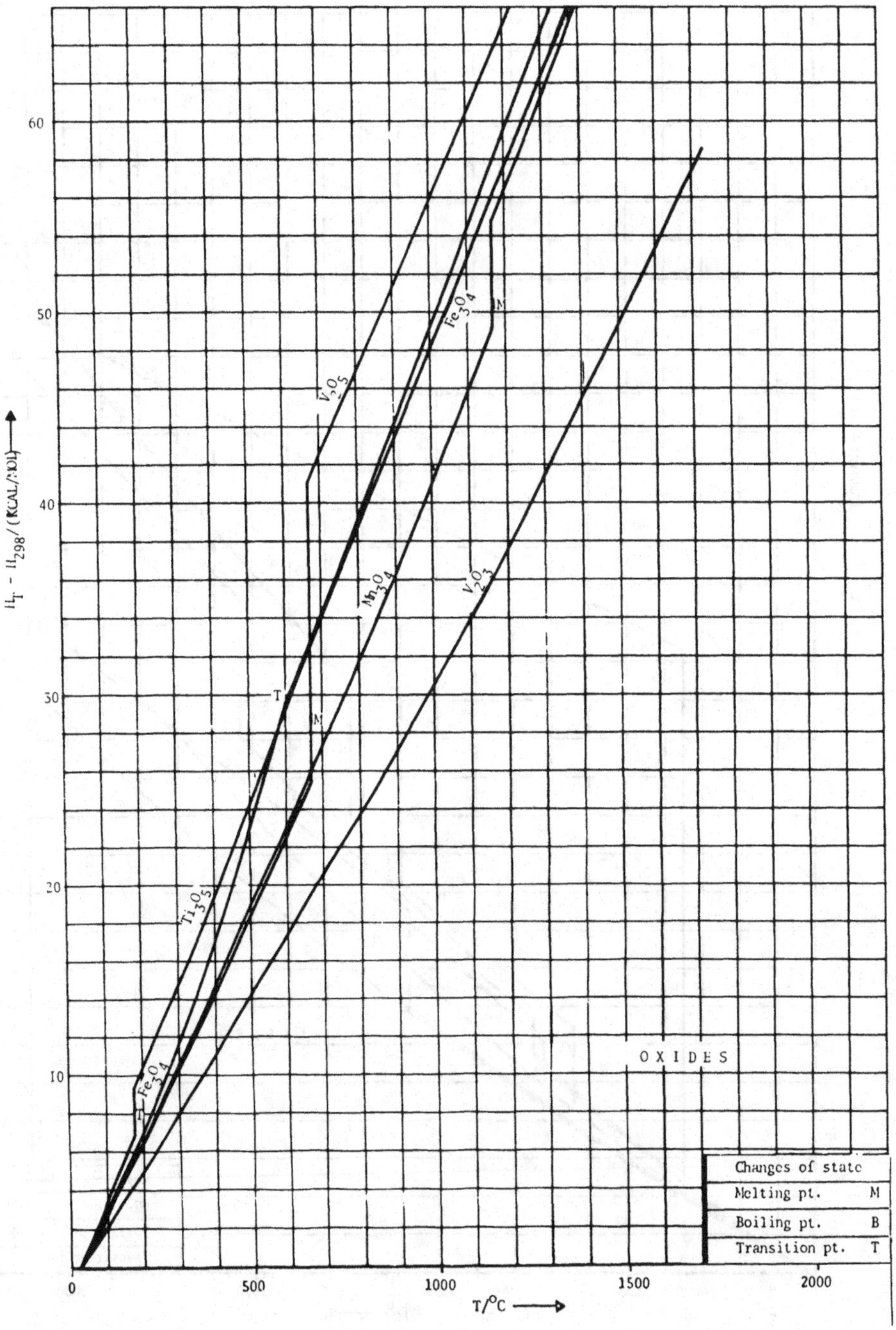

60
50
40
30
20
10
0
500
1000
1500
2000
$H_T - H_{298}$/(KCAL/MOL) ⟶
T/°C ⟶
V_2O_5
Fe_3O_4
Mn_3O_4
V_2O_3
Ti_3O_5
Fe_3O_4
M
T
OXIDES
Changes of state
Melting pt. M
Boiling pt. B
Transition pt. T

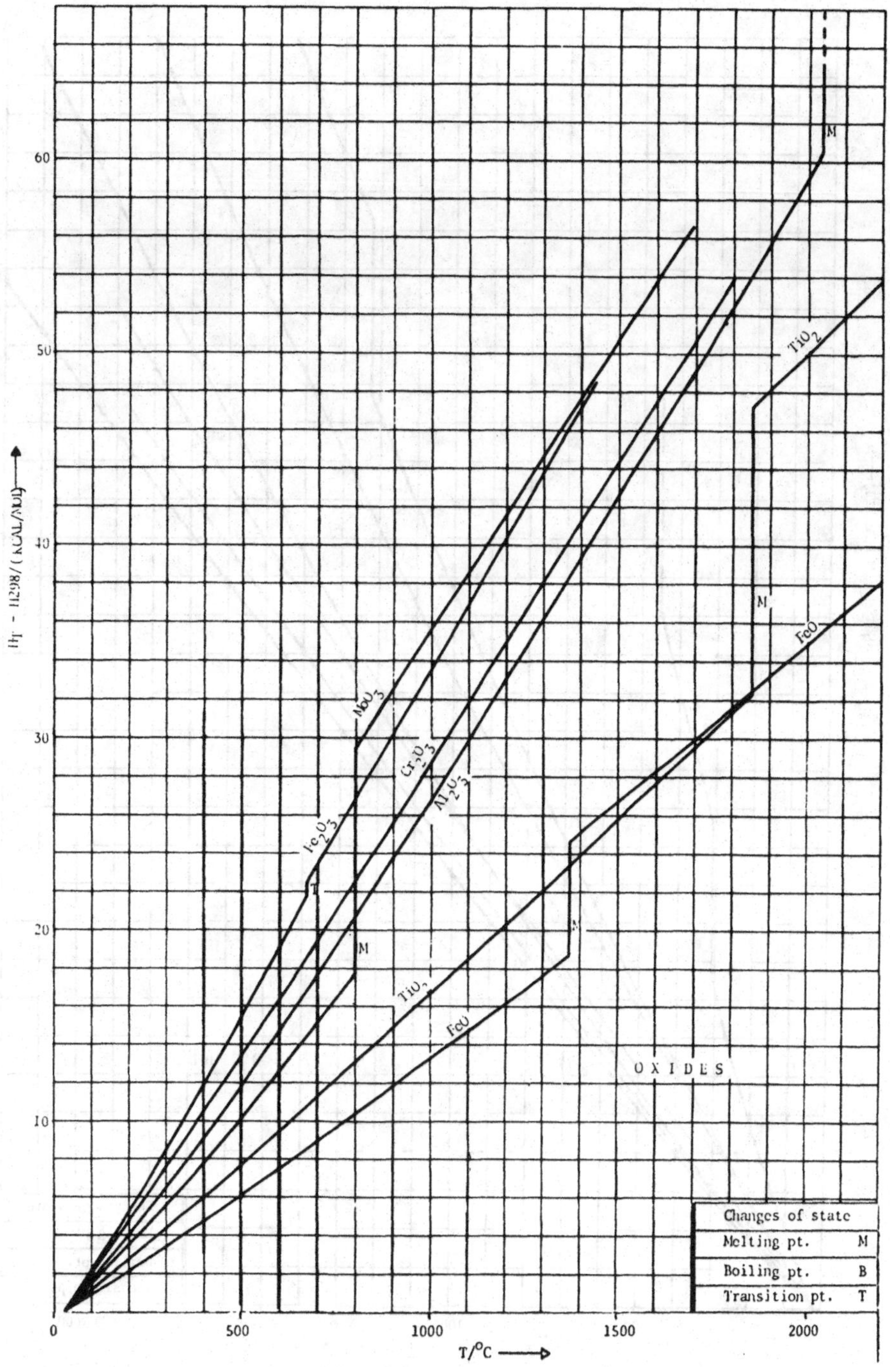

H_T - H298/(KCAL/MOL)
T/°C
0
500
1000
1500
2000
10
20
30
40
50
60
MoO_3
Cr_2O_3
Al_2O_3
Fe_2O_3
TiO_2
FeO
M
T
OXIDES
Changes of state
Melting pt. M
Boiling pt. B
Transition pt. T

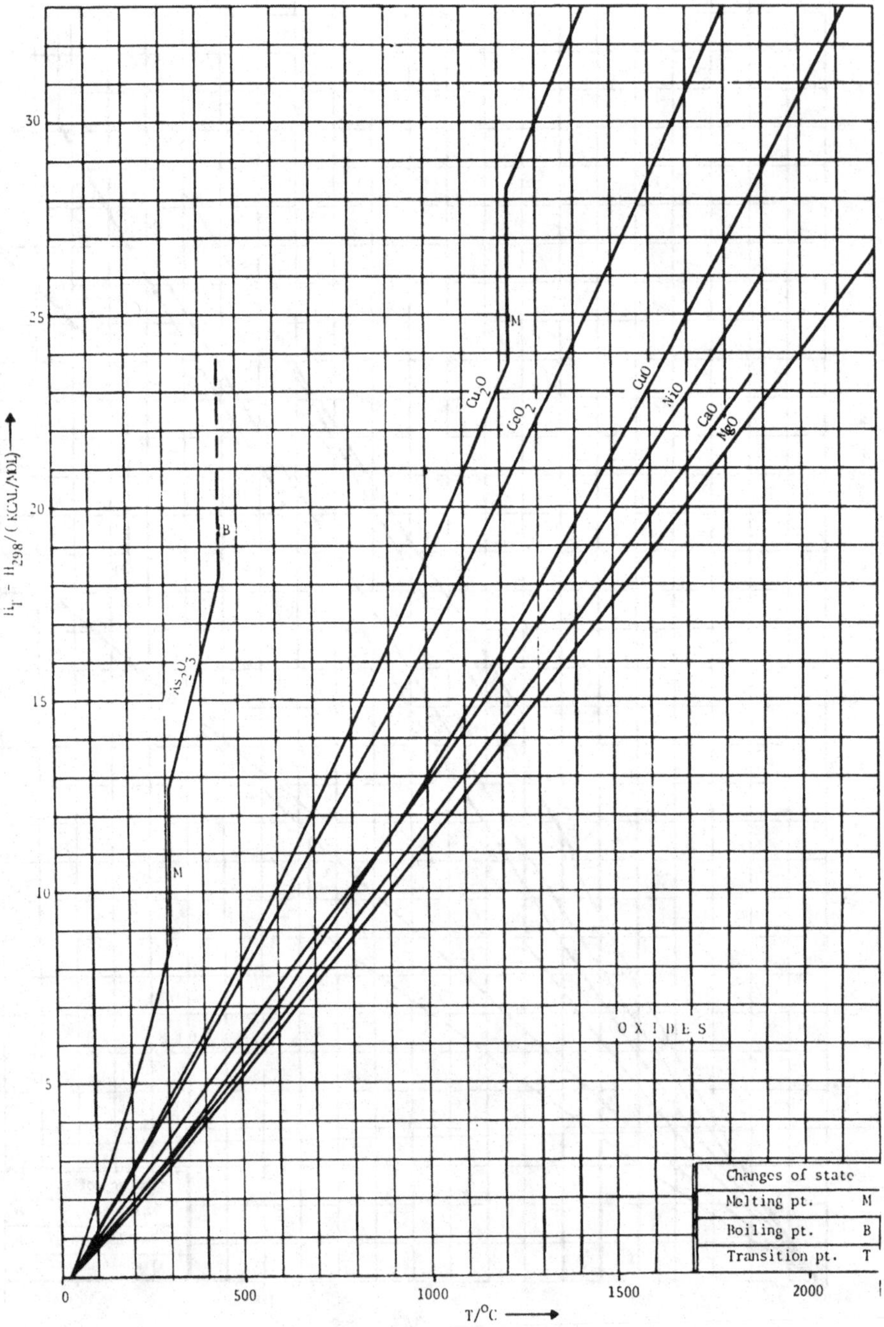

30
25
20
15
10
5
0
500
1000
1500
2000
$T/^{o}C$ ⟶
$H_T - H_{298}$/ (KCAL/MOL) ⟶
M
B
As_2O_3
Cu_2O
CoO_2
CuO
NiO
CaO
MgO
O X I D E S
Changes of state
Melting pt. M
Boiling pt. B
Transition pt. T

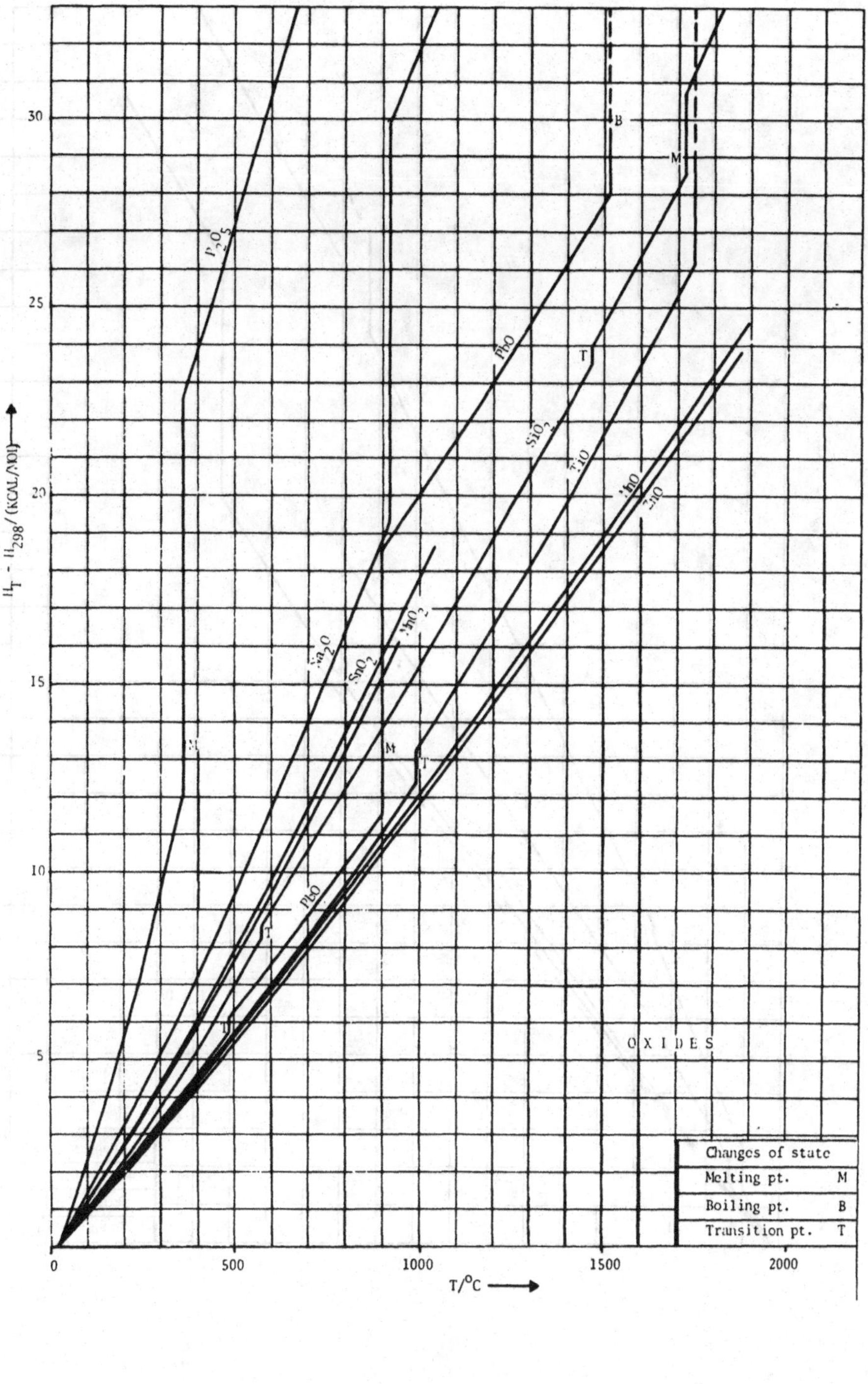

$H_T - H_{298}$/ (KCAL/MOL)
T/°C
0
500
1000
1500
2000
5
10
15
20
25
30
P_2O_5
PbO
SiO_2
TiO
MnO
ZnO
Na_2O
SnO_2
MnO_2
M
T
B
OXIDES
Changes of state
Melting pt. M
Boiling pt. B
Transition pt. T

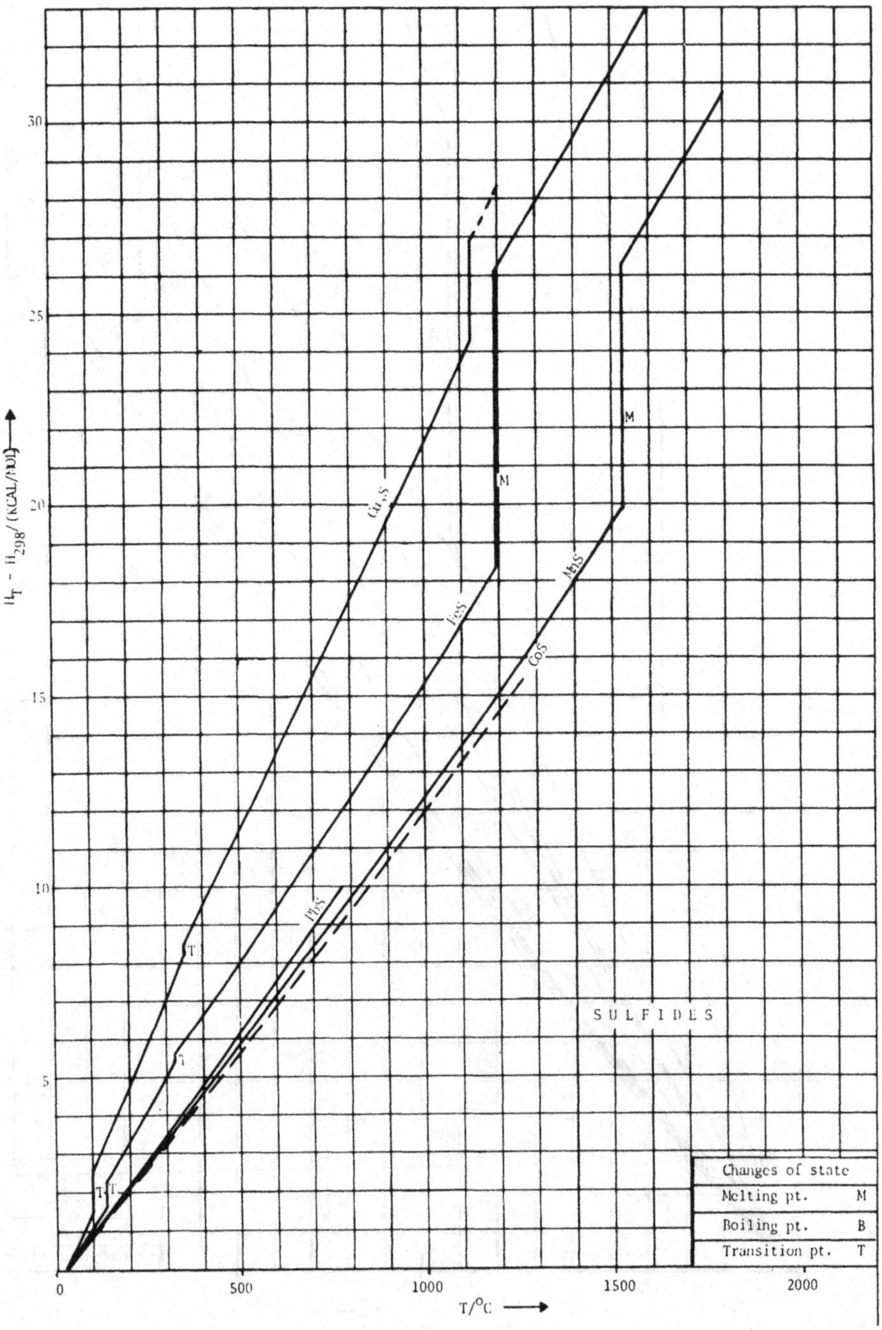

$H_T - H_{298}$/(KCAL/MOL)
T/°C
SULFIDES
Cu₂S
FeS
MnS
CoS
PbS
M
T
Changes of state
Melting pt. M
Boiling pt. B
Transition pt. T

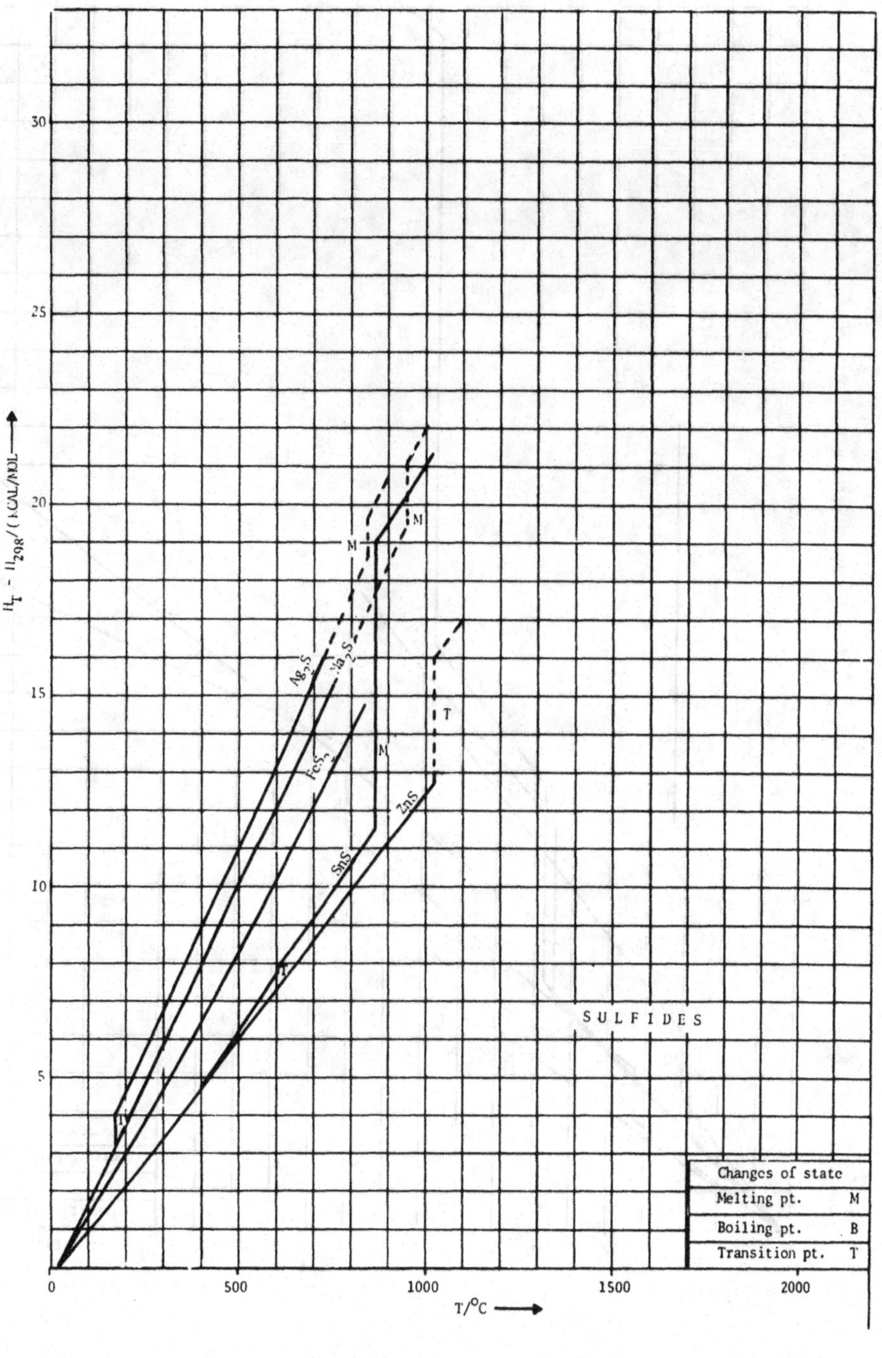
$H_T - H_{298}$/(KCAL/MOL)
T/°C
SULFIDES
Ag_2S
Na_2S
FeS_2
SnS
ZnS
M
T
Changes of state
Melting pt. M
Boiling pt. B
Transition pt. T

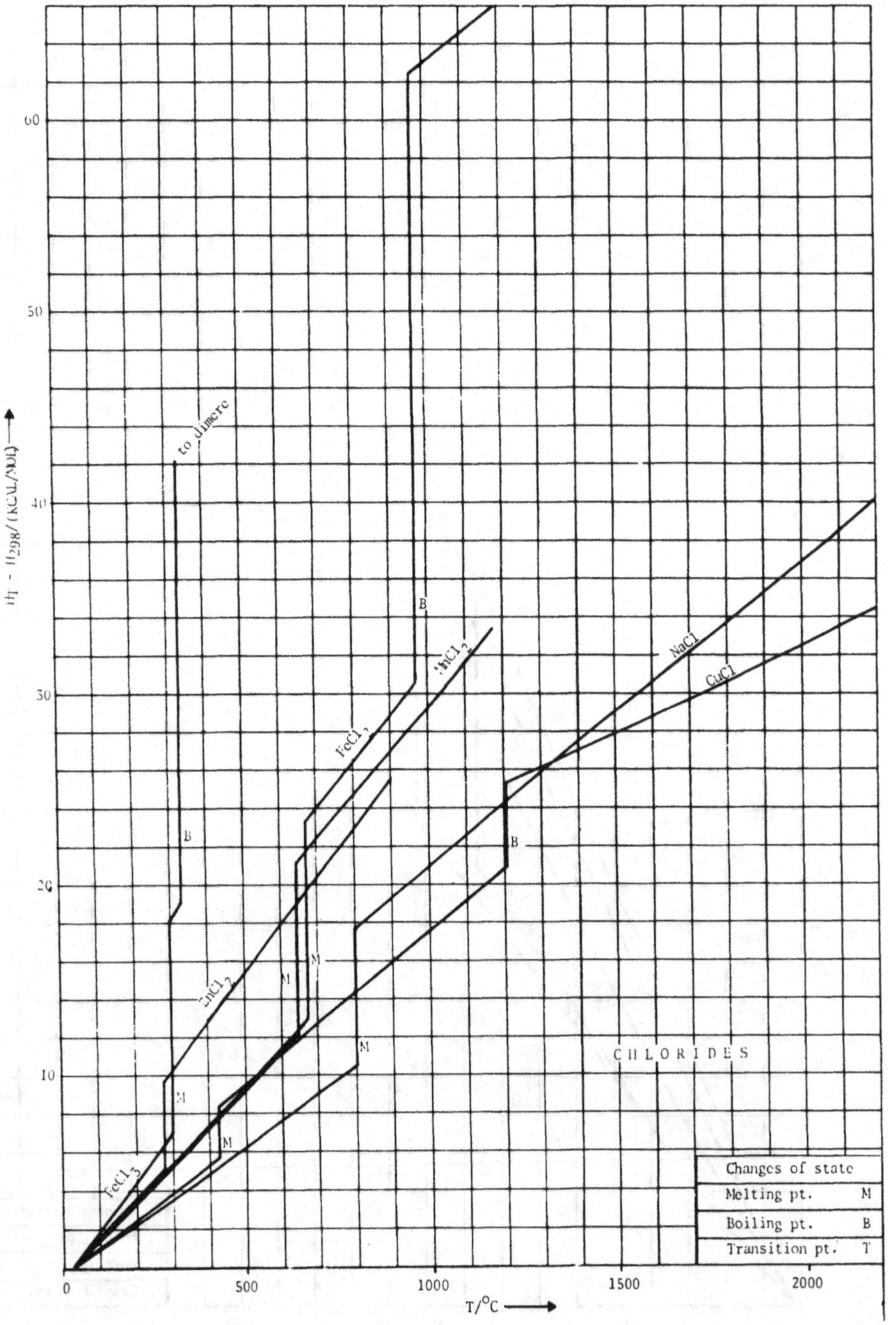

60
50
40
30
20
10
0
500
1000
1500
2000
T/°C
to dimere
B
M
FeCl3
ZnCl2
FeCl2
B
NaCl
CuCl
CHLORIDES
Changes of state
Melting pt. M
Boiling pt. B
Transition pt. T

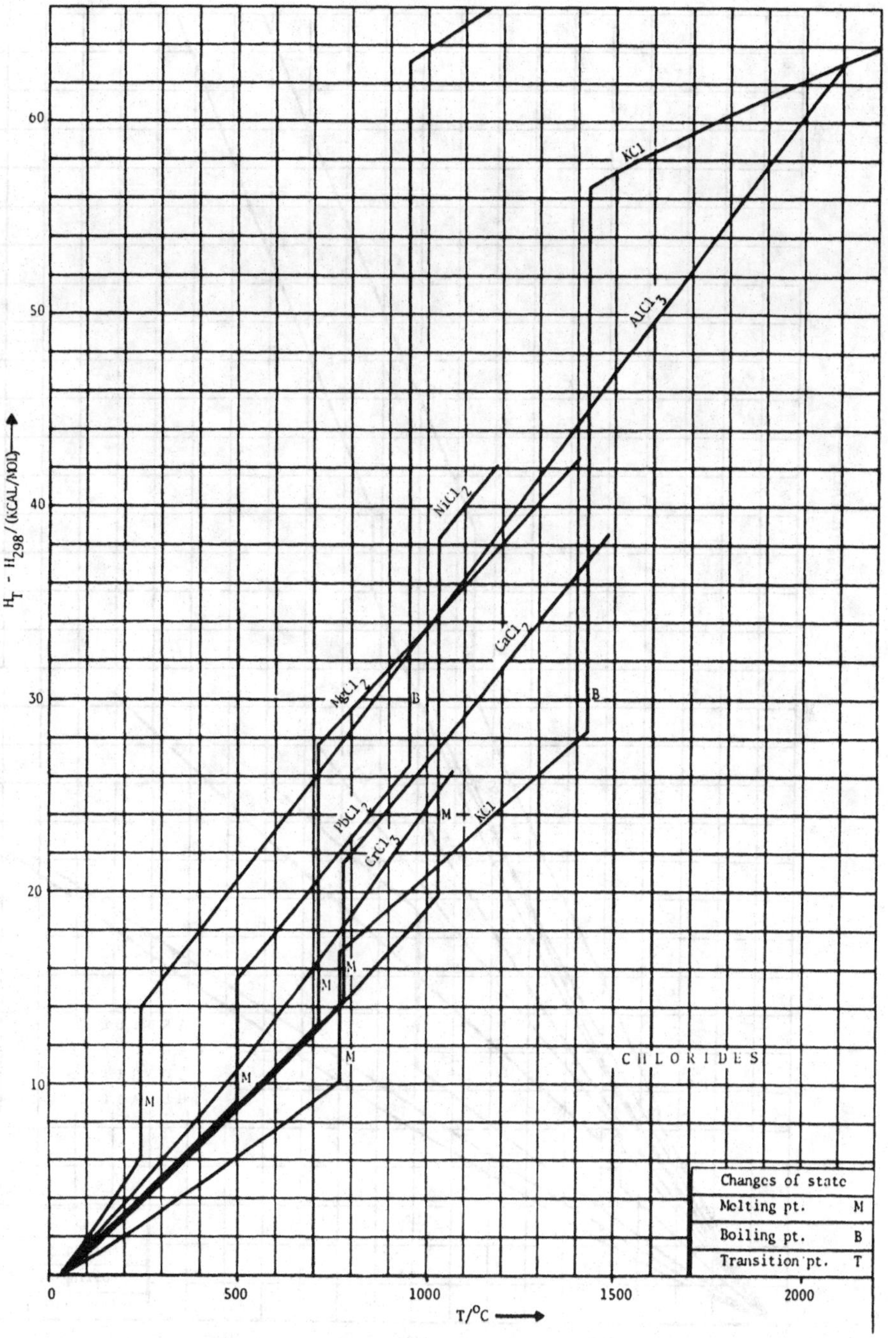

$H_T - H_{298}$/(KCAL/MOL)
T/°C
0
500
1000
1500
2000
10
20
30
40
50
60
KCl
AlCl3
NiCl2
CaCl2
MgCl2
PbCl2
CrCl3
KCl
M
B
CHLORIDES
Changes of state
Melting pt. M
Boiling pt. B
Transition pt. T

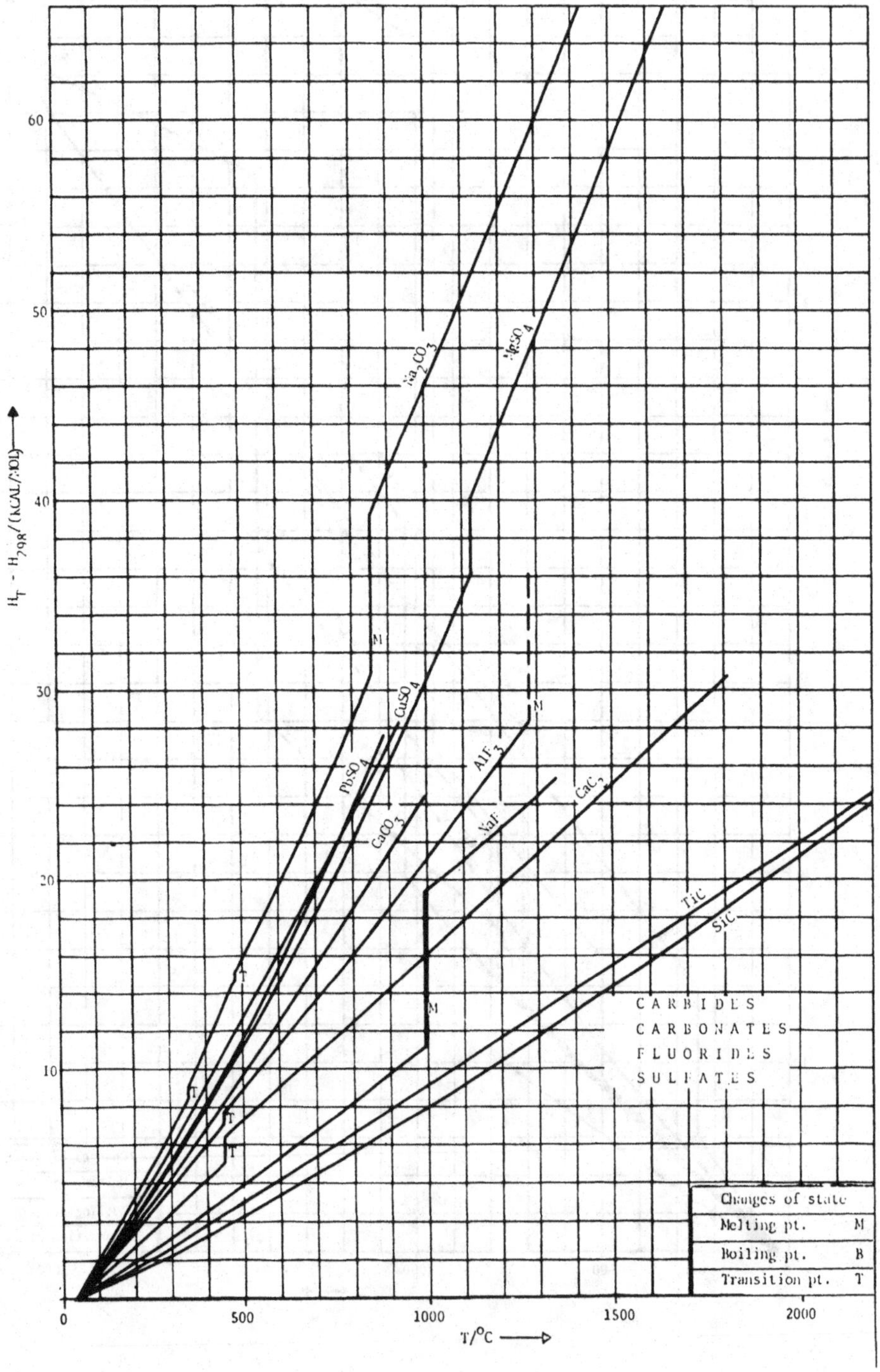

60
50
40
30
20
10
0
500
1000
1500
2000
T/°C
Na2CO3
MgSO4
CuSO4
PbSO4
CaCO3
AlF3
NaF
CaC2
TiC
SiC
M
T
CARBIDES
CARBONATES
FLUORIDES
SULFATES
Changes of state
Melting pt. M
Boiling pt. B
Transition pt. T

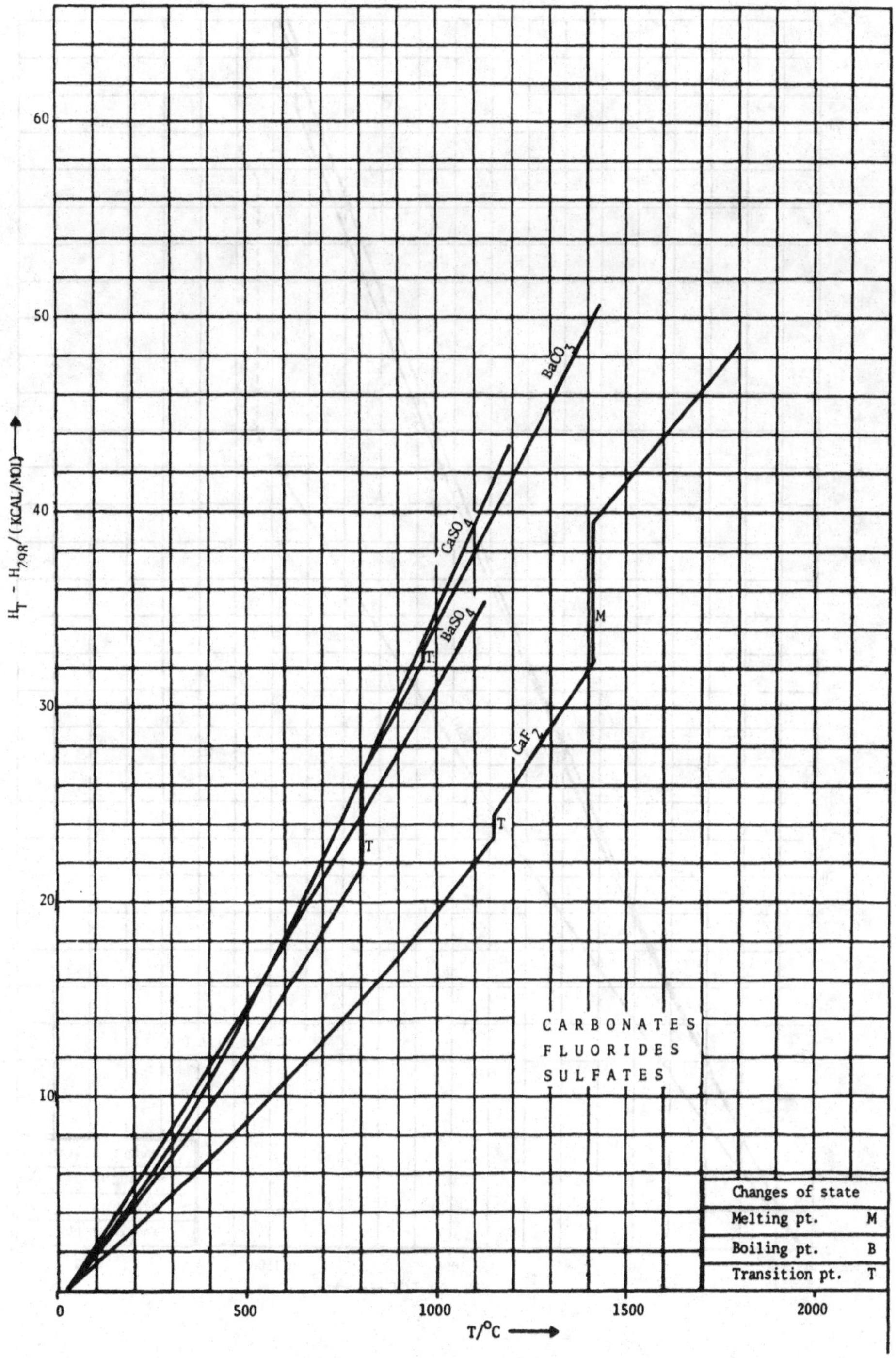

60
50
40
30
20
10
$H_T - H_{298}$/ (KCAL/MOL)
0
500
1000
1500
2000
T/°C
$BaCO_3$
$CaSO_4$
$BaSO_4$
CaF_2
T
T
T
M
CARBONATES
FLUORIDES
SULFATES
Changes of state
Melting pt. M
Boiling pt. B
Transition pt. T

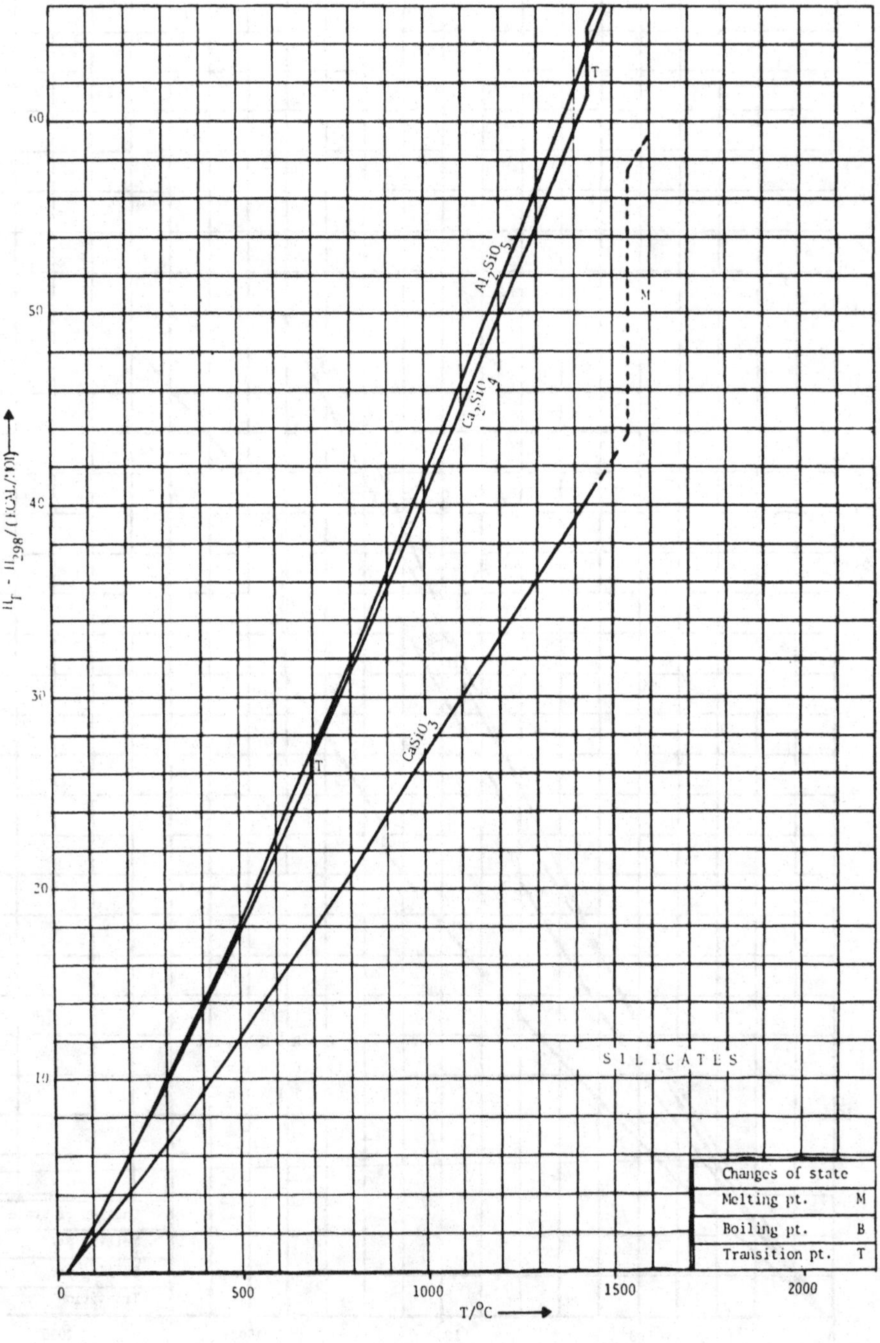

60
50
40
30
20
10
$H_T - H_{298}$/(KCAL/MOL)
0
500
1000
1500
2000
T/°C
T
M
Al_2SiO_5
Ca_2SiO_4
T
$CaSiO_3$
SILICATES
Changes of state
Melting pt. M
Boiling pt. B
Transition pt. T

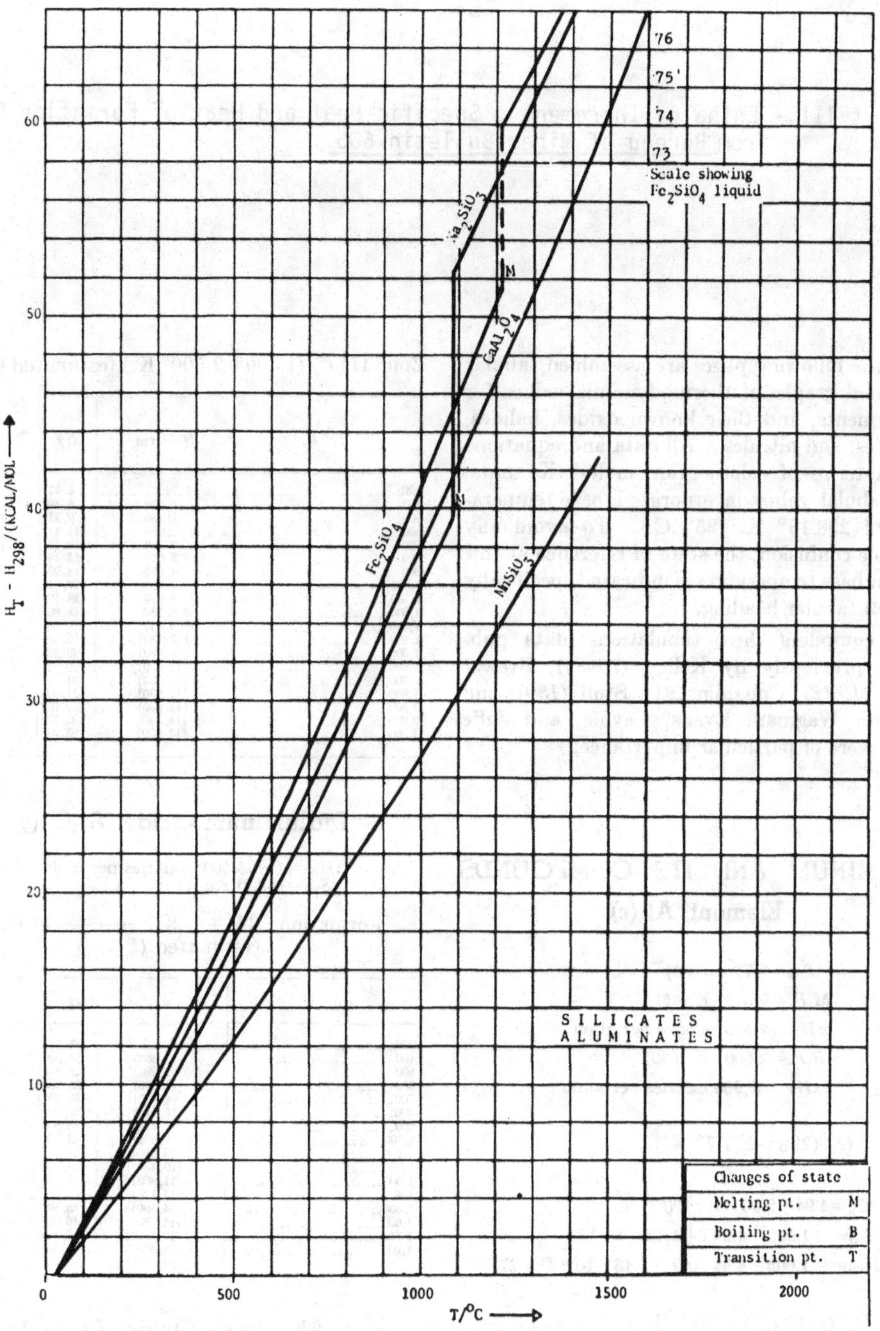
76
75
74
73
Scale showing
Fe_2SiO_4 liquid
Na_2SiO_3
M
$CaAl_2O_4$
M
Fe_2SiO_4
$MnSiO_3$
SILICATES
ALUMINATES
Changes of state
Melting pt. M
Boiling pt. B
Transition pt. T
$H_T - H_{298}$/(KCAL/MOL
T/°C
0
500
1000
1500
2000
10
20
30
40
50
60

Part III - Enthalpy Increment , Specific Heat and Heat of Formation Data from Bureau of Mines Bulletin 605.

In the following pages are assembled tabular data and graphs of thermodynamic values for 65 elements, and their known oxides, halides, carbides, and nitrides. All data and equations are in terms of calorie-gram mole-° K. units. The tabular values incorporate a base temperature of 298.15° K. (25° C.). To avoid any possible confusion, the state of reference at this chosen base temperature is indicated specifically in each tabular heading.

In compiling these tabulations, data published previously by Kelley (*79–84*), Brewer (*6–9, 11, 12*), Coughlin (*24*), Stull (*130*), and Rossini, Wagman, Evans, Levine, and Jaffe (*112*) were of particular importance.

ALUMINUM AND ITS COMPOUNDS

Element, Al (*c*)

$S_{298} = 6.77$ *e.u.*(*83*)

M.P. = 931.7° K.(*82*)

$\Delta H_M = 2{,}570$ calories per atom

B.P. = 2,600° K.(*130*)

$\Delta H_V = 67{,}950$ calories per atom

Zone I (*c*) (298°–931.7° K.)

$$C_p = 4.94 + 2.96\times10^{-3}T\ (82)$$

$$H_T - H_{298} = -1{,}605 + 4.94T + 1.48\times10^{-3}T^2$$

$$F_T - H_{298} = -1{,}605 - 4.94T\ln T - 1.48\times10^{-3}T^2 + 27.19T$$

Zone II (*l*) (931.7°–1,300° K.)

$$C_p = 7.00\ (82)$$

$$H_T - H_{298} = 330 + 7.00T$$

$$F_T - H_{298} = 330 - 7.00T\ln T + 37.83T$$

Zone III (*l*) (1,300°–2,500° K.) (estimated (*130*))

T, ° K.	$H_T - H_{298}$	S_T	$-\frac{(F_T - H_{298})}{T}$
298		6.77	6.77
400	600	8.49	6.99
500	1,230	9.91	7.45
600	1,890	11.11	7.96
700	2,580	12.17	8.48
800	3,310	13.15	9.01
900	4,060	14.03	9.53
1,000	7,330	17.53	10.20
1,100	8,030	18.19	10.89
1,200	8,730	18.80	11.52
1,300	9,430	19.36	(12.11)
1,400	(10,130)	(19.88)	(12.64)
1,500	(10,830)	(20.32)	(13.43)
1,600	(11,530)	(20.81)	(13.59)
1,700	(12,230)	(21.24)	(14.06)
1,800	(12,930)	(21.64)	(14.44)
1,900	(13,630)	(22.02)	(14.84)
2,000	(14,330)	(22.32)	(15.15)
2,500	(17,830)	(23.94)	(16.80)

Dialuminum Oxide, Al_2O (*g*)

$\Delta H^\circ_{298} = (-33{,}500)$ calories per mole (*8*)
$S_{298} = (68.2)$ *e.u.* (*24*)

Formation: $2Al + 1/2O_2 \longrightarrow Al_2O$
(estimated (*24*))

T, ° K.	$H_T - H_{298}$	ΔH°_f	ΔF°_f
298		−33,500	−42,500
400	(560)	(−34,500)	(−45,000)
500	(1,690)	(−35,000)	(−48,000)
600	(2,880)	(−35,500)	(−50,500)
700	(4,150)	(−36,000)	(−53,000)
800	(5,010)	(−37,000)	(−55,000)
900	(7,420)	(−36,500)	(−57,500)
1,000	(7,875)	(−43,000)	(−59,000)
1,100	(9,160)	(−43,500)	(−60,500)
1,200	(10,480	(−44,000)	(−62,000)
1,300	(11,800)	(−44,500)	(−63,500)
1,400	(13,120)	(−45,000)	(−65,000)
1,500	(14,500)	(−45,500)	(−66,500)
1,600	(15,780)	(−46,000)	(−68,000)
1,700	(17,130)	(−46,500)	(−69,500)
1,800	(18,470)	(−47,000)	(−70,500)

Aluminum Oxide, AlO (*g*)

$\Delta H^\circ_{298} = 10{,}000$ calories per mole (*8*)
$S_{298} = 51.40$ *e.u.* (*24*)

Formation: $Al + 1/2O_2 \longrightarrow AlO$
(estimated (*24*))

T, ° K.	H_T-H_{298}	ΔH_T°	ΔF_T°
298	----------	10,000	+4,000
400	(460)	(9,500)	(1,500)
500	(1,260)	(9,500)	(−500)
600	(2,000)	(9,000)	(−2,000)
700	(2,870)	(9,000)	(−4,000)
800	(3,700)	(8,500)	(−6,000)
900	(4,660)	(8,500)	(−8,000)
1,000	(5,550)	(5,500)	(−9,500)
1,100	(6,430)	(5,500)	(−11,000)
1,200	(7,250)	(5,000)	(−12,500)
1,300	(8,170)	(5,000)	(−14,000)
1,400	(9,000)	(4,500)	(−15,000)
1,500	(9,970)	(4,500)	(−16,500)
1,600	(10,750)	(4,000)	(−18,000)
1,700	(11,600)	(4,000)	(−19,500)
1,800	(12,500)	(3,500)	(−21,000)

Dialuminum Trioxide (Alumina), Al_2O_3 (*c*)

$\Delta H_{298}^\circ = -400{,}290$ calories per mole (*53*)
$S_{298} = 12.16$ *e.u.* (*87*)
M.P. = 2,313° K. (*112*)
$\Delta H_M = 26{,}000$ calories per mole

Zone I (*c*) (298°–1,800° K.)

$$C_p = 27.43 + 3.06\times10^{-3}T - 8.47\times10^5T^{-2} \ (82)$$
$$H_T - H_{298} = -11{,}155 + 27.43T + 1.53\times10^{-3}T^2 + 8.47\times10^5T^{-1}$$

Formation: $2Al + 3/2O_2 \longrightarrow Al_2O_3$

Zone I (298°–931.7° K.)

$$\Delta C_p = 6.81 - 4.3\times610^{-3}T - 7.87\times10^5T^{-2}$$
$$\Delta H_T = -405{,}200 + 6.81T - 2.18\times10^{-3}T^2 + 7.87\times10^5T^{-1}$$
$$\Delta F_T = -405{,}200 - 6.81T\ln T + 2.18\times10^{-3}T^2 + 3.93\times10^5T^{-1} + 123.58T$$

Zone II (931.7°–1,300° K.)

$$\Delta C_p = 2.69 + 1.56\times10^{-3}T - 7.87\times10^5T^{-2}$$
$$\Delta H_T = -408{,}660 + 2.69T + 0.78\times10^{-3}T^2 + 7.87\times10^5T^{-1}$$
$$\Delta F_T = -408{,}660 - 2.69T\ln T - 0.78\times10^{-3}T^2 + 3.93\times10^5T^{-1} + 102.38T$$

T, ° K.	H_T-H_{298}	S_T	ΔH_T°	ΔF_T°
298	----------	12.16	−400,300	−378,000
400	2,200	18.48	−400,400	−370,300
500	4,600	23.83	−400,300	−362,800
600	7,220	28.60	−400,200	−355,300
700	9,990	32.86	−399,900	−347,800
800	12,840	36.67	−399,700	−340,400
900	15,750	40.08	−399,500	−332,900
1,000	18,710	43.22	−404,400	−325,200
1,100	21,710	46.07	−404,000	−317,200
1,200	24,740	48.71	−403,600	−309,400
1,300	27,790	51.15	−403,200	−301,500
1,400	30,850	53.42	(−402,800)	(−293,800)
1,500	33,920	55.54	(−402,500)	(−286,100)
1,600	37,000	57.52	(−402,000)	(−278,100)
1,700	40,090	59.39	(−401,700)	(−270,600)
1,800	43,190	61.17	(−401,300)	(−263,100)

Aluminum Fluoride, AlF (*c*)

$\Delta H_{298}^\circ = (-84{,}000)$ calories per mole (*11*)
$S_{298} = (12)$ *e.u.* (*11*)
M.P. = (1,100°) K. (*6*)
$\Delta H_M = 5{,}000$ calories per mole
B.P. = 1,650° K. (*6*)
$\Delta H_V = 38{,}000$ calories per mole

Zone I (*g*) (298°–2,000° K.)

$$C_p = 8.9 - 1.45\times10^5T^{-2} \ (82)$$

Aluminum Trifluoride, AlF_3 (*c*)

$\Delta H_{298}^\circ = -323{,}000$ calories per mole (*11*)
$S_{298} = 23.8$ *e.u.* (*11*)
S.P. = 1,545° K. (*6*)
$\Delta H_{subl} = 77{,}000$ calories per mole

Zone I (*c*) (298°–1,100° K.)

$$C_p = 15.64 + 11.28\times10^{-3}T \ (82)$$
$$H_T - H_{298} = -5{,}164 + 15.64T + 5.64\times10^{-3}T^2$$

Formation: $Al + 3/2F_2 \longrightarrow AlF_3$

Zone I (298°–931.7° K.)

$$\Delta C_p = -1.73 + 7.66\times10^{-3}T + 1.20\times10^5T^{-2}$$
$$\Delta H_T = -322{,}400 - 1.73T + 3.83\times10^{-3}T^2 - 1.20\times10^5T^{-1}$$
$$\Delta F_T = -322{,}400 + 1.73T\ln T - 3.83\times10^{-3}T^2 - 0.60\times10^5T^{-1} + 45.83T$$

Zone II (931.7°–1,100° K.)

$$\Delta C_p = -3.79 + 10.62\times10^{-3}T + 1.20\times10^5T^{-2}$$
$$\Delta H_T = -323{,}500 - 3.79T + 5.31\times10^{-3}T^2 - 1.20\times10^5T^{-1}$$
$$\Delta F_T = -323{,}500 + 3.79T\ln T - 5.31\times10^{-3}T^2 - 0.60\times10^5T^{-1} + 34.32T$$

T, ° K.	H_T-H_{298}	S_T	ΔH_T°	ΔF_T°
298	----------	23.8	−323,000	−306,400
400	1,980	29.51	−322,800	−300,700
500	4,050	34.12	−322,550	−295,200
600	6,250	38.12	−322 400	−289,900
700	8,620	41.78	−321,850	−284,400
800	11,060	45.04	−321,450	−279,100
900	13,510	47.93	−321,000	−273,800
1,000	15,980	50.52	−322,100	−268,300
1,100	18,500	52.92	−322,600	−252,900

Aluminum Trichloride, $AlCl_3$ (*c*)

$\Delta H_{298} = -166{,}800$ calories per mole (*11*)
$S_{298} = 40.5$ *e.u.* (*11*)
M.P. = 465.6° K. (*6*)
$\Delta H_M = 8{,}500$ calories per mole
B.P. = 720° K. (*6*)
$\Delta H_V = 15{,}610$ calories per mole

Zone I (*c*) (298°–465.6° K.)

$$C_p = 13.25 + 28.00\times10^{-3}T \ (82)$$
$$H_T - H_{298} = -5{,}195 + 13.25T + 14.00\times10^{-3}T^2$$

Zone II (*l*) (465.6°–720° K.)

$$C_p = 31.2 \ (82)$$
$$H_T - H_{298} = -2{,}020 + 31.2T$$

Zone III (*g*) (720°–1,800° K.)

$$C_p = 19.8 - 2.69\times10^5T^{-2} \ (94)$$
$$H_T - H_{298} = 20{,}320 + 19.8T + 2.69\times10^5T^{-1}$$

Formation: $Al + 3/2Cl_2 \longrightarrow Al\ Cl_3$

Zone I (298°–466° K.)

$$\Delta C_p = -4.92 + 24.95\times10^{-3}T + 1.02\times10^5T^{-2}$$
$$\Delta H_T = -166{,}100 - 4.92T + 12.47\times10^{-3}T^2 - 1.02\times10^5T^{-1}$$
$$\Delta F_T = -166{,}100 + 4.92T\ln T - 12.47\times10^{-3}T^2 - 0.51\times10^5T^{-1} + 20.31T$$

Zone II (466°–720° K.)

$$\Delta C_p = 13.03 - 3.05\times10^{-3}T + 1.02\times10^5T^{-2}$$
$$\Delta H_T = -163{,}000 + 13.03T - 1.52\times10^{-3}T^2 - 1.02\times10^5T^{-1}$$

$$\Delta F_T = -163{,}000 - 13.03 T \ln T + 1.52\times10^{-3}T^2 - 0.51 \times10^5 T^{-1} + 118.0T$$

Zone III (720°–931.7° K.)

$$\Delta C_p = 1.63 - 3.05\times10^{-3}T - 1.67\times10^5 T^{-2}$$
$$\Delta H_T = -140{,}500 + 1.63T - 1.52\times10^{-3}T^2 + 1.65\times10^5 T^{-1}$$
$$\Delta F_T = -140{,}500 - 1.63 T \ln T + 1.52\times10^{-3}T^2 + 0.83 \times10^5 T^{-1} + 9.48T$$

T, ° K.	H_T-H_{298}	S_T	ΔH_T°	ΔF_T°
298	....	40.5	−166,800	−153,000
400	2,340	47.23	−166,300	−148,300
500	13,580	71.58	−157,000	−144,500
600	16,700	75.3	−155,800	−141,000
700	19,820	80.0	−154,700	−138,600
800	36,490	105.0	−140,000	−140,700
900	38,440	107.3	−140,200	−141,000
1,000	40,390	109.5	−142,400	−141,200
1,100	42,340	111.3	−142,700	−141,800
1,200	44,300	112.9	−143,000	−142,300
1,300	46,275	114.5	−142,900	−143,000
1,400	48,230	115.9	(−143,100)	(−143,800)
1,500	50,200	117.3	(−143,400)	(−144,600)
1,600	52,170	118.5	(−142,700)	(−145,300)
1,700	54,140	119.7	(−143,900)	(−146,000)
1,800	56,110	120.9	(−144,100)	(−146,500)

Aluminum Tribromide, $AlBr_3$ (*c*)

$\Delta H_{298}^\circ = -127{,}000$ calories per mole (*11*)
$S_{298} = (49)$ *e.u.* (*11*)
M.P. = 370.6° K. (*82*)
$\Delta H_M = 2{,}710$ calories per mole
B.P. = 739° K. (*6*)
$\Delta H_V = 16{,}080$ calories per mole

Zone I (*c*) (298°–370.6° K.)

$$C_p = 18.74 + 18.66\times10^{-3}T \quad (82)$$
$$H_T - H_{298} = -6{,}420 + 18.74T + 9.33\times10^{-3}T^2$$

Zone II (*l*) (370.6°–500° K.)

$$C_p = 29.5 \quad (82)$$
$$H_T - H_{298} = -6{,}410 + 29.5T$$

Formation: $Al + 3/2Br_2 \longrightarrow AlBr_3$

Zone I (298°–331° K.)

$$\Delta C_p = -11.85 + 15.7\times10^{-3}T$$
$$\Delta H_T = -124{,}200 - 11.85T + 7.85\times10^{-3}T^2$$
$$\Delta F_T = -124{,}200 + 11.85 T \ln T - 7.85\times10^{-3}T^2 - 62.35T$$

Zone II (331°–500° K.)

$$\Delta C_p = 11.0 - 2.96\times10^{-3}T + 0.55\times10^5 T^{-2}$$
$$\Delta H_T = -139{,}200 + 11.0T - 1.48\times10^{-3}T^2 - 0.55\times10^5 T^{-1}$$
$$\Delta F_T = -139{,}200 - 11.0 T \ln T + 1.48\times10^{-3}T^2 - 0.27 \times10^5 T^{-1} + 114.2T$$

T, ° K.	H_T-H_{298}	S_T	ΔH_T	ΔF_T
298	....	(49.0)	−127,000	(−123,300)
400	5,390	(64.0)	−135,200	(−119,800)
500	8,340	(70.58)	−134,200	(−116,000)

Aluminum Triiodide, AlI_3 (*c*)

$\Delta H_{298}^\circ = -74{,}400$ calories per mole (*21*)
$S_{298} = (46)$ *e.u.* (*21*)
M.P. = 464° K. (*82*)
$\Delta H_M = 3{,}980$ calories per mole
B.P. = 695° K. (*6*)
$\Delta H_V = 18{,}500$ calories per mole

Zone I (*c*) (298°–464° K.)

$$C_p = 16.88 + 22.66\times10^{-3}T \quad (82)$$
$$H_T - H_{298} = -6{,}040 + 16.88T + 11.33\times10^{-3}T^2$$

Zone II (*l*) (464°–500° K.)

$$C_p = 29.0 \quad (82)$$
$$H_T - H_{298} = -5{,}250 + 29.0T$$

Formation: $Al + 3/2I_2 \longrightarrow AlI_3$

Zone I (298°–386.8° K.)

$$\Delta C_p = -2.44 + 1.85\times10^{-3}T$$
$$\Delta H_T = -73{,}800 - 2.44T + 0.92\times10^{-3}T^2$$
$$\Delta F_T = -73{,}800 + 2.44 T \ln T - 0.92\times10^{-3}T^2 - 13.24T$$

Zone II (386.8°–464° K.)

$$\Delta C_p = -16.86 + 19.7\times10^{-3}T$$
$$\Delta H_T = -68{,}600 - 16.86T + 9.85\times10^{-3}T^2$$
$$\Delta F_T = -68{,}600 + 16.86 T \ln T - 9.85\times10^{-3}T^2 - 19.76T$$

Zone III (464°–500° K.)

$$\Delta C_p = 10.73 - 2.96\times10^{-3}T$$
$$\Delta H_T = -73{,}000 + 10.73T - 1.48\times10^{-3}T^2$$
$$\Delta F_T = -73{,}000 - 10.73 T \ln T + 1.48\times10^{-3}T^2 + 65.13T$$

T, ° K.	H_T-H_{298}	S_T	ΔH_T	ΔF_T
298	....	(46.0)	−74,400	(−73,600)
400	2,525	(53.26)	−73,500	(−73,400)
500	9,250	(67.97)	−68,200	(−73,400)

Tetraaluminum Tricarbide, Al_4C_3 (*c*)

$\Delta H^\circ_{298} = -39{,}900$ calories per mole (*9*)
$S_{298} = 25.2$ *e.u.* (*9*)

Zone I (*c*) (298°–600° K.)

$$C_p = 24.08 + 31.6\times10^{-3}T \quad (82)$$
$$H_T - H_{298} = -8{,}585 + 24.08T + 15.8\times10^{-3}T^2$$

Formation: $4Al + 3C \longrightarrow Al_4C_3$

Zone I (298°–600° K.)

$$\Delta C_p = -8.02 + 16.7\times10^{-3}T + 6.30\times10^5 T^{-2}$$
$$\Delta H_T = -36{,}150 - 8.02T + 8.35\times10^{-3}T^2 - 6.30\times10^5 T^{-1}$$
$$\Delta F_T = -36{,}150 + 8.02 T \ln T - 8.35\times10^{-3}T^2 - 3.15 \times10^5 T^{-1} - 46.2T$$

T, ° K.	H_T-H_{298}	S_T	ΔH_T°	ΔF_T°
298	....	25.2	−39,900	−38,100
400	3,580	35.5	−39,500	−37,600
500	7,430	44.1	39,100	37,150
600	11,530	51.55	38,800	36,800

Aluminum Nitride, AlN (*c*)

$\Delta H^{\circ}_{298} = -64{,}000$ calories per mole (*9*)
$S_{298} = 3.8$ *e.u.* (*9*)
M.P. = 2,500° K. (*9*)

Zone I (*c*) (298°–900° K.)

$C_p = 5.47 + 7.80 \times 10^{-3}T$ (*82*)
$H_T - H_{298} = -1{,}980 + 5.47T + 3.90 \times 10^{-3}T^2$

Formation: $Al + 1/2\ N_2 \longrightarrow AlN$

Zone I (298°–900° K.)

$\Delta C_p = -2.8 + 4.33 \times 10^{-3}T$
$\Delta H_T = -63{,}400 - 2.8T + 2.16 \times 10^{-3}T^2$
$\Delta F_T = -63{,}400 + 2.8T\ln T - 2.16 \times 10^{-3}T^2 + 8.4T$

T, ° K.	$H_T - H_{298}$	S_T	ΔH°_T	ΔF°_T
298	-------	3.8	−64, 000	−56, 300
400	600	6.19	−64, 100	−53, 600
500	1, 230	8.24	−64, 200	−51, 000
600	1, 890	10.04	−64, 200	−48, 300
700	2, 580	11.64	−64, 200	−45, 700
800	3, 310	13.1	−64, 200	−43, 200
900	4, 060	14.46	−64, 200	−40, 400

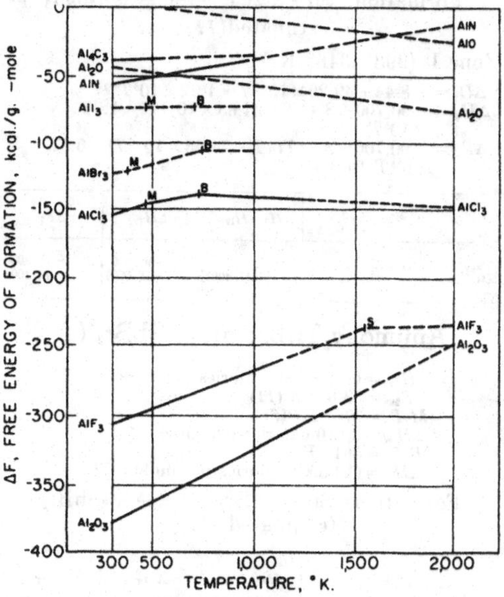

FIGURE 1.—Aluminum.

ANTIMONY AND ITS COMPOUNDS

Element, Sb (*c*)

$S_{298} = 10.5$ *e.u.* (*83*)
M.P. = 903° K. (*82*)
$\Delta H_M = 4{,}740$ calories per atom
B.P. = 1,713° K. (*112*)
$\Delta H_V = (46{,}700)$ calories per atom (*94*)

647940 O—63——2

Zone I (*c*) (298°–903° K.)

$C_p = 5.51 + 1.74 \times 10^{-3}T$ (*82*)
$H_T - H_{298} = -1{,}720 + 5.51T + 0.87 \times 10^{-3}T^2$
$F_T - H_{298} = -1{,}720 - 5.51T\ln T - 0.87 \times 10^{-3}T^2 + 26.93T$

Zone II (*l*) (903°–1,300° K.)

$C_p = 7.50$ (*82*)
$H_T - H_{298} = +1{,}940 + 7.50T$
$F_T - H_{298} = +1{,}940 - 7.50T\ln T + 35.62T$

T, ° K.	$H_T - H_{298}$	S_T	$-\frac{(F_T - H_{298})}{T}$
298	-------	10.5	10.5
400	625	12.3	10.75
500	1, 250	13.69	11.20
600	1, 890	14.86	11.71
700	2, 550	15.88	12.24
800	3, 240	16.80	12.75
900	3, 950	17.64	13.26
1,000	9, 440	23.68	14.24
1,100	10, 190	24.39	15.12
1,200	10, 940	25.04	15.92
1,300	11, 690	25.64	16.65
1,400	(12, 440)	(26.2)	(17.26)
1,500	(13, 190)	(26.8)	(17.91)
1,600	(13, 940)	(27.3)	(18.47)
1,700	(14, 690)	(27.8)	(19.01)
1,800	-------	(28.9)	(20.35)
1,900	-------	(29.3)	(20.80)
2,000	-------	(39.3)	(21.43)

Diantimony Trioxide (Orthorhombic), Sb_2O_3 (*c*)

$\Delta H^{\circ}_{298} = -168{,}500$ calories per mole (*113*)
$S_{298} = 29.4$ *e.u.* (*83*)
M.P. = 928° K. (*24*)
$\Delta H_M = 13{,}500$ calories per mole
B.P. = 1,698° K. (*24*)
$\Delta H_V = 8{,}910$ calories per mole

Zone I (*c*) (298°–929° K.)

Estimated equation:
$C_p = 19.1 + 17.1 \times 10^{-3}T$ (*82*)
$H_T - H_{298} = -6{,}450 + 19.1T + 8.55 \times 10^{-3}T^2$

Formation: $2Sb + 3/2O_2 \longrightarrow Sb_2O_3$

(estimated (*24*))

T, ° K.	$H_T - H_{298}$	ΔH°_T	ΔF°_T
298	-------	−168, 500	−149, 100
400	(2, 220)	(−168, 300)	(−142, 500)
500	(5, 180)	(−168, 000)	(−136, 100)
600	(7, 800)	(−167, 600)	(−129, 700)
700	(10, 980)	(−167, 100)	(−123, 500)
800	(14, 260)	(−166, 400)	(−117, 300)
900	(17, 700)	(−165, 600)	(−111, 200)
1,000	(35, 020)	(−160, 500)	(−105, 200)
1,100	(38, 990)	(−159, 200)	(−99, 800)
1,200	(42, 940)	(−158, 000)	(−94, 400)
1,300	(46, 990)	(−156, 700)	(−89, 200)
1,400	(50, 950)	(−155, 500)	(−84, 000)
1,500	(55, 100)	(−154, 300)	(−79, 000)
1,600	(58, 940)	(−153, 100)	(−74, 000)
1,700	(61, 880)	(−153, 000)	(−69, 100)
1,800	-------	(−236, 600)	(−60, 000)
1,900	-------	(−236, 900)	(−50, 200)
2,000	-------	(−237, 200)	(−40, 300)

Diantimony Trioxide (Cubic, Orthorhombic), Sb_2O_3 (c)

$\Delta H^\circ_{298} = -169{,}900$ calories per mole (111)
$S_{298} = 27.7$ e.u. (24)
$T.P. = 842°$ K. (24)
$\Delta H_T = 1{,}390$ calories per mole
$M.P. = 928°$ K. (24)
$\Delta H_M = 13{,}500$ calories per mole
$B.P. = 1{,}698°$ K. (24)
$\Delta H_V = 8{,}910$ calories per mole

Zone I (298°–842° K.)

Estimated equation:

$$C_p = 19.1 + 17.1 \times 10^{-3} T \quad (82)$$
$$H_T - H_{298} = -6{,}450 + 19.1T + 8.55 \times 10^{-3} T^2 \quad (82)$$

Formation: $2Sb + 3/2O_2 \longrightarrow Sb_2O_3$
(estimated (24))

T, ° K.	$H_T - H_{298}$	ΔH°_T	ΔF°_T
298	....	−169,900	−150,000
400	(2,220)	(−169,700)	(−143,200)
500	(5,180)	(−169,400)	(−136,600)
600	(7,800)	(−169,000)	(−130,100)
700	(11,080)	(−168,400)	(−123,700)
800	(14,260)	(−167,800)	(−117,400)
900	(19,100)	(−165,600)	(−111,200)
1,000	(36,420)	(−160,500)	(−105,200)
1,100	(40,390)	(−159,200)	(−99,800)
1,200	(44,340)	(−158,000)	(−94,400)
1,300	(48,390)	(−156,700)	(−89,200)
1,400	(52,350)	(−155,500)	(−84,000)
1,500	(56,500)	(−154,300)	(−79,000)
1,600	(60,340)	(−153,100)	(−74,000)
1,700	(63,280)	(−143,000)	(−69,100)
1,800	....	(−236,600)	(−60,000)
1,900	....	(−236,900)	(−50,200)
2,000	....	(−237,200)	(−40,300)

Diantimony Tetraoxide, Sb_2O_4 (c)

$\Delta H^\circ_{298} = (-209{,}000)$ calories per mole (24)
$S_{298} = 30.4$ e.u. (24)
Decomposes $= 1{,}203°$ K. (24)

Formation: $2Sb + 2O_2 \longrightarrow Sb_2O_4$
(estimated (24))

T, ° K.	$H_T - H_{298}$	ΔH°_T	ΔF°_T
298	....	(−209,000)	(−182,500)
400	(2,700)	(−209,000)	(−173,500)
500	(5,900)	(−208,500)	(−164,500)
600	(9,200)	(−208,000)	(−156,000)
700	(12,600)	(−207,500)	(−147,500)
800	(16,000)	(−207,000)	(−138,500)
900	(19,600)	(−206,500)	(−130,000)
1,000	(23,200)	(−215,500)	(−120,500)
1,100	(27,400)	(−214,500)	(−111,500)
1,200	(31,400)	(−213,500)	(−102,000)

Diantimony Pentaoxide, Sb_2O_5 (c)

$\Delta H_{298} = (-229{,}000)$ calories per mole (24)
$S_{298} = 31.3$ e.u. (24)
Decomposes $= 673°$ K. (24)

Formation: $2Sb + 5/2O_2 \longrightarrow Sb_2O_5$
(estimated (24))

T, ° K	$H_T - H_{298}$	ΔH°_T	ΔF°_T
298	....	(−229,000)	(−195,500)
400	(3,050)	(−229,000)	(−183,500)
500	(6,650)	(−228,500)	(−172,500)
600	(10,800)	(−227,500)	(−161,500)

Antimony Trifluoride, SbF_3 (c)

$\Delta H^\circ_{298} = -216{,}600$ calories per mole (11)
$S_{298} = (38)$ e.u. (11)
$M.P. = 565°$ K. (6)
$B.P. = 649°$ K. (6)
$\Delta H_V = (14{,}000)$ calories per mole

Formation: $Sb + 3/2F_2 \longrightarrow SbF_3$
(estimated (11))

T, ° K.	$H_T - H_{298}$	ΔH°_T	ΔF°_T
298	....	−216,600	(−203,200)
500	(6,000)	(−213,600)	(−193,100)

Antimony Trichloride, $SbCl_3$ (c)

$\Delta H^\circ_{298} = -91{,}400$ calories per mole (11)
$S_{298} = 44.7$ e.u. (11)
$M.P. = 346.4°$ K. (6)
$\Delta H_M = 3{,}030$ calories per mole
$B.P. = 494°$ K. (6)
$\Delta H_V = 10{,}362$ calories per mole

Zone I (c) (298°–346° K.)

$$C_p = 10.3 + 51.1 \times 10^{-3} T \quad (82)$$
$$H_T - H_{298} = -5{,}300 + 10.3T + 25.55 \times 10^{-3} T^2$$

Formation: $Sb + 3/2Cl_2 \longrightarrow SbCl_3$
(estimated (11))

Zone I (298°–346° K.)

$$\Delta C_p = -8.44 + 49.26 \times 10^{-3} T + 1.02 \times 10^5 T^{-2}$$
$$\Delta H_T = -90{,}700 - 8.44T + 24.63 \times 10^{-3} T^2 - 1.02 \times 10^5 T^{-1}$$
$$\Delta F_T = -90{,}700 + 8.44 T \ln T - 24.63 \times 10^{-3} T^2 - 0.51 \times 10^5 T^{-1} + 3.4T$$

T, ° K.	$H_T - H_{298}$	ΔH_T	ΔF_T
298	....	−91,400	−77,800
500	(17,400)	(−78,000)	(−71,100)

Antimony Tribromide, $SbBr_3$ (c)

$\Delta H^\circ_{298} = (-59{,}900)$ calories per mole (11)
$S_{298} = 53.5$ e.u. (11)
$M.P. = 370°$ K. (6)
$\Delta H_M = 3{,}510$ calories per mole
$B.P. = 561°$ K. (6)
$\Delta H_V = (12{,}000)$ calories per mole

Formation: $Sb + 3/2Br_2 \longrightarrow SbBr_3$
(estimated (11))

T, ° K.	$H_T - H_{298}$	ΔH°_T	ΔF°_T
298	....	−59,900	(−56,000)
500	(7,000)	(−68,400)	(−51,300)

Antimony Triiodide, SbI_3 (c)

$\Delta H^\circ_{298} = (-22{,}800)$ calories per mole (11)
$S_{298} = (57)$ e.u. (11)
$M.P. = 444°$ K. (6)
$B.P. = 700°$ K. (6)
$\Delta H_V = (15{,}000)$ calories per mole

Formation: $Sb+3/2I_2 \longrightarrow SbI_3$
(estimated (*11*))

T, ° K.	H_T-H_{298}	ΔH_T°	ΔF_T°
298		(−22, 800)	(−24, 100)
500	(7, 000)	(−42, 000)	(−25, 500)

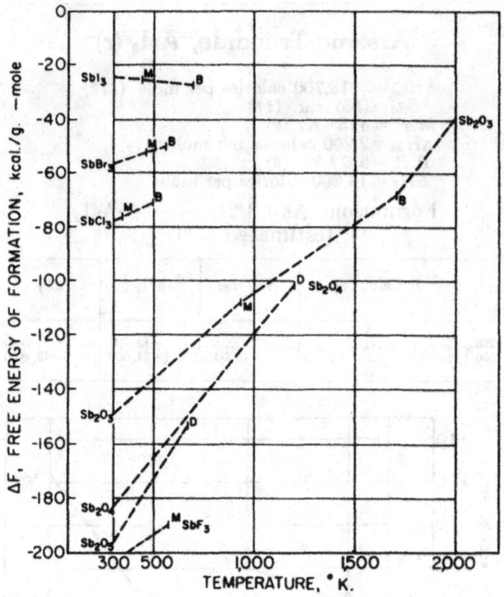

FIGURE 2.—Antimony.

ARSENIC AND ITS COMPOUNDS

Element, As (*c*)

$S_{298}=8.40$ *e.u.* (*83*)
S.P.$=886°$ K. (*130*)
$\Delta H_{subl}=7{,}630$ calories per atom

Zone I (*c*) (298°–883° K.)

$C_p=5.23+2.22\times10^{-3}T$ (*82*)
$H_T-H_{298}=-1{,}658+5.23T+1.11\times10^{-3}T^2$
$F_T-H_{298}=-1{,}658-5.23T\ln T-1.11\times10^{-3}T^2+27.28T$
Above 883° K., diatomic gas (estimated (*124*))

T, ° K.	H_T-H_{298}	S_T	$-\frac{(F_T-H_{298})}{T}$
298		8.40	8. 4
400	610	10. 16	8. 64
500	1, 240	11. 54	9. 06
600	1, 880	12. 73	9. 60
700	2, 540	13. 75	10. 12
800	3, 230	14. 68	10. 64
900	(26, 630)	(33. 39)	(3. 81)
1,000	(27, 070)	(33. 86)	(6. 79)
1,100	(27, 520)	(34. 28)	(9. 27)
1,200	(27, 960)	(34. 67)	(11. 37)
1,300	(28, 410)	(35. 03)	(13. 18)
1,400	(28, 850)	(35. 36)	(14. 76)
1,500	(29, 300)	(35. 67)	(16. 14)
1,600	(29, 740)	(35. 95)	(17. 37)
1,700	(30, 190)	(36. 22)	(18. 47)
1,800	(30, 640)	(36. 48)	(19. 46)
1,900	(31, 080)	(36. 72)	(20. 37)
2,000	(31, 530)	(36. 95)	(21. 19)

Diarsenic Trioxide (Orthorhombic), As_2O_3 (*c*)

$\Delta H_{f298}^\circ=157{,}000$ calories per mole (*114*)
$S_{298}=25.6$ *e.u.* (*112*)
T.P.$=506°$ K. (*24*)
$\Delta H_T=4{,}110$ calories per mole
M.P.$=542°$ K. (*24*)
$\Delta H_M=7{,}930$ calories per mole
B.P.$=730.3°$ K. (*24*)
$\Delta H_V=14{,}300$ calories per mole of As_4O_6

Formation: $2As+3/2O_2 \longrightarrow As_2O_3$
(estimated (*24*))

T, ° K.	H_T-H_{298}	ΔH_T°	ΔF_T°
298		−157, 000	−137, 700
400	(2, 550)	(−156, 750)	(−131, 150)
500	(5, 610)	(−156, 050)	(−124, 800)
600	(17, 225)	(−146, 850)	(−119, 500)
700	(21, 200)	(−145, 350)	(−115, 100)
800	(31, 100)	(−138, 050)	(−111, 500)
900	(36, 500)	(−154, 000)	(−107, 850)
1,000	(38, 200)	(−154, 000)	(−102, 700)
1,100	(39, 800)	(−154, 050)	(−97, 600)
1,200	(41, 500)	(−154, 050)	(−92, 450)
1,300	(43, 000)	(−154, 150)	(−87, 300)
1,400	(44, 700)	(−154, 200)	(−82, 200)
1,500	(46, 500)	(−154, 300)	(−77, 050)
1,600	(48, 000)	(−154, 400)	(−71, 900)
1,700	(49, 700)	(−154, 500)	(−66, 700)
1,800	(51, 400)	(−154, 650)	(−61, 550)
1,900	(52, 900)	(−154, 800)	(−56, 400)
2,000	(54, 500)	(−154, 950)	(−51, 200)

Diarsenic Trioxide (Monoclinic), As_2O_3 (*c*)

$\Delta H_{f298}^\circ=-152{,}900$ calories per mole (*114*)
$S_{298}=33.6$ *e.u.* (*24*)
M.P.$=586°$ K. (*24*)
$\Delta H_M=4{,}000$ calories per mole
B.P.$=730.3°$ K. (*24*)
$\Delta H_V=14{,}300$ calories per mole of As_4O_6

Formation: $2As+3/2O_2 \longrightarrow As_2O_3$
(estimated (*24*))

T, ° K.	H_T-H_{298}	ΔH_T°	ΔF_T°
298		−152, 900	−136, 000
400	(2, 600)	(−152, 600)	(−130, 300)
500	(5, 550)	(−152, 000)	(−124, 800)
600	(13, 200)	(−146, 800)	(−119, 500)
700	(17, 100)	(−145, 300)	(−115, 100)
800	(26, 900)	(−138, 100)	(−111, 500)
900	(32, 400)	(−154, 000)	(−107, 800)
1,000	(34, 100)	(−154, 000)	(−102, 700)
1,100	(35, 700)	(−154, 000)	(−97, 600)
1,200	(37, 300)	(−154, 100)	(−92, 500)
1,300	(39, 000)	(−154, 100)	(−87, 300)
1,400	(40, 600)	(−154, 200)	(−82, 200)
1,500	(42, 400)	(−154, 300)	(−77, 000)
1,600	(43, 900)	(−154, 400)	(−71, 900)
1,700	(45, 600)	(−154, 500)	(−66, 700)
1,800	(47, 200)	(−154, 600)	(−61, 600)
1,900	(48, 800)	(−154, 800)	(−56, 400)
2,000	(50, 600)	(−155, 000)	(−51, 200)

Diarsenic Tetraoxide, As_2O_4 (*c*)

$\Delta H_{f298}^\circ=-175{,}500$ calories per mole (*14*)
$S_{298}=(36)$ *e.u.* (*24*)

Formation: $2As+2O_2 \longrightarrow As_2O_4$
(estimated (*24*))

T, ° K.	H_T-H_{298}	ΔH_T°	ΔF_T°
298	....	−175,500	−149,000
400	(2,700)	(−175,500)	(−140,000)
500	(5,900)	(−175,000)	(−131,000)
600	(9,200)	(−174,500)	(−122,000)
700	(12,500)	(−174,000)	(−113,500)
800	(17,000)	(−172,500)	(−105,000)
900	(22,500)	(−186,500)	(−96,000)
1,000	(26,200)	(−184,500)	(−86,000)
1,100	(30,300)	(−182,000)	(−76,500)
1,200	(35,000)	(−179,000)	(−67,000)
1,300	(39,700)	(−176,000)	(−58,000)
1,400	(44,900)	(−172,500)	(−49,000)
1,500	(50,300)	(−169,000)	(−40,000)

Diarsenic Pentaoxide, As_2O_5 (c)

$\Delta H_{298}^\circ = -218,500$ calories per mole (*112*)
$S_{298} = 25.2$ *e.u.* (*112*)

Formation: $2As + 5/2O_2 \longrightarrow As_2O_5$
(estimated (*24*))

T, ° K.	$H-H_{298}$	ΔH_T°	ΔF_T°
298	....	−218,500	−184,500
400	(3,000)	(−218,500)	(−173,000)
500	(6,600)	(−219,000)	(−161,500)
600	(9,800)	(−219,000)	(−150,000)
700	(12,500)	(−218,500)	(−138,500)
800	(15,900)	(−218,500)	(−127,000)
900	(23,100)	(−233,500)	(−115,500)
1,000	(26,600)	(−232,500)	(−102,500)
1,100	(30,000)	(−231,500)	(−89,500)
1,200	(33,500)	(−230,500)	(−76,500)
1,300	(37,600)	(−229,000)	(−63,500)
1,400	(41,700)	(−227,500)	(−51,000)
1,500	(46,000)	(−226,000)	(−38,500)
1,600	(49,800)	(−224,500)	(−26,000)
1,700	(54,500)	(−222,500)	(−13,500)
1,800	(59,200)	(−220,500)	(−1,500)
1,900	(64,400)	(−218,000)	(+10,500)
2,000	(69,400)	(−216,000)	(+22,500)

Arsenic Trifluoride, AsF_3 (l)

$\Delta H_{298}^\circ = -218,300$ calories per mole (*112*)
$S_{298} = 69.08$ *e.u.* (*112*)
$\Delta F_{298}^\circ = -214,700$ calories per mole
M.P. = 267.2° K. (*112*)
$\Delta H_M = 2,486$ calories per mole
B.P. = 333° K. (*94*)
$\Delta H_V = 7,100$ calories per mole

Arsenic Trichloride, $AsCl_3$ (l)

$\Delta H_{298}^\circ = -80,200$ calories per mole (*112*)
$S_{298} = 55.8$ *e.u.* (*112*)
$\Delta F_{298}^\circ = -70,400$ calories per mole
M.P. = 257° K. (*112*)
$\Delta H_M = 2,420$ calories per mole
B.P. = 403° K. (*112*)
$\Delta H_V = 7,500$ calories per mole

Arsenic Tribromide, $AsBr_3$ (c)

$\Delta H_{298}^\circ = -46,610$ calories per mole (*112*)
$S_{298} = (53)$ *e.u.* (*112*)
M.P. = 304° K. (*112*)
$\Delta H_M = 2,810$ calories per mole
B.P. = 494° K. (*112*)
$\Delta H_V = 10,000$ calories per mole

Formation: $As + 3/2Br_2 \longrightarrow AsBr_3$
(estimated (*11*))

T, ° K.	H_T-H_{298}	ΔH_T°	ΔF_T°
298	....	−46,600	(−43,600)
500	(10,000)	(−43,000)	(−36,800)

Arsenic Triiodide, AsI_3 (c)

$\Delta H_{298}^\circ = -13,700$ calories per mole (*112*)
$S_{298} = (55)$ *e.u.* (*11*)
M.P. = 415° K. (*6*)
$\Delta H_M = 2,200$ calories per mole
B.P. = 687° K. (*6*)
$\Delta H_V = 14,200$ calories per mole

Formation: $As + 3/2I_2 \longrightarrow AsI_3$
(estimated (*11*))

T, ° K.	H_T-H_{298}	ΔH_T°	ΔF_T°
298	....	−13,700	(−15,000)
500	(9,000)	(−31,000)	(−13,900)

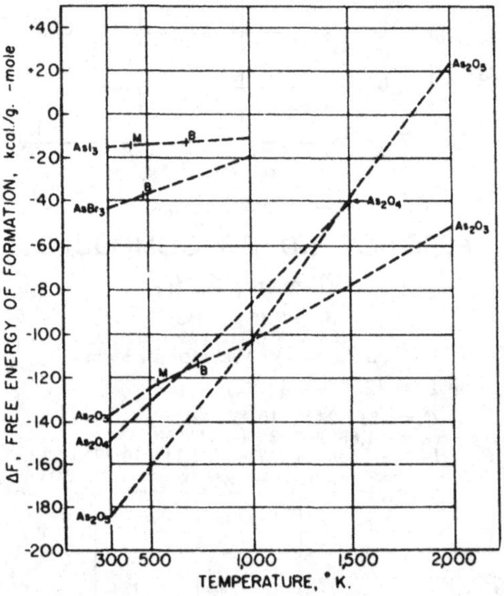

FIGURE 3.– Arsenic.

BARIUM AND ITS COMPOUNDS

Element, Ba (c)

$S_{298} = 16.0$ *e.u.* (*83*)
M.P. = 983° K. (*93*)
$\Delta H_M = 1,830$ calories per atom
B.P. = 1,911° K. (*130*)
$\Delta H_V = 35,700$ calories per atom

Zone I (c) (298°–983° K.)

$$C_p = 5.55 + 1.50 \times 10^{-3} T \ (82)$$
$$H_T - H_{298} = -1{,}720 + 5.55T + 0.75 \times 10^{-3} T^2$$
$$F_T - H_{298} = 1{,}720 - 5.55 T \ln T - 0.75 \times 10^{-3} T^2 + 21.55 T$$

Zone II (l) (983°–1,125° K.)

$$C_p = 11.5 \ (82)$$
$$H_T - H_{298} = -4{,}250 + 11.5T$$
$$F_T - H_{298} = -4{,}250 - 11.5 T \ln T + 64.6T$$

Above 1,125° K. (estimated (*130*))

T, ° K.	$H_T - H_{298}$	S_T	$-\frac{(F_T - H_{298})}{T}$
298	....	16.0	16.0
400	620	17.8	16.25
500	1,240	19.25	16.77
600	1,880	20.4	17.26
700	2,430	21.4	17.93
800	3,200	22.3	18.30
900	3,880	23.1	18.80
1,000	7,247	26.38	19.15
1,100	7,990	27.10	19.83
1,200	(8,750)	(27.75)	(20.47)
1,300	(9,500)	(28.35)	(21.05)
1,400	(10,250)	(28.91)	(21.60)
1,500	(11,000)	(20.42)	(22.09)
1,600	(11,750)	(29.91)	(22.57)
1,700	(13,000)	(30.36)	(23.01)
1,800	(13,250)	(30.79)	(23.44)
1,900	(14,000)	(31.20)	(23.84)

Barium Oxide, BaO (c)

$\Delta H^\circ_{298} = -133{,}400$ calories per mole (*50*)
$S_{298} = 16.8$ *e.u.* (*83*)
$M.P. = 2{,}196°$ K. (*8*)
$\Delta H_M = 13{,}800$ calories per mole
$B.P. = 3{,}000°$ K. (*8*)

Zone I (c) (298°–1,300° K.)

$$C_p = 12.74 + 1.040 \times 10^{-3} T - 1.984 \times 10^5 T^{-2} \ (82)$$
$$H_T - H_{298} = -4{,}500 + 12.74T + 0.52 \times 10^{-3} T^2 + 1.984 \times 10^5 T^{-1}$$

Formation: $Ba + 1/2 O_2 \longrightarrow BaO$

Zone I (298°–983° K.)

$$\Delta C_p = 3.63 - 0.96 \times 10^{-3} T - 1.78 \times 10^5 T^{-2}$$
$$\Delta H_T = -135{,}000 + 3.63T - 0.48 \times 10^{-3} T^2 + 1.78 \times 10^5 T^{-1}$$
$$\Delta F_T = -135{,}000 - 3.63 T \ln T + 0.48 \times 10^{-3} T^2 + 0.89 \times 10^5 T^{-1} + 48.77T$$

Zone II (983°–1,125° K.)

$$\Delta C_p = -2.32 + 0.54 \times 10^{-3} T - 1.78 \times 10^5 T^{-2}$$
$$\Delta H_T = -132{,}700 - 2.32T + 0.27 \times 10^{-3} T^2 + 1.78 \times 10^5 T^{-1}$$
$$\Delta F_T = -132{,}700 + 2.32 T \ln T - 0.27 \times 10^{-3} T^2 + 0.89 \times 10^5 T^{-1} + 4.8T$$

T, ° K.	$H_T - H_{298}$	S_T	ΔH°_T	ΔF°_T
298	....	16.8	−133,400	−126,300
400	1,200	20.3	−133,200	−124,000
500	2,400	22.9	−133,000	−121,650
600	3,700	25.3	−132,700	−119,400
700	4,900	27.2	−132,400	−117,100
800	6,300	29.0	−132,200	−115,000
900	7,600	30.7	−132,000	−113,100
1,000	8,950	32.0	−134,600	−111,100
1,100	10,300	33.2	−134,200	−108,500
1,200	....	....	(−133,500)	(−106,000)
1,300	....	....	(−133,500)	(−103,500)
1,400	....	....	(−133,500)	(−101,500)
1,500	(15,900)	....	(−133,500)	(−99,000)
1,600	....	....	(−133,000)	(−96,500)
1,700	....	....	(−132,500)	(−94,500)
1,800	....	....	(−132,500)	(−92,000)
1,900	....	....	(−132,000)	(−90,000)
2,000	(23,150)	....	(−167,000)	(−86,000)

Barium Dioxide, BaO_2 (c)

$\Delta H^\circ_{298} = -151{,}890 \pm 250$ calories per mole (*139*)
$S_{298} = 22.62$ *e.u.* (*24*)

Formation: $Ba + O_2 \longrightarrow BaO_2$ (estimated (*24*))

T, ° K.	$H_T - H_{298}$	ΔH°_T	ΔF°_T
298	....	−152,000	−139,500
400	(1,300)	(−152,000)	(−135,500)
500	(3,200)	(−151,500)	(−131,000)
600	(5,100)	(−151,000)	(−127,000)
700	(6,400)	(−151,000)	(−123,000)
800	(8,500)	(−150,500)	(−119,500)
900	(9,500)	(−150,000)	(−115,500)
1,000	(11,900)	(−152,000)	(−111,500)
1,100	(13,000)	(−151,500)	(−107,500)
1,200	....	(−151,500)	(−103,500)
1,300	....	(−151,000)	(−99,500)
1,400	....	(−150,500)	(−95,500)
1,500	(22,000)	(−150,500)	(−91,500)

Barium Difluoride, BaF_2 (c)

$\Delta H^\circ_{298} = -286{,}900$ calories per mole (*112*)
$S_{298} = 23.03$ *e.u.* (*83*)
$M.P. = 1{,}593°$ K. (*112*)
$\Delta H_M = 3{,}000$ calories per mole
$B.P. = 2{,}473°$ K. (*94*)
$\Delta H_V = 70{,}000$ calories per mole

Zone I (c) (298°–1,300° K.)

$$C_p = 13.98 + 10.20 \times 10^{-3} T \ (82)$$
$$H_T - H_{298} = -4{,}600 + 13.98T + 5.10 \times 10^{-3} T^2$$

Formation: $Ba + F_2 \longrightarrow BaF_2$

Zone I (298°–983° K.)

$$\Delta C_p = 0.14 + 8.26 \times 10^{-3} T + 0.80 \times 10^5 T^{-2}$$
$$\Delta H_T = -287{,}000 + 0.14T + 4.13 \times 10^{-3} T^2 - 0.80 \times 10^5 T^{-1}$$
$$\Delta F_T = -287{,}000 - 0.14 T \ln T - 4.13 \times 10^{-3} T^2 - 0.40 \times 10^5 T^{-1} + 44.48T$$

Zone II (983°–1,125° K.)

$$\Delta C_p = -5.81 + 9.76 \times 10^{-3} T + 0.80 \times 10^5 T^{-2}$$
$$\Delta H_T = -284{,}800 - 5.81T + 4.88 \times 10^{-3} T^2 - 0.80 \times 10^5 T^{-1}$$
$$\Delta F_T = -284{,}800 + 5.81 T \ln T - 4.88 \times 10^{-3} T^2 - 0.40 \times 10^5 T^{-1} + 1.91T$$

T, ° K.	$H_T - H_{298}$	S_T	ΔH°_T	ΔF°_T
298	....	23.03	−286,900	−274,500
400	1,850	28.36	−286,450	−270,350
500	3,700	32.48	−286,000	−266,300
600	5,650	36.04	−285,550	−262,450
700	7,650	39.12	−284,950	−257,550
800	9,700	41.85	−284,500	−254,900
900	11,900	44.44	−283,850	−251,250
1,000	14,200	46.86	−285,800	−247,700
1,100	16,700	49.25	−284,800	−243,900
1,200	(19,500)	....	(−283,770)	(−241,600)
1,300	(22,200)	....	(−283,110)	(−238,200)

Barium Dichloride, $BaCl_2$ (c)

$\Delta H^\circ_{298} = -205{,}300$ calories per mole (*11*)
$S_{298} = (29)$ *e.u.* (*11*)
$M.P. = 1{,}233°$ K. (*6*)
$\Delta H_M = 5{,}370$ calories per mole
$B.P. = 2{,}100°$ K. (*6*)
$\Delta H_V = (50{,}000)$ calories per mole

Zone I (*c*) (298°–1, 198° K.)

$$C_p = 17.0 + 3.34 \times 10^{-3} T \quad (82)$$
$$H_T - H_{298} = -5{,}200 + 17.0T + 1.67 \times 10^{-3} T^2$$

Formation: $Ba + Cl_2 \longrightarrow BaCl_2$

Zone I (298°–983° K.)

$$\Delta C_p = 2.63 + 1.78 \times 10^{-3} T + 0.68 \times 10^5 T^{-2}$$
$$\Delta H_T = -206{,}000 + 2.63T + 0.89 \times 10^{-3} T^2 - 0.68 \times 10^5 T^{-1}$$
$$\Delta F_T = -206{,}000 - 2.63 T \ln T - 0.89 \times 10^{-3} T^2 - 0.34 \times 10^5 T^{-1} + 58.13T$$

Zone II (983°–1,125° K.)

$$\Delta C_p = -3.32 + 3.28 \times 10^{-3} T + 0.68 \times 10^5 T^{-2}$$
$$\Delta H_T = -203{,}400 - 3.32T + 1.64 \times 10^{-3} T^2 - 0.68 \times 10^5 T^{-}$$
$$\Delta F_T = -203{,}400 + 3.32 T \ln T - 1.64 \times 10^{-3} T^2 - 0.34 \times 10^5 T^{-1} + 14.68T$$

T, ° K.	$H_T - H_{298}$	S_T	ΔH_T°	ΔF_T°
298	----------	(29.0)	−205,300	(−193,300)
400	1,900	(34.64)	−204,900	(−189,300)
500	3,700	(38.67)	−204,500	(−185,300)
600	5,600	(42.1)	−204,100	(−181,300)
700	7,500	(44.94)	−203,650	(−178,700)
800	9,500	(47.57)	−203,300	(−174,000)
900	11,500	(49.9)	−202,900	(−170,500)
1,000	13,500	(51.84)	−205,100	(−166,700)
1,100	15,500	(54.3)	−204,600	(−163,500)
1,500	(24,050)	----------	----------	(−149,000)
2,000	(35,500)	----------	----------	(−133,000)

Barium Dibromide, $BaBr_2$ (*c*)

$\Delta H^\circ_{298} = -180{,}000$ calories per mole (*11*)
$S_{298} = (35)$ *e.u.* (*11*)
M.P. = 1,120° K. (*6*)
$\Delta H_M = (6{,}000)$ calories per mole
B.P. = 2,100° K. (*6*)
$\Delta H_V = (50{,}000)$ calories per mole

Formation: $Ba + Br_2 \longrightarrow BaBr_2$
(estimated (*11*))

T, ° K.	$H_T - H_{298}$	S_T	ΔH_T°	ΔF_T°
298	----------	(35)	−180,000	(−175,000)
500	(3,800)	(45)	(−187,000)	(−167,000)
1,000	(13,700)	(58)	(−187,000)	(−148,000)
1,500	(31,400)	(73)	(−178,000)	(−131,500)

Barium Diiodide, BaI_2 (*c*)

$\Delta H^\circ_{298} = -144{,}600 \pm 1{,}000$ calories per mole (*11*)
$S_{298} = 39$ *e.u.* (*11*)
M.P. = 984° K. (*6*)
$\Delta H_M = (6{,}800)$ calories per mole
B.P. = (2,000°) K. (*6*)
$\Delta H_V = (45{,}000)$ calories per mole

Formation: $Ba + I_2 \longrightarrow BaI_2$
(estimated (*11*))

T, ° K.	$H_T - H_{298}$	S_T	ΔH_T°	ΔF_T°
298	----------	(39)	−144,600	(−143,000)
500	(3,800)	(49)	(−158,700)	(−140,000)
1,000	(20,700)	(70)	(−151,500)	(−121,500)
1,500	(32,700)	(80)	(−148,500)	(−105,500)

Tribarium Dinitride, Ba_3N_2 (*c*)

$\Delta H^\circ_{298} = -90{,}600$ calories per mole (*9*)
$S_{298} = 36.4$ *e.u.* (*9*)
Decomposes = 1,270° K. (*9*)

Formation: $3Ba + N_2 \longrightarrow Ba_3N_2$
(estimated (*9*))

T, ° K.	ΔF_T°
298	(−73,400)
500	(−61,900)
1,000	(−33,200)

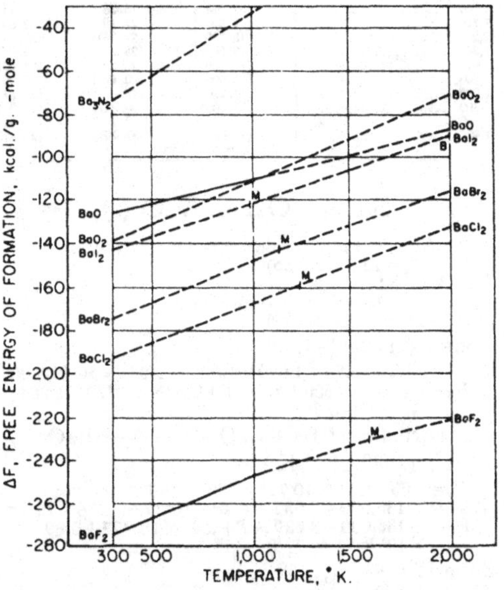

FIGURE 4.—Barium.

BERYLLIUM AND ITS COMPOUNDS

Element, Be (*c*)

$S_{298} = 2.28$ *e.u.* (*83*)
M.P. = 1,556° K. (*112*)
$\Delta H_M = 2{,}300$ calories per atom
B.P. = 3,243° K. (*138*)
$\Delta H_V = 53{,}490$ calories per atom

Zone I (*c*) (298°–1,300° K.)

$$C_p = 3.40 + 2.90 \times 10^{-3} T \quad (82)$$
$$H_T - H_{298} = -1{,}143 + 3.40T + 1.45 \times 10^{-3} T^2$$
$$F_T - H_{298} = -1{,}143 - 3.40 T \ln T - 1.45 \times 10^{-3} T^2 + 21.34T$$

T, ° K.	H_T-H_{298}	S_T	$-\frac{(F_T-H_{298})}{T}$
298	----------	2.28	2.28
400	415	3.48	2.44
500	900	4.56	2.76
600	1,415	5.50	3.14
700	1,965	6.14	3.33
800	2,535	7.10	3.93
900	3,135	7.81	4.33
1,000	3,745	8.45	4.70
1,100	4,365	9.04	5.07
1,200	4,990	9.59	5.43
1,300	5,615	10.09	5.77
1,400	(6,460)	(10.72)	(6.11)
1,500	(7,220)	(11.25)	(6.44)

Beryllium Oxide, BeO (*c*)

$\Delta H^{\circ}_{298} = -143{,}100$ calories per mole (*22*)
$S_{298} = 3.37$ *e.u.* (*83*)
M.P. = 2,823° K. (*42*)
$\Delta H_M = 17{,}000$ calories per mole
B.P. = 4,533° K. (*42*)
$\Delta H_V = 117{,}000$ calories per mole

Zone I (*c*) (298°–1,200° K.)

$$C_p = 8.45 + 4.00\times10^{-3}T - 3.17\times10^{5}T^{-2} \quad (82)$$
$$H_T - H_{298} = -3{,}760 + 8.45T + 2.00\times10^{-3}T^2 + 3.17\times10^{5}T^{-1}$$

Zone II (*c*) (1,200°–2,000° K.) (estimated (*24*))

Formation: $Be + 1/2O_2 \longrightarrow BeO$

Zone I (298°–1,000° K.)

$$\Delta C_p = +1.47 + 0.60\times10^{-3}T - 2.97\times10^{5}T^{-2}$$
$$\Delta H_T = -144{,}560 + 1.47T + 0.30\times10^{-3}T^2 + 2.97\times10^{5}T^{-1}$$
$$\Delta F_T = -144{,}560 - 1.47T\ln T - 0.30\times10^{-3}T^2 + 1.48\times10^{5}T^{-1} + 35.15T$$

T, ° K.	H_T-H_{298}	S_T	ΔH°_T	ΔF°_T
298	----------	3.37	−143,100	−136,100
400	730	5.46	−143,150	−133,700
500	1,580	7.35	−143,250	−131,400
600	2,540	9.10	−143,100	−128,800
700	3,600	10.73	−142,900	−126,800
700	4,700	12.20	−142,800	−124,350
900	5,830	13.53	−142,700	−122,000
1,000	7,010	14.77	−142,550	−119,750
1,100	8,240	15.95	−142,300	−117,450
1,200	9,510	17.05	−142,100	−115,250
1,300	(10,870)	----------	(−141,900)	(−112,850)
1,400	(12,200)	----------	(−141,700)	(−111,150)
1,500	(13,630)	----------	(−140,500)	(−109,250)
2,000	(21,300)	----------	(−142,900)	(−97,000)

Beryllium Difluoride, BeF_2 (*c*)

$\Delta H^{\circ}_{298} = (-227{,}000)$ calories per mole (*11*)
$S_{298} = (17)$ *e.u.* (*11*)
M.P. = 1,070° K. (*6*)
$\Delta H_M = (6{,}000)$ calories per mole
B.P. = (1,600°) K. (*6*)
$\Delta H_V = (40{,}000)$ calories per mole

Formation: $Be + F_2 \longrightarrow BeF_2$ (estimated (*11*))

T, ° K.	H_T-H_{298}	S_T	ΔH°_T	ΔF°_T
298	----------	(17)	−227,000	(−216,900)
500	(3,400)	(25.5)	(−226,000)	(−210,500)
1,000	(12,000)	(38)	(−224,500)	(−195,000)
1,500	(29,000)	(52.7)	(−215,500)	(−183,500)

Beryllium Dichloride, $BeCl_2$ (*c*)

$\Delta H^{\circ}_{298} = -112{,}600$ calories per mole (*11*)
$S_{298} = (23)$ *e.u.* (*11*)
M.P. = 678° K. (*6*)
$\Delta H_M = (3{,}000)$ calories per mole
B.P. = (820°) K. (*6*)
$\Delta H_V = (25{,}000)$ calories per mole

Formation: $Be + Cl_2 \longrightarrow BeCl_2$ (estimated (*11*))

T, ° K.	H_T-H_{298}	S_T	ΔH°_T	ΔF°_T
298	----------	(23)	−112,600	(−102,900)
500	(3,400)	(26.4)	(−112,000)	(−96,600)
1,000	(42,000)	(40)	(−80,500)	(−84,600)

Beryllium Dibromide, $BeBr_2$ (*c*)

$\Delta H^{\circ}_{298} = -79{,}400$ calories per mole (*11*)
$S_{298} = (29)$ *e.u.* (*11*)
M.P. = 761° K. (*6*)
$\Delta H_M = (4{,}500)$ calories per mole
B.P. = (800°) K. (*6*)
$\Delta H_V = (22{,}000)$ calories per mole

Formation: $Be + Br_2 \longrightarrow BeBr_2$ (estimated (*11*))

T, ° K.	H_T-H_{298}	S_T	ΔH°_T	ΔF°_T
298	----------	(29)	−79,400	(−76,500)
500	(3,400)	(37.5)	(−86,500)	(−70,500)
1,000	(39,000)	(58.0)	(−58,000)	(−59,000)

Beryllium Diiodide, BeI_2 (*c*)

$\Delta H^{\circ}_{298} = (-39{,}400)$ calories per mole (*11*)
$S_{298} = (31)$ *e.u.* (*11*)
M.P. = 753° K. (*6*)
$\Delta H_M = (4{,}500)$ calories per mole
B.P. = 760° K. (*6*)
$\Delta H_V = (19{,}000)$ calories per mole

Formation: $Be + I_2 \longrightarrow BeI_2$ (estimated (*11*))

T, ° K.	H_T-H_{298}	S_T	ΔH°_T	ΔF°_T
298	----------	(31)	(−39,400)	(−39,400)
500	(3,400)	(39.5)	(−53,500)	(−33,200)
1,000	(36,000)	(60)	(−28,300)	(−20,700)

Triberyllium Dinitride, Be_3N_2 (*c*)

$\Delta H^{\circ}_{298} = -133{,}500$ calories per mole (*81*)
$S_{298} = 12.0$ *e.u.* (*81*)
$M.P. = 2{,}470°$ K. (*9*)

Zone I (*c*) (298°–800° K.)

$$C_p = 7.32 + 30.8 \times 10^{-3} T \quad (82)$$
$$H_T - H_{298} = -3{,}550 + 7.32T + 15.4 \times 10^{-3} T^2$$

Formation: $3Be + N_2 \longrightarrow Be_3N_2$

Zone I (298°–800° K.)

$$\Delta C_p = -9.54 + 21.08 \times 10^{-3} T$$
$$\Delta H_T = -131{,}600 - 9.54T + 10.54 \times 10^{-3} T^2$$
$$\Delta F_T = -131{,}600 + 9.54 T \ln T - 10.54 \times 10^{-3} T^2 - 18.14T$$

T, ° K.	$H_T - H_{298}$	S_T	ΔH°_T	ΔF°_T
298	....	12.0	−133,500	−121,400
400	1,840	17.29	−133,600	−117,200
500	3,930	21.94	−133,700	−113,100
600	6,420	26.47	−133,450	−109,000
700	9,200	30.75	−133,000	−105,400
800	12,130	34.66	−132,500	−101,000
900	(15,500)	....	(−131,600)	(−98,200)
1,000	(19,200)	....	(−130,600)	(−94,500)
1,500	(42,200)	....	(−122,100)	(−77,900)
2,000	(72,700)	....	(−108,500)	(−65,100)

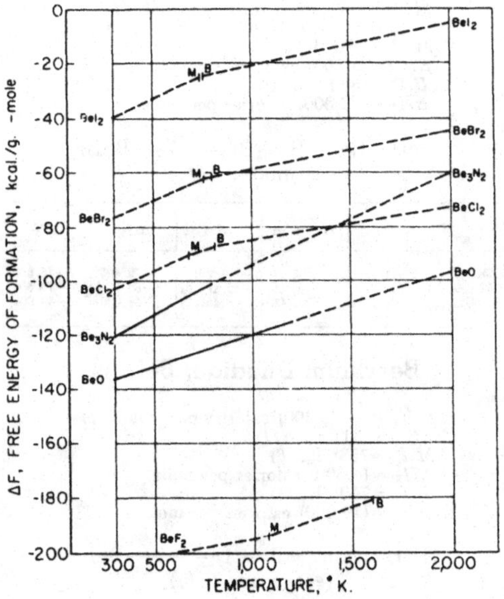

FIGURE 5.—Beryllium.

BISMUTH AND ITS COMPOUNDS

Element, Bi (*c*)

$S_{298} = 13.6$ *e.u.* (*83*)
$M.P. = 544.5°$ K. (*82*)
$\Delta H_M = 2{,}600$ calories per atom
$B.P. = 1{,}832°$ K. (*130*)
$\Delta H_V = 36{,}200$ calories per atom

Zone I (*c*) (298°–544.5° K.)

$$C_p = 4.49 + 5.40 \times 10^{-3} T \quad (82)$$
$$H_T - H_{298} = -1{,}579 + 4.49T + 2.70 \times 10^{-3} T^2$$
$$F_T - H_{298} = -1{,}579 - 4.49 T \ln T - 2.70 \times 10^{-3} T^2 + 18.08T$$

Zone II (*l*) (544.5°–1,800° K.)

$$C_p = 7.50 \quad (82)$$
$$H_T - H_{298} = 180 + 7.50T$$
$$F_T - H_{298} = 180 - 7.50 T \ln T + 32.34T$$

T, ° K.	$H_T - H_{298}$	S_T	$-\frac{(F_T - H_{298})}{T}$
298	....	13.6	13.6
400	650	15.47	13.84
500	1,340	17.01	14.33
600	4,680	23.14	15.34
700	5,430	24.30	16.54
800	6,180	25.30	17.58
900	6,930	26.18	18.48
1,000	7,680	26.97	19.29
1,100	8,430	27.67	20.01
1,200	9,180	28.32	20.67
1,300	9,930	28.92	21.29
1,400	10,680	29.47	21.85
1,500	11,430	29.99	22.37
1,600	12,180	30.48	22.87
1,700	12,930	30.93	23.32
1,800	13,680	31.36	23.76
1,900	(55,460)	(53.87)	(24.68)
2,000	(55,970)	(54.13)	(26.15)

Bismuth Oxide, BiO (*c*)

$\Delta H^{\circ}_{298} = -49{,}850$ calories per mole (*112*)
$S_{298} = (16.4)$ *e.u.* (*24*)

Formation: $Bi + 1/2O_2 \longrightarrow BiO$
(estimated (*24*))

T, ° K.	$H_T - H_{298}$	ΔH°_T	ΔF°_T
298	....	−49,850	(−43,500)
400	(1,400)	(−49,500)	(−41,000)
500	(2,400)	(−49,500)	(−39,000)
600	(4,100)	(−51,500)	(−36,500)
700	(5,300)	(−51,500)	(−34,000)
800	(6,400)	(−51,500)	(−31,500)
900	(8,100)	(−51,000)	(−29,000)
1,000	(9,200)	(−51,000)	(−26,500)
1,100	(10,900)	(−50,500)	(−24,500)
1,200	(12,000)	(−50,500)	(−22,000)
1,300	(13,200)	(−50,500)	(−19,500)
1,400	(15,000)	(−50,000)	(−17,000)
1,500	(16,300)	(−50,000)	(−15,000)
1,600	(17,400)	(−49,500)	(−12,500)

Dibismuth Trioxide, Bi_2O_3 (*c*)

$\Delta H^\circ_{298} = -137{,}900$ calories per mole (*112*)
$S_{298} = 36.2$ *e.u.* (*83*)
M.P. = 1,090° K. (*112*)
$\Delta H_M = 6{,}800$ calories per mole
B.P. = (2,160°) K. (*94*)

Zone I (*c*) (298°–800° K.)

$$C_p = 24.74 + 8.00 \times 10^{-3}T \; (82)$$
$$H_T - H_{298} = -7{,}732 + 24.74T + 4.00 \times 10^{-3}T^2$$

Formation: $2Bi + 3/2O_2 \longrightarrow Bi_2O_3$

Zone I (298°–800° K.)

$$\Delta C_p = 5.02 - 4.30 \times 10^{-3}T + 0.60 \times 10^5 T^{-2}$$
$$\Delta H_T = -139{,}000 + 5.02T - 2.15 \times 10^{-3}T^2 - 0.60 \times 10^5 T^{-1}$$
$$\Delta F_T = -139{,}000 - 5.02T\ln T + 2.15 \times 10^{-3}T^2 - 0.30 \times 10^5 T^{-1} + 96.5T$$

Above 800° K. (estimated (*24*))

T, ° K.	$H_T - H_{298}$	S_T	ΔH°_T	ΔF°_T
298	----------	36.2	−137, 900	−118, 700
400	2, 770	44.2	−137, 500	−112, 150
500	5, 630	50.56	−137, 100	−105, 800
600	8, 550	55.89	−142, 000	−99, 100
700	11, 550	60.51	−141, 700	−92, 000
800	14, 620	64.61	−141, 300	−84, 850
900	(18, 170)	(68.36)	(−140, 500)	(−77, 500)
1, 000	(21, 000)	(71.78)	(−140, 300)	(−70, 850)
1, 100	----------	----------	(−132, 500)	(−64, 500)
1, 200	----------	----------	(−131, 500)	(−58, 000)
1, 300	----------	----------	(−130, 500)	(−52, 000)
1, 400	----------	----------	(−129, 000)	(−46, 000)
1, 500	----------	----------	(−128, 000)	(−40, 000)
1, 600	----------	----------	(−126, 500)	(−34, 500)

Bismuth Trifluoride, BiF_3 (*c*)

$\Delta H^\circ_{298} = (-216{,}000)$ calories per mole (*11*)
$S_{298} = (34)$ *e.u.* (*11*)
M.P. = 1,000° K. (*6*)
$\Delta H_M = (6{,}200)$ calories per mole
B.P. = (1,300°) K. (*6*)
$\Delta H_V = (28{,}000)$ calories per mole

Formation: $Bi + 3/2F_2 \longrightarrow BiF_3$
(estimated (*11*))

T, ° K.	$H_T - H_{298}$	ΔH°_T	ΔF°_T
298	----------	(−216, 000)	(−200, 000)
500	(6, 000)	(−213, 500)	(−188, 500)

Bismuth Trichloride, $BiCl_3$ (*c*)

$\Delta H^\circ_{298} = -90{,}500$ calories per mole (*11*)
$S_{298} = 45.8$ *e.u.* (*11*)
M.P. = 502° K. (*6*)
$\Delta H_M = 2{,}600$ calories per mole
B.P. = 714° K. (*6*)
$\Delta H_V = 17{,}354$ calories per mole

Formation: $Bi + 3/2Cl_2 \longrightarrow BiCl_3$
(estimated (*11*))

T, ° K.	$H_T - H_{298}$	ΔH°_T	ΔF°_T
298	----------	−90, 500	−76, 000
500	(7, 000)	(−87, 500)	(−66, 500)

Bismuth Tribromide, $BiBr_3$ (*c*)

$\Delta H^\circ_{298} = (-60{,}000)$ calories per mole (*11*)
$S_{298} = (54)$ *e.u.* (*11*)
M.P. = 491° K. (*6*)
$\Delta H_M = (4{,}000)$ calories per mole
B.P. = 734° K. (*6*)
$\Delta H_V = 18{,}024$ calories per mole

Formation: $Bi + 3/2Br_2 \longrightarrow BiBr_3$
(estimated (*11*))

T, ° K.	$H_T - H_{298}$	ΔH°_T	ΔF°_T
298	----------	(−60, 000)	(−55, 800)
500	(7, 000)	(−68, 600)	(−47, 500)

Bismuth Triiodide, BiI_3 (*c*)

$\Delta H^\circ_{298} = -23{,}700$ calories per mole (*11*)
$S_{298} = (55)$ *e.u.* (*11*)
M.P. = 681° K. (*6*)
Decomposes = 773° K. (*6*)

Formation: $Bi + 3/2I_2 \longrightarrow BiI_3$
(estimated (*11*))

T, ° K.	$H_T - H_{298}$	ΔH°_T	ΔF°_T
298	----------	−23, 700	(−23, 500)
500	(6, 000)	(−44, 000)	(−20, 000)

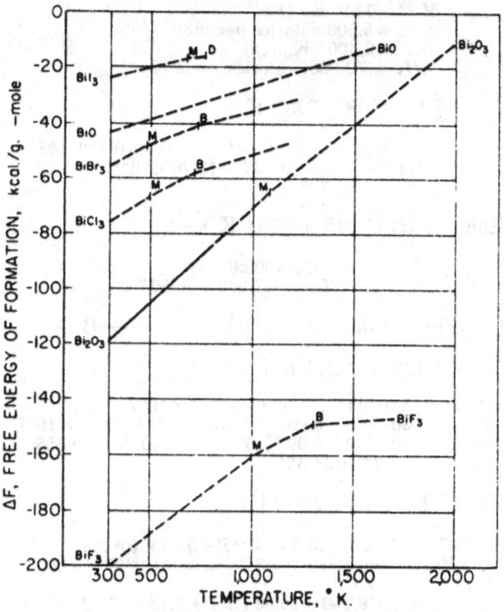

FIGURE 6.—Bismuth.

BORON AND ITS COMPOUNDS

Element, B (*c*)

$S_{298} = 1.4$ *e.u.* (*76*)
$M.P. = 2{,}300°$ K. (*24*)
$\Delta H_M = 5{,}300$ calories per atom

Zone I (*c*) (298°–1,200° K.)

$$C_p = 1.54 + 4.40 \times 10^{-3}T \quad (82)$$
$$H_T - H_{298} = -655 + 1.54T + 2.20 \times 10^{-3}T^2$$
$$F_T - H_{298} = -655 - 1.54T\ln T - 2.20 \times 10^{-3}T^2 + 10.21T$$

T, ° K.	$H_T - H_{298}$	S_T	$-\frac{(F_T - H_{298})}{T}$
298	....	1.4	1.4
400	300	2.27	1.52
500	660	3.07	1.75
600	1,080	3.83	2.03
700	1,540	4.54	2.34
800	2,040	5.20	2.65
900	2,570	5.82	2.96
1,000	3,130	6.41	3.28
1,100	3,700	6.94	3.57
1,200	4,270	7.45	3.90
1,500	(6,600)	(9.15)	(4.88)
2,000	(11,225)	(11.15)	(6.21)

Boron Oxide, BO (*g*)

$\Delta H^\circ_{298} = 5{,}300$ calories per mole (*112*)
$S_{298} = 48.60$ *e.u.* (*83*)
$\Delta F^\circ_{298} = 11{,}600$ calories per mole

Diboron Trioxide, B_2O_3 (*c*)

$\Delta H^\circ_{298} = -305{,}400$ calories per mole (*112*)
$S_{298} = 13.04$ *e.u.* (*83*)
$M.P. = 723°$ K. (*82*)
$\Delta H_M = 5{,}500$ calories per mole
$B.P. = 2{,}300°$ K. (*42*)
$\Delta H_V = (70{,}000)$ calories per mole

Zone I (*c*) (298°–723° K.)

$$C_p = 8.73 + 25.40 \times 10^{-3}T - 1.31 \times 10^5T^{-2} \quad (82)$$
$$H_T - H_{298} = -4{,}170 + 8.73T + 12.70 \times 10^{-3}T^2 + 1.31 \times 10^5T^{-1}$$

Zone II (*l*) (723°–1,800° K.)

$$C_p = 30.50 \quad (82)$$
$$H_T - H_{298} = -7{,}590 + 30.50T$$

Formation: $2B + 3/2O_2 \longrightarrow B_2O_3$

Zone I (298°–723° K.)

$$\Delta C_p = -5.09 + 15.1 \times 10^{-3}T - 0.71 \times 10^5T^{-2}$$
$$\Delta H_T = -304{,}690 - 5.09T + 7.55 \times 10^{-3}T^2 + 0.71 \times 10^5T^{-1}$$
$$\Delta F_T = -304{,}690 + 5.09T\ln T - 7.55 \times 10^{-3}T^2 + 0.355 \times 10^5T^{-1} + 34.3T$$

Zone II (723°–1,200° K.)

$$\Delta C_p = 16.68T - 10.3 \times 10^{-3}T + 0.60 \times 10^5T^{-2}$$
$$\Delta H_T = -308{,}150 + 16.68T - 5.15 \times 10^{-3}T^2 - 0.60 \times 10^5T^{-1}$$
$$\Delta F_T = -308{,}150 - 16.68T\ln T + 5.15 \times 10^{-3}T^2 - 0.30 \times 10^5T^{-1} + 173.25T$$

T, ° K.	$H_T - H_{298}$	S_T	ΔH°_T	ΔF°_T
298	....	13.04	−305,400	−286,400
400	1,640	17.75	−305,400	−279,900
500	3,700	22.33	−305,200	−273,400
600	5,860	26.27	−305,000	−267,000
700	8,350	30.10	−304,600	−261,200
800	16,810	41.65	−298,350	−254,900
900	19,860	45.24	−297,600	−249,700
1,000	22,910	48.45	−296,900	−244,500
1,100	25,960	51.36	−296,250	−239,150
1,200	29,010	54.02	−295,600	−234,000
1,300	32,060	56.45	(−295,200)	(−229,500)
1,400	35,110	58.71	(−294,900)	(−224,600)
1,500	38,160	60.82	(−294,700)	(−219,300)
1,600	41,210	62.79	(−294,500)	(−214,300)
1,700	44,260	64.64	(−294,200)	(−209,400)

Boron Trifluoride, BF_3 (*g*)

$\Delta H^\circ_{298} = -273{,}500$ calories per mole (*42*)
$S_{298} = 60.70$ *e.u.* (*112*)
$M.P. = 145°$ K. (*6*)
$\Delta H_M = 480$ calories per mole
$B.P. = 172.2°$ K. (*6*)
$\Delta H_V = 4{,}620$ calories per mole

Zone I (*g*) (298°–1,000° K.)

$$C_p = 12.44 + 6.70 \times 10^{-3}T - 2.12 \times 10^5T^{-2} \quad (82)$$
$$H_T - H_{298} = -4{,}720 + 12.44T + 3.35 \times 10^{-3}T^2 + 2.12 \times 10^5T^{-1}$$

Formation: $B + 3/2F_2 \longrightarrow BF_3$

Zone I (298°–1,000° K.)

$$\Delta C_p = -1.54 + 1.64 \times 10^{-3}T - 0.92 \times 10^5T^{-2}$$
$$\Delta H_T = -273{,}420 - 1.54T + 0.82 \times 10^{-3}T^2 + 0.92 \times 10^5T^{-1}$$
$$\Delta F_T = -273{,}420 + 1.54T\ln T - 0.82 \times 10^{-3}T^2 + 0.46 \times 10^5T^{-1} + 4.54T$$

T, ° K.	$H_T - H_{298}$	S_T	ΔH°_T	ΔF°_T
298	....	60.70	−273,500	−269,400
400	1,320	64.50	−273,660	−267,950
500	2,765	67.72	−273,780	−266,500
600	4,320	70.55	−273,890	−265,000
700	5,965	73.08	−273,970	−263,500
800	7,695	75.36	−274,050	−262,050
900	9,430	77.43	−274,110	−260,550
1,000	11,230	79.33	−274,175	−259,050
1,100	(13,240)	....	(−274,050)	(−257,600)
1,200	(15,160)	....	(−274,010)	(−256,100)
1,300	....	....	(−273,960)	(−255,200)
1,400	....	....	(−273,900)	(−253,650)
1,500	(21,640)	....	(−273,820)	(−252,050)
1,600	....	....	(−273,720)	(−250,450)
1,700	....	....	(−273,640)	(−248,900)
1,800	....	....	(−273,480)	(−247,300)
1,900	....	....	(−273,330)	(−245,700)
2,000	(33,560)	....	(−273,170)	(−244,050)

Boron Trichloride, BCl_3 (*g*)

$\Delta H^\circ_{298} = -94{,}500$ calories per mole (*112*)
$S_{298} = 69.29$ *e.u.* (*112*)
$M.P. = 166°$ K. (*6*)
$\Delta H_M = (500)$ calories per mole
$B.P. = 285.6°$ K. (*6*)
$\Delta H_V = 5{,}700$ calories per mole

Zone I (g) (298°–1,000° K.)

$$C_p = 16.86 + 2.86\times10^{-3}T - 2.44\times10^{5}T^{-2} \ (82)$$
$$H_T - H_{298} = -5{,}970 + 16.86T + 1.43\times10^{-3}T^2 + 2.44\times10^{5}T^{-1}$$

Formation: $B + 3/2Cl_2 \longrightarrow BCl_3$

Zone I (298°–1,000° K.)

$$\Delta C_p = 2.09 - 1.63\times10^{-3}T - 1.42\times10^{5}T^{-2}$$
$$\Delta H_T = -95{,}525 + 2.09T - 0.815\times10^{-3}T^2 + 1.42\times10^{5}T^{-1}$$
$$\Delta F_T = -95{,}525 - 2.09T\ln T + 0.815\times10^{-3}T^2 + 0.71\times10^{5}T^{-1} + 26.68T$$

T, ° K.	$H_T - H_{298}$	S_T	ΔH°_T	ΔF°_T
298	-----	69.29	−94,500	−90,800
400	1,610	73.93	−94,450	−89,450
500	3,295	77.68	−94,395	−88,250
600	5,065	80.91	−94,330	−87,050
700	6,885	83.71	−94,285	−85,900
800	8,745	86.19	−94,240	−84,550
900	10,630	88.41	−94,205	−83,500
1,000	12,530	90.41	−94,190	−82,150
1,100	(14,530)	-----	(−94,080)	(−81,250)
1,200	(16,500)	-----	(−94,080)	(−80,150)
1,300	-----	-----	(−94,090)	(−79,050)
1,400	-----	-----	(−94,100)	(−77,750)
1,500	(22,700)	-----	(−94,130)	(−76,650)
1,600	-----	-----	(−94,200)	(−75,550)
1,700	-----	-----	(−94,250)	(−74,300)
1,800	-----	-----	(−94,330)	(−73,100)
1,900	-----	-----	(−94,420)	(−71,850)
2,000	(33,600)	-----	(−94,530)	(−70,600)

Boron Tribromide, BBr_3 (l)

$\Delta H^\circ_{298} = -57{,}900$ calories per mole (*120*)
$S_{298} = 53.9$ *e.u.* (*11*)
M.P. = 227° K. (*6*)
$\Delta H_M = (700)$ calories per mole
B.P. = 364.4° K. (*6*)
$\Delta H_V = 7{,}298$ calories per mole

Zone I (g) (364.4°–1,000° K.)

$$C_p = 17.83 + 2.04\times10^{-3}T - 1.95\times10^{5}T^{-2} \ (82)$$
$$H_T - H_{298} = 7{,}160 + 17.83T + 1.02\times10^{-3}T^2 + 1.95\times10^{5}T^{-1}$$

Formation: $B + 3/2Br_2 \longrightarrow BBr_3$

Zone I (400°–1,000° K.)

$$\Delta C_p = 2.73 - 2.36\times10^{-3}T - 1.40\times10^{5}T^{-2}$$
$$\Delta H_T = -57{,}460 + 2.73T - 1.18\times10^{-3}T^2 + 1.40\times10^{5}T^{-1}$$
$$\Delta F_T = -57{,}460 - 2.73T\ln T + 1.18\times10^{-3}T^2 + 0.70\times10^{5}T^{-1} + 32.81T$$

T, ° K.	$H_T - H_{298}$	S_T	ΔH°_T	ΔF°_T
298	-----	53.9	−57,900	−57,200
400	14,960	82.46	−56,200	−50,500
500	16,740	86.43	−56,120	−49,100
600	18,575	89.77	−56,030	−47,700
700	20,450	92.66	−55,960	−46,300
800	22,350	95.19	−55,900	−45,000
900	24,270	97.45	−55,860	−43,600
1,000	26,200	99.49	−55,840	−42,300
1,100	(28,450)	-----	-----	(−40,900)
1,200	(30,450)	-----	-----	(−39,500)
1,300	(32,500)	-----	-----	(−38,100)
1,400	(34,700)	-----	-----	(−36,700)
1,500	(36,600)	-----	(−55,900)	(−35,400)
2,000	(46,600)	-----	(−56,850)	(−28,600)

Boron Triiodide BI_3 (c)

$\Delta H^\circ_{298} = (-27{,}600)$ calories per mole (*11*)
$S_{298} = (55)$ *e.u.* (*11*)
$\Delta F^\circ_{298} = (-31{,}100)$ calories per mole
M.P. = 316° K. (*6*)
$\Delta H_M = (1{,}000)$ calories per mole
B.P. = 483° K. (*6*)
$\Delta H_V = (10{,}000)$ calories per mole

Tetraboron Carbide, B_4C (c)

$\Delta H_{298} = -13{,}800$ calories per mole (*122*)
$S_{298} = 6.47$ *e.u.* (*83*)
M.P. = 2,623° K. (*9*)

Zone I (c) (298°–1,200° K.)

$$C_p = 22.99 + 5.40\times10^{-3}T - 10.72\times10^{5}T^{-2} \ (82)$$
$$H_T - H_{298} = -10{,}690 + 22.99T + 2.70\times10^{-3}T^2 + 10.72\times10^{5}T^{-1}$$

Formation: $4B + C \longrightarrow B_4C$

Zone I (298°–1,200° K.)

$$\Delta C_p = 12.73 - 13.22\times10^{-3}T - 8.62\times10^{5}T^{-2}$$
$$\Delta H_T = -20{,}000 + 12.73T - 6.61\times10^{-3}T^2 + 8.62\times10^{5}T^{-1}$$
$$\Delta F_T = -20{,}000 - 12.73T\ln T + 6.61\times10^{-3}T^2 + 4.31\times10^{5}T^{-1} + 87.7T$$

T, ° K.	$H_T - H_{298}$	S_T	ΔH°_T	ΔF°_T
298	-----	6.47	−13,800	−13,300
400	1,620	10.82	−13,600	−13,150
500	3,610	15.65	−13,450	−13,100
600	5,850	19.54	−13,300	−13,100
700	8,340	23.38	−13,150	−13,050
800	10,760	26.77	−13,050	−13,000
900	13,325	29.52	−13,000	−12,950
1,000	16,070	32.55	−13,000	−12,900
1,100	18,800	35.20	−12,950	−12,850
1,200	21,665	37.66	−12,900	−12,750
1,500	30,550	44.35	(−12.800)	(−12,500)
2,000	(46,550)	(53.64)	-----	-----

Boron Nitride, BN (c)

$\Delta H^\circ_{298} = -60{,}700$ calories per mole (*33*)
$S_{298} = 3.67$ *e.u.* (*33*)
S.P. = 3,270° K. (*9*)

Zone I (c) (298°–1,200° K.)

$$C_p = 3.64 + 7.24\times10^{-3}T \ (78)$$
$$H_T - H_{298} = -1{,}380 + 3.64T + 3.62\times10^{-3}T^2$$

Formation: $B + 1/2N_2 \longrightarrow BN$

Zone I (298°–1,200° K.)

$$\Delta C_p = -1.23 + 2.33\times10^{-3}T$$
$$\Delta H_T = -60{,}950 - 1.23T + 1.16\times10^{-3}T^2$$
$$\Delta F_T = -60{,}950 + 1.23T\ln T - 1.16\times10^{-3}T^2 + 9.94T$$

T, ° K.	H_T-H_{298}	S_T	ΔH°_T	ΔF°_T
298	-------	7. 34	−60, 700	−55, 650
400	630	9. 16	−60, 700	−53, 900
500	1, 340	10. 74	−60, 700	−52, 200
600	2, 070	12. 06	−60, 750	−50, 500
700	2, 900	13. 34	−60, 750	−48, 800
800	3, 840	14. 60	−60, 750	−47, 150
900	4, 840	15. 78	−60, 600	−45, 400
1, 000	5, 900	16. 90	−60, 500	−43, 750
1, 100	6, 980	17. 92	−60, 400	−42, 100
1, 200	8, 100	18. 88	−60, 250	−40, 300

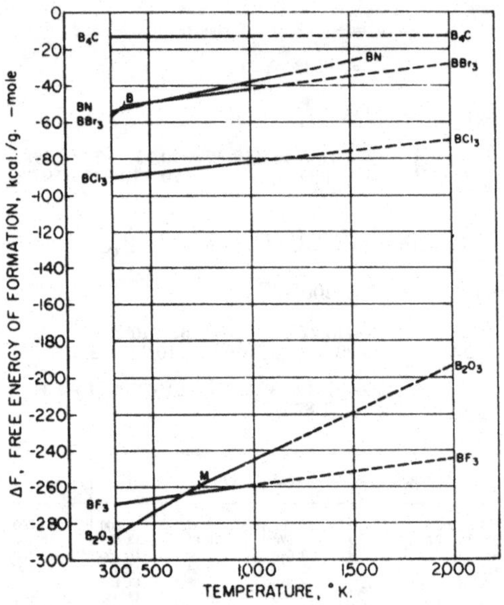

FIGURE 7.—Boron.

BROMINE

Element, Br_2 (l)

$S_{298}=36.4$ *e.u.* (*83*)
$M.P.=265.7°$ K. (*112*)
$\Delta H_M=2{,}580$ calories per atom
$B.P.=331°$ K. (*112*)
$\Delta H_V=7{,}418$ calories per atom

Zone I (l) (298°–331° K.)

$$C_p=17.1\ (82)$$
$$H_T-H_{298}=-5{,}090+17.1T$$
$$F_T-H_{298}=-5{,}090-17.1T\ln T+77.6T$$

Zone II (g) (331°–1,600° K.)

$$C_p=9.04-0.37\times10^5T^{-2}\ (82)$$
$$H_T-H_{298}=4{,}940+9.04T+0.37\times10^5T^{-1}$$
$$F_T-H_{298}=4{,}940-9.04T\ln T+0.185\times10^5T^{-1}+0.51T$$

T, ° K.	H_T-H_{298}	S_T	$-\frac{(F_T-H_{298})}{T}$
298	-------	36. 4	36. 4
400	8, 648	62. 74	41. 12
500	9, 531	64. 71	45. 65
600	10, 420	66. 33	48. 96
700	11, 313	67. 71	51. 55
800	12, 209	68. 91	53. 65
900	13, 109	69. 96	55. 39
1, 000	14, 011	70. 91	56. 90
1, 100	14, 900	71. 85	58. 30
1, 200	15, 819	72. 56	59. 37
1, 300	16, 720	73. 24	60. 38
1, 400	17, 635	73. 96	61. 36
1, 500	18, 530	74. 54	62. 19
1, 600	19, 457	75, 18	63. 02
1, 700	(20, 340)	(75. 74)	(63. 78)
1, 800	(21, 240)	(76. 29)	(64. 49)
1, 900	(22, 140)	(76. 74)	(65. 08)
2, 000	(23, 040)	(77. 24)	(65. 72)

CADMIUM AND ITS COMPOUNDS

Element, Cd (c)

$S_{298}=12.37$ *e.u.* (*28*)
$M.P.=594°$ K. (*82*)
$\Delta H_M=1{,}450$ calories per atom
$B.P.=1{,}038°$ K. (*7*)
$\Delta H_V=23{,}870$ calories per atom

Zone I (c) (298°–594° K.)

$$C_p=5.31+2.94\times10^{-3}T\ (82)$$
$$H_T-H_{298}=-1{,}714+5.31T+1.47\times10^{-3}T^2$$
$$F_T-H_{298}=-1{,}714-5.31T\ln T-1.47\times10^{-3}T^2+24.07T$$

Zone II (l) (594°–1,038° K.)

$$C_p=7.10\ (82)$$
$$H_T-H_{298}=-810+7.10T$$
$$F_T-H_{298}=-810-7.10T\ln T+32.99T$$

Zone III (g) (1,038°–2,000° K.)

$$C_p=(5.0)\ (141)$$
$$H_T-H_{298}=+25{,}370+5.0T$$
$$F_T-H_{298}=+25{,}370-5.0T\ln T-6.57T$$

T, ° K.	H_T-H_{298}	S_T	$-\frac{(F_T-H_{298})}{T}$
298	-------	12. 37	12. 37
400	645	14. 23	12. 62
500	1, 310	15. 71	13. 08
600	3, 450	19. 41	13. 67
700	4, 160	20. 50	14. 57
800	4, 870	21. 45	15. 36
900	5, 580	22. 29	16. 09
1, 000	6, 290	23. 04	16. 75
1, 100	(30, 700)	(46. 55)	(18. 62)
1, 200	(31, 200)	(46. 98)	(20. 96)
1, 300	(31, 700)	(47. 38)	(22. 98)
1, 400	(32, 200)	(47. 75)	(24. 74)
1, 500	(32, 700)	(48. 09)	(26. 28)
1, 600	(33, 200)	(48. 41)	(27. 65)
1, 700	(33, 700)	(48. 72)	(28. 90)
1, 800	(34, 200)	(49. 00)	(30. 00)
1, 900	(34, 700)	(49. 27)	(31. 01)
2, 000	(35, 200)	(49. 52)	(31. 92)

Cadmium Oxide, CdO (*c*)

$\Delta H^{\circ}_{298} = -61{,}200$ calories per mole (*98*)
$S_{298} = 13.1$ *e.u.* (*24*)

Formation: $Cd + 1/2O_2 \longrightarrow CdO$
(estimated (*24*))

T, ° K.	$H_T - H_{298}$	ΔH°_T	ΔF°_T
298	------------	−61, 200	−54, 100
400	(1, 100)	(−61, 100)	(−51, 700)
500	(2, 150)	(−61, 100)	(−47, 000)
600	(3, 250)	(−62, 500)	(−44, 400)
700	(4, 350)	(−62, 500)	(−41, 800)
800	(5, 450)	(−62, 500)	(−39, 200)
900	(6, 600)	(−62, 500)	(−36, 700)
1, 000	(7, 800)	(−62, 400)	(−32, 700)
1,100	(9, 050)	(−86, 100)	(−27, 800)
1, 500	(14, 150)	(−84, 800)	(−8, 700)

Cadmium Difluoride, CdF₂ (*c*)

$\Delta H^{\circ}_{298} = -167{,}000$ calories per mole (*11*)
$S_{298} = (22)$ *e.u.* (*11*)
M.P. = 1,383° K. (*6*)
$\Delta H_M = 5{,}400$ calories per mole
B.P. = 2,023° K. (*6*)
$\Delta H_V = 52{,}000$ calories per mole

Formation: $Cd + F_2 \longrightarrow CdF_2$
(estimated (11))

T, ° K.	$H_T - H_{298}$	ΔH°_T	ΔF°_T
298	------------	−167, 000	(−155, 400)
500	(4, 000)	(−165, 900)	(−147, 000)
1, 000	(14, 000)	(−165, 000)	(−129, 000)
1, 500	(32, 000)	------------	------------

Cadmium Dichloride, CdCl₂ (*c*)

$\Delta H^{\circ}_{298} = -93{,}000$ calories per mole (*112*)
$S_{298} = 31.2$ *e.u.* (*83*)
M.P. = 841° K. (*6*)
$\Delta H_M = 5{,}300$ calories per mole
B.P. = 1,240° K. (*6*)
$\Delta H_V = 29{,}860$ calories per mole

Zone I (298°–800° K.)

$$C_p = 14.64 + 9.60 \times 10^{-3} T \quad (82)$$
$$H_T - H_{298} = -4{,}790 + 14.64T + 4.80 \times 10^{-3} T^2$$

Formation: $Cd + Cl_2 \longrightarrow CdCl_2$

Zone I (298°–594 K.)

$$\Delta C_p = 0.51 + 6.60 \times 10^{-3} T + 0.68 \times 10^5 T^{-2}$$
$$\Delta H_T = -93{,}215 + 0.51T + 3.30 \times 10^{-3} T^2 - 0.68 \times 10^5 T^{-1}$$
$$\Delta F_T = -93{,}215 - 0.51 T \ln T - 3.30 \times 10^{-3} T^2 - 0.34 \times 10^5 T^{-1} + 39.48T$$

Zone II (594°–800° K.)

$$\Delta C_p = -1.28 + 9.54 \times 10^{-3} T + 0.68 \times 10^5 T^{-2}$$
$$\Delta H_T = -94{,}100 - 1.28T + 4.77 \times 10^{-3} T^2 - 0.68 \times 10^5 T^{-1}$$
$$\Delta F_T = -94{,}100 + 1.28 T \ln T - 4.77 \times 10^{-3} T^2 - 0.34 \times 10^5 T^{-1} + 30.5T$$

T, ° K.	$H_T - H_{298}$	S_T	ΔH°_T	ΔF°_T
298	------------	31. 2	−93, 000	−82, 700
400	1, 780	36. 33	−92, 700	−79, 200
500	3, 720	40. 66	−92, 300	−75, 900
600	5, 750	44. 36	−93, 250	−72, 650
700	7, 840	47. 58	−92, 750	−69, 300
800	9, 990	50. 45	−92, 200	−65, 900
1,000	(20, 000)	------------	(−85, 350)	(−60, 700)
1, 500	(62, 000)	------------	(−74, 000)	(−47, 900)

Cadmium Dibromide, CdBr₂ (*c*)

$\Delta H^{\circ}_{298} = -75{,}800$ calories per mole (*11*)
$S_{298} = 34.4$ *e.u.* (*83*)
M.P. = 841° K. (*6*)
$\Delta H_M = 5{,}000$ calories per mole
B.P. = 1,136° K. (*6*)
$\Delta H_V = 27{,}000$ calories per mole

Formation: $Cd + Br_2 \longrightarrow CdBr_2$
(estimated (*11*))

T, ° K.	$H_T - H_{298}$	ΔH°_T	ΔF°_T
298	------------	−75, 800	−71, 500
500	(4, 000)	(−82, 650)	(−64, 250)
1,000	(20, 000)	(−76, 100)	(−47, 000)
1,500	(59, 000	(−68, 200)	------------

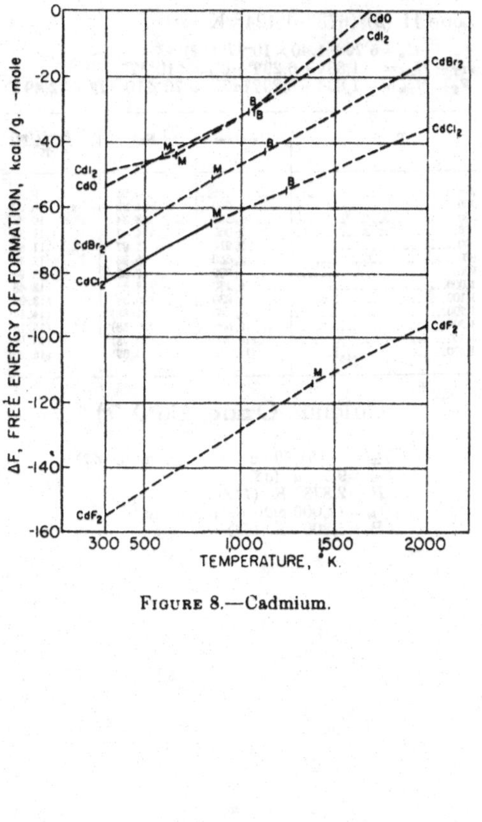

FIGURE 8.—Cadmium.

Cadmium Diiodide, CdI_2 (*c*)

$\Delta H^\circ_{298} = -48{,}750$ calories per mole (*11*)
$S_{298} = 39.5$ *e.u.* (*11*)
M.P. = 660° K. (*6*)
$\Delta H_M = 3{,}660$ calories per mole
B.P. = 1,069° K. (*6*)
$\Delta H_V = 25{,}400$ calories per mole

Formation: $Cd + I_2 \longrightarrow CdI_2$
(estimated (*11*))

T, ° K.	$H_T - H_{298}$	ΔH°_T	ΔF°_T
298	....	−48, 750	−49, 000
500	(4, 000)	(−62, 700)	(−46, 300)
1,000	(19, 000)	(−56, 200)	(−32, 000)
1,500	(56, 400)	(−50, 800)	(−13, 000)

CALCIUM AND ITS COMPOUNDS

Element, Ca (*c*)

$S_{298} = 9.95$ *e.u.* (*83*)
T.P. = 673° K. (*82*)
$\Delta H_T = 115$ calories per atom
M.P. = 1,124° K. (*80*)
$\Delta H_M = 2{,}230$ calories per atom
B.P. = 1,760° K. (*130*)
$\Delta H_V = 35{,}840$ calories per atom

Zone I (α) (298°–673° K.)

$C_p = 5.24 + 3.50 \times 10^{-3} T$ (*82*)
$H_T - H_{298} = -1{,}718 + 5.24T + 1.75 \times 10^{-3} T^2$
$F_T - H_{298} = -1{,}718 - 5.24 T \ln T - 1.75 \times 10^{-3} T^2 + 26.13T$

Zone II (β) (673°–1,124° K.)

$C_p = 6.29 + 1.40 \times 10^{-3} T$ (*82*)
$H_T - H_{298} = -1{,}834 + 6.29T + 0.70 \times 10^{-3} T^2$
$F_T - H_{298} = -1{,}834 - 6.29 T \ln T - 0.70 \times 10^{-3} T^2 + 32.49T$

T, ° K.	$H_T - H_{298}$	S_T	$-\frac{(F_T - H_{298})}{T}$
298	....	9. 95	9. 95
400	650	11. 82	10. 20
500	1, 330	13. 34	10. 88
600	2, 060	14. 67	11. 23
700	2, 910	15. 97	11. 81
800	3, 650	16. 96	12. 41
900	4, 390	17. 83	12. 94
1,000	5, 160	18. 64	13. 48
1,100	5, 930	19. 38	13. 97
1,200	(8, 880)	(21. 88)	(14. 47)
1,300	(9, 630)	(22. 58)	(15. 18)
1,400	(10, 380)	(23. 08)	(15. 58)
1,500	(11, 210)	(23. 69)	(16. 20)

Calcium Oxide, CaO (*c*)

$\Delta H^\circ_{298} = -151{,}790$ calories per mole (*57*)
$S_{298} = 9.5$ *e.u.* (*83*)
M.P. = 2,873° K. (*112*)
$\Delta H_M = 12{,}000$ calories per mole
B.P. = 3,800° K. (*94*)

Zone I (*c*) (298°–1,800° K.)

$C_p = 11.67 + 1.08 \times 10^{-3} T - 1.56 \times 10^5 T^{-2}$ (*82*)
$H_T - H_{298} = -4{,}050 + 11.67T + 0.54 \times 10^{-3} T^2 + 1.56 \times 10^5 T^{-1}$

Formation: $Ca + 1/2O_2 \longrightarrow CaO$

Zone I (298° − 673° K.)

$\Delta C_p = 2.85 - 2.92 \times 10^{-3} T - 1.36 \times 10^5 T^{-2}$
$\Delta H_T = -152{,}950 + 2.85T - 1.46 \times 10^{-3} T^2 + 1.36 \times 10^5 T^{-1}$
$\Delta F_T = -152{,}950 - 2.85 T \ln T + 1.46 \times 10^{-3} T^2 + 0.68 \times 10^5 T^{-1} + 43.87T$

Zone II (673°–1,124° K.)

$\Delta C_p = 1.80 - 0.82 \times 10^{-3} T - 1.36 \times 10^5 T^{-2}$
$\Delta H_T = -152{,}850 + 1.80T - 0.41 \times 10^{-3} T^2 + 1.36 \times 10^5 T^{-1}$
$\Delta F_T = -152{,}850 - 1.80 T \ln T + 0.41 \times 10^{-3} T^2 + 0.68 \times 10^5 T^{-1} + 37.57T$

T, ° K.	$H_T - H_{298}$	S_T	ΔH°_T	ΔF°_T
298	....	9. 5	−151, 790	−144, 350
400	1, 100	12. 67	−151, 700	−141, 850
500	2, 230	15. 19	−151, 650	−139, 400
600	3, 400	17. 32	−151, 550	−136, 950
700	4, 600	19. 17	−151, 600	−134, 500
800	5, 820	20. 80	−151, 500	−132, 050
900	7, 040	22. 23	−151, 450	−129, 650
1,000	8, 270	23. 53	−151, 400	−127, 200
1,100	9, 520	24. 72	−151, 300	−124, 700
1,200	10, 800	25. 84	(−153, 400)	(−122, 400)
1,300	12, 110	26. 88	(−153, 300)	(−119, 900)
1,400	13, 430	27. 86	(−153, 200)	(−117, 250)
1,500	14, 760	28. 78	(−153, 100)	(−114, 500)

Calcium Dioxide, CaO_2 (*c*)

$\Delta H^\circ_{298} = (-156{,}500)$ calories per mole (*24*)
$S_{298} = (15.4)$ *e.u.* (*24*)
Decomposes = 548° K. (*8*)

(estimated (*24*))

T, ° K.	$H_T - H_{298}$	ΔH°_T	ΔF°_T
298	....	(−156, 500)	(−143, 500)
400	(1, 875)	(−156, 000)	(−139, 500)
500	(3, 800)	(−155, 500)	(−135, 000)

Calcium Difluoride, CaF_2 (*c*)

$\Delta H^\circ_{298} = -290{,}200$ calories per mole (*112*)
$S_{298} = 16.48$ *e.u.* (*134*)
T.P. = 1,424° K. (*82*)
$\Delta H_T = 1{,}140$ calories per mole
M.P. = 1,691° K. (*82*)
$\Delta H_M = 6{,}780$ calories per mole
B.P. = 2,145° K. (*112*)
$\Delta H_V = 83{,}000$ calories per mole

Zone I (α) (298°–1,424° K.)

$C_p = 14.30 + 7.28 \times 10^{-3} T + 0.47 \times 10^5 T^{-2}$ (*82*)
$H_T - H_{298} = -4{,}400 + 14.30T + 3.64 \times 10^{-3} T^2 - 0.47 \times 10^5 T^{-1}$

Zone II (β) (1,424°–1,691° K.)

$$C_p = 25.81 + 2.50\times10^{-3}T \ (82)$$
$$H_T - H_{298} = -14{,}900 + 25.81T + 1.25\times10^{-3}T^2$$

Zone III (*l*) (1,691°–1,800° K.)

$$C_p = 23.90 \ (82)$$
$$H_T - H_{298} = -1{,}000 + 23.90T$$

Formation: $Ca + F_2 \longrightarrow CaF_2$

Zone I (298°–673° K.)

$$\Delta C_p = 0.77 + 3.34\times10^{-3}T + 1.27\times10^5T^{-2}$$
$$\Delta H_T = -290{,}150 + 0.77T + 1.67\times10^{-3}T^2 - 1.27\times10^5T^{-1}$$
$$\Delta F_T = -290{,}150 - 0.77T\ln T - 1.67\times10^{-3}T^2 - 0.63\times10^5T^{-1} + 47.48T$$

Zone II (673°–1,124° K.)

$$\Delta C_p = -0.28 + 5.44\times10^{-3}T + 1.27\times10^5T^{-2}$$
$$\Delta H_T = -290{,}010 - 0.28T + 2.72\times10^{-3}T^2 - 1.27\times10^5T^{-1}$$
$$\Delta F_T = -290{,}010 + 0.28T\ln T - 2.72\times10^{-3}T^2 - 0.63\times10^5T^{-1} + 41.11T$$

T, ° K.	H_T-H_{298}	S_T	ΔH_T°	ΔF_T°
298	....	16.48	−290,200	−277,700
400	1,760	21.54	−289,900	−273,500
500	3,540	25.52	−289,600	−269,400
600	5,400	28.91	−289,300	−265,300
700	7,320	31.87	−289,050	−261,400
800	9,280	34.49	−288,700	−257,500
900	11,300	36.86	−288,250	−253,500
1,000	13,380	39.06	−287,850	−249,750
1,100	15,550	41.12	−287,300	−246,000
1,200	17,850	43.12	(−288,750)	(−242,150)
1,300	20,230	45.08	(−287,800)	(−238,300)
1,400	22,680	46.84	(−286,850)	(−234,600)
1,500	26,660	49.60	(−284,990)	(−230,900)

Calcium Dichloride, $CaCl_2$ (*c*)

$\Delta H^\circ_{298} = -190{,}400$ calories per mole (*94*)
$S_{298} = 27.2$ *e.u.* (*83*)
M.P. = 1,055° K. (*82*)
$\Delta H_M = 6{,}780$ calories per mole
B.P. = (2,300°) K. (*6*)
$\Delta H_V = (55{,}000)$ calories per mole

Zone I (*c*) (298°–1,055° K.)

$$C_p = 17.18 + 3.04\times10^{-3}T - 0.60\times10^5T^{-2} \ (83)$$
$$H_T - H_{298} = -5{,}460 + 17.18T + 1.52\times10^{-3}T^2 + 0.60\times10^5T^{-1}$$

Zone II (*l*) (1,055°–1,700° K.)

$$C_p = 24.70 \ (83)$$
$$H_T - H_{298} = -4{,}880 + 24.70T$$

Formation: $Ca + Cl_2 \longrightarrow CaCl_2$

Zone I (298°–673° K.)

$$\Delta C_p = 3.12 - 0.52\times10^{-3}T + 0.08\times10^5T^{-2}$$
$$\Delta H_T = -191{,}280 + 3.12T - 0.26\times10^{-3}T^2 - 0.08\times10^5T^{-1}$$
$$\Delta F_T = -191{,}280 - 3.12T\ln T + 0.26\times10^{-3}T^2 - 0.04\times10^5T^{-1} + 56.41T$$

Zone II (673°–1,055° K.)

$$\Delta C_p = 2.07 + 1.58\times10^{-3}T + 0.08\times10^5T^{-2}$$
$$\Delta H_T = -191{,}150 + 2.07T + 0.79\times10^{-3}T^2 - 0.08\times10^5T^{-1}$$
$$\Delta F_T = -191{,}150 - 2.07T\ln T - 0.79\times10^{-3}T^2 - 0.04\times10^5T^{-1} + 50.32T$$

Zone III (1,055°–1,124° K.)

$$\Delta C_p = 9.59 - 1.46\times10^{-3}T - 0.68\times10^5T^{-2}$$
$$\Delta H_T = -190{,}500 + 9.59T - 0.73\times10^{-3}T^2 + 0.68\times10^5T^{-1}$$
$$\Delta F_T = -190{,}500 - 9.59T\ln T + 0.73\times10^{-3}T^2 + 0.34\times10^5T^{-1} + 100.69T$$

T, ° K.	H_T-H_{298}	S_T	ΔH_T°	ΔF_T°
298	....	27.2	−190,400	−179,650
400	1,850	32.53	−190,050	−175,950
500	3,700	36.66	−189,700	−172,500
600	5,540	40.02	−189,450	−169,050
700	7,400	42.88	−189,350	−165,750
800	9,290	45.4	−189,050	−162,350
900	11,230	47.69	−188,750	−159,150
1,000	13,270	49.84	−188,350	−155,700
1,100	22,340	58.44	−180,800	−152,800
1,200	24,840	60.62	(−182,300)	(−150,300)
1,300	27,320	62.6	(−181,500)	(−147,700)
1,400	29,780	64.42	(−180,600)	(−145,100)
1,500	32,210	66.10	(−179,770)	(−142,400)

Calcium Dibromide, $CaBr_2$ (*c*)

$\Delta H^\circ_{298} = -161{,}300$ calories per mole (*114*)
$S_{298} = (31)$ *e.u.* (*114*)
M.P. = 1,033° K. (*6*)
$\Delta H_M = 4{,}180$ calories per mole
B.P. = (2,100°) K. (*6*)
$\Delta H_V = (50{,}000)$ calories per mole

Formation: $Ca + Br_2 \longrightarrow CaBr_2$
(estimated (*11*))

T, ° K.	H_T-H_{298}	ΔH_T°	ΔF_T°
298	....	−161,300	(−157,500)
500	(3,500)	(−168,700)	(−150,350)
1,000	(13,900)	(−166,600)	(−133,500)
1,500	(30,100)	(−160,900)	(−119,500)

Calcium Diiodide, CaI_2 (*c*)

$\Delta H^\circ_{298} = -127{,}500$ calories per mole (*112*)
$S_{298} = (34)$ *e.u.* (*112*)
M.P. = 1,013° K. (*6*)
$\Delta H_M = (5{,}000)$ calories per mole
B.P. = (1,500°) K. (*6*)
$\Delta H_V = (35{,}000)$ calories per mole

Formation: $Ca + I_2 \longrightarrow CaI_2$
(estimated (*11*))

T, ° K.	H_T-H_{298}	ΔH_T°	ΔF_T°
298	....	−127,500	(−126,400)
500	(3,900)	(−141,600)	(−124,000)
1,000	(14,100)	(−139,700)	(−106,000)
1,500	(31,100)	(−132,200)	(−91,000)

Calcium Dicarbide, CaC_2 (*c*)

$\Delta H^\circ_{298} = -15{,}000$ calories per mole (*112*)
$S_{298} = 16.8$ *e.u.* (*83*)
$T.P. = 720°$ K. (*82*)
$\Delta H_T = 1{,}330$ calories per mole
$M.P. = 2{,}573°$ K. (*9*)

Zone I (α) (298°–720° K.)

$$C_p = 16.40 + 2.84 \times 10^{-3} T - 2.07 \times 10^5 T^{-2} \ (82)$$
$$H_T - H_{298} = -5{,}700 + 16.40T + 1.42 \times 10^{-3} T^2 + 2.07 \times 10^5 T^{-1}$$

Zone II (β) (720°–1,300° K.)

$$C_p = 15.40 + 2.00 \times 10^{-3} T \ (82)$$
$$H_T - H_{298} = -3{,}150 + 15.40T + 1.00 \times 10^{-3} T^2$$

Formation: $Ca + 2C \longrightarrow CaC_2$

Zone I (298°–673° K.)

$$\Delta C_p = 2.96 - 2.7 \times 10^{-3} T + 2.13 \times 10^5 T^{-2}$$
$$\Delta H_T = -15{,}000 + 2.96T - 1.35 \times 10^{-3} T^2 - 2.13 \times 10^5 T^{-1}$$
$$\Delta F_T = -15{,}000 - 2.96 T \ln T + 1.35 \times 10^{-3} T^2 - 1.07 \times 10^5 T^{-1} + 13.72T$$

Zone II (673°–720° K.)

$$\Delta C_p = 1.91 - 0.60 \times 10^{-3} T + 2.13 \times 10^5 T^{-2}$$
$$\Delta H_T = -14{,}700 + 1.91T - 0.30 \times 10^{-3} T^2 - 2.13 \times 10^5 T^{-1}$$
$$\Delta F_T = -14{,}700 - 1.91 T \ln T + 0.30 \times 10^{-3} T^2 - 1.06 \times 10^5 T^{-1} + 7.02T$$

Zone III (720°–1,124° K.)

$$\Delta C_p = 0.91 - 1.44 \times 10^{-3} T + 4.20 \times 10^5 T^{-2}$$
$$\Delta H_T = -12{,}320 + 0.91T - 0.72 \times 10^{-3} T^2 - 4.20 \times 10^5 T^{-1}$$
$$\Delta F_T = -12{,}320 - 0.91 T \ln T + 0.72 \times 10^{-3} T^2 - 2.10 \times 10^5 T^{-1} - 3.17T$$

T ° K.	$H_T - H_{298}$	S_T	ΔH°_f	ΔF°_f
298		16.8	−15,000	−16,200
400	1,600	21.41	−14,550	−16,700
500	3,260	25.11	−14,200	−17,300
600	5,000	28.27	−13,950	−17,950
700	6,760	30.99	−13,900	−18,600
800	9,790	35.12	−12,500	−19,450
900	11,510	37.14	−12,500	−20,300
1,000	13,250	38.98	−12,550	−21,200
1,100	15,010	40.65	−12,600	−22,100
1,200	16,780	42.19	(−14,800)	(−22,900)

Tricalcium Dinitride, Ca_3N_2 (*c*)

$\Delta H^\circ_{298} = -108{,}200$ calories per mole (*9*)
$S_{298} = 25.4$ *e.u.* (*9*)
$M.P. = 1{,}468°$ K. (*112*)

Zone I (*c*) (298°–800° K.)

$$C_p = 20.44 + 22.00 \times 10^{-3} T \ (82)$$
$$H_T - H_{298} = -7{,}100 + 20.44T + 11.00 \times 10^{-3} T^2$$

Formation: $3Ca + N_2 \longrightarrow Ca_3N_2$

Zone I (298°–673° K.)

$$\Delta C_p = -1.94 + 10.48 \times 10^{-3} T$$
$$\Delta H_T = -108{,}100 - 1.94T + 5.24 \times 10^{-3} T^2$$
$$\Delta F_T = -108{,}100 + 1.94 T \ln T - 5.24 \times 10^{-3} T^2 + 40.46T$$

Zone II (673°–800° K.)

$$\Delta C_p = -5.09 + 16.78 \times 10^{-3} T$$
$$\Delta H_T = -107{,}670 - 5.09T + 8.39 \times 10^{-3} T^2$$
$$\Delta F_T = -107{,}670 + 5.09 T \ln T - 8.39 \times 10^{-3} T^2 + 21.28T$$

T, ° K.	$H_T - H_{298}$	S_T	ΔH°_f	ΔF°_f
298		25.4	−108,200	−93,200
400	2,850	33.61	−108,100	−88,100
500	5,900	40.41	−107,700	−83,200
600	9,150	46.33	−107,350	−78,350
700	12,650	51.72	−107,100	−73,550
800	16,300	56.59	−106,500	−68,750

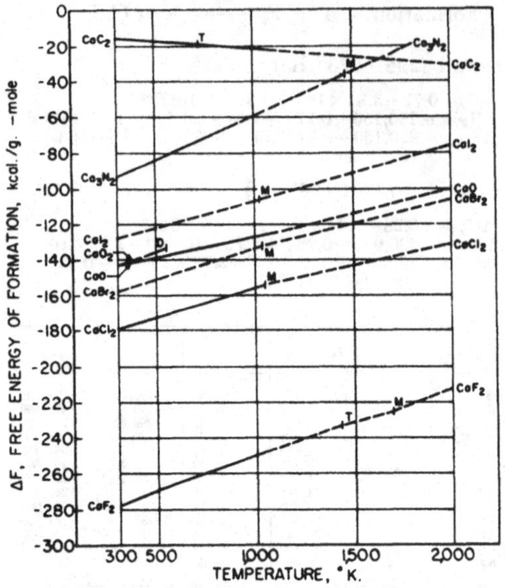

FIGURE 9.—Calcium.

CARBON AND ITS COMPOUNDS

Element, C (*c*)

$S_{298} = 1.366$ *e.u.* (*83*)
$S.P. = 4{,}620°$ K. (*130*)

Zone I (*c*) (298°–2,300° K.)

$$C_p = 4.10 + 1.02 \times 10^{-3} T - 2.10 \times 10^5 T^{-2} \ (82)$$
$$H_T - H_{298} = -1{,}972 + 4.10T + 0.51 \times 10^{-3} T^2 + 2.10 \times 10^5 T^{-1}$$
$$F_T - H_{298} = -1{,}972 - 4.10 T \ln T - 0.51 \times 10^{-3} T^2 + 1.05 \times 10^5 T^{-1} + 27.72T$$

T, ° K.	$H_T - H_{298}$	S_T	$-\frac{(F_T - H_{298})}{T}$
298		1.36	1.36
400	250	2.08	1.42
500	570	2.79	1.65
600	960	3.48	1.90
700	1,370	4.13	2.17
800	1,830	4.74	2.43
900	2,310	5.31	2.74
1,000	2,810	5.83	3.02
1,100	3,320	6.32	3.10
1,200	3,850	6.78	3.58
1,300	4,390	7.21	3.84
1,400	4,930	7.61	4.10
1,500	5,480	7.99	4.33
1,600	6,040	8.35	4.57
1,700	6,610	8.69	4.81
1,800	7,190	9.02	5.02
1,900	7,780	9.34	5.25
2,000	8,380	9.65	5.46

Carbon Monoxide, CO (*g*)

$\Delta H_{f298}^{\circ} = -26,416$ calories per mole (*112*)
$S_{298} = 47.31$ *e.u.* (*83*)
M.P. = 68.10° K. (*112*)
$\Delta H_M = 200$ calories per mole
B.P. = 81.66° K. (*112*)
$\Delta H_V = 1,444$ calories per mole

Zone I (*g*) (298°–2,500° K.)

$$C_p = 6.79 + 0.98 \times 10^{-3} T - 0.11 \times 10^5 T^{-2} \ (82)$$
$$H_T - H_{298} = -2,100 + 6.79T + 0.49 \times 10^{-3} T^2 + 0.11 \times 10^5 T^{-1}$$

Formation: $C + 1/2O_2 \longrightarrow CO$

Zone I (298°–2,000° K.)

$$\Delta C_p = -0.89 - 0.54 \times 10^{-3} T + 2.19 \times 10^5 T^{-2}$$
$$\Delta H_T = -25,380 - 0.89T - 0.27 \times 10^{-3} T^2 - 2.19 \times 10^5 T^{-1}$$
$$\Delta F_T = -25,380 + 0.89 T \ln T + 0.27 \times 10^{-3} T^2 - 1.10 \times 10^5 T^{-1} - 28.84T$$

T, ° K.	$H_T - H_{298}$	S_T	ΔH_T°	ΔF_T°
298	....	47.31	−26,400	−32,800
400	711	49.36	−26,300	−35,000
500	1,418	50.94	−26,200	−37,100
600	2,137	52.25	−26,350	−39,350
700	2,874	53.38	−26,400	−41,550
800	3,628	54.39	−26,500	−43,700
900	4,400	55.30	−26,600	−45,850
1,000	5,186	56.13	−26,750	−47,950
1,100	5,960	56.94	−26,900	−50,100
1,200	6,798	57.59	−27,000	−52,150
1,300	7,460	58.23	−27,300	−54,350
1,400	8,370	58.83	−27,350	−56,250
1,500	9,291	59.45	−27,450	−58,400
1,600	10,020	60.03	−27,650	−60,600
1,700	10,850	60.53	−27,850	−62,650
1,800	11,700	60.92	−28,000	−64,650
1,900	12,580	61.42	−28,250	−66,750
2,000	13,570	61.91	−28,450	−68,750

Carbon Dioxide, CO_2 (*g*)

$\Delta H_{f298}^{\circ} = -94,052$ calories per mole (*112*)
$S_{298} = 51.05$ *e.u.* (*83*)
S.P. = 194.7° K. (*112*)
$\Delta H_{subl} = 6,031$ calories per mole

Zone I (*g*) (298°–2,500° K.)

$$C_p = 10.55 + 2.16 \times 10^{-3} T - 2.04 \times 10^5 T^{-2} \ (82)$$
$$H_T - H_{298} = -3,926 + 10.55T + 1.08 \times 10^{-3} T^2 + 2.04 \times 10^5 T^{-1}$$

Formation: $C + O_2 \longrightarrow CO_2$

Zone I (298°–2,000° K.)

$$\Delta C_p = -0.71 + 0.14 \times 10^{-3} T + 0.46 \times 10^5 T^{-2}$$
$$\Delta H_T = -93,650 - 0.71T + 0.07 \times 10^{-3} T^2 - 0.46 \times 10^5 T^{-1}$$
$$\Delta F_T = -93,650 + 0.71 T \ln T - 0.07 \times 10^{-3} T^2 - 0.23 \times 10^5 T^{-1} - 5.56T$$

T, ° K.	$H_T - H_{298}$	S_T	ΔH_T°	ΔF_T°
298	....	51.05	−94,050	−94,250
400	958	53.76	−94,050	−94,300
500	1,987	56.10	−94,100	−94,400
600	3,088	58.11	−94,150	−94,450
700	4,248	59.89	−94,150	−94,500
800	5,458	61.51	−94,200	−94,500
900	6,708	62.98	−94,250	−94,550
1,000	7,993	64.33	−94,400	−94,600
1,100	9,308	65.59	−94,250	−94,600
1,200	10,650	66.75	−94,300	−94,650
1,300	12,010	67.84	−94,300	−94,700
1,400	13,380	68.86	−94,300	−94,750
1,500	14,780	69.82	−94,400	−94,750
1,600	15,850	70.39	−94,700	−94,800
1,700	17,240	71.34	−94,750	−94,850
1,800	18,690	72.09	−94,800	−94,900
1,900	20,100	72.85	−94,850	−94,900
2,000	21,920	73.93	−94,850	−95,000

Carbon Tetrafluoride, CF_4 (*g*)

$\Delta H_{f298}^{\circ} = -162,500$ calories per mole (*106*)
$S_{298} = 62.8$ *e.u.* (*80*)
M.P. = 89.47° K. (*106*)
$\Delta H_M = 167$ calories per mole
B.P. = 145.14° K. (*106*)
$\Delta H_V = 3,010$ calories per mole

Zone I (*g*) (298°–1,200° K.)

$$C_p = 16.64 + 7.84 \times 10^{-3} T - 4.00 \times 10^5 T^{-2} \ (79)$$
$$H_T - H_{298} = -6,650 + 16.64T + 3.92 \times 10^{-3} T^2 + 4.00 \times 10^5 T^{-1}$$

Formation: $C + 2F_2 \longrightarrow CF_4$

Zone I (298°–1,200° K.)

$$\Delta C_p = -4.04 + 5.94 \times 10^{-3} T - 0.30 \times 10^5 T^{-2}$$
$$\Delta H_T = -161,700 - 4.04T + 2.97 \times 10^{-3} T^2 + 0.30 \times 10^5 T^{-1}$$
$$\Delta F_T = -161,700 + 4.04 T \ln T - 2.97 \times 10^{-3} T^2 + 0.15 \times 10^5 T^{-1} + 10.8T$$

T, ° K.	$H_T - H_{298}$	S_T	ΔH_T°	ΔF_T°
298	....	62.8	−162,500	−151,850
400	1,615	67.47	−162,600	−148,100
500	3,430	71.49	−162,700	−144,450
600	5,410	75.10	−162,900	−140,900
700	7,520	78.35	−162,900	−137,200
800	9,720	81.28	−162,850	−133,550
900	11,995	83.96	−162,800	−129,900
1,000	14,315	86.41	−162,700	−126,250
1,100	16,760	88.72	−162,500	−122,900
1,200	19,085	90.75	−162,500	−119,000
1,500	(27,400)	....	(−160,500)	(−107,700)
2,000	(42,650)	....	(−156,900)	(−85,100)

Carbon Tetrachloride, CCl_4 (*l*)

$\Delta H_{f298}^{\circ} = -33,200$ calories per mole (*112*)
$S_{298} = 51.3$ *e.u.* (*83*)
M.P. = 249.1° K. (*112*)
$\Delta H_M = 644$ calories per mole
B.P. = 350° K. (*112*)
$\Delta H_V = 7,283$ calories per mole

647940 O—63——3

Zone I (g) (350°–1,000° K.)

$$C_p = 23.34 + 2.30\times10^{-3}T - 3.60\times10^{5}T^{-2} \quad (83)$$
$$H_T - H_{298} = -1{,}560 + 23.34T + 1.15\times10^{-3}T^2 + 3.60\times10^{5}T^{-1}$$

Formation: $C + 2Cl_2 \longrightarrow CCl_4$

Zone I (350°–1,000° K.)

$$\Delta C_p = 1.6 + 1.16\times10^{-3}T - -0.14\times10^{5}T^{-2}$$
$$\Delta H_T = -27{,}020 + 1.6T + 0.58\times10^{-3}T^2 + 0.14\times10^{5}T^{-1}$$
$$\Delta F_T = -27{,}020 - 1.6T\ln T - 0.58\times10^{-3}T^2 + 0.07\times10^{5}T^{-1} + 45.44T$$

T, ° K.	$H_T - H_{298}$	S_T	ΔH_T°	ΔF_T°
298	----------	51.3	−33,200	−16,300
400	8,845	79.98	−28,300	−12,750
500	11,100	84.99	−26,050	−9,400
600	13,450	89.29	−25,800	−6,150
700	15,860	92.99	−25,550	−2,960
800	18,310	96.26	−25,300	+450
900	20,790	99.18	−25,050	+3,500
1,000	23,290	101.82	−24,850	+6,850
1,500	(36,250)	----------	(−23,600)	(+22,300)
2,000	(49,900)	----------	(−22,350)	(+37,300)

Carbonyl Chloride (Phosgene), $COCl_2$ (g)

$\Delta H^\circ_{298} = -53{,}300$ calories per mole (*112*)
$S_{298} = 69.13$ *e.u.* (*38*)
M.P. = 145.34° K. (*38*)
$\Delta H_M = 1{,}371$ calories per mole
B.P. = 280.7° K. (*38*)
$\Delta H_V = 5{,}825$ calories per mole

Zone I (g) (298°–1,000° K.)

$$C_p = 15.60 + 3.46\times10^{-3}T - 1.91\times10^{5}T^{-2} \quad (82)$$
$$H_T - H_{298} = -5{,}446 + 15.60T + 1.73\times10^{-3}T^2 + 1.91\times10^{5}T^{-1}$$

Formation: $C + 1/2O_2 + Cl_2 \longrightarrow COCl_2$

Zone I (298°–1,000° K.)

$$\Delta C_p = -0.90 + 1.88\times10^{-3}T + 1.07\times10^{5}T^{-2}$$
$$\Delta H_T = -52{,}700 - 0.90T + 0.94\times10^{-3}T^2 - 1.07\times10^{5}T^{-1}$$
$$\Delta F_T = -52{,}700 + 0.90T\ln T - 0.94\times10^{-3}T^2 - 0.54\times10^{5}T^{-1} + 3.92T$$

T, ° K.	$H_T - H_{298}$	S_T	ΔH_T°	ΔF_T°
298	----------	69.13	−53,300	−50,300
400	1,545	73.58	−53,200	−49,300
500	3,165	77.19	−53,100	−48,300
600	4,855	80.27	−53,050	−47,300
700	6,600	82.96	−53,000	−46,400
800	8,400	85.36	−52,900	−45,450
900	10,210	87.49	−52,900	−44,500
1,000	12,060	89.44	−52,800	−43,500
1,500	(22,000)	----------	(−51,800)	(−39,400)
2,000	(32,800)	----------	(−50,600)	(−35,400)

Carbon Tetrabromide, CBr_4 (c)

$\Delta H^\circ_{298} = (-500)$ calories per mole (*11*)
$S_{298} = (56)$ *e.u.* (*11*)
T.P. = 320° K. (*82*)
$\Delta H_T = 1{,}430$ calories per mole
M.P. = 363° K. (*82*)
$\Delta H_M = 950$ calories per mole
B.P. = 463° K. (*6*)
$\Delta H_V = (9{,}700)$ calories per mole

Zone I (α) (298°–320° K.)

$$C_p = 34.5 \quad (82)$$
$$H_T - H_{298} = -10{,}287 + 34.5T$$

Zone II (β) (320°–363° K.)

$$C_p = 43.0 \quad (82)$$
$$H_T - H_{298} = -11{,}580 + 43.0T$$

Zone III (l) (363°–463° K.)

$$C_p = 36.7 \quad (82)$$
$$H_T - H_{298} = -8{,}340 + 36.7T$$

Zone IV (g) (463°–1,000° K.)

$$C_p = 25.03 + 0.60\times10^{-3}T - 3.03\times10^{5}T^{-2} \quad (82)$$
$$H_T - H_{298} = 5{,}200 + 25.03T + 0.30\times10^{-3}T^2 + 3.03\times10^{5}T^{-1}$$

Formation: $C + 2Br_2 \longrightarrow CBr_4$

Zone I (298°–320° K.)

$$\Delta C_p' = -3.8 - 1.02\times10^{-3}T + 2.10\times10^{5}T^{-2}$$
$$\Delta H_T = 1{,}877 - 3.8T - 0.51\times10^{-3}T^2 - 2.10\times10^{5}T^{-1}$$
$$\Delta F_T = 1{,}877 + 3.8T\ln T + 0.51\times10^{-3}T^2 - 1.05\times10^{5}T^{-1} - 10.1T$$

Zone II (331°–363° K.)

$$\Delta C_p = 20.86 - 1.02\times10^{-3}T + 2.84\times10^{5}T^{-2}$$
$$\Delta H_T = -20{,}000 + 20.86T - 0.51\times10^{-3}T^2 - 2.84\times10^{5}T^{-1}$$
$$\Delta F_T = -20{,}000 - 20.86T\ln T + 0.51\times10^{-3}T^2 - 1.42\times10^{5}T^{-1} + 201.0T$$

Zone III (363°–463° K.)

$$\Delta C_p = -14.55 - 1.02\times10^{-3}T + 2.84\times10^{5}T^{-2}$$
$$\Delta H_T = -16{,}730 + 14.55T - 0.51\times10^{-3}T^2 - 2.84\times10^{5}T^{-1}$$
$$\Delta F_T = -16{,}730 - 14.55T\ln T + 0.51\times10^{-3}T^2 - 1.42\times10^{5}T^{-1} + 154.2T$$

Zone IV (463°–1,000° K.)

$$\Delta C_p = 2.89 - 0.42\times10^{-3}T - 0.19\times10^{5}T^{-2}$$
$$\Delta H_T = -3{,}310 + 2.89T - 0.21\times10^{-3}T^2 + 0.19\times10^{5}T^{-1}$$
$$\Delta F_T = -3{,}310 - 2.89T\ln T + 0.21\times10^{-3}T^2 + 0.1\times10^{5}T^{-1} + 52.9T$$

T, ° K.	$H_T - H_{298}$	S_T	ΔH_T°	ΔF_T°
298	----------	(56)	(−500)	(+5,000)
400	6,340	(74.02)	(−11,700)	(9,700)
500	18,400	(100.0)	(−1,840)	(15,400)
600	20,840	(104.7)	(−1,450)	(18,300)
700	23,300	(108.5)	(−1,200)	(20,700)
800	25,790	(111.7)	(−950)	(23,750)
900	28,310	(114.8)	(−700)	(26,700)
1,000	30,840	(117.4)	(−500)	(29,700)
1,500	(42,000)	----------	(+1,750)	(46,100)
2,000	(68,100)	----------	(2,800)	(62,500)

Carbon Tetraiodide, CI_4 (c)

$\Delta H^\circ_{298} = (39{,}700)$ calories per mole (*11*)
$S_{298} = (60)$ *e.u.* (*11*)
M.P. = 444° K. (*6*)
$\Delta H_M = (1{,}150)$ calories per mole
B.P. = (580°) K. (*6*)
$\Delta H_V = (12{,}000)$ calories per mole

Formation: $C + 2I_2 \longrightarrow CI_4$
(estimated (*11*))

T, ° K.	H_T-H_{298}	ΔH_T°	ΔF_T°
298		(39,700)	(29,800)
500	(10,000)	(30,100)	(40,500)

Cyanogen, C_2N_2 (*g*)

$\Delta H^\circ_{298}=73{,}600$ calories per mole (*112*)
$S_{298}=57.86$ *e.u.* (*112*)
M.P. $=245.3°$ K. (*112*)
$\Delta H_M=1{,}938$ calories per mole
B.P. $=252°$ K. (*112*)
$\Delta H_V=5{,}576$ calories per mole

Zone I (*g*) (298°–2,000° K.)

$$C_p=14.90+3.20\times10^{-3}T-2.04\times10^{5}T^{-2}\ (82)$$
$$H_T-H_{298}=-5{,}270+14.90T+1.60\times10^{-3}T^2+2.04\times10^{5}T^{-1}$$

Formation: $2C+N_2 \longrightarrow C_2N_2$

Zone I (298°–2,000° K.)

$$\Delta C_p=0.04+0.14\times10^{-3}T+2.16\times10^{5}T^{-2}$$
$$\Delta H_T=74{,}250+0.04T+0.07\times10^{-3}T^2-2.16\times10^{5}T^{-1}$$
$$\Delta F_T=74{,}250-0.04T\ln T-0.07\times10^{-3}T^2-1.08\times10^{5}T^{-1}-10.45T$$

T, ° K.	H_T-H_{298}	S_T	ΔH_T°	ΔF_T°
298		57.86	+73,600	+71,800
400	1,445	62.02	73,800	69,800
500	2,965	65.41	74,000	68,800
600	4,560	68.32	74,100	67,700
700	6,220	70.88	74,200	66,650
800	7,930	73.16	74,250	65,550
900	9,695	75.24	74,300	64,500
1,000	11,500	77.14	74,350	63,400
1,100	13,250	78.75	74,300	62,350
1,200	15,205	80.51	74,400	61,200
1,300	16,990	82.08	74,400	60,100
1,400	19,015	83.46	74,450	59,100
1,500	20,900	84.34	74,450	57,900
1,600	22,900	86.24	74,450	56,400
1,700	24,820	87.34	74,450	55,400
1,800	26,840	88.34	74,450	54,750
1,900	28,900	89.47	74,450	53,500
2,000	30,810	90.45	74,450	52,500

Cyanogen Chloride, CNCl (*g*)

$\Delta H^\circ_{298}=34{,}500$ calories per mole (*112*)
$S_{298}=56.31$ *e.u.* (*112*)
M.P. $=266.3°$ K. (*112*)
$\Delta H_M=2{,}720$ calories per mole
B.P. $=286.1°$ K. (*112*)
$\Delta H_V=6{,}290$ calories per mole

Zone I (*g*) (298°–2,000° K.)

$$C_p=11.88+1.64\times10^{-3}T-1.49\times10^{5}T^{-2}\ (82)$$
$$H_T-H_{298}=-4{,}115+11.88T+0.82\times10^{-3}T^2+1.49\times10^{5}T^{-1}$$

Formation: $C+1/2Cl_2+1/2N_2 \longrightarrow CNCl$

Zone I (298°–2,000° K.)

$$\Delta C_p=0.04+0.08\times10^{-3}T+0.95\times10^{5}T^{-2}$$
$$\Delta H_T=34{,}800+0.04T+0.04\times10^{-3}T^2-0.95\times10^{5}T^{-1}$$
$$\Delta F_T=34{,}800-0.04T\ln T-0.04\times10^{-3}T^2-0.47\times10^{5}T^{-1}-5.66T$$

T, ° K.	H_T-H_{298}	S_T	ΔH_T°	ΔF_T°
298		56.31	34,500	32,900
400	1,135	59.58	34,600	32,350
500	2,315	62.21	34,700	31,800
600	3,540	64.45	34,750	31,150
700	4,805	66.39	34,800	30,550
800	6,100	68.12	34,800	29,800
900	7,425	69.68	34,850	29,350
1,000	8,770	71,10	34,850	28,750
1,100	10,060	72,35	34,900	28,150
1,200	11,515	73.60	34,900	27,500
1,300	12,850	74.56	34,900	27,000
1,400	14,310	75.76	34,950	26,400
1,500	15,690	76.68	34,950	25,700
1,600	17,145	77.65	34,950	25,000
1,700	18,645	78.55	34,950	24,350
1,800	20,010	79.34	34,950	23,800
1,900	21,475	80.21	34,950	23,100
2,000	22,890	80.85	34,950	22,650

Cyanogen Bromide, CNBr (*l*)

$S_{298}=59.05$ *e.u.* (*112*)
S.P. $=334°$ K. (*112*)
$\Delta H_{subl}=11{,}300$ calories per mole

Zone I (*g*) (334°–2,000° K.)

$$C_p=12.20+1.42\times10^{-3}T-1.34\times10^{5}T^{-2}\ (82)$$
$$H_T-H_{298}=-4{,}150+12.20T+0.71\times10^{-3}T^2+1.34\times10^{5}T^{-1}$$

Formation: $C+1/2Br_2+1/2N_2 \longrightarrow CNBr$

T °, K.	H_T-H_{298}	S_T
298		59.05
400	1,175	52.44
500	2,380	55.13
600	3,630	57.40
700	4,910	59.37
800	6,220	61.12
900	7,550	62.69
1,000	8,910	64.12
1,200	11,665	66.63
1,400	14,475	68.79
1,600	17,310	70.69
1,800	20,185	72.38
2,000	23,095	73.91

Cyanogen Iodide, CNI (*c*)

$\Delta H^\circ_{298}=40{,}400$ calories per mole (*112*)
$S_{298}=30.8$ *e.u.* (*112*)
S.P. $=413°$ K. (*112*)
$\Delta H_{subl}=14{,}200$ calories per mole

Zone I (*g*) (413°–2,000° K.)

$$C_p=12.30+1.38\times10^{-3}T-1.04\times10^{5}T^{-2}\ (82)$$
$$H_T-H_{298}=10{,}100+12.30T+0.69\times10^{-3}T^2+1.04\times10^{5}T^{-1}$$

Formation: $C+1/2N_2+1/2I_2 \longrightarrow CNI$

Zone I (456°–1,500° K.)

$$\Delta C_p=0.43-0.15\times10^{-3}T+1.06\times10^{5}T^{-2}$$
$$\Delta H_T=47{,}450+0.43T-0.075\times10^{-3}T^2-1.06\times10^{5}T^{-1}$$
$$\Delta F_T=47{,}450-0.43T\ln T+0.075\times10^{-3}T^2-0.53\times10^{5}T^{-1}-7.0T$$

T, ° K.	H_T-H_{298}	S_T	ΔH°_T	ΔF°_T
298	....	30.80	40,400	42,600
400	1,210	34.07	38,400	43,550
500	16,640	71.10	47,450	42,400
600	17,910	73.34	47,500	41,300
700	19,205	75.28	47,600	40,450
800	20,530	77.01	47,650	39,450
900	21,875	78.57	47,650	38,400
1,000	23,235	79.91	47,700	37,500
1,100	24,570	81.25	47,700	36,400
1,200	26,005	82.49	47,750	35,400
1,300	27,370	83.51	47,750	34,500
1,400	28,825	84.65	47,800	33,400
1,500	30,180	85.58	47,800	32,400
2,000	37,440	89.74	(47,750)	(27,150)

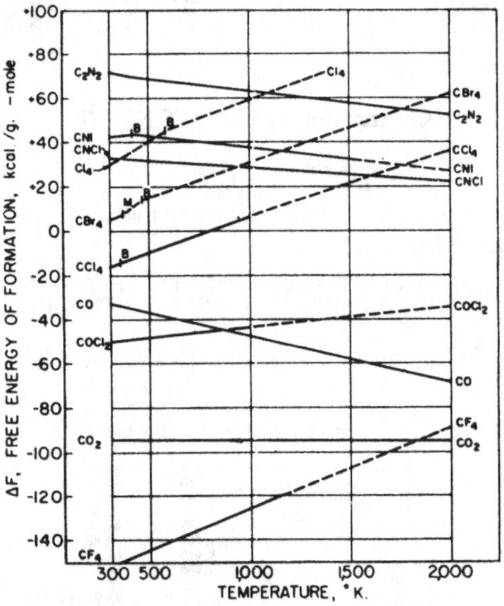

FIGURE 10.—Carbon.

CERIUM AND ITS COMPOUNDS

Element, Ce (c)

$S_{298}=13.64$ *e.u.* (*121*)
$M.P.=1{,}077°$ K. (*126*)
$\Delta H_M=2{,}120$ calories per atom (*112*)

Zone I (*c*) (298°–800° K.)

$C_p=4.40+6.00\times10^{-3}T$ (*82*)
$H_T-H_{298}=-1{,}575+4.40T+3.00\times10^{-3}T^2$
$F_T-H_{298}=-1{,}575-4.40T\ln T-3.00\times10^{-3}T^2+17.65T$

T, ° K.	H_T-H_{298}	S_T	$-\frac{(F_T-H_{298})}{T}$
298	....	13.64	13.64
400	670	15.57	13.87
500	1,380	17.15	14.38
600	2,140	18.54	14.97
700	2,970	19.82	15.57
800	3,860	21.00	16.16
900	(4,810)	(22.05)	(16.78)
1,000	(5,820)	(23.15)	(17.33)
1,500	(12,350)	(29.00)	(20.7)
2,000	(16,350)	(31.3)	(23.1)

Dicerium Trioxide, Ce_2O_3 (c)

$\Delta H^\circ_{298}=(-435{,}000)$ calories per mole (*24*)
$S_{298}=(21.81)$ *e.u.* (*24*)
$M.P.=1{,}960°$ K. (*42*)

Formation: $2Ce+3/2O_2 \longrightarrow Ce_2O_3$
(estimated (*24*))

T, ° K.	H_T-H_{298}	ΔH°_T	ΔF°_T
298	....	(−435,000)	(−411,500)
400	(2,400)	(−435,000)	(−403,500)
500	(5,500)	(−434,500)	(−395,500)
600	(8,100)	(−434,500)	(−387,500)
700	(11,400)	(−434,000)	(−380,000)
800	(14,400)	(−434,000)	(−372,000)
900	(17,500)	(−434,000)	(−364,500)
1,000	(21,300)	(−433,500)	(−356,500)
1,100	....	(−438,000)	(−348,500)
1,200	....	(−438,000)	(−340,500)
1,300	....	(−438,000)	(−332,500)
1,400	....	(−437,500)	(−324,500)
1,500	....	(−437,500)	(−316,000)
1,600	....	(−437,000)	(−308,000)
1,700	....	(−437,000)	(−300,000)
1,800	....	(−437,000)	(−292,000)
1,900	....	(−436,500)	(−284,000)

Cerium Dioxide, CeO_2 (c)

$\Delta H^\circ_{298}=-260{,}180$ calories per mole (*58*)
$S_{298}=14.88$ *e.u.* (*24*)
$M.P.=>2{,}873°$ K. (*42*)

Zone I (*c*) (298°–2,500° K.)

$C_p=15.0+2.5\times10^{-3}T$ (*94*)
$H_T-H_{298}=-4{,}580+15.0T+1.25\times10^{-3}T^2$

Formation: $Ce+O_2 \longrightarrow CeO_2$

Zone I (298°–800° K.)

$\Delta C_p=3.44-4.50\times10^{-3}T+0.40\times10^5T^{-2}$
$\Delta H_T=-259{,}500+3.44T-2.25\times10^{-3}T^2-0.40\times10^5T^{-1}$
$\Delta F_T=-259{,}500-3.44T\ln T+2.25\times10^{-3}T^2-0.20\times10^5T^{-1}+69.25T$

T, ° K.	H_T-H_{298}	S_T	ΔH°_T	ΔF°_T
298	....	14.88	−260,180	−245,940
400	1,620	19.53	−259,950	−241,250
500	3,230	23.28	−259,780	−236,480
600	4,870	26.23	−259,660	−231,360
700	6,530	28.68	−259,610	−227,110
800	8,220	30.93	−259,600	−222,460
900	9,930	32.98	(−259,600)	(−217,950)
1,000	11,680	34.83	(−259,550)	(−213,200)
1,500	20,730	42.28	(−261,500)	(−189,500)
2,000	30,400	47.73	(261,000)	(−161,000)

Cerium Trifluoride, CeF_3 (c)

$\Delta H^\circ_{298}=-391{,}000$ calories per mole (*5*)
$S_{298}=(24)$ *e.u.* (*11*)
$M.P.=(1{,}703°)$ K. (*29*)
$\Delta H_M=(9{,}000)$ calories per mole
$B.P.=(2{,}600°)$ K. (*6*)
$\Delta H_V=(62{,}000)$ calories per mole

Formation: $Ce+3/2F_2 \longrightarrow CeF_3$
(estimated (*11*))

T, °K.	H_T-H_{298}	ΔH°_T	ΔF°_T
298	....	−391,000	(−372,100)
500	(4,000)	(−391,000)	(−360,500)
1,000	(17,000)	(−388,500)	(−330,000)
1,500	(32,000)	(−386,500)	(−307,000)

Cerium Tetrafluoride, CeF_4 (*c*)

$\Delta H^\circ_{298} = -442{,}000$ calories per mole (*11*)
$S_{298} = (37)$ *e.u.* (*11*)
M.P. = (1,250°) K. (*6*)
$\Delta H_M = (10{,}000)$ calories per mole

Formation: $Ce + 2F_2 \longrightarrow CeF_4$
(estimated (*11*))

T, °K.	H_T-H_{298}	ΔH°_f	ΔF°_f
298	....	−442,000	(−420,000)
500	(6,000)	(−440,500)	(−436,000)
1,000	(23,000)	(−436,500)	(−419,000)
1,500	(52,000)	....	(−390,000)

Cerium Trichloride, $CeCl_3$ (*c*)

$\Delta H^\circ_{298} = -252{,}840$ calories per mole (*128*)
$S_{298} = 34.5$ *e.u.* (*128*)
M.P. = 1,095° K. (*29*)
$\Delta H_M = (8{,}000)$ calories per mole
B.P. = (2,000°) K. (*6*)
$\Delta H_V = (46{,}000)$ calories per mole

Formation: $Ce + 3/2Cl_2 \longrightarrow CeCl_3$
(estimated (*11*))

T, °K.	H_T-H_{298}	ΔH°_f	ΔF°_f
298	....	−252,840	−235,160
500	(5,000)	(−251,500)	(−225,300)
1,000	(19,000)	(−248,500)	(−198,800)
1,500	(43,000)	(−246,500)	(−179,300)

Cerium Tribromide, $CeBr_3$ (*c*)

$\Delta H^\circ_{298} = -192{,}000$ calories per mole (*5*)
$S_{298} = (45)$ *e.u.* (*11*)
M.P. = 1,005° K. (*29*)
$\Delta H_M = (8{,}000)$ calories per mole
B.P. = (1,830°) K. (*6*)
$\Delta H_V = (44{,}000)$ calories per mole

Formation: $Ce + 3/2Br_2 \longrightarrow CeBr_3$
(estimated (*11*))

T, °K.	H_T-H_{298}	ΔH°_f	ΔF°_f
298	....	−192,000	(−185,000)
500	(5,000)	(−202,500)	(−174,000)
1,000	(18,000)	(−200,500)	(−148,000)
1,500	(43,000)	(−189,000)	(−127,000)

Cerium Triiodide, CeI_3 (*c*)

$\Delta H^\circ_{298} = -163{,}000$ calories per mole (*5*)
$S_{298} = (50)$ *e.u.* (*11*)
M.P. = 1,038° K. (*29*)
$\Delta H_M = (8{,}000)$ calories per mole
B.P. = (1,670°) K. (*6*)
$\Delta H_V = (40{,}000)$ calories per mole

Formation: $Ce + 3/2I_2 \longrightarrow CeI_3$
(estimated (*11*))

T, °K.	H_T-H_{298}	ΔH°_f	ΔF°_f
298	....	−163,000	(−161,000)
500	(5,000)	(−184,500)	(−156,500)
1,000	(19,000)	(−181,500)	(−130,000)
1,500	(44,000)	(−169,700)	(−107,000)

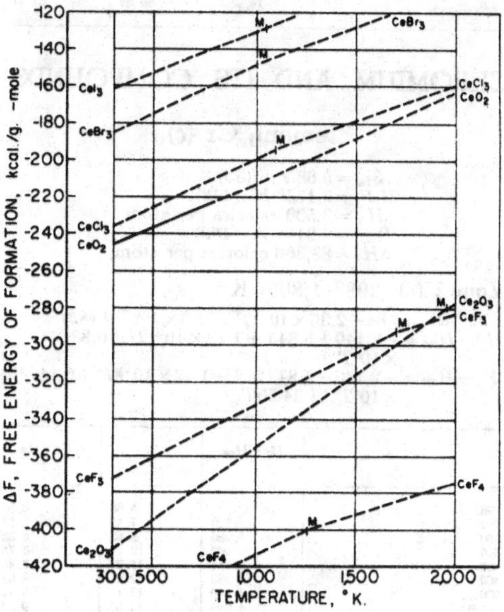

FIGURE 11.—Cerium.

CHLORINE

Element, Cl_2 (*g*)

$S_{298} = 53.31$ *e.u.* (*83*)
M.P. = 172.16° K. (*112*)
$\Delta H_M = 1{,}531$ calories per atom
B.P. = 239.1° K. (*112*)
$\Delta H_V = 4{,}878$ calories per atom

Zone I (g) (298°–3,000° K.)

$$C_p=8.82+0.06\times10^{-3}T-0.68\times10^{5}T^{-2}\ (82)$$
$$H_T-H_{298}=-2{,}861+8.82T+0.03\times10^{-3}T^2+0.68\times10^{5}T^{-1}$$
$$F_T-H_{298}=-2{,}861-8.82T\ln T-0.03\times10^{-3}T^2+0.34\times10^{5}T^{-1}+6.06T$$

T, ° K.	H_T-H_{298}	S_T	$-\frac{(F_T-H_{298})}{T}$
298	....	53.31	53.31
400	840	55.89	53.79
500	1,688	57.75	54.33
600	2,544	59.29	55.05
700	3,420	60.57	55.68
800	4,296	61.86	56.39
900	5,176	62.80	57.05
1,000	6,059	63.84	57.74
1,100	6,813	64.56	58.36
1,200	7,830	65.35	58.82
1,300	8,618	66.06	59.41
1,400	9,606	66.76	59.90
1,500	10,372	67.37	60.46
1,600	11,385	67.87	60.75
1,700	12,187	68.47	61.28
1,800	13,165	68.88	61.57
1,900	13,981	69.38	62.02
2,000	14,950	69.89	62.41

CHROMIUM AND ITS COMPOUNDS

Element, Cr (c)

$S_{298}=5.68$ *e.u.* (*83*)
$M.P.=2{,}173°$ K. (*112*)
$\Delta H_M=3{,}500$ calories per atom
$B.P.=2.915°$ K. (*130*)
$\Delta H_V=83{,}360$ calories per atom

Zone I (c) (298°–1,800° K.)

$$C_p=5.84+2.36\times10^{-3}T-0.88\times10^{5}T^{-2}\ (82)$$
$$H_T-H_{298}=-2{,}140+5.84T+1.18\times10^{-3}T^2+0.88\times10^{5}T^{-1}$$
$$F_T-H_{298}=-2{,}140-5.84T\ln T-1.18\times10^{-3}T^2+0.44\times10^{5}T^{-1}+34.56T$$

T, ° K.	H_T-H_{298}	S_T	$-\frac{(F_T-H_{298})}{T}$
298	....	5.68	5.68
400	620	7.46	5.75
500	1,280	8.93	6.38
600	1,960	10.17	6.75
700	2,660	11.25	7.45
800	3,380	12.21	8.01
900	4,140	13.11	8.51
1,000	4,940	13.95	9.04
1,100	5,770	14.74	9.49
1,200	6,630	15.49	9.96
1,300	7,520	16.20	10.41
1,400	8,430	16.88	10.84
1,500	9,350	17.51	11.27
1,600	10,290	18.12	11.70
1,700	11,250	18.70	12.09
1,800	12,230	19.27	12.44
1,900	13,260	19.78	12.79
2,000	14,300	20.41	13.26

Dichromium Trioxide Cr_2O_3 (c)

$\Delta H^{\circ}_{298}=-272{,}650$ calories per mole (*98*)
$S_{298}=19.4$ *e.u.* (*112*)
$T.P.=298.16°$ K. (*24*)
$\Delta H_T=100$ calories per mole
$M.P.=2{,}553°$ K. (*94*)
$\Delta H_M=4{,}200$ calories per mole
$B.P.=3{,}273°$ K. (*94*)

Zone I (c) (298°–1,800° K.)

$$C_p=28.53+2.20\times10^{-3}T-3.74\times10^{5}T^{-2}\ (82)$$
$$H_T-H_{298}=-9{,}760+28.53T+1.10\times10^{-3}T^2+3.74\times10^{5}T^{-1}$$

Formation: $2Cr\ 3/2O_2 \longrightarrow Cr_2O_3$

Zone I (298°–1,800° K.)

$$\Delta C_p=6.11-4.02\times10^{-3}T-1.38\times10^{5}T^{-2}$$
$$\Delta H_T=-274{,}750+6.11T-2.01\times10^{-3}T^2+1.38\times10^{5}T^{-1}$$
$$\Delta F_T=-274{,}750-6.11T\ln T+2.01\times10^{-3}T^2+0.69\times10^{5}T^{-1}+105.95T$$

T, ° K.	H_T-H_{298}	S_T	ΔH°_T	ΔF°_T
298	....	19.4	−272,650	−253,150
400	2,740	27.66	−272,650	−247,100
500	5,540	33.92	−271,850	−240,300
600	8,380	39.09	−271,500	−234,050
700	11,280	43.56	−271,150	−227,850
800	14,230	47.50	−270,850	−221,700
900	17,210	51.00	−270,600	−215,550
1,000	20,240	54.20	−270,450	−209,450
1,100	23,320	57.14	−270,200	−203,250
1,200	26,430	59.84	−270,050	−197,200
1,300	29,550	62.33	−269,950	−191,050
1,400	32,670	64.65	−269,900	−185,050
1,500	35,790	66.80	−269,950	−178,750
1,600	38,920	68.82	−270,000	−172,800
1,700	42,050	70.72	−270,100	−166,850
1,800	45,180	72.51	−270,250	−161,000

Chromium Dioxide, CrO_2 (c)

$\Delta H^{\circ}_{298}=-142{,}500$ calories per mole (*24*)
$S_{298}=(12.70)$ *e.u.* (*24*)
Disproportionates 700° K. (*8*)

Formation: $Cr+O_2 \longrightarrow CrO_2$
(estimated (*24*))

T, ° K.	H_T-H_{298}	ΔH°_T	ΔF°_T
298	....	−142,500	(−130,000)
400	(1,300)	(−142,500)	(−125,500)
500	(2,700)	(−142,500)	(−121,500)
600	(4,200)	(−142,500)	(−117,500)
700	(5,600)	(−142,500)	(−113,000)

Chromium Trioxide, CrO_3 (c)

$\Delta H^{\circ}_{298}=(-140{,}000)$ calories per mole (*24*)
$S_{298}=(24)$ *e.u.* (*24*)
$M.P.=471°$ K. (*24*)
$\Delta H_M=3{,}770$ calories per mole

Formation: $Cr+3/2O_2 \longrightarrow CrO_3$
(estimated (*24*))

T, ° K.	H_T-H_{298}	ΔH°_T	ΔF°_T
298	....	(−140,000)	(−121,000)
400	(2,700)	(−139,000)	(−114,500)
500	(9,500)	(−134,000)	(−108,500)
600	(12,300)	(−133,000)	(−103,500)

Chromium Difluoride, CrF_2 (*c*)

$\Delta H^\circ_{298} = -182{,}000$ calories per mole (*11*)
$S_{298} = (20)$ *e.u.* (*11*)
M.P. = 1,375° K. (*6*)
$\Delta H_M = (5{,}500)$ calories per mole
B.P. = (2,400°) K. (*6*)
$\Delta H_V = (60{,}000)$ calories per mole

Formation: $Cr + F_2 \longrightarrow CrF_2$
(estimated (*11*))

T, ° K.	$H_T - H_{298}$	ΔH°_T	ΔF°_T
298		−182, 000	(−172, 000)
500	(3, 000)	(−181, 900)	(−165, 500)
1,000	(13, 000)	(−179, 800)	(−148, 000)
1,500	(32, 000)	(−169, 600)	(−137, 000)

Chromium Trifluoride, CrF_3 (*c*)

$\Delta H^\circ_{298} = -266{,}000$ calories per mole (*112*)
$S_{298} = (25)$ *e.u.* (*11*)
M.P. = 1,373° K. (*6*)
$\Delta H_M = (11{,}000)$ calories per mole
B.P. = (1,700°) K. (*6*)
$\Delta H_V = (48{,}000)$ calories per mole

Formation: $Cr + 3/2F_2 \longrightarrow CrF_3$
(estimated (*11*))

T, ° K.	$H_T - H_{298}$	ΔH°_T	ΔF°_T
298		−266, 000	(−250, 000)
500	(5, 000)	(−264, 700)	(−239, 000)
1,000	(18, 000)	(−261, 100)	(−215, 000)
1,500	(42, 000)	(−248, 700)	(−197, 000)

Chromium Tetrafluoride, CrF_4 (*c*)

$\Delta H^\circ_{298} = (-286{,}500)$ calories per mole (*11*)
$S_{298} = (38)$ *e.u.* (*11*)
M.P. = (550°) K. (*6*)
$\Delta H_M = (5{,}500)$ calories per mole
B.P. = (570°) K. (*6*)
$\Delta H_V = (14{,}000)$ calories per mole

Formation: $Cr + 2F_2 \longrightarrow CrF_4$
(estimated (*11*))

T, ° K.	$H_T - H_{298}$	ΔH°_T	ΔF°_T
298		(−286, 500)	(−267, 100)
500	(6, 000)	(−284, 500)	(−254, 000)
1,000			(−241, 000)
1,500			(−220, 000)

Chromium Dichloride, $CrCl_2$ (*c*)

$\Delta H^\circ_{298} = -94{,}560$ calories per mole (*112*)
$S_{298} = 27.8$ *e.u.* (*83*)
M.P. = 1,088° K. (*112*)
$\Delta H_M = 7{,}700$ calories per mole
B.P. = 1,573° K. (*94*)
$\Delta H_V = 47{,}500$ calories per mole

Zone I (*c*) (298°–1,088° K.)

$$C_p = 15.23 + 5.30 \times 10^{-3}T \ (94)$$
$$H_T - H_{298} = -4{,}770 + 15.23T + 2.65 \times 10^{-3}T^2$$

Zone II (*l*) (1,088°–1,573° K.)

$$C_p = 24.0 \ (94)$$
$$H_T - H_{298} = -3{,}400 + 24.T$$

Formation: $Cr + Cl_2 \longrightarrow CrCl_2$

Zone I (298°–1,088° K.)

$$\Delta C_p = 0.57 + 2.88 \times 10^{-3}T + 1.56 \times 10^5 T^{-2}$$
$$\Delta H_T = -94{,}330 + 0.57T + 1.44 \times 10^{-3}T^2 - 1.56 \times 10^5 T^{-1}$$
$$\Delta F_T = -94{,}330 - 0.57T\ln T - 1.44 \times 10^{-3}T^2 - 0.78 \times 10^5 T^{-1} + 34.98T$$

Zone II (1,088°–1,573° K.)

$$\Delta C_p = 9.34 - 2.42 \times 10^{-3}T + 1.56 \times 10^5 T^{-2}$$
$$\Delta H_T = -92{,}900 + 9.34T - 1.21 \times 10^{-3}T^2 - 1.56 \times 10^5 T^{-1}$$
$$\Delta F_T = -92{,}900 - 9.34T\ln T + 1.21 \times 10^{-3}T^2 - 0.78 \times 10^5 T^{-1} + 90.87T$$

T, ° K.	$H_T - H_{298}$	S_T	ΔH°_T	ΔF°_T
298		27.8	−94, 560	−85, 250
400	1, 750	33.0	−94, 270	−82, 150
500	3, 507	36.71	−94, 020	−79, 050
600	5, 330	40.00	−93, 730	−76, 050
700	7, 215	43.02	−93, 430	−73, 050
800	9, 123	45.66	−93, 110	−70, 350
900	11, 085	48.08	−92, 780	−67, 700
1,000	13, 105	49.93	−92, 450	−65, 600
1,100	22, 985	60.4	−84, 160	−63, 400
1,200	25, 385	62.4	−83, 635	−61, 000
1,300	27, 785	64.4	−82, 910	−59, 700
1,400	30, 185	66.4	−82, 410	−58, 250
1,500	32, 585	67.8	−81, 700	−56, 100

Chromium Trichloride, $CrCl_3$ (*c*)

$\Delta H^\circ_{298} = -132{,}500$ calories per mole (*94*)
$S_{298} = (30)$ *e.u.* (*83*)
S.P. = 1,220° K. (*6*)
$\Delta H_{subl} = 56{,}800$ calories per mole

Zone I (*c*) (298°–1,200° K.)

$$C_p = 19.44 + 7.03 \times 10^{-3}T \ (94)$$
$$H_T - H_{298} = -6{,}105 + 19.44T + 3.51 \times 10^{-3}T^2$$

Formation: $Cr + 3/2Cl_2 \longrightarrow CrCl_3$

Zone I (298°–1,220° K.)

$$\Delta C_p = 0.37 + 4.58 \times 10^{-3}T + 1.90 \times 10^5 T^{-2}$$
$$\Delta H_T = -132{,}300 + 0.37T + 2.29 \times 10^{-3}T^2 - 1.90 \times 10^5 T^{-1}$$
$$\Delta F_T = -132{,}300 - 0.37T\ln T - 2.29 \times 10^{-3}T^2 - 0.95 \times 10^5 T^{-1} + 58.90T$$

T, ° K.	$H_T - H_{298}$	S_T	ΔH°_T	ΔF°_T
298		(30.0)	−132, 500	(−115, 900)
400	2, 200	(36.23)	−132, 180	(−110, 200)
500	4, 475	(41.44)	−131, 840	(−104, 800)
600	6, 830	(45.74)	−131, 450	(−99, 450)
700	9, 130	(49.24)	−131, 160	(−94, 150)
800	11, 710	(52.75)	−130, 610	(−88, 800)
900	13, 240	(55.55)	−130, 160	(−83, 650)
1,000	16, 850	(58.35)	−129, 680	(−78, 300)
1,100	19, 540	(60, 95)	−128, 950	(−73, 200)
1,200	22, 290	(63.45)	−128, 580	(−68, 550)

Chromium Tetrachloride, $CrCl_4$ (*l*)

$\Delta H^{\circ}_{298} = -110{,}000$ calories per mole (*11*)
$S_{298} = (61)$ *e.u.* (*11*)
M.P. = (245°) K. (*6*)
$\Delta H_M = (2{,}000)$ calories per mole
B.P. = (430°) K. (*6*)
$\Delta H_V = (9{,}000)$ calories per mole

Formation: $Cr + 2Cl_2 \longrightarrow CrCl_4$
(estimated (11))

T, ° K.	$H_T - H_{298}$	ΔH°_T	ΔF°_T
298		−110, 000	(−95, 000)
500	(17, 000)	(−97, 500)	(−85, 500)

Chromium Dibromide, $CrBr_2$ (*c*)

$\Delta H^{\circ}_{298} = (-74{,}000)$ calories per mole (*11*)
$S_{298} = (30)$ *e.u.* (*11*)
M.P. = 1,115° K. (*6*)
$\Delta H_M = (6{,}500)$ calories per mole
B.P. = (1,400°) K. (*6*)
$\Delta H_V = (35{,}000)$ calories per mole

Formation: $Cr + Br_2 \longrightarrow CrBr_2$
(estimated (11))

T, ° K.	$H_T - H_{298}$	ΔH°_T	ΔH°_T
298		(−74, 000)	(−70, 000)
500	(4, 000)	(−80, 500)	(−63, 500)
1,000	(14, 000)	(−79, 000)	(−48, 000)
1,500	(69, 000)	(−33, 000)	(−36, 000)

Chromium Tribromide, $CrBr_3$ (*c*)

$\Delta H^{\circ}_{298} = -91{,}000$ calories per mole (*11*)
$S_{298} = (44)$ *e.u.* (*11*)
S.P. = (1,200°) K. (*6*)
$\Delta H_{subl} = (54{,}000)$ calories per mole

Formation: $Cr + 3/2Br_2 \longrightarrow CrBr_3$
(estimated (11))

T, ° K.	$H_T - H_{298}$	ΔH°_T	ΔF°_T
298		−91, 000	(−86, 000)
500	(5, 000)	(−101, 600)	(−77, 000)

Chromium Diiodide, CrI_2 (*c*)

$\Delta H^{\circ}_{298} = (-43{,}000)$ calories per mole (*11*)
$S_{298} = (34)$ *e.u.* (*11*)
M.P. = 1,066° K. (*6*)
$\Delta H_M = (6{,}000)$ calories per mole
B.P. = (1,100°) K. (*6*)
$\Delta H_V = (24{,}000)$ calories per mole

Formation: $Cr + I_2 \longrightarrow CrI_2$
(estimated (11))

T, ° K.	$H_T - H_{298}$	ΔH°_T	ΔF°_T
298		(−43, 000)	(−43, 000)
500	(4, 000)	(−57, 000)	(−41, 000)
1,000	(14, 000)	(−55, 100)	(−26, 000)
1,500	(58, 000)	(−20, 000)	(−16, 000)

Tetrachromium Carbide, Cr_4C (*c*)

$\Delta H^{\circ}_{298} = -16{,}400$ calories per mole (*112*)
$S_{298} = 25.3$ *e.u.* (*112*)
M.P. = 1,793° K. (*94*)

Zone I (*c*) (298°-1700° K.)

$C_p = 29.35 + 7.40 \times 10^{-3}T - 5.02 \times 10^5 T^{-2}$ (*82*)
$H_T - H_{298} = -10{,}764 + 29.35T + 3.70 \times 10^{-3}T^2 + 5.02 \times 10^5 T^{-1}$

Formation: $4Cr + C \longrightarrow Cr_4C$

Zone I (298°–1,700° K.)

$\Delta C_p = 1.89 - 3.06 \times 10^{-3}T + 0.60 \times 10^5 T^{-2}$
$\Delta H_T = -16{,}620 + 1.89T - 1.53 \times 10^{-3}T^2 - 0.60 \times 10^5 T^{-1}$
$\Delta F_T = -16{,}620 - 1.89 T \ln T + 1.53 \times 10^{-3}T^2 - 0.30 \times 10^5 T^{-1} + 10.19T$

T, ° K.	$H_T - H_{298}$	S_T	ΔH°_T	ΔF°_T
298		25. 3	−16, 400	−16, 750
400	2, 800	33. 36	−16, 350	−16, 900
500	5, 850	40. 16	−16, 250	−17, 050
600	9, 010	45. 92	−16, 200	−17, 250
700	12, 290	50. 97	−16, 100	−17, 400
800	15, 700	55. 52	−16, 050	−17, 600
900	19, 200	59. 64	−16, 050	−17, 750
1, 000	22, 770	63. 41	−16, 200	−17, 950
1, 100	26, 420	66. 88	−16, 400	−18, 150
1, 200	30, 160	70. 14	−16, 600	−18, 150
1, 300	34, 000	73. 12	−16, 850	−18, 450
1, 400	37, 950	76. 14	−17, 100	−18, 500
1, 500	42, 010	78. 94	−17, 250	−18, 650
1, 600	46, 180	81. 63	−17, 550	−18, 700
1, 700	50, 480	84. 23	−17, 850	−18, 750

Heptachromium Tricarbide, Cr_7C_3 (*c*)

$\Delta H^{\circ}_{298} = -42{,}600$ calories per mole (*112*)
$S_{298} = 48.0$ *e.u.* (*112*)
Disproportionates 1,940° K. (*8*)

Zone I (*c*) (298°–1,500° K.)

$C_p = 56.96 + 14.54 \times 10^{-3}T - 10.12 \times 10^5 T^{-2}$ (*82*)
$H_T - H_{298} = -21{,}010 + 56.96T + 7.27 \times 10^{-3}T^2 + 10.12 \times 10^5 T^{-1}$

Formation: $7Cr + 3C \longrightarrow Cr_7C_3$

Zone I (298°–1,500° K.)

$\Delta C_p = 3.78 - 5.04 \times 10^{-3}T + 2.34 \times 10^5 T^{-2}$
$\Delta H_T = -42{,}600 + 3.78T - 2.52 \times 10^{-3}T^2 - 2.34 \times 10^5 T^{-1}$
$\Delta F_T = -42{,}600 - 3.78 T \ln T + 2.52 \times 10^{-3}T^2 - 1.17 \times 10^5 T^{-1} + 18.30T$

T, ° K.	H_T-H_{298}	S_T	ΔH°_T	ΔF°_T
298		48.0	−42,600	−43,840
400	5,440	63.66	−42,250	−43,350
500	11,320	76.77	−41,950	−44,900
600	17,450	87.94	−41,720	−45,500
700	23,860	97.81	−41,470	−46,150
800	30,480	106.65	−41,270	−46,850
900	37,240	114.60	−41,270	−47,500
1,000	44,230	121.97	−41,380	−48,200
1,100	51,360	128.76	−41,590	−48,850
1,200	58,600	135.06	−41,960	−49,500
1,300	66,000	140.98	−42,410	−50,150
1,400	73,700	146.69	−42,700	−50,700
1,500	81,750	152.24	−42,840	−51,300

Trichromium Dicarbide, Cr_3C_2 (*c*)

$\Delta H^\circ_{298}=-21{,}000$ calories per mole (*112*)
$S_{298}=20.42$ *e.u.* (*31*)
M.P. $=2{,}163°$ K. (*112*)

Zone I (*c*) (298°–1,700° K.)

$$C_p=26.19+9.48\times10^{-3}T-4.72\times10^{5}T^{-2}\ (84)$$
$$H_T-H_{298}=-9{,}790+26.19T+4.74\times10^{-3}T^2+4.72\times10^{5}T^{-1}$$

Formation: $3Cr+2C \longrightarrow Cr_3C_2$

Zone I (298°–1,700° K.)

$$\Delta C_p=0.47+0.36\times10^{-3}T+2.12\times10^{5}T^{-2}$$
$$\Delta H_T=-20{,}450+0.47T+0.18\times10^{-3}T^2-2.12\times10^{5}T^{-1}$$
$$\Delta F_T=-20{,}450-0.47T\ln T-0.18\times10^{-3}T^2-1.06\times10^{5}T^{-1}+1.40T$$

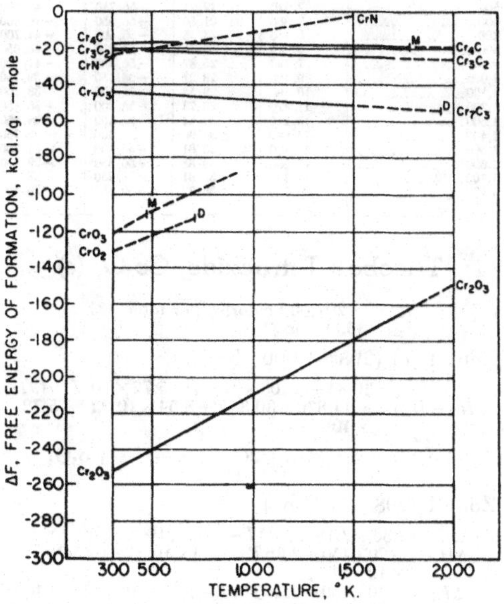

FIGURE 12.—Chromium (a).

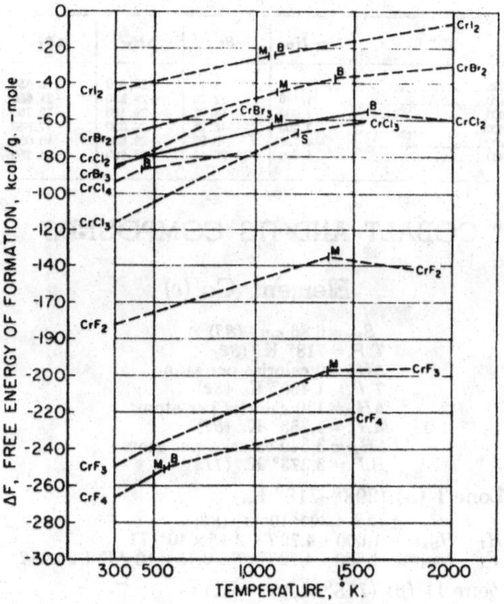

FIGURE 13.—Chromium (b).

T, ° K.	H_T-H_{298}	S_T	ΔH°_T	ΔF°_T
298		20.4	−21,000	−21,200
400	2,610	28.09	−20,750	−21,350
500	5,420	34.37	−20,560	−21,550
600	8,350	39.78	−20,430	−21,800
700	11,510	44.40	−20,210	−21,900
800	14,780	48.89	−20,020	−22,250
900	18,130	52.71	−19,910	−22,400
1,000	21,580	56.36	−19,860	−22,700
1,100	25,140	59.71	−19,810	−22,950
1,200	28,790	63.08	−19,800	−23,450
1,300	32,540	66.16	−19,800	−23,900
1,400	36,400	69.04	−19,750	−24,200
1,500	40,340	71.62	−19,670	−24,350
1,600	44,400	74.32	−19,550	−24,650
1,700	48,600	76.72	−19,370	−24,850
1,800	(52,800)	(79.0)	(−19,250)	(−24,950)
1,900	(57,150)	(81.32)	(−19,200)	(−25,450)
2,000	(61,650)	(83.92)	(−19,050)	(−25,850)

Chromium Nitride, CrN (*c*)

$\Delta H^\circ_{298}=-29{,}500$ calories per mole (*81*)
$S_{298}=8.9$ *e.u.* (*81*)
Disproportionates (1,800° K.) (*94*)

Zone I (*c*) (298°–800° K.)

$$C_p=9.84+3.9\times10^{-3}T\ (82)$$
$$H_T-H_{298}=-3{,}110+9.84T+1.95\times10^{-3}T^2$$

Formation: $Cr+1/2N_2 \longrightarrow CrN$

Zone I (298°–800° K.)

$$\Delta C_p=0.67+1.03\times10^{-3}T+0.88\times10^{5}T^{-2}$$
$$\Delta H_T=-29.450+0.67T+0.51\times10^{-3}T^2-0.88\times10^{5}T^{-1}$$
$$\Delta F_T=-29{,}450-0.67T\ln T-0.51\times10^{-3}T^2-0.44\times10^{5}T^{-1}+23.96T$$

T, ° K.	H_T-H_{298}	S_T	ΔH_T°	ΔF_T°
298	....	8.9	−29,500	−23,650
400	1,140	12.20	−29,330	−21,650
500	2,300	14.79	−29,180	−19,750
600	3,500	16.98	−29,020	−17,850
700	4,730	18.92	−28,850	−16,050
800	5,990	20.56	−28,690	−14,250

COBALT AND ITS COMPOUNDS

Element, Co (*c*)

$S_{298}=6.86$ *e.u.* (*83*)
$T.P.=718°$ K. (*82*)
$\Delta H_T=0$ calories per atom
$T.P.=1{,}400°$ K. (*82*)
$\Delta H_T=130$ calories per atom
$M.P.=1{,}763°$ K. (*82*)
$\Delta H_M=3{,}640$ calories per atom
$B.P.=3{,}373°$ K. (*112*)

Zone I (α) (298°–718° K.)

$C_p=4.72+4.30\times10^{-3}T$ (*82*)
$H_T-H_{298}=-1{,}600+4.72T+2.15\times10^{-3}T^2$
$F_T-H_{298}=-1{,}600-4.72T\ln T-2.15\times10^{-3}T^2+25.94T$

Zone II (β) (718°–1,400° K.)

$C_p=3.30+5.86\times10^{-3}T$ (*82*)
$H_T-H_{298}=-979+3.30T+2.93\times10^{-3}T^2$
$F_T-H_{298}=-979-3.30T\ln T-2.93\times10^{-3}T^2+16.36T$

Zone III (γ) (1,400°–1,763° K.)

$C_p=9.60$ (*82*)
$H_T-H_{298}=-3{,}920+9.60T$
$F_T-H_{298}=-3{,}920-9.60T\ln T+60.0T$

Zone IV (*l*) (1,763°–1,900° K.)

$C_p=8.30$ (*82*)
$H_T-H_{298}=+2{,}010+8.30T$
$F_T-H_{298}=+2{,}010-8.30T\ln T+47.07T$

T, ° K.	H_T-H_{298}	S_T	$-\frac{(F_T-H_{298})}{T}$
298	....	6.86	6.86
400	640	8.70	7.10
500	1,300	10.17	7.56
600	2,010	11.47	8.12
700	2,760	12.62	8.67
800	3,550	13.67	9.24
900	4,380	14.65	9.78
1,000	5,250	15.57	10.32
1,100	6,180	16.45	10.83
1,200	7,180	17.32	11.33
1,300	8,250	18.18	11.83
1,400	9,390	19.02	12.31
1,500	10,480	19.78	12.80
1,600	11,440	20.40	13.25
1,700	12,400	20.98	13.69
1,800	16,950	23.56	14.14
1,900	17,780	24.01	14.65
2,000	(18,610)	....	(14.60)

Cobalt Oxide, CoO (*c*)

$\Delta H_{298}^\circ=-57{,}300$ calories per mole (*4*)
$S_{298}=12.63$ *e.u.* (*88*)
$M.P.=2{,}078°$ K. (*112*)

Zone I (*c*) (298–1,800° K.)

$C_p=11.54+2.04\times10^{-3}T+0.40\times10^5T^{-2}$ (*82*)
$H_T-H_{298}=-3{,}400+11.54T+1.02\times10^{-3}T^2-0.40\times10^5T^{-1}$

Formation: $Co+1/2O_2 \longrightarrow CoO$

Zone I (298°–718° K.)

$\Delta C_p=3.24-2.76\times10^{-3}T+0.6\times10^5T^{-2}$
$\Delta H_T=-57{,}940+3.24T-1.38\times10^{-3}T^2-0.60\times10^5T^{-1}$
$\Delta F_T=-57{,}940-3.24T\ln T+1.38\times10^{-3}T^2-0.30\times10^5T^{-1}+39.28T$

Zone II (718°–1,400° K.)

$\Delta C_p=4.66-4.32\times10^{-3}T+0.60\times10^5T^{-2}$
$\Delta H_T=-58{,}590+4.66T-2.16\times10^{-3}T^2-0.60\times10^5T^{-1}$
$\Delta F_T=-58{,}590-4.66T\ln T+2.16\times10^{-3}T^2-0.30\times10^5T^{-1}+48.9T$

Zone III (1,400°–1,673° K.)

$\Delta C_p=1.64+1.54\times10^{-3}T+0.60\times10^5T^{-2}$
$\Delta H_T=-55{,}750-1.64T+0.77\times10^{-3}T^2-0.60\times10^5T^{-1}$
$\Delta F_T=-55{,}750+1.64T\ln T-0.77\times10^{-3}T^2-0.30\times10^5T^{-1}+5.47T$

Zone IV (1,673°–1,800° K.)

$\Delta C_p=-0.34+1.54\times10^{-3}T+0.60\times10^5T^{-2}$
$\Delta H_T=-61{,}480-0.34T+0.77\times10^{-3}T^2-0.60\times10^5T^{-1}$
$\Delta F_T=-61{,}480+0.34T\ln T-0.77\times10^{-3}T^2-0.30\times10^5T^{-1}+18.27T$

T, ° K.	H_T-H_{298}	S_T	ΔH_T°	ΔF_T°
298	....	12.63	−57,300	−51,700
400	1,290	16.35	−57,000	−49,850
500	2,570	19.21	−56.750	−48,100
600	3,860	21.56	−56,550	−46,400
700	5,160	23.56	−56,400	−44,700
800	6,470	25.31	−56,250	−43,050
900	7,790	26.87	−56,200	−41,100
1,000	9,120	28.27	−56,150	−39,750
1,100	10,460	29.55	−56,100	−38,100
1,200	11,820	30.73	−56,200	−36,450
1,300	13,210	31.84	−56,150	−34,650
1,400	14,640	32.90	−56,400	−33,150
1,500	16,100	33.81	−56,500	−31,300
1,600	17,600	34.88	−56,350	−29,750
1,700	19,140	35.81	−56,250	−28,150
1,800	20,750	36.72	−59,650	−26,500

Tricobalt Tetraoxide, Co_3O_4 (*c*)

$\Delta H_{298}^\circ=-207{,}000$ calories per mole (*24*)
$S_{298}=35.66$ *e.u.* (*24*)

Zone I (*c*) (298°–1,000° K.)

$C_p=30.84+17.08\times10^{-3}T-5.72\times10^5T^{-2}$ (*91*)
$H_T-H_{298}=-11{,}870+30.84T+8.54\times10^{-3}T^2+5.72\times10^5T^{-1}$

Formation: $3Co+2O_2 \longrightarrow Co_3O_4$

Zone I (298°–718° K.)

$\Delta C_p=2.36+2.18\times10^{-3}T-4.92\times10^5T^{-2}$
$\Delta H_T=-209{,}450+2.36T+1.09\times10^{-3}T^2+4.92\times10^5T^{-1}$
$\Delta F_T=-209{,}450-2.36T\ln T-1.09\times10^{-3}T^2+2.46\times10^5T^{-1}+102.16T$

Zone II (718°–1,000° K.)

$\Delta C_p = 6.62 - 2.50 \times 10^{-3} T - 4.92 \times 10^5 T^{-2}$

$\Delta H_T = -211{,}220 + 6.62T - 1.25 \times 10^{-3} T^2 - 4.92 \times 10^5 T^{-1}$

$\Delta F_T = -211{,}220 - 6.62 T \ln T + 1.25 \times 10^{-3} T^2 + 2.46 \times 10^5 T^{-1} + 131.03T$

T, ° K.	$H_T - H_{298}$	S_T	ΔH°_T	ΔF°_T
298		35.66	−207,000	−182,300
400	3,270	45.06	−207,100	−173,800
500	6,850	53.04	−206,950	−165,500
600	10,660	59.98	−206,800	−157,200
700	14,640	66.11	−206,600	−149,000
800	18,820	71.69	−206,400	−140,750
900	23,300	76.96	−206,050	−132,550
1,000	28,250	82.18	−205,350	−124,650
1,500				(−82,500)

Cobalt Difluoride, CoF_2 (*c*)

$\Delta H^\circ_{298} = -158{,}000$ calories per mole (*11*)
$S_{298} = (21)$ *e.u.* (*11*)
M.P. = 1,475° K. (*6*)
$\Delta H_M = (9{,}000)$ calories per mole
B.P. = (2,000°) K. (*6*)
$\Delta H_V = (48{,}000)$ calories per mole

Formation: $Co + F_2 \rightarrow CoF_2$
(estimated (*11*))

T, ° K.	$H_T - H_{298}$	ΔH°_T	ΔF°_T
298		−158,000	(−147,900)
500	(3,500)	(−157,400)	(−141,000)
1,000	(13,000)	(−156,100)	(−125,000)
1,500	(34,000)	(−144,700)	(−108,500)

Cobalt Trifluoride, CoF_3 (*c*)

$\Delta H^\circ_{298} = (-190{,}000)$ calories per mole (*11*)
$S_{298} = (27)$ *e.u.* (*11*)
M.P. = (1,300°) K. (*6*)
$\Delta H_M = (12{,}000)$ calories per mole
B.P. = (1,600°) K. (*6*)
$\Delta H_V = (40{,}000)$ calories per mole

Formation: $Co + 3/2F_2 \rightarrow CoF_3$
(estimated (*11*))

T, ° K.	$H_T - H_{298}$	ΔH°_T	ΔF°_T
298		(−190,000)	(−174,000)
500	(5,000)	(−188,700)	(−163,500)
1,000	(19,000)	(−185,000)	(−140,000)

Cobalt Dichloride, $CoCl_2$ (*c*)

$\Delta H^\circ_{298} = -77{,}800$ calories per mole (*112*)
$S_{298} = 25.4$ *e.u.* (*83*)
M.P. = 997° K. (*112*)
$\Delta H_M = 7{,}400$ calories per mole
B.P. = 1,323° K. (*112*)
$\Delta H_V = 27{,}200$ calories per mole

Zone I (*c*) (298°–997° K.)

$C_p = 14.41 + 14.60 \times 10^{-3} T$ (*82*)

$H_T - H_{298} = -4.945 + 14.41T + 7.30 \times 10^{-3} T^2$

Formation: $Co + Cl_2 \rightarrow CoCl_2$

Zone I (298°–718° K.)

$\Delta C_p = 0.87 + 10.24 \times 10^{-3} T + 0.68 \times 10^5 T^{-2}$

$\Delta H_T = -78{,}300 + 0.87T + 5.12 \times 10^{-3} T^2 - 0.68 \times 10^5 T^{-1}$

$\Delta F_T = -78{,}300 - 0.87 T \ln T - 5.12 \times 10^{-3} T^2 - 0.34 \times 10^5 T^{-1} + 43.3T$

Zone II (718°–997° K.)

$\Delta C_p = 2.29 + 8.68 \times 10^{-3} T + 0.68 \times 10^5 T^{-2}$

$\Delta H_T = -79{,}000 + 2.29T + 4.34 \times 10^{-3} T^2 - 0.68 \times 10^5 T^{-1}$

$\Delta F_T = -79{,}000 - 2.29 T \ln T - 4.34 \times 10^{-3} T^2 - 0.34 \times 10^5 T^{-1} + 53.02T$

T, ° K.	$H_T - H_{298}$	S_T	ΔH°_T	ΔF°_T
298		25.4	−77,800	−67,430
400	2,020	31.22	−77,260	−63,900
500	4,120	35.9	−76,670	−60,650
600	6,340	39.94	−76,010	−57,500
700	8,720	43.61	−75,260	−54,550
800	11,260	47.0	−74,400	−51,600
900	13,920	50.12	−73,440	−48,850
1,000	24,100	60.45	−65,000	−46,000

Cobalt Dibromide, $CoBr_2$ (*c*)

$\Delta H^\circ_{298} = -50{,}600$ calories per mole (*11*)
$S_{298} = (33)$ *e.u.* (*11*)
M.P. = 951° K. (*6*)
$\Delta H_M = (8{,}000)$ calories per mole
B.P. = (1,200°) K. (*6*)
$\Delta H_V = (25{,}000)$ calories per mole

Formation: $Co + Br_2 \rightarrow CoBr_2$
(estimated (*11*))

T, ° K.	$H_T - H_{298}$	ΔH°_T	ΔF°_T
298		−50,600	(−47,500)
500	(4,000)	(−57,000)	(−42,000)
1,000	(24,000)	(−45,900)	(−29,000)
1,500	(61,000)	(−18,500)	(−20,500)

Cobalt Diiodide, CoI_2 (*c*)

$\Delta H^\circ_{298} = -21{,}000$ calories per mole (*11*)
$S_{298} = (37)$ *e.u.* (*11*)
M.P. = 790° K. (*6*)
$\Delta H_M = (6{,}000)$ calories per mole
B.P. = (1,100°) K. (*6*)
$\Delta H_V = (24{,}000)$ calories per mole

Formation: $Co + I_2 \rightarrow CoI_2$
(estimated (*11*))

T, ° K.	$H_T - H_{298}$	ΔH°_T	ΔF°_T
298		−21,000	(−21,900)
500	(4,000)	(−35,000)	(−20,000)
1,000	(24,000)	(−23,400)	(−7,000)
1,500	(60,000)	(+3,000)	(+1,000)

Tricobalt Carbide, Co_3C (c)

$\Delta H^{\circ}_{298} = +9{,}330$ calories per mole (*81*)
$S_{298} = 22.9$ *e.u.* (*9*)
$\Delta F_{298} = 9{,}000$ calories per mole

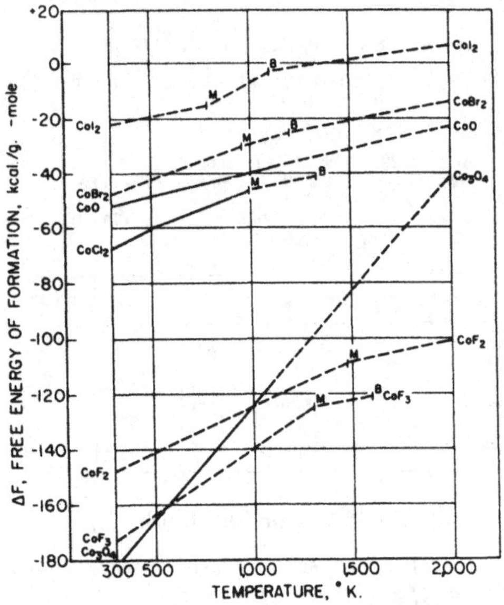

FIGURE 14.—Cobalt.

COLUMBIUM AND ITS COMPOUNDS

Element, Cb (c)

$S_{298} = 8.7$ *e.u.* (*7*)
M.P. $= 2{,}770°$ K. (*7*)
$\Delta H_M = (6{,}500)$ calories per atom
B.P. $= (5{,}400°)$ K. (*7*)
$\Delta H_V = (155{,}000)$ calories per atom

Zone I (*c*) (298°–1,900° K.)

$C_p = 5.66 + 0.96 \times 10^{-3}T$ (*82*)
$H_T - H_{298} = -1{,}730 + 5.66T + 0.48 \times 10^{-3}T^2$
$F_T - H_{298} = -1{,}730 - 5.66T\ln T - 0.48 \times 10^{-3}T^2 + 29.45T$

T, ° K.	$H_T - H_{298}$	S_T	$-\frac{(F_T - H_{298})}{T}$
298	....	8.7	8.7
400	610	10.46	8.91
500	1,215	11.81	9.36
600	1,835	12.94	9.86
700	2,470	13.92	10.40
800	3,110	14.77	10.86
900	3,750	15.53	11.39
1,000	4,400	16.21	11.81
1,100	5,070	16.85	12.23
1,200	5,760	17.45	12.67
1,300	6,450	18.00	13.04
1,400	7,160	18.53	13.44
1,500	7,870	19.02	13.78
1,600	8,580	19.48	14.11
1,700	9,300	19.91	14.44
1,800	10,020	20.34	14.77
1,900	10,760	20.72	15.05
(2,000)	(11,410)	(21.13)	(15.42)

Columbium Dioxide, CbO_2 (c)

$\Delta H^{\circ}_{298} = (-190{,}400)$ calories per mole (*24*)
$S_{298} = 13.03$ *e.u.* (*91*)

Formation: $Cb + O_2 \longrightarrow CbO_2$
(estimated (*24*))

T, ° K.	$H_T - H_{298}$	$\Delta H°_T$	$\Delta F°_T$
298	....	(190,400)	(−177,200)
400	(1,500)	(−190,200)	(−172,800)
500	(3,100)	(−190,000)	(−168,500)
600	(4,700)	(−189,700)	(−164,200)
700	(6,400)	(−189,500)	(−160,000)
800	(8,000)	(−189,300)	(−155,700)
900	(9,600)	(−189,100)	(−151,600)
1,000	(10,300)	(−188,900)	(−147,400)
1,100	(13,000)	(−188,700)	(−143,300)
1,200	(14,700)	(−188,500)	(−139,100)
1,300	(16,400)	(−188,300)	(−135,000)
1,400	(18,200)	(−188,100)	(−130,900)
1,500	(20,100)	(−187,900)	(−126,900)
1,600	(21,700)	(−187,700)	(−122,800)
1,700	(23,500)	(−187,500)	(−118,800)
1,800	(25,300)	(−187,300)	(−114,700)
1,900	(27,200)	(−187,100)	(−110,700)
2,000	(29,200)	(−186,800)	(−106,700)

Dicolumbium Pentaoxide, Cb_2O_5 (c)

$\Delta H^{\circ}_{298} = -455{,}000$ calories per mole (*67*)
$S_{298} = 32.8$ *e.u.* (*90*)
M.P. $= 1{,}785°$ K. (*107*)
$\Delta H_M = 24{,}200$ calories per mole
B.P. $= >2{,}500°$ K. (*42*)

Zone I (*c*) (298°–1,785° K.)

$C_p = 36.23 + 5.54 \times 10^{-3}T - 4.88 \times 10^{5}T^{-2}$ (*107*)
$H_T - H_{298} = -12{,}680 + 36.23T + 2.77 \times 10^{-3}T^2 + 4.88 \times 10^{5}T^{-1}$

Zone II (*l*) (1,785°–1,810° K.)

$C_p = 57.90$ (*107*)
$H_T - H_{298} = -17{,}255 + 57.90T$

Formation: $2Cb + 5/2O_2 \longrightarrow Cb_2O_5$

Zone I (298°–1,785° K.)

$\Delta C_p = 7.01 + 1.12 \times 10^{-3}T - 3.88 \times 10^{5}T^{-2}$
$\Delta H_T = -458{,}440 + 7.01T + 0.56 \times 10^{-3}T^2 + 3.88 \times 10^{5}T^{-1}$
$\Delta F_T = -458{,}440 - 7.01T\ln T - 0.56 \times 10^{-3}T^2 + 1.94 \times 10^{5}T^{-1} + 156.52T$

Zone II (1,785°–1,810° K.)

$\Delta C_p = 28.68 - 4.42 \times 10^{-3}T + 1.0 \times 10^{5}T^{-2}$
$\Delta H_T = -463{,}750 + 28.68T - 2.21 \times 10^{-3}T^2 - 1.0 \times 10^{5}T^{-1}$
$\Delta F_T = -463{,}750 - 28.68T\ln T + 2.21 \times 10^{-3}T^2 - 0.50 \times 10^{5}T^{-1} + 317.0T$

T, ° K.	$H_T - H_{298}$	S_T	ΔH°_T	ΔF°_T
298	....	32.8	−455,000	−423,050
400	3,500	42.91	−454,550	−412,250
500	7,100	51.00	−453,950	−401,750
600	10,880	57.85	−453,200	−391,200
700	14,500	63.82	−452,900	−381,300
800	18,700	69.08	−452,000	−370,900
900	23,000	73.82	−451,000	−360,500
1,000	26,800	78.14	−450,600	−350,800
1,100	30,940	82.11	−449,700	−340,600
1,200	35,260	85.78	−448,850	−330,850
1,300	39,470	89.20	−448,100	−321,100
1,400	43,890	92.42	−447,200	−311,100
1,500	48,070	95.56	−446,900	−301,900
1,600	52,720	98.34	−445,550	−291,650
1,700	57,150	100.68	−444,800	−281,800
1,800	86,430	117.35	−419,150	−273,150
1,900	(92,370)	(120.48)	(−416,950)	(−264,650)
2,000	(98,230)	(123.45)	(−415,000)	(−256,200)

Columbium Pentachloride, $CbCl_5$ (*c*)

$\Delta H^\circ_{298} = -190{,}600$ calories per mole (*48*)
$S_{298} = (65)$ *e.u.* (*11*)
M.P. = 485° K. (*6*)
$\Delta H_M = 8{,}400$ calories per mole
B.P. = 516° K. (*6*)
$\Delta H_V = 11{,}500$ calories per mole

Formation: $Cb + 5/2Cl_2 \longrightarrow CbCl_5$
(estimated (*11*))

T, ° K.	$H_T - H_{298}$	ΔH°_T	ΔF°_T
298		−190, 600	(−167, 500)
500	(13, 000)	(−183, 000)	(−153, 000)

Columbium Pentabromide, $CbBr_5$ (*c*)

$\Delta H^\circ_{298} = -132{,}850$ calories per mole (*48*)
$S_{298} = (78)$ *e.u.* (*11*)
M.P. = 500° K. (*6*)
$\Delta H_M = (8{,}500)$ calories per mole
B.P. = 545° K. (*6*)
$\Delta H_V = (12{,}000)$ calories per mole

Formation: $Cb + 5/2Br_2 \longrightarrow CbBr_5$
(estimated (*11*))

T, ° K.	$H_T - H_{298}$	ΔH°_T	ΔF°_T
298		−132, 850	(−126, 200)
500	(12, 000)	(−108, 500)	(−114, 500)

Columbium Nitride, CbN (*c*)

$\Delta H^\circ_{298} = -56{,}800$ calories per mole (*100*)
$S_{298} = 10.5$ *e.u.* (*9*)
M.P. = 2,372° K. (*94*)
$\Delta H_M = (14{,}500)$ calories per mole

Zone I (*c*) (298°–600° K.)

$C_p = 8.69 + 5.40 \times 10^{-3} T$ (*94*)
$H_T - H_{298} = -2{,}831 + 8.69T + 2.70 \times 10^{-3} T^2$

Formation: $Cb + 1/2N_2 \longrightarrow CbN$

Zone I (298°–600° K.)

$\Delta C_p = -0.30 + 3.93 \times 10^{-3} T$
$\Delta H_T = -56{,}900 - 0.30T + 1.96 \times 10^{-3} T^2$
$\Delta F_T = -56{,}900 + 0.30 T \ln T - 1.96 \times 10^{-3} T^2 + 20.42T$

T, ° K.	$H_T - H_{298}$	S_T	ΔH°_T	ΔF°_T
298		10. 5	−56, 800	−50, 550
400	1, 080	13. 61	−56, 700	−48, 400
500	2, 190	16. 09	−56, 550	−46, 300
600	3, 355	18. 21	−56, 350	−44, 300
700	(4, 570)		(−56, 150)	(−42, 300)
800	(5, 850)		(−55, 900)	(−40, 350)
900	(7, 170)		(−55, 600)	(−38, 400)
1,000	(8, 560)		(−55, 250)	(−36, 550)
1,100	(10, 000)		(−54, 850)	(−34, 700)
1,200	(11, 480)		(−54, 450)	(−32, 850)
1,300	(13, 030)		(−54, 000)	(−31, 050)
1,400	(14, 640)		(−53, 500)	(−29, 350)
1,500	(16, 270)		(−52, 950)	(−27, 600)
1,600	(18, 000)		(−52, 350)	(−26, 000)
1,700	(19, 670)		(−51, 750)	(−24, 350)
1,800	(21, 550)		(−51, 100)	(−22, 800)
1,900	(23, 420)		(−50, 400)	(−21, 250)
2,000	(25, 350)		(−49, 600)	(−19, 700)

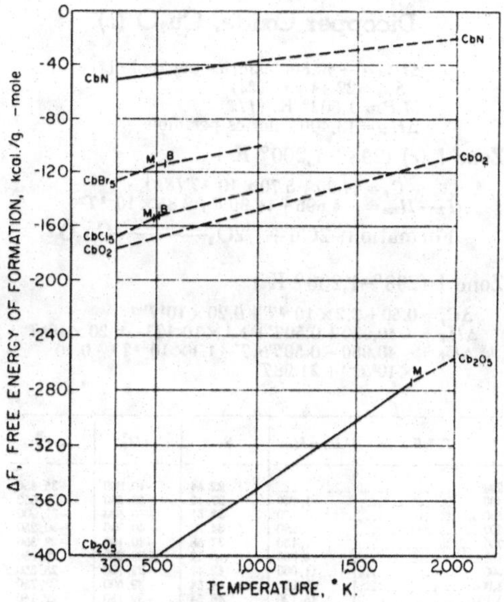

FIGURE 15.—Columbium.

COPPER AND ITS COMPOUNDS

Element, Cu (*c*)

$S_{298} = 7.97$ *e.u.* (*83*)
M.P. = 1,357° K. (*82*)
$\Delta H_M = 3{,}120$ calories per atom
B.P. = 2,855° K. (*112*)
$\Delta H_V = 72{,}800$ calories per atom

Zone I (*c*) (298°–1,357° K.)

$C_p = 5.41 + 1.50 \times 10^{-3} T$ (*82*)
$H_T - H_{298} = -1{,}680 + 5.41T + 0.75 \times 10^{-3} T^2$
$F_T - H_{298} = -1{,}680 - 5.41 T \ln T - 0.75 \times 10^{-3} T^2 + 28.7T$

Zone II (*l*) (1,357°–1,600° K.)

$C_p = 7.50$ (*82*)
$H_T - H_{298} = -20 + 7.50T$
$F_T - H_{298} = -20 - 7.50 \, T \ln T + 41.54T$

T, ° K.	$H_T - H_{298}$	S_T	$-\frac{(F_T - H_{298})}{T}$
298		7. 97	7. 97
400	600	9. 70	8. 20
500	1. 215	11. 07	8. 64
600	1. 845	12. 22	9. 14
700	2. 480	13. 20	9. 66
800	3. 130	14. 07	10. 16
900	3, 800	14. 86	10. 64
1,000	4. 490	15. 58	11. 09
1,100	5. 190	16. 25	11. 53
1,200	5, 895	16. 87	11. 96
1,300	6, 615	17. 44	12. 35
1,400	10, 480	20. 29	12. 80
1,500	11, 230	20. 81	13. 32
1,600	11, 980	21. 29	13. 80
1,700	(12, 740)	(21. 74)	(14. 24)
1,800	(13, 480)	(22. 17)	(14. 68)
1,900	(14, 230)	(22. 58)	(15. 09)
2,000	(14, 980)	(22. 96)	(15. 47)

Dicopper Oxide, Cu_2O (*c*)

$\Delta H^{\circ}_{298} = -40{,}800$ calories per mole (*2*)
$S_{298} = 22.44$ *e.u.* (*24*)
$M.P. = 1{,}502°$ K. (*112*)
$\Delta H_M = 13{,}400$ calories per mole

Zone I (*c*) (298°–1,200° K.)

$C_p = 14.90 + 5.70 \times 10^{-3} T$ (*82*)
$H_T - H_{298} = -4{,}696 + 14.90T + 2.85 \times 10^{-3} T^2$

Formation: $2Cu + 1/2O_2 \longrightarrow Cu_2O$

Zone I (298°–1,200° K.)

$\Delta C_p = 0.50 + 2.2 \times 10^{-3} T + 0.20 \times 10^5 T^{-2}$
$\Delta H_T = -40{,}980 + 0.50T + 1.1 \times 10^{-3} T^2 - 0.20 \times 10^5 T^{-1}$
$\Delta F_T = -40{,}980 - 0.50 T \ln T - 1.1 \times 10^{-3} T^2 - 0.10 \times 10^5 T^{-1} + 21.98T$

T, ° K.	$H_T - H_{298}$	S_T	ΔH°_T	ΔF°_T
298		22.44	−40,800	−35,450
400	1,720	27.40	−40,650	−33,650
500	3,470	31.31	−40,500	−31,900
600	5,280	34.70	−40,300	−30,250
700	7,150	37.58	−40,100	−28,550
800	9,050	40.12	−39,900	−26,950
900	11,000	42.42	−39,700	−25,350
1,000	13,020	44.54	−39,500	−23,750
1,100	15,120	46.54	−39,150	−22,150
1,200	17,320	48.46	−38,800	−20,650
1,300	(19,570)	(50.2)	(−38,400)	(−19,100)
1,400	(22,020)	(52.0)	(−44,100)	(−17,500)
1,500	(24,600)	(53.2)	(−43,500)	(−15,600)
1,600	(40,400)	(63.7)	(−29,600)	(−13,700)

Copper Oxide, CuO (*c*)

$\Delta H^{\circ}_{298} = -37{,}500$ calories per mole (*2*)
$S_{298} = 10.19$ *e.u.* (*56*)
$M.P. = 1{,}720°$ K. (*24*)
$\Delta H_M = 2{,}820$ calories per mole

Zone I (*c*) (298°–1,250° K.)

$C_p = 9.27 + 4.80 \times 10^{-3} T$ (*82*)
$H_T - H_{298} = -2{,}977 + 9.27T + 2.40 \times 10^{-3} T^2$

Formation: $Cu + 1/2O_2 \longrightarrow CuO$

Zone I (298°–1,250° K.)

$\Delta C_p = 0.28 + 2.8 \times 10^{-3} T + 0.20 \times 10^5 T^{-2}$
$\Delta H_T = -37{,}640 + 0.28T + 1.4 \times 10^{-3} T^2 - 0.20 \times 10^5 T^{-1}$
$\Delta F_T = -37{,}640 - 0.28 T \ln T - 1.4 \times 10^{-3} T^2 - 0.10 \times 10^5 T^{-1} + 24.93T$

T, ° K.	$H_T - H_{298}$	S_T	ΔH°_T	ΔF°_T
298		10.19	−37,500	−30,850
400	1,110	13.40	−37,350	−28,600
500	2,260	15.95	−37,200	−26,450
600	3,460	18.14	−37,000	−24,300
700	4,710	20.07	−36,750	−22,200
800	6,000	21.79	−36,500	−20,150
900	7,320	23.34	−36,300	−18,100
1,000	8,680	24.77	−36,000	−16,050
1,100	10,120	26.15	−35,700	−14,100
1,200	11,600	27.43	−35,300	−12,150
1,300	(12,860)	(28.6)	(−35,000)	(−10,250)
1,400	(14,640)	(29.9)	(−37,700)	(−8,500)
1,500	(15,870)	(31.0)	(−37,300)	(−6,400)
1,600	(18,800)	(32.0)	(−36,900)	(−4,300)

Copper Fluoride, CuF (*c*)

$\Delta H^{\circ}_{298} = -60{,}000$ calories per mole (*11*)
$S_{298} = (16)$ *e.u.* (*11*)
$M.P. = (1{,}020°)$ K. (*42*)
$\Delta H_M = (4{,}500)$ calories per mole
$B.P. = (1{,}660°)$ K. (*42*)
$\Delta H_V = (36{,}000)$ calories per mole

Formation: $Cu + 1/2F_2 \longrightarrow CuF$
(estimated (*11*))

T, ° K.	$H_T - H_{298}$	ΔH°_T	ΔF°_T
298		−60,000	(−55,200)
500	(3,000)	(−59,000)	(−52,000)
1,000	(10,000)	(−57,400)	(−46,000)
1,500	(22,000)	(−50,300)	(−39,000)

Copper Difluoride, CuF_2 (*c*)

$\Delta H^{\circ}_{298} = -128{,}000$ calories per mole (*11*)
$S_{298} = (22)$ *e.u.* (*11*)
$M.P. = (1{,}200°)$ K. (*6*)
$\Delta H_M = (6{,}000)$ calories per mole
$B.P. = (1{,}800°)$ K. (*6*)

Formation: $Cu + F_2 \longrightarrow CuF_2$
(estimated (*11*))

T, ° K.	$H_T - H_{298}$	ΔH°_T	ΔF°_T
298		−128,000	(−117,600)
500	(4,000)	(−126,800)	(−110,500)
1,000	(13,000)	(−125,300)	(−95,000)

Copper Chloride, CuCl (*c*)

$\Delta H^{\circ}_{298} = -32{,}600$ calories per mole (*11*)
$S_{298} = 20.8$ *e.u.* (*83*)
$M.P. = 703°$ K. (*82*)
$\Delta H_M = 2{,}620$ calories per mole
$B.P. = 1{,}963°$ K. (*6*)
$\Delta H_V = 39{,}600$ calories per mole

Zone I (*c*) (298°–703° K.)

$C_p = 5.87 + 19.20 \times 10^{-3} T$ (*82*)
$H_T - H_{298} = -2{,}605 + 5.87T + 9.60 \times 10^{-3} T^2$

Zone II (*l*) (703°–1,200° K.)

$C_p = 15.80$ (*79*)
$H_T - H_{298} = -2{,}220 + 15.80T$

Formation: $Cu + 1/2Cl_2 \longrightarrow CuCl$

Zone I (298°–703° K.)

$\Delta C_p = -3.95 + 17.67 \times 10^{-3} T + 0.34 \times 10^5 T^{-2}$
$\Delta H_T = -31{,}066 - 3.95T + 8.83 \times 10^{-3} T^2 - 0.34 \times 10^5 T^{-1}$
$\Delta F_T = -31{,}066 + 3.95 T \ln T - 8.83 \times 10^{-3} T^2 - 0.17 \times 10^5 T^{-1} - 7.56T$

Zone II (703°–1,200° K.)

$\Delta C_p = 6.0 - 1.53 \times 10^{-3} T + 0.34 \times 10^5 T^{-2}$
$\Delta H_T = -31{,}800 + 6.0T - 0.765 \times 10^{-3} T^2 - 0.34 \times 10^5 T^{-1}$
$\Delta F_T = -31{,}800 - 6.0 T \ln T + 0.765 \times 10^{-3} T^2 - 0.17 \times 10^5 T^{-1} + 50.5T$

T, ° K.	H_T-H_{298}	S_T	ΔH_T°	ΔF_T°
298	----------	20.8	−32,600	−28,500
400	1,280	24.48	−32,450	−27,200
500	2,720	27.68	−31,950	−25,800
600	4,385	30.71	−31,500	−24,850
700	6,210	33.52	−30,600	−23,600
800	10,420	39.37	−27,450	−22,950
900	12,000	41.23	−27,000	−22,500
1,000	13,580	42.9	−26,550	−21,950
1,100	15,160	44.41	−26,050	−21,400
1,200	16,740	45.78	−25,650	−21,150

Copper Dichloride, $CuCl_2$ (*c*)

$\Delta H^\circ_{298} = -53{,}400$ calories per mole (*11*)
$S_{298} = (27)$ *e.u.* (*11*)
Decomposes $= 810°$ K., 1 atm Cl_2 (*6*)

Zone I (*c*) (298°–800° K.)

$C_p = 15.42 + 12.00 \times 10^{-3} T$ (*82*)
$H_T - H_{298} = -5{,}131 + 15.42 T + 6.00 \times 10^{-3} T^2$

Formation: $Cu + Cl_2 \longrightarrow CuCl_2$

Zone I (298°–800° K.)

$\Delta C_p = 1.19 + 10.44 \times 10^{-3} T + 0.68 \times 10^5 T^{-2}$
$\Delta H_T = -53{,}990 + 1.19 T + 5.22 \times 10^{-3} T^2 - 0.68 \times 10^5 T^{-1}$
$\Delta F_T = -53{,}990 - 1.19 T \ln T - 5.22 \times 10^{-3} T^2 - 0.34 \times 10^5 T^{-1} + 45.0 T$

T, ° K.	H_T-H_{298}	S_T	ΔH_T°	ΔF_T°
298	----------	(27.0)	−53,400	(−43,200)
400	2,000	(32.7)	−52,850	(−39,700)
500	4,080	(37.4)	−52,200	(−36,500)
600	6,290	(41.4)	−51,500	(−33,450)
700	8,620	(45.0)	−50,700	(−30,550)
800	11,030	(49.2)	−49,800	(−28,450)

Copper Bromide, CuBr (*c*)

$\Delta H^\circ_{298} = -25{,}450$ calories per mole (*11*)
$S_{298} = 22.97$ *e.u.* (*55*)
M.P. $= 761°$ K. (*6*)
$\Delta H_M = (2{,}300)$ calories per mole
B.P. $= 1{,}591°$ K. (*6*)
$\Delta H_V = (33{,}400)$ calories per mole

Formation: $Cu + 1/2 Br_2 \longrightarrow CuBr$
(estimated (*11*))

T, ° K.	H_T-H_{298}	ΔH_T°	ΔF_T°
298	----------	−25,450	−24,400
500	(3,000)	(−28,400)	(−21,500)
1,000	(13,000)	(−24,000)	(−17,000)
1,500	(21,000)	(−24,900)	(−13,400)

Copper Dibromide, $CuBr_2$ (*c*)

$\Delta H^\circ_{298} = -33{,}200$ calories per mole (*112*)
$S_{298} = (33)$ *e.u.* (*11*)
Decomposes $= 600°$ K., 1 atm Br_2 (*6*)

Formation: $Cu + Br_2 \longrightarrow CuBr_2$
(estimated (*11*))

T, ° K.	H_T-H_{298}	ΔH_T°	ΔF_T°
298	----------	−33,200	(−30,400)
500	(4,000)	(−40,000)	(−24,300)

Copper Iodide, CuI (*c*)

$\Delta H^\circ_{298} = -16{,}500$ calories per mole (*11*)
$S_{298} = 23.1$ *e.u.* (*112*)
M.P. $= 861°$ K. (*6*)
$\Delta H_M = (2{,}600)$ calories per mole
B.P. $= 1{,}480°$ K. (*6*)
$\Delta H_V = (31{,}100)$ calories per mole

Zone I (*c*) (298°–675° K.)

$C_p = 12.1 + 2.86 \times 10^{-3} T$ (*82*)
$H_T - H_{298} = -3{,}733 + 12.1 T + 1.43 \times 10^{-3} T^2$

Formation: $Cu + 1/2 I_2 \longrightarrow CuI$

Zone I (298°–386.1° K.)

$\Delta C_p = 1.9 - 4.59 \times 10^{-3} T$
$\Delta H_T = -16{,}850 + 1.9 T - 2.29 \times 10^{-3} T^2$
$\Delta F_T = -16{,}850 - 1.9 T \ln T + 2.29 \times 10^{-3} T^2 + 9.47 T$

Zone II (386.1°–456° K.)

$\Delta C_p = -2.91 + 1.36 \times 10^{-3} T$
$\Delta H_T = -17{,}350 - 2.91 T + 0.68 \times 10^{-3} T^2$
$\Delta F_T = -17{,}350 + 2.91 T \ln T - 0.68 \times 10^{-3} T^2 - 16.1 T$

Zone III (456°–675° K.)

$\Delta C_p = 2.21 + 1.36 \times 10^{-3} T$
$\Delta H_T = -24{,}700 - 2.21 T + 0.68 \times 10^{-3} T^2$
$\Delta F_T = -24{,}700 - 2.21 T \ln T - 0.68 \times 10^{-3} T^2 + 31.3 T$

(estimated (*11*))

T, ° K.	H_T-H_{298}	S_T	ΔH_T°	ΔF_T°
298	----------	23.1	−16,500	−17,000
400	1,335	26.95	−18,400	−16,900
500	2,670	29.95	−23,400	−16,100
600	4,040	32.44	−23,100	−14,700
1,000	(13,000)	----------	(−18,500)	(−9,500)
1,500	(52,000)	----------	(+11,500)	(−5,800)

Tricopper Nitride, Cu_3N(*c*)

$\Delta H^\circ_{298} = 17{,}800$ calories per mole (*9*)
Metastable, decomposes > 740° K.

Copper Nitride, CuN (*c*)

$\Delta H^\circ_{298} = -60{,}230$ calories per mole (*131*)
$S_{298} = 39.68$ *e.u.*
$\Delta F^\circ_{298} = -62{,}850$ calories per mole

Copper Trinitride, CuN_3 (*c*)

$\Delta H^\circ_{298} = 67{,}230$ calories per mole (*43*)

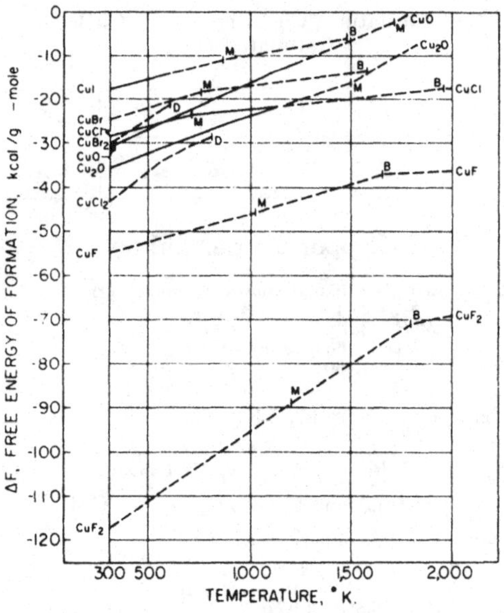

FIGURE 16.—Copper.

DYSPROSIUM AND ITS COMPOUNDS

Element, Dy (*c*)

S_{298} = 17.87 *e.u.* (*127*)
$M.P.$ = 1,673° K. (*125*)
ΔH_M = 4,100 calories per atom
$B.P.$ = 2,600° K. (*125*)
ΔH_V = 67,000 calories per atom

Data above 298° K. estimated by (*130*)

T, ° K.	H_T-H_{298}	S_T	$-\frac{(F_T-H_{298})}{T}$
298	....	17.87	17.87
400	(670)	(19.81)	(18.14)
500	(1,350)	(21.31)	(18.61)
600	(2,040)	(22.58)	(19.18)
700	(2,750)	(23.67)	(19.75)
800	(3,480)	(24.64)	(20.29)
900	(4,220)	(25.52)	(20.84)
1,000	(4,990)	(26.32)	(21.33)
1,100	(5,760)	(27.06)	(21.83)
1,200	(6,560)	(27.76)	(22.30)
1,300	(7,370)	(28.41)	(22.75)
1,400	(8,200)	(29.02)	(23.17)
1,500	(9,050)	(29.61)	(23.58)
1,600	(9,911)	(30.16)	(23.97)
1,700	(10,790)	(30.70)	(24.36)
1,800	(15,760)	(33.51)	(24.76)
1,900	(16,560)	(33.94)	(25.23)
2,000	(17,360)	(34.36)	(25.68)

Dysprosium Trifluoride, DyF_3 (*c*)

ΔH°_{298} = (−373,000) calories per mole (*5*)
S_{298} = (25) *e.u.* (*11*)
$M.P.$ = (1,427°) K. (*29*)
ΔH_M = (8,000) calories per mole
$B.P.$ = (2,500°) K. (*6*)
ΔH_V = (60,000) calories per mole

Formation: $Dy+3/2F_2 \longrightarrow DyF_3$
(estimated (*11*))

T, ° K.	H_T-H_{298}	ΔH°_T	ΔF°_T
298	....	(−373,000)	(−355,000)
500	(4,000)	(−372,500)	(−343,000)
1,000	(17,000)	(−370,000)	(−313,000)
1,500	(32,000)	(−365,500)	(−287,500)

Dysprosium Trichloride, $DyCl_3$ (*c*)

ΔH°_{298} = (−211,000) calories per mole (*5*)
S_{298} = (40) *e.u.* (*11*)
$M.P.$ = 920° K. (*29*)
ΔH_M = (7,000) calories per mole
$B.P.$ = (1,800°) K. (*6*)
ΔH_V = (45,000) calories per mole

Formation: $Dy+3/2Cl_2 \longrightarrow DyCl_3$
(estimated (*11*))

T, ° K.	H_T-H_{298}	ΔH°_T	ΔF°_T
298	....	(−211,000)	(−195,000)
500	(5,000)	(−210,000)	(−185,000)
1,000	(19,000)	(−206,000)	(−161,000)
1,500	(43,000)	(−202,500)	(−142,000)

Dysprosium Tribromide, $DyBr_3$ (*c*)

ΔH°_{298} = (−173,000) calories per mole (*5*)
S_{298} = (45) *e.u.* (*11*)
$M.P.$ = 1,152° K. (*29*)
ΔH_M = (9,000) calories per mole
$B.P.$ = (1,750°) K. (*6*)
ΔH_V = (44,000) calories per mole

Formation: $Dy+3/2Br_2 \longrightarrow DyBr_3$
(estimated (*11*))

T, ° K.	H_T-H_{298}	ΔH°_T	ΔF°_T
298	....	(−173,000)	(−166,500)
500	(5,000)	(−183,500)	(−156,000)
1,000	(18,000)	(−181,000)	(−131,000)
1,500	(43,000)	(−167,000)	(−112,000)

Dysprosium Triiodide, DyI_3 (*c*)

ΔH°_{298} = −143,700 calories per mole (*5*)
S_{298} = (47) *e.u.* (*11*)
$M.P.$ = 1,243° K. (*29*)
ΔH_M = (10,000) calories per mole
$B.P.$ = (1,590°) K. (*6*)
ΔH_V = (41,000) calories per mole

Formation: $Dy+3/2I_2 \longrightarrow DyI_3$
(estimated (*11*))

T, ° K.	H_T-H_{298}	ΔH°_T	ΔF°_T
298	....	−143,700	(−141,000)
500	(5,000)	(−165,000)	(−143,600)
1,000	(19,000)	(−161,500)	(−108,000)
1,500	(46,000)	(−145,000)	(−84,000)

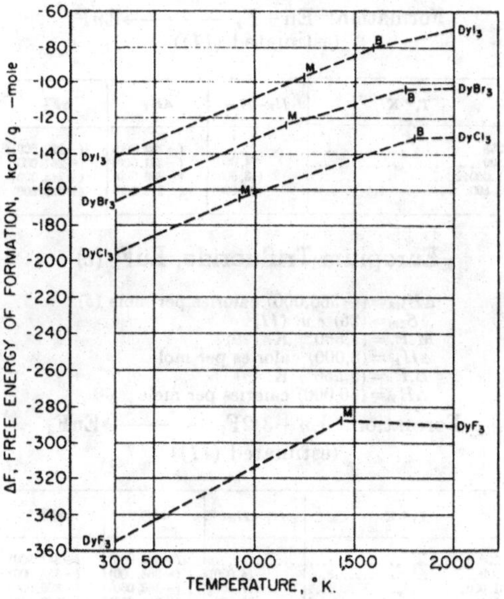

FIGURE 17.—Dysprosium.

ERBIUM AND ITS COMPOUNDS

Element, Er (*c*)

S_{298}=17.48 *e.u.* (*122*)
M.P.=1,800° K. (*125*)
ΔH_M=4,100 calories per atom
B.P.=2,900° K. (*125*)
ΔH_V=70,000 calories per atom

(estimated (*130*))

T, ° K.	H_T-H_{298}	S_T	$-\frac{(F_T-H_{298})}{T}$
298	....	17.48	17.48
400	(690)	(19.47)	(17.75)
500	(1,390)	(21.02)	(18.24)
600	(2,095)	(22.32)	(18.83)
700	(2,820)	(23.43)	(19.41)
800	(3,560)	(24.42)	(19.97)
900	(4,310)	(25.31)	(20.53)
1,000	(5,080)	(26.12)	(21.04)
1,100	(5,870)	(26.87)	(21.54)
1,200	(6,670)	(27.56)	(22.01)
1,300	(7,480)	(28.21)	(22.46)
1,400	(8,310)	(28.83)	(22.90)
1,500	(9,160)	(29.41)	(23.31)
1,600	(10,020)	(29.97)	(23.71)
1,700	(10,890)	(30.50)	(24.10)
1,800	(15,880)	(33.29)	(24.47)
1,900	(16,680)	(33.72)	(24.95)
2,000	(17,480)	(34.13)	(25.39)

Erbium Trifluoride, ErF_3 (*c*)

ΔH°_{298}=(−367,000) calories per mole (*5*)
S_{298}=(25) *e.u.* (*11*)
M.P.=(1,413°) K. (*29*)
ΔH_M=(8,000) calories per mole
B.P.=(2,500°) K. (*6*)
ΔH_V=(60,000) calories per mole

647940 O—63——4

Formation: $Er+3/2F_2 \longrightarrow ErF_3$
(estimated (*11*))

T, ° K.	H_T-H_{298}	ΔH°_T	ΔF°_T
298	....	(−367,000)	(−349,000)
500	(4,000)	(−367,000)	(−337,000)
1,000	(17,000)	(−364,000)	(−308,000)
1,500	(32,000)	(−359,500)	(−281,500)

Erbium Trichloride, $ErCl_3$ (*c*)

ΔH°_{298}=−229,070 calories per mole (*127*)
S_{298}=35.1 *e.u.* (*127*)
M.P.=(1,049°) K. (*29*)
ΔH_M=(8,000) calories per mole
B.P.=(1,770°) K. (*6*)
ΔH_V=(44,000) calories per mole

Formation: $Er+3/2Cl_2 \longrightarrow ErCl_3$
(estimated (*11*))

T, ° K.	H_T-H_{298}	ΔH°_T	ΔF°_T
298	....	−229,000	−211,400
500	(5,000)	(−228,000)	(−201,900)
1,000	(19,000)	(−224,000)	(−177,100)
1,500	(43,000)	(−210,500)	(−158,600)

Erbium Tribromide, $ErBr_3$ (*c*)

ΔH°_{298}=(−169,000) calories per mole (*5*)
S_{298}=(44) *e.u.* (*11*)
M.P.=1,196° K. (*29*)
ΔH_M=(10,000) calories per mole
B.P.=(1,730°) K. (*6*)
ΔH_V=(43,000) calories per mole

Formation: $Er+3/2Br_2 \longrightarrow ErBr_3$
(estimated (*11*))

T, ° K.	H_T-H_{298}	ΔH°_T	ΔF°_T
298	....	(−169,000)	(−162,000)
500	(5,000)	(−179,500)	(−153,000)
1,000	(18,000)	(−177,000)	(−126,000)
1,500	(43,000)	(−163,000)	(−106,500)

Erbium Triiodide, ErI_3 (*c*)

ΔH°_{298}=−140,000 calories per mole (*5*)
S_{298}=(47) *e.u.* (*11*)
M.P.=1,273° K. (*29*)
ΔH_M=(10,000) calories per mole
B.P.=(1,550°) K. (*6*)
ΔH_V=(40,000) calories per mole

Formation: $Er+3/2I_2 \longrightarrow ErI_3$
(estimated (*11*))

T, ° K.	H_T-H_{298}	ΔH°_T	ΔF°_T
298	....	−140,000	(−137,000)
500	(5,000)	(−161,500)	(−132,000)
1,000	(19,000)	(−157,500)	(−104,000)
1,500	(44,000)	(−143,000)	(−79,500)

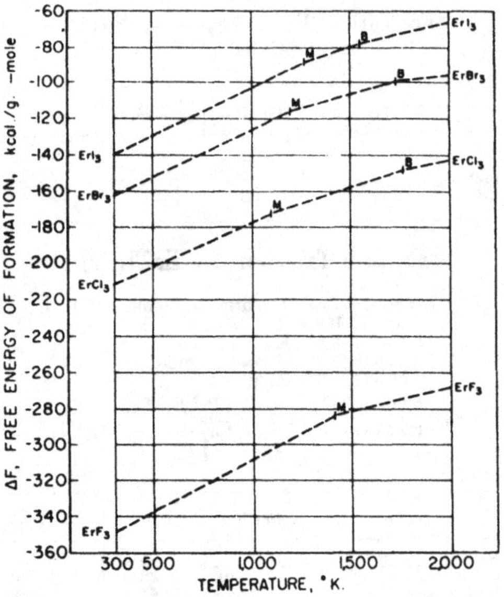

FIGURE 18.—Erbium.

EUROPIUM AND ITS COMPOUNDS

Element, Eu (c)

S_{298} = (17.0) *e.u.* (*130*)
M.P. = (1,173°) K. (*125*)
ΔH_M = 2,500 calories per atom
B.P. = (1,700°) K. (*125*)
ΔH_V = 40,000 calories per atom

(estimated (*130*))

T, ° K.	$H_T - H_{298}$	S_T	$-\frac{(F_T - H_{298})}{T}$
298		(17.0)	(17.0)
400	(660)	(18.91)	(17.26)
500	(1,330)	(20.40)	(17.74)
600	(2,020)	(21.66)	(18.30)
700	(2,730)	(22.76)	(18.86)
800	(3,460)	(23.73)	(19.41)
900	(4,210)	(24.62)	(19.95)
1,000	(4,980)	(25.43)	(20.45)
1,100	(8,270)	(28.45)	(20.94)
1,200	(9,070)	(29.15)	(21.60)
1,300	(9,870)	(29.79)	(22.20)
1,400	(10,670)	(30.38)	(22.76)
1,500	(11,470)	(30.93)	(23.29)
1,600	(12,270)	(31.45)	(23.79)
1,700	(13,070)	(31.94)	(24.26)
1,800	(48,670)	(53.49)	(25.89)
1,900	(49,170)	(53.26)	(27.38)
2,000	(49,680)	(53.57)	(28.73)

Europium Difluoride, EuF_2 (c)

ΔH°_{298} = (−282,000) calories per mole (*5*)
S_{298} = (20) *e.u.* (*11*)
M.P. = (1,571°) K. (*29*)
ΔH_M = (5,000) calories per mole
B.P. = (2,700°) K. (*6*)
ΔH_V = (78,000) calories per mole

Formation: $Eu + F_2 \longrightarrow EuF_2$
(estimated (*11*))

T, ° K.	$H_T - H_{298}$	ΔH°_T	ΔF°_T
298		(−282,000)	(−270,000)
500	(4,000)	(−281,000)	(−261,000)
1,000	(13,000)	(−280,000)	(−243,000)
1,500	(24,000)	(−279,500)	(−223,500)

Europium Trifluoride, EuF_3 (c)

ΔH°_{298} = (−366,000) calories per mole (*5*)
S_{298} = (25) *e.u.* (*11*)
M.P. = (1,560°) K. (*29*)
ΔH_M = (8,000) calories per mole
B.P. = (2,550°) K. (*6*)
ΔH_V = (60,000) calories per mole

Formation: $Eu + 3/2F_2 \longrightarrow EuF_3$
(estimated (*11*))

T, ° K.	$H_T - H_{298}$	ΔH°_T	ΔF°_T
298		(−366,000)	(−347,800)
500	(4,000)	(−366,000)	(−336,000)
1,000	(17,000)	(−363,000)	(−306,000)
1,500	(32,000)	(−361,000)	(−279,000)

Europium Dichloride, $EuCl_2$ (c)

ΔH°_{298} = (−192,000) calories per mole (*5*)
S_{298} = (30) *e.u.* (*11*)
M.P. = (1,000°) K. (*29*)
ΔH_M = (6,000) calories per mole
B.P. = (2,300°) K. (*6*)
ΔH_V = (55,000) calories per mole

Formation: $Eu + Cl_2 \longrightarrow EuCl_2$
(estimated (*11*))

T, ° K.	$H_T - H_{298}$	ΔH°_T	ΔF°_T
298		(−192,000)	(−181,000)
500	(4,000)	(−191,000)	(−173,500)
1,000	(13,000)	(−190,000)	(−158,000)
1,500	(31,000)	(−183,000)	(−145,500)

Europium Trichloride, $EuCl_3$ (c)

ΔH°_{298} = (−208,000) calories per mole (*5*)
S_{298} = (40) *e.u.* (*11*)
M.P. = 896° K. (*29*)
ΔH_M = (7,000) calories per mole
Decomposes (*6*)

Formation: $Eu + 3/2Cl_2 \longrightarrow EuCl_3$
(estimated (*11*))

T, ° K.	$H_T - H_{298}$	ΔH°_T	ΔF°_T
298		(−208,000)	(−192,000)
500	(5,000)	(−207,000)	(−181,500)
1,000	(19,000)	(−303,000)	(−158,000)
1,500	(43,000)	(−192,000)	(−139,000)

Europium Dibromide, $EuBr_2$ (*c*)

$\Delta H^{\circ}_{298} = (-162{,}000)$ calories per mole (*5*)
$S_{298} = (40)$ *e.u.* (*11*)
$M.P. = (950°)$ K. (*29*)
$\Delta H_M = (6{,}000)$ calories per mole
$B.P. = (2{,}150°)$ K. (*6*)
$\Delta H_V = (50{,}000)$ calories per mole

Formation: $Eu + Br_2 \longrightarrow EuBr_2$
(estimated (*11*))

T, ° K.	$H_T - H_{298}$	ΔH°_T	ΔF°_T
298		(−162, 000)	(−158, 000)
500	(4, 000)	(−169, 000)	(−151, 000)
1,000	(20, 000)	(−161, 000)	(−133, 000)
1,500	(32, 000)	(−160, 000)	(−122, 500)

Europium Tribromide, $EuBr_3$ (*c*)

$\Delta H^{\circ}_{298} = (-166{,}000)$ calories per mole (*5*)
$S_{298} = (46)$ *e.u.* (*11*)
$M.P. = (975°)$ K. (*6*)
$\Delta H_M = (8{,}000)$ calories per mole
Decomposes (*6*)

Formation: $Eu + 3/2Br_2 \longrightarrow EuBr_3$
(estimated (*11*))

T, ° K.	$H_T - H_{298}$	ΔH°_T	ΔF°_T
298		(−166, 000)	(−159, 000)
500	(5, 000)	(−176, 500)	(−149, 500)
1,000	(18, 000)	(−174, 000)	(−125, 000)
1,500	(43, 000)	(−162, 000)	(−106, 500)

Europium Diiodide, EuI_2 (*c*)

$\Delta H^{\circ}_{298} = (-127{,}000)$ calories per mole (*5*)
$S_{298} = (40)$ *e.u.* (*11*)
$M.P. = (800°)$ K. (*29*)
$\Delta H_M = (5{,}000)$ calories per mole
$B.P. = (1{,}850°)$ K. (*6*)
$\Delta H_V = (40{,}000)$ calories per mole

Formation: $Eu + I_2 \longrightarrow EuI_2$
(estimated (*11*))

T, ° K.	$H_T - H_{298}$	ΔH°_T	ΔF°_T
298		(−127, 000)	(−127, 000)
500	(4, 000)	(−141, 000)	(−124, 000)
1,000	(19, 000)	(−134, 000)	(−110, 000)
1,500	(31, 000)	(−133, 000)	(−99, 000)

Europium Triiodide, EuI_3 (*c*)

$\Delta H^{\circ}_{298} = (-112{,}000)$ calories per mole (*5*)
$S_{298} = (48)$ *e.u.* (*11*)
$M.P. = (1{,}150°)$ K. (*29*)
$\Delta H_M = (9{,}000)$ calories per mole
Decomposes (*6*)

Formation: $Eu + 3/2I_2 \longrightarrow EuI_3$
(estimated (*11*))

T, ° K.	$H_T - H_{298}$	ΔH°_T	ΔF°_T
298		(−112, 000)	(−110, 000)
500	(5, 000)	(−133, 000)	(−104, 000)
1,000	(19, 000)	(−129, 500)	(−77, 000)
1,500	(44, 000)	(−118, 000)	(−53, 000)

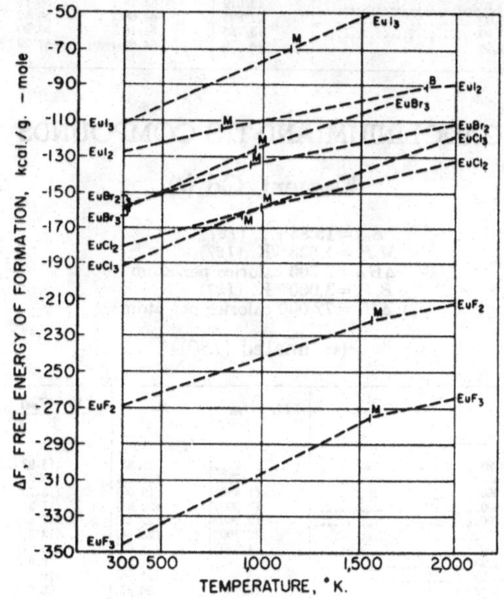

FIGURE 19.—Europium.

FLUORINE

Element, F_2 (*g*)

$S_{298} = 48.56$ *e.u.* (*83*)
$M.P. = 53.54°$ K. (*112*)
$\Delta H_M = 122$ calories per atom
$B.P. = 85.0°$ K. (*112*)
$\Delta H_V = 1{,}562$ calories per atom

Zone I (*g*) (298°–2,000° K.)

$$C_p = 8.29 + 0.44 \times 10^{-3}T - 0.80 \times 10^5 T^{-2} \quad (82)$$
$$H_T - H_{298} = -2{,}760 + 8.29T + 0.22 \times 10^{-3}T^2 + 0.80 \times 10^5 T^{-1}$$
$$F_T - H_{298} = -2{,}760 - 8.29T\ln T - 0.22 \times 10^{-3}T^2 + 0.40 \times 10^5 T^{-1} + 7.3T$$

T, ° K.	H_T-H_{298}	S_T	$-\frac{(F_T-H_{298})}{T}$
298	....	48.56	48.56
400	786	50.83	48.86
500	1,590	52.62	49.44
600	2,420	54.14	50.11
700	3,265	55.44	50.77
800	4,120	56.58	51.43
900	4,980	57.59	52.06
1,000	5,850	58.51	52.66
1,100	6,725	59.34	53.14
1,200	7,600	60.11	53.77
1,300	8,480	60.81	54.29
1,400	9,360	61.46	54.77
1,500	10,240	62.07	55.24
1,600	11,125	62.64	55.69
1,700	12,010	63.18	56.11
1,800	12,895	63.68	56.52
1,900	13,785	64.16	56.90
2,000	14,670	64.62	57.29

GADOLINIUM AND ITS COMPOUNDS

Element, Gd (c)

S_{298} = 15.83 *e.u.* (*121*)
$M.P.$ = 1,523° K. (*127*)
ΔH_M = 3,700 calories per atom
$B.P.$ = 3,000° K. (*127*)
ΔH_V = 72,000 calories per atom

(estimated (*130*))

T, ° K.	H_T-H_{298}	S_T	$-\frac{(F_T-H_{298})}{T}$
298	....	15.83	15.83
400	(780)	(18.12)	(15.93)
500	(1,480)	(19.66)	(16.70)
600	(2,200)	(20.98)	(17.31)
700	(2,940)	(22.13)	(17.93)
800	(3,700)	(23.14)	(18.51)
900	(4,480)	(24.05)	(19.07)
1,000	(5,270)	(24.89)	(19.62)
1,100	(6,080)	(25.66)	(20.13)
1,200	(6,900)	(26.37)	(20.62)
1,300	(7,740)	(27.05)	(21.09)
1,400	(8,600)	(27.68)	(21.53)
1,500	(9,480)	(28.29)	(21.97)
1,600	(14,070)	(31.17)	(22.37)
1,700	(14,870)	(31.66)	(22.91)
1,800	(15,670)	(32.12)	(23.41)
1,900	(16,470)	(32.55)	(23.88)
2,000	(17,270)	(32.96)	(24.32)

Digadolinium Trioxide, Gd_2O_3 (c)

ΔH°_{298} = −433,940 ± 860 calories per mole (*59*)

Gadolinium Trifluoride, GdF_3 (c)

ΔH°_{298} = (−379,000) calories per mole (*5*)
S_{298} = (25) *e.u.* (*11*)
$T.P.$ = 1,280° K. (*29*)
$M.P.$ = 1,650° K. (*29*)
ΔH_M = (8,000) calories per mole
$B.P.$ = (2,550°) K. (*6*)
ΔH_V = (60,000) calories per mole

Formation: $Gd + 3/2F_2 \longrightarrow GdF_3$
(estimated (*11*))

T, ° K.	H_T-H_{298}	ΔH°_T	ΔF°_T
298	....	(−379,000)	(−361,000)
500	(4,000)	(−379,000)	(−349,000)
1,000	(17,000)	(−376,000)	(−319,000)
1,500	(32,000)	(−372,000)	(−290,500)

Gadolinium Trichloride, $GdCl_3$ (c)

ΔH°_{298} = −240,080 calories per mole (*127*)
S_{298} = 34.9 *e.u.* (*127*)
$M.P.$ = 882° K. (*29*)
ΔH_M = (7,000) calories per mole
$B.P.$ = (1,850°) K. (*6*)
ΔH_V = (45,000) calories per mole

Formation: $Gd + 3/2Cl_2 \longrightarrow GdCl_3$
(estimated (*11*))

T, ° K.	H_T-H_{298}	ΔH°_T	ΔF°_T
298	....	−240,100	−222,000
500	(5,000)	(−239,000)	(−211,000)
1,000	(19,000)	(−235,500)	(−188,000)
1,500	(43,000)	(−222,000)	(−170,000)

Gadolinium Tribromide, $GdBr_3$ (c)

ΔH°_{298} = −178,000 calories per mole (*11*)
S_{298} = (46) *e.u.* (*11*)
$M.P.$ = 1,043° K. (*29*)
ΔH_M = (8,000) calories per mole
$B.P.$ = (1,760°) K. (*6*)
ΔH_V = (44,000) calories per mole

Formation: $Gd + 3/2Br_2 \longrightarrow GdBr_3$
(estimated (*11*))

T, ° K.	H_T-H_{298}	ΔH°_T	ΔF°_T
298	....	−178,000	(−171,500)
500	(5,000)	(−189,000)	(−161,500)
1,000	(18,000)	(−186,000)	(−137,000)
1,500	(43,000)	(−172,000)	(−118,500)

Gadolinium Triiodide, GdI_3 (c)

ΔH°_{298} = −147,000 calories per mole (*11*)
S_{298} = (48) *e.u.*
$M.P.$ = 1,199° K. (*29*)
ΔH_M = (10,000) calories per mole
$B.P.$ = (1,610°) K. (*6*)
ΔH_V = (40,000) calories per mole

Formation: $Gd + 3/2I_2 \longrightarrow GdI_3$
(estimated (*11*))

T, ° K.	H_T-H_{298}	ΔH°_T	ΔF°_T
298	....	−147,000	(−144,000)
500	(5,000)	(−168,000)	(−140,500)
1,000	(19,000)	(−165,000)	(−113,000)
1,500	(44,000)	(−151,000)	(−89,000)

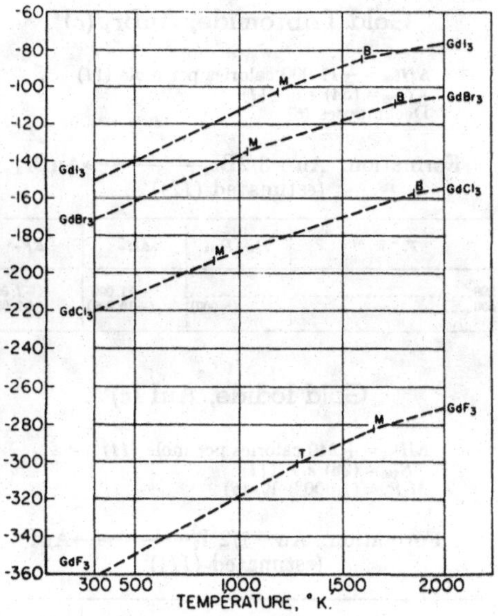

FIGURE 20.—Gadolinium.

GOLD AND ITS COMPOUNDS

Element, Au (*c*)

S_{298} = 11.32 *e.u.* (*37*)
M.P. = 1,336° K. (*82*)
ΔH_M = 2,955 calories per atom
B.P. = 2,980° K. (*130*)
ΔH_V = 77,540 calories per atom

Zone I (*c*) (298°-1,336° K.)

$$C_p = 5.66 + 1.24\times10^{-3}T \ (82)$$
$$H_T - H_{298} = -1{,}743 + 5.66T + 0.62\times10^{-3}T^2$$
$$F_T - H_{298} = -1{,}743 - 5.66T\ln T - 0.62\times10^{-3}T^2 + 26.95T$$

Zone II (*l*) (1,336°-1,600° K.)

$$C_p = 7.00 \ (82)$$
$$H_T - H_{298} = 530 + 7.00T$$
$$F_T - H_{298} = 530 - 7.00T\ln T + 34.1T$$

T, ° K.	$H_T - H_{298}$	S_T	$-\frac{(F_T - H_{298})}{T}$
298	....	11.32	11.32
400	625	13.12	11.56
500	1,245	14.51	12.02
600	1,880	15.66	12.53
700	2,530	16.67	13.06
800	3,180	17.53	13.56
900	3,850	18.32	14.04
1,000	4,530	19.04	14.51
1,100	5,220	19.70	14.95
1,200	5,930	20.32	15.37
1,300	6,660	20.90	15.78
1,400	10,330	23.64	16.26
1,500	11,030	24.12	16.77
1,600	11,730	24.57	17.24
1,700	(12,430)	(24.99)	(17.68)
1,800	(13,130)	(25.39)	(18.10)
1,900	(13,830)	(25.77)	(18.50)
2,000	(14,530)	(26.13)	(18.87)

Digold Trioxide, Au_2O_3 (*c*)

ΔH°_{298} = (−800) calories per mole (*24*)
S_{298} = (31) *e.u.* (*24*)

Formation: $2Au + 3/2O_2 \longrightarrow Au_2O_3$
(estimated (*24*))

T, ° K.	$H_T - H_{298}$	ΔH°_T	ΔF°_T
298	....	(−800)	(18,550)
400	(2,800)	(−350)	(25,100)
500	(6,000)	(+100)	(31,450)

Gold Fluoride, AuF (*g*)

ΔH°_{298} = −18,000 calories per mole (*11*)
S_{298} = (23) *e.u.* (*11*)

Formation: $Au + 1/2F_2 \longrightarrow AuF$
(estimated (*11*))

T, ° K.	$H_T - H_{298}$	ΔH°_T	ΔF°_T
298	....	−18,000	(−14,100)
500	(3,000)	(−17,000)	(−11,500)
1,000	(12,000)	(−13,500)	(−9,000)

Gold Difluoride, AuF_2 (*c*)

ΔH°_{298} = (−57,000) calories per mole (*11*)
S_{298} = (28) *e.u.* (*11*)

Formation: $Au + F_2 \longrightarrow AuF_2$
(estimated (*11*))

T, ° K.	$H_T - H_{298}$	ΔH°_T	ΔF°_T
298	....	(−57,000)	(−47,400)
500	(4,000)	(−55,800)	(−40,500)

Gold Trifluoride, AuF_3 (*c*)

ΔH°_{298} = (−100,000) calories per mole (*11*)
S_{298} = (38) *e.u.* (*11*)
M.P. = (1,000°) K. (*6*)

Formation: $Au + 3/2F_2 \longrightarrow AuF_3$
(estimated (*11*))

T, ° K.	$H_T - H_{298}$	ΔH°_T	ΔF°_T
298	....	(−100,000)	(−86,200)
500	(4,000)	(−99,600)	(−77,500)

Gold Chloride, AuCl (*c*)

ΔH°_{298} = −8,400 calories per mole (*112*)
S_{298} = (24) *e.u.* (*11*)
B.P. = (1,600°) K. (*6*)

Formation: $Au + 1/2Cl_2 \longrightarrow AuCl$
(estimated (*11*))

T, ° K.	H_T-H_{298}	ΔH_T°	ΔF_T°
298		-8,400	(-4,200)
500	(3,000)	(-7,500)	(-1,400)
1,000	(13,000)	(-2,900)	(+2,600)

Gold Dichloride, $AuCl_2$ (*c*)

$\Delta H_{298}^\circ = -18,100$ calories per mole (*112*)
$S_{298} = (36)$ *e.u.* (*11*)
Decomposes = >460° K. (*6*)

Formation: $Au + Cl_2 \longrightarrow AuCl_2$
(estimated (*11*))

T, ° K.	H_T-H_{298}	ΔH_T°	ΔF_T°
298		-18,100	(-9,500)
500	(4,000)	(-17,000)	(-3,600)

Gold Trichloride, $AuCl_3$ (*c*)

$\Delta H_{298}^\circ = -28,300$ calories per mole (*112*)
$S_{298} = (45)$ *e.u.* (*6*)
M.P. = 561° K. (*6*)
B.P. = (700°) K.

Formation: $Au + 3/2Cl_2 \longrightarrow AuCl_3$
(estimated (*11*))

T, ° K.	H_T-H_{298}	ΔH_T°	ΔF_T°
298		-28,300	(-14,600)
500	(5,000)	(-27,100)	(-5,300)

Gold Bromide, AuBr (*c*)

$\Delta H_{298}^\circ = -3,300$ calories per mole (*112*)
$S_{298} = (27)$ *e.u.* (*11*)
M.P. = (1,600°) K. (*6*)

Formation: $Au + 1/2Br_2 \longrightarrow AuBr$
(estimated (*11*))

T, ° K.	H_T-H_{298}	ΔH_T°	ΔF_T°
298		-3,300	(-2,550)
500	(3,000)	(-300)	(0)

Gold Dibromide, $AuBr_2$ (*c*)

$\Delta H_{298}^\circ = -5,550$ calories per mole (*11*)
$S_{298} = (39)$ *e.u.* (*11*)
Decomposes (*6*)

Formation: $Au + Br_2 \longrightarrow AuBr_2$
(estimated (*11*))

T, ° K.	H_T-H_{298}	ΔH_T°	ΔF_T°
298		-5,550	(-2,900)
500	(4,000)	(12,300)	(+3,100)

Gold Tribromide, $AuBr_3$ (*c*)

$\Delta H_{298}^\circ = -11,000$ calories per mole (*11*)
$S_{298} = (54)$ *e.u.* (*11*)
Decomposes (*6*)

Formation: $Au + 3/2Br_2 \longrightarrow AuBr_3$
(estimated (*11*))

T, ° K.	H_T-H_{298}	ΔH_T°	ΔF_T°
298		-11,000	(-7,600)
500	(5,000)	(-16,800)	(+2,000)

Gold Iodide, AuI (*c*)

$\Delta H_{298}^\circ = +240$ calories per mole (*11*)
$S_{298} = (28)$ *e.u.* (*11*)
M.P. = (1,600°) K. (*6*)

Formation: $Au + 1/2\ I_2 \longrightarrow AuI$
(estimated (*11*))

T, ° K.	H_T-H_{298}	ΔH_T°	ΔF_T°
298		240	(-400)
500	(3,000)	(-9,000)	(+300)

Gold Diiodide, AuI_2 (*c*)

$\Delta H_{298}^\circ = (+6,900)$ calories per mole (*11*)
$S_{298} = (39)$ *e.u.* (*11*)
Decomposes (*6*)

Formation: $Au + I_2 \longrightarrow AuI_2$
(estimated (*11*))

T, ° K.	H_T-H_{298}	ΔH_T°	ΔF_T°
298		(+6,900)	(+6,900)
500	(4,000)	(-7,000)	(+8,000

Gold Triiodide, AuI_3 (*c*)

$\Delta H_{298}^\circ = (+8,300)$ calories per mole (*11*)
$S_{298} = (50)$ *e.u.* (*11*)
Decomposes (*6*)

Formation: $Au + 3/2I_2 \longrightarrow AuI_3$
(estimated (*11*))

T, ° K.	H_T-H_{298}	ΔH_T°	ΔF_T°
298		(8,300)	(+9,200)
500	(5,000)	(-13,000)	(+9,500)

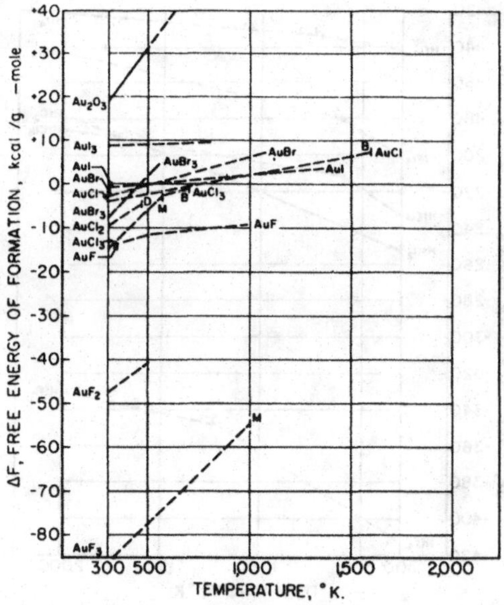

FIGURE 21.—Gold.

HAFNIUM AND ITS COMPOUNDS

Element, Hf (*c*)

$S_{298}=13.1$ *e.u.* (*83*)
M.P.$=2{,}488°$ K. (*85*)
$\Delta H_M=(6{,}000)$ calories per atom
B.P.$=5{,}500°$ K. (*7*)
$\Delta H_V=155{,}000$ calories per atom

Zone I (*c*) (298°–2,488° K.)

$$C_p=6.00+0.52\times10^{-3}T \ (82)$$
$$H_T-H_{298}=-1{,}810+6.00T+0.26\times10^{-3}T^2$$
$$F_T-H_{298}=-1{,}810-6.00T\ln T-0.26\times10^{-3}T^2+27.16T$$

T, ° K.	H_T-H_{298}	S_T	$-\frac{(F_T-H_{298})}{T}$
298	....	13.1	13.1
400	670	14.95	13.30
500	1,320	16.40	13.70
600	1,980	17.55	14.27
700	2,645	18.45	14.61
800	3,320	19.26	15.26
900	4,010	20.10	15.68
1,000	4,710	20.81	16.10
1,100	5,420	21.44	16.54
1,200	6,140	21.96	16.84
1,300	6,870	22.51	17.31
1,400	7,610	23.07	17.58
1,500	8,360	23.52	17.91
1,600	9,120	23.92	18.22
1,700	9,890	24.32	18.59
1,800	10,595	24.77	18.78
1,900	11,470	25.03	19.11
2,000	12,270	25.43	19.31
2,500	(16,440)	(28.14)	(21.56)

Hafnium Dioxide, HfO_2 (*c*)

$\Delta H^{\circ}_{298}=-266{,}050$ calories per mole (*66*)
$S_{298}=14.18$ *e.u.* (*132*)
M.P.$=3{,}063°$ K. (*8*)

Zone I (*c*) (298°–1,800° K.)

$$C_p=17.39+2.08\times10^{-3}T-3.48\times10^5T^{-2} \ (105)$$
$$H_T-H_{298}=-6{,}440+17.39T+1.04\times10^{-3}T^2+3.48\times10^5T^{-1}$$

Formation: $Hf+O_2 \longrightarrow HfO_2$

Zone I (298°–1,800° K.)

$$\Delta C_p=4.23+0.56\times10^{-3}T-3.08\times10^5T^{-2}$$
$$\Delta H_T=-268{,}400+4.23T+0.28\times10^{-3}T^2+3.08\times10^5T^{-1}$$
$$\Delta F_T=-268{,}400-4.23T\ln T-0.28\times10^{-3}T^2+1.54\times10^5T^{-1}+78.16T$$

T, ° K.	H_T-H_{298}	S_T	ΔH°_T	ΔF°_T
298	....	14.81	−266,050	−251,750
400	1,550	19.61	−265,900	−246,900
500	3,150	22.24	−265,650	−242,150
600	4,900	25.40	−265,350	−237,550
700	6,700	28.19	−264,950	−232,950
800	8,550	30.67	−264,600	−228,650
900	10,450	32.88	−264,200	−224,100
1,000	12,350	34.88	−263,850	−219,700
1,100	14,300	36.72	−263,400	−215,300
1,200	16,250	38.42	−263,100	−211,200
1,300	18,200	40.00	−262,600	−206,900
1,400	20,200	41.48	−262,200	−202,500
1,500	22,200	42.87	−261,850	−198,350
1,600	24,250	44.19	−261,350	−194,350
1,700	26,300	45.44	−260,950	−190,250
1,800	28,400	46.63	−260,350	−186,050
1,900	(30,500)	(47.76)	(−259,100)	(−181,400)
2,000	(32,600)	(48.85)	(−258,550)	(−177,600)
2,500	....	....	(−255,850)	(−159,600)

Hafnium Tetrafluoride, HfF_4 (*c*)

$\Delta H^{\circ}_{298}=(-435{,}000)$ calories per mole (*11*)
$S_{298}=(35)$ *e.u.* (*11*)
S.P.$=(1{,}200°)$ K. (*6*)
$\Delta H_{subl}=(63{,}000)$ calories per mole

Formation: $Hf+2F_2 \longrightarrow HfF_4$
(estimated (*11*))

T, ° K.	H_T-H_{298}	ΔH°_T	ΔF°_T
298	....	(−435,000)	(−412,500)
500	(6,000)	(−433,500)	(−398,500)
1,000	(22,000)	(−429,500)	(−363,000)

Hafnium Tetrachloride, $HfCl_4$ (*c*)

$\Delta H^{\circ}_{298}=(-255{,}000)$ calories per mole (*11*)
$S_{298}=45.6$ *e.u.* (*132*)
S.P.$=590°$ K. (*6*)
$\Delta H_{subl}=(24{,}000)$ calories per mole

Zone I (*c*) (298°–485° K.)

$$C_p=31.47-2.38\times10^5T^{-2} \ (105)$$
$$H_T-H_{298}=-10{,}180+31.47T+2.38\times10^5T^{-1}$$

Formation: $Hf+2Cl_2 \longrightarrow HfCl_4$

Zone I (298°–485° K.)

$\Delta C_p = 7.83 - 0.64\times10^{-3}T - 1.02\times10^{5}T^{-2}$
$\Delta H_T = -257{,}650 + 7.83T - 0.32\times10^{-3}T^2 + 1.02\times10^{5}T^{-1}$
$\Delta F_T = -257{,}650 - 7.83T\ln T + 0.32\times10^{-3}T^2 + 0.51\times10^{5}T^{-1} + 126.78T$

Zone II (500°–2,000° K.)
(estimated (*42*))

T, ° K.	H_T-H_{298}	ΔH_T°	ΔF_T°
298		(−255, 000)	(−232, 900)
400	3, 000	(−254, 350)	(−225, 350)
500	6, 000	(−253, 700)	(−218, 200
1,000			(−209, 000)
1,500			(−199, 000)
2,000			(−180, 000)

Hafnium Tetrabromide, $HfBr_4$ (*c*)

$\Delta H_{298}^\circ = (-210{,}000)$ calories per mole (*11*)
$S_{298} = (57)$ *e.u.* (*11*)
S.P. = 595° K. (*6*)
$\Delta H_{subl} = (24{,}000)$ calories per mole

Formation: $Hf+2Br_2 \longrightarrow HfBr_4$
(estimated (*11*))

T, ° K.	H_T-H_{298}	ΔH_T°	ΔF_T°
298		(−210, 000)	(−201, 750)
500	(6, 000)	(−223, 750)	(−187, 500)

Hafnium Tetraiodide, HfI_4 (*c*)

$\Delta H_{298}^\circ = (-145{,}000)$ calories per mole (*11*)
$S_{298} = (62)$ *e.u.* (*11*)
S.P. = (700°) K. (*6*)
$\Delta H_{subl} = (28{,}000)$ calories per mole

Formation: $Hf+2I_2 \longrightarrow HfI_4$
(estimated (*11*))

T, ° K.	H_T-H_{298}	ΔH_T°	ΔF_T°
298		(−145, 000)	(−142, 000)
500	(6, 000)	(−177, 700)	(−137, 500)

Hafnium Nitride, HfN (*c*)

$\Delta H_{298}^\circ = -88{,}240$ calories per mole (*66*)
$S_{298} = 13.1$ *e.u.* (*66*)
$\Delta F_{298} = 81{,}400$ calories per mole
M.P. = 3,580° K. (*9*)

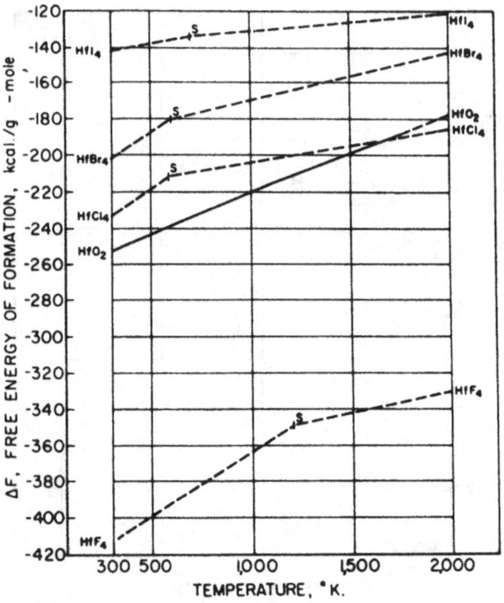

FIGURE 22.—Hafnium.

HOLMIUM AND ITS COMPOUNDS

Element, Ho (*c*)

$S_{298} = (17.77)$ *e.u.* (*121*)
M.P. = 1,773° K. (*125*)
$\Delta H_M = 4{,}100$ calories per atom
B.P. = 2,600° K. (*125*)
$\Delta H_V = 67{,}000$ calories per atom

(estimated (*130*))

T, ° K.	H_T-H_{298}	S_T	$-\frac{(F_T-H_{298})}{T}$
298		(17. 77)	(17. 77)
400	(670)	(19, 71)	(18. 04)
500	(1, 350)	(21. 21)	(18. 51)
600	(2, 040)	(22. 48)	(19. 08)
700	(2, 750)	(23. 57)	(19. 65)
800	(3, 480)	(24. 54)	(20. 19)
900	(4, 220)	(25. 42)	(20. 74)
1, 000	(4, 985)	(26. 22)	(21. 24)
1, 100	(5, 760)	(26. 96)	(21. 73)
1, 200	(6, 560)	(27. 66)	(22. 20)
1, 300	(7, 370)	(28. 31)	(22. 66)
1, 400	(8, 200)	(28. 92)	(23. 07)
1, 500	(9, 050)	(29. 51)	(23. 48)
1, 600	(9, 910)	(30. 06)	(23. 87)
1, 700	(10, 790)	(30. 60)	(24. 26)
1, 800	(15, 760)	(33. 41)	(24. 66)
1, 900	(16, 560)	(33. 84)	(25. 13)
2, 000	(17, 360)	(34. 26)	(25. 58)
2, 500	(21, 360)	(36. 04)	(27. 50)

Holmium Trifluoride, HoF_3 (*c*)

$\Delta H^\circ_{298} = -370{,}000$ calories per mole (*5*)
$S_{298} = (25)$ *e.u.* (*11*)
$M.P. = 1{,}416°$ K. (*29*)
$\Delta H_M = (8{,}000)$ calories per mole
$B.P. = (2{,}500°)$ K. (*6*)
$\Delta H_V = (60{,}000)$ calories per mole

Formation: $Ho + 3/2F_2 \longrightarrow HoF_3$
(estimated (*11*))

T, ° K.	$H_T - H_{298}$	ΔH°_T	ΔF°_T
298		−370, 000	(−352, 000)
500	(4, 000)	(−370, 000)	(−340, 000)
1, 000	(17, 000)	(−367, 000)	(−311, 000)
1, 500	(32, 000)	(−362, 500)	(−285, 000)

Holmium Trichloride, $HoCl_3$ (*c*)

$\Delta H^\circ_{298} = -233{,}000$ calories per mole (*5*)
$S_{298} = (39)$ *e.u.* (*11*)
$M.P. = 99.°$ K. (*29*)
$\Delta H_M = (8{,}000)$ calories per mole
$B.P. = (1{,}780°)$ K. (*6*)
$\Delta H_V = (44{,}000)$ calories per mole

Formation: $Ho + 3/2Cl_2 \longrightarrow HoCl_3$
(estimated (*11*))

T, ° K.	$H_T - H_{298}$	ΔH°_T	ΔF°_T
298		−233, 000	(−217, 000)
500	(5, 000)	(−232, 000)	(−206, 000)
1, 000	(19, 000)	(−228, 000)	(−181, 000)
1, 500	(43, 000)	(−224, 500)	(−162, 500)

Holmium Tribromide, $HoBr_3$ (*c*)

$\Delta H^\circ_{298} = (-171{,}000)$ calories per mole (*5*)
$S_{298} = (45)$ *e.u.* (*11*)
$M.P. = 1{,}192°$ K. (*29*)
$\Delta H_M = (10{,}000)$ calories per mole
$B.P. = (1{,}740°)$ K. (*6*)
$\Delta H_V = (43{,}000)$ calories per mole

Formation: $Ho + 3/2Br_2 \longrightarrow HoBr_3$
(estimated (*11*))

T, ° K.	$H_T - H_{298}$	ΔH°_T	ΔF°_T
298		(−171, 000)	(−164, 500)
500	(5, 000)	(−181, 500)	(−154, 000)
1, 000	(18, 000)	(−179, 000)	(−129, 000)
1, 500	(43, 000)	(−165, 000)	(−110, 000)

Holmium Triiodide, HoI_3 (*c*)

$\Delta H^\circ_{298} = (-142{,}000)$ calories per mole (*5*)
$S_{298} = (47)$ *e.u.* (*11*)
$M.P. = 1{,}262°$ K. (*29*)
$\Delta H_M = (10{,}000)$ calories per mole
$B.P. = (1{,}570°)$ K. (*6*)
$\Delta H_V = (41{,}000)$ calories per mole

Formation: $Ho + 3/2I_2 \longrightarrow HoI_3$
(estimated (*11*))

T, ° K.	$H_T - H_{298}$	ΔH°_T	ΔF°_T
298		(−142, 000)	(−139, 000)
500	(5, 000)	(−163, 000)	(−134, 500)
1, 000	(19, 000)	(−159, 500)	(−106, 000)
1, 500	(44, 000)	(−145, 500)	(−81, 500)

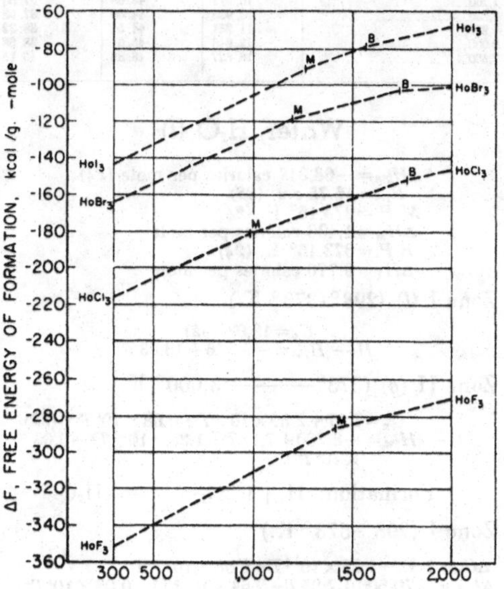

FIGURE 23.—Holmium.

HYDROGEN AND ITS COMPOUNDS

Element, H_2 (*g*)

$S_{298} = 31.22$ *e.u.* (*83*)
$M.P. = 13.96°$ K. (*130*)
$\Delta H_M = 28$ calories per atom
$B.P. = 20.39°$ K. (*130*)
$\Delta H_V = 216$ calories per atom

Zone I (*g*) (298°–3, 000° K.)

$$C_p = 6.52 + 0.78 \times 10^{-3} T + 0.12 \times 10^5 T^{-2} \ (82)$$
$$H_T - H_{298} = -1{,}939 + 6.52T + 0.39 \times 10^{-3} T^2 - 0.12 \times 10^5 T^{-1}$$
$$F_T - H_{298} = -1{,}939 - 6.52T \ln T - 0.39 \times 10^{-3} T^2 - 0.06 \times 10^5 T^{-1} + 12.7T$$

T, ° K.	H_T-H_{298}	S_T	$-\frac{(F_T-H_{298})}{T}$
298	....	31.22	31.22
400	707	33.26	31.50
500	1,406	34.82	32.01
600	2,105	36.09	32.58
700	2,808	37.18	33.17
800	3,514	38.12	33.73
900	4,224	38.96	34.27
1,000	4,942	39.71	34.77
1,100	5,681	40.32	35.15
1,200	6,422	40.99	35.64
1,300	7,180	41.67	36.15
1,400	7,937	42.15	36.49
1,500	8,670	42.73	36.95
1,600	9,571	43.21	37.23
1,700	10,271	43.68	37.63
1,800	10,935	44.06	37.98
1,900	11,851	44.54	38.32
2,000	12,648	45.01	38.68
2,500	16,827	46.88	40.15

Water, H_2O (*l*)

$\Delta H^\circ_{298} = -68,317$ calories per mole (*24*)
$S_{298} = 16.75$ *e.u.* (*83*)
M.P. $= 273.16°$ K. (*24*)
$\Delta H_M = 1,436$ calories per mole
B.P. $= 373.16°$ K. (*24*)
$\Delta H_V = 9,770$ calories per mole

Zone I (*l*) (298°–373° K.)

$$C_p = 18.03 \ (82)$$
$$H_T - H_{298} = -5,376 + 18.03T$$

Zone II (*g*) (373° ——→ 3,000° K.)

$$C_p = 7.17 + 2.56 \times 10^{-3}T + 0.08 \times 10^5 T^{-2} \ (82)$$
$$H_T - H_{298} = +8,280 + 7.17T + 1.28 \times 10^{-3}T^2 - 0.08 \times 10^5 T^{-1}$$

Formation: $H_2 + 1/2O_2 \longrightarrow H_2O$

Zone I (298°–373° K.)

$$\Delta C_p = 7.95 - 1.28 \times 10^{-3}T + 0.08 \times 10^5 T^{-2}$$
$$\Delta H_T = -70,600 + 7.95T - 0.64 \times 10^{-3}T^2 - 0.08 \times 10^5 T^{-1}$$
$$\Delta F_T = -70,600 - 7.95T \ln T + 0.64 \times 10^{-3}T^2 + 0.04 \times 10^5 T^{-1} + 91.75T$$

Zone II (373°–2,500° K.)

$$\Delta C_p = -2.91 + 1.28 \times 10^{-3}T + 0.16 \times 10^5 T^{-2}$$
$$\Delta H_T = -56,940 - 2.91T + 0.64 \times 10^{-3}T^2 - 0.16 \times 10^5 T^{-1}$$
$$\Delta F_T = -56,940 + 2.91T \ln T - 0.64 \times 10^{-3}T^2 - 0.08 \times 10^5 T^{-1} - 8.11T$$

T, ° K.	H_T-H_{298}	S_T	ΔH°_T	ΔF°_T
298	....	16.75	−68,320	−56,720
400	11,345	47.01	−58,050	−53,350
500	12,174	48.86	−58,300	−52,150
600	13,026	50.41	−58,500	−50,900
700	13,910	51.77	−58,700	−49,600
800	14,821	52.99	−58,900	−48,300
900	15,762	54.10	−59,100	−46,800
1,000	16,735	55.12	−59,250	−45,800
1,100	17,735	56.07	−59,350	−44,300
1,200	18,768	56.97	−59,500	−42,900
1,300	19,830	57.82	−59,600	−41,400
1,400	20,917	58.63	−59,700	−40,000
1,500	22,031	59.40	−59,800	−38,650
1,600	23,040	60.06	−60,050	−37,350
1,700	24,180	60.71	−60,200	−35,950
1,800	25,335	61.42	−60,050	−34,400
1,900	26,510	62.02	−60,200	−33,050
2,000	27,915	62.78	−60,250	−31,650
2,500	34,205	65.59	−60,300	−24,400

Hydrogen Peroxide, H_2O_2 (*l*)

$\Delta H^\circ_{298} = -44,750$ calories per mole (*36*)
$S_{298} = 22.35$ *e.u.* (*24*)
M.P. $= 272.5°$ K. (*94*)
$\Delta H_M = 2,920$ calories per mole
B.P. $= 425°$ K. (*94*)
$\Delta H_V = 10,530$ calories per mole

Zone I (*g*) (425°–1,500° K.)

$$C_p = 10.43 + 5.00 \times 10^{-3}T - 1.68 \times 10^5 T^{-2} \ (82)$$
$$H_T - H_{298} = 8,300 + 10.43T + 2.50 \times 10^{-3}T^2 + 1.68 \times 10^5 T^{-1}$$

Formation: $H_2 + O_2 \longrightarrow H_2O_2$

Zone I (425°–1,500° K.)

$$\Delta C_p = -3.25 + 3.22 \times 10^{-3}T - 1.4 \times 10^5 T^{-2}$$
$$\Delta H_T = -32,200 - 3.25T + 1.61 \times 10^{-3}T^2 + 1.4 \times 10^5 T^{-1}$$
$$\Delta F_T = -32,200 + 3.25T \ln T - 1.61 \times 10^{-3}T^2 + 0.7 \times 10^5 T^{-1} + 4.38T$$

T, ° K.	H_T-H_{298}	S_T	ΔH°_T	ΔF°_T
298	....	22.35	−44,750	−28,100
400	2,130	28.92	−44,750	−22,600
500	14,460	61.55	−33,150	−20,200
600	15,690	63.84	−33,400	−17,600
700	17,020	65.89	−33,450	−14,900
800	18,440	67.78	−33,600	−12,300
900	19,850	69.83	−33,650	−9,900
1,000	21,400	71.08	−33,700	−6,900
1,100	22,900	72.47	−33,700	−4,200
1,200	24,510	73.92	−33,700	−1,600
1,300	26,170	75.22	−33,650	+1,200
1,400	27,870	76.48	−33,650	3,600
1,500	29,420	77.57	−33,700	6,500
2,000	(39,150)	(83.18)	(−32,400)	(19,600)
2,500	(50,000)	(88.00)	(−30,350)	(32,350)

Hydrogen Fluoride, HF (*g*)

$\Delta H^\circ_{298} = -64,200$ calories per mole (*112*)
$S_{298} = 41.49$ *e.u.* (*83*)
M.P. $= 190.1°$ K. (*112*)
$\Delta H_M = 1,094$ calories per mole
B.P. $= 293.1°$ K. (*112*)
$\Delta H_V = 1,800$ calories per mole

Zone I (*g*) (298°–2,000° K.)

$$C_p = 6.43 + 0.82 \times 10^{-3}T + 0.26 \times 10^5 T^{-2} \ (82)$$
$$H_T - H_{298} = -1,866 + 6.43T + 0.41 \times 10^{-3}T^2 - 0.26 \times 10^5 T^{-1}$$

Formation: $1/2H_2 + 1/2F_2 \longrightarrow HF$

Zone I (298°–2,000° K.)

$$\Delta C_p = -0.98 + 0.21 \times 10^{-3}T + 0.60 \times 10^5 T^{-2}$$
$$\Delta H_T = -63,695 - 0.98T + 0.10 \times 10^{-3}T^2 - 0.60 \times 10^5 T^{-1}$$
$$\Delta F_T = -63,695 + 0.98T \ln T - 0.10 \times 10^{-3}T^2 - 0.30 \times 10^5 T^{-1} - 8.45T$$

T, ° K.	H_T-H_{298}	S_T	ΔH°_T	ΔF°_T
298	....	41.49	−64,200	−64,520
400	710	43.54	−64,250	−64,650
500	1,405	45.09	−64,300	−65,000
600	2,105	46.36	−64,350	−65,150
700	2,800	47.43	−64,450	−65,250
800	3,505	48.38	−64,500	−65,350
900	4,215	49.21	−64,600	−65,450
1,000	4,935	49.97	−64,650	−65,550
1,100	5,681	50.68	−64,700	−65,700
1,200	6,395	51.31	−64,800	−65,750
1,300	7,160	51.99	−64,900	−65,850
1,400	7,895	52.46	−64,950	−65,900
1,500	8,684	53.01	−65,000	−65,950
1,600	9,440	53.49	−65,100	−66,050
1,700	10,235	54.02	−65,100	−66,150
1,800	11,020	54.42	−65,100	−66,150
1,900	11,817	54.89	−65,200	−66,200
2,000	12,635	55.28	−65,200	−66,200
2,500	(16,761)	(55.13)	(−65,400)	(−66,450)

Hydrogen Chloride, HCl (*g*)

$\Delta H^{\circ}_{298} = -22{,}063$ calories per mole (*112*)
$S_{298} = 44.61$ *e.u.* (*83*)
$M.P. = 158.9°$ K. (*112*)
$\Delta H_M = 476$ calories per mole
$B.P. = 188.1°$ K. (*112*)
$\Delta H_V = 3{,}860$ calories per mole

Zone I (*g*) (298°–2,000° K.)

$$C_p = 6.34 + 1.10\times10^{-3}T + 0.26\times10^{5}T^{-2} \quad (82)$$
$$H_T - H_{298} = -1{,}860 + 6.34T + 0.55\times10^{-3}T^2 - 0.26\times10^{5}T^{-1}$$

Formation: $1/2H_2 + 1/2Cl_2 \longrightarrow HCl$

Zone I (298°–2,000° K.)

$$\Delta C_p = -1.33 + 0.68\times10^{-3}T + 0.54\times10^{5}T^{-2}$$
$$\Delta H_T = -21{,}500 - 1.33T + 0.34\times10^{-3}T^2 - 0.54\times10^{5}T^{-1}$$
$$\Delta F_T = -21{,}500 + 1.33T\ln T - 0.34\times10^{-3}T^2 - 0.27\times10^{5}T^{-1} - 11.39T$$

T, ° K.	$H_T - H_{298}$	S_T	ΔH°_T	ΔF°_T
298	....	44.61	−22,060	−22,750
400	710	46.66	−22,100	−22,950
500	1,405	48.21	−22,200	−23,200
600	2,105	49.49	−22,300	−23,400
700	2,815	50.58	−22,350	−23,600
800	3,535	51.54	−22,450	−23,750
900	4,265	52.40	−22,500	−23,900
1,000	5,005	53.18	−22,550	−24,000
1,100	5,765	53.84	−22,600	−24,150
1,200	6,530	54.57	−22,650	−24,400
1,300	7,315	55.16	−22,650	−24,400
1,400	8,095	55.78	−22,750	−24,650
1,500	8,800	56.28	−22,800	−24,700
1,600	9,700	56.85	−22,850	−24,900
1,700	10,520	57.30	(−22,800)	(−25,000)
1,800	11,335	57.81	(−22,800)	(−25,250)
1,900	12,170	58.32	(−22,800)	(−25,450)
2,000	12,995	58.69	(−22,850)	(−25,550)

Hydrogen Bromide, HBr (*g*)

$\Delta H^{\circ}_{298} = -8{,}660$ calories per mole (*112*)
$S_{298} = 47.63$ *e.u.* (*83*)
$M.P. = 186.24°$ K. (*112*)
$\Delta H_M = 575$ calories per mole
$B.P. = 206.4°$ K. (*112*)
$\Delta H_V = 4{,}210$ calories per mole

Zone I (*g*) (298°–1,600° K)

$$C_p = 6.25 + 1.40\times10^{-3}T + 0.26\times10^{5}T^{-2} \quad (82)$$
$$H_T - H_{298} = -1{,}838 + 6.25T + 0.70\times10^{-3}T^2 - 0.26\times10^{5}T^{-1}$$

Formation: $1/2H_2 + 1/2Br_2 \longrightarrow HBr$

Zone I (298°–331° K.)

$$\Delta C_p = -5.56 + 1.01\times10^{-3}T + 0.20\times10^{5}T^{-2}$$
$$\Delta H_T = -6{,}980 - 5.56T + 0.51\times10^{-3}T^2 - 0.20\times10^{5}T^{-1}$$
$$\Delta F_T = -6{,}980 + 5.56T\ln T - 0.51\times10^{-3}T^2 - 0.10\times10^{5}T^{-1} - 50.87T$$

Zone II (331°–1,600° K.)

$$\Delta C_p = -1.53 + 1.01\times10^{-3}T + 0.38\times10^{5}T^{-2}$$
$$\Delta H_T = -12{,}000 - 1.53T + 0.51\times10^{-3}T^2 - 0.38\times10^{5}T^{-1}$$
$$\Delta F_T = -12{,}000 + 1.53T\ln T - 0.51\times10^{-3}T^2 - 0.19\times10^{5}T^{-1} - 12.12T$$

T, ° K.	$H_T - H_{298}$	S_T	ΔH°_T	ΔF°_T
298	....	47.63	−8,660	−12,800
400	709	49.68	−12,650	−13,300
500	1,410	51.24	−12,700	−13,450
600	2,120	52, 54	−12,800	−13,600
700	2,840	53.64	−12,900	−13,750
800	3,575	54.63	−12,950	−13,850
900	4,325	55.51	−13,000	−14,000
1,000	5,090	56.31	−13,050	−14,200
1,100	5,908	57.06	−13,050	−14,150
1,200	6,665	57.75	−13,100	−14,300
1,300	7,470	58.34	−13,150	−14,400
1,400	8,285	59.00	−13,150	−14,550
1,500	9,129	59.52	−13,150	−14,600
1,600	9,945	60.11	−13,250	−14,750
2,000	(13,475)	(62.02)	(−13,100)	(−14,950)

Hydrogen Iodide, HI (*g*)

$\Delta H^{\circ}_{298} = 6{,}200$ calories per mole (*112*)
$S_{298} = 49.33$ *e.u.* (*83*)
$M.P. = 222.36°$ K. (*112*)
$\Delta H_M = 686$ calories per mole
$B.P. = 237.8°$ K. (*112*)
$\Delta H_V = 4{,}724$ calories per mole

Zone I (*g*) (298°–2,000° K.)

$$C_p = 6.29 + 1.42\times10^{-3}T + 0.22\times10^{5}T^{-2} \quad (82)$$
$$H_T - H_{298} = -1{,}865 + 6.29T + 0.71\times10^{-3}T^2 - 0.22\times10^{5}T^{-1}$$

Formation: $1/2H_2 + 1/2I_2 \longrightarrow HI$

Zone I (298°–386.8° K.)

$$\Delta C_p = -1.76 - 4.92\times10^{-3}T - 0.06\times10^{5}T^{-2}$$
$$\Delta H_T = 6{,}920 + 1.76T - 2.46\times10^{-3}T^2 + 0.06 + 10^{5}T^{-1}$$
$$\Delta F_T = 6{,}920 + 1.76T\ln T + 2.46\times10^{-3}T^2 + 0.03\times10^{5}T^{-1} - 32.9T$$

Zone II (386.8°–456° K.)

$$\Delta C_p = -6.57 + 1.03\times10^{-3}T + 0.16\times10^{5}T^{-2}$$
$$\Delta H_T = 6{,}530 - 6.57T + 0.51\times10^{-3}T^2 - 0.16\times10^{5}T^{-1}$$
$$\Delta F_T = 6{,}530 + 6.57T\ln T - 0.51\times10^{-3}T^2 - 0.08\times10^{5}T^{-1} - 58.47T$$

Zone III (456°–1,500° K.)

$$\Delta C_p = -1.41 + 1.03\times10^{-3}T + 0.16\times10^{5}T^{-2}$$
$$\Delta H_T = -800 - 1.41T + 0.52\times10^{-3}T^2 - 0.16\times10^{5}T^{-1}$$
$$\Delta F_T = -800 + 1.41T\ln T - 0.52\times10^{-3}T^2 - 0.08\times10^{5}T^{-1} - 11.76T$$

T, ° K.	$H_T - H_{298}$	S_T	ΔH°_T	ΔF°_T
298	....	49.33	+6,200	+300
400	710	51.38	+3,950	−1,750
500	1,410	52.94	−1,450	−2,450
600	2,125	54.25	−1,500	−2,650
700	2,855	55.37	−1,550	−2,800
800	3,595	56.36	−1,650	−3,000
900	4,355	57.25	−1,700	−3,150
1,000	5,180	58.07	−1,700	−3,350
1,100	5,894	58.84	−1,750	−3,550
1,200	6,715	59.51	−1,750	−3,700
1,300	7,495	60.18	−1,800	−3,850
1,400	8,345	60.77	−1,800	−4,050
1,500	9,152	61.36	−1,750	−4,150
1,600	10,000	61.87	(−1,750)	(−4,300)
1,700	10,867	62.45	(−1,750)	(−4,450)
1,800	11,685	62.87	(−1,750)	(−4,600)
1,900	12,639	63.43	(−1,750)	(−4,700)
2,000	13,385	63.76	(−1,750)	(−4,800)

Hydrogen Cyanide, HCN (*g*)

$\Delta H^{\circ}_{298}=31{,}200$ calories per mole (*112*)
$S_{298}=48.23$ *e.u.* (*83*)
M.P.$=170.4°$ K. (*112*)
$\Delta H_M=40$ calories per mole
B.P.$=298.8°$ K. (*112*)
$\Delta H_V=6{,}027$ calories per mole

Zone I (*g*) (298°–2,000° K.)

$$C_p=8.92+3.10\times10^{-3}T-1.12\times10^{5}T^{-2} \ (82)$$
$$H_T-H_{298}=-3{,}173+8.92T+1.55\times10^{-3}T^2+1.12\times10^{5}T^{-1}$$

Formation: $1/2H_2+C+1/2N_2 \longrightarrow HCN$

Zone I (298°–2,000° K.)

$$\Delta C_p=-1.77+1.18\times10^{-3}T+0.92\times10^{5}T^{-2}$$
$$\Delta H_T=31{,}980-1.77T+0.59\times10^{-3}T^2-0.92\times10^{5}T^{-1}$$
$$\Delta F_T=31{,}980+1.77T\ln T-0.59\times10^{-3}T^2-0.46\times10^{5}T^{-1}-20.43T$$

T, ° K.	H_T-H_{298}	S_T	ΔH°_T	ΔF°_T
298	....	48.23	+31,200	28,700
400	917	50.87	31,150	27,850
500	1,890	53.04	31,100	27,050
600	2,915	54.91	31,150	26,300
700	3.990	56.56	31,000	25,400
800	5,105	58.05	30,900	24,600
900	6,260	59.41	30,850	23,850
1,000	7,450	60.67	30,800	23,050
1,100	8,610	61.74	30,800	22,350
1,200	9,910	62.91	30,700	21,450
1,300	11,130	63.86	30,600	20,800
1,400	12,480	64.89	30,600	20,000
1,500	13,770	65.51	30,550	19,600
1,600	15,130	66.65	30,550	18,350
1,700	16,540	67.51	30,550	17,550
1,800	17,850	68.26	30,600	16,750
1,900	19,430	69.19	30,700	16,000
2,000	20,610	69.71	30,600	15,450
2,500	(28,860)	(73.46)	(31,200)	(11,200)

Hydroden Trinitride (Azoimide), HN_3 (*g*)

$\Delta H^{\circ}_{298}=70{,}300$ calories per mole (*112*)
$S_{298}=56.8$ *e.u.* (*83*)
M.P.$=193°$ K. (*112*)
B.P.$=309°$ K. (*112*)
$\Delta H_V=7{,}100$ calories per mole

Zone I (*g*) (309°–1,800° K.)

$$C_p=11.33+4.62\times10^{-3}T-2.38\times10^{5}T^{-2} \ (82)$$
$$H_T-H_{298}=-4{,}382+11.33T+2.31\times10^{-3}T^2+2.38\times10^{5}T^{-1}$$

Formation: $1/2H_2+3/2N_2 \longrightarrow HN_3$

Zone I (309°–1,800° K.)

$$\Delta C_p=-1.92+2.7\times10^{-3}T-2.44\times10^{5}T^{-2}$$
$$\Delta H_T=69{,}940-1.92T+1.35\times10^{-3}T^2+2.44\times10^{5}T^{-1}$$
$$\Delta F_T=69{,}940+1.92T\ln T-1.35\times10^{-3}T^2+1.22\times10^{5}T^{-1}+15.67T$$

T, ° K.	H_T-H_{298}	S_T	ΔH°_T	ΔF°_T
298	....	56.8	70,300	78,480
400	1,095	59.95	70,000	81,350
500	2,295	62.62	69,800	84,250
600	3,600	65.00	69,650	87,100
700	4,990	67.14	69,600	90,000
800	6,460	69.10	69,600	92,950
900	7,985	70.90	69.650	95,850
1,000	9,565	72.56	69,700	98,750
1,100	10,996	74.00	69,800	101,750
1,200	12,870	75.58	69,900	104,500
1,300	14,440	76.75	69,900	107,650
1,400	16,280	78.20	70,100	110,350
1,500	17,970	79.36	70,200	113,250
1,600	19,810	80.56	70,400	115,950
1,700	21,659	81.62	70,700	118,900
1,800	23,420	82.68	70,850	121,950
1,900	(25,510)	(83.79)	(71,200)	(124,650)
2,000	(27,090)	(84.61)	(70,900)	(127,400)

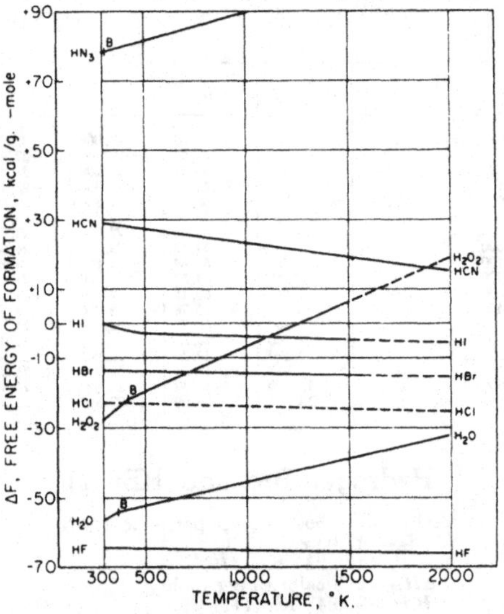

FIGURE 24.—Hydrogen.

IODINE

Element, I_2 (*c*)

$S_{298}=27.90$ *e.u.* (*112*)
M.P.$=386.1°$ K. (*112*)
B.P.$=456°$ K. (*82*)
$\Delta H_V=9{,}970$ calories per atom

Zone I (*c*) (298°–386.1° K.)

$$C_p=9.59+11.90\times10^{-3}T \ (82)$$
$$H_T-H_{298}=-3{,}388+9.59T+5.95\times10^{-3}T^2$$
$$F_T-H_{298}=-3{,}388-9.59T\ln T-5.95\times10^{-3}T^2+39.73T$$

Zone II (*l*) (386.1°–456° K.)

$$C_p = 19.20\ (82)$$
$$H_T - H_{298} = -2{,}445 + 19.20T$$
$$F_T - H_{298} = -2{,}445 - 19.20T\ln T + 92.2T$$

Zone III (*g*) (456°–1,500° K.)

$$C_p = 8.89\ (82)$$
$$H_T - H_{298} = 12{,}226 + 8.89T$$
$$F_T - H_{298} = 12{,}226 - 8.89T\ln T - 3.04T$$

T, ° K.	$H_T - H_{298}$	S_T	$-\frac{(F_T - H_{298})}{T}$
298	....	27.90	27.90
400	5,235	41.85	28.76
500	16,670	67.05	33.71
600	17,560	68.67	39.40
700	18,450	70.04	43.68
800	19,340	71.23	47.06
900	20,240	72.29	49.06
1,000	21,130	73.23	52.10
1,100	22,020	74.08	54.05
1,200	22,910	74.85	55.76
1,300	23,810	75.57	57.25
1,400	24,700	76.23	58.59
1,500	25,590	76.84	59.78
2,000	(30,000)	....	(64.42)

IRON AND ITS COMPOUNDS

Element, Fe (*c*)

$S_{298} = 6.49$ *e.u.* (*83*)
$T.P. = 1{,}033°$ K. (*82*)
$\Delta H_T = 410$ calories per atom
$T.P. = 1{,}179°$ K. (*82*)
$\Delta H_T = 210$ calories per atom
$T.P. = 1{,}674°$ K. (*82*)
$\Delta H_T = 110$ calories per atom
$M.P. = 1{,}803°$ K. (*82*)
$\Delta H_M = 3{,}700$ calories per atom
$B.P. = 3{,}008°$ K. (*8*)
$\Delta H_V = 84{,}620$ calories per atom

Zone I (α) (298°–1,033° K.)

$$C_p = 3.37 + 7.10 \times 10^{-3}T + 0.43 \times 10^5 T^{-2}\ (82)$$
$$H_T - H_{298} = -1{,}176 + 3.37T + 3.55 \times 10^{-3}T^2 - 0.43 \times 10^5 T^{-1}$$
$$F_T - H_{298} = -1{,}176 - 3.37T\ln T - 3.55 \times 10^{-3}T^2 - 0.21 \times 10^5 T^{-1} + 17.96T$$

Zone II (β) (1,033°–1,179° K.)

$$C_p = 10.40\ (82)$$
$$H_T - H_{298} = -4{,}280 + 10.40T$$
$$F_T - H_{298} = -4{,}280 - 10.40T\ln T + 66.07T$$

Zone III (γ) (1,179°–1,674° K.)

$$C_p = 4.85 + 3.00 \times 10^{-3}T\ (82)$$
$$H_T - H_{298} = 390 + 4.85T + 1.50 \times 10^{-3}T^2$$
$$F_T - H_{298} = 390 - 4.85T\ln T - 1.50 \times 10^{-3}T^2 + 24.60T$$

Zone IV (δ) (1,674°–1,803° K.)

$$C_p = 10.30\ (82)$$
$$H_T - H_{298} = -4{,}420 + 10.30T$$
$$F_T - H_{298} = -4{,}420 - 10.30T\ln T + 65.31T$$

Zone V (*l*) (1,803°–1,900° K.)

$$C_p = 10.0\ (82)$$
$$H_T - H_{298} = -180 + 10.0T$$
$$F_T - H_{298} = -180 - 10.0T\ln T + 54.4T$$

T, ° K.	$H_T - H_{298}$	S_T	$-\frac{(F_T - H_{298})}{T}$
298	....	6.49	6.49
400	640	8.34	6.75
500	1,310	9.83	7.20
600	2,050	11.17	7.75
700	2,860	12.42	8.34
800	3,720	13.57	8.92
900	4,680	14.70	9.47
1,000	5,830	15.91	10.08
1,100	7,160	17.18	10.65
1,200	8,370	18.23	11.28
1,300	9,230	18.91	11.83
1,400	10,120	19.57	12.32
1,500	11,040	20.21	12.87
1,600	11,990	20.82	13.32
1,700	13,090	21.49	13.83
1,800	14,120	22.08	14.18
1,900	18,820	24.66	14.78
2,000	(19,760)	(25.08)	(15.20)

Iron Oxide, $Fe_{0.95}O$ (*c*)

$\Delta H_{298} = -63{,}800$ calories per mole (*70*)
$S_{298} = 13.74$ *e.u.* (*70*)
$M.P. = 1{,}650°$ K. (*84*)
$\Delta H_M = 7{,}490$ calories per mole

Zone I (*c*) (298°–1,650° K.)

$$C_p = 11.66 + 2.00 \times 10^{-3}T - 0.67 \times 10^5 T^{-2}\ (84)$$
$$H_T - H_{298} = -3{,}790 + 11.66T + 1.00 \times 10^{-3}T^2 + 0.67 \times 10^5 T^{-1}$$

Zone II (*l*) (1,650°–1,800° K.)

$$C_p = 16.30\ (84)$$
$$H_T - H_{298} = -1{,}200 + 16.30T$$

Formation: $0.95Fe + 1/2O_2 \longrightarrow Fe_{0.95}O$

Zone I (298°–1,033° K.)

$$\Delta C_p = 4.71 - 5.60 \times 10^{-3}T - 0.90 \times 10^5 T^{-2}$$
$$\Delta H_T = -65{,}250 + 4.71T - 2.80 \times 10^{-3}T^2 + 0.90 \times 10^5 T^{-1}$$
$$\Delta F_T = -65{,}250 - 4.71T\ln T + 2.80 \times 10^{-3}T^2 + 0.45 \times 10^5 T^{-1} + 47.61T$$

Zone II (1,033°–1,179° K.)

$$\Delta C_p = -2.32 + 1.50 \times 10^{-3}T - 0.47 \times 10^5 T^{-2}$$
$$\Delta H_T = -62{,}200 - 2.32T + 0.75 \times 10^{-3}T^2 + 0.47 \times 10^5 T^{-1}$$
$$\Delta F_T = -62{,}200 + 2.32T\ln T - 0.75 \times 10^{-3}T^2 + 0.23 \times 10^5 T^{-1} - 0.43T$$

Zone III (1,179°–1,650° K.)

$$\Delta C_p = 3.23 - 1.50 \times 10^{-3}T - 0.47 \times 10^5 T^{-2}$$
$$\Delta H_T = -66{,}720 + 3.23T - 0.75 \times 10^{-3}T^2 + 0.47 \times 10^5 T^{-1}$$
$$\Delta F_T = -66{,}720 - 3.23T\ln T + 0.75 \times 10^{-3}T^2 + 0.23 \times 10^5 T^{-1} + 41.0T$$

Zone IV (1,674°–1,800° K.)

$$\Delta C_p = 2.42 - 0.50 \times 10^{-3}T + 0.20 \times 10^5 T^{-2}$$
$$\Delta H_T = -59{,}430 + 2.42T - 0.25 \times 10^{-3}T^2 - 0.20 \times 10^5 T^{-1}$$
$$\Delta F_T = -59{,}430 - 2.42T\ln T + 0.25 \times 10^{-3}T^2 - 0.10 \times 10^5 T^{-1} + 31.35T$$

T, ° K.	H_T-H_{298}	S_T	ΔH°_T	ΔF°_T
298	....	13.74	−63,800	−58,670
400	1,210	17.23	−63,700	−57,000
500	2,440	19.97	−63,400	−55,300
600	3,700	22.07	−63,280	−53,550
700	4,980	24.24	−63,150	−51,100
800	6,280	25.97	−63,150	−50,500
900	7,590	27.52	−63,200	−48,950
1,000	8,920	28.92	−63,400	−47,350
1,100	10,280	30.21	−63,800	−45,700
1,200	11,670	31.42	−63,900	−43,950
1,300	13,060	32.55	−63,900	−42,350
1,400	14,520	33.62	−63,750	−40,750
1,500	15,980	34.62	−63,700	−39,050
1,600	17,460	35.58	−63,550	−37,400
1,700	26,510	41.06	−56,050	−36,000
1,800	28,140	42.00	−55,900	−34,750

T, ° K.	H_T-H_{298}	S_T	ΔH°_T	ΔF°_T
298	....	35.0	−267,800	−243,200
400	3,990	46.48	−267,200	−234,900
500	8,320	56.12	−266,300	−226,900
600	13,060	64.75	−265,300	−219,100
700	18,340	72.88	−264,000	−211,500
800	24,260	80.77	−262,300	−204,200
900	30,550	88.18	−260,500	−197,000
1,000	35,350	93.24	−260,800	−189,900
1,100	40,350	97.81	−261,500	−182,600
1,200	44,950	101.99	−262,000	−175,500
1,300	49,750	105.83	−261,500	−168,300
1,400	54,550	109.39	−261,000	−161,100
1,500	59,350	112.70	−260,900	−154,000
1,600	64,150	115.80	−260,500	−146,800
1,700	68,950	118.71	−260,800	−139,800
1,800	73,750	121.45	−260,800	−133,000

Triiron Tetraoxide, Fe_3O_4 (*c*)

$\Delta H_{298}^{\circ}=-267{,}800$ calories per mole (*24*)
$S_{298}=35.0$ *e.u.* (*83*)
T.P. $=900°$ K. (*24*)
$\Delta H_T=0$ calories per mole
M.P. $=1{,}870°$ K. (*30*)
$\Delta H_M=33{,}000$ calories per mole

Zone I (α) (298°–900° K.)

$$C_p=21.88+48.20\times10^{-3}T \ (27)$$
$$H_T-H_{298}=-8{,}640+21.88T+24.10\times10^{-3}T^2$$

Zone II (β) (900°–1,800° K.)

$$C_p=48.0 \ (27)$$
$$H_T-H_{298}=-12{,}650+48.00T$$

Formation: $3Fe+2O_2 \longrightarrow Fe_3O_4$

Zone I (298°–900° K.)

$$\Delta C_p=-2.55+24.90\times24.90\times10^{-3}T-0.49\times10^5T^{-2}$$
$$\Delta H_T=-268{,}300-2.55T+12.45\times10^{-3}T^2+0.49\times10^5T^{-1}$$
$$\Delta F_T=-268{,}300+2.55T\ln T-12.45\times10^{-3}T^2+0.24\times10^5T^{-1}+73.07T$$

Zone II (900°–1,033° K.)

$$\Delta C_p=23.57-23.30\times10^{-3}T-0.49\times10^5T^{-2}$$
$$\Delta H_T=-272{,}760+23.57T-11.65\times10^{-3}T^2+0.49\times10^5T^{-1}$$
$$\Delta F_T=-272{,}760-23.57T\ln T+11.65\times10^{-3}T^2+0.24\times10^5T^{-1}+234.0T$$

Zone III (1,033°–1,179° K.)

$$\Delta C_p=2.48-2.00\times10^{-3}T+0.80\times10^5T^{-2}$$
$$\Delta H_T=-262{,}950+2.48T-1.00\times10^{-3}T^2-0.80\times10^5T^{-1}$$
$$\Delta F_T=-262{,}950-2.48T\ln T+1.00\times10^{-3}T^2-0.40\times10^5T^{-1}+89.38T$$

Zone IV (1,179°–1,674° K.)

$$\Delta C_p=19.13-11.00\times10^{-3}T+0.80\times10^5T^{-2}$$
$$\Delta H_T=-277{,}000+19.13T-5.50\times10^{-3}T^2-0.80\times10^5T^{-1}$$
$$\Delta F_T=-277{,}000-19.13T\ln T+5.50\times10^{-3}T^2-0.40\times10^5T^{-1}+162.62T$$

Zone V (1,674°–1,800° K.)

$$\Delta C_p=2.78-2.00\times10^{-3}T+0.80\times10^5T^{-2}$$
$$\Delta H_T=-262{,}500+2.78T-1.00\times10^{-3}T^2-0.80\times10^5T^{-1}$$
$$\Delta F_T=-262{,}500-2.78T\ln T+1.00\times10^{-3}T^2-0.40\times10^5T^{-1}+91.0T$$

Diiron Trioxide, Fe_2O_3 (*c*)

$\Delta H_{298}^{\circ}=-196{,}800$ calories per mole (*112*)
$S_{298}=21.5$ *e.u.* (*112*)
T.P. $=950°$ K. (*24*)
$\Delta H_T=160$ calories per mole
T.P. $=1{,}050°$ K. (*24*)
$\Delta H_T=0$ calories per mole
Decomposes $=1{,}730°$ K. (*24*)

Zone I (α) (298°–950° K.)

$$C_p=23.49+18.60\times10^{-3}T-3.55\times10^5T^{-2} \ (84)$$
$$H_T-H_{298}=-9{,}020+23.49T+9.30\times10^{-3}T^2+3.55\times10^5T^{-1}$$

Zone II (β) (950°–1,050° K.)

$$C_p=36.0 \ (84)$$
$$H_T-H_{298}=-11{,}980+36.0T$$

Zone III (γ) (1,050°–1,730° K.)

$$C_p=31.71+1.76\times10^{-3}T \ (84)$$
$$H_T-H_{298}=-8{,}450+31.71T+0.88\times10^{-3}T^2$$

Formation: $2Fe+3/2O_2 \longrightarrow Fe_2O_3$

Zone I (298°–950° K.)

$$\Delta C_p=6.01+2.90\times10^{-3}T-3.81\times10^5T^{-2}$$
$$\Delta H_T=-200{,}000+6.01T+1.45\times10^{-3}T^2+3.81\times10^5T^{-1}$$
$$\Delta F_T=-200{,}000-6.01T\ln T-1.45\times10^{-3}T^2+1.90\times10^5T^{-1}+108.4T$$

Zone II (950°–1,033° K.)

$$\Delta C_p=18.52-15.7\times10^{-3}T-0.26\times10^5T^{-2}$$
$$\Delta H_T=-203{,}300+18.52T-7.85\times10^{-3}T^2+0.26\times10^5T^{-1}$$
$$\Delta F_T=-203{,}300-18.52T\ln T+7.85\times10^{-3}T^2+0.13\times10^5T^{-1}+189.0T$$

Zone III (1,050°–1,179° K.)

$$\Delta C_p=0.17+0.26\times10^{-3}T+0.60\times10^5T^{-2}$$
$$\Delta H_T=-193{,}100+0.17T+0.13\times10^{-3}T^2-0.60\times10^5T^{-1}$$
$$\Delta F_T=-193{,}100-0.17T\ln T-0.13\times10^{-3}T^2-0.30\times10^5T^{-1}+60.07T$$

Zone IV (1,179°–1,674° K.)

$$\Delta C_p=11.27-5.74\times10^{-3}T+0.60\times10^5T^{-2}$$
$$\Delta H_T=-202{,}600+11.27T-2.87\times10^{-3}T^2-0.60\times10^5T^{-1}$$
$$\Delta F_T=-202{,}600-11.27T\ln T+2.87\times10^{-3}T^2-0.30\times10^5T^{-1}+142.29T$$

Zone V (1,674°–1,730° K.)

$\Delta C_p = 0.37 + 0.26 \times 10^{-3}T + 0.60 \times 10^5 T^{-2}$
$\Delta H_T = -192{,}400 + 0.37T + 0.13 \times 10^{-3}T^2 - 0.60 \times 10^5 T^{-1}$
$\Delta F_T = -192{,}400 - 0.37T \ln T - 0.13 \times 10^{-3}T^2 - 0.30 \times 10^5 T^{-1} + 61.3T$

T, ° K.	$H_T - H_{298}$	S_T	ΔH_T°	ΔF_T°
298	....	21.5	−196,800	−177,400
400	2,750	29.41	−196,400	−170,800
500	5,770	36.14	−195,800	−164,500
600	9,010	42.04	−195,200	−158,300
700	12,460	47.35	−194,500	−152,200
800	16,130	52.25	−193,800	−146,200
900	20,020	55.84	−193,000	−139,500
1,000	24,020	61.05	−192,600	−134,500
1,100	27,500	64.37	−192,900	−128,500
1,200	30,870	67.30	−193,200	−122,800
1,300	34,250	70.01	−192,800	−116,900
1,400	37,650	72.53	−192,500	−111,100
1,500	41,070	74.89	−192,300	−105,300
1,600	44,540	77.13	−191,900	−99,400
1,700	48,100	79.29	−191,400	−93,700

Iron Difluoride, FeF_2 (*c*)

$\Delta H_{298}^\circ = -168{,}000$ calories per mole (*112*)
$S_{298} = 20.8$ *e.u.* (*18*)
M.P. = 1,375° K. (*6*)
$\Delta H_M = (8{,}000)$ calories per mole
B.P. = (2,100°) K. (*6*)
$\Delta H_V = (50{,}000)$ calories per mole

Formation: $Fe + F_2 \longrightarrow FeF_2$
(estimated (*11*))

T, ° K.	$H_T - H_{298}$	ΔH_T°	ΔF_T°
298	....	−168,000	(−157,300)
500	(3,500)	(−167,400)	(−150,500)
1,000	(13,000)	(−166,700)	(−133,000)
1,500	(33,000)	(−156,300)	(−118,500)

Iron Trifluoride, FeF_3 (*c*)

$\Delta H_{298}^\circ = (-235{,}000)$ calories per mole (*11*)
$S_{298} = (25)$ *e.u.* (*11*)
M.P. = 1,300° K. (*6*)
$\Delta H_M = (12{,}000)$ calories per mole
B.P. = (1,600°) K. (*6*)
$\Delta H_V = (40{,}000)$ calories per mole

Formation: $Fe + 3/2F_2 \longrightarrow FeF_3$
(estimated (*11*))

T, ° K.	$H_T - H_{298}$	ΔH_T°	ΔF_T°
298	....	(−235,000)	(−219,000)
500	(5,000)	(−232,900)	(−207,500)
1,000	(19,000)	(−230,600)	(−183,000)
1,500	(46,000)	(−215,400)	(−160,000)

Iron Dichloride, $FeCl_2$ (*c*)

$\Delta H_{298}^\circ = -81{,}900$ calories per mole (*112*)
$S_{298} = 28.7$ *e.u.* (*83*)
M.P. = 950° K. (*82*)
$\Delta H_M = 10{,}280$ calories per mole
B.P. = 1,299° K. (*112*)
$\Delta H_V = 30{,}210$ calories per mole

Zone I (*c*) 298°–950° K.)

$C_p = 18.94 + 2.08 \times 10^{-3}T - 1.17 \times 10^5 T^{-2}$ (*82*)
$H_T - H_{298} = -6{,}090 + 18.94T + 1.04 \times 10^{-3}T^2 + 1.17 \times 10^5 T^{-1}$

Zone II (*l*) (950°–1,110° K.)

$C_p = 24.40$ (*82*)
$H_T - H_{298} = 81{,}100 + 24.40T$

Formation: $Fe + Cl_2 \longrightarrow FeCl_2$

Zone I (298°–950° K.)

$\Delta C_p = 6.75 - 5.08 \times 10^{-3}T - 0.92 \times 10^5 T^{-2}$
$\Delta H_T = -84{,}000 + 6.75T + 2.54 \times 10^{-3}T^2 + 0.92 \times 10^5 T^{-1}$
$\Delta F_T = -84{,}000 - 6.75T \ln T + 2.54 \times 10^{-3}T^2 + 0.46 \times 10^5 T^{-1} + 75.2T$

Zone II (950°–1,033° K.)

$\Delta C_p = 12.2 - 7.16 \times 10^{-3}T + 0.25 \times 10^5 T^{-2}$
$\Delta H_T = -77{,}980 + 12.2T - 3.58 \times 10^{-3}T^2 - 0.25 \times 10^5 T^{-1}$
$\Delta F_T = -77{,}980 - 12.2T \ln T + 3.58 \times 10^{-3}T^2 - 0.12 \times 10^5 T^{-1} + 105.57T$

T, ° K.	$H_T - H_{298}$	S_T	ΔH_T°	ΔF_T°
298	....	28.7	−81,900	−72,600
400	1,930	34.27	−81,450	−69,500
500	3,870	38.59	−81,030	−66,500
600	5,820	42.15	−80,670	−63,700
700	7,800	45.20	−80,400	−61,000
800	9,830	47.91	−80,090	−58,100
900	11,880	50.32	−79,880	−55,400
1,000	24,410	63.51	−69,380	−53,100
1,100	26,860	65.85	−69,010	−51,500

Iron Trichloride, $FeCl_3$ (*c*)

$\Delta H_{298}^\circ = -95{,}700$ calories per mole (*92*)
$S_{298} = (32.2)$ *e.u.* (*136*)
M.P. = 577° K. (*136*)
$\Delta H_M = 10{,}300$ calories per mole
B.P. = 592° K. (*136*)
$\Delta H_V = 6{,}020$ calories per mole

Zone I (*c*) (298°–577° K.)

$C_p = 29.56 - 6.11 \times 10^5 T^{-2}$ (*136*)
$H_T - H_{298} = -10{,}860 + 29.56T + 6.11 \times 10^5 T^{-1}$

Formation: $Fe + 3/2Cl_2 \longrightarrow FeCl_3$

Zone I (298°–577° K.)

$\Delta C_p = 12.96 - 7.19 \times 10^{-3}T - 5.52 \times 10^5 T^{-2}$
$\Delta H_T = -101{,}100 + 12.96T - 3.59 \times 10^{-3}T^2 + 5.52 \times 10^5 T^{-1}$
$\Delta F_T = -101{,}100 - 12.96T \ln T + 3.59 \times 10^{-3}T^2 + 2.76 \times 10^5 T^{-1} + 142.0T$

T, ° K.	H_T-H_{298}	S_T	ΔH_T°	ΔF_T°
298	....	32.2	−95,700	−79,500
400	2,500	39.38	−95,100	−74,000
500	5,140	45.27	−94,400	−68,800

Iron Dibromide, $FeBr_2$ (*c*)

$\Delta H^\circ_{298}=(-60{,}000)$ calories per mole (*112*)
$S_{298}=(32)$ *e.u.* (*11*)
M.P. = 957° K. (*6*)
$\Delta H_M=(9{,}000)$ calories per mole
B.P. = (1,200°) K. (*6*)
$\Delta H_V=(26{,}000)$ calories per mole

Formation: $Fe+Br_2 \longrightarrow FeBr_2$
(estimated (*11*))

T, ° K.	H_T-H_{298}	ΔH_T°	ΔF_T°
298	....	−60,000	(−57,700)
500	(4,000)	(−66,800)	(−52,000)
1,000	(24,000)	(−55,800)	(−38,000)
1,500	(62,000)	(−27,500)	(−29,500)

Iron Tribromide, $FeBr_3$ (*c*)

$\Delta H^\circ_{298}=(-65{,}000)$ calories per mole (*11*)
$S_{298}=(46)$ *e.u.* (*11*)
M.P. = (500°) K. (*6*)
$\Delta H_M=(5{,}000)$ calories per mole
B.P. = (900°) K. (*6*)
$\Delta H_V=(20{,}000)$ calories per mole

Formation: $Fe+3/2Br_2 \longrightarrow FeBr_3$
(estimated (*11*))

T, ° K.	H_T-H_{298}	ΔH_T°	ΔF_T°
298	....	(−65,000)	(−60,400)
500	(5,000)	(−54,500)	(−37,000)

Iron Diiodide, FeI_2 (*c*)

$\Delta H^\circ_{298}=-30{,}000$ calories per mole (*11*)
$S_{298}=(36)$ *e.u.* (*11*)
M.P. = 860° K. (*6*)
$\Delta H_M=(7{,}000)$ calories per mole
B.P. = (1,100°) K. (*6*)
$\Delta H_V=(25{,}000)$ calories per mole

Formation: $Fe+I_2 \longrightarrow FeI_2$
(estimated (*11*))

T, ° K.	H_T-H_{298}	ΔH_T°	ΔF_T°
298	....	−30,000	(−30,200)
500	(4,000)	(−44,000)	(−28,500)
1,000	(24,000)	(−33,000)	(−15,000)
1,500	(61,000)	(−5,500)	(−6,000)

Triiron Carbide, Fe_3C (*c*)

$\Delta H^\circ_{298}=5{,}780$ calories per mole (*81*)
$S_{298}=24.2$ *e.u.* (*83*)
T.P. = 463° K. (*82*)
$\Delta H_T=180$ calories per mole
M.P. = 1,500° K. (*82*)
$\Delta H_M=12{,}330$ calories per mole
Metastable above 2,000° K.

Zone I (α) (298°–463° K.)

$$C_p=19.64+20.00\times10^{-3}T \ (82)$$
$$H_T-H_{298}=-6{,}745+19.64T+10.00\times10^{-3}T^2$$

Zone II (β) (463°–1,500° K.)

$$C_p=25.62+3.00\times10^{-3}T \ (82)$$
$$H_T-H_{298}=-7{,}515+25.62T+1.50\times10^{-3}T^2$$

Zone III (l) (1,500°–1,900° K.)

$$C_p=30.60 \ (82)$$
$$H_T-H_{298}=+740+30.60T$$

Formation: $3Fe+C \longrightarrow Fe_3C$

Zone I (298°–463° K.)

$$\Delta C_p=5.43-2.32\times10^{-3}T+0.81\times10^5T^{-2}$$
$$\Delta H_T=+4{,}530+5.43T-1.16\times10^{-3}T^2-0.81\times10^5T^{-1}$$
$$\Delta F_T=+4{,}530-5.43T\ln T+1.16\times10^{-3}T^2-0.40\times10^5T^{-1}+31.98T$$

Zone II (463°–1,033° K.)

$$\Delta C_p=11.41-19.32\times10^{-3}T+0.81\times10^5T^{-2}$$
$$\Delta H_T=+3{,}850+11.41T-9.66\times10^{-3}T^2-0.81\times10^5T^{-1}$$
$$\Delta F_T=+3{,}850-11.41T\ln T+9.66\times10^{-3}T^2-0.40\times10^5T^{-1}+66.2T$$

Zone III (1,033°–1,179° K.)

$$\Delta C_p=-9.68+1.98\times10^{-3}T+2.10\times10^5T^{-2}$$
$$\Delta H_T=13{,}130-9.68T+0.99\times10^{-3}T^2-2.10\times10^5T^{-1}$$
$$\Delta F_T=13{,}130+9.68T\ln T-0.99\times10^{-3}T^2-1.05\times10^5T^{-1}-78.14T$$

Zone IV (1,179°–1,500° K.)

$$\Delta C_p=7.00-7.0\times10^{-3}T+2.10\times10^5T^{-2}$$
$$\Delta H_T=-1{,}000+7.00T-3.5\times10^{-3}T^2-2.10\times10^5T^{-1}$$
$$\Delta F_T=-1{,}000-7.00T\ln T+3.5\times10^{-3}T^2-1.05\times10^5T^{-1}+46.45T$$

Zone II (1,500°–1,674° K.)

$$\Delta C_p=11.95-10.02\times10^{-3}T+2.10\times10^5T^{-2}$$
$$\Delta H_T=7{,}340+11.95T-5.01\times10^{-3}T^2-2.10\times10^5T^{-1}$$
$$\Delta F_T=7{,}340-11.95T\ln T+5.01\times10^{-3}T^2-1.05\times10^5T^{-1}+74.62T$$

Zone III (1,674°–1,803° K.)

$$\Delta C_p=-4.4-1.02\times10^{-3}T+2.10\times10^5T^{-2}$$
$$\Delta H_T=21{,}700-4.4T-0.51\times10^{-3}T^2-2.10\times10^5T^{-1}$$
$$\Delta F_T=21{,}700+4.4T\ln T+0.51\times10^{-3}T^2-1.05\times10^5T^{-1}-47.48T$$

Zone IV (1,803°–1,900° K.)

$$\Delta C_p=-3.50-1.02\times10^{-3}T+2.10\times10^5T^{-2}$$
$$\Delta H_T=8{,}980-3.50T-0.51\times10^{-3}T^2-2.10\times10^5T^{-1}$$
$$\Delta F_T=8{,}980+3.50T\ln T+0.51\times10^{-3}T^2-1.05\times10^5T^{-1}-33.8T$$

T, ° K.	H_T-H_{298}	S_T	ΔH_T°	ΔF_T°
298	----------	24.2	+5,780	4,800
400	2,690	31.96	6,300	4,400
500	5,670	38.61	6,950	3,800
600	8,390	43.57	7,050	3,100
700	11,150	47.82	7,000	2,500
800	13,940	51.55	6,750	1,850
900	16,760	54.86	6,200	1,300
1,000	19,610	57.87	5,100	750
1,100	22,490	60.61	3,450	450
1,200	25,400	63.15	2,200	200
1,300	28,340	65.50	2,050	10
1,400	31,310	67.70	1,800	−400
1,500	46,640	77.99	13,800	−650
1,600	49,700	79.96	13,450	−1,150
1,700	52,760	81.82	12,650	−2,050
1,800	55,820	83.57	12,050	−2,900
1,900	58,880	85.23	400	−3,200

Tetrairon Nitride, Fe_4N (*c*)

$\Delta H^\circ_{298} = -2{,}550$ calories per mole (*112*)
$S_{298} = 37.3$ *e.u.* (*112*)

Zone I (*c*) (298°–1,000° K.)

$$C_p = 26.84 + 8.16\times10^{-3}T \ (82)$$
$$H_T - H_{298} = -8{,}350 + 26.84T + 4.08\times10^{-3}T^2$$

Formation: $4Fe + 1/2N_2 \longrightarrow Fe_4N$

Zone I (298°–1,000° K.)

$$\Delta C_p = 10.03 - 20.75\times10^{-3}T - 1.72\times10^5T^{-2}$$
$$\Delta H_T = -5{,}200 + 10.03T - 10.37\times10^{-3}T^2 + 1.72\times10^5T^{-1}$$
$$\Delta F_T = -5{,}200 - 10.03T\ln T + 10.37\times10^{-3}T^2 + 0.86\times10^5T^{-1} + 73.47T$$

T, ° K.	H_T-H_{298}	S_T	ΔH_T°	ΔF_T°
298	----------	37.3	−2,550	+900
400	3,030	46.4	−2,435	+1,900
500	6,090	53.0	−2,400	+3,100
600	9,230	58.8	−2,580	+4,150
700	12,450	63.8	−2,970	+5,300
800	15,650	67.8	−3,580	+6,700
900	19,120	72.2	−4,330	+7,800
1,000	22,570	75.3	−5,070	+9,500

Diiron Nitride, Fe_2N (*c*)

$\Delta H^\circ_{298} = -900$ calories per mole (*112*)
$S_{298} = 24.2$ *e.u.* (*112*)

Zone I (*c*) (298°–1,000° K.)

$$C_p = 14.91 + 6.09\times10^{-3}T \ (82)$$
$$H_T - H_{298} = -4{,}713 + 14.91T + 3.04\times10^{-3}T^2$$

Formation: $2Fe + 1/2N_2 \longrightarrow Fe_2N$

Zone I (298°–1,000° K.)

$$\Delta C_p = 4.84 - 8.62\times10^{-3}T - 0.86\times10^5T^{-2}$$
$$\Delta H_T = -2{,}250 + 4.84T - 4.31\times10^{-3}T^2 + 0.86\times10^5T^{-1}$$
$$\Delta F_T = -2{,}250 - 4.84T\ln T + 4.31\times10^{-3}T^2 + 0.43\times10^5T^{-1} + 41.7T$$

T, ° K.	H_T-H_{298}	S_T	ΔH_T°	ΔF_T°
298	----------	24.2	−900	2,500
400	1,730	29.3	−800	3,600
500	3,500	33.3	−740	4,750
600	5,320	36.7	−740	5,850
700	7,200	39.5	−830	7,000
800	9,100	42.2	−1,040	8,000
900	11,170	44.4	−1,270	9,350
1,000	13,240	46.6	−1,600	10,600

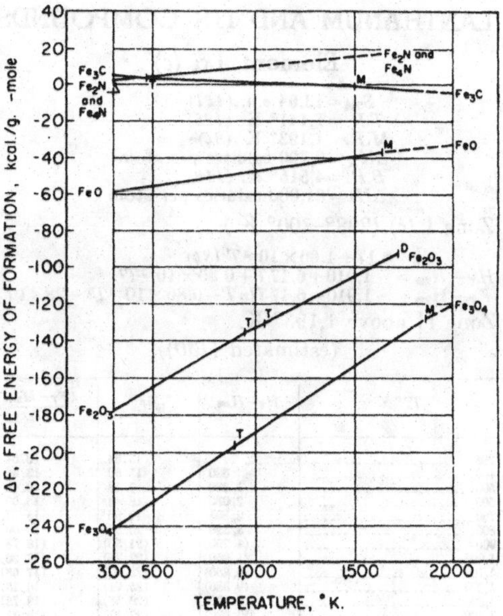

FIGURES 25.—Iron (a).

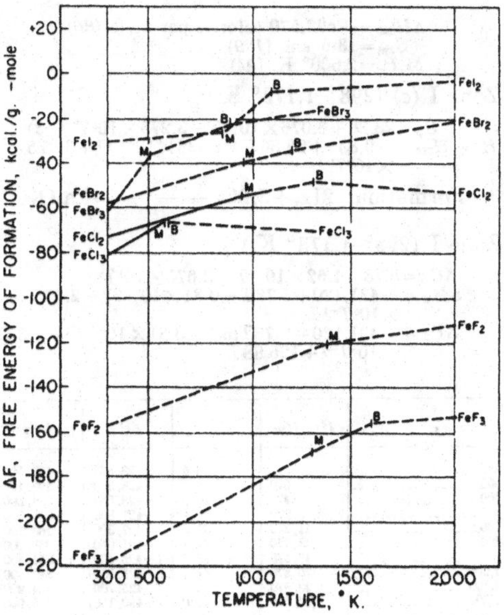

FIGURES 26.—Iron (b).

647940 O—63——5

LANTHANUM AND ITS COMPOUNDS

Element, La (c)

$S_{298} = 13.64$ *e.u.* (*121*)
$T.P. = 1{,}141°$ K. (*125*)
$M.P. = 1{,}193°$ K. (*84*)
$\Delta H_M = 2{,}790$ calories per atom
$B.P. = 4{,}515°$ K. (*112*)
$\Delta H_V = 81{,}000$ calories per atom

Zone I (*c*) (298°–800° K.)

$C_p = 6.17 + 1.60 \times 10^{-3} T$ (*84*)
$H_T - H_{298} = -1{,}910 + 6.17T + 0.80 \times 10^{-3} T^2$
$F_T - H_{298} = -1{,}910 - 6.17 T \ln T - 0.80 \times 10^{-3} T^2 + 28.11T$

Zone II above 1,193° K.
(estimated (*130*))

T, ° K.	$H_T - H_{298}$	S_T	$-\frac{(F_T - H_{298})}{T}$
298	....	13.64	13.64
400	680	15.60	13.90
500	1,380	17.16	14.40
600	2,080	18.44	15.00
700	2,805	19.56	15.57
800	3,530	20.52	16.09
900	(4,290)	(21.50)	(16.73)
1,000	(5,060)	(22.26)	(17.20)
1,100	(5,850)	(23.02)	(17.68)
1,200	(9,380)	(25.94)	(18.13)
1,300	(10,180)	(26.58)	(18.75)
1,400	(10,980)	(27.17)	(19.33)
1,500	(11,780)	(27.73)	(19.88)
1,600	(12,580)	(28.24)	(20.38)
1,700	(13,380)	(28.73)	(20.86)
1,800	(14,180)	(29.18)	(21.31)
1,900	(14,980)	(29.62)	(21.74)
2,000	(15,780)	(30.03)	(22.14)

Dilanthanum Trioxide, La_2O_3 (c)

$\Delta H^\circ_{298} = -428{,}570$ calories per mole (*60*)
$S_{298} = 13.6$ *e.u.* (*109*)
$M.P. = 2{,}600°$ K. (*94*)

Zone I (*c*) (298°–1,173° K.)

$C_p = 28.86 + 3.076 \times 10^{-3} T - 3.275 \times 10^5 T^{-2}$ (*3*)
$H_T - H_{298} = -9{,}835 + 28.86T + 1.538 \times 10^{-3} T^2 + 3.275 \times 10^5 T^{-1}$

Formation: $2La + 3/2O_2 \longrightarrow La_2O_3$

Zone I (298°–1,173° K.)

$\Delta C_p = 5.78 - 1.62 \times 10^{-3} T - 2.675 \times 10^5 T^{-2}$
$\Delta H_T = -431{,}120 + 5.78T - 0.81 \times 10^{-3} T^2 + 2.675 \times 10^5 T^{-1}$
$\Delta F_T = -431{,}120 - 5.78 T \ln T + 0.81 \times 10^{-3} T^2 + 1.337 \times 10^5 T^{-1} + 126.88T$

T, ° K.	$H_T - H_{298}$	S_T	ΔH°_T	ΔF°_T
298	....	13.6	−428,570	−402,600
400	2,670	19.7	−428,350	−393,100
500	5,470	26.5	−428,050	−384,500
600	8,270	32.6	−427,800	−376,600
700	11,370	37.5	−427,300	−368,100
800	14,170	41.7	−427,050	−360,000
900	17,570	45.7	(−426,400)	(−351,300)
1,000	20,870	48.7	(−425,950)	(−342,400)
1,100	23,870	51.9	(−425,700)	(−334,900)
1,200	(27,070)	(55.0)	(−430,150)	(−326,900)

Lanthanum Trifluoride, LaF_3 (c)

$\Delta H^\circ_{298} = (-396{,}000)$ calories per mole (*5*)
$S_{298} = (24)$ *e.u.* (*11*)
$M.P. = 1{,}766°$ K. (*29*)
$\Delta H_M = (8{,}000)$ calories per mole
$B.P. = (2{,}600°)$ K. (*6*)
$\Delta H_V = (62{,}000)$ calories per mole

Formation: $La + 3/2F_2 \longrightarrow LaF_3$
(estimated (*11*))

T, ° K.	$H_T - H_{298}$	ΔH°_T	ΔF°_T
298	....	(−396,000)	(−377,000)
500	(4,000)	(−395,800)	(−365,000)
1,000	(17,000)	(−392,800)	(−336,000)
1,500	(32,000)	(−391,250)	(−307,000)

Lanthanum Trichloride, $LaCl_3$ (c)

$\Delta H^\circ_{298} = -255{,}910$ calories per mole (*127*)
$S_{298} = 34.5$ *e.u.* (*127*)
$M.P. = 1{,}135°$ K. (*29*)
$\Delta H_M = (9{,}000)$ calories per mole
$B.P. = (2{,}020°)$ K. (*6*)
$\Delta H_V = (44{,}000)$ calories per mole

Formation: $La + 3/2Cl_2 \longrightarrow LaCl_3$
(estimated (*11*))

T, ° K.	$H_T - H_{298}$	ΔH°_T	ΔF°_T
298	....	−255,900	−238,300
500	(5,000)	(−254,700)	(−227,400)
1,000	(19,000)	(−250,900)	(−200,900)
1,500	(43,000)	(−237,800)	(−180,900)

Lanthanum Tribromide, $LaBr_3$ (c)

$\Delta H^\circ_{298} = (-197{,}000)$ calories per mole (*5*)
$S_{298} = (45)$ *e.u.* (*11*)
$M.P. = 1{,}062°$ K. (29)
$\Delta H_M = (8{,}000)$ calories per mole
$B.P. = (1{,}850°)$ K. (*6*)
$\Delta H_V = (45{,}000)$ calories per mole

Formation: $La + 3/2Br_2 \longrightarrow LaBr_3$
(estimated (*11*))

T, ° K.	$H_T - H_{298}$	ΔH°_T	ΔF°_T
298	....	(−197,000)	(−191,000)
500	(5,000)	(−213,500)	(−180,000)
1,000	(18,000)	(−217,200)	(−154,000)
1,500	(43,000)	(−209,900)	(−133,000)

Lanthanum Triiodide, LaI_3 (c)

$\Delta H^\circ_{298} = (-166{,}700)$ calories per mole (*5*)
$S_{298} = (49)$ *e.u.* (*11*)
$M.P. = 1{,}045°$ K. (*29*)
$\Delta H_M = (8{,}000)$ calories per mole
$B.P. = (1{,}675°)$ K. (*6*)
$\Delta H_V = (40{,}000)$ calories per mole

Formation: $La + 3/2I_2 \longrightarrow LaI_3$
(estimated (*11*))

T, °K.	H_T-H_{298}	ΔH°_T	ΔF°_T
298	----------	(−166, 700)	(−164, 800)
500	(5, 000)	(−188, 100)	(−160, 000)
1,000	(19, 000)	(−184, 400)	(−133, 000)
1,500	(44, 000)	(−173, 500)	(−109, 500)

Lanthanum Nitride, LaN (*c*)

$\Delta H^\circ_{298} = -72{,}100$ calories per mole (*112*)
$S_{298} = 11.5$ *e.u.* (*9*)
$\Delta F^\circ_{298} = -64{,}700$ calories per mole

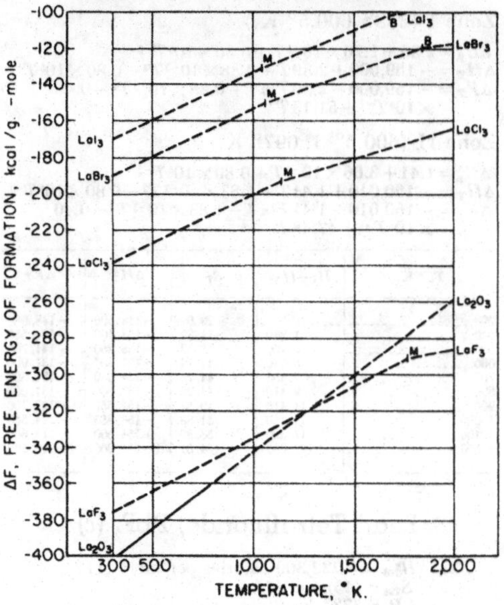

FIGURE 27.—Lanthanum.

LEAD AND ITS COMPOUNDS

Element, Pb (*c*)

$S_{298} = 15.49$ *e.u.* (*83*)
M.P. = 600.5° K. (*82*)
$\Delta H_M = 1{,}225$ calories per atom
B.P. = 2,024° K. (*130*)
$\Delta H_V = 42{,}880$ calories per atom

Zone I (*c*) (298°–600.5° K.)

$$C_p = 5.82 + 1.90 \times 10^{-3}T \ (82)$$
$$H_T - H_{298} = -1{,}820 + 5.82T + 0.95 \times 10^{-3}T^2$$
$$F_T - H_{298} = -1{,}820 - 5.82T\ln T - 0.95 \times 10^{-3}T^2 + 24.04T$$

Zone II (*l*) (600.5°–1,300° K.)

$$C_p = 6.80 \ (82)$$
$$H_T - H_{298} = -838 + 6.80T$$
$$F_T - H_{298} = -838 - 6.80T\ln T + 28.15T$$

Zone III (1,300°–2,000° K.)
(estimated (*130*))

T, °K.	H_T-H_{298}	S_T	$-\frac{(F_T-H_{298})}{T}$
298	----------	15.49	15.49
400	655	17.38	15.75
500	1, 335	18.90	15.83
600	2, 015	20.14	16.78
700	3, 920	23.23	17.63
800	4, 600	24.14	18.39
900	5, 280	24.94	19.08
1,000	5, 960	25.65	19.69
1,100	6, 640	26.30	20.26
1,200	7, 320	26.89	20.79
1,300	8, 000	27.44	21.28
1,400	(8, 780)	(28.02)	(21.75)
1,500	(9, 450)	(28.48)	(22.18)
1,600	(10, 110)	(28.91)	(22.60)
1,700	(10, 760)	(29.30)	(22.98)
1,800	(11, 410)	(29.67)	(23.34)
1,900	(12, 050)	(30.02)	(23.68)
2,000	(12, 680)	(30.34)	(24.00)

Lead Oxide (Yellow), PbO (*c*)

$\Delta H^\circ_{298} = -52{,}070$ calories per mole (*112*)
$S_{298} = 16.1$ *e.u.* (*89*)
M.P. = 1,159° K. (*112*)
$\Delta H_M = 2{,}800$ calories per mole
B.P. = 1,745° K. (*112*)
$\Delta H_V = 51{,}000$ calories per mole

Zone I (*c*) (298°–1,000° K.)

$$C_p = 9.05 + 6.40 \times 10^{-3}T \ (82)$$
$$H_T - H_{298} = -2{,}983 + 9.05T + 3.20 \times 10^{-3}T^2$$

Formation: $Pb + 1/2O_2 \longrightarrow PbO$

Zone I (298°–600.5° K.)

$$\Delta C_p = -0.35 + 4.0 \times 10^{-3}T + 0.20 \times 10^5 T^{-2}$$
$$\Delta H_T = -52{,}070 - 0.35T + 2.0 \times 10^{-3}T^2 - 0.20 \times 10^5 T^{-1}$$
$$\Delta F_T = -52{,}070 + 0.35T\ln T - 2.0 \times 10^{-3}T^2 - 0.10 \times 10^5 T^{-1} + 23.35T$$

Zone II (600.5°–1,000° K.)

$$\Delta C_p = -1.33 + 5.90 \times 10^{-3}T + 0.20 \times 10^5 T^{-2}$$
$$\Delta H_T = -53{,}070 - 1.33T + 2.95 \times 10^{-3}T^2 - 0.20 \times 10^5 T^{-1}$$
$$\Delta F_T = -53{,}070 + 1.33T\ln T - 2.95 \times 10^{-3}T^2 - 0.10 \times 10^5 T^{-1} + 18.57T$$

T, °K.	H_T-H_{298}	S_T	ΔH°_T	ΔF°_T
298	----------	16.1	−52, 070	−44, 950
400	1, 150	19.42	−51, 950	−42, 750
500	2, 340	22.07	−51, 800	−40, 200
600	3. 600	24.36	−51, 600	−37, 900
700	4, 920	26.4	−52, 600	−35, 450
800	6, 310	28.25	−52, 250	−32, 900
900	7, 760	29.96	−51, 450	−30, 600
1,000	9, 260	31.54	−51, 450	−28, 250
1,100	(10, 800)	----------	(−51, 000)	(−26, 450)
1,200	(15, 200)	----------	(−47, 700)	(−24, 400)
1,300	(16, 750)	----------	(−47, 250)	(−22, 450)
1,400	(18, 450)	----------	(−46. 750)	(−20, 550)
1,500	(20, 100)	----------	(−46. 250)	(−18, 700)
1,600	(21, 650)	----------	(−45, 750)	(−16, 850)
1,700	(23. 200)	----------	(−45, 300)	(−15, 100)

Lead Oxide (Red), PbO (*c*)

$\Delta H^\circ_{298} = -52{,}400$ calories per mole (*112*)
$S_{298} = 15.6$ *e.u.* (*89*)
T.P. = 762° K. (red ⟶ yellow) (*24*)
$\Delta H_T = 250$ calories per mole

Zone I (β) (298°–762° K.)

$$C_p = 10.60 + 4.00 \times 10^{-3} T \quad (82)$$
$$H_T - H_{298} = -3{,}338 + 10.60T + 2.00 \times 10^{-3} T^2$$

Formation: $Pb + 1/2O_2 \longrightarrow PbO$

Zone I (298°–600.5° K.)

$$\Delta C_p = 1.20 + 1.60 \times 10^{-3} T + 0.20 \times 10^5 T^{-2}$$
$$\Delta H_T = -52{,}770 + 1.20T + 0.80 \times 10^{-3} T^2 - 0.20 \times 10^5 T^{-1}$$
$$\Delta F_T = -52{,}770 - 1.20T \ln T - 0.80 \times 10^{-3} T^2 - 0.10 \times 10^5 T^{-1} + 32.77T$$

Zone II (600.5°–762° K.)

$$\Delta C_p = 0.22 + 3.50 \times 10^{-3} T + 0.20 \times 10^5 T^{-2}$$
$$\Delta H_T = -53.730 + 0.22T + 1.75 \times 10^{-3} T^2 - 0.20 \times 10^5 T^{-1}$$
$$\Delta F_T = -53{,}730 - 0.22T \ln T - 1.75 \times 10^{-3} T^2 - 0.10 \times 10^5 T^{-1} + 28.72T$$

T, ° K.	$H_T - H_{298}$	S_T	ΔH°_T	ΔF°_T
298	....	15.6	−52,400	−45,130
400	1,220	19.12	−52,200	−42,700
500	2,460	21.88	−52,000	−40,300
600	3,740	24.22	−51,900	−38,100
700	5,060	26.25	−52,750	−35,500
762	....	....	....	−34,250

Trilead Tetraoxide, Pb_3O_4 (*c*)

$\Delta H^\circ_{298} = -175{,}500$ calories per mole (*112*)
$S_{298} = 50.5$ *e.u.* (*83*)

Formation: $3Pb + 2O_2 \longrightarrow Pb_3O_4$
(estimated (*24*))

T, ° K.	$H_T - H_{298}$	ΔH°_T	ΔF°
298	....	−175,500	−147,500
400	(3,390)	(−175,500)	(−138,000)
500	(7,410)	(−175,000)	(−129,000)
600	(11,470)	(−174,500)	(−119,500)
700	(15,740)	(−177,500)	(−110,000)
800	(20,370)	(−176,500)	(−100,500)
900	(25,370)	(−175,000)	(−91,000)
1,000	(31,180)	(−173,000)	(−81,500)

Lead Dioxide, PbO_2 (*c*)

$\Delta H^\circ_{298} = -66{,}120$ calories per mole (*112*)
$S_{298} = 18.3$ *e.u.* (*83*)

Formation: $Pb + O_2 \longrightarrow PbO_2$
(estimated (*24*))

T, ° K.	$H_T - H_{298}$	ΔH°_T	ΔF°_T
298	....	−66,100	−52,300
400	(1,580)	(−65,900)	(−47,700)
500	(3,190)	(−65,700)	(−43,100)
600	(4,825)	(−65,500)	(−38,600)
700	(6,610)	(−66,400)	(−34,000)
800	(8,490)	(−66,000)	(−29,400)
900	(10,480)	(−65,500)	(−24,800)
1,000	(12,460)	(−65,000)	(−20,300)
1,100	(14,550)	(−64,400)	(−15,900)

Lead Difluoride, PbF_2 (*c*)

$\Delta H^\circ_{298} = -158{,}500$ calories per mole (*112*)
$S_{298} = (29)$ *e.u.* (*112*)
M.P. = 1,097° K. (*6*)
$\Delta H_M = 1{,}860$ calories per mole
B.P. = 1,566° K. (*6*)
$\Delta H_V = 38{,}340$ calories per mole

Zone I (298°–1,097° K.)

$$C_p = 16.50 + 4.10 \times 10^{-3} T \quad (15)$$
$$H_T - H_{298} = -5{,}100 + 16.50T + 2.05 \times 10^{-3} T^2$$

Formation: $Pb + F_2 \longrightarrow PbF_2$

Zone I (298°–600.5° K.)

$$\Delta C_p = 2.39 + 1.76 \times 10^{-3} T + 0.80 \times 10^5 T^{-2}$$
$$\Delta H_T = -159{,}000 + 2.39T + 0.88 \times 10^{-3} T^2 - 0.80 \times 10^5 T^{-1}$$
$$\Delta F_T = -159{,}000 - 2.39T \ln T - 0.88 \times 10^{-3} T^2 - 0.40 \times 10^5 T^{-1} + 51.13T$$

Zone II (600.5°–1,097° K.)

$$\Delta C_p = 1.41 + 3.66 \times 10^{-3} T + 0.80 \times 10^5 T^{-2}$$
$$\Delta H_T = -160{,}010 + 1.41T + 1.83 \times 10^{-3} T^2 - 0.80 \times 10^5 T^{-1}$$
$$\Delta F_T = -160{,}010 - 1.41T \ln T - 1.83 \times 10^{-3} T^2 - 0.40 \times 10^5 T^{-1} + 47.48T$$

T, ° K.	$H_T - H_{298}$	S_T	ΔH°_T	ΔF°_T
298	....	29.0	−158,500	−148,060
400	1,830	34.32	−158,100	−144,550
500	3,160	38.43	−158,250	−141,700
600	5,540	41.74	−157,400	−137,900
700	7,450	44.75	−158,250	−134,500
800	9,410	47.26	−157,800	−131,000
900	11,410	49.57	−157,350	−127,700
1,000	13,450	51.88	−156,850	−124,550
1,100	17,390	53.89	−154,500	−119,600
1,500	(27,000)	(64.0)	(−151,200)	(−112,100)

Lead Tetrafluoride, PbF_4 (*c*)

$\Delta H^\circ_{298} = -222{,}300$ calories per mole (*112*)
$S_{298} = (45)$ *e.u.* (*11*)
S.P. = 773° K. (*6*)

Formation: $Pb + 2F_2 \longrightarrow PbF_4$
(estimated (*11*))

T, ° K.	$H_T - H_{298}$	ΔH°_T	ΔF°_T
298	....	−222,300	(−202,000)
500	(6,000)	(−220,800)	(−189,300)

Lead Dichloride, $PbCl_2$ (*c*)

$\Delta H^\circ_{298} = -85{,}850$ calories per mole (*112*)
$S_{298} = 32.6$ *e.u.* (*83*)
M.P. = 771° K. (*82*)
$\Delta H_M = 5{,}800$ calories per mole
B.P. = 1,227° K. (*6*)
$\Delta H_V = 29{,}604$ calories per mole

Zone I (*c*) (298°–771° K.)

$$C_p = 15.96 + 8.00 \times 10^{-3} T \quad (82)$$
$$H_T - H_{298} = -5{,}115 + 15.96T + 4.00 \times 10^{-3} T^2$$

Zone II (*l*) (771°–900° K.)

$$C_p = 27.20 \quad (82)$$
$$H_T - H_{298} = -5{,}600 + 27.20T$$

Formation: $Pb+Cl_2 \longrightarrow PbCl_2$

Zone I (298°–600.5° K.)

$\Delta C_p = 1.32+6.04\times10^{-3}T+0.68\times10^5T^{-2}$
$\Delta H_T = -86,280+1.32T+3.02\times10^{-3}T^2-0.68\times10^5T^{-1}$
$\Delta F_T = -86,280-1.32T\ln T-3.02\times10^{-3}T^2-0.34\times10^5T^{-1}+46.26T$

Zone II (600.5°–771° K.)

$\Delta C_p = 0.34+7.94\times10^{-3}T+0.68\times10^5T^{-2}$
$\Delta H_T = -87,240+0.34T+3.97\times10^{-3}T^2-0.68\times10^5T^{-1}$
$\Delta F_T = -87,240-0.34T\ln T-3.97\times10^{-3}T^2-0.34\times10^5T^{-1}+42.4T$

Zone III (771°–900° K.)

$\Delta C_p = 11.58-0.06\times10^{-3}T+0.68\times10^5T^{-2}$
$\Delta H_T = -87,750+11.58T-0.03\times10^{-3}T^2-0.68\times10^5T^{-1}$
$\Delta F_T = -87,750-11.58T\ln T+0.03\times10^{-3}T^2-0.34\times10^5T^{-1}+113.74T$

T, ° K.	H_T-H_{298}	S_T	ΔH_T°	ΔF_T°
298	....	32.6	−85,850	−75,060
400	1,920	38.14	−85,450	−71,400
500	3,830	42.40	−85,050	−67,900
600	5,890	46.15	−84,500	−64,550
700	8,040	49.46	−85,150	−61,100
800	16,160	60.07	−78,590	−57,850
900	18,880	63.27	−77,430	−55,550
1,000	(21,400)	(66.1)	(−76,450)	(−53,000)

Lead Dibromide, $PbBr_2$ (*c*)

$\Delta H_{298}^\circ = -66,210$ calories per mole (*112*)
$S_{298} = 38.6$ *e.u.* (*83*)
M.P. = 761° K. (*82*)
$\Delta H_M = 4,430$ calories per mole
B.P. = 1,187° K. (*6*)
$\Delta H_V = 27,694$ calories per mole

Zone I (*c*) (298°–761° K.)

$C_p = 18.59+2.20\times10^{-3}T$ (*82*)
$H_T-H_{298} = -5,640+18.59T+1.10\times10^{-3}T^{-2}$

Zone II (*l*) (761°–900° K.)

$C_p = 27.60$ (*82*)
$H_T-H_{298} = -7,435+27.60T$

Formation: $Pb+Br_2 \longrightarrow PbBr_2$

Zone I (298°–331° K.)

$\Delta C_p = -4.33+0.30\times10^{-3}T$
$\Delta H_T = -64,930-4.33T+0.15\times10^{-3}T^2$
$\Delta F_T = -64,930+4.33T\ln T-0.15\times10^{-3}T^2-15.62T$

Zone II (331°–600.5° K.)

$\Delta C_p = 3.73+0.30\times10^{-3}T+0.37\times10^5T^{-2}$
$\Delta H_T = -74,975+3.73T+0.15\times10^{-3}T^2-0.37\times10^5T^{-1}$
$\Delta F_T = -74,975-3.73T\ln T-0.15\times10^{-3}T^2-0.18\times10^5T^{-1}+61.94T$

Zone III (600.5°–761° K.)

$\Delta C_p = 2.75+2.20\times10^{-3}T+0.37\times10^5T^{-2}$
$\Delta H_T = -75,940+2.75T+1.10\times10^{-3}T^2-0.37\times10^5T^{-1}$
$\Delta F_T = -75,940-2.75T\ln T-1.10\times10^{-3}T^2-0.18\times10^5T^{-1}+57.85T$

Zone IV (761°–900° K.)

$\Delta C_p = 11.74+0.37\times10^5T^{-2}$
$\Delta H_T = -77,730+11.74T-0.37\times10^5T^{-1}$
$\Delta F_T = -77,730-11.74T\ln T-0.18\times10^5T^{-1}+118.0T$

T, ° K.	H_T-H_{298}	S_T	ΔH_T°	ΔF_T°
298	....	38.6	−66,210	−62,250
400	1,970	44.29	−73,500	−59,150
500	3,930	48.66	−73,150	−55,650
600	5,900	52.25	−72,750	−52,200
700	7,910	55.35	−73,550	−48,600
800	14,640	64.22	−68,400	−45,300
900	17,400	67.47	−67,200	−42,500
1,000	(19,800)	(69.80)	−66,400	(−39,650)

Lead Diiodide, PbI_2 (*c*)

$\Delta H_{298}^\circ = -41,850$ calories per mole (*112*)
$S_{298} = 42.3$ *e.u.* (*112*)
M.P. = 685° K. (*82*)
$\Delta H_M = 6,010$ calories per mole
B.P. = 1,145° K. (*6*)
$\Delta H_V = 24,846$ calories per mole

Zone I (*c*) (298°–685° K.)

$C_p = 18.00+4.70\times10^{-3}T$ (*82*)
$H_T-H_{298} = -5,576+18.00T+2.35\times10^{-3}T^2$

Zone II (*l*) (685°–800° K.)

$C_p = 32.40$ (*82*)
$H_T-H_{298} = -8,325+32.40T$

Formation: $Pb+I_2 \longrightarrow PbI_2$

Zone I (298°–386.1° K.)

$\Delta C_p = 2.59-9.10\times10^{-3}T$
$\Delta H_T = -42,220+2.59T-4.55\times10^{-3}T^2$
$\Delta F_T = -42,220-2.59T\ln T+4.55\times10^{-3}T^2+15.72T$

Zone II (386.1°–456° K.)

$\Delta C_p = -7.02+2.80\times10^{-3}T$
$\Delta H_T = -43,150-7.02T+1.40\times10^{-3}T^2$
$\Delta F_T = -43,150+7.02T\ln T-1.40\times10^{-3}T^2-36.83T$

Zone III (456°–600° K.)

$\Delta C_p = 3.29+2.80\times10^{-3}T$
$\Delta H_T = -57,850+3.29T+1.40\times10^{-3}T^2$
$\Delta F_T = -57,850-3.29T\ln T-1.40\times10^{-3}T^2+58.55T$

Zone IV (685°–800° K.)

$\Delta C_p = 16.71$
$\Delta H_T = -61,550+16.71T$
$\Delta F_T = -61,550-16.71T\ln T+150.87T$

T, ° K.	H_T-H_{298}	S_T	ΔH_T°	ΔF_T°
298	....	42.3	−41,850	−41,550
400	2,010	48.1	−45,730	−41,280
500	4,000	52.54	−55,850	−39,150
600	6,070	56.31	−55,350	−36,850
700	14,360	68.58	−49,850	−32,570
800	17,600	72.9	−48,200	−30,200

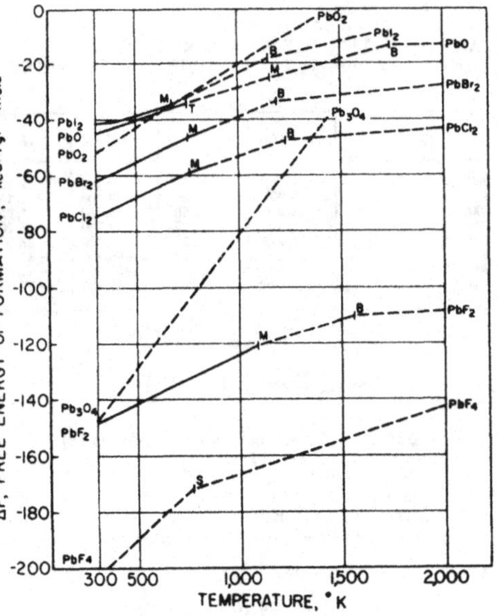

FIGURE 28.—Lead.

LITHIUM AND ITS COMPOUNDS

Element, Li (*c*)

$S_{298}=6.75$ *e.u.* (*34*)
$M.P.=453.7°$ K. (*34*)
$\Delta H_M=723$ calories per atom
$B.P.=1{,}604°$ K. (*130*)
$\Delta H_V=32{,}190$ calories per atom

Zone I (*c*) (298°–452° K.)

$$C_p=3.15+8.40\times10^{-3}T \; (82)$$
$$H_T-H_{298}=-1{,}313+3.15T+4.20\times10^{-3}T^2$$
$$F_T-H_{298}=-1{,}313-3.15T\ln T-4.20\times10^{-3}T^2+16.84T$$

Zone II (*l*) (452°-1,604° K.)

$$C_p=6.935-0.078\times10^{-3}T+0.36\times10^5T^{-2} \; (34)$$
$$H_T-H_{298}=-1{,}324+6.935T-0.039\times10^{-3}T^2-0.36\times10^5T^{-1}$$
$$F_T-H_{298}=-1{,}324-6.935T\ln T+0.039\times10^{-3}T^2-0.18\times10^5T^{-1}+38.19T$$

Zone III (*g*) (1,640°–2,500° K.)

$$C_p=3.93+0.364\times10^{-3}T+12.94\times10^5T^{-2} \; (34)$$
$$H_T-H_{298}=+38{,}956+3.93T+0.182\times10^{-3}T^2-12.94\times10^5T^{-1}$$
$$F_T-H_{298}=+38{,}956-3.93T\ln T-0.182\times10^{-3}T^2-6.47\times10^5T^{-1}-8.23T$$

T, ° K.	H_T-H_{298}	S_T	$-\frac{(F_T-H_{298})}{T}$
298		6.75	6.75
400	630	8.57	7.0
500	2,049	11.71	7.62
600	2,763	13.01	8.4
700	3,462	14.09	9.14
800	4,155	15.01	9.82
900	4,846	15.83	10.44
1,000	5,536	16.55	11.01
1,100	6,224	17.21	11.55
1,200	6,912	17.81	12.05
1,300	7,598	18.36	12.51
1,400	8,284	18.86	12.94
1,500	8,967	19.34	13.36
1,600	9,648	19.78	13.75
1,700	45,404	41.79	15.08
1,800	45,901	42.08	16.58
1,900	46,399	42.35	17.93
2,000	46,897	42.60	19.15

Dilithium Oxide, Li_2O (*c*)

$\Delta H^{\circ}_{298}=-142{,}570$ calories per mole (*75*)
$S_{298}=9.06$ *e.u.* (*75*)
$M.P.=(2{,}000)°$ K. (*42*)

Zone I (*c*) (298°–1500° K.)

$$C_p=14{,}939+6.08\times10^{-3}T-3.38\times10^5T^{-2} \; (116)$$
$$H_T=H_{298}=5{,}858+14.939T+3.04\times10^{-3}T^2+3.38\times10^5T^{-1}$$

Formation: $2Li+1/2O_2 \longrightarrow Li_2O$

Zone I (298°–452° K.)

$$\Delta C_p=5.06-11.22\times10^{-3}T-3.18\times10^5T^{-2}$$
$$\Delta H_T=144{,}648+5.06T-5.61\times10^{-3}T^2+3.18\times10^5T^{-1}$$
$$\Delta F_T=144{,}648-5.06T\ln T+5.61\times10^{-3}T^2+1.59\times10^5T^{-1}+61.28T$$

Zone II (452°-1,500° K.)

$$\Delta C_p=2.51+5.74\times10^{-3}T-3.90\times10^5T^{-2}$$
$$\Delta H_T=138{,}800-2.51T+2.87\times10^{-3}T^2+3.90\times10^5T^{-1}$$
$$\Delta F_T=138{,}800+2.51T\ln T-2.87\times10^{-3}T^2+1.95\times10^5T^{-1}+18.6T$$

T, ° K.	H_T-H_{298}	S_T	ΔH°_T	ΔF°_T
298		9.06	−142,570	−133,950
400	7,306	13.22	−136,900	−125,100
500	8,905	16.79	−138,500	−122,000
600	10,620	19.91	−138,600	−118,700
700	12,430	22.70	−138,550	−115,350
800	14,320	25.22	−138,450	−112,050
900	16,270	27.53	−138,300	−108,800
1,000	18,317	29.68	−138,050	−105,500
1,100	20,418	31.68	−137,700	−102,200
1,200	22,586	33.56	−137,350	−99,050
1,300	24,818	35.35	−136,900	−95,850
1,400	27,050	37.05	−136,550	−92,900
1,500	30,233	38.67	−136,100	−89,550

Lithium Fluoride, LiF (*c*)

$\Delta H^{\circ}_{298}=-146{,}300$ calories per mole (*112*)
$S_{298}=8.57$ *e.u.* (*112*)
$M.P.=1{,}120°$ K. (*6*)
$\Delta H_M=2{,}360$ calories per mole
$B.P.=1{,}954°$ K. (*6*)
$\Delta H_V=50{,}970$ calories per mole

Zone I (*c*) (298°–1,120° K.)

$$C_p = 9.14 + 5.19 \times 10^{-3} T \ (82)$$
$$H_T - H_{298} = -2{,}954 + 9.14T + 2.59 \times 10^{-3} T^2$$

Formation: $Li + 1/2F_2 \longrightarrow LiF$

Zone I (298°–452° K.)

$$\Delta C_p = 1.84 - 3.43 \times 10^{-3} T + 0.40 \times 10^5 T^{-2}$$
$$\Delta H_T = -146{,}550 + 1.84T - 1.71 \times 10^{-3} T^2 - 0.40 \times 10^5 T^{-1}$$
$$\Delta F_T = -146{,}550 - 1.84 T \ln T + 1.71 \times 10^{-3} T^2 - 0.20 \times 10^5 T^{-1} + 33.53T$$

Zone II (452°–1,120° K.)

$$\Delta C_p = -1.94 + 5.05 \times 10^{-3} T + 0.04 \times 10^5 T^{-2}$$
$$\Delta H_T = -146{,}400 - 1.94T + 2.52 \times 10^{-3} T^2 - 0.04 \times 10^5 T^{-1}$$
$$\Delta F_T = -146{,}400 + 1.94 T \ln T - 2.52 \times 10^{-3} T^2 - 0.02 \times 10^5 T^{-1} + 12.23T$$

T, ° K.	$H_T - H_{298}$	S_T	ΔH_T°	ΔF_T°
298	....	8.57	−146,300	−139,650
400	1,116	11.79	−146,150	−137,250
500	2,263	14.34	−146,800	−134,950
600	3,462	16.53	−146,700	−132,550
700	4,713	18.46	−146,550	−130,200
800	6,016	20.20	−146,350	−127,850
900	7,370	21.79	−146,100	−125,550
1,000	8,776	23.28	−145,800	−123,300
1,100	10,234	24.66	−145,400	−120,950
1,500	(18,200)	(30.57)	(−141,900)	(−112,200)

Lithium Chloride, LiCl (*c*)

$\Delta H_{298}^\circ = -97{,}700$ calories per mole (*112*)
$S_{298} = 13.9$ *e.u.* (*83*)
M.P. = 887° K. (*6*)
$\Delta H_M = 3{,}200$ calories per mole
B.P. = 1,653° K. (*6*)
$\Delta H_V = 35{,}960$ calories per mole

Zone I (*c*) (298°–887° K.)

$$C_p = 11.0 + 3.40 \times 10^{-3} T \ (74)$$
$$H_T - H_{298} = -3{,}429 + 11.0T + 1.70 \times 10^{-3} T^2$$

Formation: $Li + 1/2Cl_2 \longrightarrow LiCl$

Zone I (298°–452° K.)

$$\Delta C_p = 3.44 - 5.03 \times 10^{-3} T + 0.34 \times 10^5 T^{-2}$$
$$\Delta H_T = -98{,}400 + 3.44T - 2.51 \times 10^{-3} T^2 - 0.34 \times 10^5 T^{-1}$$
$$\Delta F_T = -98{,}400 - 3.44 T \ln T + 2.51 \times 10^{-3} T^2 - 0.17 \times 10^5 T^{-1} + 38.76T$$

Zone II (452°–887° K.)

$$\Delta C_p = -0.345 + 3.45 \times 10^{-3} T - 0.02 \times 10^5 T^{-2}$$
$$\Delta H_T = -98{,}360 - 0.345T + 1.72 \times 10^{-3} T^2 - 0.02 \times 10^5 T^{-1}$$
$$\Delta F_T = -98{,}360 + 0.345 T \ln T - 1.72 \times 10^{-3} T^2 - 0.01 \times 10^5 T^{-1} + 17.02T$$

T, ° K.	$H_T - H_{298}$	S_T	ΔH_T°	ΔF_T°
298	....	13.9	−97,700	−92,500
400	1,243	17.48	−97,500	−90,850
500	2,496	20.28	−98,100	−89,200
600	3,783	22.63	−97,450	−87,000
700	5,104	24.66	−97,750	−85,600
800	6,459	26.47	−97,850	−84,450
900	11,050	31.71	−94,100	−82,600
1,000	(12,700)	(33.6)	(−93,550)	(−81,500)
1,500	(20,700)	(40.1)	(−90,800)	(−75,800)

Lithium Bromide, LiBr (*c*)

$\Delta H_{298}^\circ = -83{,}720$ calories per mole (*11*)
$S_{298} = 19$ *e.u.* (*11*)
M.P. = 825° K. (*6*)
$\Delta H_M = 2{,}900$ calories per mole
B.P. = 1,583° K. (*6*)
$\Delta H_V = 35{,}420$ calories per mole

Zone I (*c*) (298°–825° K.)

$$C_p = 11.5 + 3.02 \times 10^{-3} T \ (74)$$
$$H_T - H_{298} = -3{,}560 + 11.5T + 1.51 \times 10^{-3} T^2$$

Formation: $Li + 1/2Br_2 \longrightarrow LiBr$

Zone I (298°–331° K.)

$$\Delta C_p = -0.2 - 5.38 \times 10^{-3} T$$
$$\Delta H_T = -83{,}420 - 0.2T - 2.69 \times 10^{-3} T^2$$
$$\Delta F_T = -83{,}420 + 0.2 T \ln T + 2.69 \times 10^{-3} T^2 + 3.01T$$

Zone II (331°–452° K.)

$$\Delta C_p = 3.83 - 5.38 \times 10^{-3} T + 0.19 \times 10^5 T^{-2}$$
$$\Delta H_T = -88{,}650 + 3.83T - 2.69 \times 10^{-3} T^2 - 0.19 \times 10^5 T^{-1}$$
$$\Delta F_T = -88{,}650 - 3.83 T \ln T + 2.69 \times 10^{-3} T^2 - 0.09 \times 10^5 T^{-1} + 42.83T$$

Zone III (452°–825° K.)

$$\Delta C_p = 0.05 + 3.1 \times 10^{-3} T + 0.17 \times 10^5 T^{-2}$$
$$\Delta H_T = -88{,}400 + 0.05T + 1.55 \times 10^{-3} T^2 + 0.17 \times 10^5 T^{-1}$$
$$\Delta F_T = -88{,}400 - 0.05 T \ln T - 1.55 \times 10^{-3} T^2 + 0.08 \times 10^5 T^{-1} + 20.45T$$

T, ° K.	$H_T - H_{298}$	S_T	ΔH_T°	ΔF_T°
298	....	19.0	−83,720	−81,950
400	1,281	22.69	−87,400	−80,500
500	2,567	25.56	−87,950	−78,700
600	3,884	27.95	−87,800	−76,850
700	5,240	30.03	−87,600	−75,050
800	7,606	31.87	−87,400	−73,300
1,000	(12,700)	(38.9)	(−83,550)	(−70,450)
1,500	(20,700)	(45.4)	(−81,250)	(−64,600)

Lithium Iodide, LiI (*c*)

$\Delta H_{298}^\circ = -64{,}790$ calories per mole (*112*)
$S_{298} = (21)$ *e.u.* (*11*)
M.P. = 713° K. (*6*)
$\Delta H_M = 1{,}420$ calories per mole
B.P. = 1,440° K. (*6*)
$\Delta H_V = 40{,}772$ calories per mole

Zone I (*c*) (298°–713° K.)

$$C_p = 12.3 + 2.44 \times 10^{-3} T \ (74)$$
$$H_T - H_{298} = -3{,}773 + 12.3T + 1.22 \times 10^{-3} T^2$$

Formation $Li + 1/2I_2 \longrightarrow LiI$

Zone I (298°–386.8° K.)

$$\Delta C_p = 4.36 - 11.91 \times 10^{-3} T$$
$$\Delta H_T = -65{,}550 + 4.36T - 5.95 \times 10^{-3} T^2$$
$$\Delta F_T = -65{,}550 - 4.36 T \ln T + 5.95 \times 10^{-3} T^2 + 34.3T$$

Zone II (386.8°–452° K.)

$$\Delta C_p = -0.45 - 5.96 \times 10^{-3} T$$
$$\Delta H_T = -63{,}750 - 0.45T - 2.98 \times 10^{-3} T^2$$
$$\Delta F_T = -63{,}750 + 0.45 T \ln T + 2.98 \times 10^{-3} T^2 + 2.0T$$

Zone II (456°–713° K.)

$$\Delta C_p = -0.93 + 2.52\times 10^{-3}T - 0.36\times 10^{5}T^{-2}$$
$$\Delta H_T = -65{,}600 + 0.93T + 1.26\times 10^{-3}T^2 + 0.36\times 10^{5}T^{-1}$$
$$\Delta F_T = -65{,}600 - 0.93T\ln T - 1.26\times 10^{-3}T^2 + 0.18\times 10^{5}T^{-1} + 15.54T$$

T, ° K.	H_T-H_{298}	S_T	ΔH°_T	ΔF°_T
298	....	(21.0)	−64,790	(−62,200)
400	1,342	(24.9)	−64,400	(−61,350)
500	2,682	(27.9)	−64,750	(−61,000)
600	4,046	(30.3)	−66,000	(−60,800)
700	5,435	(32.5)	−65,450	(−60,200)

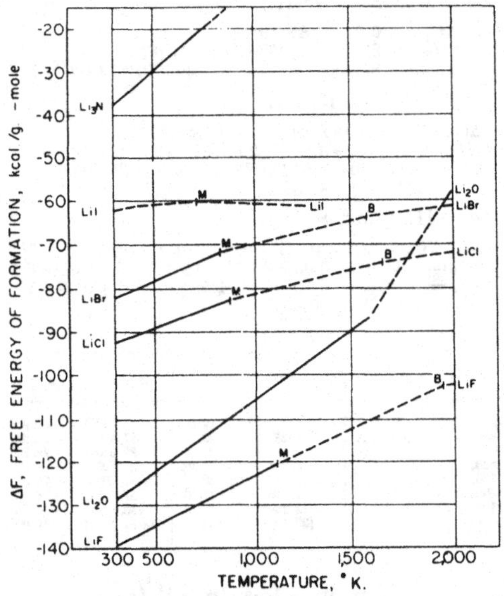

FIGURE 29.—Lithium.

Trilithium Nitride, Li_3N (*c*)

$\Delta H^\circ_{298} = -47{,}500$ calories per mole (*9*)
$S_{298} = 9$ *e.u.* (*9*)
Decomposes (*9*)

Zone I (*c*) (298°–800° K.)

$$C_p = 11.73 + 23.00\times 10^{-3}T \ (82)$$
$$H_T - H_{298} = -4{,}520 + 11.73T + 11.5\times 10^{-3}T^2$$

Formation: $3Li + 1/2N_2 \longrightarrow Li_3N$

Zone I (298°–452° K.)

$$\Delta C_p = -1.05 - 2.71\times 10^{-3}T$$
$$\Delta H_T = -47{,}050 - 1.05T - 1.35\times 10^{-3}T^2$$
$$\Delta F_T = -47{,}050 + 1.05T\ln T + 1.35\times 10^{-3}T^2 + 26.35T$$

Zone II (452°–800° *K.*)

$$\Delta C_p = -12.40 + 22.72\times 10^{-3}T - 1.08\times 10^{5}T^{-2}$$
$$\Delta H_T = -47{,}000 - 12.40T + 11.36\times 10^{-3}T^2 + 1.08\times 10^{5}T^{-1}$$
$$\Delta F_T = -47{,}000 + 12.40T\ln T - 11.36\times 10^{-3}T^2 + 0.54\times 10^{5}T^{-1} - 37.8T$$

T, ° K.	H_T-H_{298}	S_T	ΔH°_T	ΔF°_T
298	....	9.0	−47,500	−37,300
400	2,000	14.7	−48,100	−34,150
500	4,200	19.7	−50,150	−30,100
600	6,680	24.2	−50,170	−26,050
700	9,360	28.3	−49,950	−22,000
800	12,190	32.1	−49,570	−18,100

LUTETIUM AND ITS COMPOUNDS

Element, Lu (*c*)

$S_{298} = (11.79)$ *e.u.* (*121*)
M.P. = (2,000°) K. (*125*)
$\Delta H_M = (4{,}600)$ calories per atom
B.P. = (2,200°) K. (*125*)
$\Delta H_V = 59{,}000$ calories per atom
(estimated (*130*))

T, ° K.	H_T-H_{298}	S_T	$-\frac{(F_T-H_{298})}{T}$
298	....	(11.79)	(11.79)
400	(665)	(13.66)	(12.00)
500	(1,330)	(15.15)	(12.49)
600	(2,015)	(16.40)	(13.05)
700	(2,710)	(17.47)	(13.60)
800	(3,425)	(18.42)	(14.14)
900	(4,150)	(19.28)	(14.67)
1,000	(4,890)	(20.06)	(15.17)
1,100	(5,650)	(20.78)	(15.65)
1,200	(6,420)	(21.46)	(16.11)
1,300	(7,210)	(22.09)	(16.55)
1,400	(8,010)	(22.68)	(16.96)
1,500	(8,830)	(23.25)	(17.37)
1,600	(9,660)	(23.78)	(17.75)
1,700	(10,510)	(24.30)	(18.12)
1,800	(11,370)	(24.79)	(18.48)
1,900	(12,250)	(25.26)	(18.82)
2,000	(17,740)	(28.04)	(19.17)

Lutetium Trifluoride, LuF_3 (*c*)

$\Delta H^\circ_{298} = (-367{,}000)$ calories per mole (*5*)
$S_{298} = (26)$ *e.u.* (*11*)
T.P. = 927° K. (*29*)
M.P. = (1,455°) K. (*29*)
$\Delta H_M = (8{,}000)$ calories per mole
B.P. = (2,500°) K. (*6*)
$\Delta H_V = (60{,}000)$ calories per mole

Formation: $Lu + 3/2F_2 \longrightarrow LuF_3$
(estimated (*11*))

T, ° K.	H_T-H_{298}	ΔH°_T	ΔF°_T
298	....	(−367,000)	(−349,000)
500	(4,000)	(−366,700)	(−338,000)
1,000	(17,000)	(−363,700)	(−309,000)
1,500	(32,000)	(−359,200)	(−283,000)

Lutetium Trichloride, $LuCl_3$ (*c*)

$\Delta H^\circ_{298} = -228{,}000$ calories per mole (*5*)
$S_{298} = (37)$ *e.u.* (*11*)
M.P. = 1,178° K. (*29*)
$\Delta H_M = (9{,}000)$ calories per mole
B.P. = (1,750°) K. (*6*)
$\Delta H_V = (43{,}000)$ calories per mole

Formation: $Lu + 3/2Cl_2 \longrightarrow LuCl_3$
(estimated (*11*))

T, ° K.	$H_T - H_{298}$	ΔH°_T	ΔF°_T
298		−228,000	(−211,500)
500	(5,000)	(−226,100)	(−200,000)
1,000	(19,000)	(−223,000)	(−174,000)
1,500	(43,000)	(−219,400)	(−154,500)

Lutetium Tribromide, $LuBr_3$ (*c*)

$\Delta H^\circ_{298} = (-164{,}000)$ calories per mole (*5*)
$S_{298} = (44)$ *e.u.* (*11*)
$M.P. = (1{,}298°)$ K. (*29*)
$\Delta H_M = (10{,}000)$ calories per mole
$B.P. = (1{,}680°)$ K. (*6*)
$\Delta H_V = (42{,}000)$ calories per mole

Formation: $Lu + 3/2Br_2 \longrightarrow LuBr_3$
(estimated (*11*))

T, ° K.	$H_T - H_{298}$	ΔH°_T	ΔF°_T
298		(−164,000)	(−157,000)
500	(5,000)	(−174,600)	(−146,000)
1,000	(18,000)	(−171,900)	(−121,000)
1,500	(43,000)	(−157,600)	(−104,500)

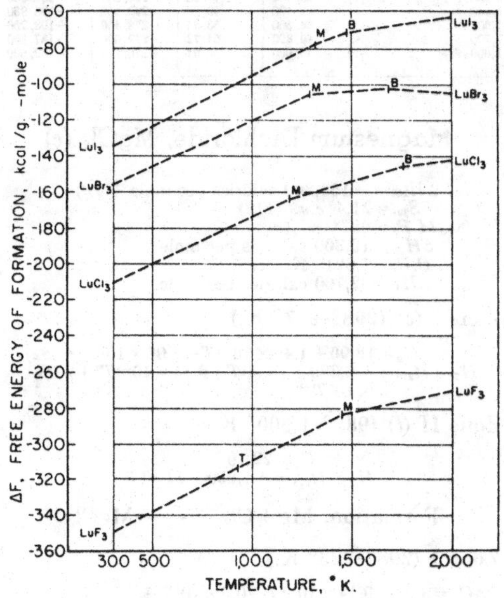

FIGURE 30.—Lutetium.

Lutetium Triiodide, LuI_3 (*c*)

$\Delta H^\circ_{298} = -133{,}000$ calories per mole (*5*)
$S_{298} = (46)$ *e.u.* (*11*)
$M.P. = 1{,}323°$ K. (*29*)
$\Delta H_M = (11{,}000)$ calories per mole
$B.P. = (1{,}480°)$ K. (*6*)
$\Delta H_V = (38{,}000)$ calories per mole

Formation: $Lu + 3/2I_2 \longrightarrow LuI_3$
(estimated (*11*))

T, ° K.	$H_T - H_{298}$	ΔH°_T	ΔF°_T
298		−133,000	(−131,000)
500	(5,000)	(−154,000)	(−125,000)
1,000	(19,000)	(−150,500)	(−96,000)
1,500	(82,000)	(−98,000)	(−71,000)

MAGNESIUM AND ITS COMPOUNDS

Element, Mg (*c*)

$S_{298} = 7.77$ *e.u.* (*83*)
$M.P. = 923°$ K. (*82*)
$\Delta H_M = 2{,}160$ calories per atom
$B.P. = 1{,}393°$ K. (*112*)
$\Delta H_V = 31{,}500$ calories per atom

Zone I (*c*) (298°–923° K.)

$$C_p = 6.14 + 1.50 \times 10^{-3}T - 0.78 \times 10^5 T^{-2} \ (82)$$
$$H_T - H_{298} = -2{,}160 + 6.14T + 0.75 \times 10^{-3}T^2 + 0.78 \times 10^5 T^{-1}$$
$$F_T - H_{298} = -2{,}160 - 6.14T\ln T - 0.75 \times 10^{-3}T^2 + 0.39 \times 10^5 T^{-1} + 33.08T$$

Zone II (*l*) (923°–1,393° K.)

$$C_p = 7.4 \ (82)$$
$$H_T - H_{298} = -440 + 7.40T$$
$$F_T - H_{298} = -440 - 7.40T\ln T + 40.2T$$

Zone III (*g*) (1,393–1,800° K.)

$$C_p = 4.97 \ (84)$$
$$H_T - H_{298} = 34{,}440 + 4.97T$$
$$F_T - H_{298} = 34{,}440 - 4.97T\ln T - 2.4T$$

T, ° K.	$H_T - H_{298}$	S_T	$-\frac{(F_T - H_{298})}{T}$
298		7.77	7.77
400	615	9.52	7.97
500	1,255	10.95	8.44
600	1,920	12.16	8.97
700	2,615	13.23	9.49
800	3,330	14.19	10.01
900	4,060	15.04	10.52
1,000	6,960	18.16	11.20
1,100	7,700	18.87	11.68
1,200	8,430	19.47	12.43
1,300	8,980	20.07	13.09
1,400	41,400	43.27	13.53
1,500	41,900	43.67	15.73
1,600	42,390	43.99	17.58
1,700	42,890	44.37	19.12
1,800	43,390	44.67	20.62
1,900	(43,890)	(44.87)	(21.79)
2,000	(44,390)	(45.17)	(22.92)

Magnesium Oxide, MgO (*c*)

$\Delta H^\circ_{298} = -143{,}700$ calories per mole (*117*)
$S_{298} = 6.40$ *e.u.* (*83*)
$M.P. = 3{,}173°$ K. (*112*)
$\Delta H_M = 18{,}500$ calories per mole

Zone I (298°–2,100° K.)

$$C_p = 10.18 + 1.74\times10^{-3}T - 1.48\times10^{5}T^{-2} \quad (82)$$
$$H_T - H_{298} = -3{,}609 + 10.18T + 0.87\times10^{-3}T^2 + 1.48\times10^{5}T^{-1}$$

Formation: $Mg + 1/2O_2 \longrightarrow MgO$

Zone I (298°–923° K.)

$$\Delta C_p = 0.46 - 0.26\times10^{-3}T - 0.50\times10^{5}T^{-2}$$
$$\Delta H_T = -144{,}000 + 0.46T - 0.13\times10^{-3}T^2 + 0.50\times10^{5}T^{-1}$$
$$\Delta F_T = -144{,}000 - 0.46T\ln T + 0.13\times10^{-3}T^2 + 0.25\times10^{5}T^{-1} + 28.73T$$

Zone II (923°–1,393° K.)

$$\Delta C_p = -0.80 + 1.24\times10^{-3}T - 1.28\times10^{5}T^{-2}$$
$$\Delta H_T = -145{,}750 - 0.80T + 0.62\times10^{-3}T^2 + 1.28\times10^{5}T^{-1}$$
$$\Delta F_T = -145{,}750 + 0.80T\ln T - 0.62\times10^{-3}T^2 + 0.64\times10^{5}T^{-1} + 22.71T$$

Zone III (1,393°–1,800° K.)

$$\Delta C_p = 1.63 + 1.24\times10^{-3}T - 1.28\times10^{5}T^{-2}$$
$$\Delta H_T = -180{,}500 + 1.63T + 0.62\times10^{-3}T^2 + 1.28\times10^{5}T^{-1}$$
$$\Delta F_T = -180{,}500 - 1.63T\ln T - 0.62\times10^{-3}T^2 + 0.64\times10^{5}T^{-1} + 65.4T$$

T, ° K.	$H_T - H_{298}$	S_T	ΔH_T°	ΔF_T°
298	....	6.40	−143,700	−136,100
400	965	9.18	−143,700	−133,500
500	1,975	11.43	−143,700	−131,000
600	3,020	13.34	−143,700	−128,400
700	4,100	15.0	−143,700	−125,900
800	5,225	16.5	−143,700	−123,300
900	6,390	17.87	−143,650	−120,750
1,000	7,580	19.13	−145,800	−118,050
1,100	8,800	20.29	−145,700	−115,200
1,200	10,050	21.38	−145,600	−112,600
1,300	11,310	22.38	−145,300	−109,500
1,400	12,570	23.32	−176,850	−106,850
1,500	13,830	24.19	−176,600	−101,700
1,600	15,090	25.0	−176,200	−96,700
1,700	16,350	25.76	−175,900	−91,600
1,800	17,610	26.48	−175,600	−86,600
1,900	(18,870)	(27.16)	(−175,200)	(−81,800)
2,000	(20,130)	(27.81)	(−175,050)	(−76,950)

Magnesium Difluoride, MgF_2 (*c*)

$\Delta H_{298}^\circ = -263{,}500$ calories per mole (*112*)
$S_{298} = 13.68$ *e.u.* (*112*)
M.P. = 1,536° K. (*82*)
$\Delta H_M = 13{,}900$ calories per mole
B.P. = 2,500° K. (*112*)
$\Delta H_V = 65{,}000$ calories per mole

Zone I (*c*) (298°–1,536° K.)

$$C_p = 16.93 + 2.52\times10^{-3}T - 2.20\times10^{5}T^{-2} \quad (82)$$
$$H_T - H_{298} = -5{,}898 + 16.93T + 1.26\times10^{-3}T^2 + 2.20\times10^{5}T^{-1}$$

Zone II (*l*) (1,536°–1,800° K.)

$$C_p = 22.60 \quad (82)$$
$$H_T - H_{298} = 2{,}400 + 22.60T$$

Formation: $Mg + F_2 \longrightarrow MgF_2$

Zone I (298°–923° K.)

$$\Delta C_p = 2.50 + 0.58\times10^{-3}T - 0.62\times10^{5}T^{-2}$$
$$\Delta H_T = -264{,}500 + 2.50T + 0.29\times10^{-3}T^2 + 0.62\times10^{5}T^{-1}$$
$$\Delta F_T = -264{,}500 - 2.50T\ln T - 0.29\times10^{-3}T^2 + 0.31\times10^{5}T^{-1} + 59.87T$$

Zone II (923°–1,393° K.)

$$\Delta C_p = 1.24 + 2.08\times10^{-3}T - 1.4\times10^{5}T^{-2}$$
$$\Delta H_T = -266{,}220 + 1.24T + 1.04\times10^{-3}T^2 + 1.4\times10^{5}T^{-1}$$
$$\Delta F_T = -266{,}220 - 1.24T\ln T - 1.04\times10^{-3}T^2 + 0.7\times10^{5}T^{-1} + 53.81T$$

Zone III (1,393°–1,536° K.)

$$\Delta C_p = 3.67 + 2.08\times10^{-3}T - 1.40\times10^{5}T^{-2}$$
$$\Delta H_T = -301{,}050 + 3.67T + 1.04\times10^{-3}T^2 + 1.40\times10^{5}T^{-1}$$
$$\Delta F_T = -301{,}050 - 3.67T\ln T - 1.04\times10^{-3}T^2 + 0.70\times10^{5}T^{-1} + 96.44T$$

Zone IV (1,536°–1,800° K.)

$$\Delta C_p = \quad 34 - 0.44\times10^{-3}T + 0.80\times10^{5}T^{-2}$$
$$\Delta H_T = -292{,}760 + 9.34T - 0.22\times10^{-3}T^2 - 0.80\times10^{5}T^{-1}$$
$$\Delta F_T = -292{,}760 - 9.34T\ln T + 0.22\times10^{-3}T^2 - 0.40\times10^{5}T^{-1} + 130.8T$$

T, ° K.	$H_T - H_{298}$	S_T	ΔH_T°	ΔF_T°
298	....	13.68	−263,500	−250,800
400	1,645	18.42	−263,250	−246,500
500	3,320	22.15	−263,000	−242,300
600	5,080	25.36	−262,750	−238,200
700	6,890	28.15	−262,600	−234,200
800	8,720	30.60	−262,250	−230,150
900	10,590	32.80	−261,950	−226,150
1,000	12,510	34.82	−263,800	−221,950
1,100	14,450	36.67	−263,450	−216,750
1,200	16,430	38.39	−263,100	−213,700
1,300	18,440	40.00	−262,500	−209,500
1,400	20,460	41.50	−293,800	−205,300
1,500	22,490	42.90	−293,150	−198,950
1,600	38,560	53.35	−278,450	−193,250
1,700	40,820	54.72	−277,600	−187,800
1,800	43,080	56.01	−276,700	−182,600

Magnesium Dichloride, $MgCl_2$ (*c*)

$\Delta H_{298}^\circ = -153{,}200$ calories per mole (*112*)
$S_{298} = 21.4$ *e.u.* (*112*)
M.P. = 987° K. (*82*)
$\Delta H_M = 10{,}300$ calories per mole
B.P. = 1,691° K. (*112*)
$\Delta H_V = 32{,}700$ calories per mole

Zone I (*c*) (298°–987° K.)

$$C_p = 18.90 + 1.42\times10^{-3}T - 2.06\times10^{5}T^{-2} \quad (82)$$
$$H_T - H_{298} = -6{,}389 + 18.90T + 0.71\times10^{-3}T^2 + 2.06\times10^{5}T^{-1}$$

Zone II (*l*) (987°–1,500° K.)

$$C_p = 22.10 \quad (82)$$
$$H_T - H_{298} = +1{,}650 + 22.10T$$

Formation: $Mg + Cl_2 \longrightarrow MgCl_2$

Zone I (298°–923° K.)

$$\Delta C_p = 3.94 - 0.14\times10^{-3}T - 0.6\times10^{5}T^{-2}$$
$$\Delta H_T = -154{,}600 + 3.94T - 0.07\times10^{-3}T^2 + 0.6\times10^{5}T^{-1}$$
$$\Delta F_T = -154{,}600 - 3.94T\ln T + 0.07\times10^{-3}T^2 - 0.3\times10^{5}T^{-1} + 66.56T$$

Zone II (923°–987° K.)

$$\Delta C_p = 2.68 + 1.36\times10^{-3}T - 1.38\times10^{5}T^{-2}$$
$$\Delta H_T = -154{,}200 + 2.68T + 0.68\times10^{-3}T^2 + 1.38\times10^{5}T^{-1}$$
$$\Delta F_T = -154{,}200 - 2.68T\ln T - 0.68\times10^{-3}T^2 - 0.69\times10^{5}T^{-1} + 55.47T$$

Zone III (987°–1,393° K.)

$\Delta C_p = 5.88 - 0.06\times10^{-3}T + 0.68\times10^{5}T^{-2}$
$\Delta H_T = -148{,}150 + 5.88T - 0.03\times10^{-3}T^2 - 0.68\times10^{5}T^{-1}$
$\Delta F_T = -148{,}150 - 5.88T\ln T + 0.03\times10^{-3}T^2 + 0.34\times10^{5}T^{-1} + 73.54T$

Zone IV (1,393°–1,500° K.)

$\Delta C_p = 8.31 - 0.06\times10^{-3}T + 0.68\times10^{5}T^{-2}$
$\Delta H_T = -183{,}100 + 8.31T - 0.03\times10^{-3}T^2 - 0.68\times10^{5}T^{-1}$
$\Delta F_T = -183{,}100 - 8.31T\ln T + 0.03\times10^{-3}T^2 - 0.34\times10^{5}T^{-1} + 116.34T$

T, ° K.	H_T-H_{298}	S_T	ΔH°_T	ΔF°_T
298	----------	21.40	−153,200	−141,400
400	1,800	26.59	−152,850	−137,300
500	3,650	30.71	−152,500	−133,500
600	5,555	34.19	−152,100	−129,700
700	7,480	37.15	−151,750	−126,100
800	9,420	39.74	−151,400	−122,400
900	11,380	42.05	−151,050	−118,850
1,000	23,750	54.67	−142,450	−115,150
1,100	25,960	56.78	−141,750	−112,450
1,200	28,170	58.70	−141,300	−110,000
1,300	30,380	60.47	−141,450	−107,050
1,400	32,590	62.10	−171,600	−104,600
1,500	34,800	63.67	−170,650	−99,650

Magnesium Dibromide, $MgBr_2$ (*c*)

$\Delta H^\circ_{298} = (-123{,}900)$ calories per mole (*11*)
$S_{298} = (30)$ *e.u.* (*11*)
M.P. = 984° K. (*6*)
$\Delta H_M = 8{,}300$ calories per mole
B.P. = (1,500°) K. (*6*)
$\Delta H_V = (35{,}000)$ calories per mole

Formation: $Mg + Br_2 \longrightarrow MgBr_2$
(estimated (*11*))

T, ° K.	H_T-H_{298}	ΔH°_T	ΔF°_T
298	----------	(−123,900)	(−120,000)
500	(4,000)	(−130,700)	(−111,000)
1,000	(22,700)	(−122,200)	(−92,450)
1,500	(34,800)	(−149,500)	(−69,600)

Magnesium Diiodide, MgI_2 (*c*)

$\Delta H^\circ_{298} = -86{,}800$ calories per mole (*11*)
$S_{298} = (33)$ *e.u.* (*11*)
M.P. = 923° K. (*6*)
$\Delta H_M = (5{,}300)$ calories per mole
B.P. = (1,200°) K. (*6*)
$\Delta H_V = (25{,}000)$ calories per mole

Formation: $Mg + I_2 \longrightarrow MgI_2$
(estimated (*11*))

T, ° K.	H_T-H_{298}	ΔH°_T	ΔF°_T
298	----------	−86,800	(−86,000)
500	(4,000)	(−100,700)	(−83,000)
1,000	(19,900)	(−95,000)	(−69,500)
1,500	(57,000)	(−97,500)	(−58,000)

Trimagnesium Dinitride, Mg_3N_2 (*c*)

$\Delta H^\circ_{298} = -110{,}200$ calories per mole (*102*)
$S_{298} = 21.8$ *e.u.* (*102*)
T.P. = 823° K. (*82*)
$\Delta H_T = 110$ calories per mole
T.P. = 1,061° K. (*82*)
$\Delta H_T = 220$ calories per mole
Decomposes = 1,300° K. (*9*)

Zone I (α) (298°–823° K.)

$C_p = 20.77 + 11.20\times10^{-3}T$ (*82*)
$H_T - H_{298} = -6{,}691 + 20.77T + 5.60\times10^{-3}T^2$

Zone II (β) (823°–1,061° K.)

$C_p = 20.07 + 10.66\times10^{-3}T$ (*82*)
$H_T - H_{298} = -5{,}830 + 20.07T + 5.33\times10^{-3}T^2$

Zone III (γ) (1,061°–1,300)° K.)

$C_p = 28.50$ (*79*)
$H_T - H_{298} = -8{,}560 + 28.50T$

Formation: $3Mg + N_2 \longrightarrow Mg_3N_2$

Zone I (298°–823° K.)

$\Delta C_p = -4.31 + 5.68\times10^{-3}T + 2.34\times10^{5}T^{-2}$
$\Delta H_T = -108{,}400 - 4.31T + 2.84\times10^{-3}T^2 - 2.34\times10^{5}T^{-1}$
$\Delta F_T = -108{,}400 + 4.31T\ln T - 2.84\times10^{-3}T^2 - 1.17\times10^{5}T^{-1} + 18.05T$

Zone II (823°–923° K.)

$\Delta C_p = -5.01 + 5.14\times10^{-3}T + 2.34\times10^{5}T^{-2}$
$\Delta H_T = -107{,}500 - 5.01T + 2.57\times10^{-3}T^2 - 2.34\times10^{5}T^{-1}$
$\Delta F_T = -107{,}500 + 5.01T\ln T - 2.57\times10^{-3}T^2 - 1.17\times10^{5}T^{-1} + 12.70T$

Zone III (923°–1,061° K.)

$\Delta C_p = -8{,}79 + 9.64\times10^{-3}T$
$\Delta H_T = -112\ 700 - 8.79T + 4.82\times10^{-3}T^2$
$\Delta F_T = -112{,}700 + 8.79T\ln T - 4.82\times10^{-3}T^2 - 5.38T$

Zone IV (1,061°–1,300° K.)

$\Delta C_p = 0.32 - 1.02\times10^{-3}T$
$\Delta H_T = -115{,}430 - 0.32T - 0.51\times10^{-3}T^2$
$\Delta F_T = -115{,}430 + 0.32T\ln T + 0.51\times10^{-3}T^2 + 50.4T$

T, ° K.	H_T-H_{298}	S_T	ΔH°_T	ΔF°_T
298	----------	21.8	−110,200	−96,100
400	2,510	29.04	−110,250	−91,300
500	5,100	34.81	−110,300	−86,600
600	7,790	39.72	−110,300	−81,900
700	10,590	44.03	−110,300	−77,100
800	13,510	47.93	−110,250	−72,400
900	16,550	51.50	−110,200	−67,700
1,000	19,570	54.69	−116,650	−62,350
1,100	22,790	57.76	−116,400	−56,900
1,200	25,640	60.24	−116,350	−51,350
1,300	28,490	62.52	−116,150	−45,650

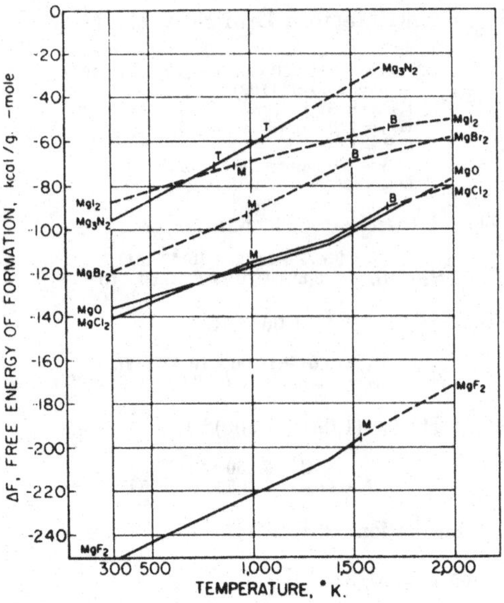

FIGURE 31.—Magnesium.

MANGANESE AND ITS COMPOUNDS

Element, Mn (c)

$S_{298}=7.59$ *e.u.* (*83*)
$T.P.=1{,}000°$ K. (*82*)
$\Delta H_T=535$ calories per atom
$T.P.=1{,}374°$ K. (*82*)
$\Delta H_T=545$ calories per atom
$T.P.=1{,}410°$ K. (*82*)
$\Delta H_T=430$ calories per atom
$M.P.=1{,}517°$ K. (*82*)
$\Delta H_M=3{,}500$ calories per atom
$B.P.=2{,}368°$ K. (*82*)
$\Delta H_V=53{,}700$ calories per atom

Zone I (α) (298°–1,000° K.)

$$C_p=5.70+3.38\times10^{-3}T-0.37\times10^{5}T^{-2}\ (82)$$
$$H_T-H_{298}=-1{,}974+5.70T+1.69\times10^{-3}T^{2}+0.37\times10^{5}T^{-1}$$
$$F_T-H_{298}=-1{,}974-5.70T\ln T-1.69\times10^{-3}T^{2}+0.18\times10^{5}T^{-1}+31.74T$$

Zone II (β) (1,000°–1,374° K.)

$$C_p=8.33+0.66\times10^{-3}T$$
$$H_T-H_{298}=-2{,}675+8.33T+0.33\times10^{-3}T^{2}$$
$$F_T-H_{298}=-2{,}675-8.33T\ln T-0.33\times10^{-3}T^{2}+49.27T$$

Zone III (γ) (1,374°–1,410° K.)

$$C_p=10.70$$
$$H_T-H_{298}=-4{,}760+10.70T$$
$$F_T-H_{298}=-4{,}760-10.70T\ln T+67.5T$$

Zone IV (δ) (1,410°–1,517° K.)

$$C_p=11.30\ (82)$$
$$H_T-H_{298}=-1{,}517+11.30T$$
$$F_T-H_{298}=-1{,}517-11.30T\ln T+69.7T$$

Zone V (l) (1,517°–2,368° K.)

$$C_p=11.0$$
$$H_T-H_{298}=-1{,}220+11.0T$$
$$F_T-H_{298}=-1{,}220-11.0T\ln T+67.2T$$

Zone VI (g) (2,368°–5,000° K.)

$$C_p=6.26\ (82)$$
$$H_T-H_{298}=63{,}710+6.26T$$
$$F_T-H_{298}=63{,}710-6.26T\ln T+4.26T$$

T, ° K.	H_T-H_{298}	S_T	$-\frac{(F_T-H_{298})}{T}$
298	----------	7.59	7.59
400	690	9.58	7.85
500	1,385	11.13	8.36
600	2,210	12.47	8.78
700	2,895	13.66	9.53
800	3,715	14.75	10.12
900	4,570	15.76	10.67
1,000	5,450	16.69	11.24
1,100	6,890	18.09	11.82
1,200	7,795	18.87	12.37
1,300	8,715	19.61	12.91
1,400	10,220	20.72	13.56
1,500	11,780	21.80	13.95
1,600	16,380	24.82	14.58
1,700	17,480	25.49	15.21
1,800	18,580	26.12	15.80
1,900	19,680	26.71	16.35
2,000	20.780	27.28	16.89
2,500	79,190	52.09	20.41

Manganese Oxide, MnO (c)

$\Delta H_{298}^{\circ}=-92{,}050$ calories per mole (*124*)
$S_{298}=14.27$ *e.u.* (*135*)
$M.P.=2{,}058°$ K. (*94*)
$\Delta H_M=13{,}000$ calories per mole
$B.P.=3{,}400°$ K. (*8*)

Zone I (c) (298°–1,800° K.)

$$C_p=11.11+1.94\times10^{-3}T-0.88\times10^{5}T^{-2}\ (82)$$
$$H_T-H_{298}=-3{,}690+11.11T+0.97\times10^{-3}T^{2}+0.88\times10^{5}T^{-1}$$

Formation: $Mn+1/2O_2 \longrightarrow MnO$

Zone I (298°–1,000° K.)

$$\Delta C_p=1.83-1.94\times10^{-3}T-0.31\times10^{5}T^{-2}$$
$$\Delta H_T=-92{,}600+1.83T-0.97\times10^{-3}T^{2}+0.31\times10^{5}T^{-1}$$
$$\Delta F_T=-92{,}600-1.83T\ln T+0.97\times10^{-3}T^{2}+0.15\times10^{5}T^{-1}+29.6T$$

Zone II (1,000°–1,374° K.)

$$\Delta C_p=-0.80+0.78\times10^{-3}T-0.68\times10^{5}T^{-2}$$
$$\Delta H_T=-92{,}950-0.80T+0.39\times10^{-3}T^{2}+0.68\times10^{5}T^{-1}$$
$$\Delta F_T=-92{,}950+0.80T\ln T-0.39\times10^{-3}T^{2}+0.34\times10^{5}T^{-1}+13.15T$$

Zone III (1,374°–1,410° K.)

$$\Delta C_p=-3.17+1.44\times10^{-3}T-0.68\times10^{5}T^{-2}$$
$$\Delta H_T=-89{,}800-3.17T+0.72\times10^{-3}T^{2}+0.68\times10^{5}T^{-1}$$
$$\Delta F_T=-89{,}800+3.17T\ln T-0.72\times10^{-3}T^{2}+0.34\times10^{5}T^{-1}-5.97T$$

Zone IV (1,410°–1,517° K.)

$$\Delta C_p=-3.77+1.44\times10^{-3}T-0.68\times10^{5}T^{-2}$$
$$\Delta H_T=-89{,}480-3.77T+0.72\times10^{-3}T^{2}+0.68\times10^{5}T^{-1}$$
$$\Delta F_T=-89{,}480+3.77T\ln T-0.72\times10^{-3}T^{2}+0.34\times10^{5}T^{-1}-10.63T$$

Zone V (1,517°–1,800° K.)

$\Delta C_p = -3.47 + 1.44\times10^{-3}T - 0.68\times10^5T^{-2}$
$\Delta H_T = -93,400 - 3.47T + 0.72\times10^{-3}T^2 + 0.68\times10^5T^{-1}$
$\Delta F_T = -93,400 + 3.47T\ln T - 0.72\times10^{-3}T^2 + 0.34\times10^5T^{-1} - 5.79T$

T, ° K.	$H_T - H_{298}$	S_T	ΔH_T°	ΔF_T°
298		14.27	−92,050	−86,750
400	1,130	17.53	−92,000	−84,950
500	2,280	20.09	−91,900	−83,200
600	3,470	22.26	−91,900	−81,550
700	4,680	24.13	−91,750	−79,700
800	5,900	25.76	−91,750	−78,000
900	7,150	27.23	−91,750	−76,300
1,000	8,430	28.54	−91,800	−74,550
1,100	8,750	29.83	−93,300	−72,750
1,200	11,100	31.01	−92,250	−71,000
1,300	12,470	32.01	−92,250	−69,100
1,400	13,840	33.12	−92,800	−67,450
1,500	15,210	34.07	−93,450	−65,600
1,600	16,590	34.96	−97,050	−63,500
1,700	17,970	35.79	−97,250	−61,450
1,800	19,350	36.58	−97,400	−59,450

Trimanganese Tetraoxide, Mn_3O_4 (*c*)

$\Delta H^\circ_{298} = -331,400$ calories per mole (*115*)
$S_{298} = 35.5$ *e.u.* (*83*)
T.P. = 1,445° K. (*82*)
$\Delta H_T = 4,970$ calories per mole
M.P. = 1,863° K. (*8*)
$\Delta H_M = (39,000)$ calories per mole (*42*)

Zone I (α) (298°–1,445° K.)

$C_p = 34.64 + 10.82\times10^{-3}T - 2.20\times10^5T^{-2}$ (*82*)
$H_T - H_{298} = -11,550 + 34.64T + 5.41\times10^{-3}T^2 + 2.20\times10^5T^{-1}$

Zone II (β) (1,445°–1,800° K.)

$C_p = 50.20$ (*82*)
$H_T - H_{298} = -17,600 + 50.20T$

Formation: $3Mn + 2O_2 \longrightarrow Mn_3O_4$

Zone I (298°–1,000° K.)

$\Delta C_p = 3.22 - 1.32\times10^{-3}T - 0.29\times10^5T^{-2}$
$\Delta H_T = -332,400 + 3.22T - 0.66\times10^{-3}T^2 + 0.29\times10^5T^{-1}$
$\Delta F_T = -332,400 - 3.22T\ln T + 0.66\times10^{-3}T^2 + 0.15\times10^5T^{-1} + 106.75T$

Zone II (1,000°–1,374° K.)

$\Delta C_p = -4.67 + 6.84\times10^{-3}T - 1.40\times10^5T^{-2}$
$\Delta H_T = -330,600 - 4.67T + 3.42\times10^{-3}T^2 + 1.40\times10^5T^{-1}$
$\Delta F_T = -330,600 + 4.67T\ln T - 3.42\times10^{-3}T^2 + 0.70\times10^5T^{-1} + 54.40T$

Zone III (1,374°–1,410° K.)

$\Delta C_p = -11.78 + 8.82\times10^{-3}T - 1.40\times10^5T^{-2}$
$\Delta H_T = -324,130 - 11.78T + 4.41\times10^{-3}T^2 + 1.40\times10^5T^{-1}$
$\Delta F_T = -324,130 + 11.78T\ln T - 4.41\times10^{-3}T^2 + 0.70\times10^5T^{-1} - 0.14T$

Zone IV (1,410°–1,445° K.)

$\Delta C_p = -13.58 + 8.82\times10^{-3}T - 1.40\times10^5T^{-2}$
$\Delta H_T = -321,600 - 13.58T + 4.41\times10^{-3}T^2 + 1.40\times10^5T^{-1}$
$\Delta F_T = -321,600 + 13.58T\ln T - 4.41\times10^{-3}T^2 + 0.70\times10^5T^{-1} - 15.43T$

Zone V (1,445°–1,517° K.)

$\Delta C_p = 2.0 - 2.00\times10^{-3}T + 0.80\times10^5T^{-2}$
$\Delta H_T = -329,100 + 2.0T - 1.00\times10^{-3}T^2 - 0.80\times10^5T^{-1}$
$\Delta F_T = -329,100 - 2.0T\ln T + 1.00\times10^{-3}T^2 - 0.40\times10^5T^{-1} + 95.46T$

Zone VI (1,517°–1,800° K.)

$\Delta C_p = 2.88 - 2.00\times10^{-3}T + 0.80\times10^5T^{-2}$
$\Delta H_T = -340,700 + 2.88T - 1.00\times10^{-3}T^2 - 0.80\times10^5T^{-1}$
$\Delta F_T = -340,700 - 2.88T\ln T + 1.00\times10^{-3}T^2 - 0.40\times10^5T^{-1} + 109.88T$

T, ° K.	$H_T - H_{298}$	S_T	ΔH_T°	ΔF_T°
298		35.5	−331,400	−308,100
400	3,730	46.25	−331,200	−297,300
500	7,590	54.86	−330,900	−288,850
600	11,590	62.15	−330,850	−280,800
700	15,740	68.54	−330,300	−272,200
800	19,980	74.20	−330,150	−263,900
900	24,250	79.23	−330,050	−255,600
1,000	28,570	83.78	−330,050	−247,350
1,100	33,020	88.02	−331,450	−238,750
1,200	37,650	92.05	−331,200	−230,400
1,300	42,510	95.93	−330,800	−221,950
1,400	47,620	99.72	−331,900	−213,650
1,500	57,690	106.68	−328,400	−205,300
1,600	62,710	109.92	−338,600	−196,250
1,700	67,730	112.96	−338,800	−187,600
1,800	72,750	115.84	−338,850	−179,100

Dimanganese Trioxide, Mn_2O_3 (*c*)

$\Delta H^\circ_{298} = -229,200$ calories per mole (*8*)
$S_{298} = 26.4$ *e.u.* (*91*)
Decomposes = 1,620° K. (*42*)

Zone I (*c*) (298°–1,350° K.)

$C_p = 24.73 + 8.38\times10^{-3}T - 3.23\times10^5T^{-2}$ (*106*)
$H_T - H_{298} = -8,830 + 24.73T + 4.19\times10^{-3}T^2 + 3.23\times10^5T^{-1}$

Formation: $2Mn + 3/2O_2 \longrightarrow Mn_2O_3$

Zone I (298°–1,000° K.)

$\Delta C_p = 2.59 + 0.12\times10^{-3}T - 1.89\times10^5T^{-2}$
$\Delta H_T = -230,600 + 2.59T + 0.06\times10^{-3}T^2 + 1.89\times10^5T^{-1}$
$\Delta F_T = -230,600 - 2.59T\ln T - 0.06\times10^{-3}T^2 + 0.94\times10^5T^{-1} + 80.7T$

Zone II (1,000°–1,350° K.)

$\Delta C_p = -2.67 + 5.56\times10^{-3}T - 2.63\times10^5T^{-2}$
$\Delta H_T = -229,210 - 2.67T + 2.78\times10^{-3}T^2 + 2.63\times10^5T^{-1}$
$\Delta F_T = -229,210 + 2.67T\ln T - 2.78\times10^{-3}T^2 + 1.31\times10^5T^{-1} + 50.84T$

T, ° K.	H_T-H_{298}	S_T	ΔH°_T	ΔF°_T
298		26.4	−229,200	−210,650
300	2,550	33.73	−229,200	−204,350
400	5,220	39.68	−228,900	−198,050
500	8,040	44.82	−228,900	−192,150
600	10,990	49.37	−228,500	−185,850
700	14,040	53.44	−228,300	−179,800
800	17,190	57.15	−228,050	−173,900
900	20,420	60.55	−227,800	−167,650
1,000	23,740	63.71	−228,550	−161,450
1,100	27,150	66.68	−228,200	−155,450
1,200	30,650	69.48	−227,900	−149,450
1,300			(−228,700)	(−143,400)
1,400			(−229,500)	(−137,300)
1,500			(−236,300)	(−130,700)

Manganese Dioxide, MnO_2 (*c*)

$\Delta H^\circ_{298} = -124{,}450$ calories per mole (*24*)
$S_{298} = 12.68$ *e.u.* (*83*)
Decomposes = 1,120° K. (*8*)

Zone I (*c*) (298°–800° K.)

$$C_p = 16.60 + 2.44 \times 10^{-3}T - 3.88 \times 10^5 T^{-2} \ (82)$$
$$H_T - H_{298} = -6{,}360 + 16.60T + 1.22 \times 10^{-3}T^2 + 3.88 \times 10^5 T^{-1}$$

Formation: $Mn + O_2 \longrightarrow MnO_2$

Zone I (298°–800° K.)

$$\Delta C_p = 3.74 - 1.94 \times 10^{-3}T - 3.11 \times 10^5 T^{-2}$$
$$\Delta H_T = -126{,}620 + 3.74T - 0.97 \times 10^{-3}T^2 + 3.11 \times 10^5 T^{-1}$$
$$\Delta F_T = -126{,}620 - 3.74 T\ln T + 0.97 \times 10^{-3}T^2 + 1.55 \times 10^5 T^{-1} + 70.21T$$

T, ° K.	H_T-H_{298}	S_T	ΔH°_T	ΔF°_T
298		12.68	−124,450	−111,350
400	1,450	16.84	−124,400	−106,900
500	3,020	20.35	−124,350	−102,600
600	4,690	23.38	−124,200	−98,250
700	6,410	26.04	−124,000	−93,900
800	8,190	28.41	−123,850	−89,650
900	(10,000)		(−123,700)	(−85,350)
1,000	(11,850)		(−123,550)	(−81,150)
1,100	(13,730)		(−123,200)	(−77,000)

Manganese Difluoride, MnF_2 (*c*)

$\Delta H^\circ_{298} = -190{,}000$ calories per mole (*11*)
$S_{298} = 22.3$ *e.u.* (*11*)
M.P. = 1,129° K. (*6*)
$\Delta H_M = (5{,}500)$ calories per mole
B.P. = (2,300°) K. (*6*)
$\Delta H_V = (57{,}000)$ calories per mole

Formation: $Mn + F_2 \longrightarrow MnF_2$
(estimated (*11*))

T, ° K.	H_T-H_{298}	ΔH°_T	ΔF°_T
298		−190,000	−180,000
500	(3,500)	(−189,500)	(−173,000)
1,000	(13,000)	(−188,300)	(−157,000)
1,500	(30,000)	(−182,000)	(−146,500)

Manganese Trifluoride, MnF_3 (*c*)

$\Delta H^\circ_{298} = -238{,}000$ calories per mole (*11*)
$S_{298} = (28)$ *e.u.* (*11*)
M.P. = (1,350°) K. (*6*)
$\Delta H_M = (11{,}000)$ calories per mole
B.P. = (1,600°) K. (*6*)
$\Delta H_V = (42{,}000)$ calories per mole

Formation: $Mn + 3/2F_2 \longrightarrow MnF_3$
(estimated (*11*))

T, ° K.	H_T-H_{298}	ΔH°_T	ΔF°_T
298		−238,000	(−222,200)
500	(5,000)	(−236,800)	(−210,500)
1,000			(−191,000)

Manganese Dichloride, $MnCl_2$ (*c*)

$\Delta H^\circ_{298} = -115{,}190$ calories per mole (*92*)
$S_{298} = 28$ *e.u.* (*83*)
M.P. = 923° K. (*82*)
$\Delta H_M = 8{,}970$ calories per mole
B.P. = 1,463° K. (*6*)
$\Delta H_V = 29{,}600$ calories per mole

Zone I (*c*) (298°–923° K.)

$$C_p = 18.04 + 3.16 \times 10^{-3}T - 1.37 \times 10^5 T^{-2} \ (82)$$
$$H_T - H_{298} = -6{,}000 + 18.04T + 1.58 \times 10^{-3}T^2 + 1.37 \times 10^5 T^{-1}$$

Zone II (*l*) (923°–1,200° K.)

$$C_p = 22.60 \ (82)$$
$$H_T - H_{298} = +280 + 22.60T$$

Formation: $Mn + Cl_2 \longrightarrow MnCl_2$

Zone I (298°–923° K.)

$$\Delta C_p = 3.52 - 0.28 \times 10^{-3}T - 0.32 \times 10^5 T^{-2}$$
$$\Delta H_T = -116{,}350 + 3.52T - 0.14 \times 10^{-3}T^2 + 0.32 \times 10^5 T^{-1}$$
$$\Delta F_T = -116{,}350 - 3.52 T\ln T + 0.14 \times 10^{-3}T^2 + 0.16 \times 10^5 T^{-1} + 58.5T$$

Zone II (923°–1,000° K.)

$$\Delta C_p = 8.08 - 3.44 \times 10^{-3}T + 1.05 \times 10^5 T^{-2}$$
$$\Delta H_T = -110{,}100 + 8.08T - 1.72 \times 10^{-3}T^2 - 1.05 \times 10^5 T^{-1}$$
$$\Delta F_T = -110{,}100 - 8.08 T\ln T + 1.72 \times 10^{-3}T^2 - 0.52 \times 10^5 T^{-1} + 79.81T$$

Zone III (1,000°–1,200° K.)

$$\Delta C_p = 5.45 - 0.72 \times 10^{-3}T + 0.68 \times 10^5 T^{-2}$$
$$\Delta H_T = -109{,}280 + 5.45T - 0.36 \times 10^{-3}T^2 - 0.68 \times 10^5 T^{-1}$$
$$\Delta F_T = -109{,}280 - 5.45 T\ln T + 0.36 \times 10^{-3}T^2 - 0.34 \times 10^5 T^{-1} + 62.33T$$

T, ° K.	H_T-H_{298}	S_T	ΔH°_T	ΔF°_T
298		28.0	−115,190	−105,100
400	1,850	33.33	−114,900	−102,050
500	3,730	37.52	−114,550	−98,850
600	5,640	41.0	−114,300	−95,850
700	7,590	44.01	−113,900	−92,750
800	9,600	46.69	−113,600	−89,650
900	11,680	49.14	−113,250	−86,800
1,000	22,880	61.21	−103,850	−84,500
1,100	25,140	63.36	−103,750	−82,550
1,200	27,400	65.33	−103,400	−80,750
1,300	(29,400)	(67.2)	(−103,150)	(−79,150)
1,400	(31,650)	(68.8)	(−102,400)	(−76,250)

Manganese Trichloride, $MnCl_3$ (*c*)

$\Delta H^\circ_{298} = -110{,}000$ calories per mole (*11*)
$S_{298} = (39)$ *e.u.* (*11*)
B.P. = (900°) K. (*6*)
$\Delta H_V = (21{,}000)$ calories per mole

Formation: $Mn + 3/2Cl_2 \longrightarrow MnCl_3$
(estimated (*11*))

T, ° K.	$H_T - H_{298}$	ΔH°_T	ΔF°_T
298	....	−110, 000	(−95, 400)
500	(5, 000)	(−108, 900)	(−85, 500)

Manganese Dibromide, $MnBr_2$ (*c*)

$\Delta H^\circ_{298} = -88{,}700$ calories per mole (*11*)
$S_{298} = (32)$ *e.u.* (*11*)
M.P. = 971° K. (*6*)
$\Delta H_M = (7{,}000)$ calories per mole
B.P. = (1,300°) K. (*6*)
$\Delta H_V = (27{,}000)$ calories per mole

Formation: $Mn + Br_2 \longrightarrow MnBr_2$
(estimated (*11*))

T, ° K.	$H_T - H_{298}$	ΔH°_T	ΔF°_T
298	....	−88, 700	(−85, 100)
500	(4, 000)	(−95, 600)	(−79, 000)
1,000	(23, 000)	(−85, 100)	(−64, 000)
1,500	(62, 000)	(−57, 000)	(−54, 000)

Manganese Diiodide, MnI_2 (*c*)

$\Delta H^\circ_{298} = -57{,}100$ calories per mole (*11*)
$S_{298} = (35)$ *e.u.* (*11*)
M.P. = 911° K. (*6*)
$\Delta H_M = (6{,}500)$ calories per mole
B.P. = (1,100°) K. (*6*)
$\Delta H_V = (23{,}000)$ calories per mole

Formation: $Mn + I_2 \longrightarrow MnI_2$
(estimated (*11*))

T, ° K.	$H_T - H_{298}$	ΔH°_T	ΔF°_T
298	....	−57, 100	(−57, 000)
500	(4, 000)	(−69, 100)	(−54, 500)
1,000	(23, 000)	(−60, 700)	(−40, 000)
1,500	(58, 000)	(−36, 500)	(−30, 000)

Trimanganese Carbide, Mn_3C (*c*)

$\Delta H^\circ_{298} = -3{,}600$ calories per mole (*89*)
$S_{298} = 23.7$ *e.u.* (*81*)
T.P. = 1,310° K. (*82*)
$\Delta H_T = 3{,}570$ calories per mole
M.P. = 1,480° K. (*9*)

Zone I (α) (298°–1,310° K.)

$$C_p = 25.26 + 5.60\times10^{-3}T - 4.07\times10^{5}T^{-2} \quad (82)$$
$$H_T - H_{298} = -9{,}145 + 25.26T + 2.80\times10^{-3}T^2 + 4.07\times10^{5}T^{-1}$$

Zone II (β) (1,310°–1,480° K.)

$$C_p = 38.00 \quad (82)$$
$$H_T - H_{298} = -17{,}150 + 38.00T$$

Formation: $3Mn + C \longrightarrow Mn_3C$

Zone I (298°–1,000° K.)

$$\Delta C_p = 4.06 - 5.56\times10^{-3}T - 0.86\times10^{5}T^{-2}$$
$$\Delta H_T = -4{,}840 + 4.06T - 2.78\times10^{-3}T^2 + 0.86\times10^{5}T^{-1}$$
$$\Delta F_T = -4{,}840 - 4.06T\ln T + 2.78\times10^{-3}T^2 + 0.43\times10^{5}T^{-1} + 26.42T$$

Zone II (1,000–1,310° K.)

$$\Delta C_p = -3.83 + 2.60\times10^{-3}T - 2.0\times10^{5}T^{-2}$$
$$\Delta H_T = -2{,}740 - 3.83T + 1.30\times10^{-3}T^2 + 2.0\times10^{5}T^{-1}$$
$$\Delta F_T = -2{,}740 + 3.83T\ln T - 1.30\times10^{-3}T^2 + 1.0\times10^{5}T^{-1} - 26.15T$$

Zone III (1,310–1,374° K.)

$$\Delta C_p = 9.0 - 3.00\times10^{-3}T + 2.10\times10^{5}T^{-2}$$
$$\Delta H_T = -12{,}600 + 9.0T - 1.50\times10^{-3}T^2 + 2.10\times10^{5}T^{-1}$$
$$\Delta F_T = -12{,}600 - 9.0T\ln T + 1.50\times10^{-3}T^2 - 1.05\times10^{5}T^{-1} + 69.4T$$

Zone IV (1,374°–1,410° K.)

$$\Delta C_p = 1.8 - 1.02\times10^{-3}T + 2.10\times10^{5}T^{-2}$$
$$\Delta H_T = -4{,}550 + 1.8T - 0.51\times10^{-3}T^2 - 2.10\times10^{5}T^{-1}$$
$$\Delta F_T = -4{,}550 - 1.8T\ln T + 0.51\times10^{-3}T^2 - 1.05\times10^{5}T^{-1} + 13.4T$$

Zone V (1,410°–1,480° K.)

$$\Delta C_p = -1.02\times10^{-3}T + 2.10\times10^{5}T^{-2}$$
$$\Delta H_T = -3{,}300 - 0.51\times10^{-3}T^2 - 2.10\times10^{5}T^{-1}$$
$$\Delta F_T = -3{,}300 + 0.51\times10^{-3}T^2 - 1.05\times10^{5}T^{-1} - 0.65T$$

T, ° K.	$H_T - H_{298}$	S_T	ΔH°_T	ΔF°_T
298	....	23. 7	−3, 600	−3, 500
400	2, 450	30. 75	−3, 450	−3, 450
500	5, 020	36. 48	−3, 300	−3, 450
600	7, 700	41. 37	−3, 500	−3, 600
700	10, 490	45. 66	−3, 150	−3, 600
800	13, 350	49. 48	−3, 200	−3, 600
900	16, 300	52. 95	−3, 300	−3, 600
1, 000	19, 320	56. 14	−3, 450	−3, 600
1, 100	22, 400	59. 07	−5, 200	−3, 600
1, 200	25, 540	61. 8	−5, 300	−3, 400
1, 300	28, 740	64. 36	−5, 400	−3, 250
1, 400	36, 050	69. 86	−3, 150	−3, 200
1, 500	39, 850	72. 48	−4, 600	−3, 200

Tetramanganese Nitride, Mn_4N (*c*)

$\Delta H^\circ_{298} = -30{,}300$ calories per mole (*97*)

Zone I (*c*) (298°–800° K.)

$$C_p = 21.15 + 30.50\times10^{-3}T \quad (82)$$
$$H_T - H_{298} = -7{,}661 + 21.15T + 15.25\times10^{-3}T^2$$

Formation: $4Mn + 1/2N_2 \longrightarrow Mn_4N$

T, ° K.	H_T-H_{298}	S_T-S_{298}	ΔH_T°
298			−30, 300
400	3, 250	9. 36	−30, 150
500	6, 720	17. 10	−29, 800
600	10, 520	24. 02	−29, 700
700	14, 640	30. 35	−28, 700
800	19, 000	36. 17	−27, 950

Pentamanganese Dinitride, Mn_5N_2 (*c*)

$\Delta H_{298}^\circ = -48{,}200$ calories per mole (*97*)
$S_{298} = 47.3$ *e.u.* (*9*)

Zone I (*c*) (298°–800° K.)

$C_p = 30.55 + 38.40 \times 10^{-3} T$ (*82*)
$H_T - H_{298} = -10{,}800 + 30.55T + 19.20 \times 10^{-3} T^2$

Formation: $5Mn + N_2 \longrightarrow Mn_5N_2$

Zone I (298°–800° K.)

$\Delta C_p = -4.61 + 20.48 \times 10^{-3} T + 1.85 \times 10^5 T^{-2}$
$\Delta H_T = -56{,}800 - 4.61T + 10.24 \times 10^{-3} T^2 - 1.85 \times 10^5 T^{-1}$
$\Delta F_T = -56{,}800 + 4.61 T\ln T - 10.24 \times 10^{-3} T^2 - 0.92 \times 10^5 T^{-1} + 10.67T$

T, ° K.	H_T-H_{298}	S_T	ΔH_T°	ΔF_T°
298		47. 3	−48, 200	−37, 350
400	4, 480	60. 2	−47, 900	−33, 700
500	9, 240	70. 81	−47, 300	−30, 200
600	14, 460	80. 32	−46, 900	−27, 250
700	20, 040	88. 91	−45, 500	−23, 650
800	25, 840	97. 65	−44, 100	−20, 600

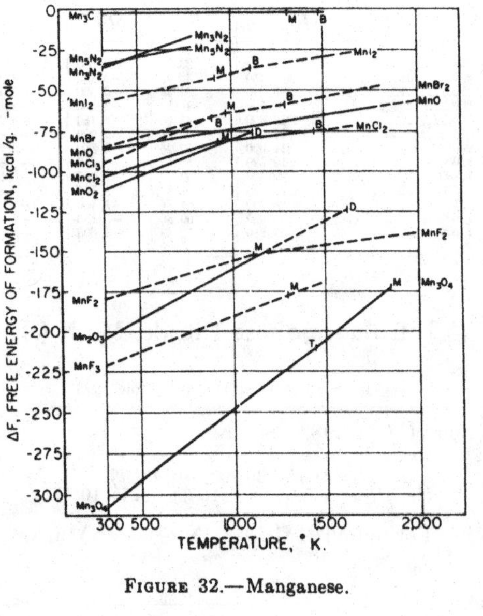

FIGURE 32.—Manganese.

Trimanganese Dinitride, Mn_3N_2 (*c*)

$\Delta H_{298}^\circ = -45{,}800$ calories per mole (*9*)
$S_{298} = 32.7$ *e.u.* (*9*)

Zone I (*c*) (298°–800° K.)

$C_p = 22.32 + 22.40 \times 10^{-3} T$ (*82*)
$H_T - H_{298} = -7{,}650 + 22.32T + 11.20 \times 10^{-3} T^2$

Formation: $3Mn + N_2 \longrightarrow Mn_3N_2$

Zone I (298°–800° K.)

$\Delta C_p = -1.44 + 11.24 \times 10^{-3} T + 1.11 \times 10^5 T^{-2}$
$\Delta H_T = -45{,}500 + 1.44T + 5.62 \times 10^{-3} T^2 - 1.11 \times 10^5 T^{-1}$
$\Delta F_T = -45{,}500 + 1.44 T\ln T - 5.62 \times 10^{-3} T^2 - 0.55 \times 10^5 T^{-1} + 28.86T$

T, ° K.	H_T-H_{298}	S_T	ΔH_T°	ΔF_T°
298		32. 7	−45, 800	−35, 150
400	3, 070	41. 55	−45, 500	−31, 500
500	6, 300	48. 75	−45, 050	−28, 050
600	9, 750	55. 03	−44, 800	−24, 950
700	13, 470	60. 76	−43, 900	−21, 500
800	17, 350	65. 94	−43, 200	−18, 300

MERCURY AND ITS COMPOUNDS

Element, Hg (*l*)

$S_{298} = 18.19$ *e.u.* (*130*)
M.P. = 234.29° K. (*130*)
$\Delta H_M = 549$ calories per atom
B.P. = 629.88° K. (*130*)
$\Delta H_V = 14{,}137$ calories per atom

Zone I (*l*) (298°–630° K.)

$C_p = 6.61$ (*82*)
$H_T - H_{298} = -1{,}971 + 6.61T$
$F_T - H_{298} = -1{,}971 - 6.61 T\ln T + 26.08T$

Zone II (*g*) (630°–3,000° K.)

$C_p = 4.969$ (*82*)
$H_T - H_{298} = 13{,}055 + 4.969T$
$F_T - H_{298} = 13{,}055 - 4.969 T\ln T - 8.21T$

T, ° K.	H_T-H_{298}	S_T	$-\frac{(F_T-H_{298})}{T}$
298		18. 19	18. 19
400	673	20. 13	18. 45
500	1, 335	21. 61	18. 94
600	1, 995	22. 81	19. 48
700	16, 535	45. 73	22. 10
800	17, 030	46. 39	25. 10
900	17, 525	46. 97	27. 50
1,000	18, 025	47. 50	29. 47
1,100	18, 521	47. 97	31. 13
1,200	19, 020	48. 41	32. 56
1,300	19, 515	48. 81	34. 57
1,400	20, 010	49. 18	35. 19
1,500	20, 509	49. 52	35. 85
1,600	21, 005	49. 84	36. 71
1,700	21, 502	50. 14	37. 49
1,800	22, 000	50. 43	38. 21
1,900	22, 496	50. 70	38. 86
2,000	22, 995	50. 95	39. 45

Dimercury Oxide, Hg_2O (*c*)

$\Delta H_{298}^\circ = -21{,}800$ calories per mole (*112*)
$S_{298} = (31.4)$ *e.u.* (*24*)

Formation: $2Hg + 1/2O_2 \longrightarrow Hg_2O$
(estimated (*24*))

T, ° K.	H_T-H_{298}	ΔH°_T	ΔF°_T
298	----------	−21, 800	(−13, 000)
400	(2, 000)	(−21, 500)	(−10, 000)
500	(3, 700)	(−21, 500)	(−7, 000)
600	(5, 900)	(−21, 000)	(−4, 000)
700	(7, 350)	(−49, 000)	(2, 000)
800	(9. 750)	(−48, 000)	(9, 000)
900	(11, 650)	(−47, 500)	(16, 000)
1,000	(13, 550)	(−47, 000)	(23, 000)

Mercury Oxide, HgO (*c*)

$\Delta H^\circ_{298} = -21{,}680$ calories per mole (*112*)
$S_{298} = 17.2$ *e.u.* (*112*)

Formation: $Hg + 1/2O_2 \longrightarrow HgO$
(estimated (*24*))

T, ° K.	H_T-H_{298}	ΔH°_T	ΔF°_T
298	----------	−21, 680	−14, 000
400	(1, 150)	(−21, 550)	(−11, 350)
500	(2, 380)	(−21, 350)	(−8, 850)
600	(3, 660)	(−21, 100)	(−6, 350)
700	(4, 870)	(−34, 850)	(−2, 400)
800	(6, 250)	(−34, 350)	(+2, 250)
900	(7, 700)	(−33, 800)	(+6, 800)
1,000	(9, 200)	(−33, 200)	(+11, 250)
1,100	(10, 700)	(−32, 600)	(+15, 650)
1,200	(12, 300)	(−31, 900)	(+20, 050)
1,300	(14, 050)	(−31, 150)	(+24, 350)
1,400	(15, 680)	(−30, 350)	(+28, 600)
1,500	(17, 490)	(−29, 550)	(+32, 750)

Mercury Fluoride, HgF (*c*)

$\Delta H^\circ_{298} = -46{,}000$ calories per mole (*11*)
$S_{298} = 22$ *e.u.* (*11*)
M.P. = 843° K. (*6*)
Decomposes to $Hg + HgF_2$ (*6*)

Formation: $Hg + 1/2F_2 \longrightarrow HgF$
(estimated (*11*))

T, ° K.	H_T-H_{298}	ΔH°_T	ΔF°_T
298	----------	−46, 000	−40, 000
500	(3, 000)	(−45, 000)	(−35, 000)

Mercury Difluoride, HgF_2 (*c*)

$\Delta H^\circ_{298} = -95{,}000$ calories per mole (*11*)
$S_{298} = (28)$ *e.u.* (*11*)
M.P. = 918° K. (*6*)
$\Delta H_M = (5{,}500)$ calories per mole
B.P. = 920° K. (*6*)
$\Delta H_V = (22{,}000)$ calories per mole

Formation: $Hg + F \longrightarrow HgF_2$
(estimated (*11*))

T, ° K.	H_T-H_{298}	ΔH°_T	ΔF°_T
298	----------	−95, 000	(−83, 000)
500	(4, 000)	(−94, 000)	(−75, 000)

647940 O—63——6

Mercury Chloride, HgCl (*c*)

$\Delta H^\circ_{298} = -31{,}600$ calories per mole (*11*)
$S_{298} = 23.5$ *e.u.* (*83*)
M.P. = 816° K. (*6*)
Decomposes to $Hg + HgCl_2$

Zone I (*c*) (298°–800° K.)

$$C_p = 11.05 + 3.70 \times 10^{-3} T \ (82)$$
$$H_T - H_{298} = -3{,}457 + 11.05T + 1.85 \times 10^{-3} T^2$$

Formation: $Hg + 1/2Cl_2 \longrightarrow HgCl$

Zone I (298°–630° K.)

$$\Delta C_p = +0.03 + 3.67 \times 10^{-3} T + 0.34 \times 10^5 T^{-2}$$
$$\Delta H_T = -31{,}650 + 0.03T + 1.83 \times 10^{-3} T^2 - 0.34 \times 10^5 T^{-1}$$
$$\Delta F_T = -31{,}650 - 0.03 T \ln T - 1.83 \times 10^{-3} T^2 - 0.17 \times 10^5 T^{-1} + 22.44T$$

Zone II (630°–800° K.)

$$\Delta C_p = 1.67 + 3.67 \times 10^{-3} T + 0.34 \times 10^5 T^{-2}$$
$$\Delta H_T = -46{,}670 + 1.67T + 1.83 \times 10^{-3} T^2 - 0.34 \times 10^5 T^{-1}$$
$$\Delta F_T = -46{,}670 - 1.67 T \ln T - 1.83 \times 10^{-3} T^2 - 0.17 \times 10^5 T^{-1} + 56.73T$$

T, ° K.	H_T-H_{298}	S_T	ΔH°_T	ΔF°_T
298	----------	23. 5	−31, 600	−25, 250
400	1, 260	27. 13	−31, 350	−22, 950
500	2, 530	29. 97	−31, 250	−21, 000
600	3, 840	32. 36	−31, 000	−19, 009
700	5, 185	34. 43	−45, 650	−16, 550
800	6, [illegible]65	36. 27	−44, 200	−11, 350

Mercury Dichloride, $HgCl_2$ (*c*)

$\Delta H^\circ_{298} = -53{,}400$ calories per mole (*11*)
$S_{298} = 34.5$ *e.u* (*83*)
M.P. = 550° K. (*6*)
$\Delta H_M = 4{,}150$ calories per mole
B.P. = 557° K. (*6*)
$\Delta H_V = 14{,}080$ calories per mole

Zone I (*c*) (298°–550° K.)

$$C_p = 15.28 + 10.4 \times 10^{-3} T \ (110)$$
$$H_T - H_{298} = -5{,}015 + 15.28T + 5.2 \times 10^{-3} T^2$$

Zone II (*g*) (557°–1,000° K.)

$$C_p = 14.66 + 0.26 \times 10^{3} T - 0.75 \times 10^5 T^{-2} \ (110)$$
$$H_T - H_{298} = 15{,}220 + 14.66T + 0.13 \times 10^{-3} T^2 + 0.75 \times 10^5 T^{-1}$$

Formation: $Hg + Cl_2 \longrightarrow HgCl_2$

Zone I (298°–550° K.)

$$\Delta C_p = -0.15 + 10.34 \times 10^{-3} T + 0.68 \times 10^5 T^{-2}$$
$$H_T = -53{,}600 - 0.15T + 5.17 \times 10^{-3} T^2 - 0.68 \times 10^5 T^{-1}$$
$$\Delta F_T = -53{,}600 + 0.15 T \ln T - 5.17 \times 10^{-3} T^2 - 0.34 \times 10^5 T^{-1} + 38.70T$$

Zone II (557°–630° K.)

$$\Delta C_p = -0.77 + 0.2 \times 10^{-3} T - 0.07 \times 10^5 T^{-2}$$
$$\Delta H_T = -33{,}335 - 0.77T + 0.10 \times 10^{-3} T^2 + 0.07 \times 10^5 T^{-1}$$
$$\Delta F_T = -33{,}335 + 0.77 T \ln T - 0.10 \times 10^{-3} T^2 + 0.035 \times 10^5 T^{-1} - 4.88T$$

Zone III (630°–1,000° K.)

$$\Delta C_p = +0.87 + 0.20 \times 10^{-3} T - 0.07 \times 10^5 T^{-2}$$
$$\Delta H_T = -48{,}350 + 0.87T + 0.10 \times 10^{-3} T^2 + 0.07 \times 10^5 T^{-1}$$
$$\Delta F_T = -48{,}350 - 0.87 T \ln T - 0.10 \times 10^{-3} T^2 + 0.035 \times 10^5 T^{-1} + 29.45T$$

T, ° K.	$H_T - H_{298}$	S_T	ΔH_T°	ΔF_T°
298	----------	34.5	−53,400	−42,370
400	1,825	40.05	−53,100	−38,700
500	3,825	44.51	−52,600	−35,200
600	24,190	81.41	−33,750	−33,350
700	25,650	83.67	−47,500	−31,650
800	27,125	85.64	−47,600	−29,500
900	28,600	87.37	−47,500	−27,350
1,000	30,085	88.94	−47,400	−25,000

Mercury Bromide, HgBr (*c*)

$\Delta H_{298}^\circ = -24,470$ calories per mole (*83*)
$S_{298} = 26.7$ *e.u.* (*83*)
M.P. = 680° K. (*6*)
Decomposes to $Hg + HgBr_2$
(estimated (*11*))

T, ° K.	$H_T - H_{298}$	ΔH_T°	ΔF_T°
298	----------	−24,470	−21,200
500	(3,000)	(−27,450)	(−16,600)

Mercury Dibromide, $HgBr_2$ (*c*)

$\Delta H_{298}^\circ = -40,500$ calories per mole (*106*)
$S_{298} = 38.9$ *e.u.* (*80*)
M.P. = 514° K. (*6*)
$\Delta H_M = 3,960$ calories per mole
B.P. = 592° K. (*6*)
$\Delta H_V = 14,080$ calories per mole

Formation: $Hg + Br_2 \longrightarrow HgBr_2$
(estimated (*11*))

T, ° K.	$H_T - H_{298}$	ΔH_T°	ΔF_T°
298	----------	−40,500	−35,900
500	(4,000)	(−47,350)	(−34,000)

Mercury Iodide, HgI (*c*)

$\Delta H_{298}^\circ = -14,455$ calories per mole (*112*)
$S_{298} = 28.6$ *e.u.* (*112*)
M.P. = 563° K. (*6*)
Decomposes to $Hg + HgI_2$

Zone I (*c*) (298°–563° K.)

$$C_p = 11.40 + 4.61 \times 10^{-3} T \quad (110)$$
$$H_T - H_{298} = -3,600 + 11.40T + 2.30 \times 10^{-3} T^2$$

Formation: $Hg + 1/2 I_2 \longrightarrow HgI$

Zone I (298°–386.8° K.)

$$\Delta C_p = -1.34 \times 10^{-3} T$$
$$\Delta H_T = -14,400 - 0.67 \times 10^{-3} T^2$$
$$\Delta F_T = -14,400 + 0.67 \times 10^{-3} T^2 + 8.14T$$

Zone II (386.8°–456° K.)

$$\Delta C_p = -4.81 + 4.61 \times 10^{-3} T$$
$$\Delta H_T = -14,910 - 4.81T + 2.30 \times 10^{-3} T^2$$
$$\Delta F_T = -14,910 + 4.81 T \ln T - 2.30 \times 10^{-3} T^2 - 23.16T$$

Zone III (456°–563° K.)

$$\Delta C_p = 0.35 + 4.61 \times 10^{-3} T$$
$$\Delta H_T = -22,200 + 0.35T + 2.30 \times 10^{-3} T^2$$
$$\Delta F_T = -22,200 - 0.35 T \ln T - 2.30 \times 10^{-3} T^2 + 13.52T$$

T, ° K.	$H_T - H_{298}$	S_T	ΔH_T°	ΔF_T°
298	----------	28.6	−14,450	−11,900
400	1,328	32.42	−16,450	−13,000
500	2,675	35.43	−21,450	−17,100

Mercury Diiodide, HgI_2 (*c*)

$\Delta H_{298}^\circ = -25,200$ calories per mole (*112*)
$S_{298} = 40.8$ *e.u.* (*83*)
T.P. = 403° K. (*82*)
$\Delta H_T = 650$ calories per mole
M.P. = 523° K. (*82*)
$\Delta H_M = 4,500$ calories per mole
B.P. = 627° K. (*82*)
$\Delta H_V = 14,263$ calories per mole

Zone I (α) (298°–403° K.)

$$C_p = 18.50 \quad (82)$$
$$H_T - H_{298} = -5,516 + 18.50T$$

Zone II (β) (403°–523° K.)

$$C_p = 20.20 \quad (82)$$
$$H_T - H_{298} = 5,550 + 20.20T$$

Zone III (*l*) (523°–627° K.)

$$C_p = 25.0 \quad (82)$$
$$H_T - H_{298} = 3,560 + 25.0T$$

Zone IV (*g*) (627°–1,000° K.)

$$C_p = 14.90 - 0.27 \times 10^5 T^{-2} \quad (82)$$
$$H_T - H_{298} = +16,993 + 14.90T + 0.27 \times 10^5 T^{-1}$$

Formation: $Hg + I_2 \longrightarrow HgI_2$

Zone I (298°–386° K.)

$$\Delta C_p = 2.30 - 11.90 \times 10^{-3} T$$
$$\Delta H_T = -25,357 + 2.30T - 5.95 \times 10^{-3} T^2$$
$$\Delta F_T = -25,357 - 2.30 T \ln T + 5.95 \times 10^{-3} T^2 + 17.14T$$

Zone II (386.1°–403° K.)

$$\Delta C_p = -7.31$$
$$\Delta H_T = -26,300 - 7.31T$$
$$\Delta F_T = -26,300 + 7.31 T \ln T - 35.41T$$

Zone III (456°–523° K.)

$$\Delta C_p = 4.7$$
$$\Delta H_T = -41,000 + 4.7T$$
$$\Delta F_T = -41,000 - 4.7 T \ln T + 70.22T$$

Zone IV (523°–627° K.)

$$\Delta C_p = 9.5$$
$$\Delta H_T = -39.015 + 9.5T$$
$$\Delta F_T = -39,015 - 9.5 T \ln T + 96.45T$$

Zone V (630°–1,000° K.)

$$\Delta C_p = 1.04 - 0.27 \times 10^5 T^{-2}$$
$$\Delta H_T = -33,500 + 1.04T + 0.27 \times 10^5 T^{-1}$$
$$\Delta F_T = -33,500 - 1.04 T \ln T + 0.135 \times 10^5 T^{-1} + 32.88T$$

T, ° K.	$H_T - H_{298}$	S_T	ΔH°_T	ΔF_T°
298	----------	40.8	−25,200	−23,600
400	1,885	46.28	−29,200	−22,950
500	4,550	52.35	−38,650	−20,500
600	11,440	65.3	−33,300	−17,600
700	27,460	90.78	−32,700	−15,200
800	28,950	92.76	−32,600	−12,750
900	30,430	94.50	−32,500	−10,250
1,000	31,920	96.08	−32,400	−7,800

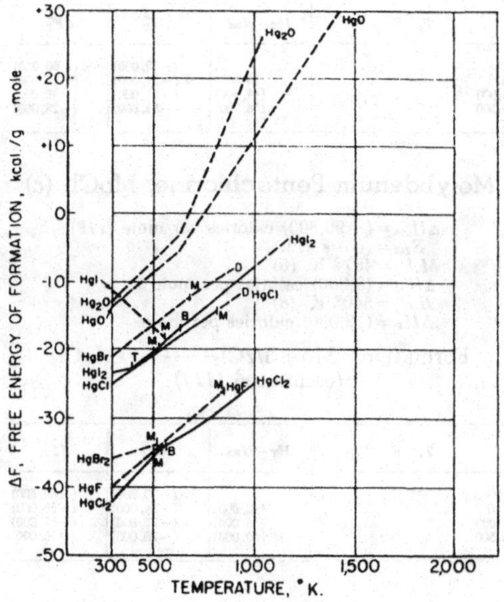

FIGURE 33.—Mercury.

MOLYBDENUM AND ITS COMPOUNDS

Element, Mo (*c*)

$S_{298}=6.83$ *e.u.* (*83*)
M.P. = 2,883° K. (*112*)

Zone I (*c*) (298°–1,800° K.)

$$C_p=5.48+1.30\times10^{-3}T$$
$$H_T-H_{298}=-1{,}690+5.48T+0.65\times10^{-3}T^2$$
$$F_T-H_{298}=-1{,}690-5.48T\ln T-0.65\times10^{-3}T^2+30.24T$$

T, ° K.	H_T-H_{298}	S_T	$-\frac{(F_T-H_{298})}{T}$
298		6.83	6.83
400	610	8.59	7.06
500	1,215	9.95	7.52
600	1,830	11.07	8.02
700	2,460	12.04	8.53
800	3,105	12.90	9.00
900	3,765	13.68	9.48
1,000	4,440	14.39	9.95
1,100	5,125	15.04	10.36
1,200	5,825	15.65	10.80
1,300	6,530	16.21	11.19
1,400	7,250	16.74	11.53
1,500	7,985	17.25	11.91
1,600	8,740	17.74	12.29
1,700	9,510	18.21	12.64
1,800	10,300	18.66	12.96
1,900	(11,075)		(13.27)
2,000	(11,890)		(13.65)
2,500	(16,070)		(14.86)

Molybdenum Dioxide, MoO_2 (*c*)

$\Delta H^\circ_{298}=(-131{,}000)$ calories per mole (*8*)
$S_{298}=11.06$ *e.u.* (*24*)
M.P. = >2,500° K. (*42*)

Formation: $Mo+O_2\longrightarrow MoO_2$
(estimated (*24*))

T, ° K.	H_T-H_{298}	ΔH°_T	ΔF°_T
298		(−131,000)	(−118,300)
400	(4,000)	(−130,500)	(−115,000)
500	(5,800)	(−130,500)	(−111,500)
600	(7,900)	(−130,500)	(−108,000)
700	(10,200)	(−130,000)	(−104,500)
800	(12,400)	(−130,000)	(−101,000)
900	(14,600)	(−130,000)	(−97,500)
1,000	(16,900)	(−129,500)	(−94,000)
1,100	(19,100)	(−129,500)	(−91,000)
1,200	(21,300)	(−129,500)	(−87,500)
1,300	(24,000)	(−129,000)	(−84,000)
1,400	(26,800)	(−129,000)	(−80,500)
1,500	(29,500)	(−129,000)	(−77,000)
1,600	(32,500)	(−128,500)	(−73,500)
1,700	(35,300)	(−128,500)	(−70,000)
1,800	(38,100)	(−128,500)	(−67,000)
1,900	(41,200)	(−128,000)	(−63,500)
2,000	(44,400)	(−128,000)	(−60,000)

Molybdenum Trioxide, MoO_3 (*c*)

$\Delta H^\circ_{298}=-180{,}330$ calories per mole (*112*)
$S_{298}=18.68$ *e.u.* (*112*)
M.P. = 1,068° K. (*112*)
$\Delta H_M=12{,}540$ calories per mole
B.P. = 1,428° K. (*112*)
$\Delta H_V=33{,}000$ calories per mole

Zone I (*c*) (298°–1,068° K.)

$$C_p=20.07+5.90\times10^{-3}T-3.68\times10^5T^{-2}\ (23)$$
$$H_T-H_{298}=-7{,}480+20.07T+2.95\times10^{-3}T^2+3.68\times10^5T^{-1}$$

Formation: $Mo+3/2O_2\longrightarrow MoO_3$

Zone I (298°–1,068° K.)

$$\Delta C_p=3.85+3.10\times10^{-3}T-3.08\times10^5T^{-2}$$
$$\Delta H_T=-182{,}600+3.85T+1.55\times10^{-3}T^2+3.08\times10^5T^{-1}$$
$$\Delta F_T=-182{,}600-3.85T\ln T-1.55\times10^{-3}T^2+1.54\times10^5T^{-1}+89.7T$$

T, ° K.	H_T-H_{298}	S_T	ΔH°_T	ΔF°_T
298		18.68	−180,330	−162,030
400	1,970	24.77	−180,050	−155,500
500	4,020	28.95	−179,700	−149,700
600	6,270	33.30	−179,200	−143,900
700	8,680	36.76	−178,700	−137,900
800	10,960	41.06	−178,150	−132,250
900	13,430	43.00	−177,550	−126,550
1,000	16,550	45.73	−176,850	−120,900
1,100	(31,070)		(−163,700)	(−115,300)
1,200	(33,770)		(−162,950)	(−109,500)
1,300	(36,430)		(−162,250)	(−105,200)
1,400	(39,310)		(−161,350)	(−101,000)
1,500	(75,040)		(−127,650)	(−96,000)
2,000	(90,800)		(−122,600)	(−90,000)

Molybdenum Hexafluoride, MoF_6 (*l*)

$\Delta H^{\circ}_{298} = -405{,}000$ calories per mole (*112*)
$S_{298} = (77)$ *e.u.* (*42*)
$M.P. = 290°$ K. (*6*)
$\Delta H_M = 2{,}500$ calories per mole
$B.P. = 309°$ K. (*6*)
$\Delta H_V = 6{,}000$ calories per mole

Formation: $Mo + 3F_2 \longrightarrow MoF_6$
(estimated (*42*))

T, ° K.	ΔF°_T
298	(−383, 000)
500	(−368, 000)
1,000	(−332, 000)
1,500	(−297, 000)
2,000	(−264, 000)

Molybdenum Dichloride, $MoCl_2$ (*c*)

$\Delta H^{\circ}_{298} = (-44{,}000)$ calories per mole (*12*)
$S_{298} = (29)$ *e.u.* (*12*)
$M.P. = 1{,}000°$ K. (*12*)
$\Delta H_M = 6{,}000$ calories per mole
$B.P. = 1{,}700°$ K. (*12*)
$\Delta H_V = 36{,}000$ calories per mole

Formation: $Mo + Cl_2 \longrightarrow MoCl_2$
(estimated (*12*))

T, ° K.	$H_T - H_{298}$	ΔH°_T	ΔF°_T
298		(−44, 000)	(−35, 000)
500	(4, 000)	(−43, 000)	(−29, 000)
1,000	(15, 000)	(−39, 000)	(−16, 000)
1,500	(34, 000)	(−28, 000)	(−8, 000)

Molybdenum Trichloride, $MoCl_3$ (*c*)

$\Delta H^{\circ}_{298} = (-65{,}000)$ calories per mole (*12*)
$S_{298} = 37.8$ *e.u.* (*12*)
$S.P. = 1{,}300°$ K. (*12*)
$\Delta H_{subl} = 52{,}000$ calories per mole

Formation: $Mo + 3/2Cl_2 \longrightarrow MoCl_3$
(estimated (*12*))

T, ° K.	$H_T - H_{298}$	ΔH°_T	ΔF°_T
298		(−65, 000)	(−50, 000)
500	(5, 000)	(−64, 000)	(−40, 000)
1,000	(20, 000)	(−58, 000)	(−20, 000)
1,500	(83, 500)	(−5, 000)	(−6, 000)

Molybdenum Tetrachloride, $MoCl_4$ (*c*)

$\Delta H^{\circ}_{298} = (-79{,}000)$ calories per mole (*112*)
$S_{298} = 47.4$ *e.u.* (*12*)
$S.P. = 595°$ K. (*12*)
$\Delta H_{subl} = 25{,}000$ calories per mole

Formation: $Mo + 2Cl_2 \longrightarrow MoCl_4$
(estimated (*12*))

T, ° K.	$H_T - H_{298}$	ΔH°_T	ΔF°_T
298		(−79, 000)	(−60, 000)
500	(6, 600)	(−77, 000)	(−47, 000)
1,000	(44, 500)	(−51, 000)	(−36, 000)
1,500	(56, 700)	(−51, 000)	(−28, 000)

Molybdenum Pentachloride, $MoCl_5$ (*c*)

$\Delta H^{\circ}_{298} = (-90{,}800)$ calories per mole (*112*)
$S_{298} = (65)$ *e.u.* (*94*)
$M.P. = 467°$ K. (6)
$\Delta H_M = (8{,}000)$ calories per mole
$B.P. = 540°$ K. (*6*)
$\Delta H_V = (12{,}000)$ calories per mole

Formation: $Mo + 5/2Cl_2 \longrightarrow MoCl_5$
(estimated (*11*))

T, ° K.	$H_T - H_{298}$	ΔH°_T	ΔF°_T
298		(−90, 800)	(−68, 500)
500	(12, 000)	(−84, 000)	(−56, 000)
1,000	(35, 000)	(−75, 000)	(−31, 000)
1,500	(50, 000)	(−75, 000)	(−10, 000)

Molybdenum Hexachloride, $MoCl_6$ (*c*)

$\Delta H^{\circ}_{298} = (-90{,}000)$ calories per mole (*112*)
$S_{298} = 72.3$ *e.u.* (*11*)
$S.P. = 630°$ K. (*6*)
$\Delta H_{subl} = (19{,}000)$ calories per mole

Formation: $Mo + 3Cl_2 \longrightarrow MoCl_6$
(estimated (*11*))

T, ° K.	$H_T - H_{298}$	ΔH°_T	ΔF°_T
298		(−90, 000)	(−62, 000)
500	(9, 300)	(−87, 000)	(−43, 000)
1,000	(45, 600)	(−67, 000)	(−13, 000)
1,500	(62, 100)	(−67, 000)	(+14, 000)

Molybdenum Dibromide, $MoBr_2$ (*c*)

$\Delta H^{\circ}_{298} = (-28{,}500)$ calories per mole (*12*)
$S_{298} = 34.5$ *e.u.* (*12*)
$M.P. = (1{,}000°)$ K. (*12*)
$\Delta H_M = (6{,}000)$ calories per mole
$B.P. = (1{,}500°)$ K. (*12*)
$\Delta H_V = (31{,}000)$ calories per mole

Formation: $Mo + Br_2 \longrightarrow MoBr_2$
(estimated (*12*))

T, ° K.	$H_T - H_{298}$	ΔH°_T	ΔF°_T
298		(−28, 500)	(−26, 000)
500	(4, 000)	(−35, 000)	(−21, 000)
1,000	(16, 500)	(−30, 000)	(−8, 000)
1,500	(30, 000)	(−19, 000)	(+1, 000)
2,000	(75, 000)	(+12, 000)	(+4, 000)

Molybdenum Tribromide, $MoBr_3$ (*c*)

$\Delta H^\circ_{298} = (-40{,}000)$ calories per mole (*12*)
$S_{298} = 43.8$ *e.u.* (*12*)
S.P. = (1,250°) K. (*12*)
$\Delta H_{subl} = (50{,}000)$ calories per mole

Formation: $Mo + 3/2Br_2 \longrightarrow MoBr_3$
(estimated (*12*))

T, ° K.	$H_T - H_{298}$	ΔH°_T	ΔF°_T
298		(−40, 000)	(−35, 000)
500	(5, 000)	(−50, 000)	(−25, 500)
1,000	(19, 000)	(−46, 000)	(−3, 000)
1,500	(81, 000)	(+6, 000)	(+7, 000)

Molybdenum Tetrabromide, $MoBr_4$ (*c*)

$\Delta H^\circ_{298} = (-45{,}300)$ calories per mole (*12*)
$S_{298} = (59)$ *e.u.* (*12*)
S.P. = 620° K. (*12*)
$\Delta H_{subl} = (26{,}000)$ calories per mole

Formation: $Mo + 2Br_2 \longrightarrow MoBr_4$
(estimated (*12*))

T, ° K.	$H_T - H_{298}$	ΔH°_T	ΔF°_T
298		(−45, 300)	(−39, 500)
500	(7, 300)	(−58, 400)	(−28, 000)
1, 000	(46, 300)	(−31, 500)	(−17, 000)
1, 500	(58, 800)	(−31, 500)	(−4, 000)

Molybdenum Pentabromide, $MoBr_5$ (*c*)

$\Delta H^\circ_{298} = (-50{,}000)$ calories per mole (*12*)
$S_{298} = (77)$ *e.u.* (*12*)
M.P. = < 500° K. (*12*)
$\Delta H_M = (8{,}000)$ calories per mole
B.P. = (600°) K. (*12*)
$\Delta H_V = (14{,}000)$ calories per mole

Formation: $Mo + 5/2Br_2 \longrightarrow MoBr_5$
(estimated (*12*))

T, ° K.	$H_T - H_{298}$	ΔH°_T	ΔF°_T
298		(−50, 000)	(−43, 000)
500	(13, 700)	(−61, 400)	(−31, 000)
1, 000	(44, 000)	(−45, 500)	(−12, 000)
1, 500	(59, 000)	(−45, 500)	(+4, 000)

Molybdenum Diiodide, MoI_2 (*c*))

$\Delta H^\circ_{298} = (-12{,}000)$ calories per mole (*112*)
$S_{298} = (36)$ *e.u.* (*12*)
M.P. = (1,000°) K. (*12*)
$\Delta H_M = (6{,}000)$ calories per mole
B.P. = (1,200°) K. (*12*)
$\Delta H_V = (25{,}000)$ calories per mole

Formation: $Mo + I_2 \longrightarrow MoI_2$
(estimated (*12*))

T, ° K.	$H_T - H_{298}$	ΔH°_T	ΔF°_T
298		(−12, 000)	(−12, 500)
500	(4, 000)	(−26, 000)	(−11, 000)
1, 000	(16, 500)	(−21, 500)	(+3, 000)
1, 500	(57, 500)	(+12, 000)	(+6, 000)

Molybdenum Triiodide, MoI_3 (*c*)

$\Delta H^\circ_{298} = (-15{,}000)$ calories per mole (*112*)
$S_{298} = (48)$ *e.u.* (*12*)
S.P. = (1,200°) K. (*12*)
$\Delta H_{subl} = (48{,}000)$ calories per mole

Formation: $Mo + 3/2I_2 \longrightarrow MoI_3$
(estimated (*12*))

T, ° K.	$H_T - H_{298}$	ΔH°_T	ΔF°_T
298		(−15, 000)	(−15, 000)
500	(5, 000)	(−36, 000)	(−11, 000)
1, 000	(19, 000)	(−32, 000)	(+12, 000)
1, 500	(67, 000)	(+18, 000)	(+21, 000)

Molybdenum Tetraiodide, MoI_4 (*c*)

$\Delta H^\circ_{298} = (-18{,}000)$ calories per mole (*112*)
$S_{298} = (64)$ *e.u.* (*12*)
S.P. = 695° K. (12)
$\Delta H_{subl} = (29{,}000)$ calories per mole

Formation: $Mo + 2I_2 \longrightarrow MoI_4$
(estimated (*12*))

T, ° K.	$H_T - H_{298}$	ΔH°_T	ΔF°_T
298		(−18, 000)	(−18, 500)
500	(6, 000)	(−46, 000)	(−15, 000)
1, 000	(50, 000)	(−15, 000)	(+4, 000)
1, 500	(59, 000)	(−15, 000)	(+13, 000)

Molybdenum Pentaiodide, MoI_5 (*c*)

$\Delta H^\circ_{298} = (-18{,}000)$ calories per mole (*112*)
$S_{298} = 81.5$ *e.u.* (*12*)
S.P. = 650° K. (*12*)
$\Delta H_{subl} = (26{,}000)$ calories per mole

Formation: $Mo + 5/2I_2 \longrightarrow MoI_5$
(estimated (*12*))

T, ° K.	$H_T - H_{298}$	ΔH°_T	ΔF°_T
298		(−18, 000)	(−19, 500)
500	(8, 000)	(−52, 000)	(−15, 000)
1, 000	(51, 000)	(−24, 000)	(+10, 000)
1, 500	(62, 000)	(−24, 000)	(+25, 000)

Dimolybdenum Carbide, Mo_2C (*c*)

$\Delta H^\circ_{298} = 4{,}200$ calories per mole (*9*)
$S_{298} = 19.1$ *e.u* (*9*)
M.P. = 2,965° K. (*9*)

Formation: $2Mo + C \longrightarrow Mo_2C$

Zone I (300°–3,000° K.)

$$\Delta F_T = 4{,}200 - 4.8T \ (81)$$

T, ° K.	ΔF°_T	T, ° K.	ΔF°_T
298	2,970	1,200	(−1,550)
400	(2,300)	1,300	(−2,050)
500	(1,800)	1,400	(−2,500)
600	(1,300)	1,500	(−3,000)
700	(800)	1,600	(−3,500)
800	(350)	1,700	(−3,950)
900	(−100)	1,800	(−4,450)
1,000	(−600)	1,900	(−4,900)
1,100	(−1,100)	2,000	(−5,400)

Dimolybdenum Nitride, Mo_2N (*c*)

$\Delta H^\circ_{298} = -16{,}600$ calories per mole (*81*)
$S_{298} = 21$ *e. u.* (*81*)

Zone I (*c*) (298°–800° K.)

$$C_p = 11.19 + 13.80 \times 10^{-3}T \ (82)$$
$$H_T - H_{298} = -3{,}950 + 11.19T + 6.90 \times 10^{-3}T^2$$

Formation: $2Mo + 1/2N_2 \longrightarrow Mo_2N$

Zone I (298°–800° K.)

$$\Delta C_p = -3.10 + 10.69 \times 10^{-3}T$$
$$\Delta H_T = -16{,}150 - 3.10T + 5.34 \times 10^{-3}T^2$$
$$\Delta F_T = -16{,}150 + 3.10T\ln T - 5.34 \times 10^{-3}T^2 - 1.98T$$

T, ° K.	$H_T - H_{298}$	S_T	ΔH°_T	ΔF°_T
298		21.0	−16,600	−11,970
400	1,610	25.64	−16,550	−10,400
500	3,360	29.54	−16,400	−8,850
600	5,280	33.03	−16,050	−7,400
700	7,290	36.13	−15,650	−5,950
800	9,370	38.90	−15,050	−4,400

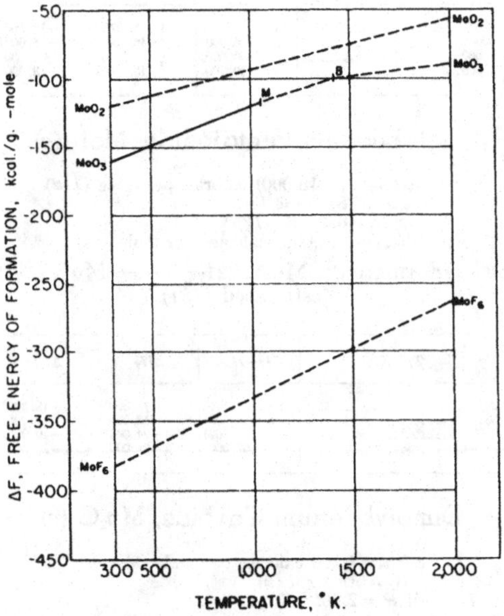

FIGURE 34.—Molybdenum (a).

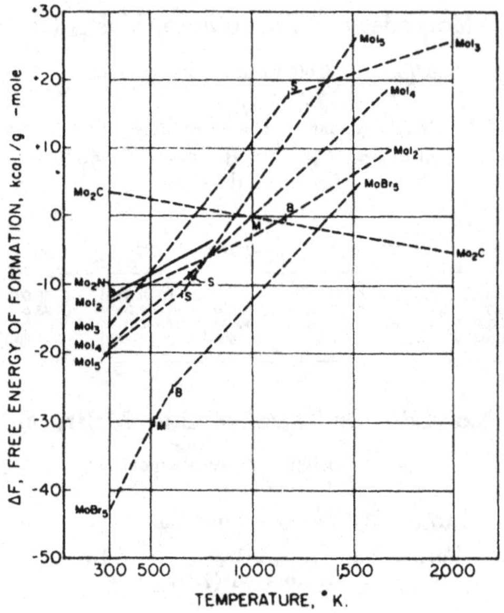

FIGURE 35.—Molybdenum (b).

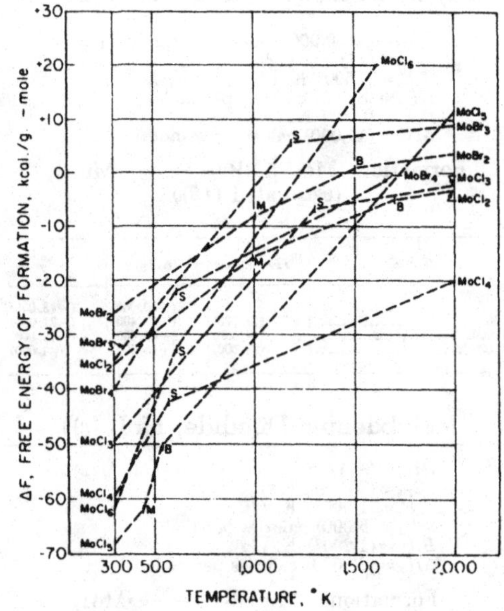

FIGURE 36.—Molybdenum (c).

NEODYMIUM AND ITS COMPOUNDS

Element, Nd (*c*)

$S_{298}=17.54$ *e.u.* (*121*)
M.P. = 1,297° K. (*125*)
$\Delta H_M=2{,}600$ calories per atom
B.P. = 3,450° K. (*125*)
$\Delta H_V=69{,}000$ calories per atom

Zone I (*c*) (298°–900° K.)

$C_p=5.61+5.34\times10^{-3}T$ (*82*)
$H_T-H_{298}=-1{,}910+5.61T+2.67\times10^{-3}T^2$
$F_T-H_{298}=-1{,}910-5.51T\ln T-2.67\times10^{-3}T^2+21.61T$

T, ° K	H_T-H_{298}	S_T	$-\frac{(F_T-H_{298})}{T}$
298	----------	17.54	17.54
400	755	19.72	17.85
500	1,560	21.51	18.38
600	2,420	23.08	19.05
700	3,330	24.48	19.70
800	4,290	25.76	20.40
900	5,300	26.95	21.07
1,000	(6,370)	(28.1)	(21.83)
1,100	(7,490)	(29.2)	(22.39)
1,200	(8,670)	(30.2)	(22.97)
1,300	(12,280)	----------	(23.56)
1,400	(13,080)	----------	(24.25)
1,500	(13,880)	----------	(24.90)
1,600	(14,680)	----------	(25.47)
1,700	(15,480)	----------	(26.04)
1,800	(16,300)	----------	(26.55)
1,900	(17,100)	----------	(27.03)
2,000	(17,900)	----------	(27.48)

Dineodymium Trioxide, Nd_2O_3 (*c*)

$\Delta H^\circ_{298}=-432{,}150$ calories per mole (*61*)
$S_{298}=(41.6)$ *e.u.* (*24*)

Zone I (*c*) (298°–1,175° K.)

$C_p=28.99+5.76\times10^{-3}T-4.159\times10^5T^{-2}$ (*3*)
$H_T-H_{298}=-10{,}290+28.99T+2.88\times10^{-3}T^2+4.159\times10^5T^{-1}$

Formation: $2Nd+3/2O_2\longrightarrow Nd_2O_3$

Zone I (298°–900° K.)

$\Delta C_p=7.03-6.42\times10^{-3}T-3.559\times10^5T^{-2}$
$\Delta H_T=-435{,}150+7.03T-3.21\times10^{-3}T^2+3.559\times10^5T^{-1}$
$\Delta F_T=-435{,}150-7.03T\ln T+3.21\times10^{-3}T^2+1.78\times10^5T^{-1}+115.1T$

T, ° K.	H_T-H_{298}	S_T	ΔH°_T	ΔF°_T
298	----------	(41.6)	−432,150	(−412,500)
400	3,100	(48.8)	−432,000	(−405,000)
500	6,050	(55.3)	−431,700	(−398,300)
600	9,150	(61.1)	−431,500	(−391,700)
700	12,350	(66.0)	−431,250	(−385,100)
800	15,600	(70.4)	−431,000	(−378,600)
900	18,950	(74.0)	−431,000	(−372,100)
1,000	22,300	(77.7)	(−431,000)	(−365,200)
1,100	25,750	(81.0)	(−431,000)	(−359,000)
1,200	29,300	(84.3)	(−436,900)	(−352,250)

Neodymium Trifluoride, NdF_3 (*c*)

$\Delta H^\circ_{298}=(-385{,}000)$ calories per mole (*5*)
$S_{298}=(24)$ *e.u.* (*11*)
M.P. = 1,647° K. (*29*)
$\Delta H_M=(8{,}000)$ calories per mole
B.P. = (2,600°) K. (*6*)
$\Delta H_V=(62{,}000)$ calories per mole

Formation: $Nd+3/2F_2\longrightarrow NdF_3$
(estimated (*11*))

T, ° K.	H_T-H_{298}	ΔH°_T	ΔF°_T
298	----------	(−385,000)	(−366,000)
500	(4,000)	(−385,000)	(−354,500)
1,000	(17,000)	(−383,100)	(−324,000)
1,500	(32,000)	(−382,200)	(−296,500)

Neodymium Trichloride, $NdCl_3$ (*c*)

$\Delta H^\circ_{298}=-245{,}600$ calories per mole (*128*)
$S_{298}=34.6$ *e.u.* (*128*)
M.P. = 1,031° K. (*29*)
$\Delta H_M=(8{,}000)$ calories per mole
B.P. = (1,940°) K. (*6*)
$\Delta H_V=(46{,}000)$ calories per mole

Formation: $Nd+3/2Cl_2\longrightarrow NdCl_3$
(estimated (*11*))

T, ° K.	H_T-H_{298}	ΔH°_T	ΔF°_T
298	----------	−245,600	−227,930
500	(5,000)	(−244,000)	(−217,600)
1,000	(19,000)	(−241,100)	(−192,600)
1,500	(43,000)	(−232,000)	(−173,600)

Neodymium Tribromide, $NdBr_3$ (*c*)

$\Delta H^\circ_{298}=(-187{,}000)$ calories per mole (*5*)
$S_{298}=(47)$ *e.u.* (*11*)
M.P. = 955° K. (*29*)
$\Delta H_M=(8{,}000)$ calories per mole
B.P. = (1,810°) K. (*6*)
$\Delta H_V=(45{,}000)$ calories per mole

Formation: $Nd+3/2Br_2\longrightarrow NdBr_3$
(estimated (*11*))

T, ° K.	H_T-H_{298}	ΔH°_T	ΔF°_T
298	----------	(−187,000)	(−180,600)
500	(5,000)	(−197,800)	(−170,000)
1,000	(18,000)	(−196,400)	(−145,000)
1,500	(43,000)	(−185,700)	(−126,500)

Neodymium Triiodide, NdI_3 (*c*)

$\Delta H^\circ_{298}=-158{,}000$ calories per mole (*5*)
$S_{298}=(49)$ *e.u.* (*11*)
T.P. = 827° K. (*29*)
M.P. = 1,048° K. (*29*)
$\Delta H_M=(8{,}000)$ calories per mole
B.P. = (1,640°) K. (*6*)
$\Delta H_V=(41{,}000)$ calories per mole

Formation: $Nd+3/2I_2\longrightarrow NdI_3$
(estimated (*11*))

T, ° K.	H_T-H_{298}	ΔH°_T	ΔF°_T
298	----------	−158,000	(−156,000)
500	(5,000)	(−179,500)	(−151,000)
1,000	(19,000)	(−177,000)	(−124,000)
1,500	(44,000)	(−166,000)	(−100,000)

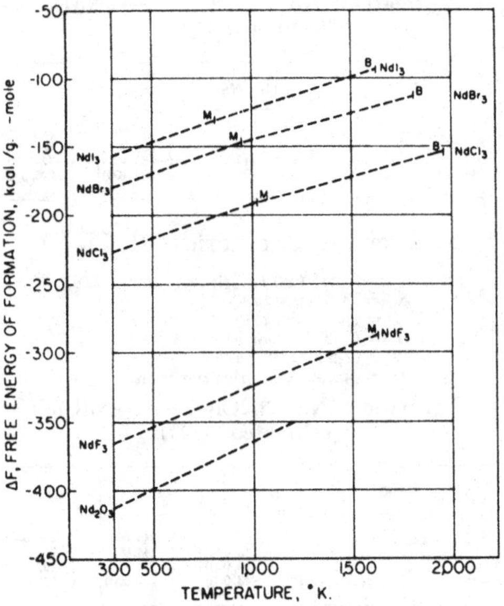

FIGURE 37.—Neodymium.

NICKEL AND ITS COMPOUNDS

Element, Ni (*c*)

$S_{298}=7.12$ *e.u.* (*83*)
$T.P.=633°$ K. (*82*)
$\Delta H_T=0$ calories per atom
$M.P.=1{,}725°$ K. (*82*)
$\Delta H_M=4{,}210$ calories per atom
$B.P.=3{,}073°$ K. (*112*)
$\Delta H_V=91{,}000$ calories per atom

Zone I (α) (298°–633° K.)

$$C_p=4.06+7.04\times10^{-3}T\ (82)$$
$$H_T-H_{298}=-1{,}525+4.06T+3.52\times10^{-3}T^2$$
$$F_T-H_{298}=-1{,}525-4.06T\ln T-3.52\times10^{-3}T^2+22.16T$$

Zone II (β) (633°–1,725° K.)

$$C_p=6.00+1.80\times10^{-3}T\ (82)$$
$$H_T-H_{298}=-1{,}700+6.00T+0.90\times10^{-3}T^2$$
$$F_T-H_{298}=-1{,}700-6.00T\ln T-0.90\times10^{-3}T^2+33.3T$$

Zone III (*l*) (1,725°–1,900° K.)

$$C_p=9.20\ (82)$$
$$H_T-H_{298}=-330+9.20T$$
$$F_T-H_{298}=-330-9.20T\ln T+54.85T$$

T, ° K.	H_T-H_{298}	S_T	$-\frac{(F_T-H_{298})}{T}$
298	....	7.12	7.12
400	665	9.03	7.37
500	1,380	10.63	7.87
600	2,180	12.08	8.44
700	2,940	13.23	9.06
800	3,690	14.26	9.65
900	4,445	15.15	10.21
1,000	5,210	15.96	10.75
1,100	5,985	16.70	11.28
1,200	6,780	17.39	11.74
1,300	7,600	18.05	12.20
1,400	8,450	18.68	12.64
1,500	9,320	19.28	13.07
1,600	10,210	19.85	13.47
1,700	11,110	20.40	13.86
1,800	16,230	23.35	14.33
1,900	17,150	23.85	14.82
2,000	(18,070)	(24.32)	(15.30)

Nickel Oxide, NiO (*c*)

$\Delta H^{\circ}_{298}=-57{,}300$ calories per mole (*4*)
$S_{298}=9.08$ *e.u.* (*88*)
$T.P.=525°$ K. (*82*)
$\Delta H_T=0$ calories per mole
$T.P.=565°$ K. (*82*)
$\Delta H_T=0$ calories per mole
$M.P.=2{,}233°$ K. (*42*)

Zone I (α) (298°–525° K.)

$$C_p=-4.99+37.58\times10^{-3}T+3.89\times10^5T^{-2}\ (82)$$
$$H_T-H_{298}=+1{,}122-4.99T+18.79\times10^{-3}T^2-3.89\times10^5T^{-1}$$

Zone II (β) (525°–565° K.)

$$C_p=13.88\ (82)$$
$$H_T-H_{298}=-4{,}347+13.88T$$

Zone III (γ) (565°–1,800° K.)

$$C_p=11.18+2.02\times10^{-3}T\ (82)$$
$$H_T-H_{298}=-3{,}140+11.18T+1.01\times10^{-3}T^2$$

Formation: $Ni+1/2O_2 \longrightarrow NiO$

Zone I (298°–525° K.)

$$\Delta C_p=-12.63+30.04\times10^{-3}T+4.09\times10^5T^{-2}$$
$$\Delta H_T=-53{,}500-12.63T+15.02\times10^{-3}T^2-4.09\times10^5T^{-1}$$
$$\Delta F_T=-53{,}500+12.63T\ln T-15.02\times10^{-3}T^2-2.04\times10^5T^{-1}-55.39T$$

Zone II (525°–565° K.)

$$\Delta C_p=6.24-7.54\times10^{-3}T+0.20\times10^5T^{-2}$$
$$\Delta H_T=-58{,}970+6.24T-3.77\times10^{-3}T^2-0.20\times10^5T^{-1}$$
$$\Delta F_T=-58{,}970-6.24T\ln T+3.77\times10^{-3}T^2-0.10\times10^5T^{-1}+63.65T$$

Zone III (565°–633° K.)

$$\Delta C_p=3.54-5.52\times10^{-3}T+0.20\times10^5T^{-2}$$
$$\Delta H_T=-57{,}808+3.54T-2.76\times10^{-3}T^2-0.20\times10^5T^{-1}$$
$$\Delta F_T=-57{,}808-3.54T\ln T+2.76\times10^{-3}T^2-0.10\times10^5T^{-1}+43.98T$$

Zone IV (633°–1,725° K.)

$\Delta C_p = 1.6 - 0.28\times10^{-3}T + 0.20\times10^5T^{-2}$
$\Delta H_T = -57,600 + 1.6T - 0.14\times10^{-3}T^2 - 0.20\times10^5T^{-1}$
$\Delta F_T = -57,600 - 1.6T\ln T + 0.14\times10^{-3}T^2 - 0.10\times10^5T^{-1} + 32.82T$

T, ° K.	H_T-H_{298}	S_T	ΔH_T°	ΔF_T°
298	......	9.08	−57,300	−50,600
400	1,165	12.43	−57,150	−48,300
500	2,535	15.47	−56,850	−46,100
600	3,940	18.05	−56,650	−44,000
700	5,220	20.02	−56,500	−41,900
800	6,500	21.73	−56,400	−39,800
900	7,780	23.24	−56,250	−37,750
1,000	9,070	24.60	−56,150	−35,700
1,100	10,370	25.84	−56,000	−33,600
1,200	11,700	26.97	−55,900	−31,600
1,300	13,060	28.08	−55,800	−29,550
1,400	14,450	29.11	−55,650	−27,550
1,500	15,860	30.08	−55,600	−25,550
1,600	17,300	31.01	−55,450	−23,500
1,700	18,770	31.90	−55,300	−21,550
1,800	20,260	32.76	−59,400	−19,550

Nickel Difluoride, NiF_2 (*c*)

$\Delta H^\circ_{298} = -158,000$ calories per mole (*11*)
$S_{298} = 17.69$ *e.u.* (*18*)
M.P. = (1,300°) K. (*6*)
$\Delta H_M = (8,000)$ calories per mole
B.P. = (1,900°) K. (*6*)
$\Delta H_V = (48,000)$ calories per mole

Formation: $Ni + F_2 \longrightarrow NiF_2$
(estimated (*11*))

T, ° K.	H_T-H_{298}	ΔH_T°	ΔF_T°
298	......	−158,000	−146,700
500	(4,000)	(−156,900)	(−140,000)
1,000	(13,000)	(−156,000)	(−124,000)
1,500	(34,000)	(−143,600)	(−108,500)

Nickel Dichloride, $NiCl_2$ (*c*)

$\Delta H^\circ_{298} = -73,000$ calories per mole (*11*)
$S_{298} = 23.3$ *e.u.* (*16*)
M.P. = 1,303° K. (*25*)
$\Delta H_M = 18,470$ calories per mole

Zone I (*c*) (298°–1,303° K.)

$C_p = 17.50 + 3.16\times10^{-3}T - 1.19\times10^5T^{-2}$ (*25*)
$H_T - H_{298} = -5,750 + 17.50T + 1.58\times10^{-3}T^2 + 1.19\times10^5T^{-1}$

Formation: $Ni + Cl_2 \longrightarrow NiCl_2$

Zone I (298°–633° K.)

$\Delta C_p = 4.62 - 3.94\times10^{-3}T - 0.51\times10^5T^{-2}$
$\Delta H_T = -74,375 + 4.62T - 1.97\times10^{-3}T^2 + 0.51\times10^5T^{-1}$
$\Delta F_T = -74,375 - 4.62T\ln T + 1.97\times10^{-3}T^2 + 0.25\times10^5T^{-1} + 67.13T$

Zone II (633°–1,303° K.)

$\Delta C_p = 2.68 + 1.30\times10^{-3}T - 0.51\times10^5T^{-2}$
$\Delta H_T = -74,200 + 2.68T + 0.65\times10^{-3}T^2 + 0.51\times10^5T^{-1}$
$\Delta F_T = -74,200 - 2.68T\ln T - 0.65\times10^{-3}T^2 + 0.25\times10^5T^{-1} + 56.05T$

T, ° K.	H_T-H_{298}	S_T	ΔH_T°	ΔF_T°
298	......	23.3	−73,000	−61,900
400	1,800	28.46	−72,700	−58,100
500	3,650	32.61	−72,400	−54,500
600	5,550	36.06	−72,150	−51,000
700	7,465	39.02	−71,900	−47,500
800	9,400	41.60	−71,600	−44,000
900	11,360	43.91	−71,250	−40,600
1,000	13,350	46.01	−70,900	−37,100
1,100	15,390	47.95	−70,400	−33,800
1,200	17,510	49.80	−70,100	−30,600
1,300	19,750	51.59	−69,450	−25,900

Nickel Dibromide, $NiBr_2$ (*c*)

$\Delta H^\circ_{298} = -51,700$ calories per mole (*11*)
$S_{298} = (30)$ *e.u.* (*11*)
S.P. = (1,150°) K. (*6*)
$\Delta H_{subl} = (36,000)$ calories per mole

Formation: $Ni + Br_2 \longrightarrow NiBr_2$
(estimated (*11*))

T, ° K.	H_T-H_{298}	ΔH_T°	ΔF_T°
298	......	−51,700	(−47,000)
500	(4,000)	(−58,600)	(−41,500)
1,000	(14,000)	(−56,900)	(−24,000)
1,500	(72,000)	(−7,500)	(−17,000)

Nickel Diiodide, NiI_2 (*c*)

$\Delta H^\circ_{298} = -23,100$ calories per mole (*11*)
$S_{298} = (34)$ *e.u.* (*11*)
S.P. = (1,020°) K. (*6*)
$\Delta H_{subl} = (32,000)$ calories per mole

Formation: $Ni + I_2 \longrightarrow NiI_2$
(estimated (*11*))

T, ° K.	H_T-H_{298}	ΔH_T°	ΔF_T°
298	......	−23,100	(−23,000)
500	(4,000)	(−37,000)	(−20,500)
1,000	(14,000)	(−35,300)	(−2,000)
1,500	(68,000)	(+10,000)	(+4,000)

Trinickel Carbide, Ni_3C (*c*)

$\Delta H^\circ_{298} = 9,200$ calories per mole (*81*)
$S_{298} = (23.8)$ *e.u.* (*78*)

Formation: $3Ni + C \longrightarrow Ni_3C$
(estimated (*81*))

T, ° K.	ΔF_T°
298	(+8,900)
400	(4,800)
500	(3,700)
600	(2,600)
700	(1,500)

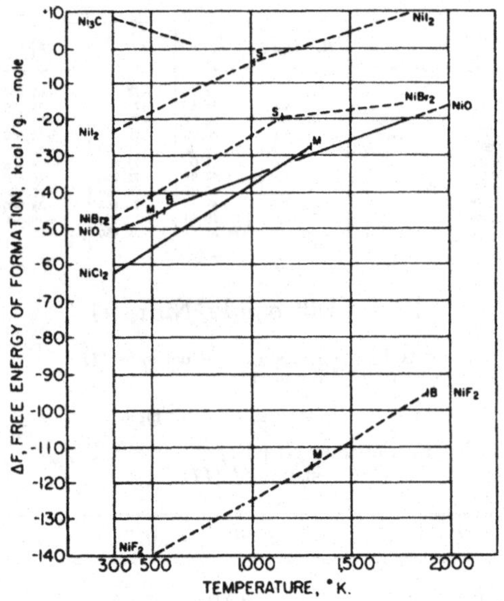

FIGURE 38.—Nickel.

NITROGEN AND ITS COMPOUNDS

Element, N_2 (*g*)

S_{298} = 45.77 *e.u.* (*83*)
M.P. = 63.18° K. (*112*)
ΔH_M = 172 calories per atom
B.P. = 77.36° K. (*112*)
ΔH_V = 1,335 calories per atom

Zone I (*g*) (298°–2,500° K.)

$$C_p = 6.66 + 1.02 \times 10^{-3} T \ (82)$$
$$H_T - H_{298} = -2{,}031 + 6.66T + 0.51 \times 10^{-3} T^2$$
$$F_T - H_{298} = -2{,}031 - 6.66 T \ln T - 0.51 \times 10^{-3} T^2 - 0.87T$$

T, ° K.	$H_T - H_{298}$	S_T	$-\frac{(F_T - H_{298})}{T}$
298	----------	45.77	45.77
400	710	47.82	46.05
500	1,413	49.39	46.48
600	2,126	50.69	47.19
700	2,854	51.81	47.70
800	3,598	52.80	48.25
900	4,358	53.70	48.86
1,000	5,132	54.51	49.38
1,100	5,916	55.25	49.87
1,200	6,723	55.96	50.31
1,300	7,500	56.65	50.72
1,400	8,288	57.26	51.25
1,500	9,186	57.79	51.60
1,600	9,943	58.26	52.08
1,700	10,750	58.76	52.40
1,800	11,620	59.37	52.78
1,900	12,470	59.77	53.14
2,000	13,433	60.23	53.51

Dinitrogen Oxide, N_2O (*g*)

ΔH_{298}° = 19,490 calories per mole (*112*)
S_{298} = 52.8 *e.u.* (*112*)
M.P. = 182.30° K. (*112*)
ΔH_M = 1,563 calories per mole
B.P. = 184.68° K. (*112*)
ΔH_V = 3,956 calories per mole

Zone I (*g*) (298°–2,000° K.)

$$C_p = 10.92 + 2.06 \times 10^{-3} T - 2.04 \times 10^{5} T^{-2} \ (82)$$
$$H_T - H_{298} = -4{,}032 + 10.92T + 1.03 \times 10^{-3} T^2 + 2.04 \times 10^{5} T^{-1}$$

Formation: $N_2 + 1/2O_2 \longrightarrow N_2O$

Zone I (298°–2,000° K.)

$$\Delta C_p = 0.68 + 0.54 \times 10^{-3} T - 1.84 \times 10^{5} T^{-2}$$
$$\Delta H_T = 18{,}650 + 0.68T + 0.27 \times 10^{-3} T^2 + 1.84 \times 10^{5} T^{-1}$$
$$\Delta F_T = 18{,}650 - 0.68 T \ln T - 0.27 \times 10^{-3} T^2 + 0.92 \times 10^{5} T^{-1} + 23.2T$$

T, ° K.	$H_T - H_{298}$	S_T	ΔH_T°	ΔF_T°
298	----------	52.8	+19,500	24,700
400	990	55.65	19,400	26,100
500	2,055	58.02	19,400	28,250
600	3,175	60.07	19,450	30,050
700	4,360	61.89	19,500	31,800
800	5,585	63.53	19,600	33,550
900	6,855	65.02	19,700	35,300
1,000	8,145	66.38	19,800	37,050
1,100	9,318	67.66	19,900	38,700
1,200	10,815	68.88	20,000	40,450
1,300	12,062	69.87	20,100	42,150
1,400	13,555	70.92	20,300	43,900
1,500	14,801	71.82	20,500	45,700
1,600	16,345	72.78	20,600	47,200
1,700	17,629	73.58	20,700	48,800
1,800	19,170	74.45	20,800	50,450
1,900	20,541	75.22	20,900	52,000
2,000	22,030	75.59	21,000	54,500

Nitrogen Oxide, NO (*g*)

ΔH_{298}° = 21,600 calories per mole (*112*)
S_{298} = 50.34 *e.u.* (*83*)
M.P. = 109.5° K. (*112*)
ΔH_M = 550 calories per mole
B.P. = 121.4° K. (*112*)
ΔH_V = 3,293 calories per mole

Zone I (*g*) (298°–2,500° K.)

$$C_p = 7.03 + 0.92 \times 10^{-3} T - 0.14 \times 10^{5} T^{-2} \ (82)$$
$$H_T - H_{298} = -2{,}184 + 7.03T + 0.46 \times 10^{-3} T^2 + 0.14 \times 10^{5} T^{-1}$$

Formation: $1/2N_2 + 1/2O_2 \longrightarrow NO$

Zone I (298°–2,500° K.)

$$\Delta C_p = 0.12 - 0.08 \times 10^{-3} T + 0.06 \times 10^{5} T^{-2}$$
$$\Delta H_T = 21{,}590 + 0.12T - 0.04 \times 10^{-3} T^2 - 0.06 \times 10^{5} T^{-1}$$
$$\Delta F_T = 21{,}590 - 0.12 T \ln T + 0.04 \times 10^{-3} T^2 - 0.03 \times 10^{5} T^{-1} - 2.2T$$

T, ° K.	$H_T - H_{298}$	S_T	ΔH°_T	ΔF°_T
298		50.34	+21,600	20,700
400	727	52.44	21,600	20,400
500	1,450	54.05	21,600	20,100
600	2,189	55.40	21,600	19,800
700	2,946	56.57	21,600	19,500
800	3,718	57.60	21,600	19,200
900	4,532	58.55	21,600	18,850
1,000	5,318	59.38	21,600	18,650
1,100	6,119	60.13	21,600	18,300
1,200	6,926	60.84	21,650	18,050
1,300	7,743	61.5	21,650	17,800
1,400	8,570	62.11	21,650	17,500
1,500	9,508	62.77	21,650	17,100
1,600	10,251	63.24	21,700	16,900
1,700	11,104	63.75	21,700	16,550
1,800	11,968	64.26	21,700	16,250
1,900	12,841	64.73	21,700	15,950
2,000	13,844	65.26	21,700	15,650

Dinitrogen Trioxide, N_2O_3 (g)

$\Delta H^\circ_{298} = 17{,}500$ calories per mole (*24*)
$S_{298} = (63.9)$ *e.u.* (*24*)
M.P. $= 162°$ K. (*112*)
B.P. $= 275°$ K. (*112*)
$\Delta H_V = 9{,}400$ calories per mole

Formation: $N_2 + 3/2O_2 \longrightarrow N_2O_3$ (estimated (*24*))

T, ° K.	$H_T - H_{298}$	ΔH°_T	ΔF°_T
298		17,500	(33,500)
400	(1,800)	(17,500)	(39,000)
500	(3,600)	(17,500)	(44,000)

Nitrogen Dioxide, NO_2 (g)

$\Delta H^\circ_{298} = 8{,}091$ calories per mole (*112*)
$S_{298} = 57.46$ *e.u.* (*83*)

Zone I (g) (298°–2,000° K.)

$$C_p = 10.26 + 2.04 \times 10^{-3} T - 1.61 \times 10^5 T^{-2} \quad (82)$$
$$H_T - H_{298} = -3{,}690 + 10.26T + 1.02 \times 10^{-3} T^2 + 1.61 \times 10^5 T^{-1}$$

Formation: $1/2N_2 + O_2 \longrightarrow NO_2$

Zone I (298°–2,000° K.)

$$\Delta C_p = -0.23 - 0.54 \times 10^{-3} T - 1.21 \times 10^5 T^{-2}$$
$$\Delta H_T = 7{,}780 - 0.23T - 0.27 \times 10^{-3} T^2 + 1.21 \times 10^5 T^{-1}$$
$$\Delta F_T = 7{,}780 + 0.23T \ln T + 0.27 \times 10^{-3} T^2 + 0.60 \times 10^5 T^{-1} + 13.41T$$

T, ° K.	$H_T - H_{298}$	S_T	ΔH°_T	ΔF°_T
298		57.46	8,100	12,400
400	960	60.23	8,000	13,900
500	1,975	62.49	7,900	15,350
600	3,055	64.46	7,800	16,750
700	4,190	66.2	7,750	18,250
800	5,370	67.78	7,750	19,750
900	6,585	69.21	7,900	21,350
1,000	7,830	70.52	7,900	22,850
1,100	8,976	71.65	7,900	24,400
1,200	10,375	72.84	8,050	25,900
1,300	11,500	73.77	8,000	27,400
1,400	12,975	74.84	8,000	28,700
1,500	14,092	75.64	7,900	30,300
1,600	15.625	76.61	8,200	31,750
1,700	16,795	77.83	8,200	33,300
1,800	18,300	78.19	8,300	34,600
1,900	19,571	78.88	8.300	36,150
2,000	20,990	79.60	8,200	37,700

Dinitrogen Tetraoxide, N_2O_4 (g)

$\Delta H^\circ_{298} = 2{,}309$ calories per mole (*112*)
$S_{298} = 72.73$ *e.u.* (*112*)
M.P. $= 261.96°$ K. (*112*)
$\Delta H_M = 3{,}502$ calories per mole
B.P. $= 294°$ K. (*112*)
$\Delta H_V = 9{,}101$ calories per mole

Zone I (g) (298°–1,000° K.)

$$C_p = 20.05 + 9.50 \times 10^{-3} T - 3.56 \times 10^5 T^{-2} \quad (82)$$
$$H_T - H_{298} = -7{,}594 + 20.05T + 4.75 \times 10^{-3} T^2 + 3.56 \times 10^5 T^{-1}$$

Formation: $N_2 + 2O_2 \longrightarrow N_2O_4$

Zone I (298°–1,000° K.)

$$\Delta C_p = -0.93 + 6.48 \times 10^{-3} T - 2.76 \times 10^5 T^{-2}$$
$$\Delta H_T = 1{,}372 - 0.93T + 3.24 \times 10^{-3} T^2 + 2.76 \times 10^5 T^{-1}$$
$$\Delta F_T = 1{,}372 + 0.93T \ln T - 3.24 \times 10^{-3} T^2 + 1.38 \times 10^5 T^{-1} + 68.31T$$

T, ° K.	$H_T - H_{298}$	S_T	ΔH°_T	ΔF°_T
298		72.73	2,300	+23,500
400	2,060	78.66	2,200	30,750
500	4,310	83.67	2,300	37,600
600	6,740	88.09	2,500	44,000
700	9,300	92.03	2,850	52,100
800	11,980	95.61	3,100	59,050
900	14,730	98.85	3,450	66,000
1,000	17,560	101.83	3,900	73,000

Nitrosyl Chloride, NOCl (g)

$\Delta H^\circ_{298} = 12{,}570$ calories per mole (*112*)
$S_{298} = 63$ *e.u.* (*112*)
M.P. $= 211.7°$ K. (*112*)
B.P. $= 267.4°$ K. (*112*)
$\Delta H_V = 6{,}000$ calories per mole

Zone I (g) (298°–2,000° K.)

$$C_p = 10.73 + 1.84 \times 10^{-3} T - 1.66 \times 10^5 T^{-2} \quad (82)$$
$$H_T - H_{298} = -3{,}838 + 10.73T + 0.92 \times 10^{-3} T^2 + 1.66 \times 10^5 T^{-1}$$

Formation: $1/2N_2 + 1/2O_2 + 1/2Cl_2 \longrightarrow NOCl$

Zone I (298°–2,000° K.)

$$\Delta C_p = -0.59 + 0.81 \times 10^{-3} T - 1.12 \times 10^5 T^{-2}$$
$$\Delta H_T = 12{,}335 - 0.59T + 0.40 \times 10^{-3} T^2 + 1.12 \times 10^5 T^{-1}$$
$$\Delta F_T = 12{,}335 + 0.59T \ln T - 0.40 \times 10^{-3} T^2 + 0.56 \times 10^5 T^{-1} + 7.96T$$

T, ° K.	$H_T - H_{298}$	ΔH°_T	ΔF°_T
298		12,550	15,850
400	1,000	12,450	17,050
500	2,055	12,350	18,200
600	3,175	12,300	19,350
700	4,340	12,300	20,500
800	5,540	12,250	21,750
900	6,770	12,250	22,900
1,000	8,030	12,300	24,100
1,100	9,229	12,350	25,300
1,200	10,600	12,350	26,350
1,300	11,794	12,350	27,700
1,400	13,215	12,400	28,900
1,500	14,434	12,400	30,000
1,600	15,875	12,550	31,100
1,700	17,160	12,600	32,300
1,800	18,555	12,600	33,350
1,900	19,957	12,700	34,600
2,000	21,255	12,600	35,600

Ammonia, NH_3 (*g*)

$\Delta H^\circ_{298} = -11{,}040$ calories per mole (*112*)
$S_{298} = 45.96$ *e.u.* (*83*)
$M.P. = 195.40°$ K. (*112*)
$\Delta H_M = 1{,}350$ calories per mole
$B.P. = 239.73°$ K. (*112*)
$\Delta H_V = 5{,}580$ calories per mole

Zone I (*g*) (298°–2,000° K.)

$C_p = 7.11 + 6.00 \times 10^{-3}T - 0.37 \times 10^5 T^{-2}$ (*83*)
$H_T - H_{298} = -2{,}510 + 7.11T + 3.00 \times 10^{-3}T^2 + 0.37 \times 10^5 T^{-1}$

Formation: $1/2N_2 + 3/2H_2 \longrightarrow NH_3$

Zone I (298°–2,000° K.)

$\Delta C_p = -6.0 + 4.32 \times 10^{-3}T - 0.55 \times 10^5 T^{-2}$
$\Delta H_T = -9{,}630 - 6.0T + 2.16 \times 10^{-3}T^2 + 0.55 \times 10^5 T^{-1}$
$\Delta F_T = -9{,}630 + 6.0TlnT - 2.16 \times 10^{-3}T^2 + 0.27 \times 10^5 T^{-1} - 14.93T$

T, ° K.	$H_T - H_{298}$	S_T	ΔH°_T	ΔF°_T
298	----------	45.96	−11,050	−4,000
400	895	48.54	−11,650	−1,600
500	1,845	50.66	−12,000	+1,200
600	2,885	52.55	−12,400	+3,700
700	3,975	54.23	−12,800	+6,350
800	5,145	55.79	−12,950	9,200
900	6,380	57.28	−13,150	12,000
1,000	7,680	58.61	−13,350	14,800
1,100	8,975	59.88	−13,500	16,400
1,200	10,450	61.14	−13,600	20,300
1,300	11,830	62.28	−13,700	23,200
1,400	13,420	63.42	−13,700	26,000
1,500	14,930	64.47	−13,700	28,900
1,600	16,565	65.52	−13,800	31,500
1,700	18,269	66.58	−13,550	34,400
1,800	19,820	67.44	−13,400	37,400
1.900	21,848	68.58	−13,200	40,050
2,000	23,195	69.21	−13,500	43,150

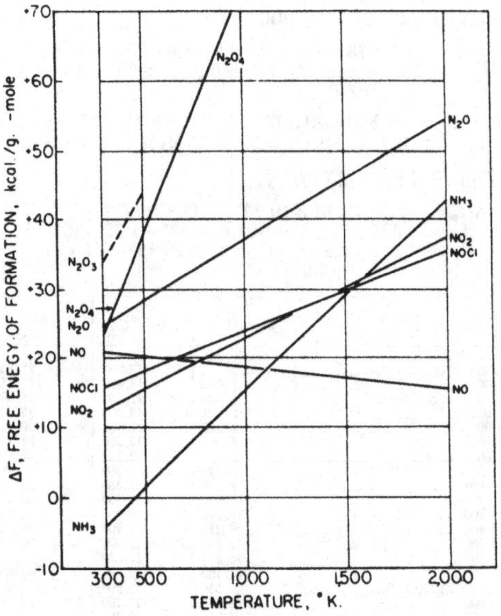

FIGURE 39.—Nitrogen.

OXYGEN

Element, O_2 (*g*)

$S_{298} = 49.01$ *e.u.* (*83*)
$M.P. = 54.36°$ K. (*112*)
$\Delta H_M = 106$ calories per atom
$B.P. = 90.19°$ K. (*112*)
$\Delta H_V = 1{,}630$ calories per atom

Zone I (*g*) (298°–3,000° K.)

$C_p = 7.16 + 1.00 \times 10^{-3}T - 0.40 \times 10^5 T^{-2}$ (*82*)
$H_T - H_{298} = -2{,}313 + 7.16T + 0.50 \times 10^{-3}T^2 + 0.40 \times 10^5 T^{-1}$
$F_T - H_{298} = -2{,}313 - 7.16TlnT - 0.50 \times 10^{-3}T^2 + 0.20 \times 10^5 T^{-1} - 0.55T$

T, ° K.	$H_T - H_{298}$	S_T	$-\frac{(F_T - H_{298})}{T}$
298	----------	49.01	49.01
400	723	51.1	49.29
500	1,455	52.73	49.82
600	2,210	54.10	50.42
700	2,989	55.30	51.03
800	3,786	56.37	51.64
900	4,602	57.33	52.21
1,000	5,430	58.20	52.77
1,100	6,208	59.01	53.27
1,200	7,040	59.72	53.83
1,300	7,873	60.41	54.35
1,400	8,716	61.01	54.78
1,500	9,711	61.67	55.21
1,600	10,442	62.21	55.68
1,700	11,334	62.71	56.04
1,800	12,219	63.11	56.32
1,900	13,113	63.66	56.77
2,000	14,155	64.22	57.14

PHOSPHORUS AND ITS COMPOUNDS

Element (White), P_4 (*c*)

$S_{298} = 42.4$ *e.u.* (*83*)
$M.P. = 317.4°$ K. (*82*)
$\Delta H_M = 601$ calories per atom
$B.P. = 553°$ K. (*112*)
$\Delta H_V = 11{,}880$ calories per atom

Zone I (*c*) (298°–317.4° K.)

$C_p = 22.50$ (*82*)
$H_T - H_{298} = -6{,}709 + 22.50T$
$F_T - H_{298} = -6{,}709 - 22.50TlnT + 108.28T$

Zone II (*l*) (317.4°–553° K.)

$C_p = 23.50$ (*82*)
$H_T - H_{298} = -6{,}435 + 23.50T$
$F_T - H_{298} = -6{,}435 - 23.50TlnT + 113.15T$

Zone III (*g*) (553°–1,500° K.)

$C_p = 18.93 + 0.86 \times 10^{-3}T - 2.81 \times 10^5 T^{-2}$ (*82*)
$H_T - H_{298} = 7{,}343 + 18.93T + 0.43 \times 10^{-3}T^2 + 2.81 \times 10^5 T^{-1}$
$F_T - H_{298} = 7{,}343 - 18.93TlnT - 0.43 \times 10^{-3}T^2 + 1.40 \times 10^5 T^{-1} + 59.17T$

T, ° K.	$H_T - H_{298}$	S_T	$-\frac{(F_T - H_{298})}{T}$
298	----------	42.4	42.4
400	2,975	51.14	43.7
500	5,325	56.40	45.75
600	19,324	81.76	49.55
700	21,206	84.66	54.36
800	23,110	87.21	58.32
900	25,040	89.48	61.66
1,000	26,980	91.52	64.5
1,100	28,940	93.39	67.08
1,200	30,900	95.10	69.35
1,300	32,860	96.66	71.39
1,400	34,820	98.12	73.25
1,500	36,790	99.48	74.95

Element (Red), P_4 (*c*)

$S_{298}=21.84$ *e.u.* (*130*)
S.P. $=870°$ K. (*112*)
$\Delta H_{subl}=7,350$ calories per atom

Zone I (*c*) (298°–870° K.)

$C_p=18.96+15.60\times10^{-3}T$ (*82*)
$H_T-H_{298}=-6,348+18.96T+7.80\times10^{-3}T^2$
$F_T-H_{298}=-6,348-18.96T\ln T-7.80\times10^{-3}T^2+109.78T$

Zone II (*g*) (870°–1,500° K.)

$C_p=18.93+0.86\times10^{-3}T-2.81\times10^5T^{-2}$ (*82*)
$H_T-H_{298}=6,293+18.93T+0.43\times10^{-3}T^2+2.81\times10^5T^{-1}$
$F_T-H_{298}=6,293-18.93T\ln T-0.43\times10^{-3}T^2+1.40\times10^5T^{-1}+89.47T$

T, ° K.	H_T-H_{298}	S_T	$-\frac{(F_T-H_{298})}{T}$
298		21.84	21.84
400	2,480	29.00	22.80
500	5,080	34.80	24.64
600	7,880	39.88	26.75
700	10,760	44.32	28.95
800	13,760	48.32	31.12
900	23,990	59.19	32.53
1,000	25,934	61.23	35.29
1,100	27,890	63.10	37.75
1,200	29,850	64.81	39.94
1,300	31,810	66.37	41.90
1,400	33,770	67.83	43.71
1,500	35,740	69.19	45.36

Phosphorus Oxide, PO (*g*)

$\Delta H^\circ_{298}=-9,500$ calories per mole (*112*)
$S_{298}=(53.6)$ *e.u.* (*24*)

Formation: $1/4P_4+1/2O_2\longrightarrow PO$
(estimated (*24*))

T, ° K.	H_T-H_{298}	ΔH°_T	ΔF°_T
298		−9,500	(−15,000)
400	(600)	(−10,000)	(−17,000)
500	(1,550)	(−10,000)	(−18,500)
600	(2,450)	(−13,000)	(−20,000)
700	(3,000)	(−13,500)	(−21,000)
800	(3,650)	(−13,500)	(−22,000)
900	(4,550)	(−13,500)	(−23,500)
1,000	(5,450)	(−13,500)	(−24,500)
1,100	(6,350)	(−13,500)	(−25,500)
1,200	(7,000)	(−14,000)	(−26,500)
1,300	(7,650)	(−14,000)	(−27,500)
1,400	(8,550)	(−14,000)	(−28,500)
1,500	(9,550)	(−14,000)	(−29,500)

Tetraphosphorus Decaoxide, P_4O_{10} (*c*)

$\Delta H_{298}=-720,000$ calories per mole (*112*)
$S_{298}=(67.4)$ *e.u.* (*24*)
S.P. $=631°$ K. (*82*)
$\Delta H_{subl}=17,600$ calories per mole

Zone I (*c*) (298°–631° K.)

$C_p=16.75+108.0\times10^{-3}T$ (*82*)
$H_T-H_{298}=-9,795+16.75T+54.0\times10^{-3}T^2$

Zone II (*g*) (631°–1,400° K.)

$C_p=73.60$ (*82*)
$H_T-H_{298}=-6,570+73.60T$

Formation: $P_4+5O_2\longrightarrow P_4O_{10}$

Zone I (298°–317.4° *K.*)

$\Delta C_p=-41.55+103.00\times10^{-3}T+2.0\times10^5T^{-2}$
$\Delta H_T=-711,525-41.55T+51.5\times10^{-3}T^2-2.0\times10^5T^{-1}$
$\Delta F_T=-711,525+41.55T\ln T-51.5\times10^{-3}T^2-1.0\times10^5T^{-1}-28.68T$

Zone II (317.4°–553° *K.*)

$\Delta C_p=-42.55+103.0\times10^{-3}T+2.0\times10^5T^{-2}$
$\Delta H_T=-711,720-42.55T+51.5\times10^{-3}T^2-2.0\times10^5T^{-1}$
$\Delta F_T=-711,720+42.55T\ln T-51.5\times10^{-3}T^2-1.0\times10^5T^{-1}-33.72T$

Zone III (553°–631° K.)

$\Delta C_p=-38.0+102.14\times10^{-3}T+4.81\times10^5T^{-2}$
$\Delta H_T=-725,450-38.0T+51.07\times10^{-3}T^2-4.81\times10^5T^{-1}$
$\Delta F_T=-725,450+38.0T\ln T-51.07\times10^{-3}T^2-2.40\times10^5T^{-1}+1.05T$

Zone IV (631°–1,400° K.)

$\Delta C_p=18.87-5.86\times10^{-3}T+4.81\times10^5T^{-2}$
$\Delta H_T=-722,540+18.87T-2.93\times10^{-3}T^2-4.81\times10^5T^{-1}$
$\Delta F_T=-722,540-18.87T\ln T+2.93\times10^{-3}T^2-2.40\times10^5T^{-1}+347.93T$

(estimated (*24*))

T, ° K.	H_T-H_{298}	S_T	ΔH°_T	ΔF°_T
298		(67.4)	−720,000	(−654,400)
400	5,550	(83.37)	−721,100	(−631,800)
500	12,080	(97.87)	−720,500	(−609,400)
600	19,700	(110.74)	−730,700	(−597,800)
700	44,950	(151.45)	−711,200	(−564,400)
800	52,300	(161.26)	−709,700	(−543,700)
900	59,650	(169.91)	−708.400	(−522,800)
1,000	67,050	(177.71)	−707,100	(−502,300)
1,100	74,400	(184.72)	−705,600	(−481,500)
1,200	81,750	(191.12)	−704,300	(−461,300)
1,300	89,100	(196.99)	−703,100	(−440,900)
1,400	96,450	(202.44)	−702,000	(−420,900)

Phosphorus Trifluoride, PF_3 (*g*)

$\Delta H^\circ_{298}=(-170,000)$ calories per mole (*42*)
$S_{298}=64.1$ *e.u.* (*83*)
M.P. $=122°$ K. (*112*)
B.P. $=172°$ K. (*112*)
$\Delta H_V=3,700$ calories per mole

Zone I (*g*) (298°–2,000° K.)

$C_p=17.18+1.92\times10^{-3}T-3.88\times10^5T^{-2}$ (*82*)
$H_T-H_{298}=-6,509+17.18T+0.96\times10^{-3}T^2+3.88\times10^5T^{-1}$

Formation: $1/4P_4+3/2F_2\longrightarrow PF_3$

Zone I (298°–317.4° K.)

$\Delta C_p=-0.88+1.26\times10^{-3}T-2.68\times10^5T^{-2}$
$\Delta H_T=-170,700-0.88T+0.63\times10^{-3}T^2+2.68\times10^5T^{-1}$
$\Delta F_T=-170,700+0.88T\ln T-0.63\times10^{-3}T^2+1.34\times10^5T^{-1}+15.33T$

Zone II (317.4°–553° K.)

$\Delta C_p=-1.13+1.26\times10^{-3}T-2.68\times10^5T^{-2}$
$\Delta H_T=-170,770-1.13T+0.63\times10^{-3}T^2+2.68\times10^5T^{-1}$
$\Delta F_T=-170,770+1.13T\ln T-0.63\times10^{-3}T^2+1.34\times10^5T^{-1}+14.13T$

Zone III (553°--1,500° K.)

$\Delta C_p = 0.01 + 1.05 \times 10^{-3} T - 1.98 \times 10^5 T^{-2}$
$\Delta H_T = -174,100 + 0.01 T + 0.525 \times 10^{-3} T^2 + 1.98 \times 10^5 T^{-1}$
$\Delta F_T = -174,100 - 0.01 T \ln T - 0.52 \times 10^{-3} T^2 + 0.99 \times 10^5 T^{-1} + 27.5 T$

T, ° K.	$H_T - H_{298}$	S_T	ΔH_T°	ΔF_T°
298	....	64.1	(−170,000)	(−164,250)
400	1,470	68.33	(−170,500)	(−162,200)
500	3,075	71.91	(−170,600)	(−160,000)
600	4,790	75.03	(−173,600)	(−157,600)
700	6,565	77.77	(−173,600)	(−155,000)
800	8,390	80.2	(−173,500)	(−152,400)
900	10,245	82.38	(−173,500)	(−149,800)
1,000	12,125	84.37	(−173,400)	(−147,100)
1,100	13,804	86.06	(−173,500)	(−144,600)
1,200	15,940	87.84	(−173,200)	(−141,900)
1,300	17,745	89.28	(−173,200)	(−139,200)
1,400	19,795	90.81	(−173,000)	(−136,700)
1,500	21,680	92.11	(−172,900)	(−134,100)

Phosphorus Trichloride, PCl_3 (*l*)

$\Delta H_{298}^\circ = -76,900$ calories per mole (*11*)
$S_{298} = 52.2$ *e.u.* (*11*)
$\Delta F_{298}^\circ = -74,500$ calories per mole
M.P. = 182° K. (*6*)
B.P. = 348° K. (*6*)
$\Delta H_V = 7,278$ calories per atom

Phosphorus Pentachloride, PCl_5 (*c*)

$\Delta H_{298}^\circ = -106,500$ calories per mole (*11*)
$S_{298} = 40.8$ *e.u.* (*11*)
$\Delta F_{298}^\circ = -75,800$ calories per mole
S.P. = 439° K. (*6*)
$\Delta H_{subl} = 14,000$ calories per mole

Phosphoryl Chloride, $POCl_3$ (*l*)

$\Delta H_{298}^\circ = -151,000$ calories per mole (*112*)
M.P. = 274.3° K. (*112*)
B.P. = 378.5° K. (*112*)
$\Delta H_V = 8,211$ calories per mole

Phosphorus Tribromide, PBr_3 (*l*)

$\Delta H_{298}^\circ = (-47,500)$ calories per mole (*112*)
$S_{298} = (59)$ *e.u.* (*11*)
M.P. = 233° K. (*6*)
B.P. = 447° K. (*6*)
$\Delta H_V = 9,500$ calories per mole

Formation: $1/4 P_4 + 3/2 Br_2 \longrightarrow PBr_3$
(estimated (*11*))

T, ° K.	$H_T - H_{298}$	ΔH_T°	ΔF_T°
298	....	(−47,500)	(−45,500)
500	(15,500)	(−44,500)	(−40,000)

Phosphorus Pentabromide, PBr_5 (*c*)

$\Delta H_{298}^\circ = -66,000$ calories per mole (*112*)
$S_{298} = (53)$ *e.u.* (*11*)
$\Delta F_{298} = (-41,500)$ calories per mole
S.P. = 379° K. (*6*)
$H_{subl} = 13,000$ calories per mole

Phosphoryl Bromide, $POBr_3$ (*c*)

$\Delta H_{298}^\circ = -114,600$ calories per mole (*112*)
M.P. = 328° K. (*112*)
B.P. = 464.9° K. (*112*)
$\Delta H_V = 9,080$ calories per mole

Phosphorus Triiodide, PI_3 (*c*)

$\Delta H_{298}^\circ = -10,900$ calories per mole (*11*)
$S_{298} = (57)$ *e.u.* (*11*)
M.P. = 334° K. (*6*)
B.P. = (500°) K. (*6*)
$\Delta H_V = 10,500$ calories per mole

Formation: $1/4 P_4 + 3/2 I_2 \longrightarrow PI_3$
(estimated (*11*))

T, ° K.	$H_T - H_{298}$	ΔH_T°	ΔF_T°
298	....	−10,900	(−12,000)
500	(7,000)	(−19,300)	(−9,600)

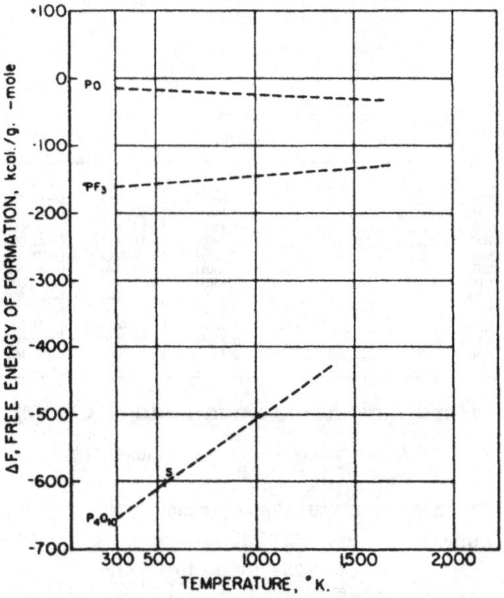

FIGURE 40.—Phosphorus.

PLATINUM AND ITS COMPOUNDS

Element, Pt (c)

$S_{298} = 10.00$ *e.u.* (*83*)
M.P. $= 2{,}042.5°$ K. (*112*)
$\Delta H_M = 5{,}200$ calories per atom
B.P. $= 4{,}100°$ K. (*112*)
$\Delta H_V = (122{,}000)$ calories per atom

Zone I (c) (298°–1, 900° K.)

$$C_p = 5.74 + 1.34 \times 10^{-3}T + 0.10 \times 10^5 T^{-2} \ (82)$$
$$H_T - H_{298} = -1{,}737 + 5.74T + 0.67 \times 10^{-3}T^2 - 0.10 \times 10^5 T^{-1}$$
$$F_T - H_{298} = -1{,}737 - 5.74T\ln T - 0.67 \times 10^{-3}T^2 - 0.05 \times 10^5 T^{-1} + 28.79T$$

T, ° K.	$H_T - H_{298}$	S_T	$-\frac{(F_T - H_{298})}{T}$
298	----------	10.0	10.0
400	645	11.86	10.25
500	1,280	13.28	10.72
600	1,920	14.44	11.24
700	2,580	15.46	11.77
800	3,260	16.37	12.29
900	3,950	17.18	12.79
1,000	4,660	17.93	13.27
1,100	5,380	18.61	13.71
1,200	6,110	19.25	14.16
1,300	6,850	19.84	14.57
1,400	7,600	20.39	14.96
1,500	8,370	20.93	15.35
1,600	9,150	21.43	15.71
1,700	9,940	21.93	16.08
1,800	10,740	22.37	16.40
1,900	11,550	22.81	16.73
2,000	(12,370)	(23.52)	(19.34)

Platinum Chloride, PtCl (c)

$\Delta H^\circ_{298} = -13{,}000$ calories per mole (*11*)
$S_{298} = (22)$ *e.u.* (*11*)
Decomposes $= 856°$ K., 1 atm Cl_2 (*6*)

Formation: $Pt + 1/2Cl_2 \longrightarrow PtCl$
(estimated (*11*))

T, ° K.	$H_T - H_{298}$	ΔH°_T	ΔF°_T
298	----------	−13,000	(−8,500)
500	(2,000)	(−13,100)	(−6,000)
1,000	(11,000)	(−9,700)	(+3,000)

Platinum Dichloride, $PtCl_2$ (c)

$\Delta H^\circ_{298} = -29{,}000$ calories per mole (*11*)
$S_{298} = (31)$ *e.u.* (*11*)
Decomposes $= 854°$ K., 1 atm Cl_2 (*6*)

Formation: $Pt + Cl_2 \longrightarrow PtCl_2$
(estimated (*11*))

T, ° K.	$H_T - H_{298}$	ΔH°_T	ΔF°_T
298	----------	−29,000	(−19,500)
500	(4,000)	(−28,000)	(−12,500)
1,000	(15,000)	(−24,700)	(+4,000)

Platinum Trichloride, $PtCl_3$ (c)

$\Delta H^\circ_{298} = -43{,}000$ calories per mole (*11*)
$S_{298} = (35)$ *e.u.* (*11*)
Decomposes $= 708°$ K., 1 atm Cl_2 (*6*)

Formation: $Pt + 2/3Cl_2 \longrightarrow PtCl_3$
(estimated (*11*))

T, ° K.	$H_T - H_{298}$	ΔH°_T	ΔH°_T
298	----------	−43,000	(−26,600)
500	(5,000)	(−41,800)	(−15,500)
1,000	(20,000)	(−36,800)	(+9,000)

Platinum Tetrachloride, $PtCl_4$ (c)

$\Delta H^\circ_{298} = -53{,}000$ calories per mole (*11*)
$S_{298} = (50)$ *e.u.* (*11*)
M.P. $= >600°$ K. (*6*)

Formation: $Pt + 2Cl_2 \longrightarrow PtCl_4$
(estimated (*11*))

T, ° K.	$H_T - H_{298}$	ΔH°_T	ΔF°_T
298	----------	−53,000	(−33,000)
500	(6,000)	(−51,700)	(−20,500)
1,000	(24,000)	(−45,000)	(+10,000)

Platinum Bromide, PtBr (c)

$\Delta H^\circ_{298} = -6{,}350$ calories per mole (*11*)
$S_{298} = (25)$ *e.u.* (*11*)
Disproportionates (*6*)

Formation: $Pt + 1/2Br_2 \longrightarrow PtBr$
(estimated (*11*))

T, ° K.	$H_T - H_{298}$	ΔH°_T	ΔF°_T
298	----------	−6,350	(−5,400)
500	(2,000)	(−10,400)	(−3,000)

Platinum Dibromide, $PtBr_2$ (c)

$\Delta H^\circ_{298} = -15{,}650$ calories per mole (*11*)
$S_{298} = (36)$ *e.u.* (*11*)
Decomposes $= 683°$ K., 1 atm Br_2 (*6*)

Formation: $Pt + Br_2 \longrightarrow PtBr_2$
(estimated (*11*))

T, ° K.	$H_T - H_{298}$	ΔH°_T	ΔF°_T
298	----------	−15,650	(−12,400)
500	(4,000)	(−22,450)	(−7,000)

Platinum Tribromide, $PtBr_3$ (*c*)

$\Delta H^{\circ}_{298} = -24{,}000$ calories per mole (*11*)
$S_{298} = (47)$ *e.u.* (*11*)
Decomposes = 678° K., 1 atm Br_2 (*6*)

Formation: $Pt + 3/2Br_2 \longrightarrow PtBr_3$
(estimated (*11*))

T, ° K.	$H_T - H_{298}$	ΔH°_T	ΔF°_T
298		−24, 000	(−18, 700)
500	(5, 000)	(−34, 500)	(−9, 500)

Platinum Tetrabromide, $PtBr_4$ (*c*)

$\Delta H^{\circ}_{298} = -32{,}300$ calories per mole (*11*)
$S_{298} = (60)$ *e.u.* (*11*)
Decomposes = 600° K., 1 atm Br_2 (*6*)

Formation: $Pt + 2Br_2 \longrightarrow PtBr_4$
(estimated (*11*))

T, ° K.	$H_T - H_{298}$	ΔH°_T	ΔF°_T
298		−32, 300	(−25, 500)
500	(6, 000)	(−46, 600)	(−14, 000)

Platinum Iodide, PtI (*c*)

$\Delta H^{\circ}_{298} = 440$ calories per mole (*11*)
$S_{298} = (26)$ *e.u.* (*11*)
Disproportionates (*6*)

Formation: $Pt + 1/2I_2 \longrightarrow PtI$
(estimated (*11*))

T, ° K.	$H_T - H_{298}$	ΔH°_T	ΔF°_T
298		+440	(−200)
500	(2, 000)	(−7, 200)	(0)

Platinum Diiodide, PtI_2 (*c*)

$\Delta H^{\circ}_{298} = -4{,}100$ calories per mole (*11*)
$S_{298} = (38)$ *e.u.* (*11*)
Decomposes = 600° K., 1 atm I_2 (*6*)

Formation: $Pt + I_2 \longrightarrow PtI_2$
(estimated (*11*))

T, ° K.	$H_T - H_{298}$	ΔH°_T	ΔF°_T
298		−4, 100	(−4, 300)
500	(4, 000)	(−18, 000)	(−1, 500)

Platinum Triiodide, PtI_3 (*c*)

$\Delta H^{\circ}_{298} = -8{,}700$ calories per mole (*11*)
$S_{298} = (50)$ *e.u.* (*11*)
Decomposes = 550° K., 1 atm I_2 (*6*)

Formation: $Pt + 3/2I_2 \longrightarrow PtI_3$
(estimated (*11*))

T, ° K.	$H_T - H_{298}$	$H\Delta^{\circ}_T$	ΔF°_T
298		−8, 700	(−8, 200)
500	(5, 000)	(−30, 000)	(−4, 000)

Platinum Tetraiodide, PtI_4 (*c*)

$\Delta H^{\circ}_{298} = -11{,}250$ calories per mole (*11*)
$S_{298} = (64)$ *e.u.* (*11*)
Decomposes = 550° K., 1 atm (*6*)

Formation: $Pt + 2I_2 \longrightarrow PtI_4$
(estimated (*11*))

T, ° K.	$H_T - H_{298}$	ΔH°_T	ΔF°_T
298		−11, 250	(−13, 900)
500		(−39, 900)	(−5, 500)

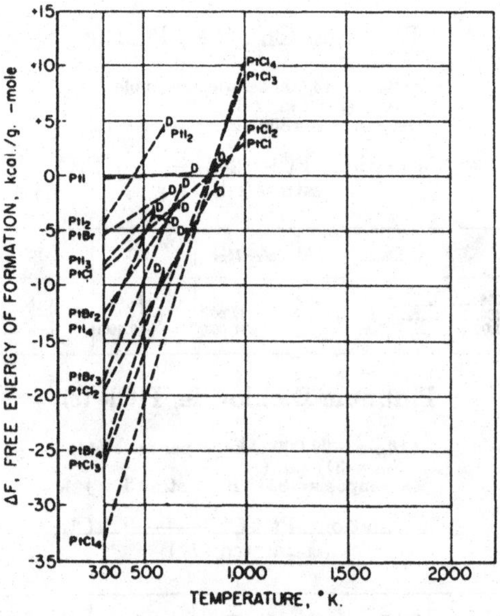

Figure 41.—Platinum.

POTASSIUM AND ITS COMPOUNDS

Element, K (*c*)

$S_{298}=15.3$ *e.u.* (*83*)
$M.P.=336.7°$ K. (*32*)
$\Delta H_M=554$ calories per atom
$B.P.=1{,}030°$ K. (*130*)
$\Delta H_V=18{,}530$ calories per atom

Zone I (*c*) (298°–336.7° K.)

$C_p=6.04+3.12\times10^{-3}T$ *34*
$H_T-H_{298}=-1{,}940+6.04T+1.56\times10^{-3}T^2$
$F_T-H_{298}=-1{,}940-6.04T\ln T-1.56\times10^{-3}T^2+26.08T$

Zone II (*l*) (336.7°–1,030° K.)

$C_p=6.03+0.992\times10^{-3}T+1.96\times10^5T^{-2}$ (*34*)
$H_T-H_{298}=-676+6.03T+0.49\times10^{-3}T^2--1.96\times10^5T^{-1}$
$F_T-H_{298}=-676-6.03T\ln T-0.496\times10^{-3}T^2-0.98\times10^5T^{-1}+22.79T$

Zone III (*g*) (1,030°–2,500° K.)

$C_p=4.90+0.054\times10^{-3}T+0.033\times10^5T^{-2}$ (*34*)
$H_T-H_{298}=+20{,}016+4.90T+0.027\times10^{-3}T^2-0.033\times10^5T^{-1}$
$F_T-H_{298}=+20{,}016-4.90T\ln T-0.027\times10^{-3}T^2-0.016\times10^5T^{-1}-5.47T$

T, ° K.	H_T-H_{298}	S_T	$-\frac{(F_T-H_{298})}{T}$
298	....	15.3	15.3
400	1,324	19.17	15.86
500	2,067	20.83	16.70
600	2,793	22.15	17.50
700	3,509	23.26	18.24
800	4,220	24.21	18.94
900	4,934	25.05	19.57
1,000	5,654	25.81	20.16
1,100	25,404	44.69	21.60
1,200	25,900	45.12	23.54
1,300	26,397	45.52	25.22
1,400	26,894	45.89	26.68
1,500	27,391	46.23	27.97
1,600	27,889	46.56	29.13
1,700	28,386	46.86	30.16
1,800	28,885	47.14	31.09
1,900	29,384	47.41	31.94
2,000	29,884	47.67	32.73

Dipotassium Oxide, K_2O (*c*)

$\Delta H^\circ_{298}=-86{,}400$ calories per mole (*112*)
$S_{298}=(20.3)$ *e.u.* (*24*)

Formation: $2K+1/2O_2 \longrightarrow K_2O$
(estimated (*24*))

T, ° K.	H_T-H_{298}	ΔH°_T	ΔF°_T
298	....	−86,400	(−76,300)
400	(1,800)	(−87,600)	(−72,600)
500	(3,650)	(−87,600)	(−68,900)
600	(5,400)	(−87,700)	(−65,100)
700	(7,200)	(−87,700)	(−61,400)
800	(8,950)	(−87,800)	(−57,600)
900	(10,700)	(−87,800)	(−53,800)
1,000	(12,500)	(−87,900)	(−50,000)
1,100	(14,900)	(−125,400)	(−44,400)
1,200	(17,000)	(−124,700)	(−37,100)
1,300	(19,100)	(−124,000)	(−29,800)
1,400	(21,250)	(−123,300)	(−22,600)
1,500	(23,400)	(−122,600)	(−15,400)

Dipotassium Dioxide, K_2O_2 (*c*)

$\Delta H^\circ_{298}=-118{,}000$ calories per mole (*112*)
$S_{298}=(26.3)$ *e.u.* (*24*)
$M.P.=763°$ K. (*106*)
$\Delta H_M=6{,}100$ calories per mole (*24*)

Formation: $2K+O_2 \longrightarrow K_2O_2$
(estimated (*24*))

T, ° K.	H_T-H_{298}	ΔH°_T	ΔF°_T
298	....	−118,000	(−102,000)
400	(2,350)	(−119,000)	(−96,500)
500	(4,600)	(−119,000)	(−91,000)
600	(7,300)	(−118,500)	(−85,500)
700	(9,500)	(−118,500)	(−80,000)
800	(18,200)	(−112,000)	(−75,000)
900	(21,450)	(−111,000)	(−70,500)
1,000	(24,750)	(−110,000)	(−66,000)
1,100	(28,000)	(−147,000)	(−59,500)
1,200	(31,350)	(−145,500)	(−51,500)
1,300	(34,150)	(−144,500)	(−44,000)
1,400	(37,500)	(−143,000)	(−36,500)
1,500	(41,950)	(−141,500)	(−28,500)

Dipotassium Trioxide, K_2O_3 (*c*)

$\Delta H^\circ_{298}=-125{,}000$ calories per mole (*112*)
$S_{298}=(28.6)$ *e.u.* (*24*)
$M.P.=703°$ K. (*112*)
$\Delta H_M=7{,}030$ calories per mole

Formation: $2K+3/2O_2 \longrightarrow K_2O_3$
(estimated (*24*))

T, ° K.	H_T-H_{298}	ΔH°_T	ΔF°_T
298	....	−125,000	−102,500
400	(3,250)	(−125,500)	(−95,000)
500	(6,300)	(−125,000)	(−87,500)
600	(9,400)	(−124,500)	(−80,000)
700	(12,500)	(−124,000)	(−72,500)
800	(23,600)	(−115,500)	(−66,000)
900	(27,250)	(−114,500)	(−60,000)
1,000	(30,950)	(−113,500)	(−54,000)

Potassium Dioxide, KO_2 (*c*)

$\Delta H^\circ_{298}=-67{,}600$ calories per mole (*40*)
$S_{298}=27.9$ *e.u.* (*133*)
$M.P.=653°$ K. (*8*)
$\Delta H_M=3.920$ calories per mole

Formation: $K+O_2 \longrightarrow KO_2$
(estimated (*24*))

T, ° K.	H_T-H_{298}	ΔH°_T	ΔF°_T
298	....	−67,600	−56,800
400	(1,950)	(−67,700)	(−53,000)
500	(3,809)	(−67,300)	(−49,400)
600	(5,800)	(−66,800)	(−45,900)
700	(11,800)	(−62,300)	(−42,700)
800	(14,100)	(−61,500)	(−40,000)
900	(16,450)	(−60,700)	(−37,400)
1,000	(18,750)	(−59,900)	(−34,800)

647940 O—63——7

Potassium Trioxide, KO_3 (c)

$\Delta H^\circ_{298} = -62{,}000$ calories per mole (*104*)
$S_{298} = (33.5)$ *e.u.* (*24*)

Formation: $K + 3/2O_2 \longrightarrow KO_3$
(estimated (*24*))

T, ° K.	$H_T - H_{298}$	ΔH°_T	ΔF°_T
298	....	−62, 000	(−45, 500)
400	(2, 400)	(−62, 000)	(−39, 500)
500	(4, 750)	(−61, 500)	(−34, 000)

Potassium Fluoride, KF (c)

$\Delta H^\circ_{298} = -134{,}500$ calories per mole (*112*)
$S_{298} = 15.91$ *e.u.* (*112*)
M.P. = 1,130° K. (*82*)
$\Delta H_M = 6{,}750$ calories per mole
B.P. = 1,775° K. (*6*)
$\Delta H_V = 41{,}275$ calories per mole

Zone I (c) (298°–1,130° K.)

$$C_p = 11.02 + 3.12\times10^{-3}T \ (82)$$
$$H_T - H_{298} = -3{,}424 + 11.02T + 1.56\times10^{-3}T^2$$

Zone II (l) (1,130°–1,200° K.)

$$C_p = 16.0 \ (82)$$
$$H_T - H_{298} = -310 + 16.0T$$

Formation: $K + 1/2F_2 \longrightarrow KF$

Zone I (298°–336. 7° K.)

$$\Delta C_p = 0.83 - 0.22\times10^{-3}T + 0.40\times10^5T^{-2}$$
$$\Delta H_T = -134{,}600 + 0.83T - 0.11\times10^{-3}T^2 - 0.40\times10^5T^{-1}$$
$$\Delta F_T = -134{,}600 - 0.83T\ln T + 0.11\times10^{-3}T^2 - 0.20\times10^5T^{-1} + 28.95T$$

Zone II (336.7°–1,030° K.)

$$\Delta C_p = 0.84 + 1.91\times10^{-3}T - 1.56\times10^5T^{-2}$$
$$\Delta H_T = -135{,}860 + 0.84T + 0.95\times10^{-3}T^2 + 1.56\times10^5T^{-1}$$
$$\Delta F_T = -135{,}860 - 0.84T\ln T - 0.95\times10^{-3}T^2 + 0.78\times10^5T^{-1} + 32.24T$$

Zone III (1,030°–1,130° K.)

$$\Delta C_p = 1.97 + 2.85\times10^{-3}T + 0.37\times10^5T^{-2}$$
$$\Delta H_T = -156{,}540 + 1.97T + 1.42\times10^{-3}T^2 - 0.37\times10^5T^{-1}$$
$$\Delta F_T = -156{,}540 - 1.97T\ln T - 1.42\times10^{-3}T^2 - 0.18\times10^5T^{-1} + 60.44T$$

Zone IV (1,130°–1,200° K.)

$$\Delta C_p = 6.95 - 0.274\times10^{-3}T + 0.37\times10^5T^{-2}$$
$$\Delta H_T = -151{,}650 + 6.95T - 0.137\times10^{-3}T^2 - 0.37\times10^5T^{-1}$$
$$\Delta F_T = -151{,}650 - 6.95T\ln T + 0.137\times10^{-3}T^2 - 0.185\times10^5T^{-1} + 90.95T$$

T, ° K.	$H_T - H_{298}$	S_T	ΔH°_T	ΔF°_T
298	....	15. 91	−134, 500	−127, 500
400	1, 230	19. 46	−134, 500	−124, 400
500	2, 470	22. 22	−134, 900	−122, 400
600	3, 760	24. 57	−134, 750	−119, 900
700	5, 070	26. 59	−134, 550	−117, 500
800	6, 400	28. 37	−134, 400	−115, 100
900	7, 770	29. 98	−134, 150	−112, 700
1,000	9, 160	31. 45	−133, 900	−110, 300
1,100	10, 580	32. 80	−152, 700	−106, 900
1,200	18, 890	40. 12	−145, 300	−103, 300
1,500	(23, 300)	(43. 2)	(−143, 700)	(−92, 600)

Potassium Chloride, KCl (c)

$\Delta H^\circ_{298} = -104{,}175$ calories per mole (*112*)
$S_{298} = 19.76$ *e.u.* (*83*)
M.P. = 1,043° K. (*82*)
$\Delta H_M = 6{,}100$ calories per mole
B.P. = 1,680° K. (*6*)
$\Delta H_V = 38{,}840$ calories per mole

Zone I (c) (298°–1,043° K.)

$$C_p = 9.89 + 5.20\times10^{-3}T + 0.77\times10^5T^{-2} \ (82)$$
$$H_T - H_{298} = -2{,}920 + 9.89T + 2.60\times10^{-3}T^2 - 0.77\times10^5T^{-1}$$

Zone II (l) (1,043°–1,200° K.)

$$C_p = 16.00 \ (82)$$
$$H_T - H_{298} = -440 + 16.00T$$

Formation: $K + 1/2Cl_2 \longrightarrow KCl$

Zone I (298°–336.7° K.)

$$\Delta C_p = -0.56 + 2.05\times10^{-3}T + 1.11\times10^5T^{-2}$$
$$\Delta H_T = -103{,}730 - 0.56T + 1.02\times10^{-3}T^2 - 1.11\times10^5T^{-1}$$
$$\Delta F_T = -103{,}730 + 0.56T\ln T - 1.02\times10^{-3}T^2 - 0.56\times10^5T^{-1} + 18.44T$$

Zone II (336.7°–1,030° K.)

$$\Delta C_p = -0.55 + 4.18\times10^{-3}T - 0.85\times10^5T^{-2}$$
$$\Delta H_T = -104{,}990 - 0.55T + 2.09\times10^{-3}T^2 + 0.85\times10^5T^{-1}$$
$$\Delta F_T = -104{,}990 + 0.55T\ln T - 2.09\times10^{-3}T^2 + 0.42\times10^5T^{-1} + 21.78T$$

Zone III (1,043°–1,200° K.)

$$\Delta C_p = 5.69 - 0.084\times10^{-3}T + 0.31\times10^5T^{-2}$$
$$\Delta H_T = -122{,}000 + 5.69T - 0.042\times10^{-3}T^2 - 0.31\times10^5T^{-1}$$
$$\Delta F_T = -122{,}000 - 5.69T\ln T + 0.042\times10^{-3}T^2 - 0.15\times10^5T^{-1} + 79.28T$$

T, ° K.	$H_T - H_{298}$	S_T	ΔH°_T	ΔF°_T
298	....	19. 76	−104, 175	−97, 550
400	1, 260	23. 40	−104, 650	−95, 200
500	2, 520	26, 21	−104, 550	−92, 800
600	3, 810	28. 56	−104, 450	−90, 500
700	5, 150	30. 62	−104, 250	−88, 200
800	6, 550	32. 49	−104, 000	−85, 900
900	8, 000	34. 20	−103. 700	−83, 650
1,000	9, 500	35. 78	−103, 350	−81, 400
1,100	17, 160	43. 12	−115, 800	−78, 600
1,200	18, 760	44. 51	−115, 250	−75. 300
(1,500)	(23, 800)	(48. 36)	(−112, 950)	(−65, 600)

Potassium Bromide, KBr (c)

$\Delta H^\circ_{298} = -93{,}730$ calories per mole (*112*)
$S_{298} = 22.6$ *e.u.* (*83*)
M.P. = 1,015° K. (*6*)
$\Delta H_M = 5{,}000$ calories per mole
B.P. = 1,656° K. (*6*)
$\Delta H_V = 37{,}060$ calories per mole

Zone I (c) (298°–1,000° K.)

$$C_p = 11.56 + 3.32\times10^{-3}T \ (20)$$
$$H_T - H_{298} = -3{,}594 + 11.56T + 1.66\times10^{-3}T^2$$

Formation: $K + 1/2Br_2 \longrightarrow KBr$

Zone I (298°–331° K.)

$$\Delta C_p = 3.03 + 0.20\times10^{-3}T$$
$$\Delta H_T = -92{,}840 - 3.03T + 0.10\times10^{-3}T^2$$
$$\Delta F_T = -92{,}840 + 3.03T\ln T - 0.10\times10^{-3}T^2 - 9.13T$$

Zone II (331°–1,015° K.)

$\Delta C_p = 1.01 + 2.33\times10^{-3}T - 1.78\times10^5T^{-2}$
$\Delta H_T = -99{,}100 + 1.01T + 1.16\times10^{-3}T^2 + 1.78\times10^5T^{-1}$
$\Delta F_T = -99{,}100 - 1.01T\ln T - 1.16\times10^{-3}T^2 + 0.89\times10^5T^{-1} + 32.75T$

T, ° K.	$H_T - H_{298}$	S_T	ΔH_T°	ΔF_T°
298		22.6	−93,750	−90,400
400	1,295	26.33	−98,100	−88,400
500	2,600	29.24	−97,950	−86,000
600	3,940	31.69	−97,800	−83,600
700	5,310	33.80	−97,600	−81,250
800	6,710	35.67	−97,350	−78,950
900	8,150	37.36	−97,100	−76,650
1,000	9,630	38.92	−96,750	−74,400
(1,500)	(22,700)	(50.5)	(−107,700)	−58,200

Potassium Iodide, KI (*c*)

$\Delta H_{298}^\circ = -78{,}310$ calories per mole (*112*)
$S_{298} = 24.9$ *e.u.* (*112*)
M.P. = 955° K. (*6*)
$\Delta H_M = 4{,}100$ calories per mole
B.P. = 1,597° K. (*6*)
$\Delta H_V = 34{,}691$ calories per mole

Zone I (*c*) (298°–950° K.)

$C_p = 11.36 + 4.00\times10^{-3}T$ (*82*)
$H_T - H_{298} = -3{,}565 + 11.36T + 2.00\times10^{-3}T^2$

Formation: $K + 1/2I_2 \longrightarrow KI$

Zone I (298°–337° K.)

$\Delta C_p = 0.53 - 5.07\times10^{-3}T$
$\Delta H_T = -77{,}260 + 0.53T - 2.53\times10^{-3}T^2$
$\Delta F_T = -77{,}260 - 0.53T\ln T + 2.53\times10^{-3}T^2 + 3.14T$

Zone II (337°–387° K.)

$\Delta C_p = 0.54 - 2.94\times10^{-3}T - 1.96\times10^5T^{-2}$
$\Delta H_T = -78{,}500 + 0.54T - 1.47\times10^{-3}T^2 + 1.96\times10^5T^{-1}$
$\Delta F_T = -78{,}500 - 0.54T\ln T + 1.47\times10^{-3}T^2 + 0.98\times10^5T^{-1} + 6.37T$

Zone III (387°–456° K.)

$\Delta C_p = -4.27 + 3.01\times10^{-3}T - 1.96\times10^5T^{-2}$
$\Delta H_T = -80{,}000 - 4.27T + 1.50\times10^{-3}T^2 + 1.96\times10^5T^{-1}$
$\Delta F_T = -80{,}000 + 4.27T\ln T - 1.50\times10^{-3}T^2 + 0.98\times10^5T^{-1} - 16.9T$

Zone IV (456°–955° K.)

$\Delta C_p = 0.89 + 3.01\times10^{-3}T - 1.96\times10^5T^{-2}$
$\Delta H_T = -87{,}350 + 0.89T + 1.50\times10^{-3}T^2 + 1.96\times10^5T^{-1}$
$\Delta F_T = -87{,}350 - 0.89T\ln T - 1.50\times10^{-3}T^2 + 0.98\times10^5T^{-1} + 31.04T$

T, ° K.	$H_T - H_{298}$	S_T	ΔH_T°	ΔF_T°
298		24.9	−78,310	−77,000
400	1,290	26.62	−80,950	−76,500
500	2,630	31.61	−86,200	−74,800
600	3,990	34.09	−85,900	−72,450
700	5,390	36.21	−85,650	−70,200
800	6,800	38.12	−85,400	−68,000
900	8,250	39.86	−85,100	−65,900

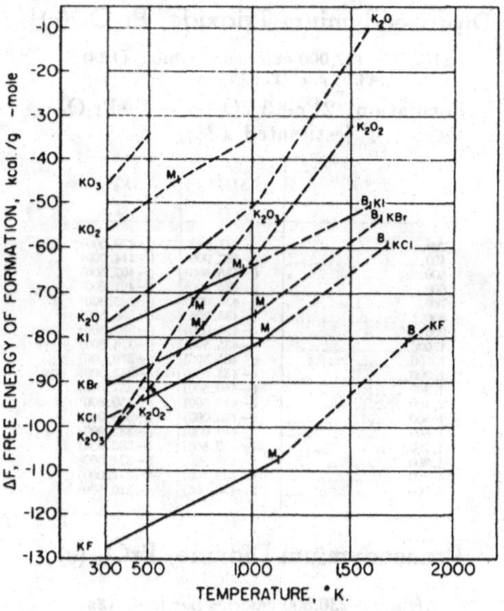

FIGURE 42.—Potassium.

PRASEODYMIUM AND ITS COMPOUNDS

Element, Pr (*c*)

$S_{298} = 17.49$ *e.u.* (*121*)
T.P. = 1,071° K. (*125*)
$\Delta H_T = (320)$ calories per atom
M.P. = 1,208° K. (*125*)
$\Delta H_M = (2{,}400)$ calories per atom
B.P. = (3,290°) K. (*125*)
$\Delta H_V = (2{,}200)$ calories per atom

T, ° K.	$H_T - H_{298}$	S_T	$-\frac{(F_T - H_{298})}{T}$
298		17.49	17.49
400	(670)	(19.39)	(17.72)
500	(1,370)	(20.94)	(18.20)
600	(2,090)	(22.26)	(18.78)
700	(2,850)	(23.43)	(19.36)
800	(3,640)	(24.48)	(19.93)
900	(4,460)	(25.45)	(20.50)
1,000	(5,320)	(26.35)	(21.03)
1,100	(6,500)	(27.47)	(21.57)
1,200	(7,300)	(28.16)	(22.08)
1,300	(8,100)	(30.79)	(24.56)
1,400	(8,900)	(31.39)	(25.04)
1,500	(9,700)	(31.94)	(25.48)
1,600	(10,500)	(32.46)	(25.90)
1,700	(11,300)	(32.94)	(26.30)
1,800	(12,100)	(33.40)	(26.68)
1,900	(12,900)	(33.83)	(27.05)
2,000	(13,700)	(34.24)	(27.39)

Dipraseodymium Trioxide, Pr_2O_3 (*c*)

$\Delta H^\circ_{298} = -437{,}000$ calories per mole (*129*)
$S_{298} = (43.5)$ *e.u.* (*24*)

Formation: $2Pr + 3/2O_2 \longrightarrow Pr_2O_3$
(estimated (*24*))

T, ° K.	ΔH°_T	ΔF°_T
298	−437,000	(−420,500)
400	(−437,000)	(−414,000)
500	(−436,500)	(−407,500)
600	(−436,500)	(−401,000)
700	(−436,000)	(−395,000)
800	(−436,000)	(−388,500)
900	(−436,000)	(−382,500)
1,000	(−435,500)	(−376,000)
1,100	(−435,500)	(−370,000)
1,200	(−435,000)	(−363,500)
1,300	(−440,500)	(−357,000)
1,400	(−440,500)	(−350,000)
1,500	(−440,000)	(−343,000)
1,600	(−440,000)	(−337,000)
1,700	(−439,500)	(−330,500)
1,800	(−439,500)	(−324,000)
1,900	(−439,500)	(−317,000)
2,000	(−439,000)	(−310,500)

Praseodymium Dioxide, PrO_2 (*c*)

$\Delta H^\circ_{298} = -230{,}500$ calories per mole (*24*)
$S_{298} = 22.9$ *e.u.* (*24*)

Formation: $Pr + O_2 \longrightarrow PrO_2$
(estimated (*24*))

T, ° K.	ΔH°_T	ΔF°_T
298	−230,500	−217,500
400	(−230,500)	(−213,000)
500	(−230,500)	(−208,500)
600	(−230,000)	(−204,000)
700	(−230,000)	(−200,000)
800	(−230,500)	(−195,500)
900	(−230,500)	(−191,000)
1,000	(−230,500)	(−187,000)
1,100	(−231,000)	(−182,500)
1,200	(−231,000)	(−178,000)

Praseodymium Trifluoride, PrF_3 (*c*)

$\Delta H^\circ_{298} = (-388{,}000)$ calories per mole (*5*)
$S_{298} = (25)$ *e.u.* (*11*)
M.P. = 1,668° K. (*29*)
$\Delta H_M = (8{,}000)$ calories per mole
B.P. = (2,600°) K. (*6*)
$\Delta H_V = (62{,}000)$ calories per mole

Formation: $Pr + 3/2F_2 \longrightarrow PrF_3$
(estimated (*11*))

T, ° K.	$H_T - H_{298}$	ΔH°_T	ΔF°_T
298		(−388,000)	(−369,500)
500	(4,000)	(−388,000)	(−358,000)
1,000	(17,000)	(−385,000)	(−328,000)
1,500	(32,000)	(−381,000)	(−301,000)

Praseodymium Trichloride, $PrCl_3$ (*c*)

$\Delta H^\circ_{298} = -252{,}090$ calories per mole (*127*)
$S_{298} = (34.5)$ *e.u.* (*127*)
M.P. = 1,059° K. (*29*)
$\Delta H_M = (8{,}000)$ calories per mole
B.P. = (1,980°) K. (*6*)
$\Delta H_V = (46{,}000)$ calories per mole

Formation: $Pr + 3/2Cl_2 \longrightarrow PrCl_3$
(estimated (*11*))

T, ° K.	$H_T - H_{298}$	ΔH°_T	ΔF°_T
298		−252,090	(−244,000)
500	(5,000)	(−251,000)	(−224,500)
1,000	(19,000)	(−247,500)	(−199,000)
1,500	(43,000)	(−233,400)	(−180,000)

Praseodymium Tribromide, $PrBr_3$ (*c*)

$\Delta H^\circ_{298} = (-189{,}000)$ calories per mole (*5*)
$S_{298} = (46)$ *e.u.* (*11*)
M.P. = 964° K. (*29*)
$\Delta H_M = (8{,}000)$ calories per mole
B.P. = (1,820°) K. (*6*)
$\Delta H_V = (45{,}000)$ calories per mole

Formation: $Pr + 3/2Br_2 \longrightarrow PrBr_3$
(estimated (*11*))

T, ° K.	$H_T - H_{298}$	ΔH°_T	ΔF°_T
298		(−189,000)	(−182,000)
500	(5,000)	(−199,700)	(−171,500)
1,000	(18,000)	(−197,300)	(−148,000)
1,500	(43,000)	(−194,500)	(−127,000)

Praseodymium Triiodide, PrI_3 (*c*)

$\Delta H^\circ_{298} = -162{,}000$ calories per mole (*5*)
$S_{298} = (50)$ *e.u.* (*11*)
M.P. = 1,010° K. (*29*)
$\Delta H_M = (8{,}000)$ calories per mole
B.P. = (1,650°) K. (*6*)
$\Delta H_V = (40{,}000)$ calories per mole

Formation: $Pr + 3/2I_2 \longrightarrow PrI_3$
(estimated (*11*))

T, ° K.	$H_T - H_{298}$	ΔH°_T	ΔF°_T
298		−162,000	(−160,000)
500	(5,000)	(−183,400)	(−148,500)
1,000	(19,000)	(−180,000)	(−122,000)
1,500	(44,000)	(−166,000)	(−99,000)

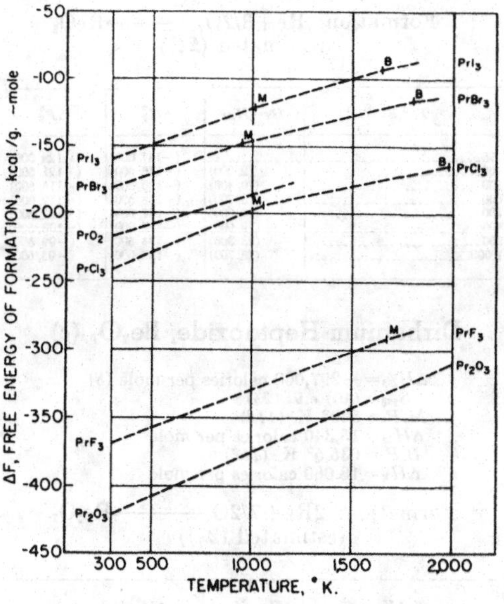

FIGURE 43.—Praseodymium.

PROMETHIUM AND ITS COMPOUNDS

Element, Pm (*c*)

S_{298} = (17.25) *e.u.* (*121*)
M.P. = 1,573° K. (*125*)
ΔH_M = (3,000) calories per atom
B.P. = (3,000°) K. (*125*)
ΔH_V = (70,000) calories per atom

(estimated (*130*))

T, °K.	H_T-H_{298}	S_T	$-\frac{(F_T-H_{298})}{T}$
298		(17.25)	(17.25)
400	(670)	(19.15)	(17.48)
500	(1,360)	(20.69)	(17.97)
600	(2,070)	(21.98)	(18.53)
700	(2,810)	(23.12)	(19.11)
800	(3,570)	(24.14)	(19.68)
900	(4,360)	(25.07)	(20.23)
1,000	(5,170)	(25.92)	(20.75)
1,100	(6,010)	(26.72)	(21.26)
1,200	(6,870)	(27.47)	(21.75)
1,300	(7,760)	(28.19)	(22.22)
1,400	(8,560)	(28.94)	(22.83)
1,500	(9,360)	(29.64)	(23.40)
1,600	(13,160)	(32.15)	(23.93)
1,700	(13,960)	(32.64)	(24.43)
1,800	(14,760)	(33.09)	(24.89)
1,900	(15,560)	(33.53)	(25.35)
2,000	(16,360)	(33.94)	(25.76)

Promethium Trifluoride, PmF_3 (*c*)

ΔH°_{298} = (−383,000) calories per mole (*5*)
S_{298} = (24) *e.u* (*11*)
M.P. = 1,680° K. (*6*)
ΔH_M = (8,000) calories per mole
B.P. = (2,600°) K. (*6*)
ΔH_V = (62,000) calories per mole

Formation: $Pm + 3/2F_2 \longrightarrow PmF_3$
(estimated (*11*))

T, °K.	H_T-H_{298}	ΔH°_T	ΔF°_T
298		(−383,000)	(−364,500)
500	(4,000)	(−383,000)	(−352,500)
1,000	(17,000)	(−380,000)	(−322,000)
1,500	(32,000)	(−379,000)	(−296,500)

Promethium Trichloride, $PmCl_3$ (*c*)

ΔH°_{298} = (−227,000) calories per mole (*5*)
S_{298} = (39) *e.u.* (*11*)
M.P. = 1,010° K. (*6*)
ΔH_M = (8,000) calories per mole
B.P. = (1,940°) K. (*6*)
ΔH_V = (46,000) calories per mole

Formation: $Pm + 3/2Cl_2 \longrightarrow PmCl_3$
(estimated (*11*))

T, °K.	H_T-H_{298}	ΔH°_T	ΔF°_T
298		(−227,000)	(−211,000)
500	(5,000)	(−225,900)	(−200,000)
1,000	(19,000)	(−222,300)	(−175,000)
1,500	(43,000)	(−209,000)	(−156,500)

Promethium Tribromide, $PmBr_3$ (*c*)

ΔH°_{298} = (−183,000) calories per mole (*5*)
S_{298} = (47) *e.u.* (*11*)
M.P. = 950° K. (*6*)
ΔH_M = (8,000) calories per mole
B.P. = (1,800°) K. (*6*)
ΔH_V = (45,000) calories per mole

Formation: $Pm + 3/2Br_2 \longrightarrow PmBr_3$
(estimated (*11*))

T, °K.	H_T-H_{298}	ΔH°_T	ΔF°_T
298		(−183,000)	(−177,000)
500	(5,000)	(−193,600)	(−167,000)
1,000	(18,000)	(−191,200)	(−142,000)
1,500	(43,000)	(−180,000)	(−123,500)

Promethium Triiodide, PmI_3 (*c*)

ΔH°_{298} = (−131,000) calories per mole (*5*)
S_{298} = (49) *e.u.* (*11*)
M.P. = 1,070° K. (*6*)
ΔH_M = (8,000) calories per mole
B.P. = (1,640°) K. (*6*)
ΔH_V = (41,000) calories per mole

Formation: $Pm + 3/2I_2 \longrightarrow PmI_3$
(estimated (*11*))

T, °K.	H_T-H_{298}	ΔH°_T	ΔF°_T
298		(−131,000)	(−129,000)
500	(5,000)	(−152,400)	(−124,000)
1,000	(19,000)	(−148,090)	(−97,000)
1,500	(44,000)	(−137,300)	(−73,500)

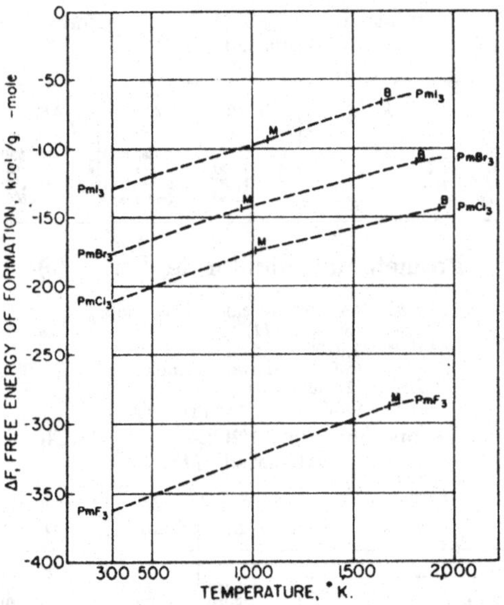

FIGURE 44.—Promethium.

RHENIUM AND ITS COMPOUNDS

Element, Re (c)

$S_{298}=8.89$ *e.u.* (*123*)
$M.P.=3{,}453°$ K. (*118*)
$\Delta H_M=(7{,}900)$ calories per atom
$B.P.=5{,}900°$ K. (*118*)

Zone I (c) (298°–1,500° K.)

$C_p=5.66+1.30\times10^{-3}T$ (*82*)
$H_T-H_{298}=-1{,}745+5.66T+0.65\times10^{-3}T^2$
$F_T-H_{298}=-1{,}745-5.66T\,ln\,T-0.65\times10^{-3}T^2+29.39T$

T, ° K.	H_T-H_{298}	S_T	$-\frac{(F_T-H_{298})}{T}$
298		8.89	8.89
400	620	10.68	9.13
500	1,240	12.06	9.58
600	1,890	13.25	10.10
700	2,550	14.26	10.62
800	3,210	15.14	11.12
900	3,880	15.93	11.62
1,000	4,570	16.66	12.09
1,100	5,270	17.33	12.54
1,200	5,980	17.94	12.96
1,300	6,710	18.53	13.36
1,400	7,460	19.08	13.75
1,500	8,220	19.61	14.13
1,600	(8,990)	(20.10)	(14.48)
1,700	(9,770)	(20.56)	(14.81)
1,800	(10,560)	(21.02)	(15.15)
1,900	(11,370)	(21.45)	(15.47)
2,000	(12,180)	(21.88)	(15.79)

Rhenium Trioxide, ReO_3 (c)

$\Delta H^\circ_{298}=(-147{,}000)$ calories per mole (*8*)
$S_{298}=(18.6)$ *e.u.* (*24*)
$M.P.=433°$ K. (*24*)
$\Delta H_M=5{,}200$ calories per mole

Formation: $Re+3/2O_2 \longrightarrow ReO_3$
(estimated (*24*))

T, ° K.	H_T-H_{298}	ΔH°_T	ΔF°_T
298		(−147,000)	(−128,000)
400	(2,200)	(−146,500)	(−121,500)
500	(10,400)	(−140,000)	(−116,500)
600	(13,700)	(−138,500)	(−112,000)
700	(17,000)	(−137,000)	(−107,500)
800	(19,900)	(−136,000)	(−103,500)
900	(23,300)	(−134,500)	(−99,500)
1,000	(26,700)	(−133,000)	(−95,500)

Dirhenium Heptaoxide, Re_2O_7 (c)

$\Delta H^\circ_{298}=-297{,}000$ calories per mole (*8*)
$S_{298}=(40)$ *e.u.* (*24*)
$M.P.=569°$ K. (*112*)
$\Delta H_M=15{,}340$ calories per mole
$B.P.=635.5°$ K. (*112*)
$\Delta H_V=18{,}060$ calories per mole

Formation: $2Re+7/2O_2 \longrightarrow Re_2O_7$
(estimated (*24*))

T, ° K.	H_T-H_{298}	ΔH°_T	ΔF°_T
298		−297,000	(−252,000)
400	(5,300)	(−295,500)	(−237,000)
500	(10,600)	(−294,000)	(−223,000)
600	(32,000)	(−276,500)	(−209,500)
700	(55,000)	(−257,500)	(−200,500)
800	(59,200)	(−257,500)	(−192,000)
900	(62,900)	(−258,000)	(−184,000)
1,000	(67,100)	(−258,000)	(−176,000)
1,100	(71,200)	(−258,000)	(−167,500)
1,200	(75,300)	(−258,500)	(−159,500)
1,300	(79,500)	(−258,500)	(−151,000)
1,400	(83,900)	(−258,500)	(−143,000)
1,500	(88,800)	(−258,500)	(−134,500)

Dirhenium Octaoxide, Re_2O_8 (c)

$\Delta H^\circ_{298}=(-308{,}500)$ calories per mole (*24*)
$S_{298}=(41)$ *e.u.* (*24*)
$M.P.=420°$ K. (*24*)
$\Delta H_M=3{,}800$ calories per mole

Formation: $2Re+4O_2 \longrightarrow Re_2O_8$
(estimated (*24*))

T, ° K.	H_T-H_{298}	ΔH°_T	ΔF°_T
298		(−308,500)	(−257,000)
400	(6,100)	(−306,500)	(−239,500)
500	(17,300)	(−299,500)	(−224,000)
600	(25,600)	(−295,500)	(−209,000)

Rhenium Trifluoride, ReF_3 (c)

$\Delta H^\circ_{298}=(-170{,}000)$ calories per mole (*42*)
$S_{298}=(26)$ *e.u.* (*11*)
$M.P.=(1{,}380°)$ K. (*42*)
$\Delta H_M=(1{,}100)$ calories per mole
$B.P.=(1{,}530°)$ K. (*42*)
$\Delta H_V=(37{,}000)$ calories per mole

Formation: $Re+3/2F_2 \longrightarrow ReF_3$
(estimated (*11*))

T, ° K.	H_T-H_{298}	ΔH_T°	ΔF_T°
298		(−170, 000)	(−153, 300)
500	(4, 000)	(−169, 500)	(−142, 500)
1,000	(17, 000)	(−166, 400)	(−116, 000)

Rhenium Tetrafluoride, ReF_4 (*c*)

$\Delta H_{298}^\circ=(-220{,}000)$ calories per mole (*42*)
$S_{298}=(36)$ *e.u.* (*11*)
M.P. = 398° K. (*6*)
$\Delta H_M=(4{,}500)$ calories per mole
B.P. = (1,070°) K. (*6*)
$\Delta H_V=(27{,}000)$ calories per mole

Formation: $Re+2F_2 \longrightarrow ReF_4$
(estimated (*11*))

T, ° K.	H_T-H_{298}	ΔH_T°	ΔF_T°
298		(−220, 000)	(−209, 300)
500	(6, 000)	(−218, 500)	(−201, 000)
1,000	(22, 000)	(−214, 300)	(−169, 000)

Rhenium Pentafluoride, ReF_5 (*c*)

$\Delta H_{298}^\circ=(-225{,}000)$ calories per mole (*11*)
$S_{298}=(59)$ *e.u.* (*11*)
M.P. = (398°) K. (*42*)
$\Delta H_M=(4{,}500)$ calories per mole
B.P. = (660°) K. (*42*)
$\Delta H_V=(15{,}000)$ calories per mole

Formation: $Re+5/2F_2 \longrightarrow ReF_5$
(estimated (*11*))

T, ° K.	H_T-H_{298}	ΔH_T°	ΔF_T°
298		(−225, 000)	(−204, 000)
500	(12, 000)	(−218, 000)	(−192, 500)

Rhenium Hexafluoride, ReF_6 (*l*)

$\Delta H_{298}^\circ=-278{,}000$ calories per mole (*112*)
$S_{298}=(78)$ *e.u.* (*11*)
$\Delta F_{298}^\circ=(-255{,}000)$ calories per mole
M.P. = 292° K. (*6*)
$\Delta H_M=5{,}000$ calories per mole
B.P. = 321° K. (*6*)
$\Delta H_V=6{,}900$ calories per mole

Rhenium Trichloride, $ReCl_3$ (*c*)

$\Delta H_{298}^\circ=(-55{,}000)$ calories per mole (*11*)
$S_{298}=(38)$ *e.u.* (*11*)
M.P. = (1,000°) K. (*6*)
$\Delta H_M=(15{,}000)$ calories per mole
B.P. = (1,100°) K. (*6*)
$\Delta H_V=(27{,}000)$ calories per mole

Formation: $Re+3/2Cl_2 \longrightarrow ReCl_3$
(estimated (*11*))

T, ° K.	H_T-H_{298}	ΔH_T°	ΔF_T°
298		(−55, 000)	(−39, 800)
500	(5, 000)	(−53, 800)	(−29, 500)
1,000	(19, 000)	(−49, 700)	(−6, 000)

Rhenium Tetrachloride, $ReCl_4$ (*c*)

$\Delta H_{298}^\circ=(-60{,}000)$ calories per mole (*11*)
$S_{298}=(50)$ *e.u.* (*11*)
M.P. = (450°) K. (*42*)
$\Delta H_M=(4{,}000)$ calories per mole
B.P. = 650° K. (*42*)
$\Delta H_V=(14{,}000)$ calories per mole

Formation: $Re+2Cl_2 \longrightarrow ReCl_4$
(estimated (*11*))

T, ° K.	H_T-H_{298}	ΔH_T°	ΔF_T°
298		(−60, 000)	(−41, 000)
500	(6, 000)	(−58, 600)	(−28, 000)

Rhenium Pentachloride, $ReCl_5$ (*c*)

$\Delta H_{298}^\circ=(-70{,}000)$ calories per mole (*11*)
$S_{298}=(66)$ *e.u.* (*11*)
M.P. = (530°) K. (*6*)
$\Delta H_M=(9{,}000)$ calories per mole
B.P. = (600°) K. (*6*)
$\Delta H_V=(14{,}000)$ calories per mole

Formation: $Re+5/2Cl_2 \longrightarrow ReCl_5$
(estimated (*11*))

T, ° K.	H_T-H_{298}	ΔH_T°	ΔF_T°
298		(−70, 000)	(−47, 400)
500	(7, 000)	(−68, 000)	(−33, 000)

Rhenium Tribromide, $ReBr_3$ (*c*)

$\Delta H_{298}^\circ=(-32{,}700)$ calories per mole (*11*)
$S_{298}=(44)$ *e.u.* (*11*)
M.P. = (900°) K. (*6*)
$\Delta H_M=(13{,}500)$ calories per mole
B.P. = (1,000°) K. (*6*)
$\Delta H_V=(25{,}000)$ calories per mole

Formation: $Re+3/2Br_2 \longrightarrow ReBr_3$
(estimated (*11*))

T, ° K.	H_T-H_{298}	ΔH_T°	ΔF_T°
298		(−32, 700)	(−23, 400)
500	(5, 000)	(−43, 200)	(−12, 700)
1,000	(18, 000)	(−40, 300)	(+10, 700)

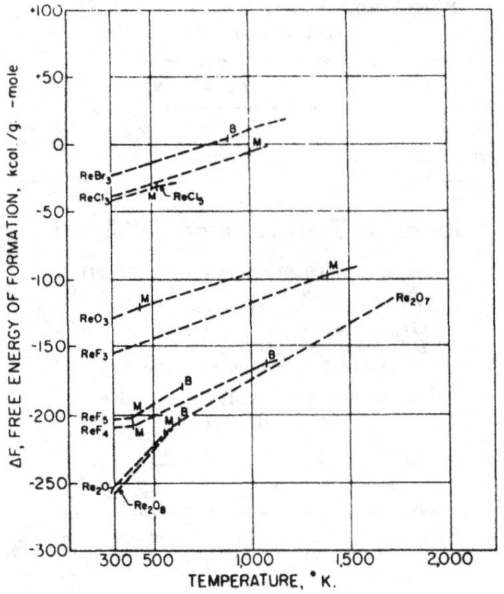

FIGURE 45.—Rhenium.

SAMARIUM AND ITS COMPOUNDS

Element, Sm (*c*)

S_{298} = (16.32) *e.u.* (*121*)
T.P. = 1,190° K. (*125*)
ΔH_T = (360) calories per atom
M.P. = (1,325°) K. (*125*)
ΔH_M = (2,650) calories per atom
B.P. = 1,860° K. (*125*)
ΔH_V = (45,800) calories per atom

(estimated (*130*))

T, ° K.	H_T-H_{298}	S_T	$-\frac{(F_T-H_{298})}{T}$
298	....	(16.32)	(16.32)
400	(675)	(18.23)	(16.55)
500	(1,370)	(19.77)	(17.03)
600	(2,090)	(21.08)	(17.60)
700	(2,835)	(22.23)	(18.18)
800	(3,610)	(23.27)	(18.76)
900	(4,415)	(24.22)	(19.32)
1,000	(5,250)	(25.09)	(19.84)
1,100	(6,110)	(25.91)	(20.36)
1,200	(7,350)	(26.98)	(20.86)
1,300	(8,150)	(27.62)	(21.36)
1,400	(11,600)	(30.21)	(21.93)
1,500	(12,400)	(30.76)	(22.50)
1,600	(13,200)	(31.28)	(23.03)
1,700	(14,000)	(31.76)	(23.53)
1,800	(14,800)	(32.22)	(24.00)
1,900	(61,340)	(57.04)	(24.76)
2,000	(61,970)	(57.36)	(26.38)

Disamarium Trioxide, Sm_2O_3 (*c*)

ΔH°_{298} = −433,890 calories per mole (*64*)
S_{298} = (41) *e.u.* (*24*)

Formation: $2Sm + 3/2O_2 \longrightarrow Sm_2O_3$
(estimated (*24*))

T, ° K.	ΔH°_T	ΔF°_T
298	−434,000	(−410,500)
400	(−434,000)	(−404,000)
500	(−433,500)	(−397,500)
600	(−433,500)	(−391,000)
700	(−433,000)	(−385,000)
800	(−433,000)	(−378,500)
900	(−433,000)	(−372,500)
1,000	(−425,500)	(−366,000)
1,100	(−425,500)	(−360,000)
1,200	(−425,000)	(−353,500)
1,300	(−425,000)	(−347,500)
1,400	(−425,000)	(−341,000)
1,500	(−424,500)	(−335,000)
1,600	(−424,500)	(−329,000)
1,700	(−438,500)	(−322,500)
1,800	(−438,500)	(−315,500)
1,900	(−438,000)	(−309,000)
2,000	(−438,000)	(−302,500)

Samarium Difluoride, SmF_2 (*c*)

ΔH°_{298} = (−272,000) calories per mole (*5*)
S_{298} = (23) *e.u.* (*11*)
M.P. = (1,603°) K. (*29*)
ΔH_M = (5,000) calories per mole
B.P. = (2,700°) K. (*6*)
ΔH_V = (78,000) calories per mole

Formation: $Sm + F_2 \longrightarrow SmF_2$
(estimated (*11*))

T, ° K.	H_T-H_{298}	ΔH°_T	ΔF°_T
298	....	(−272,000)	(−259,500)
500	(4,000)	(−271,000)	(−251,000)
1,000	(13,000)	(−270,100)	(−232,000)
1,500	(24,000)	(−270,600)	(−213,500)

Samarium Trifluoride, SmF_3 (*c*)

ΔH°_{298} = (−380,000) calories per mole (*5*)
S_{298} = 27 *e.u.* (*11*)
M.P. = (1,579°) K. (*29*)
ΔH_M = (8,000) calories per mole
B.P. = (2,600°) K. (*6*)
ΔH_V = (62,000) calories per mole

Formation: $Sm + 3/2F_2 \longrightarrow SmF_3$
(estimated (*11*))

T, ° K.	H_T-H_{298}	ΔH°_T	ΔF°_T
298	....	(−380,000)	(−361,500)
500	(4,000)	(−379,700)	(−349,500)
1,000	(17,000)	(−377,000)	(−319,000)
1,500	(32,000)	(−375,800)	(−291,500)

Samarium Dichloride, $SmCl_2$ (*c*)

ΔH°_{298} = −195,600 calories per mole (*96*)
S_{298} = 30 *e.u.* (*11*)
M.P. = 835° K. (*29*)
ΔH_M = (6,000) calories per mole
B.P. = (2,300°) K. (*6*)
ΔH_V = (55,000) calories per mole

Formation: $Sm + Cl_2 \longrightarrow SmCl_2$
(estimated (*11*))

T, ° K.	H_T-H_{298}	ΔH_T°	ΔF_T°
298	------------	−195,600	(−184,000)
500	(4,000)	(−194,500)	(−177,000)
1,000	(13,000)	(−193,900)	(−160,000)
1,500	(31,000)	(−187,400)	(−149,000)

Samarium Trichloride, $SmCl_3$ (*c*)

ΔH°_{298} = (−223,000) calories per mole (*11*)
S_{298} = (39) *e.u.* (*11*)
M.P. = 955° K. (*29*)
ΔH_M = (8,000) calories per mole
Decomposes (*6*)

Formation: $Sm + 3/2Cl_2 \longrightarrow SmCl_3$
(estimated (*11*))

T, ° K.	H_T-H_{298}	ΔH_T°	ΔF_T°
298	------------	(−223,000)	(−206,500)
500	(5,000)	(−221,900)	(−196,000)
1,000	(19,000)	(−218,700)	(−172,000)
1,500	(43,000)	(−208,000)	(−152,000)

Samarium Dibromide, $SmBr_2$ (*c*)

ΔH°_{298} = (−157,000) calories per mole (*5*)
S_{298} = (35) *e.u.* (*11*)
M.P. = 781° K. (*29*)
ΔH_M = (6,000) calories per mole
B.P. = (2,150°) K. (*6*)
ΔH_V = (50,000) calories per mole

Formation: $Sm + Br_2 \longrightarrow SmBr_2$
(estimated (*11*))

T, ° K.	H_T-H_{298}	ΔH_T°	ΔF_T°
298	------------	(−157,000)	(−152,000)
500	(4,000)	(−163,900)	(−145,000)
1,000	(20,000)	(−156,700)	(−127,000)
1,500	(32,000)	(−155,900)	(−115,000)

Samarium Tribromide, $SmBr_3$ (*c*)

ΔH°_{298} = (−180,000) calories per mole (*5*)
S_{298} = (47) *e.u.* (*11*)
M.P. = 937° K. (*6*)
ΔH_M = (8,000) calories per mole
B.P. = 1,675° K. (*51*)
ΔH_V = 46,100 calories per mole

Formation: $Sm + 3/2Br_2 \longrightarrow SmBr_3$
(estimated (*11*))

T, ° K.	H_T-H_{298}	ΔH_T°	ΔF_T°
298	------------	(−180,000)	(−173,500)
500	(5,000)	(−190,600)	(−163,000)
1,000	(18,000)	(−188,300)	(−138,000)
1,500	(43,000)	(−177,200)	(−119,500)

Samarium Diiodide, SmI_2 (*c*)

ΔH°_{298} = (−122,000) calories per mole (*5*)
S_{298} = (40) *e.u.* (*11*)
M.P. = (773°) K. (*29*)
ΔH_M = (5,000) calories per mole
B.P. = (1,850°) K. (*6*)
ΔH_V = (40,000) calories per mole

Formation: $Sm + I_2 \longrightarrow SmI_2$
(estimated (*11*))

T, ° K.	H_T-H_{298}	ΔH_T°	ΔF_T°
298	------------	(−122,000)	(−121,000)
500	(4,000)	(−136,000)	(−114,500)
1,000	(19,000)	(−129,400)	(−99,000)
1,500	(31,000)	(−128,000)	(−88,500)

Samarium Triiodide, SmI_3 (*c*)

ΔH°_{298} = (−127,000) calories per mole (*5*)
S_{298} = (49) *e.u.* (*11*)
M.P. = 1,123° K. (*5*)
ΔH_M = (9,000) calories per mole
Decomposes (*6*)

Formation: $Sm + 3/2I_2 \longrightarrow SmI_3$
(estimated (*11*))

T, ° K.	H_T-H_{298}	ΔH_T°	ΔF_T°
298	------------	(−127,000)	(−125,000)
500	(5,000)	(−148,400)	(−113,000)
1,000	(19,000)	(−144,900)	(−86,000)
1,500	(44,000)	(−133,800)	(−62,500)

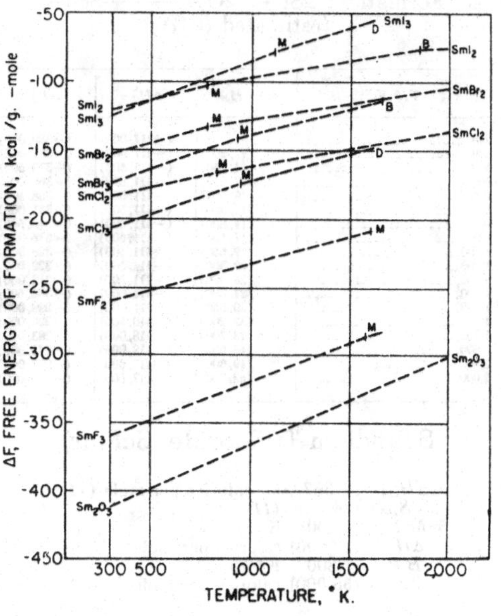

FIGURE 46.—Samarium.

SCANDIUM AND ITS COMPOUNDS

Element, Sc (*c*)

S_{298} = (9.00) *e.u.* (*7*)
M.P. = 1,673° K. (*130*)
ΔH_M = (3,850) calories per atom
B.P. = (2,750°) K. (*130*)
ΔH_V = (72,850) calories per atom
(estimated (*130*))

T, ° K.	$H_T - H_{298}$	S_T	$-\frac{(F_T - H_{298})}{T}$
298	----------	(9.00)	(9.00)
400	(617)	(10.78)	(9.24)
500	(1,235)	(12.16)	(9.69)
600	(1,860)	(13.30)	(10.20)
700	(2,500)	(14.29)	(10.72)
800	(3,150)	(15.15)	(11.22)
900	(3,850)	(15.93)	(11.70)
1,000	(4,480)	(16.64)	(12.16)
1,100	(5,170)	(17.29)	(12.59)
1,200	(5,860)	(17.89)	(13.01)
1,300	(6,560)	(18.46)	(13.42)
1,400	(7,280)	(18.99)	(13.79)
1,500	(8,010)	(19.49)	(14.15)
1,600	(8,760)	(19.96)	(14.50)
1,700	(13,350)	(22.72)	(14.87)
1,800	(14,150)	(23.18)	(15.32)
1,900	(14,950)	(23.61)	(15.75)
2,000	(15,750)	(24.02)	(16.15)
2,500	(19,750)	(25.81)	(17.91)

Discandium Trioxide, Sc_2O_3 (*c*)

ΔH°_{298} = (−411,000) calories per mole (*8*)
S_{298} = (18) *e.u.* (*24*)

Formation: $2Sc + 3/2O_2 \longrightarrow Sc_2O_3$
(estimated (*24*))

T, ° K.	$H_T - H_{298}$	ΔH°_T	ΔF°_T
298	----------	(−411,000)	(−389,000)
400	(2,300)	(−411,000)	(−381,000)
500	(4,650)	(−411,500)	(−373,500)
600	(7,050)	(−411,500)	(−366,000)
700	(9,500)	(−411,500)	(−358,500)
800	(12,000)	(−411,500)	(−351,000)
900	(14,600)	(−411,500)	(−343,500)
1,000	(17,100)	(−411,500)	(−336,000)
1,100	(19,650)	(−411,500)	(−328,000)
1,200	(22,300)	(−411,500)	(−320,000)
1,300	(24,900)	(−411,500)	(−313,000)
1,400	(27,600)	(−411,000)	(−305,500)
1,500	(30,550)	(−411,000)	(−298,000)
1,600	(33,200)	(−410,500)	(−290,000)
1,700	(43,700)	(−418,000)	(−283,000)
1,800	(46,600)	(−418,000)	(−275,000)
1,900	(49,600)	(−417,500)	(−267,000)
2,000	(52,700)	(−417,500)	(−259,000)

Scandium Trifluoride, ScF_3 (*c*)

ΔH°_{298} = (−367,000) calories per mole (*11*)
S_{298} = (24) *e.u.* (*11*)
M.P. = (1,500°) K. (*6*)
ΔH_M = (12,000) calories per mole
B.P. = (1,800°) K. (*6*)
ΔH_V = (55,000) calories per mole

Formation: $Sc + 3/2F_2 \longrightarrow ScF_3$
(estimated (*11*))

T, ° K.	$H_T - H_{298}$	ΔH°_T	ΔF°_T
298	----------	(−367,000)	(−349,700)
500	(4,000)	(−366,500)	(−338,000)
1,000	(17,000)	(−363,000)	(−311,000)
1,500	(43,000)	(−347,500)	(−287,500)

Scandium Trichloride, $ScCl_3$ (*c*)

ΔH°_{298} = −221,000 calories per mole (*11*)
S_{298} = (32) *e.u.* (*11*)
M.P. = 1,213° K. (29)
ΔH_M = (19,000) calories per mole
B.P. = 1,240° K. (*6*)
ΔH_V = 46,000 calories per mole

Formation: $Sc + 3/2Cl_2 \longrightarrow ScCl_3$
(estimated (*11*))

T, ° K.	$H_T - H_{298}$	ΔH°_T	ΔF°_T
298	----------	−221,000	(−204,000)
500	(4,000)	(−220,800)	(−193,000)
1,000	(17,000)	(−217,600)	(−166,000)
1,500	(97,000)	(−147,500)	(−146,000)

Scandium Tribromide, $ScBr_3$ (*c*)

ΔH°_{298} = −183,000 calories per mole (*11*)
S_{298} = (40) *e.u.* (*11*)
M.P. = 1,213° K. (29)
ΔH_M = (19,000) calories per mole

Formation: $Sc + 3/2Br_2 \longrightarrow ScBr_3$
(estimated (*11*))

T, ° K.	$H_T - H_{298}$	ΔH°_T	ΔF°_T
298	----------	−183,000	(−165,000)
500	(5,000)	(−193,500)	(−161,000)
1,000	(18,000)	(−190,500)	(−136,000)
1,500	(58,000)	(−160,800)	(−117,000)

Scandium Triiodide, ScI_3 (*c*)

ΔH°_{298} = (−109,000) calories per mole (*11*)
S_{298} = (44) *e.u.* (*11*)
M.P. = 1,218° K. (*6*)
ΔH_M = (18,000) calories per mole

Formation: $Sc + 3/2I_2 \longrightarrow ScI_3$
(estimated (*11*))

T, ° K.	$H_T - H_{298}$	ΔH°_T	ΔF°_T
298	----------	(−109,000)	(−102,500)
500	(5,000)	(−130,200)	(−94,000)
1,000	(19,000)	(−126,200)	(−69,000)
1,500	(52,000)	(−103,400)	(−51,000)

Scandium Nitride, ScN (*c*)

ΔH°_{298} = −68,000 calories per mole (*9*)
S_{298} = 7 *e.u.* (*9*)
ΔS_{298} = (−25) *e.u.*
ΔF_{298} = (−60,500) calories per mole
M.P. = 2,923° K. (*9*)

SILICON (REFERENCE STATE) GRAM FORMULA WEIGHT 28.086

Si: Crystals 298.15° to melting point 1685°K. Liquid 1685° to 2000°K.

TEMP. DEG K	$H_T - H_{298}$ (KCAL)	S_T (CAL/DEG-GFW)	$-(G_T - H_{298})/T$ (CAL/DEG-GFW)	FORMATION FROM THE ELEMENTS: ENTHALPY (KCAL/GFW)	FREE ENERGY (KCAL/GFW)	LOG K
298.15	0.000	4.500	4.500	0.000	0.000	0.000
UNCERTAINTY		0.020	0.020			
400	0.516	5.980	4.690	.000	.000	.000
500	1.062	7.200	5.076	.000	.000	.000
600	1.634	8.240	5.517	.000	.000	.000
700	2.225	9.150	5.971	.000	.000	.000
800	2.831	9.960	6.421	.000	.000	.000
900	3.451	10.690	6.856	.000	.000	.000
1000	4.081	11.360	7.279	.000	.000	.000
1100	4.720	11.970	7.679	.000	.000	.000
1200	5.366	12.530	8.058	.000	.000	.000
1300	6.018	13.050	8.421	.000	.000	.000
1400	6.676	13.540	8.771	.000	.000	.000
1500	7.340	13.990	9.097	.000	.000	.000
1600	8.010	14.420	9.414	.000	.000	.000
1685	8.583	14.770	9.676	.000	.000	.000
1685	20.665	21.940	9.676	.000	.000	.000
1700	20.756	21.990	9.781	.000	.000	.000
1800	21.366	22.340	10.470	.000	.000	.000
1900	21.976	22.670	11.104	.000	.000	.000
2000	22.586	22.980	11.687	.000	.000	.000

MELTING POINT 1685 DEG K BOILING POINT 3553 DEG K

HEAT OF FUSION 12.082 KCAL HEAT OF VAPOR. 93.891 KCAL

$H_{298} - H_0$ 0.7690 KCAL MOLAR VOLUME 0.28815 CAL/BAR

TRANSITIONS IN REFERENCE STATE ELEMENTS

SILICON MONOXIDE GRAM FORMULA WEIGHT 44.085

SiO: Ideal gas 298.15° to 2000°K.

TEMP. DEG K	$H_T - H_{298}$ (KCAL)	S_T (CAL/DEG-GFW)	$-(G_T - H_{298})/T$ (CAL/DEG-GFW)	FORMATION FROM THE ELEMENTS: ENTHALPY (KCAL/GFW)	FREE ENERGY (KCAL/GFW)	LOG K
298.15	0.000	50.55	50.55	-23.800	-30.226	22.156
UNCERTAINTY		0.20	0.20	1.000	1.010	0.740
400	0.742	52.69	50.83	-23.936	-32.403	17.704
500	1.501	54.38	51.38	-24.088	-34.501	15.081
600	2.288	55.82	52.00	-24.251	-36.570	13.320
700	3.096	57.06	52.64	-24.423	-38.610	12.055
800	3.922	58.16	53.26	-24.602	-40.624	11.098
900	4.761	59.15	53.86	-24.790	-42.615	10.348
1000	5.610	60.05	54.44	-24.984	-44.578	9.743
1100	6.468	60.86	54.98	-25.185	-46.528	9.244
1200	7.332	61.62	55.51	-25.391	-48.462	8.826
1300	8.201	62.31	56.00	-25.602	-50.379	8.469
1400	9.076	62.96	56.48	-25.817	-52.271	8.160
1500	9.953	63.56	56.93	-26.040	-54.166	7.892
1600	10.834	64.13	57.36	-26.267	-56.039	7.655
1700	11.718	64.67	57.78	-38.570	-57.790	7.429
1800	12.604	65.18	58.17	-38.739	-58.912	7.153
1900	13.493	65.66	58.55	-38.907	-60.027	6.905
2000	14.383	66.11	58.92	-39.077	-61.141	6.681

MELTING POINT DEG K BOILING POINT DEG K

HEAT OF FUSION KCAL HEAT OF VAPOR. KCAL

$H_{298} - H_0$ 2.0820 KCAL MOLAR VOLUME 584.727 CAL/BAR

TRANSITIONS IN REFERENCE STATE ELEMENTS

SILICON.... M. P. 1685 DEG K.

QUARTZ GRAM FORMULA WEIGHT 60.085

SiO_2: α quartz 298.15° to 848°K. β quartz 848° to 2000°K.

β quartz is metastable above 1140°K.

TEMP. DEG K	$H_T - H_{298}$ (KCAL)	S_T (CAL/DEG-GFW)	$-(G_T - H_{298})/T$ (CAL/DEG-GFW)	FORMATION FROM THE ELEMENTS: ENTHALPY (KCAL/GFW)	FREE ENERGY (KCAL/GFW)	LOG K
298.15	0.000	9.88	9.88	-217.650	-204.646	150.009
UNCERTAINTY		0.02	0.02	0.400	0.410	0.301
400	1.200	13.33	10.33	-217.690	-200.197	109.382
500	2.560	16.36	11.24	-217.607	-195.830	85.597
600	4.040	19.05	12.32	-217.454	-191.486	69.749
700	5.630	21.50	13.46	-217.233	-187.176	58.439
800	7.320	23.76	14.61	-216.947	-182.905	49.967
848	8.170	24.79	15.16	-216.787	-180.870	46.615
848	8.460	25.13	15.16	-216.497	-180.870	46.615
900	9.300	26.09	15.76	-216.401	-178.680	43.389
1000	10.920	27.80	16.88	-216.238	-174.494	38.136
1100	12.570	29.37	17.94	-216.066	-170.325	33.840
1140	13.247	29.97	18.35	-215.997	-168.662	32.334
1200	14.250	30.83	18.95	-215.880	-166.175	30.264
1300	15.940	32.18	19.92	-215.699	-162.039	27.241
1400	17.640	33.44	20.84	-215.521	-157.915	24.652
1500	19.360	34.63	21.72	-215.336	-153.824	22.412
1600	21.100	35.76	22.57	-215.143	-149.745	20.456
1700	22.860	36.82	23.37	-227.011	-145.549	18.712
1800	24.630	37.84	24.16	-226.740	-140.777	17.093
1900	26.420	38.81	24.90	-226.455	-136.013	15.645
2000	28.220	39.73	25.62	-226.165	-131.261	14.343

MELTING POINT	DEG K	BOILING POINT	DEG K
HEAT OF FUSION	KCAL	HEAT OF VAPOR.	KCAL
$H_{298} - H_0$	1.6257 KCAL	MOLAR VOLUME	0.54226 CAL/BAR

TRANSITIONS IN REFERENCE STATE ELEMENTS

SILICON.... M. P. 1685 DEG K.

CRISTOBALITE GRAM FORMULA WEIGHT 60.085

SiO_2: α cristobalite 298.15° to 523°K. β cristobalite 523° to 2000°K. Cristobalite melts at 1996°K.

TEMP. DEG K	$H_T - H_{298}$ (KCAL)	S_T (CAL/DEG-GFW)	$-(G_T - H_{298})/T$ (CAL/DEG-GFW)	FORMATION FROM THE ELEMENTS: ENTHALPY (KCAL/GFW)	FREE ENERGY (KCAL/GFW)	LOG K
298.15	0.000	10.38	10.38	-216.930	-204.075	149.591
UNCERTAINTY		0.02	0.02	0.450	0.460	0.337
400	1.210	13.86	10.83	-216.960	-199.679	109.099
500	2.560	16.86	11.74	-216.887	-195.360	85.392
523	2.910	17.54	11.97	-216.870	-194.367	81.223
523	3.110	17.92	11.97	-216.670	-194.367	81.223
600	4.310	20.06	12.88	-216.464	-191.102	69.609
700	5.850	22.43	14.07	-216.293	-186.837	58.348
800	7.460	24.58	15.25	-216.087	-182.701	49.911
900	9.090	26.50	16.40	-215.891	-178.539	43.355
1000	10.730	28.23	17.50	-215.708	-174.394	38.114
1100	12.390	29.81	18.55	-215.526	-170.269	33.829
1200	14.080	31.28	19.55	-215.330	-166.165	30.263
1300	15.790	32.65	20.50	-215.129	-162.080	27.248
1400	17.510	33.92	21.41	-214.931	-157.997	24.664
1500	19.240	35.12	22.29	-214.736	-153.959	22.432
1600	20.990	36.25	23.13	-214.533	-149.919	20.478
1700	22.750	37.31	23.93	-226.401	-145.772	18.740
1743	23.513	37.75	24.26	-226.280	-143.735	18.022
1800	24.530	38.33	24.70	-226.120	-141.039	17.124
1900	26.320	39.30	25.45	-225.835	-136.324	15.681
1996	28.050	40.19	26.14	-225.557	-131.809	14.432
2000	28.120	40.22	26.16	-225.545	-131.621	14.383

MELTING POINT	1996 DEG K	BOILING POINT	DEG K
HEAT OF FUSION	1.950 KCAL	HEAT OF VAPOR.	KCAL
$H_{298} - H_0$	1.6875 KCAL	MOLAR VOLUME	0.61518 CAL/BAR

TRANSITIONS IN REFERENCE STATE ELEMENTS

SILICON.... M. P. 1685 DEG K.

TRIDYMITE GRAM FORMULA WEIGHT 60.085

SiO_2: α tridymite 298.15° to 390°K. β tridymite 390° to 2000°K.

TEMP. DEG K	$H_T - H_{298}$ (KCAL)	S_T (CAL/DEG-GFW)	$-(G_T - H_{298})/T$ (CAL/DEG-GFW)	FORMATION FROM THE ELEMENTS: ENTHALPY (KCAL/GFW)	FREE ENERGY (KCAL/GFW)	LOG K
298.15	0.000	10.50	10.50	-216.895	-204.076	149.591
UNCERTAINTY		0.10	0.10	0.570	0.580	0.425
390	1.085	13.66	10.88	-216.920	-200.120	112.143
390	1.125	13.76	10.88	-216.880	-200.120	112.143
400	1.270	14.13	10.95	-216.865	-199.692	109.106
500	2.710	17.34	11.92	-216.702	-195.415	85.416
600	4.170	20.00	13.05	-216.569	-191.171	69.634
700	5.710	22.37	14.21	-216.398	-186.950	58.368
800	7.320	24.52	15.37	-216.192	-182.758	49.927
900	8.950	26.44	16.50	-215.996	-178.590	43.368
1000	10.590	28.17	17.58	-215.813	-174.439	38.124
1100	12.250	29.75	18.61	-215.631	-170.308	33.837
1140	12.928	30.25	19.01	-215.555	-168.662	22.334
1200	13.940	31.22	19.60	-215.435	-166.198	30.269
1300	15.650	32.59	20.55	-215.234	-162.107	27.253
1400	17.370	33.87	21.46	-215.036	-158.032	24.670
1500	19.100	35.06	22.33	-214.841	-153.974	22.434
1600	20.850	36.19	23.16	-214.638	-149.928	20.479
1700	22.610	37.25	23.95	-226.506	-145.775	18.741
1743	23.373	37.69	24.28	-226.386	-143.735	18.022
1800	24.390	38.27	24.72	-226.225	-141.036	17.124
1900	26.180	39.24	25.46	-225.940	-136.315	15.680
2000	27.980	40.16	26.17	-225.650	-131.606	14.381

MELTING POINT	DEG K	BOILING POINT	DEG K
HEAT OF FUSION	KCAL	HEAT OF VAPOR.	KCAL
$H_{298} - H_0$	KCAL	MOLAR VOLUME 0.63408	CAL/BAR

TRANSITIONS IN REFERENCE STATE ELEMENTS

SILICON.... M. P. 1685 DEG K.

Silicon Carbide, Alpha(α-SiC)

(Crystal) GFW = 40.09715

T, °K	Cp° (gibbs/mol)	S° (gibbs/mol)	$-(G°-H°_{298})/T$ (gibbs/mol)	$H°-H°_{298}$ (kcal/mol)	ΔHf° (kcal/mol)	ΔGf° (kcal/mol)	Log Kp
0	.000	.000	INFINITE	-.777	-16.856	-16.555	INFINITE
100	1.003	.338	7.855	-.757	-16.909	-16.930	36.782
200	3.583	1.931	4.449	-.515	-17.026	-16.704	18.253
298	6.396	3.940	3.940	.000	-17.100	-16.525	12.116
300	6.437	3.980	3.940	.012	-17.101	-16.525	12.039
400	8.150	6.086	4.216	.748	-17.119	-16.379	8.922
500	9.228	8.029	4.787	1.621	-17.109	-16.137	7.051
600	9.972	9.781	5.476	2.583	-17.092	-15.940	5.806
700	10.518	11.361	6.206	3.609	-17.077	-15.745	4.917
800	10.939	12.794	6.941	4.687	-17.062	-15.559	4.251
900	11.274	14.102	7.665	5.793	-17.050	-15.372	3.733
1000	11.549	15.305	8.370	6.935	-17.035	-15.187	3.319
1100	11.779	16.417	9.051	8.102	-17.029	-15.003	2.981
1200	11.976	17.450	9.709	9.290	-17.022	-14.818	2.699
1300	12.147	18.416	10.342	10.494	-17.020	-14.634	2.461
1400	12.297	19.321	10.951	11.719	-17.019	-14.450	2.256
1500	12.431	20.174	11.538	12.955	-17.021	-14.266	2.079
1600	12.552	20.981	12.103	14.208	-17.027	-14.085	1.924
1700	12.661	21.745	12.648	15.465	-29.028	-13.795	1.773
1800	12.762	22.472	13.174	16.736	-28.986	-12.899	1.566
1900	12.854	23.164	13.681	18.017	-28.937	-12.006	1.381
2000	12.940	23.826	14.172	19.307	-28.882	-11.114	1.215
2100	13.020	24.459	14.647	20.605	-28.821	-10.224	1.064
2200	13.094	25.066	15.107	21.910	-28.757	-9.336	.928
2300	13.165	25.650	15.553	23.223	-28.686	-8.464	.804
2400	13.232	26.212	15.985	24.543	-28.611	-7.587	.691
2500	13.296	26.753	16.405	25.870	-28.530	-6.711	.587
2600	13.356	27.276	16.813	27.202	-28.447	-5.842	.491
2700	13.414	27.781	17.210	28.541	-28.358	-4.976	.403
2800	13.470	28.270	17.597	29.885	-28.265	-4.109	.321
2900	13.523	28.743	17.973	31.235	-28.169	-3.247	.245
3000	13.575	29.203	18.340	32.590	-28.069	-2.392	.174
3100	13.625	29.649	18.697	33.950	-27.965	-1.539	.108
3200	13.673	30.082	19.046	35.314	-27.859	-.684	.047
3300	13.720	30.503	19.387	36.684	-27.749	.163	-.011
3400	13.765	30.914	19.720	38.059	-27.636	1.005	-.065
3500	13.810	31.313	20.046	39.437	-27.520	1.845	-.115
3600	13.853	31.703	20.364	40.820	-119.362	4.939	-.300
3700	13.895	32.083	20.676	42.209	-119.141	8.390	-.496
3800	13.937	32.454	20.981	43.599	-118.915	11.837	-.681
3900	13.977	32.817	21.280	44.995	-118.693	15.272	-.856
4000	14.017	33.171	21.572	46.395	-118.465	18.707	-1.022

Silicon Carbide, Beta (β -SiC)

(Crystal) GFW = 40.09715

T, °K	Cp° (gibbs/mol)	S° (gibbs/mol)	$-(G°-H°_{298})/T$ (gibbs/mol)	$H°-H°_{298}$ (kcal/mol)	ΔHf° (kcal/mol)	ΔGf° (kcal/mol)	Log Kp
0	.000	.000	INFINITE	-.782	-17.261	-17.261	INFINITE
100	1.019	.349	7.902	-.755	-17.312	-17.235	37.666
200	3.596	1.920	4.479	-.512	-17.428	-17.110	18.697
298	6.415	3.970	3.970	.000	-17.500	-16.937	12.415
300	6.455	4.010	3.970	.012	-17.501	-16.934	12.336
400	8.150	6.118	4.746	.749	-17.517	-16.741	9.147
500	9.235	8.061	4.818	1.629	-17.507	-16.547	7.233
600	9.985	9.815	5.507	2.585	-17.490	-16.359	5.958
700	10.542	11.398	6.238	3.613	-17.472	-16.170	5.048
800	10.965	12.835	6.974	4.689	-17.455	-15.986	4.367
900	11.301	14.146	7.699	5.803	-17.440	-15.802	3.837
1000	11.573	15.351	8.405	6.947	-17.426	-15.621	3.414
1100	11.798	16.465	9.088	8.116	-17.415	-15.442	3.068
1200	11.988	17.530	9.746	9.305	-17.407	-15.263	2.780
1300	12.151	18.456	10.380	10.512	-17.404	-15.085	2.536
1400	12.292	19.372	10.990	11.735	-17.402	-14.905	2.327
1500	12.415	20.224	11.578	12.970	-17.406	-14.726	2.146
1600	12.524	21.029	12.144	14.217	-17.414	-14.549	1.987
1700	12.621	21.791	12.689	15.474	-29.419	-14.263	1.834
1800	12.708	22.515	13.215	16.741	-29.381	-13.372	1.624
1900	12.786	23.205	13.723	18.016	-29.335	-12.485	1.436
2000	12.855	23.862	14.213	19.298	-29.291	-11.597	1.267
2100	12.923	24.491	14.688	20.587	-29.239	-10.713	1.115
2200	12.953	25.094	15.147	21.882	-29.185	-9.836	.977
2300	13.039	25.672	15.592	23.183	-29.126	-8.955	.851
2400	13.090	26.225	16.024	24.490	-29.064	-8.078	.736
2500	13.137	26.764	16.443	25.801	-28.999	-7.209	.630
2600	13.182	27.280	16.850	27.117	-28.932	-6.338	.533
2700	13.224	27.778	17.245	28.439	-28.861	-5.471	.443
2800	13.263	28.260	17.630	29.762	-28.788	-4.604	.359
2900	13.299	28.726	18.005	31.090	-28.714	-3.742	.282
3000	13.334	29.177	18.370	32.422	-28.637	-2.882	.210
3100	13.367	29.615	18.726	33.757	-28.558	-2.025	.143
3200	13.398	30.040	19.072	35.095	-28.478	-1.169	.080
3300	13.428	30.452	19.411	36.436	-28.397	-.317	.021
3400	13.456	30.854	19.742	37.781	-28.313	.532	-.034
3500	13.483	31.244	20.065	39.128	-28.229	1.377	-.086
3600	13.509	31.624	20.381	40.477	-120.105	4.480	-.272
3700	13.534	31.995	20.690	41.829	-119.920	7.935	-.469
3800	13.557	32.356	20.992	43.184	-119.733	11.394	-.655
3900	13.580	32.709	21.285	44.541	-119.547	14.839	-.832
4000	13.602	33.053	21.578	45.900	-119.360	18.284	-.999

SILVER AND ITS COMPOUNDS

Element, Ag (*c*)

$S_{298} = 10.20$ *e.u.* (*83*)
$M.P. = 1{,}234°$ K. (*82*)
$\Delta H_M = 2{,}855$ calories per atom
$B.P. = 2{,}450°$ K. (7)
$\Delta H_V = 60{,}720$ calories per atom

Zone I (*c*) (298°–1,234° K.)

$$C_p = 5.09 + 2.04\times10^{-3}T + 0.36\times10^{5}T^{-2} \quad (82)$$
$$H_T - H_{298} = -1{,}488 + 5.09T + 1.02\times10^{-3}T^2 - 0.36\times10^{5}T^{-1}$$
$$F_T - H_{298} = -1{,}488 - 5.09T\ln T - 1.02\times10^{-3}T^2 - 0.18\times10^{5}T^{-1} + 24.29T$$

Zone II (*l*) (1,234°–1,600° K.)

$$C_p = 7.30 \quad (82)$$
$$H_T - H_{298} = +160 + 7.30T$$
$$F_T - H_{298} = +160 - 7.30T\ln T + 37.42T$$

T, ° K.	$H_T - H_{298}$	S_T	$-\frac{(F_T - H_{298})}{T}$
298		10.20	10.20
400	615	11.78	10.25
500	1,240	13.37	10.90
600	1,885	14.55	11.42
700	2,535	15.55	11.93
800	3,195	16.43	12.44
900	3,880	17.24	12.93
1,000	4,585	17.98	13.40
1,100	5,210	18.67	13.84
1,200	6,060	19.32	14.27
1,300	9,650	22.22	14.80
1,400	10,380	22.76	15.34
1,500	11,110	23.26	15.85
1,600	11,840	23.74	16.33
1,700	(12,570)	(24.18)	(16.78)
1,800	(13,300)	(24.60)	(17.21)
1,900	(14,030)	(24.99)	(17.55)
2,000	(14,760)	(25.36)	(17.96)

Disilver Oxide, Ag_2O (*c*)

$\Delta H^{\circ}_{298} = -7{,}200$ calories per mole (*24*)
$S_{298} = 29.1$ *e.u.* (*24*)

Formation: $2Ag + 1/2O_2 \longrightarrow Ag_2O$
(estimated (*24*))

T, ° K.	$H_T - H_{298}$	ΔH°_T	ΔF°_T
298		−7,200	−2,500
400	(1,800)	(−7,000)	(−900)
500	(3,550)	(−6,850	(600)
600	(5,400)	(−6,650)	(2,050)
700	(7.250)	(−6,500)	(3,500)
800	(9,200)	(−6,300)	(4,900)
900	(11,150)	(−6,100	(6,300)
1,000	(13,100)	(−5,950)	(7,700)

Disilver Dioxide, Ag_2O_2 (*c*)

$\Delta H^{\circ}_{298} = -6{,}200$ calories per mole (*112*)
$S_{298} = (26.4)$ *e.u.* (*24*)

Formation: $2Ag + O_2 \longrightarrow Ag_2O_2$
(estimated (*24*))

T, ° K.	$H_T - H_{298}$	ΔH°_T	ΔF°_T
298		−6,200	(+6,600)
400	(2,050)	(−6,100)	(+10,900)
500	(4,250)	(−5,900)	(+15,100)

Silver Fluoride, AgF (*c*)

$\Delta H^{\circ}_{298} = -48{,}700$ calories per mole (*112*)
$S_{298} = (21)$ *e.u.* (*11*)
$M.P. = 708°$ K. (*6*)
$B.P. = 1{,}420°$ K. (*6*)

Formation: $Ag + 1/2F_2 \longrightarrow AgF$
(estimated (*11*))

T, ° K.	$H_T - H_{298}$	ΔH°_T	ΔF°_T
298		−48,700	(−44,500)
500	(3,000)	(−47,700)	(−41,700)
1,000	(12,000)	(−44,200)	(−38,700)
1,500	(19,000)	(−45,900)	(−35,200)

Silver Difluoride, AgF_2 (*c*)

$\Delta H^{\circ}_{298} = -83{,}000$ calories per mole (*11*)
$S_{298} = (25)$ *e.u.* (*11*)
$M.P. = >963°$ K. (*6*)

Formation: $Ag + F_2 \longrightarrow AgF_2$
(estimated (*11*))

T, ° K.	$H_T - H_{298}$	ΔH°_T	ΔF°_T
298		−83,000	(−72,900)
500	(4,000)	(−81,800)	(−66,000)
1,000	(17,000)	(−76,400)	(−51,000)

Silver Chloride, AgCl (*c*)

$\Delta H^{\circ}_{298} = -30{,}360$ calories per mole (*112*)
$S_{298} = 22.97$ *e.u.* (*83*)
$M.P. = 728°$ K. (*6*)
$\Delta H_M = 3{,}155$ calories per mole
$B.P. = 1{,}837°$ K. (*6*)
$\Delta H_V = 42{,}520$ calories per mole

Zone I (*c*) (298°–728° K.)

$$C_p = 14.88 + 1.00\times10^{-3}T - 2.70\times10^{5}T^{-2} \quad (82)$$
$$H_T - H_{298} = -5{,}390 + 14.88T + 0.50\times10^{-3}T^2 + 2.70\times10^{5}T^{-1}$$

Zone II (*l*) (728°–900° K.)

$$C_p = 16.0 \quad (82)$$
$$H_T - H_{298} = -2{,}490 + 16.0T$$

Formation: $Ag + 1/2Cl_2 \longrightarrow AgCl$

Zone I (298°–728° K.)

$$\Delta C_p = 5.38 - 1.07\times10^{-3}T - 2.72\times10^{5}T^{-2}$$
$$\Delta H_T = -32{,}830 + 5.38T - 0.535\times10^{-3}T^2 + 2.72\times10^{5}T^{-1}$$
$$\Delta F_T = -32{,}830 - 5.38T\ln T + 0.535\times10^{-3}T^2 + 1.36\times10^{5}T^{-1} + 51.2T$$

Zone II (728°–900° K.)

$\Delta C_p = 6.50 - 2.07\times10^{-3}T - 0.02\times10^5T^{-2}$
$\Delta H_T = -29{,}940 + 6.50T - 1.03\times10^{-3}T^2 + 0.02\times10^5T^{-1}$
$\Delta F_T = -29{,}940 - 6.50T\ln T + 1.03\times10^{-3}T^2 + 0.01\times10^5T^{-1} + 54.5T$

T, ° K.	$H_T - H_{298}$	S_T	ΔH_T°	ΔF_T°
298	------------	22.97	−30,350	−26,200
400	1,320	26.78	−30,100	−25,900
500	2,720	29.89	−29,700	−23,550
600	4,150	32.51	−29,350	−22,350
700	5,660	34.83	−28,950	−21,250
800	10,310	41.16	−25,400	−20,450
900	11,910	42.94	−24,900	−19,800
1,000	(13,500)	(44.64)	(−24,500)	(−19,100)
1,500	(20,200)	(50.0)	(−26,450)	(−16,000)

Silver Bromide, AgBr (*c*)

$\Delta H_{298}^\circ = -20{,}060$ calories per mole (*112*)
$S_{298} = 25.60$ *e.u.* (*83*)
M.P. = 703° K. (*82*)
$\Delta H_M = 2{,}190$ calories per mole
B.P. = (1,810°) K. (*6*)
$\Delta H_V = (37{,}000)$ calories per mole

Zone I (*c*) (298°–703° K.)

$C_p = 7.93 + 15.40\times10^{-3}T$ (*82*)
$H_T - H_{298} = -3{,}049 + 7.93T + 7.70\times10^{-3}T^2$

Zone II (*l*) (703°–900° K.)

$C_p = 14.9$ (*82*)
$H_T - H_{298} = 1{,}950 + 14.9T$

Formation: $Ag + 1/2Br_2 \longrightarrow AgBr$

Zone I (298°–331° K.)

$\Delta C_p = -5.71 + 13.36\times10^{-3}T - 0.36\times10^5T^{-2}$
$\Delta H_T = -19{,}050 - 5.71T + 6.68\times10^{-3}T^2 + 0.36\times10^5T^{-1}$
$\Delta F_T = -19{,}050 + 5.71T\ln T - 6.68\times10^{-3}T^2 + 0.18\times10^5T^{-1} - 31.2T$

Zone II (331°–703° K.)

$\Delta C_p = -1.68 + 13.36\times10^{-3}T - 0.18\times10^5T^{-2}$
$\Delta H_T = -24{,}100 - 1.68T + 6.68\times10^{-3}T^2 + 0.18\times10^5T^{-1}$
$\Delta F_T = -24{,}100 + 1.68T\ln T - 6.68\times10^{-3}T^2 + 0.09\times10^5T^{-1} + 7.35T$

Zone III (703°–900° K.)

$\Delta C_p = 5.29 - 2.04\times10^{-3}T - 0.18\times10^5T^{-2}$
$\Delta H_T = -22{,}950 + 5.29T - 1.02\times10^{-3}T^2 + 0.18\times10^5T^{-1}$
$\Delta F_T = -22{,}950 - 5.29T\ln T + 1.02\times10^{-3}T^2 + 0.09\times10^5T^{-1} + 46.22T$

T, ° K.	$H_T - H_{298}$	S_T	ΔH_T°	ΔF_T°
298	------------	25.60	−20,060	−19,200
400	1,355	29.50	−23,650	−18,200
500	2,840	32.81	−23,200	−16,750
600	4,480	35.79	−22,650	−15,500
700	6,275	38.56	−22,000	−14,400
800	9,970	43.68	−19,400	−13,650
900	11,460	45.43	−19,000	−12,900
1,000	(12,950)	(47.00)	(−18,700)	(12,300)

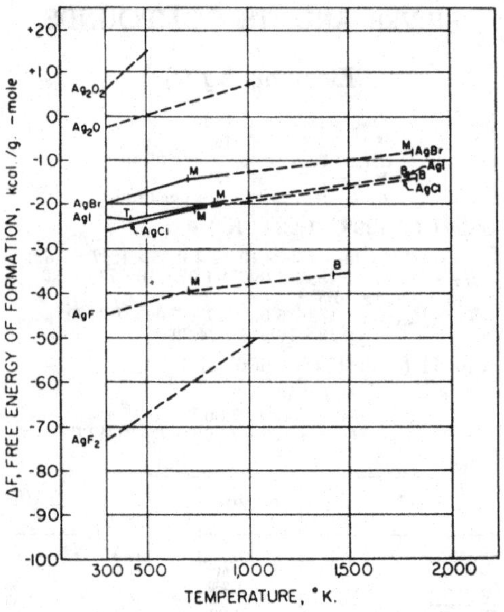

FIGURE 49.—Silver.

Silver Iodide, AgI (*c*)

$\Delta H_{298}^\circ = -22{,}300$ calories per mole (*112*)
$S_{298} = 27.6$ *e.u.* (*83*)
T.P. = 423° K. (*82*)
$\Delta H_T = 1{,}470$ calories per mole
M.P. = 830° K. (*6*)
$\Delta H_M = 2{,}250$ calories per mole
B.P. = 1,779° K. (*6*)
$\Delta H_V = 34{,}447$ calories per mole

Zone I (α) (298°–423° K.)

$C_p = 5.82 + 24.10\times10^{-3}T$ (*82*)
$H_T - H_{298} = -2{,}806 + 5.82T + 12.05\times10^{-3}T^2$

Zone II (β) (423°–600° K.)

$C_p = 13.5$ (*82*)
$H_T - H_{298} = -2{,}430 + 13.5T$

Formation: $Ag + 1/2I_2 \longrightarrow AgI$

Zone I (298°–386.8° K.)

$\Delta C_p = -4.06 + 16.11\times10^{-3}T - 0.36\times10^5T^{-2}$
$\Delta H_T = -21{,}925 - 4.06T + 8.05\times10^{-3}T^2 + 0.36\times10^5T^{-1}$
$\Delta F_T = -21{,}925 + 4.06T\ln T - 8.05\times10^{-3}T^2 + 0.18\times10^5T^{-1} - 25.44T$

Zone II (386.8°–423° K.)

$\Delta C_p = -8.87 + 22.06\times10^{-3}T - 0.36\times10^5T^{-2}$
$\Delta H_T = -22{,}390 - 8.87T + 11.03\times10^{-3}T^2 + 0.36\times10^5T^{-1}$
$\Delta F_T = -22{,}390 + 8.87T\ln T - 11.03\times10^{-3}T^2 + 0.18\times10^5T^{-1} - 52.11T$

Zone III (423°–600° K.)

$\Delta C_p = 3.97 - 2.04 \times 10^{-3} T - 0.36 \times 10^5 T^{-2}$

$\Delta H_T = -29{,}350 + 3.97T - 1.02 \times 10^{-3} T^2 + 0.36 \times 10^5 T^{-1}$

$\Delta F_T = -29{,}350 - 3.97 T \ln T + 1.02 \times 10^{-3} T^2 + 0.18 \times 10^5 T^{-1} + 36.18T$

T, ° K.	$H_T - H_{298}$	S_T	ΔH_T°	ΔF_T°
298		27.6	−22,300	−23,300
400	1,450	31.77	−24,100	−23,700
500	4,320	38.38	−27,550	−23,300
600	5,670	40.84	−27,300	−22,500
1,000	(12,900)	(49.1)	(−24,550)	(−18,950)
1,500	(20,400)	(55.2)	(−25,800)	(−16,100)

SODIUM AND ITS COMPOUNDS

Element, Na (*c*)

$S_{298} = 12.23$ *e.u.* (*83*)
M.P. = 371° K. (*41*)
$\Delta H_M = 630$ calories per atom
B.P. = 1,162° K. (*41*)
$\Delta H_V = 23{,}120$ calories per atom

Zone I (*c*) (298°–391° K.)

$C_p = 9.9261 - 28.038 \times 10^{-3} T + 5.785 \times 10^{-5} T^2$ (*41*)

$H_T - H_{298} = -2{,}235 + 9.93T - 14.02 \times 10^{-3} T^2 + 1.93 \times 10^{-5} T^3$

$F_T - H_{298} = -2{,}235 - 9.93 T \ln T + 14.02 \times 10^{-3} T^2 - 0.96 \times 10^{-5} T^3 + 48.35T$

Zone II (*l*) (391°–1,162° K.)

$C_p = 9.0696 - 45.765 \times 10^{-4} T + 2.54 \times 10^{-6} T^2$ (*41*)

$H_T - H_{298} = -1{,}960 + 9.07T - 22.88 \times 10^{-4} T^2 + 0.85 \times 10^{-6} T^3$

$F_T - H_{298} = -1{,}960 - 9.07 T \ln T + 22.88 \times 10^{-4} T^2 - 0.42 \times 10^{-6} T^3 + 45.66T$

Zone III (*g*) (1,162°–2,500° K.)

$C_p = 4.87$ (*34*)

$H_T - H_{298} = 24{,}530 + 4.87T$

$F_T - H_{298} = 24{,}530 - 4.87 T \ln T - 4.23T$

T, ° K.	$H_T - H_{298}$	S_T	$-\frac{(F_T - H_{298})}{T}$
298		12.23	12.23
400	1,355	16.08	12.69
500	2,097	17.73	13.52
600	2,819	19.05	14.33
700	3,521	20.13	15.11
800	4,218	21.06	15.79
900	4,909	21.88	16.44
1,000	5,597	22.60	17.00
1,100	6,294	23.26	17.55
1,200	30,380	43.63	18.58
1,300	30,877	44.03	20.46
1,400	31,374	44.40	22.07
1,500	31,871	44.74	23.60
1,600	32,367	45.06	24.94
1,700	32,864	45.36	26.12
1,800	33,361	45.65	27.17
1,900	33,858	45.91	28.10
2,000	34,255	46.17	28.95
2,500	36,847	47.28	32.64

Disodium Oxide, Na_2O (*c*)

$\Delta H_{298}^\circ = -99{,}400$ calories per mole (*112*)
$S_{298} = 17.4$ *e.u.* (*112*)
M.P. = 1,190° K. (*24*)
$\Delta H_M = 7{,}140$ calories per mole
B.P. = >2,500° K. (*42*)

Zone I (*c*) (298°–1,100° K.)

$C_p = 15.70 + 5.40 \times 10^{-3} T$ (*82*)

$H_T - H_{298} = -4{,}920 + 15.70T + 2.70 \times 10^{-3} T^2$

Formation: $2Na + 1/2O_2 \longrightarrow Na_2O$

Zone I (298°–391° K.)

$\Delta C_p = -7.74 + 60.98 \times 10^{-3} T - 11.57 \times 10^{-6} T^2 + 0.20 \times 10^5 T^{-2}$

$\Delta H_T = -98{,}700 - 7.74T + 30.49 \times 10^{-3} T^2 - 3.86 \times 10^{-6} T^3 - 0.2 \times 10^5 T^{-1}$

$\Delta F_T = -98{,}700 + 7.74 T \ln T - 30.49 \times 10^{-3} T^2 + 1.92 \times 10^{-6} T^3 - 0.10 \times 10^5 T^{-1} - 7.14T$

Zone II (391°–1,100° K.)

$\Delta C_p = -6.02 + 14.05 \times 10^{-3} T - 5.08 \times 10^{-6} T^2 + 0.20 \times 10^5 T^{-2}$

$\Delta H_T = -99{,}300 - 6.02T + 7.02 \times 10^{-3} T^2 - 1.69 \times 10^{-6} T^3 - 0.20 \times 10^5 T^{-1}$

$\Delta F_T = -99{,}300 + 6.02 T \ln T - 7.02 \times 10^{-3} T^2 + 0.85 \times 10^{-6} T^3 - 0.10 \times 10^5 T^{-1} - 1.61T$

T, ° K.	$H_T - H_{298}$	S_T	ΔH_T°	ΔF_T°
298		17.4	−99,400	−89,950
400	1,750	22.45	−100,700	−86,600
500	3,600	26.57	−100,700	−83,100
600	5,500	30.03	−100,650	−79,600
700	7,400	32.96	−100,550	−76,100
800	9,350	35.56	−100,400	−72,400
900	11,350	37.92	−100,150	−69,200
1,000	13,500	40.18	−99,800	−65,700
1,100	15,750	42.33	−99,350	−62,000
1,200	(25,050)		(−138,100)	(−58,200)
1,300	(27,450)		(−137,100)	(−51,600)
1,400	(29,900)		(−136,100)	(−45,100)
1,500	(32,350)		(−135,100)	(−38,600)
1,600	(34,700)		(−134,100)	(−32,200)
1,700	(37,200)		(−133,100)	(−25,900)
1,800	(39,600)		(−132,100)	(−19,600)
1,900	(42,050)		(−131,100)	(−13,400)
2,000	(44,660)		(−130,100)	(−7,200)

Disodium Dioxide, Na_2O_2 (*c*)

$\Delta H_{298}^\circ = -122{,}100$ calories per mole (*40*)
$S_{298} = 22.6$ *e.u.* (*133*)
M.P. = 733° K. (*24*)
$\Delta H_M = 5.860$ calories per mole
Decomposes = 919° K. (*3*)

Formation: $2Na + O_2 \longrightarrow Na_2O_2$ (estimated (*24*))

T, ° K.	$H_T - H_{298}$	ΔH_T°	ΔF_T°
298		−122,100	−107,000
400	(2,600)	(−123,300)	(−101,800)
500	(4,600)	(−123,100)	(−96,400)
600	(7,100)	(−122,800)	(−91,100)
700	(9,500)	(−122,600)	(−85,900)
800	(18,300)	(−116,000)	(−81,200)
900	(21,400)	(−115,100)	(−76,900)

Sodium Dioxide, NaO_2 (*c*)

$\Delta H_{298}^\circ = -62{,}100$ calories per mole (*40*)
$S_{298} = 27.7$ *e.u.* (*133*)

Formation: $Na+O_2 \longrightarrow NaO_2$
(estimated (*24*))

T, ° K.	H_T-H_{298}	ΔH_T°	ΔF_T°
298		−62, 100	−52, 100
400	(2, 200)	(−62, 000)	(−48, 500)
500	(4, 150)	(−61, 500)	(−45, 000)
600	(6, 100)	(−61, 000)	(−42, 000)
700	(8, 100)	(−60, 500)	(−39, 000)
800	(10, 100)	(−60, 000)	(−35, 500)
900	(12, 100)	(−59, 500)	(−32, 500)
1,000	(14, 100)	(−59, 000)	(−29, 500)

Sodium Fluoride, NaF (*c*)

$\Delta H^\circ_{298}=-136{,}000$ calories per mole (*112*)
$S_{298}=13.1$ *e.u.* (*11*)
$M.P.=1{,}265°$ K. (*82*)
$\Delta H_M=7{,}780$ calories per mole
$B.P.=1{,}977°$ K. (*6*)
$\Delta H_V=53{,}260$ calories per mole

Zone I (*c*) (298°–1,265° K.)

$$C_p=9.66+4.50\times10^{-3}T \quad (82)$$
$$H_T-H_{298}=-3{,}080+9.66T+2.25\times10^{-3}T^2$$

Zone II (*l*) (1,265–1,300° K.)

$$C_p=16.0 \quad (82)$$
$$H_T-H_{298}=280+16.0T$$

Formation: $Na+1/2F_2 \longrightarrow NaF$

Zone I (298°–371° K.)

$$\Delta C_p=-4.42+32.32\times10^{-3}T-5.78\times10^{-5}T^2+0.40\times10^5T^{-2}$$
$$\Delta H_T=-135{,}500-4.42T+16.16\times10^{-3}T^2-1.93\times10^{-5}T^3-0.40\times10^5T^{-1}$$
$$\Delta F_T=-135{,}500+4.42T\ln T-16.16\times10^{-3}T^2+0.96\times10^{-5}T^3-0.20\times10^5T^{-1}+0.66T$$

Zone II (391°–1,162° K.)

$$\Delta C_p=-3.56+8.86\times10^{-3}T-2.54\times10^{-6}T^2+0.40\times10^5T^{-2}$$
$$\Delta H_T=-135{,}700-3.56T+4.43\times10^{-3}T^2-0.85\times10^{-6}T^3-0.40\times10^5T^{-1}$$
$$\Delta F_T=-135{,}700+3.56T\ln T-4.43\times10^{-3}T^2+0.42\times10^{-6}T^3-0.20\times10^5T^{-1}+3.56T$$

Zone III (1,162°–1,265° K.)

$$\Delta C_p=0.64+4.28\times10^{-3}T+0.40\times10^5T^{-2}$$
$$\Delta H_T=-162{,}040+0.64T+2.14\times10^{-3}T^2-0.40\times10^5T^{-1}$$
$$\Delta F_T=-162{,}040-0.64T\ln T-2.14\times10^{-3}T^2-0.20\times10^5T^{-1}+53.19T$$

T, ° K.	H_T-H_{298}	S_T	ΔH_T°	ΔF_T°
298		13. 1	−136, 000	−129, 000
400	1, 140	16. 39	−136, 600	−126, 500
500	2, 310	19. 0	−136, 600	−124, 050
600	3, 530	21. 22	−136, 550	−121, 550
700	4, 780	23. 15	−136, 350	−119, 050
800	6, 080	24. 88	−136, 200	−116, 600
900	7, 420	26. 46	−136, 000	−114, 200
1,000	8, 810	27. 92	−135, 700	−111, 800
1,100	10, 260	29. 30	−135, 400	−109, 350
1,200	11, 760	30. 61	−158, 150	−106, 550
1,300	21, 080	37. 98	−149, 750	−105, 000

647940 O—63——S

Sodium Chloride, NaCl (*c*)

$\Delta H^\circ_{298}=-98{,}330$ calories per mole (*11*)
$S_{298}=17.3$ *e.u.* (*112*)
$M.P.=1{,}073°$ K. (*82*)
$\Delta H_M=6{,}850$ calories per mole
$B.P.=1{,}738°$ K. (*6*)
$\Delta H_V=40{,}800$ calories per mole

Zone I (*c*) (298°–1,073° K.)

$$C_p=10.98+3.90\times10^{-3}T \quad (82)$$
$$H_T-H_{298}=-3{,}447+10.98T+1.95\times10^{-3}T^2$$

Zone II (*l*) (1,073°–1,300° K.)

$$C_p=16.0 \quad (82)$$
$$H_T-H_{298}=+260+16.0T$$

Formation: $Na+1/2Cl_2 \longrightarrow NaCl$

Zone I (298°–391° K.)

$$\Delta C_p=-3.36+31.9\times10^{-3}T-5.78\times10^{-5}T^2+0.34\times10^5T^{-2}$$
$$\Delta H_T=-98{,}100-3.36T+15.95\times10^{-3}T^2-1.93\times10^{-5}T^3-0.34\times10^5T^{-1}$$
$$\Delta F_T=-98{,}100+3.36T\ln T-15.95\times10^{-3}T^2+0.96\times10^{-5}T^3-0.17\times10^5T^{-1}+5.88T$$

Zone II (391°–1,073° K.)

$$\Delta C_p=-2.5+8.45\times10^{-3}T-2.54\times10^{-6}T^2+0.34\times10^5T^{-2}$$
$$\Delta H_T=-98{,}400-2.5T+4.22\times10^{-3}T^2-0.85\times10^{-6}T^3-0.34\times10^5T^{-1}$$
$$\Delta F_T=-98{,}400+2.5T\ln T-4.22\times10^{-3}T^2+0.42\times10^{-6}T^3-0.17\times10^5T^{-1}+8.71T$$

Zone III (1,073°–1,162° K.)

$$\Delta C_p=2.52+4.55\times10^{-3}T-2.54\times10^{-6}T^2+0.34\times10^5T^{-2}$$
$$\Delta H_T=-94{,}500+2.52T+2.27\times10^{-3}T^2-0.85\times10^{-6}T^3-0.34\times10^5T^{-1}$$
$$\Delta F_T=-94{,}500-2.52T\ln T-2.27\times10^{-3}T^2+0.42\times10^{-6}T^3-0.17\times10^5T^{-1}+38.02T$$

Zone IV (1,162°–1,300° K.)

$$\Delta C_p=6.72-0.03\times10^{-3}T+0.34\times10^5T^{-2}$$
$$\Delta H_T=-120{,}900+6.72T-0.015\times10^{-3}T^2-0.34\times10^5T^{-1}$$
$$\Delta F_T=-120{,}900-6.72T\ln T+0.015\times10^{-3}T^2-0.17\times10^5T^{-1}+88.05T$$

T, ° K.	H_T-H_{298}	S_T	ΔH_T°	ΔF_T°
298		17. 30	−98, 330	−91, 900
400	1, 240	20. 88	−98, 850	−89, 600
500	2, 510	23. 71	−98, 750	−87, 500
600	3, 830	26. 12	−98, 600	−85, 050
700	5, 190	28. 21	−98, 350	−82, 850
800	6, 590	30. 08	−98, 100	−80, 600
900	8, 020	31. 76	−97, 800	−78, 400
1,000	9, 480	33. 30	−97, 500	−76, 250
1,100	17, 860	41. 14	−90, 150	−74, 350
1,200	19, 460	42. 53	−112, 900	−72, 500
1,300	21, 060	43. 81	−112, 200	−69, 000

Sodium Bromide, NaBr (*c*)

$\Delta H^\circ_{298}=-86{,}500$ calories per mole (*11*)
$S_{298}=20.1$ *e.u.* (*83*)
$M.P.=1{,}020°$ K. (*6*)
$\Delta H_M=6{,}140$ calories per mole
$B.P.=1{,}665°$ K. (*6*)
$\Delta H_V=37{,}950$ calories per mole

Zone I (*c*) (298°–550° K.)

$$C_p = 11.87 + 2.10\times 10^{-3}T \ (82)$$
$$H_T - H_{298} = -3{,}632 + 11.87T + 1.05\times 10^{-3}T^2$$

Formation: $Na + 1/2Br_2 \longrightarrow NaBr$

Zone I (298°–331° K.)

$$\Delta C_p = -6.61 + 30.14\times 10^{-3}T - 5.78\times 10^{-6}T^2$$
$$\Delta H_T = -85{,}350 - 6.61T + 15.07\times 10^{-3}T^2 - 1.93\times 10^{-6}T^3$$
$$\Delta F_T = -85{,}350 + 6.61T\ln T - 15.07\times 10^{-3}T^2 + 0.96\times 10^{-6}T^3 - 27.42T$$

Zone II (331°–391° K.)

$$\Delta C_p = -2.58 + 30.14\times 10^{-3}T - 5.78\times 10^{-6}T^2 + 0.18\times 10^5T^{-2}$$
$$\Delta H_T = -90{,}250 - 2.58T + 15.07\times 10^{-3}T^2 - 1.93\times 10^{-6}T^3 - 0.18\times 10^5T^{-1}$$
$$\Delta F_T = -90{,}250 + 2.58T\ln T - 15.07\times 10^{-3}T^2 + 0.96\times 10^{-6}T^3 - 0.09\times 10^5T^{-1} + 12.33T$$

Zone III (391°–550° K.)

$$\Delta C_p = -1.72 + 6.68\times 10^{-3}T - 2.54\times 10^{-6}T^2 + 0.18\times 10^5T^{-2}$$
$$\Delta H_T = -90{,}640 - 1.72T + 3.34\times 10^{-3}T^2 - 0.85\times 10^{-6}T^3 - 0.18\times 10^5T^{-1}$$
$$\Delta F_T = -90{,}640 + 1.72T\ln T - 3.34\times 10^{-3}T^2 + 0.42\times 10^{-6}T^3 - 0.09\times 10^5T^{-1} + 14.04T$$

T, ° K.	$H_T - H_{298}$	S_T	ΔH°_T	ΔF°_T
298	------------	20.1	−86, 500	−83, 400
400	1, 285	23.81	−90, 900	−81, 450
500	2, 565	26.66	−90, 800	−79, 100
1,000	(9, 300)	------------	(−89, 000)	(−66, 000)
1,500	(23, 400)	------------	(−104, 000)	(−61, 050)

Sodium Iodide, NaI (*c*)

$\Delta H^\circ_{298} = -70{,}650$ calories per mole (*112*)
$S_{298} = 22.50$ *e.u.* (*112*)
M.P. = 935° K. (*6*)
$\Delta H_M = 5{,}240$ calories per mole
B.P. = 1,577° K. (*6*)
$\Delta H_V = 38{,}160$ calories per mole

Formation: $Na + 1/2I_2 \longrightarrow NaI$
(estimated (*11*))

T, ° K.	$H_T - H_{298}$	ΔH°_T	ΔF°_T
298	------------	−70, 650	−69, 200
500	(2, 650)	(−78, 400)	(−64, 800)
1,000	(14, 800)	(−71, 900)	(−54, 700)
1,500	(22, 800)	(−92, 100)	(−47, 800)

Disodium Dicarbide, Na_2C_2 (*c*)

$\Delta H^\circ_{298} = -9{,}660$ calories per mole (*81*)
$S_{298} = 16.9$ *e.u.* (*81*)
$\Delta F^\circ_{298} = -6{,}570$ calories per mole
Decomposes = 1,073° K.

Sodium Trinitride, NaN_3 (*c*)

$\Delta H^\circ_{298} = +5{,}080$ calories per mole (*43*)
$S_{298} = 16.85$ *e.u.* (*43*)
$\Delta F^\circ_{298} = +24{,}180$ calories per mole
Decomposes = 548° K.

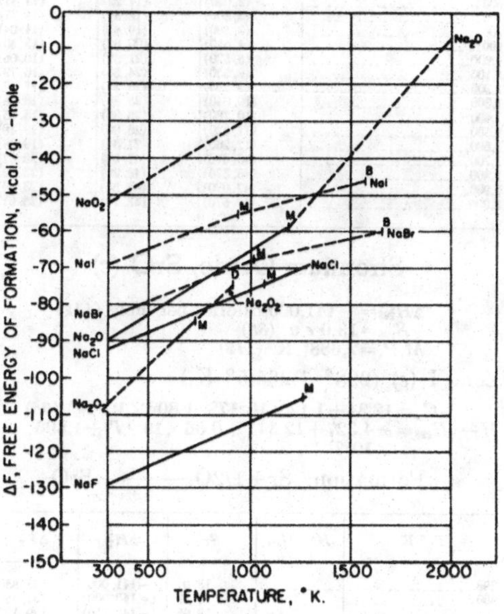

FIGURE 50.—Sodium.

STRONTIUM AND ITS COMPOUNDS

Element, Sr (*c*)

$S_{298} = 12.50$ *e.u.* (*83*)
T.P. = 862° K. (*82*)
$\Delta H_T = 200$ calories per atom
M.P. = 1,043° K. (*112*)
$\Delta H_M = 2{,}200$ calories per atom
B.P. = 1,657° K. (*79*)
$\Delta H_V = 33{,}610$ calories per atom

Zone I (α) (298°–862° K.)

$$C_p = 5.31 + 3.32\times 10^{-3}T \text{ (estimated (84))}$$
$$H_T - H_{298} = -1{,}731 + 5.31T + 1.66\times 10^{-3}T^2$$
$$F_T - H_{298} = -1{,}731 - 5.31T\ln T - 1.66\times 10^{-3}T^2 + 24.04T$$

Zone II (β) (862°–1,043° K.)

$$C_p = 9.12 \text{ (estimated (84))}$$
$$H_T - H_{298} = -3{,}582 + 9.12T$$
$$F_T - H_{298} = -3{,}582 - 9.12T\ln T + 50.54T$$

Zone III (*l*) (1,043°–1,600° K.)

$$C_p = 7.40 \text{ (estimated (84))}$$
$$H_T - H_{298} = 610 + 7.40T$$
$$F_T - H_{298} = 610 - 7.40T\ln T + 34.66T$$

T, ° K.	H_T-H_{298}	S_T	$-\frac{(F_T-H_{298})}{T}$
298	----------	12.50	12.50
400	(660)	(14.40)	(12.75)
500	(1,340)	(15.92)	(13.24)
600	(2,050)	(17.22)	(13.81)
700	(2,800)	(18.37)	(14.37)
800	(3,500)	(19.41)	(14.94)
900	(4,610)	(20.62)	(15.50)
1,000	(5,520)	(21.58)	(16.06)
1,100	(8,550)	(24.56)	(16.79)
1,200	(9,290)	(25.21)	(17.47)
1,300	(10,040)	(25.80)	(18.08)
1,400	(10,780)	(26.35)	(18.65)
1,500	(11,520)	(26.86)	(19.18)
1,600	(12,260)	(27.34)	(19.68)
1,700	(46,070)	(47.97)	(20.87)
1,800	(46,570)	(48.26)	(22.39)
1,900	(47,070)	(48.53)	(23.76)
2,000	(47,570)	(48.79)	(25.01)

Strontium Oxide, SrO (*c*)

$\Delta H^\circ_{298} = -141{,}000$ calories per mole (*112*)
$S_{298} = 13.0$ *e.u.* (*83*)
M.P. $= 2{,}688°$ K. (*112*)

Zone I (*c*) (298°–1,265.5° K.)

$$C_p = 12.34 + 1.12\times10^{-3}T - 1.806\times10^{5}T^{-2} \ (95)$$
$$H_T - H_{298} = -4{,}327 + 12.34T + 0.56\times10^{-3}T^2 + 1.806\times10^{5}T^{-1}$$

Formation: $Sr + 1/2O_2 \longrightarrow SrO$

T, ° K.	H_T-H_{298}	S_T	ΔH°_T	ΔF°_T
298	----------	13.0	−141,000	−133,850
400	1,143	16.29	(−140,900)	(−131,450)
500	2,334	18.96	(−140,750)	(−129,100)
600	3,565	21.21	(−140,600)	(−126,750)
700	4,824	23.15	(−140,400)	(−124,400)
800	6,104	24.88	(−140,400)	(−122,250)
900	7,401	26.41	(−140,600)	(−120,000)
1,000	8,714	27.80	(−140,300)	(−117,400)
1,100	9,940	29.05	(−142,700)	(−115,200)
1,200	11,380	30.24	(−141,400)	(−112,600)
1,500	(16,600)	----------	(−140,800)	(−106,000)

Strontium Dioxide, SrO₂ (*c*)

$\Delta H^\circ_{298} = -150{,}800$ calories per mole (*139*)
$S_{298} = (19.6)$ *e.u.* (*24*)

Formation: $Sr + O_2 \longrightarrow SrO_2$
(estimated (*24*))

T, ° K.	H_T-H_{298}	ΔH°_T	ΔF°_T
298	----------	−150,800	−138,300
400	(1,900)	(−150,300)	(−134,300)
500	(3,300)	(−150,300)	(−130,300)
600	(5,250)	(−149,800)	(−126,300)
700	(6,300)	(−149,300)	(−122,300)
800	(9,350)	(−148,800)	(−118,800)
900	(11,700)	(−148,300)	(−114,800)
1,000	(14,000)	(−147,800)	(−111,300)

Strontium Difluoride, SrF₂ (*c*)

$\Delta H^\circ_{298} = -290{,}300$ calories per mole (*112*)
$S_{298} = (18)$ *e.u.* (*11*)
M.P. $= 1{,}673°$ K. (*6*)
$\Delta H_M = 4{,}260$ calories per mole
B.P. $= 2{,}750°$ K. (*6*)
$\Delta H_V = 71{,}000$ calories per mole

Formation: $Sr + F_2 \longrightarrow SrF_2$
(estimated (*11*))

T, ° K.	H_T-H_{298}	ΔH°_T	ΔF°_T
298	----------	−290,300	(−277,200)
500	(3,600)	(−289,600)	(−268,500)
1,000	(13,000)	(−288,700)	(−247,900)
1,500	(23,300)	(−288,800)	(−227,800)

Strontium Dichloride, SrCl₂ (*c*)

$\Delta H^\circ_{298} = -198{,}000$ calories per mole (*112*)
$S_{298} = 28$ *e.u.* (*112*)
M.P. $= 1{,}145°$ K. (*6*)
$\Delta H_M = 4{,}100$ calories per mole
B.P. $= (2{,}300°)$ K. (*6*)
$\Delta H_V = (55{,}000)$ calories per mole

Zone I (*c*) (298°–1,145° K.)

$$C_p = 18.2 + 2.45\times10^{-3}T \ (110)$$
$$H_T - H_{298} = -5{,}533 + 18.2T + 1.225\times10^{-3}T^2$$

Formation: $Sr + Cl_2 \longrightarrow SrCl_2$

T, ° K.	H_T-H_{298}	S_T	ΔH°_T	ΔF°_T
298	----------	28.0	−198,000	−186,750
400	1,943	33.61	(−197,550)	(−182,900)
500	3,873	37.92	(−197,150)	(−179,300)
600	5,828	41.48	(−196,750)	(−175,700)
700	7,807	44.53	(−196,400)	(−172,300)
800	9,811	47.21	(−196,050)	(−168,800)
900	11,833	49.59	(−195,950)	(−165,500)
1,000	13,892	51.77	(−195,700)	(−162,000)
1,100	15,969	53.72	(−195,400)	(−158,500)
1,500	(29,600)	(65.0)	(−190,300)	(−146,500)

Strontium Dibromide, SrBr₂ (*c*)

$\Delta H^\circ_{298} = -171{,}100$ calories per mole (*112*)
$S_{298} = (34)$ *e.u.* (*11*)
M.P. $= 926°$ K. (*6*)
$\Delta H_M = 4{,}780$ calories per mole
B.P. $= (2{,}150°)$ K. (*6*)
$\Delta H_V = (50{,}000)$ calories per mole

Zone I (*c*) (298°–926° K.)

$$C_p = 18.1 + 3.15\times10^{-3}T \ (73)$$
$$H_T - H_{298} = -5{,}535 + 18.1T + 1.57\times10^{-3}T^2$$

Formation: $Sr + Br_2 \longrightarrow SrBr_2$

T, ° K.	H_T-H_{298}	S_T	ΔH°_T	ΔF°_T
298	----------	(34.0)	−171,100	(−166,700)
400	1,958	(39.6)	(−178,400)	(−163,400)
500	3,910	(44.0)	(−178,050)	(−159,700)
600	5,893	(47.6)	(−177,700)	(−156,100)
700	7,908	(50.7)	(−177,300)	(−153,200)
800	9,954	(53.5)	(−176,900)	(−149,100)
900	12,031	(55.9)	(−176,800)	(−145,600)
1,000	(19,200)	(63.0)	(−171,400)	(−141,900)
1,500	(31,200)	(73.0)	(−170,000)	(−127,400)

Strontium Diiodide, SrI₂ (*c*)

$\Delta H^\circ_{298} = -135{,}500$ calories per mole (*112*)
$S_{298} = (38.0)$ *e.u.* (*11*)
M.P. $= 788°$ K. (*6*)
$\Delta H_M = (5{,}400)$ calories per mole
B.P. $= (1{,}850°)$ K. (*6*)
$\Delta H_V = (40{,}000)$ calories per mole

Zone I (*c*) (298°–788° K.)

$$C_p = 18.6 + 3.05\times10^{-3}T \ (73)$$
$$H_T - H_{298} = -5{,}680 + 18.6T + 1.52\times10^{-3}T^2$$

Formation: $Sr + I_2 \longrightarrow SrI_2$

T, ° K.	$H_T - H_{298}$	S_T	ΔH_T°	ΔF_T°
298	----------	(38.0)	−135,500	(−134,800)
400	1,000	(43.8)	(−140,400)	(−135,400)
500	4,000	(48.2)	(−149,500)	(−132,100)
600	6,030	(51.9)	(−149,300)	(−128,900)
700	8,075	(55.1)	(−148,700)	(−125,400)
1,000	(20,400)	(60.0)	(−141,800)	(−117,000)
1,500	(32,400)	(79.0)	(−140,200)	(−103,200)

Tristrontium Dinitride, Sr_3N_2 (*c*)

$\Delta H_{298}^\circ = -92{,}200$ calories per mole* (*9*)
$S_{298} = 57.8$ *e.u.* (*9*)
$\Delta F_{298}^\circ = -77{,}000$ calories per mole
M.P. = 1,300° K. (*9*)

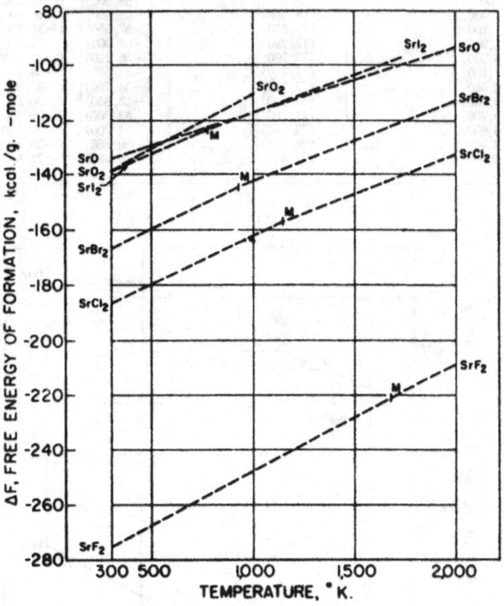

FIGURE 51.—Strontium.

TANTALUM AND ITS COMPOUNDS

Element, Ta (*c*)

$S_{298} = 9.94$ *e.u.* (*83*)
M.P. = 3,269° K. (*130*)
$\Delta H_M = (7{,}500)$ calories per mole
B.P. = 5,700° K. (*130*)
$\Delta H_V = 180{,}000$ calories per mole

Zone I (*c*) (298–1,900° K.)

$$C_p = 5.82 + 0.78\times10^{-3}T \ (82)$$
$$H_T - H_{298} = -1{,}770 + 5.82T + 0.39\times10^{-3}T^2$$
$$F_T - H_{298} = -1{,}770 - 5.82T\ln T - 0.39\times10^{-3}T^2 + 29.21T$$

T, ° K.	$H_T - H_{298}$	S_T	$-\frac{(F_T - H_{298})}{T}$
298	----------	9.94	9.94
400	620	11.73	10.2[illegible]
500	1,230	13.09	10.6[illegible]
600	1,845	14.22	11.1[illegible]
700	2,470	15.18	11.61
800	3,120	16.05	12.12
900	3,780	16.83	12.64
1,000	4,440	17.53	13.09
1,100	5,100	18.16	13.53
1,200	5,770	18.74	13.94
1,300	6,445	19.29	14.31
1,400	7,130	19.80	14.66
1,500	7,825	20.28	15.05
1,600	8,530	20.73	14.42
1,700	9,250	21.17	15.71
1,800	9,980	21.59	16.03
1,900	10,720	21.98	16.25

Ditantalum Pentaoxide, Ta_2O_5 (*c*)

$\Delta H_{298}^\circ = -488{,}800$ calories per mole (*24*)
$S_{298} = 34.2$ *e.u.* (*83*)
M.P. = 2,150° K. (*8*)
$\Delta H_M = 48{,}000$ calories per mole
B.P. = >2,500° K. (*24*)

Zone I (*c*) (298°–1,800° K.)

$$C_p = 37.00 + 6.56\times10^{-3}T - 5.92\times10^5T^{-2} \ (107)$$
$$H_T - H_{298} = -13{,}215 + 37.00T + 3.28\times10^{-3}T^2 + 5.92\times10^5T^{-1}$$

Formation: $2Ta + 5/2O_2 \longrightarrow Ta_2O_5$

Zone I (298°–1,700° K.)

$$\Delta C_p = 7.46 + 2.50\times10^{-3}T - 4.92\times10^5T^{-2}$$
$$\Delta H_T = -492{,}780 + 7.46T + 1.25\times10^{-3}T^2 + 4.92\times10^5T^{-1}$$
$$\Delta F_T = -492{,}780 - 7.46T\ln T - 1.25\times10^{-3}T^2 + 2.46\times10^5T^{-1} + 161.6T$$

T, ° K.	$H_T - H_{298}$	S_T	ΔH_T°	ΔF_T°
298	----------	34.2	−488,800	−456,500
400	3,430	44.07	−488,350	−445,650
500	7,070	52.18	−487,750	−435,000
600	10,950	59.25	−487,050	−424,400
700	14,990	66.08	−486,250	−414,250
800	19,130	71.01	−485,400	−404,150
900	23,340	75,96.	−484,500	−393,800
1,000	27,630	80.49	−483,600	−383,700
1,100	31,990	84.64	−482,600	−373,550
1,200	36,410	88.49	−481,600	−363,850
1,300	40,880	92.06	−480,600	−354,200
1,400	45,390	95.4	−479,550	−344,550
1,500	49,970	98.56	−478,450	−335,350
1,600	54,630	101.57	−477,350	−325,850
1,700	59,380	104.45	−476,300	−316,300

Tantalum Pentachloride, $TaCl_5$ (*c*)

$\Delta H_{298}^\circ = -205{,}500$ calories per mole (*48*)
$S_{298} = (66)$ *e.u.* (*11*)
M.P. = 480° K. (*6*)
$\Delta H_M = 9{,}000$ calories per mole
B.P. = 507° K. (*6*)
$\Delta H_V = 12{,}500$ calories per mole

Formation: $Ta + 5/2Cl_2 \longrightarrow TaCl_5$
(estimated (*11*))

T, ° K.	$H_T - H_{298}$	ΔH°_T	ΔF°_T
298		−205, 500	(−182, 500)
500	(13, 000)	(−199, 000)	(−168, 000)
1,000	(30, 000)	(−195, 000)	(−152, 000)

Tantalum Pentabromide, $TaBr_5$ (*c*)

$\Delta H^\circ_{298} = -142{,}900$ calories per mole (*49*)
$S_{298} = (78)$ *e.u.* (*11*)
$M.P. = 513°$ K. (*6*)
$\Delta H_M = (9{,}000)$ calories per mole
$B.P. = 622°$ K.
$\Delta H_V = 14{,}900$ calories per mole

Formation: $Ta + 5/2Br_2 \longrightarrow TaBr_5$

(estimated (*11*))

T, ° K.	$H_T - H_{298}$	ΔH°_T	ΔF°_T
298		−143, 000	(−135, 000)
500	(8, 000)	(−159, 000)	(−122, 000)

Tantalum Carbide, TaC (*c*)

$\Delta H^\circ_{298} = -38{,}500$ calories per mole (*66*)
$S_{298} = 10.1$ *e.u.* (*112*)
$M.P. = 4{,}070°$ K. (*9*)

Zone I (*c*) (298°–1,800° K.)

$$C_p = 7.28 + 1.65 \times 10^{-3} T \ (94)$$
$$H_T - H_{298} = -2{,}242 + 7.28T + 0.825 \times 10^{-3} T^2$$

Formation: $Ta + C \longrightarrow TaC$

Zone I (298°–1,800° K.)

$$\Delta C_p = -2.64 - 0.15 \times 10^{-3} T + 2.10 \times 10^5 T^{-2}$$
$$\Delta H_T = -37{,}000 - 2.64T - 0.075 \times 10^{-3} T^2 - 2.10 \times 10^5 T^{-1}$$
$$\Delta F_T = -37{,}000 + 2.64 T \ln T + 0.075 \times 10^{-3} T^2 - 1.05 \times 10^5 T^{-1} - 17.64T$$

T, ° K.	$H_T - H_{298}$	S_T	ΔH°_T	ΔF°_T
298		10. 1	−38, 500	−38, 090
400	800	12. 41	−38, 600	−38, 000
500	1, 604	14. 20	−38, 750	−37, 800
600	2, 423	15. 70	−38, 950	−37, 650
700	3, 258	16. 98	−39, 200	−37, 350
800	4, 110	18. 12	−39, 400	−37, 100
900	4, 978	19. 14	−39, 650	−36, 750
1,000	5, 863	20. 08	−39, 900	−36, 450
1,100	6, 764	20. 92	−40, 200	−36, 050
1,200	7, 682	21. 73	−40, 450	−35, 800
1,300	8, 616	22. 48	−40, 700	−35, 500
1,400	9, 567	23. 19	−41, 000	−34, 900
1,500	10, 534	23. 85	−41, 250	−34, 300
1,600	11, 516	24. 49	−41, 550	−33, 850
1,700	12, 518	25. 09	−41, 800	−33, 500
1,800	13, 535	25. 68	−42, 100	−33, 000

Tantalum Nitride, TaN (*c*)

$\Delta H^\circ_{298} = -60{,}000$ calories per mole (*100*)
$S_{298} = 12.4$ *e.u.* (*94*)
$M.P. = (3{,}360°)$ K. (*9*)

Zone I (*c*) (298°–773° K.)

$$C_p = 7.73 + 7.80 \times 10^{-3} T \ (82)$$
$$H_T - H_{298} = -2{,}652 + 7.73T + 3.90 \times 10^{-3} T^2$$

Formation: $Ta + 1/2N_2 \longrightarrow TaN$

Zone I (298°–773° K.)

$$\Delta C_p = -1.42 + 6.51 \times 10^{-3} T$$
$$\Delta H_T = -59{,}900 - 1.42T + 3.25 \times 10^{-3} T^2$$
$$\Delta F_T = -59{,}900 + 1.42 T \ln T - 3.25 \times 10^{-3} T^2 + 12.87T$$

T, ° K.	$H_T - H_{298}$	S_T	ΔH°_T	ΔF°_T
298		12. 4	−60, 000	−53, 930
400	1, 050	15. 43	−59, 900	−51, 850
500	2, 190	17. 97	−59, 750	−49, 950
600	3, 400	20. 17	−59, 500	−47, 900
700	4, 680	22. 14	−59, 200	−45, 950
800	6, 030	23. 94	−58, 900	−44, 100
900	(7, 460)		(−58, 500)	(−43, 300)
1,000	(8, 980)		(−58, 050)	(−40, 500)
1,100	(10, 570)		(−57, 550)	(−38, 750)
1,200	(12, 240)		(−56, 900)	(−37, 300)
1,300	(13, 990)		(−56, 250)	(−35, 450)
1,400	(15, 810)		(−55, 550)	(−33, 850)
1,500	(17, 720)		(−54, 750)	(−32, 300)
1,600	(19, 700)		(−53, 850)	(−30, 850)
1,700	(21, 760)		(−52, 900)	(−29, 550)
1,800	(23, 590)		(−51, 900)	(−28, 150)
1,900	(26, 110)		(−50, 850)	(−26, 850)
2,000	(28, 410)		(−49, 700)	(−25, 600)

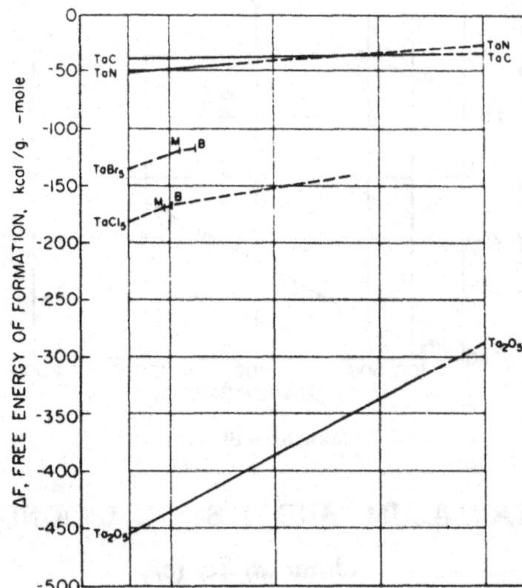

FIGURE 52.—Tantalum.

TERBIUM AND ITS COMPOUNDS

Element, Tb (*c*)

S_{298} = (17.50) *e.u.* (*121*)
M.P. = (1,638°) K. (*125*)
ΔH_M = (3,900) calories per atom
B.P. = (2,800°) K. (*125*)
ΔH_V = (70,000) calories per atom

(estimated (*130*))

T, ° K.	$H_T - H_{298}$	S_T	$-\frac{(F_T - H_{298})}{T}$
298		(17.50)	(17.50)
400	(675)	(19.41)	(17.73)
500	(1,360)	(20.92)	(18.20)
600	(2,055)	(22.20)	(18.78)
700	(2,770)	(23.30)	(19.35)
800	(3,510)	(24.28)	(19.90)
900	(4,260)	(25.17)	(20.44)
1,000	(5,030)	(25.98)	(20.95)
1,100	(5,820)	(26.74)	(21.45)
1,200	(6,630)	(27.44)	(21.92)
1,300	(7,450)	(28.10)	(22.37)
1,400	(8,300)	(28.72)	(22.80)
1,500	(9,160)	(29.32)	(23.22)
1,600	(10,040)	(29.88)	(23.61)
1,700	(14,830)	(32.72)	(24.00)
1,800	(15,630)	(33.17)	(24.49)
1,900	(16,430)	(33.61)	(25.41)
2,000	(17,230)	(34.02)	(25.83)

Diterbium Trioxide, Tb_2O_3 (*c*)

$\Delta H^\circ_{298} = -436,800 \pm 2,000$ calories per mole (*129*)

Heptaterbium Dodecaoxide, Tb_7O_{12} (*c*)

$\Delta H^\circ_{298} = -1,563,000 \pm 7,000$ calories per mole (*129*)

Pentaterbium Enneaoxide, Tb_5O_9 (*c*)

$\Delta H^\circ_{298} = -1,132,000 \pm 5,000$ calories per mole (*129*)

Terbium Trifluoride, TbF_3 (*c*)

ΔH°_{298} = (−375,000) calories per mole (*5*)
S_{298} = (25) *e.u.* (*11*)
M.P. = (1,445°) K. (*29*)
ΔH_M = (8,000) calories per mole
B.P. = (2,550°) K. (*6*)
ΔH_V = (60,000) calories per mole

Formation: $Tb + 3/2F_2 \longrightarrow TbF_3$
(estimated (*11*))

T, ° K.	$H_T - H_{298}$	ΔH°_T	ΔF°_T
298		(−375,000)	(−357,000)
500	(4,000)	(−374,700)	(−345,000)
1,000	(17,000)	(−371,800)	(−315,000)
1,500	(32,000)	(−368,000)	(−289,500)

Terbium Trichloride, $TbCl_3$ (*c*)

ΔH°_{298} = (−216,000) calories per mole (*5*)
S_{298} = (41) *e.u.* (*11*)
T.P. = 770° K. (*29*)
M.P. = 855° K. (*29*)
ΔH_M = (7,000) calories per mole
B.P. = (1,820°) K. (*6*)
ΔH_V = (45,000) calories per mole

Formation: $Tb + 3/2Cl_2 \longrightarrow TbCl_3$
(estimated (*11*))

T, ° K.	$H_T - H_{298}$	ΔH°_T	ΔF°_T
298		(−216,000)	(−200,500)
500	(5,000)	(−214,900)	(−190,000)
1,000	(19,000)	(−211,100)	(−167,000)
1,500	(43,000)	(−197,600)	(−148,500)

Terbium Tribromide, $TbBr_3$ (*c*)

ΔH°_{298} = (−175,000) calories per mole (*5*)
S_{298} = (46) *e.u.* (*11*)
M.P. = (1,100°) K. (*29*)
ΔH_M = (9,000) calories per mole
B.P. = (1,760°) K. (*6*)
ΔH_V = (44,000) calories per mole

Formation: $Tb + 3/2Br_2 \longrightarrow TbBr_3$
(estimated (*11*))

T, ° K.	$H_T - H_{298}$	ΔH°_T	ΔF°_T
298		(−175,000)	(−168,500)
500	(5,000)	(−185,600)	(−157,500)
1,000	(18,000)	(−183,000)	(−133,000)
1,500	(43,000)	(−181,000)	(−94,500)

Terbium Triiodide, TbI_3 (*c*)

ΔH°_{298} = (−122,000) calories per mole (*5*)
S_{298} = (48) *e.u.* (*11*)
M.P. = (1,219°) K. (*29*)
ΔH_M = (10,000) calories per mole
B.P. = (1,600°) K. (*6*)
ΔH_V = (40,000) calories per mole

Formation: $Tb + 3/2I_2 \longrightarrow TbI_3$
(estimated (*11*))

T, ° K.	$H_T - H_{298}$	ΔH°_T	ΔF°_T
298		(−122,000)	(−124,000)
500	(5,000)	(−143,400)	(−112,000)
1,000	(19,000)	(−139,500)	(−84,000)
1,500	(44,000)	(−125,000)	(−59,500)

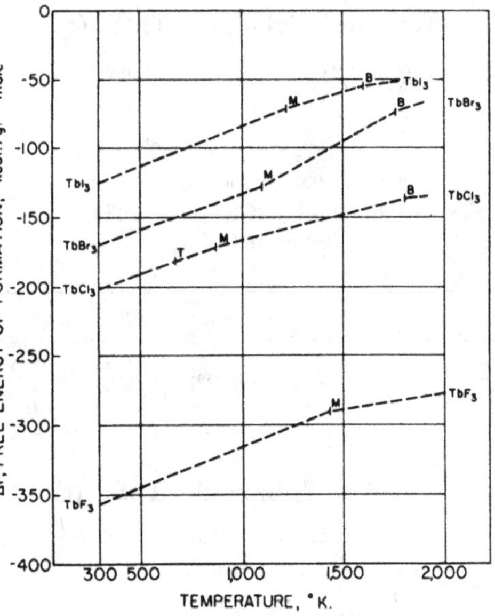

FIGURE 53.—Terbium.

THORIUM AND ITS COMPOUNDS

Element, Th (*c*)

S_{298} = 12.76 *e.u.* (*46*)
$T.P.$ = 1,673° K. (*130*)
ΔH_T = (670) calories per atom
$M.P.$ = 1,968° K. (*130*)
ΔH_M = (3,740) calories per atom
$B.P.$ = 4,500° K. (*130*)
ΔH_V = (130,000) calories per atom

Zone I (*c*) (298°–1,500° K.)

$C_p = 6.40 + 3.06 \times 10^{-3}T + 0.35 \times 10^5 T^{-2}$ (*82*)

$H_T - H_{298} = -1{,}927 + 6.40T + 1.53 \times 10^{-3}T^2 - 0.35 \times 10^5 T^{-1}$

$F_T - H_{298} = -1{,}927 - 6.40T\ln T - 1.53 \times 10^{-3}T^2 - 0.175 \times 10^5 T^{-1} + 30.71T$

T, ° K.	$H_T - H_{298}$	S_T	$-\frac{(F_T - H_{298})}{T}$
298	....	12.76	12.76
400	790	15.04	13.09
500	1,600	16.85	13.65
600	2,420	18.34	14.31
700	3,260	19.64	14.97
800	4,120	20.79	15.64
900	5,010	21.85	16.28
1,000	5,930	22.80	16.87
1,100	6,880	23.71	17.45
1,200	7,870	24.57	18.01
1,300	8,910	25.40	18.55
1,400	10,000	26.21	19.07
1,500	11,130	26.99	19.57
1,600	(12,200)	(27.59)	(19.98)
1,700	(13,200)	(28.30)	(20.54)
1,800	(14,500)	(29.06)	(21.01)

Thorium Oxide, ThO (*c*)

$\Delta H^\circ_{298} = -145{,}000$ calories per mole (*42*)
S_{298} = 16.1 *e.u.* (*42*)
$M.P.$ = >2,500° K. (*42*)

Formation: $Th + 1/2O_2 \longrightarrow ThO$
(estimated (*42*))

T, ° K.	ΔF°_T	T, ° K.	ΔF°_T
298	(−138,000)	1,500	(−113,000)
500	(−134,000)	2,000	(−102,000)
1,000	(−123,000)	2,500	(−91,000)

Thorium Dioxide, ThO_2 (*c*)

$\Delta H^\circ_{298} = -293{,}200$ calories per mole (*63*)
S_{298} = 15.59 *e.u.* (*24*)
$M.P.$ = 3,225° K. (*8*)
ΔH_M = 291,100 calories per mole
$B.P.$ = 4,670° K. (*8*)

Zone I (*c*) (298°–1,800° K.)

$C_p = 15.84 + 2.88 \times 10^{-3}T - 1.60 \times 10^5 T^{-2}$ (*82*)

$H_T - H_{298} = -5{,}388 + 15.84T + 1.44 \times 10^{-3}T^2 + 1.60 \times 10^5 T^{-1}$

Formation: $Th + O_2 \longrightarrow ThO_2$

Zone I (298°–1,500° K.)

$\Delta C_p = 2.28 - 1.18 \times 10^{-3}T - 1.55 \times 10^5 T^{-2}$

$\Delta H_T = -294{,}350 + 2.28T - 0.59 \times 10^{-3}T^2 + 1.55 \times 10^5 T^{-1}$

$\Delta F_T = -294{,}350 - 2.28T\ln T + 0.59 \times 10^{-3}T^2 + 0.775 \times 10^5 T^{-1} + 61.96T$

T, ° K.	$H_T - H_{298}$	S_T	ΔH°_T	ΔF°_T
298	....	15.59	−293,200	−279,450
400	1,600	20.20	−293,100	−274,700
500	3,210	23.79	−293,050	−270,150
600	4,890	26.85	−292,950	−265,600
700	6,620	29.91	−292,850	−261,050
800	8,390	31.88	−292,700	−256,500
900	10,200	34.01	−292,600	−251,950
1,000	12,050	35.96	−292,500	−247,450
1,100	13,940	37.76	−292,350	−242,450
1,200	15,860	39.43	−292,250	−238,400
1,300	17,800	40.98	−292,200	−233,900
1,400	19,760	42.43	−292,150	−229,450
1,500	21,740	43.80	−292,250	−225,000
1,600	23,740	45.09	(−292,100)	(−220,550)
1,700	25,750	46.31	(−292,000)	(−216,000)
1,800	27,770	47.46	(−292,150)	(−211,700)

Thorium Tetrafluoride, ThF_4 (*c*)

$\Delta H^\circ_{298} = (-477{,}000)$ calories per mole (*11*)
S_{298} = (35) *e.u.* (*11*)
$M.P.$ = (1,300°) K. (*11*)
ΔH_M = (17,000) calories per mole
$B.P.$ = (2,000°) K. (*11*)
ΔH_V = (50,000) calories per mole

Formation: $Th + 2F_2 \longrightarrow ThF_4$
(estimated (*11*))

T, ° K.	$H_T - H_{298}$	ΔH_T°	ΔF_T°
298	----------	(−477,000)	(−454,000)
500	(6,000)	(−475,800)	(−438,000)
1,000	(22,000)	(−472,600)	(−403,000)
1,500	(59,000)	(−449,600)	(−373,000)

Thorium Trichloride, $ThCl_3$ (*c*)

$\Delta H^\circ_{298} = (-242{,}000)$ calories per mole (*42*)
$S_{298} = (43.2)$ *e.u.* (*42*)
M.P. = (1,100°) K. (*6*)
$\Delta H_M = (9{,}000)$ calories per mole
B.P. = (1,890°) K. (*6*)
$\Delta H_V = (46{,}000)$ calories per mole

Formation: $Th + 3/2Cl_2 \longrightarrow ThCl_3$
(estimated (*42*))

T, ° K.	ΔF_T°	T, ° K.	ΔF_T°
298	(−227,000)	1,500	(−179,000)
500	(−218,000)	2,000	(−172,000)
1,000	(−196,000)	2,500	(−161,000)

Thorium Tetrachloride, $ThCl_4$ (*c*)

$\Delta H^\circ_{298} = -285{,}200$ calories per mole (*11*)
$S_{298} = (44)$ *e.u.* (*11*)
M.P. = 1,038° K. (*6*)
$\Delta H_M = 22{,}500$ calories per mole
B.P. = 1,195° K. (*6*)
$\Delta H_V = 36{,}500$ calories per mole

Formation: $Th + 2Cl_2 \longrightarrow ThCl_4$
(estimated (*11*))

T, ° K.	$H_T - H_{298}$	ΔH_T°	ΔF_T°
298	----------	−285,200	(−262,600)
500	(6,000)	(−284,000)	(−247,200)
1,000	(23,000)	(−280,000)	(−211,000)
1,500	(84,500)	(−232,500)	(−198,000)

Thorium Tetrabromide, $ThBr_4$ (*c*)

$\Delta H^\circ_{298} = (-230{,}300)$ calories per mole (*11*)
$S_{298} = (56)$ *e.u.* (*11*)
M.P. = 953° K. (*6*)
$\Delta H_M = 9{,}500$ calories per mole
B.P. = 1,130° K. (*6*)
$\Delta H_V = 34{,}500$ calories per mole

Formation: $Th + 2Br_2 \longrightarrow ThBr_4$
(estimated (*11*))

T, ° K.	$H_T - H_{298}$	ΔH_T°	ΔF_T°
298	----------	(−230,300)	(−221,200)
500	(6,000)	(−244,900)	(−207,000)
1,000	(33,000)	(−231,200)	(−172,500)
1,500	(88,500)	(−189,700)	----------

Thorium Tetraiodide, ThI_4 (*c*)

$\Delta H^\circ_{298} = (-161{,}200)$ calories per mole (*11*)
$S_{298} = (63)$ *e.u.* (*11*)
M.P. = 839° K. (*6*)
$\Delta H_M = 8{,}000$ calories per mole
B.P. = 1,110° K. (*6*)
$\Delta H_V = 31{,}500$ calories per mole

Formation: $Th + 2I_2 \longrightarrow ThI_4$
(estimated (*11*))

T, ° K.	$H_T - H_{298}$	ΔH_T°	ΔF_T°
298	----------	(−161,200)	(−159,000)
500	(6,000)	(−189,900)	(−154,200)
1,000	(33,000)	(−176,200)	(−118,000)
1,500	(85,500)	(−137,800)	(−72,000)

Thorium Dicarbide, ThC_2 (*c*)

$\Delta H^\circ_{298} = -45{,}600$ calories per mole (*9*)
$S_{298} = (30)$ *e.u.* (*9*)
$\Delta F^\circ_{298} = (-50{,}000)$ calories per mole (*9*)

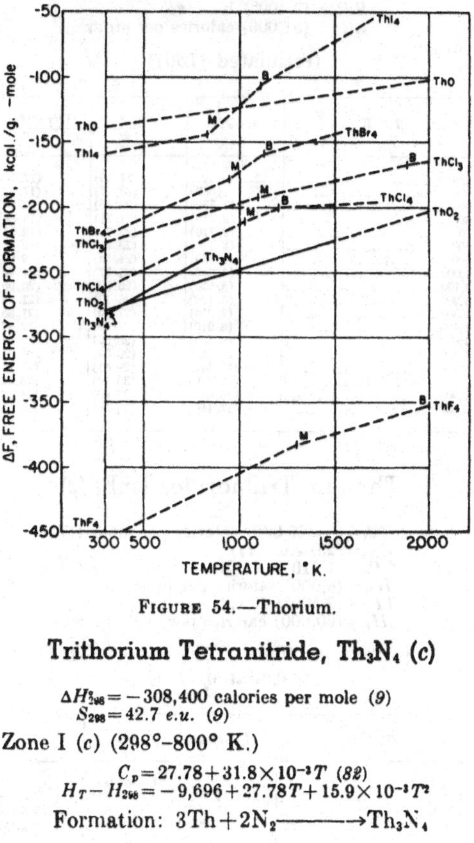

FIGURE 54.—Thorium.

Trithorium Tetranitride, Th_3N_4 (*c*)

$\Delta H^\circ_{298} = -308{,}400$ calories per mole (*9*)
$S_{298} = 42.7$ *e.u.* (*9*)

Zone I (*c*) (298°–800° K.)

$$C_p = 27.78 + 31.8 \times 10^{-3}T \ (82)$$
$$H_T - H_{298} = -9{,}696 + 27.78T + 15.9 \times 10^{-3}T^2$$

Formation: $3Th + 2N_2 \longrightarrow Th_3N_4$

Zone I (298°–800° K.)

$$\Delta C_p = -4.74 + 20.58 \times 10^{-3} T - 1.05 \times 10^5 T^{-2}$$
$$\Delta H_T = -308{,}250 - 4.74T + 10.29 \times 10^{-3} T^2 + 1.05 \times 10^5 T^{-1}$$
$$\Delta F_T = -308{,}250 + 4.74 T \ln T - 10.29 \times 10^{-3} T^2 + 0.525 \times 10^5 T^{-1} + 62.07 T$$

T, ° K.	H_T-H_{298}	S_T	ΔH°_T	ΔF°_T
298	------------	42.7	−308,400	−282,450
400	3,960	54.11	−308,250	−273,600
500	8,180	63.51	−307,850	−264,950
600	12,720	71.78	−307,200	−256,400
700	17,540	79.20	−306,350	−248,000
800	22,540	85.87	−305,400	−239,700

THULIUM AND ITS COMPOUNDS

Element, Tm (c)

S_{298} = (17.10) *e.u.* (*127*)
M.P. = (1,900°) K. (*125*)
ΔH_M = (4,400) calories per atom
B.P. = (2,400°) K. (*125*)
ΔH_V = (51,000) calories per atom

(estimated (*130*))

T, ° K.	H_T-H_{298}	S_T	$-\frac{(F_T-H_{298})}{T}$
298	------------	(17.10)	(17.10)
400	(660)	(18.97)	(17.32)
500	(1,330)	(20.46)	(17.80)
600	(2,010)	(21.71)	(18.36)
700	(2,710)	(22.78)	(18.91)
800	(3,420)	(23.73)	(19.46)
900	(4,150)	(24.59)	(19.98)
1,000	(4,890)	(25.37)	(20.48)
1,100	(5,650)	(26.09)	(20.96)
1,200	(6,420)	(26.77)	(21.42)
1,300	(7,210)	(27.40)	(21.86)
1,400	(8,010)	(27.99)	(22.27)
1,500	(8,830)	(28.56)	(22.68)
1,600	(9,660)	(29.09)	(23.06)
1,700	(10,510)	(29.61)	(23.43)
1,800	(11,370)	(30.10)	(23.79)
1,900	(16,650)	(32.89)	(24.13)
2,000	(17,450)	(33.30)	(24.58)

Thulium Trifluoride, TmF_3 (c)

ΔH°_{298} = (−366,000) calories per mole (*5*)
S_{298} = (25) *e.u.* (*11*)
M.P. = (1,610°) K. (*6*)
ΔH_M = (8,000) calories per mole
B.P. = (2,500°) K. (*6*)
ΔH_V = (60,000) calories per mole

Formation: $Tm + 3/2F_2 \longrightarrow TmF_3$
(estimated (*11*))

T, ° K.	H_T-H_{298}	ΔH°_T	ΔF°_T
298	------------	(−366,000)	(−348,000)
500	(4,000)	(−365,700)	(−336,500)
1,000	(17,000)	(−362,600)	(−308,000)
1,500	(32,000)	(−378,000)	(−282,000)

Thulium Trichloride, $TmCl_3$ (c)

ΔH°_{298} = −229,000 calories per mole (*5*)
S_{298} = (39) *e.u.* (*11*)
M.P. = 1,094° K. (*6*)
ΔH_M = (9,000) calories per mole
B.P. = (1,760°) K. (*6*)
ΔH_V = (44,000) calories mole

Formation: $Tm + 3/2Cl_2 \longrightarrow TmCl_3$
(estimated (*11*))

T, ° K.	H_T-H_{298}	ΔH°_T	ΔF°_T
298	------------	−229,000	(−212,500)
500	(5,000)	(−227,900)	(−201,500)
1,000	(19,000)	(−224,000)	(−176,000)
1,500	(43,000)	(−210,400)	(−157,000)

Thulium Tribromide, $TmBr_3$ (c)

ΔH°_{298} = (−167,000) calories per mole (*5*)
S_{298} = (44) *e.u.* (*11*)
M.P. = (1,225°) K. (*6*)
ΔH_M = (10,000) calories per mole
B.P. = (1,710°) K. (*6*)
ΔH_V = (43,000) calories per mole

Formation: $Tm + 3/2Br_2 \longrightarrow TmBr_3$
(estimated (*11*))

T, ° K.	H_T-H_{298}	ΔH°_T	ΔF°_T
298	------------	(−167,000)	(−160,000)
500	(5,000)	(−177,600)	(−148,500)
1,000	(18,000)	(−174,900)	(−123,000)
1,500	(43,000)	(−160,600)	(−103,500)

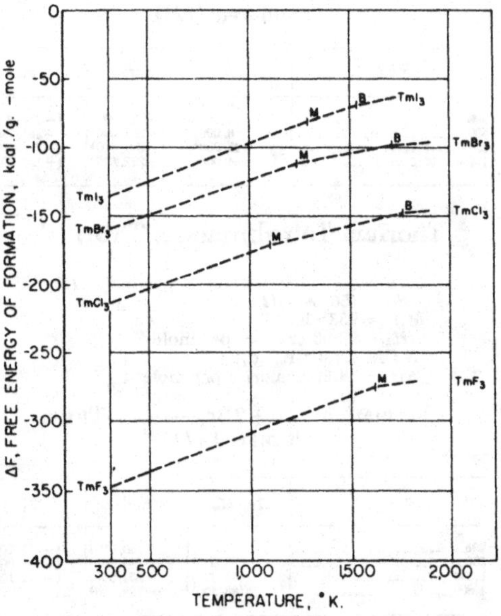

FIGURE 55.—Thulium.

Thulium Triiodide, TmI_3 (*c*)

$\Delta H^\circ_{298} = (-138{,}000)$ calories per mole (*5*)
$S_{298} = (47)$ *e.u.* (*11*)
M.P. = 1,288° K. (*6*)
$\Delta H_M = (10{,}000)$ calories per mole
B.P. = (1,530°) K. (*6*)
$\Delta H_V = (40{,}000)$ calories per mole

Formation: $Tm + 3/2 I_2 \longrightarrow TmI_3$
(estimated (*11*))

T, ° K.	$H_T - H_{298}$	ΔH°_T	ΔF°_T
298		(−138,000)	(−135,500)
500	(5,000)	(−159,300)	(−123,500)
1,000	(19,000)	(−155,600)	(−95,000)
1,500	(44,000)	(−142,000)	(−70,500)

TIN AND ITS COMPOUNDS

Element, Sn (*c*)

$S_{298} = 12.29$ *e.u.* (*83*)
M.P. = 505° K. (*82*)
$\Delta H_M = 1{,}720$ calories per atom
B.P. = 2,960° K. (*130*)
$\Delta H_V = 69{,}400$ calories per atom

Zone I (*c*) (298°–505° K.)

$$C_p = 4.42 + 6.30 \times 10^{-3} T \ (82)$$
$$H_T - H_{298} = -1{,}598 + 4.42T + 3.15 \times 10^{-3} T^2$$
$$F_T - H_{298} = -1{,}598 - 4.42 T \ln T - 3.15 \times 10^{-3} T^2 + 19.19T$$

Zone II (*l*) (505°–1,300° K.)

$$C_p = 7.30 \ (82)$$
$$H_T - H_{298} = -526 + 7.30T$$
$$F_T - H_{298} = -526 - 7.30 T \ln T + 33.41T$$

Zone III (*l*) (1,300°–2,000° K.)
(estimated (*130*))

T, ° K.	$H_T - H_{298}$	S_T	$-\frac{(F_T - H_{298})}{T}$
298		12.29	12.29
400	680	14.25	12.55
500	1,400	15.85	13.05
600	3,850	20.59	14.17
700	4,580	21.72	15.17
800	5,310	22.69	16.05
900	6,040	23.55	16.83
1,000	6,770	24.32	17.55
1,100	7,500	25.02	18.20
1,200	8,230	25.65	18.79
1,300	8,960	26.23	19.34
1,400	(9,690)	(26.77)	(19.85)
1,500	(10,420)	(27.28)	(20.34)
1,600	(11,150)	(27.75)	(20.79)
1,700	(11,880)	(28.19)	(21.21)
1,800	(12,610)	(28.61)	(21.61)
1,900	(13,340)	(29.00)	(21.98)
2,000	(14,070)	(29.38)	(22.35)

Tin Oxide, SnO (*c*)

$\Delta H^\circ_{298} = -68{,}350$ calories per mole (*71*)
$S_{298} = 13.5$ *e.u.* (*83*)
M.P. = (1,315°) K. (*42*)
$\Delta H_M = (6{,}400)$ calories per mole
B.P. = (1,800°) K. (*42*)
$\Delta H_V = (60{,}000)$ calories per mole

Zone I (*c*) (298°–1,273° K.)

$$C_p = 9.95 + 3.50 \times 10^{-3} T$$
$$H_T - H_{298} = -3{,}120 + 9.95T + 1.75 \times 10^{-3} T^2$$

Formation: $Sn + 1/2 O_2 \longrightarrow SnO$

Zone I (298°–505° K.)

$$\Delta C_p = 1.95 - 3.3 \times 10^{-3} T + 0.20 \times 10^5 T^{-2}$$
$$\Delta H_T = -68{,}720 + 1.95T - 1.65 \times 10^{-3} T^2 - 0.20 \times 10^5 T^{-1}$$
$$\Delta F_T = -68{,}720 - 1.95 T \ln T + 1.65 \times 10^{-3} T^2 - 0.10 \times 10^5 T^{-1} + 35.26T$$

Zone II (505°–1,300° K.)

$$\Delta C_p = -0.93 + 3.0 \times 10^{-3} T + 0.20 \times 10^5 T^{-2}$$
$$\Delta H_T = -69{,}800 - 0.93T + 1.5 \times 10^{-3} T^2 - 0.20 \times 10^5 T^{-1}$$
$$\Delta F_T = -69{,}800 + 0.93 T \ln T - 1.5 \times 10^{-3} T^2 - 0.10 \times 10^5 T^{-1} + 21.07T$$

T, ° K.	$H_T - H_{298}$	S_T	ΔH°_T	ΔF°_T
298		13.5	−68,350	−61,400
400	1,130	16.79	−68,250	−59,000
500	2,280	19.37	−68,200	−56,800
600	3,460	21.53	−69,850	−54,200
700	4,680	23.41	−69,750	−51,600
800	5,930	25.09	−69,700	−49,100
900	7,210	26.61	−69,500	−46,500
1,000	8,580	28.01	−69,250	−43,900
1,100	9,880	29.30	−69,050	−41,300
1,200	11,270	30.52	−68,850	−38,800
1,300	12,690	31.67	−68,550	−36,300

Tin Dioxide, SnO_2 (*c*)

$\Delta H^\circ_{298} = -138{,}820$ calories per mole (*71*)
$S_{298} = 12.5$ *e.u.* (*83*)
S.P. = 2,123° K. (*94*)

Zone I (c) (298°–1,500° K.)

$$C_p = 17.66 + 2.40 \times 10^{-3} T - 5.16 \times 10^5 T^{-2} \ (82)$$
$$H_T - H_{298} = -7{,}100 + 17.66T + 1.20 \times 10^{-3} T^2 + 5.16 \times 10^5 T^{-1}$$

Formation: $Sn + O_2 \longrightarrow SnO_2$

Zone I (298°–505° K.)

$$\Delta C_p = 6.08 - 4.90 \times 10^{-3} T - 4.76 \times 10^5 T^{-2}$$
$$\Delta H_T = -142{,}010 + 6.08T - 2.45 \times 10^{-3} T^2 + 4.76 \times 10^5 T^{-1}$$
$$\Delta F_T = -142{,}010 - 6.08 T \ln T + 2.45 \times 10^{-3} T^2 + 2.38 \times 10^5 T^{-1} + 90.74T$$

Zone II (505°–1,300° K.)

$$\Delta C_p = 3.2 + 1.4 \times 10^{-3} T - 4.76 \times 10^5 T^{-2}$$
$$\Delta H_T = -143{,}190 + 3.2T + 0.7 \times 10^{-3} T^2 + 4.76 \times 10^5 T^{-1}$$
$$\Delta F_T = -143{,}190 - 3.2 T \ln T - 0.7 \times 10^{-3} T^2 + 2.38 \times 10^5 T^{-1} + 76.58T$$

T, ° K.	$H_T - H_{298}$	S_T	ΔH°_T	ΔF°_T
298		12.5	−138,820	−124,300
400	1,510	16.84	−138,700	−119,300
500	3,100	20.38	−138,600	−114,500
600	4,780	23.45	−140,100	−109,300
700	6,550	26.18	−139,800	−104,300
800	8,390	28.63	−139,500	−99,200
900	10,280	30.85	−139,200	−94,200
1,000	12,210	32.88	−138,800	−89,200
1,100	14,190	34.77	−138,300	−84,200
1,200	16,210	36.53	−137,900	−79,300
1,300	18,260	38.17	−137,400	−74,400
1,400	20,340	39.71	−136,800	−69,500
1,500	22,440	41.16	−136,500	−64,800

Tin Difluoride, SnF_2 (*c*)

$\Delta H^\circ_{298} = (-158{,}000)$ calories per mole (*11*)
$S_{298} = 29$ *e.u.* (*11*)
M.P. = >900° K. (*6*)
B.P. = >1,500° K. (*6*)

Formation: $Sn + F_2 \longrightarrow SnF_2$
(estimated (*11*))

T, ° K.	$H_T - H_{298}$	ΔH°_T	ΔF°_T
298		(−158,000)	(−148,500)
500	(4,000)	(−157,000)	(−141,500)

Tin Dichloride, $SnCl_2$ (*c*)

$\Delta H^\circ_{298} = -81{,}100$ calories per mole (*11*)
$S_{298} = (34)$ *e.u.* (*11*)
M.P. = 500° K. (*6*)
$\Delta H_M = 3{,}050$ calories per mole
B.P. = 925° K. (*6*)
$\Delta H_V = 19{,}500$ calories per mole

Formation: $Sn + Cl_2 \longrightarrow SnCl_2$
(estimated (*11*))

T, ° K.	$H_T - H_{298}$	ΔH°_T	ΔF°_T
298		−81,000	(−71,600)
500	(4,000)	(−80,000)	(−65,500)

Tin Tetrachloride, $SnCl_4$ (*l*)

$\Delta H^\circ_{298} = -127{,}400$ calories per mole (*11*)
$S_{298} = 62.2$ *e.u.* (*83*)
$\Delta F^\circ_{298} = -16{,}900$ calories per mole
M.P. = 240° K. (*6*)
$\Delta H_M = 2{,}190$ calories per mole
B.P. = 386° K.
$\Delta H_V = 8{,}325$ calories per mole

Zone I (*g*) (298°–1,000° K.)

$$C_p = 25.57 + 0.20 \times 10^{-3}T - 1.87 \times 10^5 T^{-2} \ (82)$$
$$H_T - H_{298} = -8{,}260 + 25.57T + 0.10 \times 10^{-3}T^2 + 1.87 \times 10^5 T^{-1}$$

Formation: $Sn + 2Cl_2 \longrightarrow SnCl_4$
(estimated (*11*))

Tin Dibromide, $SnBr_2$ (*c*)

$\Delta H^\circ_{298} = -61{,}400$ calories per mole (*11*)
$S_{298} = (39)$ *e.u.* (*11*)
M.P. = 505° K. (*6*)
$\Delta H_M = 1{,}720$ calories per mole
B.P. = 912° K. (*6*)
$\Delta H_V = 23{,}500$ calories per mole

Formation: $Sn + Br_2 \longrightarrow SnBr_2$
(estimated (*11*))

T, ° K.	$H_T - H_{298}$	ΔH°_T	ΔF°_T
298		−61,400	(−58,500)
500	(4,000)	(−68,400)	(−52,000)

Tin Tetrabromide, $SnBr_4$ (*c*)

$\Delta H^\circ_{298} = -94{,}700$ calories per mole (*11*)
$S_{298} = (62)$ *e.u.* (*11*)
M.P. = 303° K. (*6*)
$\Delta H_M = 3{,}000$ calories per mole
B.P. = 480° K. (*6*)
$\Delta H_V = (10{,}500)$ calories per mole

Formation: $Sn + 2Br_2 \longrightarrow SnBr_4$
(estimated (*11*))

T, ° K.	$H_T - H_{298}$	ΔH°_T	ΔF°_T
298		−94,700	(−87,000)
500	(22,000)	(−98,000)	(−79,000)

Tin Diiodide, SnI_2 (*c*)

$\Delta H^\circ_{298} = -38{,}900$ calories per mole (*11*)
$S_{298} = (41)$ *e.u.* (*11*)
M.P. = 593° K. (*6*)
$\Delta H_M = (3{,}000)$ calories per mole
B.P. = 987° K. (*6*)
$\Delta H_V = 24{,}000$ calories per mole

Formation: $Sn + I_2 \longrightarrow SnI_2$
(estimated (*11*))

T, ° K.	$H_T - H_{298}$	ΔH°_T	ΔF°_T
298		−38,900	(−39,100)
500	(4,000)	(−53,000)	(−36,800)

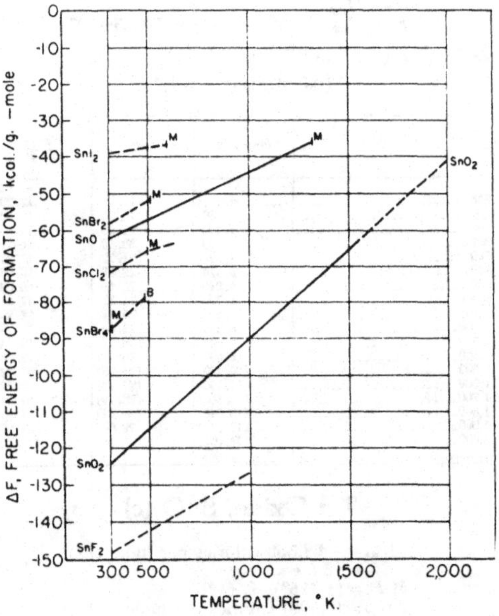

FIGURE 56.—Tin.

TITANIUM AND ITS COMPOUNDS

Element, Ti(*c*)

$S_{298}=7.24$ *e.u.* (*83*)
$T.P.=1{,}150°$ K. (*82*)
$\Delta H_T=950$ calories per atom
$M.P.=1{,}998°$ K. (*94*)
$\Delta H_M=4{,}500$ calories per atom
$B.P.=3{,}550°$ K. (*7*)
$\Delta H_V=101{,}000$ calories per atom

Zone I (α) (298°–1,150° K.)

$C_p=5.25+2.52\times10^{-3}T$ (*82*)
$H_T-H_{298}=-1{,}677+5.25T+1.26\times10^{-3}T^2$
$F_T-H_{298}=-1{,}677-5.25T\ln T-1.26\times10^{-3}T^2+28.66T$

Zone II (β) (1,150°–1,988° K.)

$C_p=7.50$ (*82*)
$H_T-H_{298}=-1{,}650+7.50T$
$F_T-H_{298}=-1{,}650-7.50T\ln T+43.1T$

T, ° K.	H_T-H_{298}	S_T	$-\frac{(F_T-H_{298})}{T}$
298	------	7.24	7.24
400	625	9.04	7.50
500	1,280	10.44	7.90
600	1,920	11.66	8.47
700	2,610	12.72	9.07
800	3,330	13.68	9.56
900	4,070	14.55	10.03
1,000	4,840	15.36	10.52
1,100	5,630	16.11	10.99
1,200	7,350	17.62	11.51
1,300	8,100	18.22	12.00
1,400	8,850	18.77	12.40
1,500	9,600	19.29	12.84
1,600	10,350	19.70	13.22
1,700	11,100	20.15	13.59
1,800	11,850	20.5	13.92
1,900	12,590	20.95	14.32
2,000	17,850	23.65	14.72

Titanium Oxide, TiO (*c*)

$\Delta H_{298}^{\circ}=-123{,}900$ calories per mole (*68*)
$S_{298}=8.31$ *e.u.* (*83*)
$T.P.=1{,}264°$ K. (*82*)
$\Delta H_T=820$ calories per mole
$M.P.=2{,}293°$ K. (*94*)
$\Delta H_M=14{,}000$ calories per mole

Zone I (*c*) (298°–1,264° K.)

$C_p=10.57+3.60\times10^{-3}T-1.86\times10^{5}T^{-2}$ (*82*)
$H_T-H_{298}=-3{,}935+10.57T+1.80\times10^{-3}T^2+1.86\times10^{5}T^{-1}$

Zone II (*l*) (1,264°–2,000° K.)

$C_p=11.85+3.00\times10^{-3}T$ (*82*)
$H_T-H_{298}=-4{,}100+11.85T+1.50\times10^{-3}T^2$

Formation: $Ti+1/2O_2\longrightarrow TiO$

Zone I (298°–1,150° K.)

$\Delta C_p=1.74+0.58\times10^{-3}T-1.66\times10^{5}T^{-2}$
$\Delta H_T=-125{,}000+1.74T+0.29\times10^{-3}T^2+1.66\times10^{5}T^{-1}$
$\Delta F_T=-125{,}000-1.74T\ln T-0.29\times10^{-3}T^2+0.83\times10^{5}T^{-1}+36.12T$

Zone II (1,150°–1,264° K.)

$\Delta C_p=-0.51+3.10\times10^{-3}T-1.66\times10^{5}T^{-2}$
$\Delta H_T=-125{,}050-0.51T+1.55\times10^{-3}T^2+1.66\times10^{5}T^{-1}$
$\Delta F_T=-125{,}050+0.51T\ln T-1.55\times10^{-3}T^2+0.83\times10^{5}T^{-1}+21.80T$

Zone III (1,264°–1,800° K.)

$\Delta C_p=0.77+2.50\times10^{-3}T+0.20\times10^{5}T^{-2}$
$\Delta H_T=-125{,}245+0.77T+1.25\times10^{-3}T^2-0.20\times10^{5}T^{-1}$
$\Delta F_T=-125{,}245-0.77T\ln T-1.25\times10^{-3}T^2-0.10\times10^{5}T^{-1}+30.8T$

T, ° K.	H_T-H_{298}	S_T	ΔH_T°	ΔF_T°
298	------	8.31	−123,900	−116,900
400	1,080	11.42	−123,800	−113,550
500	2,220	13.96	−123,650	−112,250
600	3,410	16.13	−123,500	−110,000
700	4,640	18.02	−123,350	−107,700
800	5,910	19.72	−123,200	−105,500
900	7,240	21.27	−123,050	−103,300
1,000	8,600	22.71	−122,850	−101,100
1,100	10,020	24.06	−122,600	−98,900
1,200	11,490	25.34	−123,300	−96,700
1,300	13,840	27.21	−122,100	−94,500
1,400	15,430	28.39	−121,700	−92,500
1,500	17,050	29.51	−121,300	−90,350
1,600	18,700	30.57	−120,750	−88,350
1,700	20,380	31.59	−120,300	−86,400
1,800	22,090	32.57	−119,750	−84,750

Dititanium Trioxide, Ti_2O_3 (*c*)

$\Delta H_{298}^{\circ}=-362{,}900$ calories per mole (*68*)
$S_{298}=18.83$ *e.u.* (*83*)
$T.P.=473°$ K. (*82*)
$\Delta H_T=215$ calories per mole
$M.P.=2{,}400°$ K. (*8*)
$\Delta H_M=38{,}400$ calories per mole

Zone I (α) (298°–473° K.)

$C_p=7.31+53.52\times10^{-3}T$ (*82*)
$H_T-H_{298}=-4{,}558+7.31T+26.76\times10^{-3}T^2$

Zone II (β) (473°–1,800° K.)

$C_p=34.68+1.30\times10^{-3}T-10.20\times10^{5}T^{-2}$ (*82*)
$H_T-H_{298}=-13{,}605+34.68T+0.65\times10^{-3}T^2+10.20\times10^{5}T^{-1}$

Formation: $2Ti+3/2O_2\longrightarrow Ti_2O_3$

Zone I (298°–473° K.)

$\Delta C_p=-13.93+46.98\times10^{-3}T+0.60\times10^{5}T^{-2}$
$\Delta H_T=-360{,}630-13.93T+23.49\times10^{-3}T^2-0.60\times10^{5}T^{-1}$
$\Delta F_T=-360{,}630+13.93T\ln T-23.49\times10^{-3}T^2-0.30\times10^{5}T^{-1}-17.41T$

Zone II (473°–1,150° K.)

$\Delta C_p=13.44-5.24\times10^{-3}T-9.60\times10^{5}T^{-2}$
$\Delta H_T=-369{,}620+13.44T-2.62\times10^{-3}T^2+9.60\times10^{5}T^{-1}$
$\Delta F_T=-369{,}620-13.44T\ln T+2.62\times10^{-3}T^2+4.80\times10^{5}T^{-1}+163.11T$

Zone III (1,150°–1,800° K.)

$\Delta C_p=8.94-0.2\times10^{-3}T-9.60\times10^{5}T^{-2}$
$\Delta H_T=-369{,}730+8.94T-0.10\times10^{-3}T^2+9.60\times10^{5}T^{-1}$
$\Delta F_T=-369{,}730-8.94T\ln T+0.10\times10^{-3}T^2+4.80\times10^{5}T^{-1}+133.85T$

T, ° K.	$H_T - H_{298}$	S_T	ΔH°_T	ΔF°_T
298	----------	18.83	−362,900	−344,000
400	2,610	26.32	−362,600	−335,200
500	5,940	33.70	−361,600	−328,250
600	9,140	39.54	−360,900	−322,000
700	12,440	44.62	−360,150	−315,550
800	15,930	49.14	−359,400	−309,300
900	19,270	53.19	−358,650	−303,050
1,000	22,740	56.85	−358,000	−297,800
1,100	26,260	60.21	−357,200	−290,600
1,200	29,800	63.29	−358,350	−284,350
1,300	33,360	66.13	−357,550	−278,550
1,400	36,950	68.79	−356,700	−272,300
1,500	40,560	71.29	−356,050	−266,450
1,600	44,180	73.62	−355,100	−260,600
1,700	47,830	75.83	−354,300	−254,750
1,800	51,490	77.93	−353,450	−249,450

Trititanium Pentaoxide, Ti_3O_5 (*c*)

$\Delta H^\circ_{298} = -587{,}000$ calories per mole (*68*)
$S_{298} = 30.9$ *e.u.* (*83*)
T.P. = 405° K. (*82*)
$\Delta H_T = 2{,}240$ calories per mole
M.P. = >2,500° K. (*42*)

Zone I (α) (298°–405° K.)

$$C_p = 35.47 + 29.50 \times 10^{-3}T \ (82)$$
$$H_T - H_{298} = -11{,}887 + 35.47T + 14.75 \times 10^{-3}T^2$$

Zone II (β) (405°–1, 400° K.)

$$C_p = 41.60 + 8.00 \times 10^{-3}T \ (82)$$
$$H_T - H_{298} = -10{,}230 + 41.60T + 4.00 \times 10^{-3}T^2$$

Formation: $3Ti + 5/2O_2 \longrightarrow Ti_3O_5$

Zone I (298°–450° K.)

$$\Delta C_p = 1.82 + 19.44 \times 10^{-3}T + 1.0 \times 10^5T^{-2}$$
$$\Delta H_T = -588{,}070 + 1.82T + 9.72 \times 10^{-3}T^2 - 1.0 \times 10^5T^{-1}$$
$$\Delta F_T = -588{,}070 - 1.82T\ln T - 9.72 \times 10^{-3}T^2 - 0.50 \times 10^5T^{-1} + 130.7T$$

Zone II (450°–1, 150° K.)

$$\Delta C_p = 7.95 - 2.06 \times 10^{-3}T + 1.0 \times 10^5T^{-2}$$
$$\Delta H_T = -586{,}330 + 7.95T - 1.03 \times 10^{-3}T^2 - 1.0 \times 10^5T^{-1}$$
$$\Delta F_T = -586{,}330 - 7.95T\ln T + 1.03 \times 10^{-3}T^2 - 0.50 \times 10^5T^{-1} + 159.57T$$

Zone III (1, 150°–1, 400° K.)

$$\Delta C_p = 1.20 + 5.50 \times 10^{-3}T + 1.0 \times 10^5T^{-2}$$
$$\Delta H_T = -586{,}460 + 1.20T + 2.75 \times 10^{-3}T^2 - 1.0 \times 10^5T^{-1}$$
$$\Delta F_T = -586{,}460 - 1.20T\ln T - 2.75 \times 10^{-3}T^2 - 0.50 \times 10^5T^{-1} + 116.2T$$

T, ° K.	$H_T - H_{298}$	S_T	ΔH°_T	ΔF°_T
298	----------	30.9	−587,000	−553,200
400	4,660	44.33	−586,000	−541,900
500	11,570	59.75	−582,800	−531,100
600	16,220	68.22	−582,050	−520,650
700	20,880	75.40	−581,400	−510,700
800	25,550	81.64	−580,900	−500,700
900	30,290	87.21	−580,400	−491,600
1,000	35,030	92.42	−580,050	−480,900
1,100	40,270	97.42	−579,150	−471,050
1,200	45,510	101.79	−581,150	−460,850
1,300	50,660	105.91	−580,300	−450,500
1,400	55,810	109.73	−579,550	−440,750

Titanium Dioxide (Rutile), TiO_2 (*c*)

$\Delta H^\circ_{298} = -225{,}600$ calories per mole (*68*)
$S_{298} = 12.01$ *e.u.* (*83*)
M.P. = 2,123° K. (*94*)
$\Delta H_M = 15{,}500$ calories per mole
B.P. = 3,273° K. (*94*)

Zone I (*c*) (298°–1,800° K.)

$$C_p = 17.97 + 0.28 \times 10^{-3}T - 4.35 \times 10^5T^{-2} \ (83)$$
$$H_T - H_{298} = -6{,}829 + 17.97T + 0.14 \times 10^{-3}T^2 + 4.35 \times 10^5T^{-1}$$

Formation: $Ti + O_2 \longrightarrow TiO_2$

Zone I (298°–1,150° K.)

$$\Delta C_p = 5.56 - 3.24 \times 10^{-3}T - 3.95 \times 10^5T^{-2}$$
$$\Delta H_T = -228{,}520 + 5.56T - 1.62 \times 10^{-3}T^2 + 3.95 \times 10^5T^{-1}$$
$$\Delta F_T = -228{,}520 - 5.56T\ln T + 1.62 \times 10^{-3}T^2 + 1.97 \times 10^5T^{-1} + 82.64T$$

Zone II (1,150°–1,800° K.)

$$\Delta C_p = 3.31 - 0.72 \times 10^{-3}T - 3.95 \times 10^5T^{-2}$$
$$\Delta H_T = -228{,}570 + 3.31T - 0.36 \times 10^{-3}T^2 + 3.95 \times 10^5T^{-1}$$
$$\Delta F_T = -228{,}570 - 3.31T\ln T + 0.36 \times 10^{-3}T^2 + 1.95 \times 10^5T^{-1} + 68.47T$$

T, ° K.	$H_T - H_{298}$	S_T	ΔH°_T	ΔF°_T
298	----------	12.01	−225,600	−212,400
400	1,540	16.44	−225,400	−207,900
500	3,100	19.92	−225,200	−203,600
600	4,735	22.90	−225,000	−199,300
700	6,440	25.52	−224,750	−194,950
800	8,160	27.82	−224,550	−190,750
900	9,900	29.87	−224,350	−186,550
1,000	11,650	31.71	−224,200	−182,350
1,100	13,420	33.40	−224,000	−178,000
1,200	15,200	34.95	−224,800	−173,890
1,300	17,000	36.39	−224,550	−169,550
1,400	18,820	37.74	−224,350	−165,450
1,500	20,660	39.01	−224,200	−161,200
1,600	22,540	40.22	−223,850	−157,250
1,700	24,340	41.37	−223,600	−153,100
1,800	26,340	42.46	−223,350	−149,350

Titanium Difluoride, TiF_2 (*c*)

$\Delta H^\circ_{298} = (-198{,}000)$ calories per mole (*11*)
$S_{298} = (18)$ *e.u.* (*11*)

Formation: $Ti + F_2 \longrightarrow TiF_2$
(estimated (*11*))

T, ° K.	$H_T - H_{298}$	ΔH°_T	ΔF°_T
298	----------	(−198,000)	(−187,000)
500	(4,000)	(−197,000)	(−179,000)
1,000	(14,000)	(−195,000)	(−160,000)
1,500	(25,000)	(−193,000)	(−142,000)

Titanium Trifluoride, TiF_3 (*c*)

$\Delta H^\circ_{298} = (-315{,}000)$ calories per mole (*11*)
$S_{298} = (28)$ *e.u.* (*11*)
M.P. = (1,500°) K. (*6*)
$\Delta H_M = (12{,}000)$ calories per mole
B.P. = (1,700°) K. (*6*)
$\Delta H_V = (49{,}000)$ calories per mole

Formation: $Ti+3/2F_2 \longrightarrow TiF_3$
(estimated (*11*))

T, ° K.	H_T-H_{298}	ΔH°_T	ΔF°_T
298		(−315, 000)	(−299, 500)
500	(4, 000)	(−314, 600)	(−289, 000)
1, 000	(17, 000)	(−309, 000)	(−262, 000)
1, 500	(32, 000)	(−303, 000)	(−237, 000)

Titanium Dichloride, $TiCl_2$ (*c*)

$\Delta H^\circ_{298} = -123{,}700$ calories per mole (*86*)
$S_{298} = (24.3)$ *e.u.* (*86*)

Formation: $Ti+Cl_2 \longrightarrow TiCl_2$
(estimated (*86*))

T, ° K.	H_T-H_{298}	ΔH°_T	ΔF°_T
298		−123, 000	(−112, 150)
400	(1, 790)	(−122, 700)	(−108, 500)
500	(3, 600)	(−122, 350)	(−105, 000)
600	(5, 450)	(−122, 050)	(−101, 550)
700	(7, 340)	(−121, 700)	(−98, 200)
800	(9, 280)	(−121, 400)	(−94, 850)
900	(11, 270)	(−121, 000)	(−91, 550)
1,000	(13, 300)	(−120, 650)	(−88, 300)
1,100	(15, 380)	(−120, 050)	(−85, 100)
1,200	(17, 500)	(−120, 750)	(−81, 850)

Titanium Trichloride, $TiCl_3$ (*c*)

$\Delta H^\circ_{298} = -172{,}000$ calories per mole (*86*)
$S_{298} = (33.4)$ *e.u.* (*86*)
Disproportionates (1,200°) K. (*6*)

Formation: $Ti+3/2Cl_2 \longrightarrow TiCl_3$
(estimated (*86*))

T, ° K.	H_T-H_{298}	ΔH°_T	ΔF°_T
298		−172, 000	(−155, 950)
400	(2, 320)	(−171, 550)	(−150, 500)
500	(4, 660)	(−171, 150)	(−145, 300)
600	(7, 070)	(−170, 700)	(−140, 150)
700	(9, 560)	(−170, 200)	(−135, 100)
800	(12, 110)	(−169, 700)	(−130, 150)
900	(14, 740)	(−169, 150)	(−125, 200)
1,000	(17, 430)	(−168, 600)	(−120, 350)
1,100	(20, 190)	(−167, 950)	(−115, 600)
1,200	(23, 030)	(−168, 200)	(−110, 750)

Titanium Tetrachloride, $TiCl_4$ (*l*)

$\Delta H^\circ_{298} = -192{,}100$ calories per mole (*86*)
$S_{298} = 59.50$ *e.u.* (*86*)
M.P. = 250° K. (*6*)
$\Delta H_M = 2{,}240$ calories per mole
B.P. = 409° K. (*6*)
$\Delta H_V = 8{,}346$ calories per mole

Zone I (*g*) (409°–2,000° K.)

$$C_p = 25.45 + 0.24\times10^{-3}T - 2.36\times10^{5}T^{-2} \quad (82)$$
$$H_T - H_{298} = -8{,}390 + 25.45T + 0.12\times10^{-3}T^2 + 2.36\times10^{5}T^{-1}$$

Formation: $Ti+2Cl_2 \longrightarrow TiCl_4$

Zone I (298°–409° K.)

Zone II (409°–1,150° K.)

$$\Delta C_p = 2.56 - 2.4\times10^{-3}T - 1.0\times10^{5}T^{-2}$$
$$\Delta H_T = -183{,}300 + 2.56T - 1.2\times10^{-3}T^2 + 1.0\times10^{5}T^{-1}$$
$$\Delta F_T = -183{,}300 - 2.56T\ln T + 1.2\times10^{-3}T^2 + 0.5\times10^{5}T^{-1} + 46.78T$$

Zone III (1,150°–1,900° K.)

$$\Delta C_p = 0.31 + 0.12\times10^{-3}T - 1.0\times10^{5}T^{-2}$$
$$\Delta H_T = -183{,}300 + 0.31T + 0.06\times10^{-3}T^2 + 1.0\times10^{5}T^{-1}$$
$$\Delta F_T = -183{,}300 - 0.31T\ln T - 0.06\times10^{-3}T^2 + 0.5\times10^{5}T^{-1} + 32.41T$$

(estimated (86))

T, ° K.	H_T-H_{298}	S_T	ΔH°_T	ΔF°_T
298		59. 50	−192, 100	−175, 900
400	3, 820	70. 52	−190, 600	−170, 550
500	14, 670	96. 94	−182, 100	−167, 450
600	17, 150	101. 16	−181, 950	−164, 500
700	19, 655	105. 02	−181, 900	−161, 600
800	22, 180	108. 39	−181, 850	−158, 700
900	24, 720	111. 38	−181, 800	−155, 850
1,000	27, 265	114. 06	−181, 800	−152, 950
1,100	29, 805	116. 49	−181, 750	−150, 050
1,200	32, 375	118. 72	−182, 750	−147, 100
1,300	34, 915	120. 77	−182, 750	−144, 150
1,400	37, 505	122. 67	−182, 650	−141, 150
1,500	40, 150	124. 44	−182, 600	−138, 150
1,600	42, 640	126. 10	−182, 600	−135, 200
1,700	45, 200	127. 66	−182, 600	−132, 250
1,800	47, 785	129. 13	−182, 600	−129, 250
1,900	50, 360	130. 52	−182, 600	−126, 250

Titanium Dibromide, $TiBr_2$ (*c*)

$\Delta H^\circ_{298} = (-95{,}000)$ calories per mole (*11*)
$S_{298} = (30)$ *e.u.* (*11*)
M.P. = (900°) K. (*6*)
$\Delta H_M = (6{,}000)$ calories per mole
B.P. = (1,500°) K. (*6*)
$\Delta H_V = (33{,}000)$ calories per mole

Formation: $Ti+Br_2 \longrightarrow TiBr_2$
(estimated (*11*))

T, ° K.	H_T-H_{298}	ΔH°_T	ΔF°_T
298		(−95, 000)	(−91, 000)
500	(4, 000)	(−102, 000)	(−84, 000)
1,000	(21, 000)	(−93, 000)	(−65, 000)
1,500	(33, 000)	(−90, 000)	(−53, 000)

Titanium Tribromide, $TiBr_3$ (*c*)

$\Delta H^\circ_{298} = (-132{,}000)$ calories per mole (*11*)
$S_{298} = (43)$ *e.u.* (*11*)
Disproportionates (1,200°) K. (*6*)

Formation: $Ti+3/2Br_2 \longrightarrow TiBr_3$
(estimated (*11*))

T, ° K.	H_T-H_{298}	ΔH°_T	ΔF°_T
298		(−132, 000)	(−126, 500)
500	(5, 000)	(−142, 500)	(−117, 000)
1,000	(19, 000)	(−139, 000)	(−91, 000)

Titanium Tetrabromide, $TiBr_4$ (*c*)

$\Delta H^\circ_{298} = -148{,}200$ calories per mole (*86*)
$S_{298} = 58.0$ *e.u.* (*86*)
$M.P. = 311°$ K. (*6*)
$\Delta H_M = 2{,}060$ calories per mole
$B.P. = 503°$ K. (*6*)
$\Delta H_V = (11{,}000)$ calories per mole

Formation: $Ti + 2Br_2 \longrightarrow TiBr_4$
(estimated (*86*))

T, ° K.	$H_T - H_{298}$	ΔH°_T	ΔF°_T
298		−148, 200	−142, 000
500	(11, 000)	(−159, 000)	(−132, 000)

Titanium Diiodide, TiI_2 (*c*)

$\Delta H^\circ_{298} = (-61{,}100)$ calories per mole (*11*)
$S_{298} = (33)$ *e.u.* (*11*)
$M.P. = (900°)$ K. (*6*)
$\Delta H_M = (6{,}000)$ calories per mole
$B.P. = (1{,}300°)$ K. (*6*)
$\Delta H_V = (27{,}000)$ calories per mole

Formation: $Ti + I_2 \longrightarrow TiI_2$
(estimated (*11*))

T, ° K.	$H_T - H_{298}$	ΔH°_T	ΔF°_T
298		(−61, 100)	(−60, 500)
500	(4, 000)	(−75, 000)	(−57, 000)
1,000	(21, 000)	(−66, 000)	(−39, 000)
1,500	(60, 000)	(−36, 000)	(−25, 000)

Titanium Triiodide, TiI_3 (*c*)

$\Delta H^\circ_{298} = (-80{,}000)$ calories per mole (*11*)
$S_{298} = (47)$ *e.u.* (*11*)
Disproportionates $>1{,}200°$ K. (*6*)

Formation: $Ti + 3/2I_2 \longrightarrow TiI_3$
(estimated (*11*))

T, ° K.	$H_T - H_{298}$	ΔH°_T	ΔF°_T
298		(−80, 000)	(−79, 000)
500	(5, 000)	(−102, 000)	(−75, 000)
1,000	(20, 000)	(−97, 500)	(−48, 000)

Titanium Tetraiodide, TiI_4 (*c*)

$\Delta H^\circ_{298} = (-101{,}000)$ calories per mole (*11*)
$S_{298} = (64)$ *e.u.* (*11*)
$M.P. = 423°$ K. (*6*)
$\Delta H_M = (3{,}000)$ calories per mole
$B.P. = 650°$ K. (*6*)
$\Delta H_V = 13{,}500$ calories per mole

Formation: $Ti + 2I_2 \longrightarrow TiI_4$
(estimated (*11*))

T, ° K.	$H_T - H_{298}$	ΔH°_T	ΔF°_T
298		(−101, 000)	(−101, 000)
500	(11, 000)	(−124, 600)	(−89, 500)

Titanium Carbide, TiC (*c*)

$\Delta H^\circ_{298} = -44{,}100$ calories per mole (*86*)
$S_{298} = 5.79$ *e.u.* (*88*)
$M.P. = 3{,}450°$ K. (*9*)

Zone I (*c*) (298°–1,800° K.)

$$C_p = 11.83 + 0.80 \times 10^{-3}T - 3.58 \times 10^5 T^{-2} \ (82)$$
$$H_T - H_{298} = -4{,}764 + 11.83T + 0.40 \times 10^{-3}T^2 + 3.58 \times 10^5 T^{-1}$$

Formation: $Ti + C \longrightarrow TiC$

Zone I (298°–1,150° K.)

$$\Delta C_p = 2.48 - 2.74 \times 10^{-3}T - 1.48 \times 10^5 T^{-2}$$
$$\Delta H_T = -45{,}100 + 2.48T - 1.37 \times 10^{-3}T^2 + 1.48 \times 10^5 T^{-1}$$
$$\Delta F_T = -45{,}100 - 2.48T\ln T + 1.37 \times 10^{-3}T^2 + 0.74 \times 10^5 T^{-1} + 19.41T$$

Zone II (1,150°–1,800° K.)

$$\Delta C_p = 0.23 - 0.22 \times 10^{-3}T - 1.48 \times 10^5 T^{-2}$$
$$\Delta H_T = -45{,}200 + 0.23T - 0.11 \times 10^{-3}T^2 + 1.48 \times 10^5 T^{-1}$$
$$\Delta F_T = -45{,}200 - 0.23T\ln T + 0.11 \times 10^{-3}T^2 + 0.74 \times 10^5 T^{-1} + 4.96T$$

T, ° K.	$H_T - H_{298}$	S_T	ΔH°_T	ΔF°_T
298		5. 79	−44, 100	−43, 300
400	945	8. 51	−44, 050	−43, 000
500	1, 957	10. 8	−43, 950	−42, 700
600	3, 085	12. 82	−43, 900	−42, 500
700	4, 225	14. 58	−43, 850	−42, 250
800	5, 395	16. 14	−43, 850	−42, 050
900	6, 600	17. 56	−43, 900	−41, 800
1,000	7, 830	18. 86	−43, 900	−41, 600
1,100	9, 080	20. 05	−43, 950	−41, 350
1,200	10, 330	21. 04	−44, 950	−40, 950
1,300	11, 590	22. 14	−45, 000	−40, 700
1,400	12, 860	23. 08	−45, 000	−40, 400
1,500	14, 130	23. 96	−45, 050	−40, 050
1,600	15, 400	24. 78	−45, 100	−39, 850
1,700	16. 670	25. 55	−45, 150	−39, 550
1,800	17, 940	26. 28	−45, 150	−39, 400

Titqnium Nitride, TiN (*c*)

$\Delta H^\circ_{298} = -80{,}700$ calories per mole (*86*)
$S_{298} = 7.24$ *e.u.* (*88*)
$M.P. = 3{,}200°$ K. (*9*)

Zone I (*c*) (298°–1,800° K.)

$$C_p = 11.91 + 0.94\times10^{-3}T - 2.96\times10^{5}T^{-2}\ (82)$$
$$H_T - H_{298} = -4{,}586 + 11.91T + 0.47\times10^{-3}T^2 + 2.96\times10^{5}T^{-1}$$

Formation: $Ti + 1/2N_2 \longrightarrow TiN$

Zone I (298°–1,150° K.)

$$\Delta C_p = 3.33 - 2.09\times10^{-3}T - 2.96\times10^{5}T^{-2}$$
$$\Delta H_T = -82{,}590 + 3.33T - 1.04\times10^{-3}T^2 + 2.96\times10^{5}T^{-1}$$
$$\Delta F_T = -82{,}590 - 3.33T\ln T + 1.04\times10^{-3}T^2 + 1.48\times10^{5}T^{-1} + 46.13T$$

Zone II (1,150°–1,800° K.)

$$\Delta C_p = 1.08 + 0.43\times10^{-3}T - 2.96\times10^{5}T^{-2}$$
$$\Delta H_T = -82{,}650 + 1.08T + 0.21\times10^{-3}T^2 + 2.96\times10^{5}T^{-1}$$
$$\Delta F_T = -82{,}650 - 1.08T\ln T - 0.21\times10^{-3}T^2 + 1.48\times10^{5}T^{-1} + 31.86T$$

T, ° K.	H_T-H_{298}	S_T	ΔH°_T	ΔF°_T
298	....	7.24	−80,700	−73,870
400	1,000	10.12	−80,700	−71,550
500	2,090	12.54	−80,600	−69,100
600	3,230	14.63	−80,450	−65,700
700	4,400	16.43	−80,350	−64,300
800	5,590	18.02	−80,250	−61,900
900	6,810	19.45	−80,150	−59,550
1,000	8,050	20.76	−80,050	−58,200
1,100	9,310	21.96	−80,000	−55,000
1,200	10,600	23.08	−80,800	−53,800
1,300	11,910	24.13	−80,650	−51,450
1,400	13,230	25.11	−80,450	−49,200
1,500	14,550	26.02	−80,350	−47,250
1,600	15,870	26.87	−80,150	−44,950
1,700	17,190	27.67	−80,000	−42,800
1,800	18,510	28.43	−79,850	−40,650

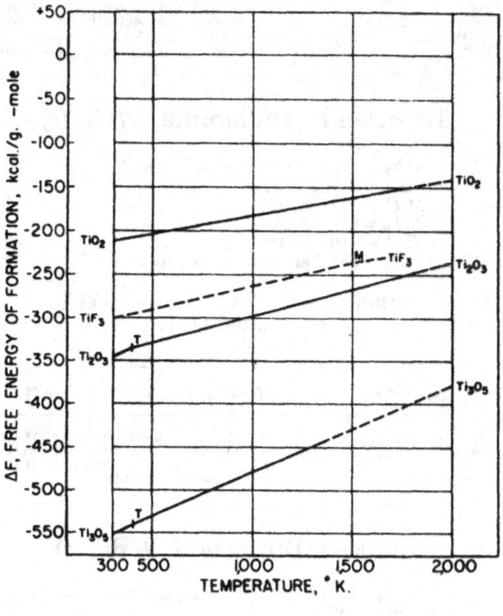

FIGURE 57.—Titanium (a).

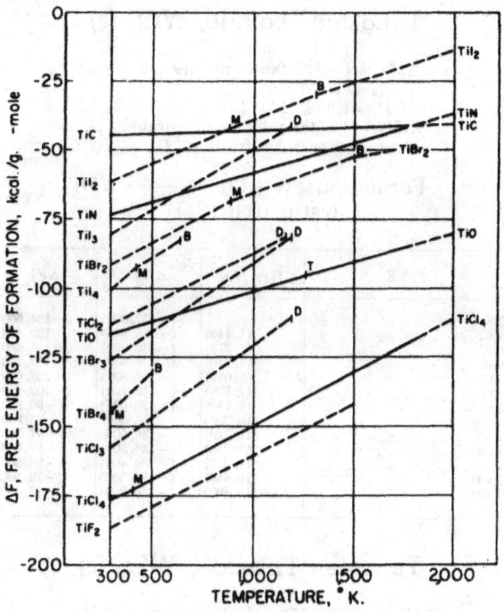

FIGURE 58.—Titanium (b).

TUNGSTEN AND ITS COMPOUNDS

Element, W (*c*)

$S_{298} = 8.0 \pm 0.2$ *e.u.* (*83*)
M.P. = 3,650° K. (*7*)
ΔH_M = 8,420 calories per atom
B.P. = 5,950° K. (*7*)
ΔH_V = 184,580 calories per atom

Zone I (*c*) (298°–2,000° K.)

$$C_p = 5.74 + 0.76\times10^{-3}T\ (82)$$
$$H_T - H_{298} = -1{,}745 + 5.74T + 0.38\times10^{-3}T^2$$
$$F_T - H_{298} = -1{,}745 - 5.74T\ln T - 0.38\times10^{-3}T^2 + 30.61T$$

T, ° K.	H_T-H_{298}	S_T	$-\frac{(F_T-H_{298})}{T}$
298	....	8.0	8.0
400	615	9.77	8.23
500	1,220	11.12	8.68
600	1,830	12.23	9.18
700	2,450	13.19	9.69
800	3,080	14.03	10.18
900	3,710	14.77	10.65
1,000	4,360	15.46	11.10
1,100	5,010	16.08	11.53
1,200	5,670	16.65	11.92
1,300	6,340	17.19	12.31
1,400	7,030	17.70	12.68
1,500	7,730	18.18	13.03
1,600	8,430	18.63	13.36
1,700	9,130	19.06	13.69
1,800	9,840	19.47	14.00
1,900	10,550	19.85	14.30
2,000	11,2[illegible]0	20.21	14.58

Tungsten Dioxide, WO_2 (*c*)

$\Delta H^{\circ}_{298} = (-137,000)$ calories per mole (*24*)
$S_{298} = (15.5)$ *e.u.* (*24*)
M.P. = 1,543° K. (*42*)
$\Delta H_M = 11,500$ calories per mole
Decomposes = 2,125° K. (*42*)

Formation: $W + O_2 \longrightarrow WO_2$
(estimated (*24*))

T, ° K.	$H_T - H_{298}$	ΔH°_T	ΔF°_T
298		(−137, 000)	(−124, 600)
400	(1, 440)	(−136, 900)	(−120, 400)
500	(2, 775)	(−136, 900)	(−116, 300)
600	(4, 240)	(−136, 800)	(−112, 100)
700	(5, 640)	(−136, 800)	(−108, 000)
800	(7, 170)	(−136, 700)	(−103, 900)
900	(8, 710)	(−136, 600)	(−99, 800)
1,000	(10, 200)	(−136, 600)	(−95, 800)
1,100	(11, 720)	(−136, 500)	(−91, 700)
1,200	(13, 210)	(−136, 500)	(−87, 600)
1,300	(14, 810)	(−136, 400)	(−83, 500)
1,400	(16, 450)	(−136, 300)	(−79, 500)
1,500	(18, 100)	(−136, 300)	(−75, 400)

Tungsten Trioxide, WO_3 (*c*)

$\Delta H^{\circ}_{298} = -200,850$ calories per mole (*65*)
$S_{298} = 19.9$ *e.u.* (*83*)
M.P. = 1,743° K. (*24*)
$\Delta H_M = 13,940$ calories per mole
B.P. = >2,500° K. (*42*)

Formation: $W + 3/2O_2 \longrightarrow WO_3$
(estimated (*24*))

T, ° K.	$H_T - H_{298}$	ΔH°_T	ΔF°_T
298		−200, 850	−182, 450
400	(3, 000)	(−197, 850)	(−176, 250)
500	(4, 050)	(−196, 800)	(−170, 200)
600	(6, 150)	(−194, 700)	(−164, 200)
700	(8, 750)	(−192, 100)	(−158, 300)
800	(10, 450)	(−190, 400)	(−152, 450)
900	(12, 800)	(−188, 050)	(−146, 650)
1, 000	(15, 200)	(−185, 650)	(−140, 900)
1, 100	(17, 500)	(−183, 350)	(−135, 200)
1, 200	(20, 000)	(−180, 850)	(−129, 550)
1, 300	(22, 500)	(−178, 350)	(−123, 900)
1, 400	(25, 050)	(−175, 800)	(−118, 350)
1, 500	(27, 850)	(−173, 000)	(−112, 850)
1, 600	(30, 400)	(−170, 450)	(−107, 400)
1, 700	(33, 200)	(−167, 650)	(−101, 950)
1, 800	(48, 100)	(−152, 750)	(−97, 050)
1, 900	(53, 300)	(−147, 550)	(−92, 500)
2, 000	(56, 700)	(−144, 150)	(−88, 050)

Tungsten Dichloride, WCl_2 (*c*)

$\Delta H^{\circ}_{298} = (-36,000)$ calories per mole (*11*)
$S_{298} = (31)$ *e.u.* (*11*)

Formation: $W + Cl_2 \longrightarrow WCl_2$
(estimated (*11*))

T, ° K.	$H_T - H_{298}$	ΔH°_T	ΔF°_T
298		(−36, 000)	(−27, 000)
500	(4, 000)	(−35, 000)	(−21, 000)
1, 000	(15, 000)	(−31, 000)	(−9, 000)

647940 O—63——9

Tungsten Tetrachloride, WCl_4 (*c*)

$\Delta H^{\circ}_{298} = (-69,000)$ calories per mole (*11*)
$S_{298} = (50)$ *e.u.* (*11*)
M.P. = (600°) K. (*6*)
$\Delta H_M = (6,000)$ calories per mole
B.P. = (605°) K. (*6*)
$\Delta H_V = (17,000)$ calories per mole

Formation: $W + 2Cl_2 \longrightarrow WCl_4$
(estimated (*11*))

T, ° K.	$H_T - H_{298}$	ΔH°_T	ΔF°_T
298		(−69, 000)	(−50, 000)
500	(6, 000)	(−68, 000)	(−37, 000)
1, 000	(41, 000)	(−44, 000)	(−21, 000)
1, 500	(53, 000)	(−44, 000)	(−10, 000)

Tungsten Pentachloride, WCl_5 (*c*)

$\Delta H^{\circ}_{298} = (-82,000)$ calories per mole (*11*)
$S_{298} = (66)$ *e.u.* (*11*)
M.P. = 517° K. (*6*)
$\Delta H_M = (8,000)$ calories per mole
B.P. = 549° K. (*6*)
$\Delta H_V = (12,500)$ calories per mole

Formation: $W + 5/2Cl_2 \longrightarrow WCl_5$
(estimated (*11*))

T, ° K.	$H_T - H_{298}$	ΔH°_T	ΔF°_T
298		(−82, 000)	(−60, 000)
500	(7, 000)	(−80, 000)	(−45, 000)
1, 000	(44, 000)	(−57, 000)	(−30, 000)
1, 500	(59, 000)	(−57, 000)	(−17, 000)

Tungsten Hexachloride, WCl_6 (*c*)

$\Delta H^{\circ}_{298} = (-96,900)$ calories per mole (*11*)
$S_{298} = (75)$ *e.u.* (*11*)
M.P. = 548° K. (*6*)
$\Delta H_M = (5,700)$ calories per mole
B.P. = 610° K. (*6*)
$\Delta H_V = (15,200)$ calories per mole

Formation: $W + 3Cl_2 \longrightarrow WCl_6$
(estimated (*11*))

T, ° K.	$H_T - H_{298}$	ΔH°_T	ΔF°_T
298		(−96, 900)	(−74, 000)
500	(9, 000)	(−94, 000)	(−50, 600)
1, 000	(50, 000)	(−69, 000)	(−24, 000)
1, 500	(67, 000)	(−69, 000)	(−2, 000)

Tungsten Dibromide, WBr_2 (*c*)

$\Delta H^{\circ}_{298} = (-18,700)$ calories per mole (*11*)
$S_{298} = (36)$ *e.u.* (*11*)
M.P. = (1,000°) K. (*6*)
$\Delta H_M = (6,000)$ calories per mole
B.P. = (1,500°) K. (*6*)
$\Delta H_V = (33,000)$ calories per mole

Formation: $W + Br_2 \longrightarrow WBr_2$
(estimated (*11*))

T, ° K.	$H_T - H_{298}$	ΔH_T°	ΔF_T°
298	------	(−18, 700)	(−16, 000)
500	(4, 000)	(−25, 000)	(−11, 000)
1,000	(16, 000)	(−21, 000)	(+1, 000)
1,500	(68, 000)	(+23, 000)	(+8, 000)

Tungsten Tetrabromide, WBr_4 (*c*)

$\Delta H_{298}^\circ = (-35{,}000)$ calories per mole (*11*)
$S_{298} = (59)$ *e.u.* (*11*)
S.P. = (600°) K. (*6*)
$\Delta H_{subl} = (24{,}000)$ calories per mole

Formation: $W + 2Br_2 \longrightarrow WBr_4$
(estimated (*11*))

T, ° K.	$H_T - H_{298}$	ΔH_T°	ΔF_T°
298	------	(−35, 000)	(−28, 500)
500	(5, 000)	(−50, 000)	(−18, 000)
1,000	(42, 000)	(−25, 000)	(−5, 000)
1,500	(55, 000)	(−25, 000)	(+4, 000)
2,000	------	------	(+14, 000)

Tungsten Pentabromide, WBr_5 (*c*)

$\Delta H_{298}^\circ = (42{,}000)$ calories per mole (*11*)
$S_{298} = (78)$ *e.u.* (*11*)
M.P. = 549° K. (*6*)
$\Delta H_M = (8{,}000)$ calories per mole
B.P. = 606° K. (*6*)
$\Delta H_V = (14{,}000)$ calories per mole

Formation: $W + 5/2Br_2 \longrightarrow WBr_5$
(estimated (*11*))

T, ° K.	$H_T - H_{298}$	ΔH_T°	ΔF_T°
298	------	(−42, 000)	(−35, 000)
500	(7, 000)	(−60, 000)	(−29, 000)
1,000	(43, 000)	(−38, 000)	(1, 000)
1,500	(58, 000)	(−38, 000)	(19, 000)

Tungsten Hexabromide, WBr_6 (*c*)

$\Delta H_{298}^\circ = (-44{,}000)$ calories per mole (*11*)
$S_{298} = (89)$ *e.u.* (*11*)

Formation: $W + 3Br_2 \longrightarrow WBr_6$
(estimated (*11*))

T, ° K.	$H_T - H_{298}$	ΔH_T°	ΔF_T°
298	------	(−44, 000)	(−36, 000)
500	(35, 000)	(−39, 000)	(−14, 000)
1,000	(51, 000)	(−39, 000)	(+11, 000)

Tungsten Diiodide, WI_2 (*c*)

$\Delta H_{298}^\circ = (-1{,}000)$ calories per mole (*11*)
$S_{298} = (38)$ *e.u.* (*11*)
M.P. = (1,000°) K. (*6*)
$\Delta H_M = (6{,}000)$ calories per mole
B.P. = (1,260°) K. (*6*)
$\Delta H_V = (27{,}000)$ calories per mole

Formation: $W + I_2 \longrightarrow WI_2$
(estimated (*11*))

T, ° K.	$H_T - H_{298}$	ΔH_T°	ΔH_T°
298	------	(−1, 000)	(−1, 500)
500	(4, 000)	(−15, 000)	(0)
1,000	(17, 000)	(−11, 000)	(+13, 000)
1,500	(59, 000)	(+25, 000)	(+16, 000)

Tungsten Tetraiodide, WI_4 (*c*)

$\Delta H_{298}^\circ = (-500)$ calories per mole (*11*)
$S_{298} = (65)$ *e.u.* (*11*)
S.P. = (690°) K. (*6*)
$\Delta H_{subl} = (20{,}000)$ calories per mole

Formation: $W + 2I_2 \longrightarrow WI_4$
(estimated (*11*))

T, ° K.	$H_T - H_{298}$	ΔH_T°	ΔF_T°
298	------	(−500)	(−500)
500	(6, 000)	(−29, 000)	(+3, 000)
1,000	(47, 000)	(0)	(+26, 000)
1,500	(59, 000)	(0)	(+39, 000)

Tungsten Carbide, WC (*c*)

$\Delta H_{298} = -9{,}100$ calories per mole (*112*)
$S_{298} = 8.5$ *e.u.* (*94*)
Decomposes = 2,900° K. (*9*)

Zone I (*c*) (298°–2,000° K.)

$$C_p = 7.98 + 2.17 \times 10^{-3} T$$
$$H_T - H_{298} = -2{,}470 + 7.98T + 1.08 \times 10^{-3} T^2$$

Formation: $W + C \longrightarrow WC$

Zone I (298°–2,000° K.)

$$\Delta C_p = -1.86 + 0.39 \times 10^{-3} T + 2.10 \times 10^5 T^{-2}$$
$$\Delta H_T = -7{,}860 - 1.86T + 0.20 \times 10^{-3} T^2 - 2.10 \times 10^5 T^{-1}$$
$$\Delta F_T = -7{,}860 + 1.86 T \ln T - 0.20 \times 10^{-3} T^2 - 1.05 \times 10^5 T^{-1} - 12.63T$$

T, ° K.	$H_T - H_{298}$	S_T	ΔH_T°	ΔF_T°
298	------	8. 5	−9, 100	−8, 800
400	920	11. 12	−9, 050	−8, 800
500	1, 810	13. 14	−9, 080	−8, 700
600	2, 730	15. 75	−9, 150	−8, 600
700	3, 670	16. 12	−9, 250	−8, 500
800	4, 630	17. 49	−9, 380	−8, 400
900	5, 610	18. 61	−9, 510	−8, 200
1,000	6, 610	19. 72	−9, 660	−8, 000
1,100	7, 650	20. 64	−9, 780	−7, 800
1,200	8, 690	21. 66	−9, 930	−7, 600
1,300	9, 760	22. 37	−10, 070	−7, 400
1,400	10, 850	23. 29	−10, 210	−7, 300
1,500	11, 970	24. 01	−10, 340	−7, 100
1,600	13, 100	24. 83	−10, 470	−7, 000
1,700	14, 260	25. 55	−10, 580	−6, 800
1,800	15, 440	26. 16	−10, 690	−6, 500
1,900	16, 630	26. 78	−10, 800	−6, 200
2,000	17, 830	27. 40	−10, 910	−6, 000

Ditungsten Nitride, W_2N (*c*)

$\Delta H_{298}^\circ = -17{,}000$ (*9*)
$S_{298} = (18.0)$ *e.u.* (*9*)
$\Delta F_{298}^\circ = (-11{,}000)$ calories per mole

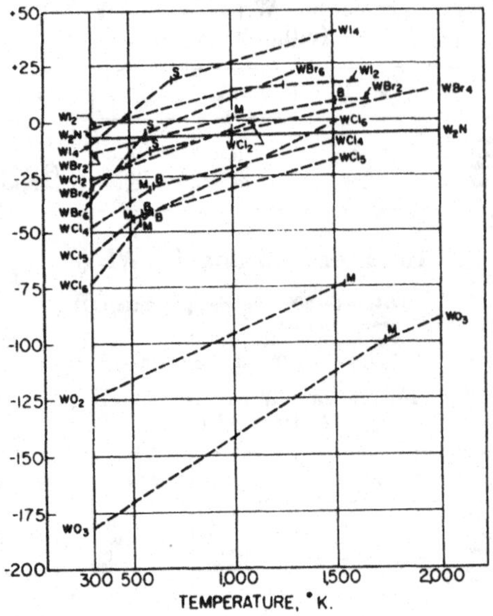

FIGURE 59.—Tungsten.

URANIUM AND ITS COMPOUNDS

Element, U (*c*)

$S_{298}=12.03$ *e.u.* (*77*)
$T.P.=935°$ K. (*82*)
$\Delta H_T=700$ calories per atom
$T.P.=1{,}045°$ K. (*82*)
$\Delta H_T=1{,}145$ calories per atom
$M.P.=1{,}405°$ K. (*24*)
$\Delta H_M=3{,}200$ calories per atom
$B.P.=3{,}800°$ K. (*8*)
$\Delta H_V=110{,}000$ calories per atom

Zone I (α) (298°–935° K.)

$$C_p=3.39+8.02\times10^{-3}T+0.70\times10^5T^{-2}\ (82)$$
$$H_T-H_{298}=-1{,}132+3.39T+4.01\times10^{-3}T^2-0.70\times10^5T^{-1}$$
$$F_T-H_{298}=-1{,}132-3.39T\ln T-4.01\times10^{-3}T^2-0.35\times10^5T^{-1}+12.67T$$

Zone II (β) (935°–1,045° K.)

$$C_p=10.18\ (82)$$
$$H_T-H_{298}=-3{,}348+10.18T$$
$$F_T-H_{298}=-3{,}348-10.18T\ln T+57.69T$$

Zone III (γ) (1,045°–1,300° K.)

$$C_p=9.20\ (82)$$
$$H_T-H_{298}=-1{,}180+9.20T$$
$$F_T-H_{298}=-1{,}180-9.2T\ln T+48.76T$$

T, ° K.	H_T-H_{298}	S_T	$-\frac{(F_T-H_{298})}{T}$
298	------	12.03	12.03
400	690	14.02	12.29
500	1,430	15.67	12.81
600	2,230	17.12	13.40
700	3,100	18.47	14.04
800	4,050	19.74	14.68
900	5,090	20.96	15.30
1,000	6,830	22.81	15.98
1,100	8,940	24.84	16.71
1,200	9,860	25.64	17.42
1,300	10,780	26.37	18.08
1,500	(15,700)	(30.52)	(20.05)
2,000	(20,000)	(32.88)	(22.80)

Uranium Dioxide, UO_2 (*c*)

$\Delta H^0_{298}=-259{,}200$ calories per mole (*63*)
$S_{298}=18.63$ *e.u.* (*52*)
$M.P.=3{,}000°$ K. (*8*)

Zone I (*c*) (298°–1,500° K.)

$$C_p=19.20+1.62\times10^{-3}T-3.96\times10^5T^{-2}\ (82)$$
$$H_T-H_{298}=-7{,}125+19.20T+0.81\times10^{-3}T^2+3.96\times10^5T^{-1}$$

Formation: $U+O_2 \longrightarrow UO_2$

Zone I (298°–935° K.)

$$\Delta C_p=8.65-7.4\times10^{-3}T-4.26\times10^5T^{-2}$$
$$\Delta H_T=-262{,}880+8.65T-3.7\times10^{-3}T^2+4.26\times10^5T^{-1}$$
$$\Delta F_T=-262{,}880-8.65T\ln T+3.7\times10^{-3}T^2+2.13\times10^5T^{-1}+100.55T$$

Zone II (935°–1,045° K.)

$$\Delta C_p=1.86+0.62\times10^{-3}T-3.56\times10^5T^{-2}$$
$$\Delta H_T=-260{,}700+1.86T+0.31\times10^{-3}T^2+3.56\times10^5T^{-1}$$
$$\Delta F_T=-260{,}700-1.86T\ln T-0.31\times10^{-3}T^2+1.78\times10^5T^{-1}+55.55T$$

Zone III (1,045°–1,300° K.)

$$\Delta C_p=2.84+0.62\times10^{-3}T-3.56\times10^5T^{-2}$$
$$\Delta H_T=-262{,}820+2.84T+0.31\times10^{-3}T^2+3.56\times10^5T^{-1}$$
$$\Delta F_T=-262{,}820-2.84T\ln T-0.31\times10^{-3}T^2+1.78\times10^5T^{-1}+64.45T$$

T, ° K.	H_T-H_{298}	S_T	ΔH°_T	ΔF°_T
298	------	18.63	−259,200	−246,550
400	1,680	23.47	−258,900	−242,300
500	3,470	27.46	−258,600	−238,100
600	5,340	30.86	−258,300	−234,100
700	7,280	33.85	−258,000	−230,100
800	9,250	36.48	−257,800	−226,100
900	11,250	38.86	−257,600	−222,100
1,000	13,280	40.97	−258,200	−218,100
1,100	15,340	42.94	−259,000	−214,000
1,200	17,420	44.75	−258,700	−210,000
1,300	19,510	46.42	−258,300	−205,900
1,400	21,620	47.98	------	------
1,500	23,750	49.45	(−260,800)	(−196,700)

Triuranium Octaoxide, U_3O_8 (*c*)

$\Delta H^0_{298}=-853{,}500$ calories per mole (*52*)
$S_{298}=66$ *e.u.* (*52*)
Decomposes = 1,950° K. (*10*)

Formation: $3U+4O_2 \longrightarrow U_3O_8$
(estimated (*24*))

T, °K.	H_T-H_{298}	ΔH_T°	ΔF_T°
298		−853, 500	−804, 000
400	(5, 860)	(−852, 600)	(−787, 200)
500	(11, 910)	(−851, 700)	(−770, 900)
600	(18, 330)	(−850, 700)	(−754, 800)
700	(24, 860)	(−849, 900)	(−738, 900)
800	(31, 750)	(−849, 200)	(−723, 100)
900	(38, 380)	(−848, 800)	(−707, 400)
1,000	(45, 410)	(−850, 300)	(−691, 500)
1,100	(52, 150)	(−853, 000)	(−675, 500)
1,200	(59, 240)	(−852, 000)	(−659, 400)
1,300	(66, 330)	(−851, 000)	(−643, 400)
1,400	(73, 450)	(−850, 000)	(−627, 500)
1,500	(80, 530)	(−858, 800)	(−611, 000)

Uranium Trioxide, UO_3 (*c*)

$\Delta H_{298}^\circ = -291{,}600$ calories per mole (*52*)
$S_{298} = 23.57$ *e.u.* (*52*)
Decomposes = 925° K. (*10*)

Zone I (*c*) (298°–925° K.)

$C_p = 22.09 + 2.54\times10^{-3}T - 2.97\times10^{5}T^{-2}$ (*82*)
$H_T - H_{298} = -7{,}695 + 22.09T + 1.27\times10^{-3}T^2 + 2.97\times10^{5}T^{-1}$

Formation: $U + 3/2O_2 \longrightarrow UO_3$

Zone I (298°–925° K.)

$\Delta C_p = 7.96 - 6.98\times10^{-3}T - 3.07\times10^{5}T^{-2}$
$\Delta H_T = -294{,}690 + 7.96T - 3.49\times10^{-3}T^2 + 3.07\times10^{5}T^{-1}$
$\Delta F_T = -294{,}690 - 7.96T\ln T + 3.49\times10^{-3}T^2 + 1.53\times10^{5}T^{-1} + 114.92T$

T, °K.	H_T-H_{298}	S_T	ΔH_T°	ΔF_T°
298		23. 57	−291, 600	−273, 200
400	2, 090	29. 59	−291, 300	−266, 800
500	4, 260	34. 43	−290, 950	−260, 800
600	6, 510	38. 53	−290, 650	−254, 800
700	8, 820	42. 09	−290, 350	−249, 800
800	11, 160	45. 21	−290, 150	−242, 900
900	13, 540	48. 01	−290, 050	−237, 000

Uranium Trifluoride, UF_3 (*c*)

$\Delta H_{298}^\circ = (-357{,}000)$ calories per mole (*10*)
$S_{298} = (26)$ *e.u.* (*10*)
M.P. = (1,700°) K. (*10*)
$\Delta H_M = (8{,}500)$ calories per mole
B.P. = (2,550°) K. (*10*)
$\Delta H_V = (61{,}000)$ calories per mole

Formation: $U + 3/2F_2 \longrightarrow UF_3$
(estimated (*10*))

T, °K.	H_T-H_{298}	ΔH_T°	ΔF_T°
298		(−357, 000)	(−339, 500)
500	(4, 000)	(−356, 800)	(−328, 000)
1,000	(17, 000)	(−355, 600)	(−299, 000)
1,500	(32, 000)	(−356, 000)	(−281, 000)

Uranium Tetrafluoride, UF_4 (*c*)

$\Delta H_{298}^\circ = (-443{,}000)$ calories per mole (*10*)
$S_{298} = 36.13$ *e.u.* (*10*)
M.P. = 1,309° K. (*10*)
$\Delta H_M = 5{,}700$ calories per mole
B.P. = 1,690° K. (*10*)
$\Delta H_V = 57{,}500$ calories per mole

Formation: $U + 2F_2 \longrightarrow UF_4$
(estimated (*10*))

T, °K.	H_T-H_{298}	ΔH_T°	ΔF_T°
298		(−443, 000)	(−421, 200)
500	(6, 300)	(−441, 300)	(−406, 500)
1,000	(25, 500)	(−436, 000)	(−375, 000)
1,500	(57, 000)	(−422, 200)	(−347, 000)

Uranium Pentafluoride, UF_5 (*c*)

$\Delta H_{298}^\circ = (-488{,}000)$ calories per mole (10)
$S_{298} = (43)$ *e.u.* (*10*)
M.P. = (600°) K. (*10*)
$\Delta H_M = (8{,}000)$ calories per mole
B.P. = (1,000°) K. (*10*)
$\Delta H_V = (23{,}000)$ calories per mole

Formation: $U + 5/2F_2 \longrightarrow UF_5$
(estimated (*10*))

T, °K.	H_T-H_{298}	ΔH_T°	ΔF_T°
298		(−488, 000)	(−461, 200)
500	(7, 000)	(−486, 400)	(−443, 500)

Uranium Hexafluoride, UF_6 (*c*)

$\Delta H_{298}^\circ = -517{,}000$ calories per mole (*52*)
$S_{298} = 54.45$ *e.u.* (*52*)
S.P. = 337° K. (*101*)
$\Delta H_{subl} = 11{,}430$ calories per mole

Formation: $U + 3F_2 \longrightarrow UF_6$
(estimated (*10*))

T, °K.	H_T-H_{298}	ΔH_T°	ΔF_T°
298		−517, 000	−486, 300
500	(18, 600)	(−516, 000)	(−471, 750)
1,000		(−515, 400)	(−440, 200)

Uranyl Fluoride, UO_2F_2

$\Delta H_{298}^\circ = -64{,}500$ calories per mole (*140*)
$S_{298} = 32.4$ *e.u.* (*52*)
$\Delta F_{298}^\circ = -41{,}500$ calories per mole

Uranium Trichloride, UCl_3 (*c*)

$\Delta H^\circ_{298} = -213{,}000$ calories per mole (*52*)
$S_{298} = 37.99$ *e.u.* (*52*)
M.P. = 1,108° K. (*10*)
$\Delta H_M = 9{,}000$ calories per mole
B.P. = 2,000° K. (*10*)
$\Delta H_V = 41{,}000$ calories per mole

Zone I (*c*) (298°–1,000° K.)

$C_p = 20.98 + 7.44 \times 10^{-3}T + 1.16 \times 10^5 T^{-2}$ (*82*)
$H_T - H_{298} = -6{,}200 + 20.98T + 3.72 \times 10^{-3}T^2 - 1.16 \times 10^5 T^{-1}$

Formation: $U + 3/2Cl_2 \longrightarrow UCl_3$

Zone I (298°–935° K.)

$\Delta C_p = 4.36 - 0.67 \times 10^{-3}T + 1.48 \times 10^5 T^{-2}$
$\Delta H_T = -213{,}770 + 4.36T - 0.33 \times 10^{-3}T^2 - 1.48 \times 10^5 T^{-1}$
$\Delta F_T = -213{,}770 - 4.36T\ln T + 0.33 \times 10^{-3}T^2 - 0.74 \times 10^5 T^{-1} + 82.17T$

Zone II (935°–1,045° K.)

$\Delta C_p = -2.43 + 7.35 \times 10^{-3}T + 2.18 \times 10^5 T^{-2}$
$\Delta H_T = -211{,}900 - 2.43T + 3.67 \times 10^{-3}T^2 - 2.18 \times 10^5 T^{-1}$
$\Delta F_T = -211{,}900 + 2.43T\ln T - 3.67 \times 10^{-3}T^2 - 1.09 \times 10^5 T^{-1} + 37.35T$

T, ° K.	$H_T - H_{298}$	S_T	ΔH°_T	ΔF°_T
298		37.99	−213,000	−196,900
400	2,500	45.21	−212,450	−191,400
500	5,000	50.78	−212,000	−186,200
600	7,540	55.41	−211,500	−181,100
700	10,140	59.42	−211,100	−176,150
800	12,810	62.98	−210,700	−171,050
900	15,570	66.23	−210,300	−166,250
1,000	18,430	69.24	−210,900	−161,600
1,100	(21,280)		(−210,800)	(−156,200)

Uranium Tetrachloride, UCl_4 (*c*)

$\Delta H^\circ_{298} = -251{,}200$ calories per mole (*52*)
$S_{298} = 62$ *e.u.* (*52*)
M.P. = 863° K. (*10*)
$\Delta H_M = 10{,}300$ calories per mole
B.P. = 1,060° K. (*10*)
$\Delta H_V = 33{,}000$ calories per mole

Zone I (*c*) (298°–700° K.)

$C_p = 26.64 + 9.60 \times 10^{-3}T$ (*82*)
$H_T - H_{298} = -8{,}370 + 26.64T + 4.80 \times 10^{-3}T^2$

Formation: $U + 2Cl_2 \longrightarrow UCl_4$

Zone I (298°–700° K.)

$\Delta C_p = 5.61 + 1.46 \times 10^{-3}T + 0.66 \times 10^5 T^{-2}$
$\Delta H_T = -252{,}700 + 5.61T + 0.73 \times 10^{-3}T^2 - 0.66 \times 10^5 T^{-1}$
$\Delta F_T = -252{,}700 - 5.61T\ln T - 0.73 \times 10^{-3}T^2 - 0.33 \times 10^5 T^{-1} + 94.27T$

T, ° K.	$H_T - H_{298}$	S_T	ΔH°_T	ΔF°_T
298		62.0	−251,200	−234,300
400	3,030	70.74	−250,550	−228,550
500	6,150	77.70	−249,850	−223,100
600	9,330	83.50	−249,200	−217,900
700	12,630	88.58	−248,500	−212,800

Uranium Pentachloride, UCl_5 (*c*)

$\Delta H^\circ_{298} = (-262{,}100)$ calories per mole (*10*)
$S_{298} = (62)$ *e.u.* (*10*)
M.P. = 600° K. (*10*)
$\Delta H_M = 8{,}500$ calories per mole
B.P. = 800° K. (*10*)
$\Delta H_V = (18{,}000)$ calories per mole

Formation: $U + 5/2Cl_2 \longrightarrow UCl_5$
(estimated (*10*))

T, ° K.	$H_T - H_{298}$	ΔH°_T	ΔF°_T
298		(−262,[illegible]	(−237,400)
500	(7,000)	(−260[illegible]	(−221,600)

Uranium Hexachloride, UCl_6 (*c*)

$\Delta H^\circ_{298} = (-272{,}400)$ calories per mole (*10*)
$S_{298} = 68.3$ *e.u.* (*10*)
M.P. = 452° K. (*10*)
$\Delta H_M = (5{,}000)$ calories per mole
B.P. = 550° K. (*10*)
$\Delta H_V = (11{,}000)$ calories per mole

Formation: $U + 3Cl_2 \longrightarrow UCl_6$
(estimated (*10*))

T, ° K.	$H_T - H_{298}$	ΔH°_T	ΔF°_T
298		(−272,400)	(−241,500)
500	(14,000)	(−264,900)	(−221,400)

Uranium (IV) Oxychloride, $UOCl_2$

$\Delta H^\circ_{298} = -261{,}700$ calories per mole (*44*)
$S_{298} = 33.06$ *e.u.* (*44*)
$\Delta F^\circ_{298} = -244{,}800$ calories per mole

Uranyl Chloride, UO_2Cl_2 (*c*)

$\Delta H^\circ_{298} = -300{,}000$ calories per mole (*45*)
$S_{298} = 35.98$ *e.u.* (*45*)
$\Delta F^\circ_{298} = -276{,}700$ calories per mole

Uranium Tribromide, UBr_3 (*c*)

$\Delta H^\circ_{298} = (-170{,}100)$ calories per mole (*10*)
$S_{298} = (49)$ *e.u.* (*10*)
M.P. = 1,025° K. (*10*)
$\Delta H_M = 11{,}000$ calories per mole
B.P. = (1,840°) K. (*10*)
$\Delta H_V = 45{,}000$ calories per mole

Formation: $U + 3/2Br_2 \longrightarrow UBr_3$
(estimated (*10*))

T, ° K.	$H_T - H_{298}$	ΔH°_T	ΔF°_T
298		(−170,000)	(−164,900)
500	(5,000)	(−180,800)	(−155,500)
1,000	(18,000)	(−179,900)	(−133,000)
1,500	(43,000)	(−170,600)	(−112,000)

Uranium Tetrabromide, UBr_4 (*c*)

$\Delta H^\circ_{298} = (-196{,}600)$ calories per mole (*10*)
$S_{298} = (58)$ *e.u.* (*10*)
M.P. = 792° K. (*10*)
$\Delta H_M = 7{,}200$ calories per mole
B.P. = 1,039° K. (*10*)
$\Delta H_V = 31{,}000$ calories per mole

Formation: $U + 2Br_2 \longrightarrow UBr_4$
(estimated (*10*))

T, ° K.	$H_T - H_{298}$	ΔH°_T	ΔF°_T
298		(−196, 600)	(−188, 600)
500	(6, 000)	(−206, 300)	(−175, 800)
1,000	(34, 300)	(−196, 400)	(−146, 500)

Uranium (IV) Oxybromide, $UOBr_2$

$\Delta H^\circ_{298} = -246{,}900$ calories per mole (*44*)
$S_{298} = 37.66$ *e.u.* (*44*)
$\Delta F^\circ_{298} = -236{,}400$ calories per mole

Uranium Triiodide, UI_3 (*c*)

$\Delta H^\circ_{298} = (-114{,}700)$ calories per mole (*10*)
$S_{298} = (56)$ *e.u.* (*10*)
M.P. = (1,030°) K. (*10*)
$\Delta H_M = (7{,}500)$ calories per mole
B.P. = (1,700°) K. (*10*)
$\Delta H_V = (40{,}800)$ calories per mole

Formation: $U + 3/2I_2 \longrightarrow UI_3$
(estimated (*10*))

T, ° K.	$H_T - H_{298}$	ΔH°_T	ΔF°_T
298		(−114, 700)	(−115, 100)
500	(5, 000)	(−136, 100)	(−112, 000)
1,000	(20, 000)	(−133, 200)	(−90, 000)
1,500	(43, 000)	(−103, 600)	(−68, 000)

Uranium Tetraiodide, UI_4 (*c*)

$\Delta H^\circ_{298} = (-127{,}000)$ calories per mole (*10*)
$S_{298} = (65)$ *e.u.* (*10*)
M.P. = 779° K. (*10*)
$\Delta H_M = 15{,}000$ calories per mole
B.P. = 1,032° K. (*10*)
$\Delta H_V = 30{,}700$ calories per mole

Formation: $U + 2I_2 \longrightarrow UI_4$
(estimated (*10*))

T, ° K.	$H_T - H_{298}$	ΔH°_T	ΔF°_T
298		(−127, 000)	(−125, 900)
500	(6, 000)	(−145, 800)	(−121, 000)
1,000	(34, 000)	(−142, 100)	(−93, 500)

Uranium Carbide, UC (*c*)

$\Delta H^\circ_{298} = -43{,}000$ calories per mole (*10*)
$S_{298} = (15.4)$ *e.u.* (*10*)
$\Delta F^\circ_{298} = (-43{,}600)$ calories per mole
M.P. = 2,550° K. (*10*)

Diuranium Tricarbide, U_2C_3 (*c*)

$\Delta H_{298} = -76{,}000$ calories per mole (*10*)
$S_{298} = (24)$ *e.u.* (*10*)
$\Delta F^\circ_{298} = (-78{,}400)$ calories per mole
M.P. = 2,700° K. (*10*)

Uranium Dicarbide, UC_2 (*c*)

$\Delta H^\circ_{298} = -36{,}000$ calories per mole (*10*)
$S_{298} = (20)$ *e.u.* (*10*)
$\Delta F^\circ_{298} = (-37{,}500)$ calories per mole
M.P. = 2,700° K. (*10*)

Uranium Nitride, UN (*c*)

$\Delta H^\circ_{298} = -80{,}000$ calories per mole (*10*)
$S_{298} = (18.0)$ *e.u.* (*10*)
$\Delta F^\circ_{298} = (-74{,}900)$ calories per mole
M.P. = 2,900° K. (*10*)

Diuranium Trinitride, U_2N_3 (*c*)

$\Delta H^\circ_{298} = -213{,}000$ calories per mole (*10*)
$S_{298} = (29)$ *e.u.* (*10*)
$\Delta F^\circ_{298} = (-193{,}900)$ calories per mole

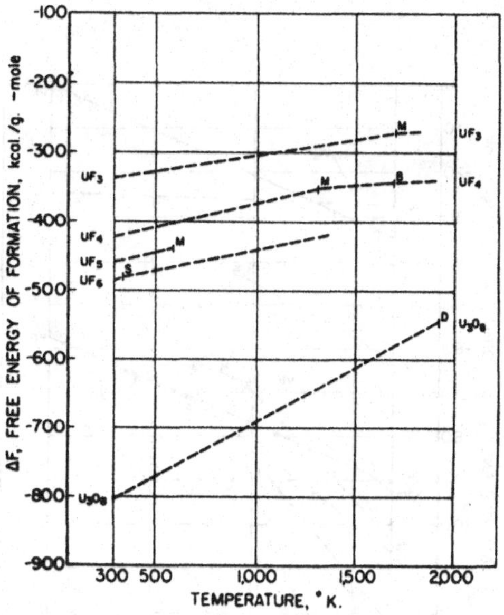

FIGURE 60.—Uranium (a).

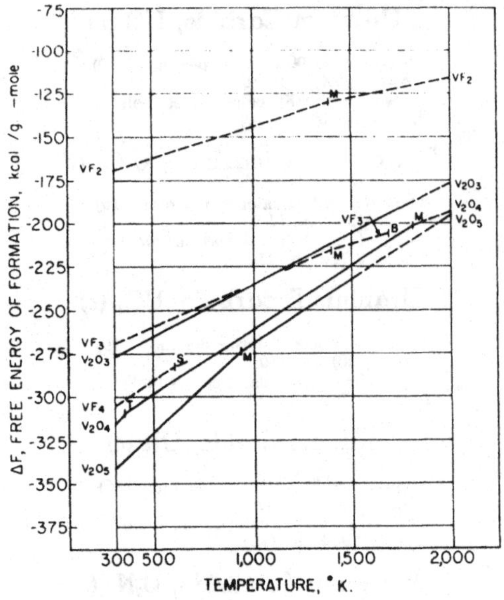

FIGURE 62.—Vanadium (a).

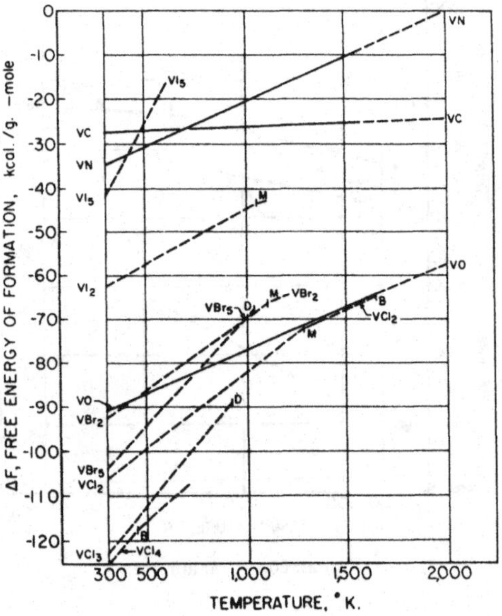

FIGURE 63.—Vanadium (b).

YTTERBIUM AND ITS COMPOUNDS

Element, Yb (*c*)

S_{298} = 15.0 *e.u.* (*130*)
M.P. = 1,097° K. (*125*)
ΔH_M = (2,200) calories per atom
B.P. = 1,800° K. (*125*)
ΔH_V = 37,100 calories per atom

(estimated (*124*))

T, ° K.	$H_T - H_{298}$	S_T	$-\frac{(F_T - H_{298})}{T}$
298	-----------	15.0	15.0
400	(620)	(16.79)	(15.24)
500	(1,250)	(18.19)	(15.69)
600	(1,900)	(19.38)	(16.22)
700	(2,570)	(20.41)	(16.74)
800	(3,260)	(21.33)	(17.26)
900	(3,970)	(22.17)	(17.76)
1,000	(4,700)	(22.94)	(18.24)
1,100	(7,950)	(25.94)	(18.72)
1,200	(8,700)	(26.59)	(19.34)
1,300	(9,450)	(27.19)	(19.93)
1,400	(10,200)	(27.75)	(20.47)
1,500	(10,950)	(28.27)	(20.97)
1,600	(11,700)	(28.75)	(21.44)
1,700	(12,450)	(29.20)	(21.88)
1,800	(13,200)	(29.64)	(22.31)
1,900	(50,860)	(50.55)	(23.79)
2,000	(51,360)	(50.81)	(25.13)

Ytterbium Difluoride, YbF_2 (*c*)

ΔH°_{298} = (−262,000) calories per mole (*5*)
S_{298} = (20) *e.u.* (*11*)
M.P. = 1,325° K. (*29*)
ΔH_M = 5,000 calories per mole
B.P. = (2,650°) K. (*6*)
ΔH_V = (75,000) calories per mole

Formation: $Yb + F_2 \longrightarrow YbF_2$
(estimated (*11*))

T, ° K.	$H_T - H_{298}$	ΔH°_T	ΔF°_T
298	-----------	(−262,000)	(−250,000)
500	(4,000)	(−260,800)	(−241,000)
1,000	(13,000)	(−259,500)	(−223,000)
1,500	(24,000)	(−259,200)	(−203,000)

Ytterbium Trifluoride, YbF_3 (*c*)

ΔH°_{298} = (−351,000) calories per mole (*5*)
S_{298} = (26) *e.u.* (*11*)
M.P. = 1,430° K. (*29*)
ΔH_M = 8,000 calories per mole
B.P. = (2,500°) K. (*6*)
ΔH_V = (60,000) calories per mole

Formation: $Yb + 3/2F_2 \longrightarrow YbF_3$
(estimated (*11*))

T, ° K.	$H_T - H_{298}$	ΔH°_T	ΔF°_T
298	-----------	(−351,000)	(−333,000)
500	(4,000)	(−350,600)	(−321,000)
1,000	(17,000)	(−347,500)	(−293,000)
1,500	(32,000)	(−345,300)	(−367,000)

Ytterbium Dichloride, $YbCl_2$ (*c*)

ΔH°_{298} = (−162,000) calories per mole (*5*)
S_{298} = (30) *e.u.* (*11*)
M.P. = 975° K. (*29*)
ΔH_M = 6,000 calories per mole
B.P. = (2,200°) K. (*6*)
ΔH_V = (50,000) calories per mole

Formation: $Yb + Cl_2 \longrightarrow YbCl_2$
(estimated (*11*))

T, ° K.	$H_T - H_{298}$	ΔH°_T	ΔF°_T
298		(−162, 000)	(−161, 000)
500	(4, 000)	(−160, 900)	(−153, 500)
1,000	(13, 000)	(−159, 700)	(−128, 000)
1,500	(31, 000)	(−152, 300)	(−115, 500)

Ytterbium Trichloride, $YbCl_3$ (*c*)

ΔH°_{298} = (−189,000) calories per mole (*5*)
S_{298} = (38) *e.u.* (*11*)
M.P. = 1,138° K. (*29*)
ΔH_M = (9,000) calories per mole
Decomposes above 1,500° K.

Formation: $Yb + 3/2Cl_2 \longrightarrow YbCl_3$
(estimated (*11*))

T, ° K.	$H_T - H_{298}$	ΔH°_T	ΔF°_T
298		(−189, 000)	(−173, 000)
500	(5, 000)	(−187, 800)	(−161, 500)
1,000	(19, 000)	(−184, 800)	(−136, 000)
1,500	(43, 000)	(172, 500)	(−117, 000)

Ytterbium Dibromide, $YbBr_2$ (*c*)

ΔH°_{298} = (−132,000) calories per mole (*5*)
S_{298} = (36) *e.u.* (*11*)
M.P. = 945° K. (*29*)
ΔH_M = 6,000 calories per mole
B.P. = (2,100°) K. (*6*)
ΔH_V = (48,000) calories per mole

Formation: $Yb + Br_2 \longrightarrow YbBr_2$
(estimated (*11*))

T, ° K.	$H_T - H_{298}$	ΔH°_T	ΔF°_T
298		(−132, 000)	(−127, 000)
500	(4, 000)	(−150, 000)	(−120, 500)
1,000	(20, 000)	(−140, 000)	(−104, 000)
1,500	(32, 000)	(−129, 500)	(−93, 000)

Ytterbium Tribromide, $YbBr_3$ (*c*)

ΔH°_{298} = (−149,000) calories per mole (*5*)
S_{298} = (44) *e.u.* (*11*)
M.P. = 1,227° K. (*29*)
ΔH_M = 10,000 calories per mole
Decomposes above 1,500° K. (*6*)

Formation: $Yb + 3/2Br_2 \longrightarrow YbBr_3$
(estimated (*11*))

T, ° K.	$H_T - H_{298}$	ΔH°_T	ΔF°_T
298		(−149, 000)	(−142, 000)
500	(5, 000)	(−170, 800)	(−132, 500)
1,000	(18, 000)	(−166, 700)	(−106, 000)
1,500	(43, 000)	(−144, 700)	(−86, 500)

Ytterbium Diiodide, YbI_2 (*c*)

ΔH°_{298} = (−102,000) calories per mole (*5*)
S_{298} = (40) *e.u.* (*11*)
M.P. = 1,045° K. (*29*)
ΔH_M = 5,000 calories per mole
B.P. = (1,600°) K. (*6*)
ΔH_V = (37,000) calories per mole

Formation: $Yb + I_2 \longrightarrow YbI_2$
(estimated (*11*))

T, ° K.	$H_T - H_{298}$	ΔH°_T	ΔF°_T
298		(−102, 000)	(−102, 000)
500	(4, 000)	(−115, 900)	(−99, 500)
1,000	(19, 000)	(−108, 800)	(−85, 000)
1,500	(31, 000)	(−107, 500)	(−73, 500)

Ytterbium Triiodide, YbI_3 (*c*)

ΔH°_{298} = (−96,000) calories per mole (*5*)
S_{298} = (47) *e.u.* (*11*)
M.P. = (1,300°) K. (*29*)
ΔH_M = (10,000) calories per mole
Decomposes above 1,500° K. (*6*)

Formation: $Yb + 3/2I_2 \longrightarrow YbI_3$
(estimated (*11*))

T, ° K.	$H_T - H_{298}$	ΔH°_T	ΔF°_T
298		(−96, 000)	(−93, 500)
500	(5, 000)	(−120, 200)	(−88, 500)
1,000	(19, 000)	(−116, 200)	(−60, 000)
1,500	(44, 000)	(−95, 500)	(−35, 500)

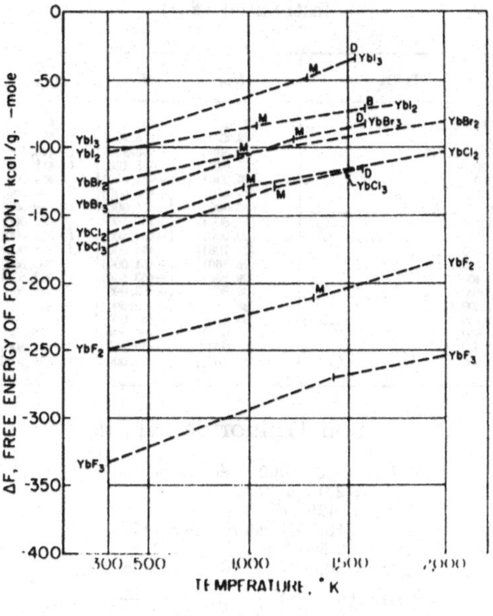

FIGURE 64. Ytterbium.

YTTRIUM AND ITS COMPOUNDS

Element, Y (*c*)

S_{298} = 11.3 *e.u.* (*127*)
M.P. = (1,773°) K. (*125*)
ΔH_M = (4,100) calories per atom
B.P. = (3,500°) K. (*125*)
ΔH_V = (94,000) calories per atom
(estimated (*130*))

T, ° K.	$H_T - H_{298}$	S_T	$-\frac{(F_T - H_{298})}{T}$
298	....	11.30	11.30
400	(617)	(13.08)	(11.52)
500	(1,233)	(14.45)	(11.98)
600	(1,859)	(15.59)	(12.50)
700	(2,495)	(16.57)	(13.00)
800	(3,141)	(17.44)	(13.51)
900	(3,798)	(18.21)	(13.99)
1,000	(4,465)	(18.91)	(14.45)
1,100	(5,142)	(19.56)	(14.88)
1,200	(5,829)	(20.16)	(15.40)
1,300	(6,527)	(20.71)	(15.70)
1,400	(7,235)	(21.24)	(16.07)
1,500	(7,935)	(21.73)	(16.44)
1,600	(8,681)	(22.20)	(16.77)
1,700	(9,419)	(22.65)	(17.11)
1,800	(14,280)	(25.40)	(17.46)
1,900	(15.080)	(25.83)	(17.90)
2,000	(15,880)	(26.24)	(18.30)

Diyttrium Trioxide, Y_2O_3 (*c*)

ΔH°_{298} = −455,450 calories per mole (*62*)
S_{298} = 27.1 *e.u.* (*8*)
M.P. = 2,500° K. (*42*)
ΔH_M = 25,000 calories per mole
B.P. = 4,570 ± 300° K. (*42*)

Formation: $2Y + 3/2O_2 \longrightarrow Y_2O_3$
(estimated (*24*))

T, ° K.	$H_T - H_{298}$	ΔH°_T	ΔF°_T
298	....	−455,450	−433,450
400	(2,320)	(−455,400)	(−425,500)
500	(4,800)	(−455,400)	(−418,500)
600	(7,030)	(−455.400)	(−411,500)
700	(9,930)	(−455,000)	(−404,000)
800	(12,460)	(−455,000)	(−396,500)
900	(15,000)	(−455,000)	(−389,500)
1,000	(17,800)	(−454,500)	(−382,000)
1,100	(20,900)	(−454,500)	(−375,000)
1,200	(23,700)	(−454,000)	(−368,000)
1,300	(26,360)	(−454,000)	(−360,500)
1,400	(29,550)	(−453,500)	(−353,500)
1,500	(32,800)	(−453,000)	(−346,500)
1,600	(36,030)	(−452,500)	(−339,000)
1,700	(39,340)	(−452,000)	(−332,000)
1,800	(46,000)	(−460,000)	(−325,000)
1,900	(49,000)	(−459,500)	(−317,500)
2,000	(52,700)	(−459,000)	(−309,500)

Yttrium Trifluoride, YF_3 (*c*)

ΔH°_{298} = (−372,000) calories per mole (*5*)
S_{298} = (23) *e.u.* (*11*)
M.P. = 1,425° K. (*29*)
ΔH_M = (13,000) calories per mole
B.P. = (2,500°) K. (*6*)
ΔH_V = (60,000) calories per mole

Formation: $Y + 3/2F_2 \longrightarrow YF_3$
(estimated (*11*))

T, ° K.	$H_T - H_{298}$	ΔH°_T	ΔF°_T
298	....	(−372,000)	(−353,800)
500	(4,000)	(−371,000)	(−342,000)
1,000	(17,000)	(−369,300)	(−314,000)
1,500	(32,000)	(−364,900)	(−288,000)

Yttrium Trichloride, YCl_3 (*c*)

ΔH°_{298} = −232,690 calories per mole (*127*)
S_{298} = 32.7 *e.u.* (*127*)
M.P. = 982° K. (*29*)
ΔH_M = (9,000) calories per mole
B.P. = (1,780°) K. (*6*)
ΔH_V = (45,000) calories per mole

Formation: $Y + 3/2Cl_2 \longrightarrow YCl_3$
(estimated (*11*))

T, ° K.	$H_T - H_{298}$	ΔH°_T	ΔF°_T
298	....	−232,700	−215,200
500	(5,000)	(−231,800)	(−206,200)
1,000	(28,000)	(−229,600)	(−183,700)
1,500	(44,000)	(−214,000)	(−163,700)

Yttrium Tribromide, YBr_3 (*c*)

ΔH°_{298} = (−172,000) calories per mole (*5*)
S_{298} = (42) *e.u.* (*11*)
M.P. = 1,186° K. (*29*)
ΔH_M = (9,000) calories per mole
B.P. = (1,740°) K. (*6*)
ΔH_V = (44,000) calories per mole

Formation: $Y + 3/2Br_2 \longrightarrow YBr_3$
(estimated (*11*))

T, ° K.	$H_T - H_{298}$	ΔH°_T	ΔH°_T
298	....	(−172,000)	(−166,000)
500	(5,000)	(−183,000)	(−155,000)
1,000	(18,000)	(−180,000)	(−129,000)
1,500	(44,000)	(−165,300)	(−108,000)

Yttrium Triiodide, YI_3 (*c*)

ΔH°_{298} = −143,000 calories per mole (*5*)
S_{298} = (45) *e.u.* (*11*)
M.P. = 1,238° K. (*29*)
ΔH_M = (12,000) calories per mole
B.P. = (1,580°) K. (*6*)
ΔH_V = (41,000) calories per mole

Formation: $Y + 3/2I_2 \longrightarrow YI_3$
(estimated (*11*))

T, ° K.	$H_T - H_{298}$	ΔH°_T	ΔF°_T
298	....	−143,000	(−140,500)
500	(5,000)	(−164,300)	(−135,000)
1,000	(19,000)	(−161,200)	(−109,000)
1,500	(46,000)	(−145,000)	(−87,000)

Yttrium Nitride, YN (*c*)

$\Delta H^{\circ}_{298} = -71{,}500$ calories per mole (*9*)
$S_{298} = (14.2)$ *e.u.* (*9*)
$\Delta F^{\circ}_{298} = -64{,}000$ calories per mole

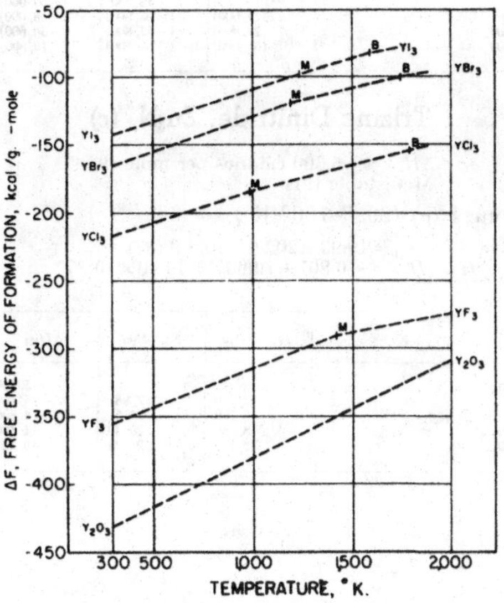

FIGURE 65.—Yttrium.

ZINC AND ITS COMPOUNDS

Element, Zn (*c*)

$S_{298} = 9.95$ *e.u.* (*83*)
M.P. = 692.7° K. (*82*)
$\Delta H_M = 1{,}765$ calories per atom
B.P. = 1,181° K. (*130*)
$\Delta H_V = 27{,}560$ calories per atom

Zone I (*c*) (298°–692.7° K.)

$$C_p = 5.35 + 2.40 \times 10^{-3} T \ (82)$$
$$H_T - H_{298} = -1{,}702 + 5.35T + 1.20 \times 10^{-3} T^2$$
$$F_T - H_{298} = -1{,}702 - 5.35 T \ln T - 1.20 \times 10^{-3} T^2 + 26.56T$$

Zone II (*l*) (692.7°–1,181° K.)

$$C_p = 7.50 \ (82)$$
$$H_T - H_{298} = -850 + 7.50T$$
$$F_T - H_{298} = -850 - 7.50 T \ln T + 38.57T$$

Zone III (*g*) (1,181°–2,000° K.)

T, ° K.	$H_T - H_{298}$	S_T	$-\frac{(F_T - H_{298})}{T}$
298	....	9.95	9.95
400	625	11.75	10.19
500	1,270	13.19	10.65
600	1,940	14.41	11.18
700	4,400	18.03	11.74
800	5,150	19.03	12.59
900	5,900	19.91	13.36
1,000	6,650	20.70	14.05
1,100	7,400	21.42	14.69
1,200	(35,660)	(45.37)	(15.66)
1,300	(36,160)	(45.77)	(17.96)
1,400	(36,650)	(46.14)	(19.96)
1,500	(37,150)	(46.48)	(21.72)
1,600	(37,650)	(46.80)	(23.28)
1,700	(38,140)	(47.10)	(24.67)
1,800	(38,640)	(47.38)	(25.92)
1,900	(39,140)	(47.65)	(27.06)
2,000	(39,630)	(47.91)	(28.10)
2,500	(42,120)	(49.02)	(32.18)

Zinc Oxide, ZnO (*c*)

$\Delta H^{\circ}_{298} = -83{,}250$ calories per mole (*24*)
$S_{298} = 10.43$ *e.u.* (*83*)
M.P. = 2,248° K. (*112*)

Zone I (*c*) (298°–1,600° K.)

$$C_p = 11.71 + 1.22 \times 10^{-3} T - 2.18 \times 10^{5} T^{-2} \ (82)$$
$$H_T - H_{298} = -4{,}280 + 11.71T + 0.61 \times 10^{-3} T^2 + 2.18 \times 10^{5} T^{-1}$$

Formation: $Zn + 1/2O_2 \longrightarrow ZnO$

Zone I (298°–692.7° K.)

$$\Delta C_p = 2.78 - 1.68 \times 10^{-3} T - 1.98 \times 10^{5} T^{-2}$$
$$\Delta H_T = -84{,}670 + 2.78T - 0.84 \times 10^{-3} T^2 + 1.98 \times 10^{5} T^{-1}$$
$$\Delta F_T = -84{,}670 - 2.78 T \ln T + 0.84 \times 10^{-3} T^2 + 0.99 \times 10^{5} T^{-1} + 43.24T$$

Zone II (692.7–1,181° K.)

$$\Delta C_p = 0.63 + 0.72 \times 10^{-3} T - 1.98 \times 10^{5} T^{-2}$$
$$\Delta H_T = -85{,}600 + 0.63T + 0.36 \times 10^{-3} T^2 + 1.98 \times 10^{5} T^{-1}$$
$$\Delta F_T = -85{,}600 - 0.63 T \ln T - 0.36 \times 10^{-3} T^2 + 0.99 \times 10^{5} T^{-1} + 31.28T$$

T, ° K.	$H_T - H_{298}$	S_T	ΔH°_T	ΔF°_T
298	....	10.43	−83,250	−76,100
400	1,070	13.51	−83,200	−74,650
500	2,190	16.01	−83,050	−71,300
600	3,350	18.12	−82,950	−68,950
700	4,530	19.94	−84,600	−66,600
800	5,740	21.56	−84,550	−64,000
900	6,970	23.00	−84,500	−61,500
1,000	8,220	24.32	−84,400	−59,000
1,100	9,500	25.54	−84,000	−56,100
1,200	10,800	26.67	(−111,600)	(−53,400)
1,300	12,120	27.72	(−111,200)	(−50,700)
1,400	13,450	28.71	(−110,800)	(−47,900)
1,500	14,800	29.64	(−110,400)	(−45,100)
1,600	16,160	30.52	(−110,000)	(−42,200)

Zinc Difluoride, ZnF_2 (*c*)

$\Delta H^\circ_{298} = -176,000$ calories per mole (*11*)
$S_{298} = (24)$ *e.u.* (*11*)
M.P. = 1,145° K. (*6*)
$\Delta H_M = (7,000)$ calories per mole
B.P. = 1,775° K. (*6*)
$\Delta H_V = 44,000$ calories per mole

Formation: $Zn + F_2 \longrightarrow ZnF_2$
(estimated (*11*))

T, ° K.	$H_T - H_{298}$	ΔH°_T	ΔF°_T
298		−176,000	(−165,600)
500	(4,000)	(−174,900)	(−158,500)
1,000	(14,000)	(−174,500)	(−142,000)
1,500	(32,000)	(−191,400)	(−123,000)

Zinc Dichloride, $ZnCl_2$ (*c*)

$\Delta H^\circ_{298} = -99,600$ calories per mole (*11*)
$S_{298} = 25.9$ *e.u.* (*11*)
M.P. = 556° K. (*6*)
$\Delta H_M = 5,540$ calories per mole
B.P. = 1,005° K. (*6*)
$\Delta H_V = 28,700$ calories per mole

Formation: $Zn + Cl_2 \longrightarrow ZnCl_2$
(estimated (*11*))

T, ° K.	$H_T - H_{298}$	ΔH°_T	ΔH°_T
298		−99,600	−88,450
500	(4,000)	(−98,600)	(−81,100)
1,000	(21,000)	(−91,300)	(−70,000)
1,500	(61,000)	(−86,000)	(−53,600)

Zinc Dibromide, $ZnBr_2$ (*c*)

$\Delta H^\circ_{298} = -78,200$ calories per mole (*11*)
$S_{298} = (33)$ *e.u.* (*11*)
M.P. = 665° K. (*6*)
$\Delta H_M = 4,000$ calories per mole
B.P. = 975° K. (*6*)
$\Delta H_V = 24,250$ calories per mole

Formation: $Zn + Br_2 \longrightarrow ZnBr_2$
(estimated (*11*))

T, ° K.	$H_T - H_{298}$	ΔH°_T	ΔF°_T
298		−78,200	(−74,400)
500	(4,000)	(−71,400)	(−68,000)
1,000	(21,000)	(−78,500)	(−56,800)
1,500	(56,000)	(−77,700)	(−39,000)

Zinc Diiodide, ZnI_2 (*c*)

$\Delta H^\circ_{298} = -49,980$ calories per mole (*112*)
$S_{298} = (38)$ *e.u.* (*112*)
M.P. = 719° K. (*6*)
$\Delta H_M = 4,500$ calories per mole
B.P. = 1,000° K. (*6*)
$\Delta H_V = 23,000$ calories per mole

Formation: $Zn + I_2 \longrightarrow ZnI_2$
(estimated (*11*))

T, ° K.	$H_T - H_{298}$	ΔH°_T	ΔF°_T
298		−49,980	(−50,000)
500	(4,000)	(−63,900)	(−48,000)
1,000	(20,000)	(−57,800)	(−36,600)
1,500	(54,000)	(−58,500)	(−19,800)

Trizinc Dinitride, Zn_3N_2 (*c*)

$\Delta H^\circ_{298} = -5,300$ calories per mole (*9*)
Metastable (*9*)

Zone I (*c*) (298°–700° K.)

$$C_p = 19.93 + 20.80 \times 10^{-3}T \ (82)$$
$$H_T - H_{298} = -6,867 + 19.93T + 10.40 \times 10^{-3}T^2$$

T, ° K.	$H_T - H_{298}$	$S_T - S_{298}$	ΔH_{298}
298			−5,300
400	2,770	7.98	−5,100
500	5,700	14.51	−4,800
600	8,880	20.30	−4,400
700	12,180	25.38	−9,150

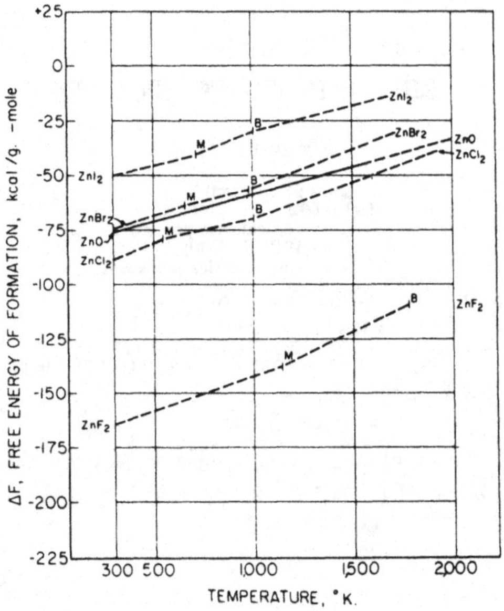

FIGURE 66.—Zinc.

ZIRCONIUM AND ITS COMPOUNDS

Element, Zr (*c*)

$T.P. = 1{,}135°$ K. (*26*)
$\Delta H_T = 920$ calories per atom
$M.P. = 2{,}125°$ K. (*130*)
$\Delta H_M = (5{,}500)$ calories per atom

Zone I (α) (298°–1,135° K.)

$$C_p = 6.83 + 1.12\times10^{-3}T - 0.87\times10^5T^{-2} \ (26)$$
$$H_T - H_{298} = -2{,}380 + 6.83T + 0.56\times10^{-3}T^2 + 0.87\times10^5T^{-1}$$
$$F_T - H_{298} = -2{,}380 - 6.83T\ln T - 0.56\times10^{-3}T^2 + 0.43\times10^5T^{-1} + 37.25T$$

Zone II (β) (1,135°–2,133° K.)

$$C_p = 7.27 \ (26)$$
$$H_T - H_{298} = -1{,}170 + 7.27T$$
$$F_T - H_{298} = -1{,}170 - 7.27T\ln T + 38.67T$$

T, ° K.	$H_T - H_{298}$	S_T	$-\frac{(F_T - H_{298})}{T}$
298	....	9.29	9.29
400	665	11.20	9.57
500	1,350	12.73	10.02
600	2,065	14.03	10.61
700	2,800	15.17	11.14
800	3,550	16.17	11.71
900	4,315	17.07	12.29
1,000	5,095	17.89	12.80
1,100	5,895	18.65	13.28
1,200	7,560	20.11	13.82
1,300	8,290	20.69	14.28
1,400	9,015	21.23	14.85
1,500	(9,730)	(21.70)	(15.18)
1,600	(10,450)	(22.15)	(15.67)
1,700	(11,210)	(22.65)	(16.16)
1,800	(11,930)	(23.10)	(16.46)
1,900	(12.640)	(23.50)	(16.72)
2,000	(13,380)	(23.85)	(17.13)

Zirconium Dioxide, ZrO_2 (*c*)

$\Delta H^\circ_{298} = -261{,}500$ calories per mole (*24*)
$S_{298} = 12.12$ *e.u.* (*83*)
$T.P. = 1{,}478°$ K. (*26*)
$\Delta H_T = 1{,}420$ calories per mole
$M.P. = 2{,}950°$ K. (*42*)
$\Delta H_M = 20{,}800$ calories per mole
$B.P. = 4{,}570°$ K.

Zone I (α) (298°–1,478° K.)

$$C_p = 16.64 + 1.80\times10^{-3}T - 3.36\times10^5T^{-2} \ (26)$$
$$H_T - H_{298} = -6{,}160 + 16.64T + 0.90\times10^{-3}T^2 + 3.36\times10^5T^{-1}$$

Zone II (β) (1,478°–2,100° K.)

$$C_p = 17.80 \ (26)$$
$$H_T - H_{298} = -4{,}267 + 17.80T$$

Formation: $Zr + O_2 \longrightarrow ZrO_2$

Zone I (298°–1,135° K.)

$$\Delta C_p = 2.65 - 0.32\times10^{-3}T - 2.09\times10^5T^{-2}$$
$$\Delta H_T = -262{,}960 + 2.65T - 0.16\times10^{-3}T^2 + 2.09\times10^5T^{-1}$$
$$\Delta F_T = -262{,}960 - 2.65T\ln T + 0.16\times10^{-3}T^2 + 1.04\times10^5T^{-1} + 65.0T$$

Zone II (1,135°–1,478° K.)

$$\Delta C_p = 2.21 + 0.80\times10^{-3}T - 2.96\times10^5T^{-2}$$
$$\Delta H_T = -264{,}360 + 2.21T + 0.40\times10^{-3}T^2 + 2.96\times10^5T^{-1}$$
$$\Delta F_T = -264{,}360 - 2.21T\ln T - 0.40\times10^{-3}T^2 + 1.48\times10^5T^{-1} + 63.5T$$

Zone III (1,478°–2,100° K.)

$$\Delta C_p = 3.37 - 1.00\times10^{-3}T + 0.40\times10^5T^{-2}$$
$$\Delta H_T = -262{,}400 + 3.37T - 0.50\times10^{-3}T^2 - 0.40\times10^5T^{-1}$$
$$\Delta F_T = -262{,}400 - 3.37T\ln T + 0.50\times10^{-3}T^2 - 0.20\times10^5T^{-1} + 69.44T$$

T, ° K.	$H_T - H_{298}$	S_T	ΔH°_T	ΔF°_T
298	....	12.12	−261,500	−247,750
400	1,475	16.36	−261,400	−243,100
500	3,050	19.87	−261,250	−239,450
600	4,690	22.86	−261,100	−233,900
700	6,380	25.46	−260,900	−229,400
800	8,120	27.80	−260,700	−224,900
900	9,990	29.91	−260,450	−220,450
1,000	11,730	31.82	−260,300	−216,000
1,100	13,570	33.58	−260,150	−211,550
1,200	15,420	35.19	−260,900	−207,400
1,300	17,280	36.67	−260.400	−202,800
1,400	19,150	38.06	−260,100	−198,350
1,500	22,430	40.30	(−258,500)	(−193,900)
1,600	24,210	41.45	(−258,200)	(−189,700)
1,700	25,990	42.53	(−258,050)	(−185,250)
1,800	27,770	43.55	(−257,850)	(−180,950)

Zirconium Difluoride, ZrF_2 (*c*)

$\Delta H^\circ_{298} = (-230{,}000)$ calories per mole (*11*)
$S_{298} = (21)$ *e.u.* (*11*)
$M.P. = 1{,}800°$ K. (*6*)
$\Delta H_M = 14{,}500$ calories per mole
$B.P. = >2{,}500°$ K. (*6*)

Formation: $Zr + F_2 \longrightarrow ZrF_2$
(estimated (*11*))

T, ° K.	$H_T - H_{298}$	ΔH°_T	ΔH°_T
298	....	(−230,000)	(−219,000)
500	(4,000)	(−229,000)	(−211,500)
1,000	(14,000)	(−227,500)	(−196,000)
1,500	(25,000)	(−225,000)	(−182,000)

Zirconium Trifluoride, ZrF_3 (*c*)

$\Delta H^\circ_{298} = (-350{,}000)$ calories per mole (*11*)
$S_{298} = (24)$ *e.u.* (*11*)
$M.P. = (1{,}600°)$ K. (*6*)
$\Delta H_M = (13{,}000)$ calories per mole
$B.P. = (2{,}400°)$ K. (*6*)
$\Delta H_V = (58{,}000)$ calories per mole

Formation: $Zr + 3/2F_2 \longrightarrow ZrF_3$
(estimated (*11*))

T, ° K.	$H_T - H_{298}$	ΔH°_T	ΔF°_T
298	....	(−350.000)	(−333.000)
500	(4.000)	(−349.500)	(−321.000)
1,000	(17,000)	(−347.000)	(−293.000)
1,500	(32,000)	(−344.000)	(−269.000)

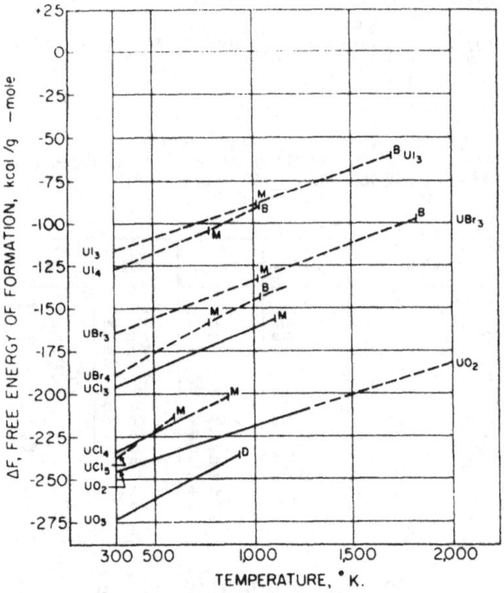

FIGURE 61.—Uranium (b).

VANADIUM AND ITS COMPOUNDS

Element, V (*c*)

$S_{298}=7.01$ *e.u.* (*83*)
$M.P.=2{,}190°$ K. (*130*)
$\Delta H_M=(4{,}200)$ calories per atom
$B.P.=3{,}650°$ K. (*130*)
$\Delta H_V=109{,}600$ calories per atom

Zone I (*c*) (298°–1,900° K.)

$C_p=5.40+2.00\times10^{-3}T$ (*82*)
$H_T-H_{298}=-1{,}699+5.40T+1.00\times10^{-3}T^2$
$F_T-H_{298}=-1{,}699-5.40T\ln T-1.00\times10^{-3}T^2+29.73T$

T, ° K.	H_T-H_{298}	S_T	$-\frac{(F_T-H_{298})}{T}$
298	----------	7.01	7.01
400	630	8.83	7.25
500	1,270	10.25	7.71
600	1,920	11.44	8.24
700	2,580	12.46	8.77
800	3,260	13.36	9.28
900	3,960	14.19	9.80
1,000	4,680	14.95	10.27
1,100	5,410	15.64	10.72
1,200	6,160	16.29	11.15
1,300	6,930	16.91	11.58
1,400	7,740	17.51	11.98
1,500	8,600	18.10	12.37
1,600	9,510	18.69	12.74
1,700	10,450	19.26	13.11
1,800	11,420	19.82	13.48
1,900	12,420	20.36	13.82
2,000	13,140	----------	----------

Vanadium Oxide, VO (*c*)

$\Delta H^\circ_{298}=-98{,}000$ calories per mole (*112*)
$S_{298}=9.3$ *e.u.* (*135*)
$M.P.=2{,}350°$ K. (*42*)
$\Delta H_M=15{,}000$ calories per mole
$B.P.=3{,}400°$ K. (*42*)
$\Delta H_V=70{,}000$ calories per mole

Zone I (*c*) (298°–1,700° K.)

$C_p=11.32+3.22\times10^{-3}T-1.26\times10^5T^{-2}$ (*106*)
$H_T-H_{298}=-3{,}940+11.32T+1.61\times10^{-3}T^2+1.26\times10^5T^{-1}$

Formation: $V+1/2O_2 \longrightarrow VO$

Zone I (298°–1,700° K.)

$\Delta C_p=2.34+0.72\times10^{-3}T-1.06\times10^5T^{-2}$
$\Delta H_T=-99{,}100+2.34T+0.36\times10^{-3}T^2+1.06\times10^5T^{-1}$
$\Delta F_T=-99{,}100-2.34T\ln T-0.36\times10^{-3}T^2+0.53\times10^5T^{-1}+38.64T$

T, ° K.	H_T-H_{298}	S_T	ΔH°_T	ΔF°_T
298	----------	9.3	−98,000	−91,400
400	1,160	12.64	−97,800	−89,100
500	2,380	15.36	−97,600	−87,000
600	3,640	17.65	−97,400	−84,900
700	4,940	19.66	−97,150	−82,800
800	6,280	21.45	−96,850	−80,800
900	7,660	23.07	−96,600	−78,800
1,000	9,090	24.58	−96,300	−76,800
1,100	10,560	25.98	−95,950	−74,900
1,200	12,070	27.29	−95,600	−73,000
1,300	13,610	28.52	−95,250	−71,100
1,400	15,170	29.68	−94,950	−69,300
1,500	16,760	30.78	−94,700	−67,500
1,600	18,370	31.82	−94,350	−65,600
1,700	20,000	32.80	−94,100	−63,900

Divanadium Trioxide, V_2O_3 (*c*)

$\Delta H^\circ_{298}=-296{,}000$ calories per mole (*8*)
$S_{298}=23.58$ *e.u.* (*83*)
$M.P.=2{,}250°$ K. (*112*)
$\Delta H_M=(28{,}000)$ calories per mole

Zone I (*c*) (298°–1,800° K.)

$C_p=29.35+4.76\times10^{-3}T-5.42\times10^5T^{-2}$
$H_T-H_{298}=-10{,}780+29.35T+2.38\times10^{-3}T^2+5.42\times10^5T^{-1}$

Formation: $2V+3/2O_2 \longrightarrow V_2O_3$

Zone I (298°–1,800° K.)

$\Delta C_p=7.81-0.74\times10^{-3}T-4.82\times10^5T^{-2}$
$\Delta H_T=-299{,}900+7.81T-0.37\times10^{-3}T^2+4.82\times10^5T^{-1}$
$\Delta F_T=-299{,}900-7.81T\ln T+0.37\times10^{-3}T^2+2.41\times10^5T^{-1}+118.8T$

T, ° K.	H_T-H_{298}	S_T	ΔH°_T	ΔF°_T
298	----------	23.58	−296,000	−276,970
400	2,720	31.41	−295,600	−268,150
500	5,990	37.81	−294,700	−263,850
600	8,600	43.30	−294,550	−258,150
700	11,700	48.07	−293,950	−252,150
800	14,870	52.30	−293,350	−246,250
900	18,100	56.11	−292,700	−240,250
1,000	21,370	59.55	−292,150	−234,500
1,100	24,660	62.69	−291,450	−229,950
1,200	27,960	65.56	−290,900	−223,700
1,300	31,360	68.28	−290,300	−217,900
1,400	34,940	70.93	−289,600	−212,300
1,500	38,940	73.50	−288,800	−205,900
1,600	42,480	75.96	−288,200	−201,800
1,700	46,370	78.31	−287,550	−196,050
1,800	50,350	80.59	−286,800	−191,100

Divanadium Tetraoxide, V_2O_4 (*c*)

$\Delta H^{\circ}_{298} = -342,000$ calories per mole (*24*)
$S_{298} = 24.5$ *e.u.* (*8*)
T.P. $= 345°$ K. (*82*)
$\Delta H_T = 2,050$ calories per mole
M.P. $= 1,818°$ K. (*82*)
$\Delta H_M = 27,210$ calories per mole

Zone I (α) (298°–345° K.)

$$C_p = 29.91 \ (82)$$
$$H_T - H_{298} = -8,918 + 29.91T$$

Zone II (β) (345°–1,818° K.)

$$C_p = 35.70 + 3.40\times10^{-3}T - 7.89\times10^{5}T^{-2} \ (82)$$
$$H_T - H_{298} = -11,355 + 35.70T + 1.70\times10^{-3}T^2 + 7.89\times10^{5}T^{-1}$$

Zone III (*l*) (1,818°–1,900° K.)

$$C_p = 51.0 \ (82)$$
$$H_T - H_{298} = -5,910 + 51.00T$$

Formation: $2V + 2O_2 \longrightarrow V_2O_4$

Zone I (298°–345° K.)

$$\Delta C_p = 4.79 - 6.00\times10^{-3}T + 0.80\times10^{5}T^{-2}$$
$$\Delta H_T = -342,900 + 4.79T - 3.00\times10^{-3}T^2 - 0.80\times10^{5}T^{-1}$$
$$\Delta F_T = -342,900 - 4.79T\ln T + 3.00\times10^{-3}T^2 - 0.40\times10^{5}T^{-1} + 117.46T$$

Zone II (345°–1,818° K.)

$$\Delta C_p = 10.58 - 2.60\times10^{-3}T - 7.09\times10^{5}T^{-2}$$
$$\Delta H_T = -345,270 + 10.58T - 1.30\times10^{-3}T^2 + 7.09\times10^{5}T^{-1}$$
$$\Delta F_T = -345,270 - 10.58T\ln T + 1.30\times10^{-3}T^2 + 3.54\times10^{5}T^{-1} + 155.21T$$

Zone III (1,818°–1,900° K.)

$$\Delta C_p = 25.88 - 6.00\times10^{-3}T + 0.80\times10^{5}T^{-2}$$
$$\Delta H_T = -339,820 + 25.88T - 3.00\times10^{-3}T^2 - 0.80\times10^{5}T^{-1}$$
$$\Delta F_T = -339,820 - 25.88T\ln T + 3.00\times10^{-3}T^2 - 0.40\times10^{5}T^{-1} + 264T$$

T, ° K.	$H_T - H_{298}$	S_T	ΔH°_T	ΔF°_T
298		24.5	−342,000	−315,950
400	5,270	39.7	−339,500	−307,400
500	8,600	47.12	−338,850	−299,450
600	12,000	53.29	−338,250	−291,450
700	15,560	58.79	−337,600	−283,900
800	19,230	63.69	−336,850	−276,250
900	22,990	68.11	−336,150	−268,750
1,000	26,830	72.16	−335,400	−261,250
1,100	30,730	75.88	−334,500	−253,800
1,200	34,670	79.31	−333,750	−245,650
1,300	38,630	82.47	−333,000	−239,250
1,400	42,600	85.42	−332,300	−232,100
1,500	46,590	88.17	−331,950	−225,050
1,600	50,620	90.77	−331,300	−218,400
1,700	54,710	93.25	−330,850	−211,000
1,800	58,850	95.62	−330,450	−203,850
1,900	91,000	113.25	−301,500	−199,000

Divanadium Pentaoxide, V_2O_5 (*c*)

$\Delta H^{\circ}_{298} = -372,500$ calories per mole (*8*)
$S_{298} = 31.3$ *e.u.* (*83*)
M.P. $= 943°$ K. (*24*)
$\Delta H_M = 15,560$ calories per mole
B.P. $= 2,325°$ K. (*42*)
$\Delta H_V = 63,000$ calories per mole

Zone I (*c*) (298°–943° K.)

$$C_p = 46.54 = 3.90\times10^{-3}T - 13.22\times10^{5}T^{-2} \ (82)$$
$$H_T - H_{298} = -18,137 + 46.54T - 1.95\times10^{-3}T^2 + 13.22\times10^{5}T^{-1}$$

Zone II (*l*) (943°–1,500° K.)

$$C_p = 45.60 \ (82)$$
$$H_T - H_{298} = -2,020 + 45.60T$$

Formation: $2V + 5/2O_2 \longrightarrow V_2O_5$

Zone I (298°–943° K.)

$$\Delta C_p = 17.84 - 10.40\times10^{-3}T - 12.22\times10^{5}T^{-2}$$
$$\Delta H_T = -381,450 + 17.84T - 5.20\times10^{-3}T^2 + 12.22\times10^{5}T^{-1}$$
$$\Delta F_T = -381,450 - 17.84T\ln T + 5.20\times10^{-3}T^2 + 6.11\times10^{5}T^{-1} + 228.56T$$

Zone II (943°–1,500° K.)

$$\Delta C_p = 16.90 - 6.50\times10^{-3}T + 1.0\times10^{5}T^{-2}$$
$$\Delta H_T = -365,400 + 16.90T - 3.25\times10^{-3}T^2 - 1.0\times10^{5}T^{-1}$$
$$\Delta F_T = -365,400 - 16.90T\ln T + 3.25\times10^{-3}T^2 - 0.50\times10^{5}T^{-1} + 207.16T$$

T, ° K.	$H_T - H_{298}$	S_T	ΔH°_T	ΔF°_T
298		31.3	−372,500	−341,250
400	3,650	41.79	−371,900	−330,500
500	7,400	50.15	−371,300	−320,200
600	11,290	57.24	−370,550	−310,500
700	15,290	63.40	−369,850	−300,000
800	19,390	68.88	−369,100	−290,100
900	23,590	73.82	−368,350	−280,300
1,000	43,580	94.99	−351,850	−271,400
1,100	48,140	99.34	−351,700	−264,300
1,200	52,700	103.30	−349,700	−255,300
1,300	57,260	106.95	−348,800	−247,800
1,400	61,820	110.33	−347,950	−239,900
1,500	66,380	113.48	−347,500	−232,500

Vanadium Difluoride, VF_2 (*c*)

$\Delta H^{\circ}_{298} = (-180,000)$ calories per mole (*11*)
$S_{298} = (19)$ *e.u.* (*11*)
M.P. $= (1,400°)$ K. (*6*)
$\Delta H_M = (6,000)$ calories per mole
B.P. $= (2,500°)$ K. (*6*)
$\Delta H_V = (65,000)$ calories per mole

Formation: $V + F_2 \longrightarrow VF_2$
(estimated (*11*))

T, ° K.	$H_T - H_{298}$	ΔH°_T	ΔF°_T
298		(−180,000)	(−169,000)
500	(3,000)	(−179,900)	(−162,000)
1,000	(12,000)	(−178,500)	(−144,000)
1,500	(29,000)	(−169,800)	(−127,500)

Vanadium Trifluoride, VF_3 (*c*)

$\Delta H^{\circ}_{298} = (-285,000)$ calories per mole (*11*)
$S_{298} = (28)$ *e.u.* (*11*)
M.P. $= (1,400°)$ K. (*6*)
$\Delta H_M = (11,000)$ calories per mole
B.P. $= (1,700°)$ K. (*6*)
$\Delta H_V = (49,000)$ calories per mole

Formation: $V + 3/2F_2 \longrightarrow VF_3$
(estimated (*11*))

T, ° K.	H_T-H_{298}	ΔH_T°	ΔF_T°
298	....	(−285, 000)	(−269, 500)
500	(4, 000)	(−284, 700)	(−259, 500)
1,000	(17, 000)	(−281, 500)	(−234, 000)
1,500	(32, 000)	(−277, 000)	(−212, 000)

Vanadium Tetrafluoride, VF_4 (*c*)

$\Delta H_{298}^\circ = (-325{,}000)$ calories per mole (*11*)
$S_{298} = (38)$ *e.u.* (*11*)
S.P. = (600°) K. (*6*)
$\Delta H_{subl} = (23{,}000)$ calories per mole

Formation: $V + 2F_2 \longrightarrow VF_4$
(estimated (*11*))

T, ° K.	H_T-H_{298}	ΔH_T°	ΔF_T°
298	....	(−325, 000)	(−305, 000)
500	(6, 000)	(−323, 500)	(−291, 000)

Vanadium Pentafluoride, VF_5 (*c*)

$\Delta H_{298}^\circ = (-335{,}000)$ calories per mole (*11*)
$S_{298} = (50)$ *e.u.* (*11*)
M.P. = (375°) K. (*6*)
B.P. = 384° K. (*6*)
$\Delta H_V = 8{,}500$ calories per mole

Formation: $V + 5/2F_2 \longrightarrow VF_5$
(estimated (*11*))

T, ° K.	H_T-H_{298}	ΔH_T°	ΔF_T°
298	....	(−335, 000)	(−312, 000)
500	(10, 000)	(−330, 200)	(−298, 000)

Vanadium Dichloride, VCl_2 (*c*)

$\Delta H_{298}^\circ = (-117{,}000)$ calories per mole (*11*)
$S_{298} = 23.2$ *e.u.* (*83*)
M.P. = 1,300° K. (*6*)
$\Delta H_M = 8{,}000$ calories per mole
B.P. = (1,650°) K. (*6*)
$\Delta H_V = (35{,}000)$ calories per mole

Zone I (*c*) (298°–1,300° K.)

$$C_p = 17.25 + 2.72\times10^{-3}T - 0.71\times10^{5}T^{-2} \ (82)$$
$$H_T - H_{298} = -5{,}500 + 17.25T + 1.36\times10^{-3}T^2 + 0.71\times10^{5}T^{-1}$$

Formation: $V + Cl_2 \longrightarrow VCl_2$

Zone I (298°–1,300° K.)

$$\Delta C_p = 3.03 + 0.66\times10^{-3}T - 0.03\times10^{5}T^{-2}$$
$$\Delta H_T = -117{,}950 + 3.03T + 0.33\times10^{-3}T^2 + 0.03\times10^{5}T^{-1}$$
$$\Delta F_T = -117{,}950 - 3.03T\ln T - 0.33\times10^{-3}T^2 + 0.015\times10^{5}T^{-1} + 57.61T$$

T, ° K.	H_T-H_{298}	S_T	ΔH_T°	ΔF_T°
298	....	23. 2	(−117, 000)	(−105, 900)
400	1, 840	28. 5	(−116, 650)	(−102, 100)
500	3, 620	32. 47	(−116, 350)	(−98, 600)
600	5, 450	35. 81	(−116, 000)	(−95, 000)
700	7, 330	38. 7	(−115, 650)	(−91, 700)
800	9, 250	41. 27	(−115, 300)	(−88, 200)
900	11, 200	43. 56	(−114, 950)	(−84, 800)
1, 000	13, 180	45. 65	(−114, 550)	(−81, 400)
1, 100	15, 190	47. 56	(−114, 150)	(−78, 200)
1, 200	17, 220	49. 33	(−113, 750)	(−75, 000)
1, 300	19, 270	50. 97	(−113, 350)	(−71, 800)
1, 500	....	....	(−84, 000)	(−67, 000)

Vanadium Trichloride, VCl_3 (*c*)

$\Delta H_{298}^\circ = (-139{,}000)$ calories per mole (*11*)
$S_{298} = 31.3$ *e.u.* (*83*)
Disproportionates < 1,000° K. (*6*)

Zone I (*c*) (298°–900° K.)

$$C_p = 22.99 + 3.92\times10^{-3}T - 1.68\times10^{5}T^{-2} \ (82)$$
$$H_T - H_{298} = -7{,}592 + 22.99T + 1.96\times10^{-3}T^2 + 1.68\times10^{5}T^{-1}$$

Formation: $V + 3/2Cl_2 \longrightarrow VCl_3$

Zone I (298°–900° K.)

$$\Delta C_p = 4.36 + 1.83\times10^{-3}T - 0.64\times10^{5}T^{-2}$$
$$\Delta H_T = -140{,}600 + 4.36T + 0.915\times10^{-3}T^2 + 0.64\times10^{5}T^{-1}$$
$$\Delta F_T = -140{,}600 - 4.36T\ln T - 0.915\times10^{-3}T^2 + 0.32\times10^{5}T^{-1} + 85.82T$$

T, ° K.	H_T-H_{298}	S_T	ΔH_T°	ΔF_T°
298	....	31. 3	(−139, 000)	(−123, 400)
400	2, 360	38. 1	(−138, 500)	(−116, 700)
500	4, 730	43. 39	(−138, 100)	(−111, 200)
600	7, 180	47. 85	(−137, 500)	(−106, 100)
700	9, 700	51. 73	(−137, 000)	(−100, 900)
800	12, 270	55. 15	(−136, 400)	(−95, 600)
900	14, 860	58. 20	(−135, 900)	(−90, 700)

Vanadium Tetrachloride, VCl_4 (*l*)

$\Delta H_{298}^\circ = (-141{,}000)$ calories per mole (*11*)
$S_{298} = (61)$ *e.u.* (*11*)
M.P. = 247° K. (*6*)
$\Delta H_M = (2{,}200)$ calories per mole
B.P. = 437° K. (*6*)
$\Delta H_V = 7{,}700$ calories per mole

Formation: $V + 2Cl_2 \longrightarrow VCl_4$
(estimated (*11*))

T, ° K.	H_T-H_{298}	ΔH_T°	ΔF_T°
298	....	(−141, 000)	(−125, 000)
500	(14, 700)	(−131, 000)	(−115, 000)

Vanadium Dibromide, VBr_2 (*c*)

$\Delta H^{\circ}_{298} = (-97{,}000)$ calories per mole (*11*)
$S_{298} = (30)$ *e.u.* (*11*)
M.P. = (1,100°) K. (*6*)
$\Delta H_M = (7{,}000)$ calories per mole
B.P. = (1,500°) K. (*6*)
$\Delta H_V = (32{,}000)$ calories per mole

Formation: $V + Br_2 \longrightarrow VBr_2$
(estimated (*11*))

T, ° K.	$H_T - H_{298}$	ΔH°_T	ΔF°_T
298		(−97, 000)	(−93, 000)
500	(4, 000)	(−103, 800)	(−86, 000)
1, 000	(14, 000)	(−101, 700)	(−70, 000)

Vanadium Tribromide, VBr_3 (*c*)

$\Delta H^{\circ}_{298} = (-109{,}000)$ calories per mole (*11*)
$S_{298} = (43)$ *e.u.* (*11*)
Decomposes to VBr_2

Formation: $V + 3/2Br_2 \longrightarrow VBr_3$
(estimated (*11*))

T, ° K.	$H_T - H_{298}$	ΔH°_T	ΔF°_T
298		(−109, 000)	(−103, 500)
500	(5, 000)	(−119, 600)	(−94, 000)
1, 000	(20, 000)	(−114, 700)	(−70, 000)

Vanadium Diiodide, VI_2 (*c*)

$\Delta H^{\circ}_{298} = (-62{,}000)$ calories per mole (*11*)
$S_{298} = (33)$ *e.u.* (*11*)
M.P. = (1,050°) K. (*6*)
$\Delta H_M = (6{,}000)$ calories per mole
B.P. = (1,200°) K. (*6*)
$\Delta H_V = (25{,}000)$ calories per mole

Formation: $V + I_2 \longrightarrow VI_2$
(estimated (*11*))

T, ° K.	$H_T - H_{298}$	ΔH°_T	ΔF°_T
298		(−62, 000)	(−62, 000)
500	(4, 000)	(−75, 900)	(−59, 000)
1, 000	(14, 000)	(−73, 800)	(−44, 000)

Vanadium Pentaiodide, VI_5 (*c*)

$\Delta H^{\circ}_{298} = (-42{,}000)$ calories per mole (*11*)
$S_{298} = (78)$ *e.u.* (*11*)

Formation: $V + 5/2I_2 \longrightarrow VI_5$
(estimated (*11*))

T, ° K.	$H_T - H_{298}$	ΔH°_T	ΔF°_T
298		(−42, 000)	(−42, 000)
500	(8, 000)	(−77, 000)	(−25, 500)

Vanadium Carbide, VC (*c*)

$\Delta H^{\circ}_{298} = -28{,}000$ calories per mole (*9*)
$S_{298} = 6.77$ *e.u.*
M.P. = 3,100° K. (*9*)

Zone I (*c*) (298°–1,600° K.)

$$C_p = 9.18 + 3.30 \times 10^{-3}T - 1.95 \times 10^5 T^{-2} \ (82)$$
$$H_T - H_{298} = -3{,}725 + 9.18T + 1.65 \times 10^{-3}T^2 + 1.95 \times 10^5 T^{-1}$$

Formation: $V + C \longrightarrow VC$

Zone I (298°–1,600° K.)

$$\Delta C_p = -0.32 + 0.28 \times 10^{-3}T + 0.15 \times 10^5 T^{-2}$$
$$\Delta H_T = -27{,}870 - 0.32T + 0.14 \times 10^{-3}T^2 - 0.15 \times 10^5 T^{-1}$$
$$\Delta F_T = -27{,}870 + 0.32T\ln T - 0.14 \times 10^{-3}T^2 - 0.075 \times 10^5 T^{-1} - 0.53T$$

T, ° K.	$H_T - H_{298}$	S_T	ΔH°_T	ΔF°_T
298		6. 77	−28, 000	−27, 525
400	990	9. 32	−28, 000	−27, 350
500	1, 850	11. 47	−28, 000	−27, 200
600	2, 870	13. 32	−28, 000	−27, 050
700	3. 950	14. 99	−28. 000	−26, 900
800	5, 090	16. 51	−28, 000	−26, 750
900	6, 280	17. 91	−28, 000	−26, 550
1,000	7, 510	19. 20	−28, 000	−26, 400
1,100	8, 770	20. 41	−27, 950	−26, 250
1,200	10, 060	21. 53	−27, 950	−26, 100
1,300	11, 380	22. 58	−27, 950	−25, 950
1,400	12, 720	23. 57	−27, 950	−25, 800
1,500	14, 080	24. 51	−27, 900	−25, 650
1,600	15, 450	25. 40	−27, 800	−25, 600

Vanadium Nitride, VN (*c*)

$\Delta H^{\circ}_{298} = -40{,}800$ calories per mole (*94*)
$S_{298} = 8.9$ *e.u.* (*83*)
M.P. = 2,320° K. (*9*)

Zone I (*c*) (298°–1, 600° K.)

$$C_p = 10.94 + 2.10 \times 10^{-3}T - 2.21 \times 10^5 T^{-2} \ (82)$$
$$H_T - H_{298} = -4{,}096 + 10.94T + 1.05 \times 10^{-3}T^2 + 2.21 \times 10^5 T^{-1}$$

Formation: $V + 1/2N_2 \longrightarrow VN$

Zone I (298°–1, 600° K.)

$$\Delta C_p = 2.21 - 0.41 \times 10^{-3}T - 2.21 \times 10^5 T^{-2}$$
$$\Delta H_T = -42{,}180 + 2.21T - 0.205 \times 10^{-3}T^2 + 2.21 \times 10^5 T^{-1}$$
$$\Delta F_T = -42{,}180 - 2.21T\ln T + 0.205 \times 10^{-3}T^2 + 1.105 \times 10^5 T^{-1} + 36.91T$$

T, ° K.	$H_T - H_{298}$	S_T	ΔH°_T	ΔF°_T
298		8. 9	−40, 800	−34, 550
400	1, 010	11. 81	−40, 750	−32, 400
500	2, 080	14. 20	−40, 700	−30, 300
600	3, 200	16. 24	−40, 600	−28, 250
700	4, 370	18. 04	−40, 450	−26, 200
800	5, 590	19. 66	−40, 250	−24, 150
900	6, 750	21. 15	−40, 100	−22, 200
1, 000	8, 130	22. 50	−39, 900	−20, 200
1, 100	9, 430	23. 74	−39, 750	−18, 250
1, 200	10, 750	24. 89	−39, 550	−16, 350
1, 300	12, 090	25. 96	−39, 400	−14, 350
1, 400	13, 450	26. 97	−39, 250	−12. 450
1, 500	14, 820	27. 91	−39. 150	−10. 600
1, 600	16, 200	28. 80	−39. 100	−8. 650

Zirconium Tetrafluoride, ZrF_4 (*c*)

$\Delta H^\circ_{298} = (-445{,}000)$ calories per mole (*11*)
$S_{298} = (33)$ *e.u.* (*11*)
$S.P. = (1{,}200^\circ)$ K. (*6*)
$\Delta H_{subl} = (45{,}000)$ calories per mole

Formation: $Zr + 2F_2 \longrightarrow ZrF_4$
(estimated (*11*))

T, ° K.	$H_T - H_{298}$	ΔH°_T	ΔF°_T
298	....	(−445,000)	(−423,000)
500	(6,000)	(−443,500)	(−408,000)
1,000	(22,000)	(−439,800)	(−375,000)

Zirconium Dichloride, $ZrCl_2$ (*c*)

$\Delta H^\circ_{298} = (-145{,}000)$ calories per mole (*11*)
$S_{298} = (27)$ *e.u.* (*11*)
$M.P. = (1{,}000^\circ)$ K. (*6*)
$\Delta H_M = 7{,}300$ calories per mole
$B.P. = (1{,}750^\circ)$ K. (*6*)
$\Delta H_V = (35{,}000)$ calories per mole

Formation: $Zr + Cl_2 \longrightarrow ZrCl_2$
(estimated (*11*))

T, ° K.	$H_T - H_{298}$	ΔH°_T	ΔF°_T
298	....	(−145,000)	(−134,000)
500	(4,000)	(−144,000)	(−127,000)
1,000	(21,000)	(−135,000)	(−112,000)

Zirconium Trichloride, $ZrCl_3$ (*c*)

$\Delta H^\circ_{298} = (-208{,}000)$ calories per mole (*11*)
$S_{298} = (40)$ *e.u.* (*11*)
$M.P. = (900^\circ)$ K. (*6*)
Disproportionates above 1,000° K. (*6*)

Formation: $Zr + 3/2Cl_2 \longrightarrow ZrCl_3$
(estimated (*11*))

T, ° K.	$H_T - H_{298}$	ΔH°_T	ΔF°_T
298	....	(−208,000)	(−193,000)
500	(5,000)	(−207,000)	(−183,000)
1,000	(20,000)	(−202,000)	(−162,000)

Zirconium Tetrachloride, $ZrCl_4$ (*c*)

$\Delta H^\circ_{298} = (-230{,}000)$ calories per mole (*11*)
$S_{298} = 44.5$ *e.u.* (*83*)
$S.P. = 604^\circ$ K. (*6*)
$\Delta H_{subl} = 25{,}290$ calories per mole

Zone I (*c*) (298°–604° K.)

$$C_p = 31.92 - 2.91 \times 10^5 T^{-2} \quad (26)$$
$$H_T - H_{298} = -10{,}495 + 31.92T + 2.91 \times 10^5 T^{-1}$$

Formation: $Zr + 2Cl_2 \longrightarrow ZrCl_4$

Zone I (298°–604° K.)

$$\Delta C_p = 7.45 - 1.24 \times 10^{-3} T - 0.68 \times 10^5 T^{-2}$$
$$\Delta H_T = -232{,}400 + 7.45T - 0.62 \times 10^{-3} T^2 + 0.68 \times 10^5 T^{-1}$$
$$\Delta F_T = -232{,}400 - 7.45 T \ln T + 0.62 \times 10^{-3} T^2 + 0.34 \times 10^5 T^{-1} + 121.27T$$

T, ° K.	$H_T - H_{298}$	S_T	ΔH°_T	ΔF°_T
298	....	44.5	(−230,000)	(−208,750)
400	3,000	53.15	(−229,360)	(−201,400)
500	6,050	59.94	(−228,670)	(−194,550)
600	9,120	65.60	(−227,560)	(−187,450)
1,000	....	....	....	(−177,000)
1,500	....	....	....	(−164,000)
2,000	....	....	....	(−149,000)

Zirconium Dibromide, $ZrBr_2$ (*c*)

$\Delta H^\circ_{298} = (-120{,}000)$ calories per mole (*11*)
$S_{298} = (32)$ *e.u.* (*11*)
$M.P. = (900^\circ)$ K. (*6*)
$B.P. = (1{,}500^\circ)$ K. (*6*)
$\Delta H_V = (33{,}000)$ calories per mole

Formation: $Zr + Br_2 \longrightarrow ZrBr_2$
(estimated (*11*))

T, ° K.	$H_T - H_{298}$	ΔH°_T	ΔF°_T
298	....	(−120,000)	(−116,000)
500	(4,000)	(−127,000)	(−109,500)
1,000	(22,000)	(−117,000)	(− 94,000)

Zirconium Tribromide, $ZrBr_3$ (*c*)

$\Delta H^\circ_{298} = (-174{,}000)$ calories per mole (*11*)
$S_{298} = (42)$ *e.u.* (*11*)
Disproportionates above 1,100° K. (*6*)

Formation: $Zr + 3/2Br_2 \longrightarrow ZrBr_3$
(estimated (*11*))

T, ° K.	$H_T - H_{298}$	ΔH°_T	ΔF°_T
298	....	(−174,000)	(−168,000)
500	(5,000)	(−185,000)	(−157,000)
1,000	(19,000)	(−181,000)	(−132,000)

Zirconium Tetrabromide, $ZrBr_4$ (*c*)

$\Delta H^\circ_{298} = (-192{,}300)$ calories per mole (*11*)
$S_{298} = (54)$ *e.u.* (*11*)
$S.P. = 595^\circ$ K.
$\Delta H_{subl} = 24{,}000$ calories per mole

Formation: $Zr + 2Br_2 \longrightarrow ZrBr_4$
(estimated (*11*))

T, ° K.	$H_T - H_{298}$	ΔH°_T	ΔF°_T
298	....	(−192,300)	(−183,000)
500	(6,000)	(−206,400)	(−171,500)

647940 O—63——10

Zirconium Diiodide, ZrI_2 (*c*)

$\Delta H^\circ_{298} = (-90{,}000)$ calories per mole (*11*)
$S_{298} = (35)$ *e.u.* (*11*)
M.P. = (700°) K. (*6*)
B.P. = (1,300°) K. (*6*)
$\Delta H_V = (27{,}000)$ calories per mole

Formation: $Zr + I_2 \longrightarrow ZrI_2$
(estimated (*11*))

T, ° K.	$H_T - H_{298}$	ΔH°_T	ΔF°_T
298		(−90,000)	(−89,500)
500	(4,000)	(−97,000)	(−87,000)
1,000	(22,000)	(−87,000)	(−71,000)

Zirconium Triiodide, ZrI_3 (*c*)

$\Delta H^\circ_{298} = (-128{,}000)$ calories per mole (*11*)
$S_{298} = (45)$ *e.u.* (*11*)
Disproportionates above 1,200° K. (*6*)

Formation: $Zr + 3/2I_2 \longrightarrow ZrI_3$
(estimated (*11*))

T, ° K.	$H_T - H_{298}$	ΔH°_T	ΔF°_T
298		(−128,000)	(−126,000)
500	(5,000)	(−149,000)	(−121,000)
1,000	(19,000)	(−146,000)	(− 95,000)

Zirconium Tetraiodide, ZrI_4 (*c*)

$\Delta H^\circ_{298} = (-130{,}000)$ calories per mole (*11*)
$S_{298} = (60)$ *e.u.* (*11*)
S.P. = 704° K. (*6*)
$\Delta H_{subl} = 29{,}000$ calories per mole

Formation: $Zr + 2I_2 \longrightarrow ZrI_4$
(estimated (*11*))

T, ° K.	$H_T - H_{298}$	ΔH°_T	ΔF°_T
298		(−130,000)	(−129,000)
500	(6,000)	(−158,700)	(−124,000)

Zirconium Carbide, ZrC (*c*)

$\Delta H^\circ_{298} = -44{,}100$ calories per mole (*99*)
$S_{298} = (8.5)$ *e.u.* (*94*)
$\nabla F^\circ_{298} = (-43{,}450)$ calories per mole
M.P. = 3,805° K. (*9*)

Zirconium Nitride, ZrN (*c*)

$\Delta H^\circ_{298} = -87{,}300$ calories per mole (*100*)
$S_{298} = 9.29$ *e.u.* (*83*)
M.P. = 3,255° K. (*9*)

Zone I (*c*) (298°–1,700° K.)

$$C_p = 11.0 + 1.68 \times 10^{-3}T - 1.72 \times 10^5 T^{-2} \ (26)$$
$$H_T - H_{298} = -3{,}930 + 11.0T + 0.84 \times 10^{-3}T^2 + 1.72 \times 10^5 T^{-1}$$

Formation: $Zr + 1/2N_2 \longrightarrow ZrN$

Zone I (298°–1,135° K.)

$$\Delta C_p = 0.94 + 0.05 \times 10^{-3}T - 0.85 \times 10^5 T^{-2}$$
$$\Delta H_T = -87{,}870 + 0.94T + 0.025 \times 10^{-3}T^2 + 0.85 \times 10^5 T^{-1}$$
$$\Delta F_T = -87{,}870 - 0.94T\ln T - 0.025 \times 10^{-3}T^2 + 0.42 \times 10^5 T^{-1} + 28.77T$$

Zone II (1,135°–1,700° K.)

$$\Delta C_p = 0.40 + 1.17 \times 10^{-3}T - 1.72 \times 10^5 T^{-2}$$
$$\Delta H_T = -89{,}100 + 0.40T + 0.58 \times 10^{-3}T^2 + 1.72 \times 10^5 T^{-1}$$
$$\Delta F_T = -89{,}100 - 0.40T\ln T - 0.58 \times 10^{-3}T^2 + 0.86 \times 10^5 T^{-1} + 28.0T$$

T, ° K.	$H_T - H_{298}$	S_T	ΔH°_T	ΔF°_T
298		9.29	−87,300	−80,500
400	1,040	12.29	−87,300	−78,150
500	2,120	14.69	−87,250	−75,900
600	3,260	16.77	−87,150	−73,600
700	4,450	18.60	−87,100	−71,350
800	5,670	20.23	−87,000	−69,100
900	6,920	21.70	−86,850	−66,850
1,000	8,190	23.04	−86,750	−64,650
1,100	9,470	24.26	−86,700	−62,500
1,200	10,660	25.39	−87,550	−60,350
1,300	12,060	26.43	−87,300	−58,000
1,400	13,370	27.40	−87,100	−55,600
1,500	14,690	28.31	(−86,950)	(−53,550)
1,600	16,020	29.17	(−86,700)	(−51,300)
1,700	17,360	29.98	(−86,500)	(−49,100)

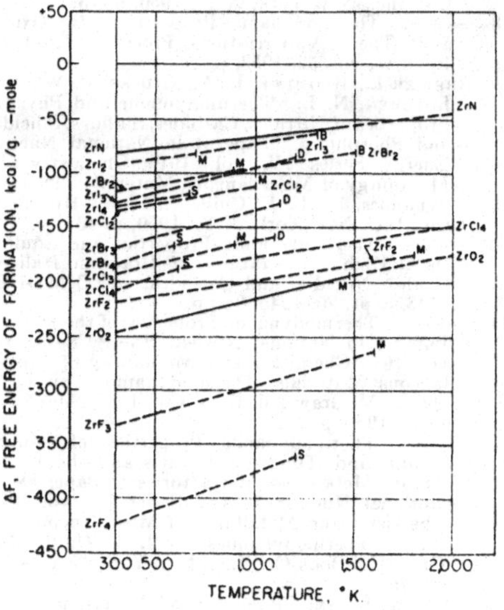

FIGURE 67.—Zirconium.

REFERENCES

1. Altman, D., Farber, M., and Mason, D. M. Thermodynamic Properties of the Titanium Chlorides. Jour. Chem. Phys., vol. 25, 1956, p. 531.
2. Balesdent, D. Détermination of the Heats of Formation of the Oxides of Copper. Compt. rend., vol. 240, 1955, p. 1884.
3. Blomeke, J. O., and Ziegler, W. T. Heat Content, Specific Heat and Entropy of La_2O_3, Pr_6O_{11} and Nd_2O_3 Between 30° and 900°. Jour. Am. Chem. Soc., vol. 73, 1951, p. 5099.
4. Boyle, B. J., King, E. G., and Conway, K. C. Heats of Formation of Nickel and Cobalt Oxides (NiO and CoO) of Combustion Calorimetry. Jour. Am. Chem. Soc., vol. 76, 1954, p. 3835.
5. Brewer, L. Private Communications, 1958. Available for inspection at the Albany Metallurgy Research Center, Bureau of Mines, Albany, Oreg.
6. ———. Fusion and Vaporization Data of the Halides. Paper 7 in National Nuclear Energy Series IV, vol. 19B, Chemistry and Metallurgy of Miscellaneous Materials: Thermodynamics, by L. L. Quill. McGraw-Hill Book Co., Inc., New York, N.Y., 1950, p. 193.
7. ———. Thermodynamic and Physical Properties of the Elements. Paper 3 in National Nuclear Energy Series IV, vol. 19B, Chemistry and Metallurgy of Miscellaneous Materials: Thermodynamics, by L. L. Quill. McGraw-Hill Book Co., Inc., New York, N.Y., 1950, p. 13.
8. ———. Thermodynamic Properties of the Oxides and Their Vaporization Processes. Chem. Reviews, vol. 52, 1953, p. 1.
9. Brewer L., Bromley, L. A., Gilles, P. W., and Lofgren, N. L. Thermodynamic and Physical Properties of Nitrides, Carbides, Sulfides, Silicides and Phosphides. Paper 4 in National Nuclear Energy Series IV, vol. 19B, Chemistry and Metallurgy of Miscellaneous Materials: Thermodynamics, by L. L. Quill. McGraw-Hill Book Co., Inc., New York, N.Y., 1950, p. 40.
10. ———. Thermodynamic Properties and Equilibria at High Temperatures of Uranium Halides, Oxides, Nitrides, and Carbides. AEC MDDC-1543, Sept. 20, 1945, 84 pp.
11. ———. Thermodynamic Properties of the Halides. Paper 6 in National Nuclear Energy Series IV, vol. 19B, Chemistry and Metallurgy of Miscellaneous Materials: Thermodynamics, by L. L. Quill. McGraw-Hill Book Co., Inc., New York, N.Y., 1950, p. 76.
12. ———. Thermodynamic Properties of Molybdenum and Tungsten Halides and the Use of These Metals as Refractories. Paper 8 in National Nuclear Energy Series IV, vol. 19B, Chemistry and Metallurgy of Miscellaneous Materials: Thermodynamics, by L. L. Quill. McGraw-Hill Book Co., Inc., New York, N.Y., 1950, p. 276.
13. Brewer, L., and Searcy, A. W. Gaseous Species of the Al-Al_2O_3 System. Jour. Am. Chem. Soc., vol. 73, 1951, p. 5308.
14. Britzke, E. V., Kapustinskii, A. F., and Chenzova, L. G. Die Affintat von Metallen zu Schwefel, III Mitteilung. Verbrennungs und Bildungswarmen von Sulfide von Arsen und der Verbindungen As_2O_3, As_2O_5 and As_2O_3, SO_3. Ztschr. anorg. allgem. Chem., vol. 213, 1933, p. 58.
15. Brönsted, J. N. Untersuchungen über die Spezifiche Wärme. I. Ztschr. Elektrochem., vol. 18, 1912, p. 714.
16. Busey, R. H., and Giauque, W. F. Heat Capacity of Anhydrous $NiCl_2$ From 15° to 300° K. The Antiferromagnetic Anomaly Near 52° K. Entropy and Free Energy. Jour. Am. Chem. Soc., vol. 74, 1952, p. 4443.
17. ———. Heat Capacity of Nickel From 15° to 300° K. Entropy and Free Energy Functions. Jour. Am. Chem. Soc., vol. 74, 1952, p. 3157.
18. Catalano, E., and Stout, J. W. Heat Capacity and Entropy of FeF_2 and CoF_2 From 11° to 300° K. The Thermal Anomalies Associated With Antiferromagnetic Ordering. Jour. Chem. Phys., vol. 23, 1955, p. 1803.
19. ———. Heat Capacity of NiF_2 From 12° to 300° K. Thermodynamic Functions of NiF_2. The Thermal Anomaly Associated With the Antiferromagnetic Ordering. Jour. Chem. Phys., vol. 23, 1955, p. 1284.
20. Cooper, C. B. Precision Measurements of the Enthalpy of KBr and KI at High Temperatures. Jour. Chem Phys., vol. 21, 1953, p. 777.
21. Corbett, J. D., and Gregory, N. W. AlI_3-HCl-$AlCl_3$-HI System. The Free Energy of Formation of Aluminum Iodide. Jour. Am. Chem. Soc., vol. 76, 1954, p. 1446.
22. Cosgrove, L. A., and Snyder, P. E. Heat of Formation of Beryllium Oxide. Jour. Am. Chem. Soc., vol. 75, 1953, p. 3102.
23. ———. High Temperature Thermodynamic Properties of Molybdenum Trioxide. Jour. Am. Chem. Soc., vol. 75, 1953, p. 1227.
24. Coughlin, J. P. Contributions to the Data on Theoretical Metallurgy. XII. Heats and Free Energies of Formation of Inorganic Oxides. Bureau of Mines Bull. 542, 1954, 80 pp.
25. ———. High Temperature Heat Contents of Nickel Chloride. Jour. Am. Chem. Soc., vol. 73, 1951, p. 5314.
26. Coughlin, J. P., and King, E. G. High-Temperature Heat Contents of Some Zirconium-Containing Substances. Jour. Am. Chem. Soc., vol. 72, 1950, p. 2262.
27. Coughlin, J. P., King, E. G., and Bonnickson, K. R. High Temperature Heat Contents of Ferrous Oxide, Magnetite, and Ferric Oxide. Jour. Am. Chem. Soc., vol. 73, 1951, p. 3891.
28. Craig, R. S., Krier, C. A., Coffer, L. W., Bates, E. A., and Wallace, W. E. Magnesium-Cadmium Alloys. VI. Heat Capacities Between 12 and 320° K. and the Entropies at 25° C. of Magnesium and Cadmium. Jour. Am. Chem. Soc., vol. 76, 1954, 238.

29. Daane, A. H. Private Communications, 1959. Available for inspection at the Albany Metallurgy Research Center, Bureau of Mines, Albany, Oreg.
30. Darken, L. S., and Gurry, R. W. System Iron-Oxygen. II. Equilibrium and Thermodynamics of Liquid Oxide and Other Phases. Jour. Am. Chem. Soc., vol. 68, 1946, p. 799.
31. De Sorbo, W. Heat Capacity of Chromium Carbide (Cr_3C_2) From 13° to 300° K. Jour. Am. Chem. Soc., vol. 75, 1953, p. 1825.
32. Douglas, T. B., Ball, A. F., Ginnings, D. C., and Davis, W. D. Heat Capacity of Potassium and Three Potassium-Sodium Alloys Between 0° and 800°, the Triple Point and Heat of Fusion of Potassium. Jour. Am. Chem. Soc., vol. 74, 1952, p. 2472.
33. Dworkin, A. S., Sasmor, D. J., and Van Artsdalen, E. R. Thermodynamics of Boron Nitride; Low-Temperature Heat Capacity and Entropy; Heats of Combustion and Formation. Jour. Chem. Phys., vol. 22, 1954, p. 837.
34. Evans, W. H., Jacobson, R., Munson, T. R., and Wagman, D. D. Thermodynamic Properties of the Alkali Metals. Nat. Bureau of Standards Res. Jour., vol. 55, 1955, RP 2608, p. 83.
35. Farber, M., and Darnell, A. J. Heat of Formation and Entropy of Titanium Tetrachloride From an Investigation of the Equilibrium: $TiO_2(s) + 4HCl(g) = TiCl_4(g) + 2H_2O(g)$. Jour. Chem. Phys., vol. 23, 1955, p. 1460.
36. Foley, W. T., and Giguere, P. A. Hydrogen Peroxide and Its Analogs. IV. Some Thermal Properties of Hydrogen Peroxide. Canadian Jour. Chem., vol. 29, 1951, p. 895.
37. Geballe, T. H., and Giauque, W. F. Heat Capacity and Entropy of Gold From 15° to 300° K. Jour. Am. Chem. Soc., vol. 74, 1952, p. 2368.
38. Giauque, W. F., and Jones, W. M. Carbonyl Chloride, Entropy, Heat Capacity. Vapor Pressure, Heats of Fusion and Vaporization. Comments on Solid Sulfur Dioxide Structure. Jour. Am. Chem. Soc., vol. 70, 1948, p. 120.
39. Giguere, P. A., Liu, I. D., Dugdale, J. S., and Morrison, J. A. Hydrogen Peroxide: The Low-Temperature Heat Capacity of the Solid and the Third Law Entropy. Canadian Jour. Chem., vol. 32, 1954, p. 117.
40. Gilles, P. W., and Margrave, J. L. Heats of Formation of Na_2O_2, NaO_2 and KO_2. Jour. Phys. Chem., vol. 60, 1956, p. 1333.
41. Ginnings, D. C., Douglas, T. B., and Ball, A. F. Heat Capacity of Sodium Between 0° and 900° C., the Triple Point and Heat of Fusion. Nat. Bureau of Standards Res. Jour., vol. 45, 1950, RP 2110, p. 23.
42. Glassner, A. Thermochemical Properties of the Oxides, Fluorides, and Chlorides to 2500° K. Argonne Nat. Lab., ANL–5750, 1957, 70 pp.
43. Gray, P., and Waddington, T. C. Thermochemistry and Reactivity of the Azides. I. Thermochemistry of the Inorganic Azides. Proc. Royal Soc. (London), Math. and Phys. Sci., vol. 235A, 1956, p. 106.
44. Greenberg, E., and Westrum, E. F., Jr. Heat Capacity and Thermodynamic Functions of Uranium (IV) Oxychloride and Uranium (IV) Oxybromide From 10° to 350° K. Jour. Am. Chem. Soc., vol. 78, 1956, p. 5144.
45. ———. Heat Capacity and Thermodynamic Functions of Uranyl Chloride From 6° to 350° K. Jour. Am. Chem. Soc., vol. 78, 1956, p. 4526.
46. Griffel, M., and Skochdopole, R. E. The Heat Capacity and Entropy of Thorium From 18° to 300° K. Jour. Am. Chem. Soc., vol. 75, 1953, p. 5250.
47. Griffel, M., Skochdopole, R. E., and Spedding, F. H. The Heat Capacity of Dysprosium From 15° to 300° K. Jour. Chem. Phys., vol. 25, 1956, p.75.
48. Gross, P., Hayman, C., and Levi, D. L. Heats of Formation of Metallic Halides. Titanium Tetrachloride. Trans. Faraday Soc., vol. 51, 1955, p. 626.
49. Gross, P., Hayman, C., Levi, D. L., and Wilson, G. Studies and Experimental Investigations in the Field of Measurements of the Heat or Free Energy of Formation of the Chlorides and Bromides of Niobium and Tantalum. Fulmer Research Inst., Ltd., Bucks, England, R–115/5/23, NP–7847, 1959, 44 pp.
50. Guntz, M. Heats of Formation of Barium Compounds. Compt. rend., vol. 136, 1903, p. 1071.
51. Harrison, E. R. Vapor Pressures of Some Rare-Earth Halides. Jour. Applied Chem. (London), vol. 2, 1952, p. 601.
52. Hoekstra, H. R., and Katz, J. J. Chemistry of Uranium. Ch. 6 in The Actinide Elements, by G. T. Seaborg and J. J. Katz. McGraw-Hill Book Co., Inc., New York, N.Y., 1954, p. 130.
53. Holley, C. E., Jr., and Huber, E. J., Jr. Heats of Combustion of Magnesium and Aluminum. Jour. Am. Chem. Soc., vol. 73, 1951, p. 5577.
54. Horning, R. Private Communications, 1957. Available for inspection at the Albany Metallurgy Research Center, Bureau of Mines, Albany, Oreg.
55. Hu, J. H., and Johnston, H. L. Low Temperature Heat Capacities of Inorganic Solids. XII. Heat Capacity and Thermodynamic Properties of Cuprous Bromide From 16° to 300° K. Jour. Am. Chem. Soc., vol. 74, 1952, p. 4771.
56. ———. Low Temperature Heat Capacities of Inorganic Solids. XVI. Heat Capacity of Cupric Oxide From 15° to 300° K. Jour. Am. Chem. Soc., vol. 75, 1953, p. 2471.
57. Huber, E. J., Jr., and Holley, C. E., Jr. Heat of Combustion of Calcium. Jour. Phys. Chem., vol. 60, 1956, p. 498.
58. ———. Heat of Combustion of Cerium. Jour. Am. Chem. Soc., vol. 75, 1953, p. 5645.
59. ———. Heat of Combustion of Gadolinium. Jour. Am. Chem. Soc., vol. 75, 1953, p. 1444.
60. ———. Heat of Combustion of Lanthanum. Jour. Am. Chem. Soc., vol. 75, 1953, p. 3594.
61. ———. Heat of Combustion of Neodymium. Jour. Am. Chem. Soc., vol. 74, 1952, p. 5530.
62. ———. Heat of Combustion of Yttrium. Jour. Phys. Chem., vol. 61, 1957, p. 497.
63. Huber, E. J., Jr., Holley, C. E., Jr., and Meierkord, E. H. Heats of Combustion of Thorium and Uranium. Jour. Am. Chem. Soc., vol. 74, 1952, p. 3406.
64. Huber, E. J., Jr., Matthews, C. O., and Holley, C. E., Jr. Heat of Combustion of Samarium. Jour. Am. Chem. Soc., vol. 77, 1955, p. 6493.
65. Huff, G., Squitiere, E., and Snyder, P. E. Heat of Formation of Tungstic Oxide, WO_3. Jour. Am. Chem. Soc., vol. 70, 1948, p. 3380.
66. Humphrey, G. L. Heats of Formation of Hafnium Oxide and Hafnium Nitride. Jour. Am. Chem. Soc., vol. 75, 1953, p. 2806.
67. ———. Heats of Formation of Tantalum, Niobium and Zirconium Oxides, and Tantalum Carbide. Jour. Am. Chem. Soc., vol. 76, 1954, p. 978.
68. ———. Heats of Formation of TiO, Ti_2O_3, Ti_3O_5, and TiO_2 From Combustion Calorimetry. Jour. Am. Chem. Soc., vol. 73, 1951, p. 1587.
69. Humphrey, G. L., and King, E. G. Heats of Formation of Quartz and Cristobalite. Jour. Am. Chem. Soc., vol. 74, 1952, p. 2041.

70. HUMPHREY, G. L., KING, E. G., AND KELLEY, K. K. Some Thermodynamic Values of Ferrous Oxide. Bureau of Mines Rept. of Investigations 4870, 1952, 16 pp.

71. HUMPHREY, G. L., AND O'BRIEN, C. J. Heats of Formation of Stannic and Stannous Oxides From Combustion Calorimetry. Jour. Am. Chem. Soc., vol. 75, 1953, p. 2805.

72. HUMPHREY, G. L., TODD, S. S., COUGHLIN, J. P., AND KING, E. G. Some Thermodynamic Properties of Silicon Carbide. Bureau of Mines Rept. of Investigations 4888, 1952, 23 pp.

73. HÜTTIG, G. F., AND SLONIM, CH. Die Spezifischen Wärmen, Bildungswärmen und Zersetzungsdrucke der Strontiumhalogenidhydrate. Ztsch. anorg. allgem. Chem., vol. 181, 1929, p. 65.

74. HÜTTIG, G. F., AND WEHLING, W. Zur Kenntnis der Spezisfichen Wärmen Homogener Phasen, an deren Aufbau Wasser beteiligtist. Kolloidchem. Beihefte, vol. 23, 1926, p. 354.

75. JOHNSTON, H. L., AND BAUER, T. W. Low Temperature Heat Capacities of Inorganic Solids. VII. Heat Capacity and Thermodynamic Functions of Li_2O. Thermodynamics of the Li_2O-H_2O System. Jour. Am. Chem. Soc., vol. 73, 1951, p. 1119.

76. JOHNSTON, H. L., HERSCH, H. N., AND KERR, E. C. Low-Temperature Heat Capacities of Inorganic Solids. V. The Heat Capacity of Pure Elementary Boron in Both Amorphous and Crystalline Conditions Between 13° and 305° K. Some Free Energies of Formation. Jour. Am. Chem. Soc., vol. 73, 1951, p. 1112.

77. JONES, W. M., GORDON, J., AND LONG, E. A. The Heat Capacities of Uranium, Uranium Trioxide, and Uranium Dioxide From 15° K to 300° K. Jour. Chem. Phys., vol. 20, 1952, p. 695.

78. KELLEY, K. K. Private Communication, 1959. Available for inspection at the Albany Metallurgy Research Center, Bureau of Mines, Albany, Oreg.

79. ———. Contributions to the Data on Theoretical Metallurgy. III. The Free Energies of Vaporization and Vapor Pressures of Inorganic Substances. Bureau of Mines Bull. 383, 1935, 132 pp.

80. ———. Contributions to the Data on Theoretical Metallurgy. V. Heats of Fusion of Inorganic Substances. Bureau of Mines Bull. 393, 1936, 166 pp.

81. ———. Contributions to the Data on Theoretical Metallurgy. VIII. The Thermodynamic Properties of Metal Carbides and Nitrides. Bureau of Mines Bull. 407, 1937, 66 pp.

82. ———. Contributions to the Data on Theoretical Metallurgy. X. High Tempertaure Heat-Content, Heat-Capacity, and Entropy Data for Inorganic Compounds. Bureau of Mines Bull. 476, 1949, 241 pp.

83. ———. Contributions to the Data on Theoretical Metallurgy. XI. Entropies of Inorganic Substances. Revision (1948) of Data and Methods of Calculation. Bureau of Mines Bull. 477, 1950, 147 pp.

84. ———. Contributions to the Data on Theoretical Metallurgy. XIII. High-Temperature Heat-Content, Heat-Capacity, and Entropy Data for the Elements and Inorganic Compounds. Bureau of Mines Bull. 584, 1960, 232 pp.

85. ———. Thermodynamic Properties of Zirconium Compounds. Ch. 4 in National Nuclear Energy Series 7, vol. 4, Metallurgy of Zirconium, by B. Lustman and F. Kerze, Jr. McGraw-Hill Book Co., Inc., New York, N.Y.. 1955, p. 59.

86. KELLEY, K. K., AND MAH, A. D. Metallurgical Thermochemistry of Titanium. Bureau of Mines Rept. of Investigations 5490, 1959, 48 pp.

87. KERR, E. C., JOHNSTON, H. L., AND HALLETT, N. C. Low Temperature Heat Capacities of Inorganic Solids. III. Heat Capacity of Aluminum Oxide (Synthetic Sapphire) From 19° to 300° K. Jour. Am. Chem. Soc., vol. 72, 1950, p. 4740.

88. KING, E. G. Heat Capacities at Low Temperature and Entropies at 298.15° K. of Nickelous Oxide, Cobaltous Oxide and Cobalt Spinel. Jour. Am. Chem. Soc., vol. 79, 1957, p. 2399.

89. ———. Low Temperature Heat Capacities and Entropies at 298.15° K. of Lead Sesquioxide and Red and Yellow Lead Monoxide. Jour. Am. Chem. Soc., vol. 80, 1958, p. 2400.

90. ———. Low Temperature Heat Capacities and Entropies at 298.16° K. of Manganese Sesquioxide and Niobium Pentoxide. Jour. Am. Chem. Soc., vol. 76, 1954, p. 3289.

91. KING, E. G., AND CHRISTENSEN, A. U., JR. Heat Contents Above 298.15° K. of Oxides of Cobalt and Nickel. Jour. Am. Chem. Soc., vol. 80, 1958, p. 1800.

92. KOEHLER, M. F., AND COUGHLIN, J. P. Heats of Formation of Ferrous Chloride, Ferric Chloride and Manganous Chloride. Jour. Phys. Chem., vol. 63, 1959, p. 605.

93. KUBASCHEWSKI, O., BRIZGYS, P., HUCHLER, O., JAUCH, R., AND REINARTZ, K. Heats of Fusion and Transformation of Metals. Ztschr. Elektrochem., vol. 54, 1950, p. 275.

94. KUBASCHEWSKI, O., AND EVANS, E. L. International Series of Monographs on Metal Physics and Physical Metallurgy. Metallurgical Thermochemistry. Pergamon Press, New York, N.Y., vol. 1, 3d ed., 1958, 426 pp.

95. LANDER, J. J. Experimental Heat Contents of SrO, BaO, CaO, $BaCO_3$, and $SrCO_3$ at High Temperatures. Dissociation Pressures of $BaCO_3$ and $SrCO_3$. Jour. Am. Chem. Soc., vol. 73, 1951, p. 5794.

96. MACHLAN, G. R., STUBBLEFIELD, C. T., AND EYRING L. Heats of Reaction of the Dichlorides of Samarium and Ytterbium With Hydrochloric Acid. A Microcalorimeter. Jour. Am. Chem. Soc., vol. 77, 1955, p. 2975.

97. MAH, A. D. Heats of Combustion and Formation of Two Manganese Nitrides, Mn_5N_2 and Mn_4N. Jour. Am. Chem. Soc., vol. 80, 1958, p. 2954.

98. ———. Heats of Formation of Chromium Oxide and Cadmium Oxide From Combustion Calorimetry. Jour. Am. Chem. Soc., vol. 76, 1954, p. 3363.

99. MAH, A. D., AND BOYLE, B. J. Heats of Formation of Niobium Carbide and Zirconium Carbide From Combustion Calorimetry. Jour. Am. Chem. Soc., vol. 77, 1955, p. 6512.

100. MAH, A. D., AND GELLERT, N. L. Heats of Formation of Niobium Nitride, Tantalum Nitride and Zirconium Nitride From Combustion Calorimetry. Jour. Am. Chem. Soc., vol. 78, 1956, p. 3261.

101. MASI, J. F. Heats of Vaporization of Uranium Hexafluoride. Jour. Chem. Phys., vol. 17, 1949, p. 755.

102. MITCHELL, D. W. Heat Contents and Heat of Formation of Magnesium Nitride. High-Temperature Measurements. Ind. Eng. Chem., vol. 41, 1949, p. 2027.

103. MURPHY, G. M., AND RUBIN, E. L. Thermodynamic Calculations on Some Equilibria Between Alkaline Earth Fluorides and Hydrogen Fluoride. AEC NYO-3505, July 24, 1952, 21 pp.

104. NIKOL'SKI, G. P., KAZARNOVSKAYA, L. I., BAGDASAR'YAN, Z. A., AND KAZARNOVSKII, I. A. Heat of Formation of Potassium Ozonide and Electron Affinity of the Ozone Molecule. Doklady Akad. Nauk U.S.S.R., vol. 72, 1950, p. 713.

105. Orr, R. L. High Temperature Heat Contents of Hafnium Dioxide and Hafnium Tetrachloride. Jour. Am. Chem. Soc., vol. 75, 1953, p. 1231.

106. ———. High Temperature Heat Contents of Manganese Sesquioxide and Vanadium Monoxide. Jour. Am. Chem. Soc., vol. 76, 1954, p. 857.

107. Orr, R. L. High-Temperature Heat Contents of Tantalum and Niobium Oxides. Jour. Am. Chem. Soc., vol. 75, 1953, p. 2808.

108. Osborne, D. W., Westrum, E. F., Jr., and Lohr, H. R. Heat Capacity of Uranium Tetrafluoride From 5° to 300° K. Jour. Am. Chem. Soc., vol. 77, 1955, p. 2737.

109. Parkinson, D. H., Simon F. E., and Spedding, F. H. The Atomic Heats of the Rare Earth Elements. Proc. Roy Soc. (London), vol. 207–A, 1951, p. 137.

110. Regnault, V. Untersuchungen über die spezifische Wärme einfacher und zusammengesetzter Körper. Poggendorf's Ann., vol. 53, 1841, p. 61.

111. Roberts, E. J., and Fenwick, F. Antimony-Antimony Trioxide Electrode and Its Use as a Measure of Acidity. Jour. Am. Chem. Soc., vol. 50, 1928, p. 2125.

112. Rossini, F. D., Wagman, D. D., Evans, W. H., Levine, S., and Jaffe, I. Selected Values of Chemical Thermodynamic Properties. Nat. Bureau of Standards Circ. 500, 1952, 1266 pp.

113. Schuhmann, R. Free Energy of Antimony Trioxide and the Reduction Potential of Antimony. Jour. Am. Chem. Soc., vol. 46, 1924, p. 52.

114. ———. Free Energy and Heat Content of Arsenic Trioxide and the Reduction Potential of Arsenic. Jour. Am. Chem. Soc., vol. 46, 1924, p. 1444.

115. Shomate, C. H. Heats of Formation of Manganomanganic Oxide and Manganese Dioxide. Jour. Am. Chem. Soc., vol. 65, 1943, p. 785.

116. Shomate, C. H., and Cohen, A. J. High-Temperature Heat Content and Entropy of Lithium Oxide and Lithium Hydroxide. Jour. Am. Chem. Soc., vol. 77, 1955, p. 285.

117. Shomate, C. H., and Huffman, E. H. Heats of Formation of MgO, $MgCl_2$, $MgCl_2{\cdot}H_2O$, $MgCl_2{\cdot}2H_2O$, $MgCl_2{\cdot}4H_2O$ and $MgCl_2{\cdot}6H_2O$. Jour. Am. Chem. Soc., vol. 65, 1943, p. 1625.

118. Sims, C. T., Craighead, C. M., and Jaffe, R. I. Physical and Mechanical Properties of Rhenium. Jour. Metals, vol. 7, 1955, p. 168.

119. Skinner, G. B., and Johnston, H. L. Low Temperature Heat Capacities of Inorganic Solids. VIII. Heat Capacity of Zirconium From 14° to 300°K. Jour. Am. Chem. Soc., vol. 73, 1951, p. 4549.

120. Skinner, H. A., and Smith, N. B. Heat of Hydrolysis of Boron Tribromide. Trans. Faraday Soc., vol. 51, 1955, p. 19.

121. Skochdopole, R. E., Griffel, M., and Spedding, F. H. Heat Capacity of Erbium From 15° to 320° K. Jour. Chem. Phys., vol. 23, 1955, p. 2258.

122. Smith, D., Dworkin, A. S., and Van Artsdalen, E. R. Heats of Combustion and Formation of Boron Carbide. Jour. Am. Chem. Soc., vol. 77, 1955, p. 2654.

123. Smith, W. T., Jr., Oliver, G. D., and Cobble, J. W. Thermodynamic Properties of Technetium and Rhenium Compounds. IV. Low Temperature Heat Capacity and Thermodynamics of Rhenium. Jour. Am. Chem. Soc., vol. 75, 1953, p. 5785.

124. Southard, J. C., and Shomate, C. H. Heats of Formation and High-Temperature Heat Contents of Manganous Oxide and Manganous Sulfate. High-Temperature Heat Content of Manganese. Jour. Am. Chem. Soc., vol. 64, 1942, p. 1770.

125. Spedding, F. H. Private Communications, 1956. Available for inspection at the Albany Metallurgy Research Center, Bureau of Mines, Albany, Oreg.

126. Spedding, F. H., and Daane, A. H. Production of Rare Earth Metals in Quantity Allows Testing of Physical Properties. Jour. Metals, vol. 6, 1954, p. 504.

127. Spedding, F. H., and Flynn, J. P. Thermochemistry of the Rare Earths. II. Lanthanum, Praseodymium, Samarium, Gadolinium, Erbium, Ytterbium and Yttrium. Jour. Am. Chem. Soc., vol. 76, 1959, p. 1474.

128. Spedding, F. H., and Miller, C. F. Thermochemistry of Rare Earths. I. Cerium and Neodymium. Jour. Am. Chem. Soc., vol. 74, 1952, p. 4195.

129. Stubblefield, C. T., Eick, H., and Eyring, L. Praseodymium Oxides. III. Heats of Formation of Several Oxides. Jour. Am. Chem. Soc., vol. 78, 1956, p. 3018.

130. Stull, D. R., and Sinke, G. C. Thermodynamic Properties of the Elements. Am. Chem. Soc., Washington, D.C., 1956, 233 pp.

131. Suzuki, S. Thermodynamic Studies of Cuprous Azide. Jour. Chem. Soc. (Japan), vol. 74, 1953, p. 269.

132. Todd, S. S. Heat Capacities at Low Temperatures and Entropies at 298.16° K. of Hafnium Dioxide and Hafnium Tetrachloride. Jour. Am. Chem. Soc., vol. 75, 1953, p. 3055.

133. ———. Heat Capacities at Low-Temperatures and Entropies at 298.16° K. of Sodium Superoxide, Potassium Superoxide, and Sodium Peroxide. Jour. Am. Chem. Soc., vol. 75, 1953, p. 1229.

134. ———. Heat Capacities at Low-Temperatures and Entropies of Magnesium and Calcium Fluorides. Jour. Am. Chem. Soc., vol. 71, 1949, p. 4115.

135. Todd, S. S., and Bonnickson, K. R. Low-Temperature Heat Capacities and Entropies at 298.16° K. of Ferrous Oxide, Manganous Oxide, and Vanadium Monoxide. Jour. Am. Chem. Soc., vol. 73, 1951, p. 3894.

136. Todd, S. S., and Coughlin, J. P. Low-Temperature Heat Capacity, Entropy at 298.16° K, and High-Temperature Heat Content of Ferric Chloride. Jour. Am. Chem. Soc., vol. 73, 1951, p. 4184.

137. Udy, M. C., and Boulger, F. W. Properties of Beryllium Oxide. AEC BMI–T–18, 1949, 28, pp.

138. Udy, M. C., Shaw, H. L., and Boulger, F. W. Properties of Beryllium. Nucleonics, vol. 11, May 1953, p. 52.

139. Vedeneev, A. V., Kazarnovskaya, L. I., and Kazarnovskii, I. A. Heats of Formation of Barium, Strontium, and Calcium Peroxides, and the Affinity of the Oxygen Molecule for Two Electrons. Zhur. Fiz. Khim., vol. 26, 1952, p. 1808.

140. Wacker, P. F., and Cheney, R. K. Specific Heat, Enthalpy, and Entropy of Uranyl Fluoride. Nat. Bureau of Standards Res. Jour., vol. 39, 1947, RP 1832, p. 317.

141. White, J. L., Orr, R. L., and Hultgren, R. Selected Values for the Thermodynamic Properties of Metals and Alloys. Minerals Research Laboratory Reports. Institute of Eng. Res., Berkeley, Calif., 1956 and supplements.

U.S. GOVERNMENT PRINTING OFFICE [illegible]

APPENDIX C

BIBLIOGRAPHY OF MATERIAL AND ENERGY BALANCES ON METALLURGICAL, CEMENT AND LIME PROCESSES

C.0 INTRODUCTION

The following bibliography was developed to give an indication of what is available in the published literature on material and energy balances of modern processes for the production and processing of metals, cement, and lime. The Chemical Abstracts were searched from 1960 to 1975, and, in addition, all available books on the subject were reviewed.

Brief annotations are provided which indicate whether or not the paper gives actual plant data or hypothetical calculations. Obviously, the authors cannot vouch for the accuracy of the papers themselves, nor the abstracts from which the annotations were obtained in many cases.

C.1 METALLURGICAL PROCESSES

1. Optimization of Metallurgical Processes, Proceedings of a Conference, No. 3, Y. Solntsev, (Russ).

2. Process Engineering Calculations - Material and Energy Balances, M. Tyner, The Ronald Press Company, New York, 1960.
   Energy and material balances for combustion processes are described.

3. "Fuel Economy of Furnaces", W. Trinks and M. Mawhinney, Industrial Furances, 5th Edition, 109-204, (1961).

4. "Plant Material Balance", F. Tingey, Ind. Eng. Chem. 54, No. 4, 36-41 (1962).
   A mathematical procedure is described, which is of help in making decisions on whether or not to shut down a process in which excessive product losses are suspected.

5. "Necessity of External Electrical Heating of Furnaces for the Investigation of Exothermic Processes", I. Rafalovich, Tsvetn. Metal. 36 (11), 32-4 (1963) (Russ).
   Equations and calculations are given for the determination of heat balances.

6. "Thermal Balance Measurements for Production Processes. II", M. Kwiatkowski, T. Missala, and W. Szacki, Przemysl Chem. 42 (10), 576-80 (1963).
   Effect of measurement errors on heat balances is discussed.

7. "Determination of the Gas Utilization Factor in Industrial Furnaces During Carbon Dioxide Evolution", M. Ravich, Gaz. Delo, Nauchn. - Tekhn. Sb. 1963(3), 33-6.
Energy balances, fuel efficiency, combustion.

8. "Adjustment of Measurements in Chemical Industry", A. Swenker, Proc. Intern. Meas. Conf., 3rd, Budapest 1964(1), 29-48 (Ger).
Measurement errors.

9. "New Method of Heat Calculation for Shaft Furnaces", V. Romanovskii, Tsvetn. Metal. 37(8), 41-7 (1964) (Russ).

10. "Determining Heat Losses on the Outer Surface of Electrical Furnaces", T. Schwartz and Z. Czajczynski, Przegl. Elektrotech. 1964(7), 341-3 (Pol).
A simple method for determining heat losses which eliminates the use of complicated formulas is presented.

11. "Thermal Balance Measurement of Production Processes. III. Conclusion", M. Kwiztkowski, T. Missala, and W. Szacki, Przemysl Chem. 43(4), 221-3 (1964).
Energy balance for a new plant.

12. "Factors Affecting Thermal Efficiency in Steelmaking Processes", J. Edwards, J. Inst. Fuel 38(297), 443-50 (1965).
General discussion and review.

13. "Direct Measurement of Heat Losses through the Walls of Industrial Furnaces by Fluxmeters", J. Deliere and S. Glaverbel, Silicates Ind. 30, 233-42 (1965).
The preparation of fluxmeter systems is accurately described.

14. "Use of an Actinometer for Determining Heat Losses into a Surrounding Medium", L. Krigman and I. Polosin, Gaz. Prom. 11(8), 37-8 (1966) (Russ).
Heat loss measurement.

15. "The Recovery of Heat in Industry", R. Zimmermann, Metallurgie (Paris) 98(7-8), 535-6 (1966).
A description is given of what is done and might be done in the steel and nonferrous metal industries to conserve heat.

16. "The Heat Economy of Metallurgical Processes", A. Lange, Nehezip, Musz, Egyet. Kozlem. 1967, 13, 197-216 (Hung).
A review is given of the relation of winning processes, reaction kinetics, and energy requirements in nonferrous metallurgy, together with a qualitative evaluation of various types of metallurgical furnaces.

17. "Calculation of the Yield of Serviceable Metal", A. Bigeev, P. Perchatkin, A. Millyaev, Y. Kolesnikov, S. Milyukov, N. Urtsev, and A. Lapin, Intensifikatsiya Avtomat. Martenovskogo Protessessa, Tr. Vses. Mezhvus. Konf. 1967 (Pub. 1969) 59-62 (Russ). Edited by M. Glinkov, "Metallurgiya": Moscow,

USSR.
Balances of Fe and other elements during melting are used to improve the calculation of the yield.

18. "Thermodynamics of Recuperation of Waste Heat from the Combustion Products of Industrial Furnaces", S. Cernoch, J. Iron Steel Inst., London 1969, 207 (Pt. 12), 1578-90.
Thermal work of a furnace is analyzed graphically.

19. "Ecological and Technological Aspects of Utilization of Waste Disposal Heat", C. Thomas, AIChE, Symp. Ser. 1972, 68(122). 58-62.
Evaluation of heat transfer surfaces.

20. "Recuperative Waste Heat Recovery", D. Larson, Precis. Metal 1974, 32(6), 28-30.
General quantitative discussion of energy recovery methods.

21. "Energy Efficiency in the Age of Scarcity", H. Kellogg, J. Metals 1974, 26(6), 25-9.
Process Fuel Equivalent and Material Fuel Equivalent are defined and used to compare processes.

22. Energy - Use and Conservation in the Metals Industry, ed. by Y. Chang, W. Danver, and J. Cigan, A.I.M.E., New York (1975).

C.2 METAL PROCESSING

C.2.1 CASTING

1. "Heat Flow in the Continuous Casting of Steel", J. Donaldson, R. Easton, and R. Kraus, Giessereipraxis 1966(24), 465-71 (1965) (Ger).
Experimental heat loss data.

2. "Combined Radiative and Convective Heat Loss from the Surface of a Solidifying Melt", J. Ludley and J. Szekely, J. Iron Steel Inst. 204(1), 12-15 (1966).
Theoretical calculations: heat losses to surroundings.

3. "Thermal Investigation of the Feedhead of Small Steel Ingots", G. Csabalik, Magy. Tud. Akad. Musz. Tud. Oszt. Kozlem. 37(3-4), 409-26 (1966 (Hung).
Experimental heat losses from hot-top.

4. "Change in the Heat Content in Hardened Ingots", V. Temnik, Izv. Vyssh, Ucheb. Zaved., Chern. Met. 1973, (3), 169-71 (Russ).
Heat content measured as f(Temp.).

5. "Thermal Balance of the Cylinder-Metal Mold System", W. Przybytniowski and W. Longa, Zesz. Nauk. Akad. Gorn.-Hutn., Krakow, Zesz. Spec. 1973, 44, 30-50 (Pol).
Heat balances on commercial ingots.

6. "Heat Balance and Temperature Stratification of Liquid Steel in Ladles", H. Verhoogs, S. Rosier, H. Den Hartog, A. Snoeyer, and P. Kreyger, Hoesch, Ber. Forsch. Entwickl. Unserer Werke 1974, 9(3), 114-20.
Math models for the heat balance and measurements of the steel temperature.

C.2.2 COMBUSTION CALCULATIONS

1. "Nomograph for Determining Heat Loss from Discharging Gases", E. Kleinfel'der and Y. Shchelokov, Gazov. Prom. 1969, 14(2), 35-6 (Russ).
Heat loss nomograph.

2. "Determination of the Effectiveness of Using Natural Gas in Multistage Apparatus"-G. Klimov, Gazov. Prom. 1972, (10), 43 (Russ).
Heat losses.

3. "Operational Calculations of Heat Losses with Exhaust Gases of Boiler Units Operating on Gaseous and Liquid Fuels", B. Syutkin, Tr. Tashkent. Politekh. Inst. 1972, No. 87, 117-22 (Russ). From Ref. Ah., Teploeberg, 1973, Abstr. No. 1T27.
Heat losses.

4. "Mercury Balance of a Large Pulverized Coal-Fired Furnace", C. Billings, A. Sacco, W. Matson, R. Griffin, W. Coniglio, and R. Harley, J. Air Pollut. Contr., Ass. 1973, 23(9), 773-7.
Mercury balance.

C.2.3 DRYERS

1. "Engineering Calculation of Thermal Equilibrium of Dryers and Determination of Moisture Loss in Drying", V. Kurochkin, Inzh.-Fiz. Zh. 1975, 28(1), 143-5 (Russ).
Energy and material balances, theoretical model.

2. Drying Principles and Practice, R. Keey, 6-18, Pergamon Press (1972).
Energy and material balances, example problems.

C.2.4 FORGING

1. "Optimal Conditions for Metal Heating", S. Malyi, Izv. Akad. Nauk SSR, Tekh, Kibern. 1966(5), 172-6 (Russ).
Math expressions developed for optimal conditions for heating metals.

2. "Optimal Evenness of Drop Forging", Y. Okhrimenko, Izv. Vyssh. Ucheb. Zaved., Chern. Met. 1971, 14(5), 84-8 (Russ).
Heat balance on drop forging die.

C.2.5 HEAT TREATING

1. "Thermal Operation of 5-Zone Holding Furnaces for Heating Cast Slabs", G. Shchapov. Stal' 1972, (10), 957-8.

Heat balance on plant measured.

2. "Effectiveness of the Regeneration of Heat in Heat-Treatment Furnaces with Stagewise Use of Natural Gas", V. Grigor'ev, Gazov. Prom. 1973, (4), 32-7 (Russ).
   Problems of effective utilization of heat in the effluent gases from natural-gas furnaces are discussed. Equations for calculating heat losses, efficiencies, and overall economy of the process are given.

C.2.6 MINERAL PROCESSING

1. "Advances in Mineral Processing Material Balances", R. Wiegel, Can. Met. Quart. 1972, 11(2), 413-24.
   Material balances, review.

C.2.7 SINTERING

1. "Material and Heat Balances of Sintering Agglomeration", P. Grekov, Trudy Ural. Politekh. Inst. im. S.M. Kirova No. 105, 5-17 (1960).
   Material and heat balances were determined from the gas analyses.

2. "Calculation Method for Sintering Charge", E. Vegman, Izv. Vysshikh Uchebn. Zavedenii, - Chern. Met. 7(5), 28-32 (1964).
   Heat balance and material balance equations for the sintering process.

3. "Self-Fluxing Sinters and Possibilities of Improving them by Considering Thermal Balances", F. Asensio Gonzalo and J. Boned Sopena, Inst. Hierro Acero 17(93), 467-99; (94), 533-51; (95), 619-62 (1964).
   The heat balance was made on fines of limonite.

4. "Experience in the Computation of Sintering Charges Using the 'Ural-2' Electronic Computer", E. Volkova, Kh. Zaitsev, I. Medvedev, and V. Vikhrov, Met. i Koksokhim., Mezhvedomstv. Resp. Nauchn.-Tekhn. Sb. 1965 (1), 191-200 (Russ).
   Computer program equations are based on O balance, basicity balance, Mn balance, and Fe balance.

5. "Sintering Equipment at Tarente (ITALSIDER) Steelworks. Analysis of Results Obtained form December 1965 to September 1966", M. Carignani and A. Chiaverini, Centre Doc. Siderurg., Circ. Inform. Tech. 24 (7-8), 1741-65 (1967) (Fr).
   Material balances in terms of sinter and production yields are discussed. Heat balances comparing mixtures with zero and 15% magnetite confirm the reduced need for coke as more magnetite is used.

6. "Calculation of the Composition of the Gas Phase Formed During the Combustion of Carbon in Sintering and Roasting Processes", V. Shurkhal, Izv. Vyssh. Ucheb. Zaved., Chern. Met. 1969, 12(8), 27-31.
   Heat balances taking into account incomplete combustion.

7. "Temperature-Heating System for the Sintering of Kursk Magnetic Anomaly Ores", V. Shurkhal, V. Yakubovskii, E. Nevmerzhitskii, Teor. Prakt. Avtomat, Aglom, Proizvod. 1971, 133-45 (Russ). Edited by N. Fedorovskii, Inst. Avtomat.: Kiev, USSR.
Large-scale agglomeration processes were studied. Heat input as well as output was calculated. A heat balance of the sintering process was presented.

8. "Constructing Balances of the Process of Agglomeration and Calculation of Charges by Linear Programming on a Computer", A. Pokhvisnev, E. Vegman, M. Frenkel, M. Vol'pert, and A. Grishkova, Sb., Mosk. Inst. Stali Splavov, 1971, No. 69, 3-15 (Russ).
Calculation of agglomeration of ore charges is accomplished by using material and heat-balance equations.

9. Agglomeration of Iron Ores, D. Ball, et al, American Elsevier, New York, pp. 95-97 (1973).
Material and energy balances summarized for sinter process.

C.2.8 WELDING

1. "Temperature of Aluminothermal (Thermite) Processes", A. Dubrovin and Y. Pliner, Izv. Sibirski. Otd. Akad. Nauk SSSR 1962 (12), 9-15 (Russ).
Energy balance, heat losses.

2. "Heat Balances of Submerged-Arc Welds Dependent on Some Powders", G. Becker and L. Rink, ZIS (Zentralinst. Schweisstech.) -Mitt. 6, 134-57 (1964).
Energy consumption.

3. "Heat Balance in UP (Submerged Arc) Welding with MnO-Containing Welding Powders and the Properties of the Powders", G. Becker, L. Rink, and G. Hesse, ZIS (Zentralinst. Schweisstech.)-Mitt. 7(2), 236-57 (1965) (Ger).
Energy balance.

4. "Heat Balance of Co_2- Welding", G. Becker, ZIS (Zentralinst. Schweisstech.)-Mitt. 9(7), 1088-1103 (1967) (Ger).
Energy balance.

5. "Thermal Regimes in Electroslag Welding and its Metallurgical Consequences", M. LeFevre, Arcos (Brussels) 1968, No. 157, 4205-16 (French).
Energy Balance.

C.3 METAL PRODUCTION

C.3.1 ALUMINUM

1. "Method of Calculating Energy Balances of Electrolytic Aluminum Cells", V. Krivoruchenko and B. Gulyanitskii, Legkie Metally (Leningrad) Sbornik 1957, No. 2, 9-18.
Energy balance, theory.

2. "Problems in the Electrolytic Production of Metals from Melts. The Example of the Molten-Bath Electrolysis of Aluminum", H. Ginsberg, Chem.-Ingr.-Tech. 33, 80-4 (1961).
Material and energy balances, review.

3. "Relaxation Technique for Calculating Heat Loss from an Aluminum Smelting Cell", W. Haupin, Ext. Met. Aluminum 2, 139-51 (1963) AIME.
Experimental heat losses, theoretical calculation.

4. "Discussion of Heat Balance in the Digestion Process of Bauxite", Extractive Metallurgy of Aluminum, Volume 1, Alumina, ed. by G. Gerard and P. Stroup, Interscience Publishers, pp. 71-82, (1963).

5. "Determination of the Dependence of Heat Loss of Aluminum Electrolysis Cells on the Temperature of the Melt", V. Dmitriev, Tsvetn. Metal. 38(2), 49-52 (1965).
Heat losses, review.

6. "A Material Balance for Leaching Bauxite Clinker", K. Tomko, Neue Huette 11(1), 33-7 (1966) (Ger).
Material balances.

7. "Level of Metal in Heat Transfer and its Effect on the Electrolysis of Aluminum", M. Korobov and E. Yanko, Tsvet. Metal. 1968, 41(2), 51-3 (Russ).
Energy balance.

8. "Oxidation and Other Losses of Metal during Melting of Aluminum and its Alloys in Reverberatory Furnaces", A. Radin, Tr., Mosk. Aviats. Tekhnol. Inst. 1969, No. 70, 53-71 (Russ).
Material balance, plant data.

9. "Effect of the Technological Parameters of Electrolytic Aluminum Cells on Electrolyte Temperature", A. Smorodinov, Tsvet. Metal. 1972, 45(6), 30-3.
Energy balance.

10. "External Heating of the Bottom of a Cathode as a Regulator of an Electrolytic Aluminum Cell", G. Potylitsyn, A. Bazhenov, M. Korobov, and N. Grechukhin, Tsvet. Metal. 1972, 45(2), 30-3 (Russ).
Experimental heat losses, commercial cells.

11. "Heat Losses from Electrolytic Aluminum Cells", M. Korobov and A. Smorodinov, Tsvet. Metal. 1972, 45(3), 20-3 (Russ).
Heat losses, theory and plant data.

12. "Concentration Fields of Alumina in Electrolytes of Industrial Electrolytic Cells", V. Kryukovskii, P. Polyakov, A. Tsyplakov, and V. Yagodzinskii, Tsvet. Metal. 1973, (8), 18-21 (Russ).
Material balance, source of error.

13. "Behavior of Calcium in Aluminum Electrolysis", J. Thonstad, A. Stattavik, and J. Abrahamsen, Aluminum (Duesseldorf) 1973, 49(10), 674-7.
Ca balance, laboratory and plant data.

14. "Effect of the Service Life of Electrolytic Cells on Aluminum Electrolysis Indicators", A. Dmitriev, M. Kulesh, and A. Smorodinov, Tsvet. Metal. 1973, (5), 34-8 (Russ).
Heat losses, plant data.

15. "Fluid-Bed Calcination of Alumina. New Heat-Saving Process", L. Reh, Keram. Z. 1974, 26(6), 336-7 (Ger).
Energy balance.

16. "Energy Balance of Aluminum Reduction Cell", C. McMinn, Light Metals Vol. 1, pp. 65-77 (1975), A.I.M.E.
Energy balance, theoretical model, heat loss operating data.

17. "Heat Balance and Thermal Losses in Advanced Prebaked Anode Cells", K. Yamazaki and K. Arai, Light Metals Vol. 1, pp. 193-214 (1975), A.I.M.E.
Energy balances, heat losses, experimental data.

C.3.2 ANTIMONY

1. "Oxidizing Roasting in Fluidized Bed of Antimony and Mercury-Antimony Flotation-Concentrate", A. Rozlovskii, V. Til'ga, and A. Ustinov, Tsvetn. Metal. 36(12), 34-7 (1963).
Material balance of large-scale laboratory-scale installation.

2. "Design of Industrial Cyclone Reactor by Using Data Obtained in Pilot-Plant Installation", V. Holy, Hutnicke Listy 4(19), 262-8 (1964).
Energy balances, pilot plant.

C.3.3 CALCIUM

1. "Behavior of Calcium in Aluminum Electrolysis", J. Thonstad, A. Stattavik, and J. Abrahamsen, Aluminum (Duesseldorf) 1973, 49(10), 674-7.
Ca balance, laboratory and plant data.

C.3.4 CESIUM

1. "Material and Heat Balances for Metallothermic Melts", V. Bogolyubov, Fiz.-Khim. Osnovy Met. Protsessov, Komis. po Fiz.-Khim. Osnovam Proisz. Stali, Sb. Statei 1964, 72-6.
Energy and material balances.

C.3.5 CHROMIUM

1. "Material and Heat Balances for Metallothermic Melts", V. Bogolyubov, Fiz.-Khim. Osnovy Met. Protsessov, Komis, po Fiz.-Khim. Osnovam Proisz. Stali, Sb. Statei 1964, 72-6.

Energy and material balances.

C.3.6 COBALT

1. Extractive Metallurgy of Copper, Nickel and Cobalt, ed. P. Queneau, A.I.M.E. (1961).
   Material and energy data.

2. "Reductional Electric Smelting of Converter Slags in Nickel Production", L. Pimenov and A. Zyazev, Tsvetn. Metal. 38(1), 34-6 (1965) (Russ).
   Energy and material balances, plant data.

C.3.7 COPPER

A. *Hydrometallurgy:*

1. Hydrometallurgy of Base Metals, G. Van Arsdale, ed., McGraw-Hill Book Co., Inc., pp. 167-262 (1953).
   Material balances for several plants. Discusses copper and zinc industries utilizing various processes.

2. "Process for the Recovery of Copper from Oxide Copper-Bearing Ores by Leaching, Liquid Ion Exchange and Electrowinning at Ranchers Bluebird Mine, Miami, Arizona", A. Miller, in The Design of Metal Producing Processes, ed. Kibby, Publ. A.I.M.E., New York, 1967.
   Design and plant data.

B. *Refining:*

1. "Charge Calculations", Extractive Metallurgy, J. Newton, John Wiley and Sons, pp. 383-7 (1959).
   Calculating charge to yield proper reverberatory slag.

2. "New Tough Pitch Continuous Copper Melting and Casting Unit at Asarco's Perth Amboy Plant", G. Storm and J. Stone, Trans. A.I.M.E. 218, 584-91 (1960).
   Design features and operating methods.

3. "Mathematical Model for Copper Converter Control", J. Foreman, International Symposium on Statistics, Operations Research and Computers, Colorado School of Mines, 431-9 (1964).
   Fundamental equations, process material and energy measurments.

4. "Converting of High-Grade Copper Matte", J. Bystron and M. Malusecki, Rudy Metale Niezelazne 10(6), 314-18 (1965) (Pol).
   Energy balances, theoretical model, plant data.

5. "The Arizona Chemcopper Plant - A Chemical Copper Refinery", W. Yurko, Met. Soc. Conf. 49, 55-91 (1966) (Publ. 1968).
   Operating data.

C. *Smelting:*

1. "Copper Matte Smelting: the Influence of Charge Composition upon Heat Requirement", G. Evans, Bull. Inst. Mining Met. No. 639, 201-9 (1960).
   Energy balance, data estimates.

2. "Cyclone Smelting of Copper-Zinc Concentrates", P. Myasnikov, A. Okunev, and D. Lutokhin, Trudy Inst. Energet., Akad. Nauk Kazakh. S.S.R. 2, 274-84 (1960).
   Energy and material balances, review.

3. "Some Physicochemical Properties of Charges and Products of Electric Smelting of Dzhezkazgan Copper Concentrates", V. Shchurovskii, V. Vladimirov, G. Gnatyshenko, A. Kurochkin, Y. Shchurovskii, N. Adson, and V. Golovko, Izvest. Akad. Nauk Kazakh. S.S.R., Ser. Met., Obogashchen. i Ogneuporov, 1961, No. 1, 8-13.
   Energy balance, experimental data.

4. "Electric Smelting of Sulphide Ores", Extractive Metallurgy of Copper, Nickel and Cobalt, ed. by P. Queneau, A.I.M.E. 252-3 (1961).
   Energy balance from plant data.

5. "The Smelting of Copper and Nickel Sulfide Ores and Concentrates in the Electric Furnace", O. Barth, Freiberger Forschungsh. B67, 35-55 (1962).
   Energy balances.

6. "Stoichiometric and Heat Calculations in the Oxygen Smelting of Copper Sulfide Concentrates", F. Egorov, Y. Bykhovskii, and L. Bochkarev, Tsvetn. Metal. 36(10), 30-4 (1963).
   Material and energy balances, theory.

7. "Design of Industrial Cyclone Reactor by Using Data Obtained in Pilot-Plant Installation", V. Holy, Hutnicke Listy 4(19), 262-8 (1964).
   Energy balances, pilot plant.

8. "Material and Heat Balances of Electric Smelting of Copper-Nickel-Sulfide Ores in a 30,000 kv.-amp. Electric Furnace", V. Brovkin, M. Palysaev, Y. Slobodin, and M. Chetvertkov, Tsvetn. Metal. 38(12), 34-40 (1965) (Russ).
   Energy and material balances, plant data.

9. "Copper Blast-Furnace Practice at Union Miniere de Haut-Katanga", B. Claus and A. Guebels, Met. Soc. Conf. 39, 93-114 (1965) (Publ. 1967).
   Operating data.

10. "The Development of Flash Smelting Process at Ashico Copper Smelter, Furukawa Mining Co., Ltd.", T. Okazoe, T. Kato, and K. Murao, Met. Soc. Conf. 39, 175-95 (1965) (Publ. 1967).
   Energy and material balances, operating data.

11. "Mathematical Model of a Metallurgical Conversion Process in a Copper Smelter", I. Rabinovich, Tsvet. Metal. 40(3), 46-51 (1967).
Material balances, theoretical model.

12. "Material and Heat Balances of Cu-S Smelting with Air and with an Oxygen-Enriched Blast", R. Fel'man, Sb. Nauch. Tr., Gos. Nauch.-Issled. Inst. Tsvet. Metal. 1969, No. 29, 66-79.
Material and energy balances.

13. "Technical Thermal Parameters of the Reverberatory Process Using Oxygen-Enriched Air", Y. Kupryakov, V. Chakhotin, N. Artem'ev, V. Galaktionov, and V. Makarov, Tsvet. Metal. 1969, 42(6), 32-5 (Russ).
Energy balance, operating data.

14. "Operations at Kennecott's Utah Copper Division Smelter Reverberatory Department", R. J. Anderson in Copper Metallurgy, ed. Ehrlich, publ. A.I.M.E., New York, 1970.
Complete material and energy balance on reverberatory furnace.

15. "Physical Chemistry of 'One-Step' Copper Production from a Chalcopyrite Concentrate", J. Jeffes and C. Diaz, Inst. Mining Met., Trans., Sect. C 1971, 80(March), 1-6.
Energy balances, review.

16. The Future of Copper Pyrometallurgy, C. Diaz, Chilean Inst. of Mining Eng. (1973).
Operating data.

17. "Modern Flash Smelting Technology", A. Lauria and B. Anderson, 4th BNF Int. Conf., BNF Metal Technology Center (1974).

C.3.8 FERROALLOYS

A. *General:*

1. Production of Ferroalloys, 2nd Edition, V. Elyutin, Y. Pavlov, B. Levin, and E. Alekseev, State Scientific and Tech. Publ. House for Literature on Ferrous and Nonferrous Metallurgy, Moscow, 1957: Publ. for NSF by Israel Program for Scientific Translations, 1961; available from OTS, Dept. of Commerce, Washington, D.C.
Contains practical data on material and energy balances for submerged arc smelting of ferrosilicon, pp. 65-73; ferromanganese, pp. 121-125; ferrochromium, pp. 185-190, 208-212; ferrotungsten, pp. 238-244; ferromolybdenum, pp. 273-280; vanadium (by aluminothermic method), pp. 307-310; (by silicothermic method), pp. 310-318; pig iron, pp. 393-404.

2. Physical Chemistry of Production or Use of Alloy Additions, ed. J. Farrell, Proc. of Conf., A.I.M.E., 1974.
Many papers on ferroalloy production and utilization, including energy balances and heat effects of additions.

B. *Ferroboron:*

1. "Material Balance of High-Grade Ferroboron Production", N. Chirkov, Sb. Tr. Chelyabinsk. Elektromet. Komb. 1970, No. 2, 150-4 from Ref. Zh., Met. 1971, Abstr. No. 8V273.
   Plant scale study.

C. *Ferrochromium:*

1. "Balance of Basic Elements in Smelting of Carbon Ferrochromium", Kh. Kadarmetov, Teoriya i Prakt. Met., Chelyabinsk, Sb. 1963(6), 83-6.
   Material balance in a 7500-kv.-amp. electric furnace.

2. "Material and Energy Balances of Ferrochromium Smelting in a Large Furnace", V. Shmel'kov, Ya. Shchedrovitskii, Kh. Kadarmetov, O. Brikova, Yu. Shiryaev, N. Agarkova, R. Kravchinskii, and V. Tambovtsev, Stal' 24(12), 1094-6 (1964) (Russ).
   Operating data.

3. "Mass and Energy Balances During Smelting of a High-Carbon Ferrochromium in a Closed (Electric) Furnace", Kh. Kadarmetov, Stal' 27(8), 707-10 (1967).
   Comprehensive data on plant.

4. "Thermal Operation of Rotary Drum-Type Furnaces for Roasting Chromite Charges", V. Sherman, R. Sukhanova, Tr. Ural. Mauch.-Issled. Khim. Inst. 1967, No. 16, 175-85 (Russ).
   Data on the mineral and heat balances on the furnace are given.

5. "Sanitary-Engineering Characteristics of Closed Ferroalloy Furnaces", P. Alikin, B. Velichkovskii, and L. Pokrovskaya, Vop. Gig. Prof. Patol. Tsvet. Chern. Met. 1971, 207-13 (Russ). Edited by S. Miller, S. V. Sverdlovsk. Nauch.-Issled. Inst. Gig. Tr. Profzabol.: Sverdlovsk. USSR.
   Effects of furnace hoods on heat losses; quantitative.

6. "Material and Heat Balance in Furnaces During Melting of Refined Ferrochromium", T. Petrova, V. Vasil'ev, L. D'yakonova, and V. Zaiko, Sb. Tr. Chelyabinsk, Elektromet. Komb. 1972, No. 3, 38-43 (Russ). From Ref. Zn., Met. 1972, Abstr. No. 10V241.
   Plant scale study. Complete balances.

7. Electric Smelting Processes, A. Robiette, John Wiley & Sons, New York (1973) pp. 160-161.
   Material balance.

D. *Ferromanganese:*

1. Electric Smelting Processes, A. Robiette, John Wiley & Sons, New York (1973) p. 139.
   Material balance.

E. *Ferromolybdenum:*

1. "Material and Thermal Equilibria of the Alumino-Silicothermic Smelting of Ferromolybdenum", K. Rispel and L. Klokotina, Izv. Akad. Nauk SSSR, Metally 1966(3), 34-41 (Russ).
   Thermal equilibrium determined by measuring actual temperatures of the smelting process.

F. *Ferronickel:*

1. "Utilization of Nickeliferous Serpentine", Extractive Metallurgy of Copper, Nickel and Cobalt, ed. P. Queneau, A.I.M.E. (1961) 302-7.
   Material and energy data from test.

G. *Ferrosilicochromium:*

1. "Material and Thermal Balances of a Heavy-Duty Furnace During Smelting of Ferrosilicochromium", G. Belyaev, Stal' 1968, 28(4), 335-7.
   Plant scale study; detailed material and energy balances.

H. *Ferrosilicon:*

1. "Material and Energy Balances in Making 75% Ferrosilicon in a Closed Furnace", Ya. Shchedrovitskii, Stal' 27(11), 2112-14 (1967).
   Plant scale study; material and energy balances.

2. "Sanitary-Engineering Characteristics of Closed Ferroalloy Furnaces", P. Alikin, B. Velichkovskii, and L. Pokrovskaya, Vop. Gig. Prof. Patol. Tsvet. Chern. Met. 1971, 207-13 (Russ). Edited by Miller, S. V. Sverdlovsk. Nauch.-Issled. Inst. Gig. Tr. Profzabol.: Sverdlovsk. USSR.
   Effects of furnace hoods on heat losses; quantitative.

3. "Energy Balance of a Furnace During Ferrosilicon Melting", B. Semenovich, Stal' 1972, (1), 51-3.
   Plant scale study.

4. "Material Balances of the Melting of Different Brands of Ferrosilicon", I. Volkova, Stal' 1973, (12), 1096-8.
   Plant scale study; complete material and energy balances.

5. Electric Smelting Processes, A. Robiette, John Wiley & Sons, New York (1973) p. 116.
   Material balance for silicon alloys.

I. *Ferrosiliconcalcium:*

1. "Material and Heat Balances of a Furnace for Smelting Ferrosilicocalcium", G. Belyaev, R. Kravchinskii, M. Ryss, A. Mikulinskii, G. Kozhevnikov, V. Zaiko, A. Pereverzev, V. Vasil'ev, and L. D'yakonova, Tr. Inst. Met., Sverdlovsk 1970 No. 19, 86-93 (Russ).
   Plant scale study; complete material and energy balances.

2. "Material and Heat Balance in Furnaces for Melting Calcium Ferrosilicon", M. Ryss, R. Kravchinskii, A. Colev, V. Zaiko, V. Vasil'ev, E. Popova, G. Belyaev, V. Gusarov, and S. Pigasov, Sb. Tr., Chelyabinski. Elektromet. Komb. 1972, No. 3, 66-73 (Russ). From Ref. Zh., Met. 1972, Abstr. No. 10V251.
Plant scale study; complete material and energy balance.

J. *Ferrosiliconzirconium:*

1. "Thermal Calculations of the Aluminothermic Smelting of Ferrosilicozirconium", Yu. Pliner, Sb. Tr. Klyuchevsk. Zavoda Ferrosplavov No. 2, 72-8 (1965). Rev. Zh., Met., v. 1966, Abstract No. 6V154.
Plant scale calculations.

K. *Ferrotungsten:*

1. "Material and Heat Balance in Furnaces for Melting Ferro-Tungsten", V. Zaiko, Sb. Tr., Chelyabinsk. Elecktromet. Komb. 1972, No. 3, 93-100. From Ref. Zh., Met. 1972, Abstr. No. 10V246.
Plant scale study; effects of furnace charges on heat losses reported.

L. *Silicomanganese:*

1. "Sanitary-Engineering Characteristics of Closed Ferroalloy Furnaces", P. Alikin, B. Velichkovskii, and L. Pokrovskaya, Vop. Gig. Prof. Patol. Tsvet. Chern. Met. 1971, 207-13 (Russ). Edited by Miller, S. V. Sverdlovsk. Nauch.-Issled. Inst. Gig. Tr. Profzabol.: Sverdlovsk. USSR.
Effects of furnace hoods on heat losses; quantitative.

C.3.9 IRON, CAST

1. "Developments in the Melting of Metals for Foundries", The British Foundryman, March, 1961, p. 103.
Detailed material and energy balances of cupola.

2. "Material Balance and Heat Balance of the Cold-Blast Cupola", W. Patterson, H. Siepmann, and H. Pacyna, Intern. Giesserei-Kongr. Vortr., 28, Vienna 1961, 29-52.
Plant data. Material and heat balances.

3. Fundamentals of Metal Casting, R. Flinn, Addison-Wesley, Reading, Mass., pp. 288-293 (1963).
Example of mass balance for cupola charge.

4. "Calculation of Metal, Coke, and Gas Temperature Distributions in the Cupola Furnace", J. Breen, Australasian Inst. Mining Met. Proc. No. 208, 25-42 (1963).
Experimental results on small cupola.

5. "Thermochemical Model for Computer Prediction of Cupola Performance", R. Pehlke, Mod. Castings 44(5), 580-7 (1963).
A thermochemical model.

6. "Possibility of Dissolving Magnesium in Cast Iron", V. Bedarev and A. Khrapov, Izv. Vysshikh Uchebn. Zavedenii, Chern. Met. 6(12), 171-6 (1963).
Plant study of thermal effect of Mg addition.

7. "Thermochemical Model of a Basic Hot-Blast Cupola", R. Shields, K. Roessing, and H. Bishop, Met. Soc. Conf. 32, 167-95 (1964) (Pub. 1966).
Energy and material balances, operating data.

8. "Cupola Performance: the Potential for Oil Injection", R. Pehlke, Mod. Castings, New Technol. Sect. 48(1), 59-63 (1965).
A thermochemical model.

9. "Computer Prediction of Cupola Performance Utilizing O and Natural Gas in the Blast", R. Pehlke, Mod. Castings, New Technol. Sect. 47(1), 806-12 (1965).
A thermochemical model.

10. "The Performance of a Cupola Furnace", L. Bos, Metalen 20(4), 96-104 (1965) (Dut).
Detailed heat and mass balances of the melting tests are given.

11. "Characteristics of the Cocombustion of Two Types of Fuel in Cupolas", A. Poplavskii, Tr. Vses. Nauch.-Issled. Inst. Zheleznodorozh. Transp. 1969, No. 376, 118-27 (Russ).
Energy balance, operating data.

12. "When is a Cubic Foot Not a Cubic Foot?" F. Ekman, Mod. Cast., 1970, 57(4), 33-4, 36.
Heat balances are calculated for blast air, including dust loadings.

13. "Effect of a Blast Temperature of 500° on the Melting Product, Heat Balance, and Some Processes in a Cupola", W. Patterson and B. Koehler, Giessereiforschung 1971, 23(3), 109-22(Ger).
Experimental data.

14. "Material and Heat Balances of Blowing Vanadium Cast Iron by Oxygen with Fuel Additions", V. Zonov, Tr. Ural. Politekn. Inst. 1971, No. 202, 19-22 (Russ). From Ref. Zh. Met. 1972, Abstr. No. 4V369.
Material and heat balances of melts with flame blowing of cast iron compared with those for melts made in 100-ton converters.

15. "Partial Substitution of Coke by Natural Gas or Liquified Petroleum Gases in Cast Iron Smelting Cupolas", J. Leveque, J. Pouriel, F. Rey, and G. Ulmer, World Gas Conf., (Proc.), 12th, 1973 (Pub. 1973). IGU/E 32-73, 20 pp. (Fr). Int. Gas Union: London, Engl.
Operating data on energy requirements.

16. "Petroleum Coke in Cupola", P. Mukherjee and S. Pinge, Met. Eng. - IIT (Indian Inst. Technol.) Bombay 1972-1973 (Pub. 1973). 4, 28-34.
Material and heat balances are calculated.

17. "Calculation of Heat Exchange in a Radiation Preheater", A. Ardeleanu, I. Sporea, and H. Gutmayer, Metalurgia (Bucharest) 1973, 25(6), 315-18 (Rom).
Heat-transfer and hydrodynamic design calculations are given for a cupola furnace air preheater.

18. "Heat Balance of a Cupola Using Anthracite", C. Kang. Kumsok Hakhoe Chi 1974, 12(1), 21-6 (Korean).
Heat balance calculated with practical operating data.

C.3.10 IRON, PIG

A. *General:*

1. "Energy Requirements for Iron-Ore Reduction", R. Wild, J. Inst. Fuel 34, 381-92 (1961).
The basic ore-reduction reactions are considered with a view toward methods to supply the energy deficit when carbonatious reducing agents are used.

2. "Preliminary Reduction of Fe Sand in a Rotary Furnace. V. Heat Balance in the Reduction Furnace", H. Arakawa, Tetsue to Hagane 51, 2301-9 (1965).

B. *Blast Furnace:*

1. "Metallurgical Evaluation of Iron Ores of Eastern Siberia", A. Pokhvisnev, Razvitie Proizvoditel. Sil Vostock Sibiri, Chernaya Met., Trudy Konf. 1958, 46-53 (Pub. 1960).
Relates ore composition to coke requirements.

2. "Charge Calculations", Extractive Metallurgy, J. Newton, John Wiley & Sons (1959), 383-5, 387-8.
Calculating blast furnace burden, given compositions.
Heat balance sheet of an iron blast furnace.

3. "Material and Heat Balances of Blast-Furnace Operation with a Blow of Reducing Gases", A. Ramm, Trudy Leningrad. Politekh. Inst. im. M. I. Kalinina, No. 212, 24-39 (1960).
Detailed equations are given for setting up the material and heat balances of blast furnace operations with mixed blow.

4. "Heat Transfer in the Blast Furnace", N. Makhanek, Trudy Ural. Politekh. Inst. im. S. M. Kirova No. 105, 78-89 (1960).
Theoretical energy balances by zones.

5. "Oxygen-Enriched Blast Furnace Operation of Higashida No. 5 Blast Furnace", K. Tsujihata, K. Kodama, N. Tsuboi, K. Kato, and S. Hashimoto, Tetsu to Hagane 46, 955-61 (1960).

Operating data of the furnace, with an O-enriched blast were analyzed. A set of material balances and heat balances was worked out.

6. "Predicting Effects of Oxygen, Moisture, and Fuel Additions on Blast-Furnace Operation with Electronic Computers", A. Hodge, Yearbook Am. Iron Steel Inst. 1960, 75-106.
   Theoretical heat and material balances.

7. "Carbon and Heat Consumption in a Blast Furnace. I.", T. Yatsuzaka, J. Sawamura, S. Ota, and T. Fukuda, Tetsu to Hagane 46, 643-52 (1960).
   C consumption and heat distribution in each part of a blast furnace were examined. The results were applied to blast furnace operation.

8. "Carbon and Heat Consumption in the Blast Furnaces. II. Application of Results", T. Yatsuzaka, J. Sawamura, S. Ota, and T. Fukuda, Tetsu tc Hagane 46, 741-7 (1960).
   Heat required for control of the Si in pig iron was calculated. Formulas to show the blast required and the blast temperature differential accompanying a 100-kg. increase or decrease of charged ore were established.

9. "Thermochemical Calculations of the Blast-Furnace Process", R. Linder, Jernkontorets Ann. 144, 859-967 (1960).
   Theoretical staged heat and material balances used to evaluate blast furnace operations.

10. "Effect of Natural Gas and Oxygen Added During Blowing on the Size of the Combustion Zone and the Temperature of the Hearth of a Forced-Draft Blast Furnace", Yu. Borisov and L. Tsylev, Izvest. Akad. Nauk S.S.S.R., Otdel. Tekh. Nauk, Met. i Toplivo 1960, No. 1, 9-20.
   Analysis of material and thermal balances in the plant working with a blast containing air, O, and natural gas.

11. "Methods of Making Up Material and Heat Balances for Blast-Furnace Smelting", A. Shur, Stal', Sb. Statei 1961, 13-23.
   To improve the accuracy of blast-furnace heat and material balances, an analysis of possible sources of error is made on the basis of equations given by Ramm (Material'nyi i Teplobye Balansy Domennoi Plavki, Leningrad. Politekhn. Inst. 1950.

12. "Thermal Balance of the Iron Blast Furnace", H. Bell, J. Iron Steel Inst. (London) 199, 285-7 (1961).

13. "Thermal Aspects of Blast Furnace Fuel Injection", A. Decker, J. Metals 13, 41-4 (1961).
   Heat balances from low-shaft furnace operations.

14. "Relations Among Heat and Material Balances and Reduction for Different Operating Conditions", E. Schuermann and D. Buelter, Journees Intern. Siderurgie, 3e, Luxemburg 1962, 89-112 (Ger).

15. "Fuel-Oil Injection: Comparison of Actual Performance with Predictions Made by Using a Mathematical Model", R. Trense and D. Rosborough, Iron Steel Inst. (London) Spec. Rept. No. 72, 53-8 (1962).
Mathematical model.

16. "A Critical Analysis of Methods for Blast-Furnace Heat Balance Investigations", N. Makhanek, Forsirovanie Domennoi Plavki, Tr. Nauchn. Konf po Tecr. Vopr, Met. Chuguna, Dnepropetrovsk 1961, 159-68 (Pub. 1963).
Evaluation of heat loss and consumption into heat balances.

17. "Blast-Furnace Heat Balance in Stages: Development of a Computer Program", J. Ridgion, J. Iron Steel Inst. (London) 200, 389-94 (1962).
Theoretical model.

18. "Thermotechnical Characteristics of Coke-Oven and Blast-Furnace Gases", G. Nikolaenko, Tr. Vses. Zaochn. Energ. Inst. No. 20, 76-82 (1962) (Russ).
Simplified formulas are given for the calculation of thermal losses associated with the combustion of coke-oven and blast-furnace gases.

19. "The Application of a Digital Computer to the Calculation of Material and Heat Balances in the Blast Furnace", R. Sevrin, J. Iron Steel Inst. (London) 200, 34-6 (1962).

20. "Calculation of the Heat Balance of Blast-Furnace Smelting", N. Makhanek, Izv. Vysshikh Uchebn. Zavedenii, Chernaya Met. 5, No. 11, 30-6 (1962).
A new way of calculating the heat balance of blast-furnace smelting is suggested, in which the actual amounts of heat according to the reactions which occur in the furnace are considered.

21. "The Significance of Exergy for the Thermodynamic Study of the Blast-Furnace Process", H. Brauer and R. Jeschar, Arch. Eisenhuettenw. 34, 9-16 (1963).
Calculated energy balances on thirty furnaces.

22. "Automation of Blast Furnaces: Thermal Control by Composition of Waste Gases", Ch. Thibaut, Centre Doc. Siderrurg., Circ. Inform. Tech. 21(12), 2633-9 (1964) (Fr).
Data for an 11-month run were used.

23. "Control of Heat Balance of a Blast Furnace", J. Czernek, Hutnicke Listy 19(11), 769-74 (1964) (Czech).
The optimum heat balance of the hearth was determined.

24. "Fuel Injection Blast Furnaces: Sharon Steel Corporation", J. Walsh, Reg. Tech. Meetings Am. Iron Steel Inst. 1964, 95-109.
Tables show the material balance and operating data.

25. "Determination of Material and Heat Transfer in the Blast Furnace in Two States", B. Marincek, Arch. Eisenhuettenw. 35 (11), 1029-38 (1964).
Coke consumption and mass and heat transfer in a modern blast furnace were determined mathematically.

26. "Significance of Heat Balance in Blast-Furnace Operation. I. Two-Zone Heat Balance for Blast Furnace Operation with Hot Blast", B. Gerstenberg and T. Kootz, Stahl Eisen 84(18), 1105-20 (1974).
The two-zone heat balances of two domestic blast furnaces were compared with two foreign furnaces.

27. "Distribution of Sulfur in a Blast Furnace During Operation with a High-Iron Charge", Z. Nekrasov, G. Volovik, and V. Pokryshkin, Izv. Vysshikh Uchebn. Zavedenii, Chern. Met. 7(2), 26-33 (1974).
Plant data sulfur balance.

28. "Determination of Oxygen Removal Rate by Heat Balance and Waste Gas Analysis in the Reduction of Iron Ore and Sinter", E. Schuermann, J. Willems, and G. Sommer, Arch.Eisenhuettenw. 35(3), 169-71 (1974).
Calculations showing use of heat balance.

29. "Calculation of Coke Consumption in the Blast Furnace Based on Two-Stage Heat, Material, and Reduction Balances. I. Mathematical Relations of Material and Heat Balances of the Blast Furnace", E. Schuermann and D. Buelter, Arch. Eisenhuettenw. 35(6), 475-83 (1964): "II. Effect of Lime on the Reduction Process and Coke Consumption of the Blast Furnace", Ibid. 484-6.
Equations are presented for calculating coke and heat requirements and for calculating the effect of CaO and humidity of the blast on coke consumption.

30. "Material and Heat Balances of the Cherepovets Plant Blast Furnaces", A. Shur, L. Byalyi, and P. Rusakov, Stal' 25(4), 301-6 (1965) (Russ).
Plant data shows a substantial discrepancy in heat balance.

31. "Study of the Hearth of the Blast Furnace No. 1 of the Wendel Works in Hayange Saint-Jacques", S. Sayegh and R. Baro, Rev. Met. (Paris) 62(12), 1181-6 (1965) (Fr).
Heat balance of the hearth of the blast furnace is given.

32. "Study of Hot Stoves of a 2000-m^3 Blast Furnace", I. Vas'ko, E. Gol'dfarb, M. Kruskal, P. Netrebko, G. Rabinovich, N. Taits, and I. Fainshtein, Stal' 26(9), 781-5 (1966) (Russ).
Pressure and temperature measurements made it possible to determine thermal balances and fuel efficiency.

33. "Theory of Fuel Economy with Air Heating", A. Evdokimenko, Tsvetn. Metal. 39(1), 33-7 (1966) (Russ).
Mathematical analysis of the heat balance of a blast furnace showed that preheating the blast results in heat saving.

34. "Test Carried out on B-F 5 at the Rhenon (Providence) Works", I. Renaurt, Rev. Met. (Paris) 63(10), 765-77 (1966).
Plant data. Heat and material balance.

35. "Thermal Behavior of a Group of Hot Blast Chambers for Highest Blast Temperatures in 2- and 3-Chamber Operation", E. Hofmann, Stahl Eisen 86(24), 1594-1601 (1966).
Heat balances measured and calculated.

36. "A Dual Graphic Representation of the B-F Mass and Heat Balances", A. Rist and N. Meysson, J. Metals 19(4), 50-9 (1967).
Mathematical model using mass and heat balances.

37. "Heat Balance for Ironmaking in Experimental Low-Shaft Furnace", S. Prasad and A. Chatterjea, NML (Nat. Met. Lab., Jamshedpur, India), Tech. J. 1969, 11(4), 68-82.
Comparison of the heat balances of the conventional blast furnace and of a low-shaft furnace, based on the operational parameters.

38. "Thermal Operation of the Lower Part of a Blast Furnace", M. Laverent'ev, Stal' 1969, 29(2), 105-10 (Russ).
Theoretical analysis of published data.

39. "Chronological Sequence of the Cooling Losses in the Various Zones of a Blast Furnace", F. Hillnhuetter, H. Kister, and U. Pueckoff, Stahl Eisen 1969, 89(23), 1306-9 (Ger).
Operating data on heat losses to cooling water.

40. "Importance of Thermodynamic Data for Solving Problems in Blast-Furnace Production", M. Stefanovich and S. Sibagatullin, Zh. Fiz. Khim. 1970, 44(1), 213-15 (Russ).
Some thermodynamic calculations are discussed, knowledge of which is very important for the theoretical development of pig-iron production in blast furnaces: errors of 5-20% can be made in the heat balance of the furnace zone, where the temperature reaches $\leqq 850°$, by using the rule of additivity of the specific heats, because the dependence of the specific heats on the chemical state and on the process conditions is not yet fully known, and the values of the specific heats from the literature are very different.

41. "Material and Thermal Balances of Blast-Furnace Smelting with Blowing of Pulverized Coal Fuel", A. Akberdin and B. Plastinin, Tr. Khim. - Met. Inst., Akad. Nauk Kaz. SSR 1972, No. 13, 91-100 (Russ). From Ref. Zh., Met. 1972, Abstr. No. 11V123.
From the heat and material balances of a blast furnace, the degree of the gas consumption and the degree of indirect reduction were improved by blowing with coal.

42. "Heat Losses and Thermal Operation of Blast Furances", I. Semikin, G. Tsygankov, A. Borodulin, I. Vas'ko, M. Kruskal, and V. Mirshavka, Izv. Vyssh. Ucheb. Zaved., Chern. Met. 1972, (8), 159-63 (Russ).

Plant study of heat losses to cooling water. Complete heat balance.

43. "Measurement of the Efficiency of Preheating of Air Blown in Blast Furnaces", P. Sanna, Cent. Doc. Siderurg., Circ. Inform. Tech. 1973, 30(7-8), 1731-45.
    Thermal balance of a Cowper stove.

44. "Influence of Thermodynamic Factors on the Heat Balance in the Bosh and the Hearth of a Blast Furnace", R. Benesch and P. Mandelka, Hutnik 1973, 40(11), 492-7 (Pol).
    Operating data analyzed.

45. "Application of the Material and Energy Balances for Determining the Effect of Blast Temperature on the Heat Characteristics of the Blast-Furnace Process", J. Szargut and A. Ziebik, Arch. Hutn. 1974, 19(4), 395-414 (Pol).
    Theoretical material and energy balances.

C. *Electric Smelter:*

1. Iron Ore Reduction, Proceedings of a Symposium of the Electrothermics and Metallurgy Division of the Electrochemical Society, held in Chicago, 3-5 May, 1960, ed. R. R. Rogers, Pergamon Press Ltd., (1962) p. 107, 248.
    Energy balance of 20 MW electric furnace. Power requirements for Strategic-Udy process compared with conventional.

2. "Balance of Energy in Electric Reduction Furnaces", K. Fritzsche and R. Sroka, Intern. Electrowaerme-Kongr., 5, Wiesbaden, Ger, 1963 (106), 4 pp.
    The difference between the heat balance and electro-heat balance in electric reduction furnaces is shown.

3. "Thermal Balance in an Electric Reduction Furnace for the Production of Pig Iron", E. Scharle, Met. ABM (Ass. Brasil. Metais) 1969, 25(140), 523-31.
    Theoretical calculations.

4. Electric Smelting Processes, A. Robiette, John Wiley & Sons, New York (1973) p. 94.
    Heat balance of electric pig iron furnace with kiln pretreatment.

D. *Krupp-Renn:*

1. "Reduction of Iron Ore in the Rotary Furnace (Krupp Sponge Iron Process)", F. Lucke, H. Serbert, and G. Meyer, Stahl Eisen 82, 1222-32 (1962).
    The material and heat balances of the Krupp process are discussed.

E. *Pyrite Roasting:*

1. "The Stürzelberger Process", F. Giménez Blasco, Inst. hierro y acero 13, 242-5 (1960).

Heat balance is given.

2. "Fluid Bed Roasting - Principles and Practice", Extractive Metallurgy of Copper, Nickel and Cobalt, ed. P. Queneau, A.I.M.E. (1961) 15-21.
   Operating data for material balances.

3. "Design of Industrial Cyclone Reactor by Using Data Obtained in Pilot-Plant Installation", V. Holy, Hutnicke Listy 4 (19), 262-8 (1964).
   Energy balances, pilot plant.

4. "Heat Utilization from the Roasting of Pyrite in DKSM Furnaces (Furnace-Boilers with Two Double-Decked Fluidized Beds)", Ya. Korenberg, A. Ternovskaya, B. Vasil'ev, V. Anurov, and E. Shipov, Khim. Prom. 1968, 44 (6), 466-8 (Russ).

C.3.11 LEAD

1. "Comparison of Technological Factors Affecting Balances of Heat and Materials in Shaft Smelting of Lead Agglomerates", I. Reznik, Byull. Tsvetnoi Met. 1957, No. 22, 15-23.
   Energy and material balances, plant data.

2. "Certain Characteristics of Lead Refining in Sulfamic Electrolyte. II. Prevention of Decrease of Lead Content in Electrolyte", V. Chernenko and M. Loshkarev, Trudy Dnepropetrovsk. Khim. - Tekhnol. Inst. 1958, No. 6, 202-7.
   Material balance.

3. "Coke Consumption in Lead Blast Furnaces", F. Johannsen and G. Waechter, Z. Erzbergbau u. Metallhüttenw. 14, No. 2, 53-63 (1961).
   Heat balance, experimental design.

4. "Testing on Commercial Scale of Processing Altai Polymetallic Ores Following the Flow Sheet of Magnitogorsk Combine", I. Tsygoda, V. Kazakov, N. Kolesnikov, N. Bryukhanov, A. Burba, V. Sadykov, and A. Pigarev, Tsvetn. Metal. 36(12), 12-15 (1963).
   Material balances, plant data.

5. "Heat Balance of a Short Rotary Furnace for the Production of Pb.", G. Pungartnik, D. Pavko, and D. Kozelj, Rudarsko-Met. Zb. 1965 (2), 195-207 (Slovenian).
   Energy and material balances, plant data.

6. "Lead Smelting Improvements at La Oroya", L. Harris, Met. Soc. Conf. 39, 197-223 (1965) (Pub. 1967).
   Energy and material balances, operating data.

7. "The Boliden Lead Process", H. Elvander, Met. Soc. Conf. 39, 225-45 (1965) (Pub. 1967).
   Energy and material balances, operating data.

8. "Flash Smelting of Lead Concentrates", P. Bryk, R. Malmstrom, and E. Nyholm, J. Metals 18 (12), 1298-302 (1966).
Heat balance.

9. "Simultaneous Reduction of Metal Oxides by Gases", W. Ptak and M. Sukiennik, Zesz. Mauk. Akad. Gorn. - Hutn., Cracow, Met Odlew. 1973, No. 50, 17-30 (Pol).
Energy and mass balances, theoretical model.

10. "Heat Conservation in Zinc and Lead Extraction and Refining", S. Woods, Metals & Materials 1974, 8(3), 187-9.
Suggestions for conserving energy.

11. "Design of Industrial Cyclone Reactor by Using Data Obtained in Pilot-Plant Installation", V. Holy, Hutnicke Listy 4(19), 262-8 (1964).
Energy balances, pilot plant.

C.3.12 MAGNESIUM

1. "The Loss of Magnesium in its Electrolytic Preparation", E. Zhemchuzhina and A. Belyaev, Nauch. Doklayd Vysshei Shkoly, Met. 1959, No. 2, 61-4.
Material balance.

2. "Energy Balance of an Experimental Industrial Electrolysis Cell used in the Triple-Layer Refining of Mg Alloy Scrap and Waste Products", O. Lebedev, A. Tatakin, and G. Svalov, Tsvetn. Metal. 38(7), 62-6 (1965) (Russ).
Energy balance, plant data.

3. "Heat Balance of a Large Diaphram-Type Mg Electrolytic Cell with Overhead Anodes", G. Olyunin, Tsvet. Metal. 1971, 44(3), 62-4.
Energy balance, plant data.

4. "Heat Losses of Magnesium Electrolysis Cells with the Overhead Introduction of Anodes", V. Devyatkin and V. Gribov, Tsvet. Metal. 1972, (8), 48-50 (Russ).
Heat losses, review.

5. "Material Balance of Commercial Magnesium Electrolyzer with Anhydrous Carnallite", Z. Yastrebova, Yu. Ryabukhin, E. Chukal'skii, A. Bogdanov, and L. Davydova, Nauchn. Tr., Veses. Nauchn. - Issled. Proektn. Inst. Titana 1973, No. 9, 34-7 (Russ). From Ref. Zh., Met. 1974, Abstr. No. 4G200.
Material balances, plant data.

6. "Heat Losses from an Electrolytic Magnesium Cell", V. Gribov, V. Devyatking, R. Usenov, and A. Chesnokov, Tsvet. Metal. 1973, (7), 36-7 (Russ).
Heat losses.

C.3.13 MERCURY

1. "Heat Balance in the Production of Mercury in Rotary Furnaces", F. Pavlin and N. Medved, Rud. Met. Zb. 1966 (2), 195-202 (Slovenian).
   Heat balance.

2. "Oxidizing Roasting in Fluidized Bed of Antimony and Mercury-Antimony Flotation-Concentrate", A. Rozlovskii, V. Til'ga, and A. Ustinov, Tsvetn. Metal. 36(12), 34-7 (1963).
   Large-scale laboratory-scale installation; material balance.

C.3.14 MOLYBDENUM

1. "Calculations for Furnaces for Fluidized-Bed Roasting", A. Zelikman and G. Vol'dman, Tsvetn. Metal. 37(5), 23-9 (1964).
   Heat balance of a furnace operating on Mo sulfide.

2. "Calculation of the Heat Balance of a Furnace for Chlorinating a Scheelite-Molybdenite Intermediate Product", V. Kremnev, O. Krichevskaya, and P. Yakovlev, Tr. Proekt. Nauch.-Issled. Inst. "Gipronikel" (Gos. Inst. Proekt. Predpr. Nikelevoi Prom.) 1969, No. 42, 63-7 (Russ).

C.3.15 NICKEL

1. "Material and Heat Balances of Electric Ore-Smelting Furnace at 'Pechenganikel' Combine", Ya. Osipov, G. Talovikov, Ya. Serebryannyi, and V. Sudarkina, Tsvetnye Metally 33, No. 10, 35-8 (1960).
   Material and heat balances, plant data.

2. "Utilization of Nickeliferous Serpentine", Extractive Metallurgy of Copper, Nickel and Cobalt, ed. P. Queneau, A.I.M.E., (1961) 302-7.
   Material and energy data from test.

3. "The Smelting of Copper and Nickel Sulfide Ores and Concentrates in the Electric Furnace", O. Barth, Freiberger Forschungsh. B67, 35-55 (1962).
   Energy balances.

4. "Cyclone Smelting of Nickel Concentrates", R. Khmel'nitskii, Tsiklonnye Plavil'nye Energo-Tekhnol. Protessy, Tr. Nauchn.-Tekhn. Soveshch., Provedennogo Mosk. Energ. Inst. 1962, 111-20 (Pub. 1963).
   Energy balance, laboratory experiments.

5. "Direct Production of Metallic Nickel from the Liquid Intermediate Sulfide Product of Matte Converting", A. Okunev, P. Kusakin, N. Vatolin, B. Kolmogorov, and L. Zamorin, Tr. Inst. Met., Akad. Nauk SSSR, Ural'sk. Filial No. 8, 75-82 (1963).
   Energy balance, laboratory experiment.

6. "Heat Balance in Shaft-Furnace Smelting of Nickel Sinter with Oxygen-Enriched Blast and Hot Blast", I. Reznik and M. Kruglyakova, Tsvetn. Metal. 37(7), 33-9 (1964) (Russ).
   Energy balance, plant data.

7. "Reductional Electric Smelting of Converter Slags in Nickel Production", L. Pimenov and A. Zyazev, Tsvetn. Metal. 38(1), 34-6 (1965) (Russ).
Energy and material balances, plant data.

8. "Replacing of Coke by Natural Gas in the Shaft-Furnace Smelting of Oxidized Nickel Ores", I. Reznik and A. Evdokimenko, Tsvetn. Metal. 38(7), 36-40 (1965) (Russ).
Energy balance, theoretical model.

9. "Material and Heat Balances of Electric Smelting of Copper-Nickel Sulfide Ores in a 30,000 kv.-amp. Electric Furnace", V. Brovkin, M. Palysaev, Yu. Slobodin, and M. Chetvertkov, Tsvetn. Metal. 38(12), 34-40 (1965) (Russ).
Energy and material balances, plant data.

10. "Thermal Constraints on the Segregation of Nickel from Lateritic Ores", J. Warner, R. Sridhar, and H. Bakker, Nickel Segregation Proc. Panel Discuss. 1972 (Pub. 1973), 241-64. Edited by A. Dor, A.I.M.E.: New York, N.Y.
Calculations and laboratory tests.

11. "Modern Flash Smelting Technology", A. Lauria and B. Anderson, 4th BNF Int. Conf. (1974), BNF Metal Technology Center.

C.3.16 NIOBIUM

1. "Material and Heat Balances for Metallothermic Melts", V. Bogolyubov, Fiz.-Khim. Osnovy Met. Protesessov, Komis, po. Fiz.-Khim. Osnovam Proisz. Stali, Sb. Statei 1964, 72-6.
Energy and material balances.

C.3.17 PHOSPHOROUS

1. Electric Smelting Processes, A. Robiette, John Wiley & Sons, New York, (1973) p. 269.
Material balance of elemental phosphorous.

C.3.18 SODIUM

1. "Thermal Balance of Sodium Electrolyzers and Effective Use of the Results", I. Veneraki, Izv. Vyssh. Ucheb. Zaved., Energ. 1971, 14(2), 64-7.
Plant scale study of heat losses.

C.3.19 STAINLESS STEEL

1. "Metallic Oxidation in Chromium Steel Melting", D. Hilty, G. Healy and W. Crafts, Trans. A.I.M.E. 197, (1953), p. 649.
Material balance on chromium in Appendix.

2. (Article on Development of Chromium Oxidation Model), G. Healy, A.I.M.E. Electric Furnace Conf. Proceedings, 1958, 16, p. 252.
Includes details of heat balance development based on actual furnace data.

3. "Calculation Method and Results of Material Balance in Smelting of Steel 1Kh18N9T", E. Kadinov, A. Rabinovich, and S. Khitrik, Izvest. Vysshikh Ucheb. Zavedenii, Chernaya Met. 1961, No. 8, 56-71.

4. "Material Balance of Titanium During Electromelting of Stainless Steel", V. Kamardin, E. Kadinov, and E. Moshkevich, Izv. Vysshikih Uchebn. Zavedenii, Chern. Met. 9(6), 80-7 (1966) (Russ).
   Material balances of two heats were made. Total nonreversible losses of Ti were determined.

5. "Material Balance of Titanium During Electromelting of Stainless Steel", V. Kamardin, E. Kadinov, and E. Moshkevich, Izv. Vyssh. Ucheb. Zaved., Chern. Met. 1968, 11(6), 57-9 (Russ).
   Plant scale study. Material balance on Ti only.

C.3.20 STEEL

A. *General:*

1. "The Effect of the Various Steelmaking Processes on the Energy Balances of Integrated Iron and Steelworks", W. Walker, J. Iron Steel Inst. (London) 200, 349 (1962).
   A review of the Institute's Special Report No. 71.

2. "Material Balances in an Integrated Steel Plant", J. Irvin Stanford Univ. Publ., Univ. Ser., Geol. Sci. 9(1), 237-53 (1964).
   Theoretical material balances.

3. Development of Fuel-Energy Principles and the Efficiency of Fuel Use in Ferrous Metallurgy, N. Kalita, Kiev: Navkova Dumka (1965) 266 pp.
   Fuel efficiency.

4. "Energy of Typical Raw Materials and Products of the Iron and Steel Industry", J. Szargut, Neue Huette 10(5), 266-75 (1965) (Ger).
   Formulas are derived for calculating the energy balances of Fe and steel melting processes.

5. "Application of Energy Balances in Iron and Steel Works", R. Jeschar and R. Goergen, Stahl Eisen 85(12), 724-30 (1965) (Ger).
   The results of thermal and of energy balances made on soaking pits, arc-melting furnaces, open hearths, and blast furnaces are compared to determine in which cases the thermal balance is sufficient to evaluate the performance of a furnace and when the additional energy balance is needed. The exergy concept is reviewed briefly.

6. "Importance of Material and Heat in Steel Processing. Utilization of Scrap", T. Kootz, Freiberger Forschungsh 106B, 5-21 (1965).
   A review with 17 references.

7. "Development of a Continuous Melt-Down Process", H. Langhammer and H. Geck. Stahl Eisen 1972, 92(11), 501-18 (Ger/Fr).

B. *Ajax Process:*

1. "Energy Requirements of the Ajax Steel-Making Process", A. Jackson and S. Brooks, J. Inst. Fuel 33, 580-4 (1960).
   Discussion of thermal efficiency and general performance.

C. *Bessemer Process:*

1. "Material and Heat Balances of the Bessemer Process", L. Tsykin, Sh. Bektursunov, G. Rekhlis, M. Kuznetsov, D. Ul'yanov, G. Oiks, and V. Yavoiskii, Sb. Nauchn. Tr. Zhdanovsk. Met. Inst. 1961, No. 7, 95-107.
   On the basis of plant data, the material and heat balances of seven heats were compared.

D. *BOF Process:*

1. "Iron Balance in Oxygen Top-Blowing Process", S. Lifshits, Stal' 21, 109-12 (161).
   Material balances of several heats of a basic top-blown 26-ton vessel, a 275-ton open-hearth furnace, and a 19-ton bottom-blown acid vessel are given in detail.

2. "Fuel and Power Required for Making Steel in the LD (Linz-Donawitz) Process", K. Rosner and F. Dobrowsky, J. Inst. Fuel 34, 3-7 (1961).
   A heat balance is presented.

3. "Oxygen Steelmaking Processes. Controlled Heat Balance in the LD Process", R. Rinesch, J. Metals 14, 497-501 (1962).

4. "Energy Balances for the Open-Hearth Process with and without O, and for the LDAC and LD Processes with Waste-Heat Boilers", R. Aspland and P. Tidy, Iron Steel Inst. (London) Spec. Rept. No. 71, 4-23 (1962).
   Breakdowns of thermal balance are given in detail.

5. "Energy Balances for LD (Linz-Donawitz), Kaldo, and Oxy/Steam Processes", A. Raper, A. Collinson, and S. Desai, Iron Steel Inst. (London) Spec. Rept. No. 71, 24-44 (1962).
   Heat balances are illustrated graphically by Sankey diagrams; materials requirements for the plants are presented as flow sheets.

6. "Comparison of Material Balances of Oxygen Converter and Large-Sized Siemens Martin Furnace with the Use of Oxygen", N. Korkoshko, G. Kolganov, Yu. Krivchenko, and V. Servetnik, Stal' 23 (9), 788-91 (1963) (Russ).

7. "Steelmaking Heat Potentialities Point Toward Continuous Processing", G. Alexandrovsky, J. Metals 15 (8), 585-92 (1963).
   Theoretical calculations.

8. "Material and Heat Balance in Blowing Basic Bessemer Pig Iron by the LDAC Process in a 50-ton Converter"- H. Voigt and G. Mahn, Stahl Eisen 84(18), 1120-8 (1964).
Plant data used.

9. "Development of Nomograms for the Calculation of Amount of Additives During the Smelting of Converter Steel", S. Zaikov, B. Nikiforov, V. Koval, and P. Rubinskii, Met. i Gornorudn. Prom., Inform. Nauchn. - Tekhn. Sb. 1965(4), 25-9 (Russ).
An average heat balance of 0 converter smelting is shown.

10. "Material and Heat Balances in Oxygen Converters", A. Kranjc, Met. ABM (Assoc. Brasil Metais) 22 (107), 815-28 (1966).
The substitution of scrap iron for pig iron was studied. Heat losses were examined and the results plotted.

11. "Operating Performance and Improvements to Membrane Hoods for Basic Oxygen Furnaces", T. Hurst, Iron Steel Eng. 43(12), 101-7 (1966).
Partial heat recovery from a simple hood gives a better return on capital investment than full heat recovery from a hood followed by a complete waste-heat boiler unit.

12. "Material and Heat Balances of an Oxygen Conversion Melt", V. Kocho, L. Paizanskii, Yu. Reshetnyak, and B. Boichenko, Izv. Vysshikh Uchebn. Zavedenii, Chern. Met. 9(8), 50-5 (1966) (Russ).
Pig iron (27.9 tons) was introduced into the converter and blown to a steel. Material and heat balances were calculated for the whole operation as well as for the following four periods: 0 to 2 min. 15 sec.; 2 min. 15 sec. to 7 min.; 7 to 10 min.; 10 to 17 min.

13. "Consumption of Coolants (Scrap, Ore, Sinter) During the Oxygen Converter Process and Evaluation of the Influence of Various Factors on it", V. Romenets and S. Kremenevskii, Sb., Mosk. Inst. Stali Splavov 1969, No. 55, 91-107 (Russ).
Based on industrial experience, total heat balance was established.

14. "Use of Top Gas Analysis for Predicting Metal Carbon Levels", E. Bicknese, Proc., Nat. Open Hearth, Basic Oxygen Steel Conf. 1970 (Pub. 1971), 53, 172-81.
Studies were carried out on a commercial vessel.

15. "Material and Heat Balances of 140-ton Oxygen-Converter Melts", E. Zarvin, V. Maron, A. Nikolaev, Yu. Nikitin, and B. Sel'skii, Izv. Vyssh. Ucheb. Azved., Chern. Met. 1970, 13(10), 44-7 (Russ).
Energy and material balances, operating data.

16. "Procedures for Improving the Heat Balance of Pure-Oxygen Converters", C. Roederer, Cent. Doc. Siderurg., Circ. Inform. Tech. 1970, 27(12) 2669-88 (Fr).
Energy balance, scrap preheating.

17. "Mathematical Model for the Calculation of the Load, and Application to the Control, of a Large Scale Oxygen Converter", G. Bozza, A. Cecere, B. Costa, G. Violi, Met. Ital. 1971, 63(11), 569-76 (Ital).
Material and energy balances, operating data, model for control.

18. "Technological Features During the Conversion of Pig Iron with Varying Manganese Content", V. Didkovskii, Met. Koksokhim 1971, No. 25, 23-7.
Material balance on converter for yield and Mn recovery.

19. "Utilization of Heat of Converter Gases", V. Bondarenko, F. Belin, I. Gritsyuk, and M. Rimer, Prom. Energ. 1972, (6), 36-40 (Russ).
Waste heat in O steel converters was utilized.

20. BOF Steelmaking Vol. 4, Chpt. 13, ed. R. Pehlke et al, A.I.M.E. (1975).
Energy and material balances, operating data.

21. "Heat and Material Balances", G. Healy, Chapter in Basic Oxygen Steelmaking, ed. R. Pehlke, Pub. A.I.M.E., New York, (1975).
Example of development of balances.

E. *Electric Arc Process :*

1. Electric Melting and Smelting Practice, A. Robiette, Charles Griffin and Co., Ltd., London, 1955, pp. 39-40.
Heat balance in arc furnace.

2. "Energy Balances for Electric-Arc Processes", G. Ovens, Iron Steel Inst. (London) Spec. Rept. No. 71, 45-51 (1962).
A hypothetical energy balance.

3. "Electrical Aspects of Arc Furnaces", K. Kukan, Iron and Steel Engr., 40, Oct., 1963 P. 137.
Energy requirements for steel as function of furnace size.

4. "Energy Balance in the Arc Furnace", Electric Furnace Steelmaking, Volume II, Chpt. 19, Ed. C. Sims, A.I.M.E. (1963) pp. 283-315.

5. "Materials Usage Optimization in Electric Furnace Steel Production", B. Bernacchi, Metals Eng. Quart. 4(4), 57-62 (1964).
Solution of material balance by linear programming.

6. "Losses of Tungsten During the Production of High-Speed Steel", A. Stroganov, Yu. Pyl'nev, E. Chernyshev, N. Keis, V. Pakuleva, I. Donets, Yu. Kholodov, and F. Germelin, Metallurg)(Moscow) 1971, 16(1), 21-3 (Russ).
Material balance on W. Plant scale study.

7. Electric Melting Practice, A. Robiette, Griffin & Co. Ltd., London, 1972, pp. 89-91.
Heat balance for 120-ton arc furnace.

8. "Energy Balances for Conventional Versus Pretreated Smelting of Pig Iron in Electric Furnace", S. Ghorpade, Trans. Indian Inst. Metals 1972, 25(4), 103-11.
Energy balances, theoretical and operating data.

F. *Electroslag Refining:*

1. "Examination of Electrode-Change Practice in Electroslag Melting", R. Jackson, J. Vac. Sci. Technol. 1972, 9(6), 1301-5.
Unsteady-state energy balance.

2. "Thermal Characteristics of the Electroslag Process", A. Mitchell and S. Joshi, Met. Trans. 1973, 4(3), 631-42.
Energy balance, laboratory model.

G. *Kaldo Process:*

1. "Fuel and Energy Required for the Manufacture of Steel by the Kaldo Process", F. Johansson and B. Kalling, J. Inst. Fuel 34, 172-6 (1961).

H. *Ladle Treatments:*

1. "Preliminary Refining of High Phosphorus Iron in the Ladle with the Aid of Oxygen", I. Zaitsev and V. Kovraiskii, Byull. Nauch. - Tekh. Inform., Ukr. Nauch - Issledovatel. Inst. Metal. 1958, No. 5, 29-41.
Energy balance.

2. "Change of Metal Temperature During Deoxidation and Alloying (of Steel)", R. Savel'eva, Sb. Nauch. Tr., Manitogorsk. Gornomet. Inst. 1972, No. 115, 31-7. From Ref. Zh., Met. 1973, Abstr. No. 3V356.
Effect of alloying additions on metal temperature reported.

I. *LWS Process:*

1. Oxygen Bottom Blowing by the LWS Process", P. Leroy, Iron & Steel Eng. 1972, 49(10), 51-5.
Heat balance for 30-ton vessel.

J. *Open-Hearth Process:*

1. "Material and Heat Balances of Meltings in Open-Hearth Furnaces", I. Kobeza and Yu. Kiselev, Voprosy Proizvodstva Stali, Akad. Nauk Ukr. S.S.R., Otdel. Tekh. Nauk 1958, No. 6, 20-6.
Effect of roof materials on heat losses.

2. "Use of Oxygen in the Open-Hearth Furnace and its Effect on the Heat Utilization", L. Efimov, Freiberger Forschungsh. 40B, 5-24 (1959).
Energy balance, operating data.

3. "Calculation of Heat Absorption by the Charge of the Open-Hearth Furnace by Means of a Reverse Heat Balance on the Working Space", A. Voitov, Sbornik Trudov Tsentral, Nauch. - Issledovatel. Inst. Chernoi Met. No. 21, 266-80 (1960.

4. "Use of Forsterite Checkers", F. Volovik, P. Gorshtein, V. Zelenskii and A. Poyarkov, Stal' 20, 125-7 (19600.
The reduction of thermal efficiency of the open-hearth checkers is produced by the divergence of the travel of gases and air in the checkers and not by the forsterite use.

5. "Use of a Computer for Controlling Thermal Balance of Open-Hearth Furnaces", M. Korobko and V. Artynskii, Stal' 20, 981-4 (1960).
Data collected in application to a 430-ton open-hearth furnace.

6. "Study of Charge Melting in Open-Hearth Furnaces with Various Methods of Oxygen Delivery", K. Trubetskov, V. Kornfel'd, E. Grekov, A. Voitov, L. Shteinberg, and G. Lomtatidze, Stal' 21, 214-21 (1961).

7. "Fuel and Energy Required for Steelmaking in the Open-Hearth Furnace", R. Mayorcas and I. McGregor, J. Inst. Fuel 34, 153-6 (1961).
Material, O, and heat balances are presented.

8. Steelmaking: The Chipman Conference, Proceedings of the Conference on the Physical Chemistry and Technology of Steelmaking, The M.I.T. Press, Cambridge, Mass., June 1962, pp. 244, 245.
Heat Balances for several open-hearth practices. Approximate high-temperature heat balances, hot metal to tap period.

9. "Heat Balances of a Recuperative Recirculation Furance", G. Demin and A. Pluzhnikov, Izv. Vysshikh Uchebn. Zavedenii, Chernaya Met. 5, No. 9, 188-92 (1962).
Heat balances in tabular form.

10. "Material Balance of Scrap-Ore Process", V. Grigor'ev, V. Luzgin, E. Abrosimov, V. Orlov, V. Yavoiskii, G. Gurskii, I. Goncharov, and P. Starkov, Izv. Vysshikh Uvhebn. Zavedenii, Chernaya Met. 5, No. 5, 63-8 (1962).
Complete material balance was performed with 15 melts in a 185-ton open-hearth furnace.

11. "Heat Generation in Steel Melting Baths of 500-ton Open-Hearth Furnaces During Melting", M. Glinkov and E. Stul'pin, Izv. Vysshikh Uchebn. Zavedenii, Chern. Met. 6(11), 223-9 (1963).
Plant data. Heat balances.

12. "Thermotechnical Investigation of Open-Hearth Furnaces", D. Pavko and B. Sicherl, Rudarsko-Met. Zbornik 1963(2), 111-23.
Experimental results.

13. "A Thermochemical Approach to Prediction of Open Hearth Variables", J. Koros and F. Altimore, Met. Soc. Conf. 32, 267-93 (1964) (Pub. 1966).
Energy and material balances, error analysis, operating data.

14. "Basic Open-Hearth Process Operated with Gas or Gas Producer", J. Menendez Alvarez, Hierro Acero 17(91), 229-52; (93), 399-432 (1964).
Study of charge and thermal balance.

15. "Thermal and Economic Indexes of the Open-Hearth Furnace Operating with Pure Oxygen", R. Goergen, Stahl Eisen 84(6), 350-61 (1964).
Thermal balances are discussed.

16. "Heat Requirement During the Melting Period in the Scrap-Ore Open-Hearth Process", T. Sabirzyanov and E. Abrosimov, Izv. Vysshikh Uchebn. Zavedenii, Chern. Met. 8(1), 26-31 (1965) (Russ).
Material and thermal balances were calculated on the basis of production data.

17. "Heat and Oxygen Balances of an Open-Hearth Furnace Blown with Oxygen", V. Pogorelyi, N. Korkoshko, M. Babenyshev, E. Grekov, and L. Shteinberg, Stal' 26(2), 120-3 (1966) (Russ).
Plant data analyzed.

18. "Obtaining Data on the Progress of an Open Hearth Heat", A. Zakurdaev, G. Sharonov, V. Gogenko, and A. Kunakhovich, Stal' 26(7), 603-8 (1966) (Russ).
Problems involved in development of time-dependent heat balance.

19. "Heat Absorption by the Bath and Thermal Efficiency of Very Large Open-Hearth Furnaces", N. Korneva, Stal' 27(2), 178-80 (1967).
Heat balances on full-scale furnaces.

20. "Thermal Operation of O-H Furnaces Viewed from Thermodynamic Aspects of Energy Efficiency", B. Stjepovic, Tehnika (Belgrade) 1969, 24(12), 1967-72.
Energy balance, operating data.

21. "Material Balance of a Process in Two-Bath and Open-Hearth Furnaces", K. Trubetskov, V. Tarasov, I. Konovalov, A. Alymov, K. Mokrushin, A. Tat'yanshchikov, V. Chizhova, and V. Yakushin, Sb. Tr. Tsent. Nauch.-Issled. Inst. Chern. Met. 1970, No. 75, 68-73 (Russ).
Fe losses determined by plant experiments.

22. "Material and Heat Balances of Open-Hearth Melting", Ah. Portnaya and M. Chelyadin, Met. Koksokhim. 1970, No. 20, 60-4 (Russ).
Energy and material balances .

23. "Oxidation of Carbon During the Automatic Control of an Open-Hearth Furnace by an Air System", B. Litvinov, L. Gol'bers, and I. Semikin, Met. Gornorud. Prom. 1970, (4), 8-11 (Russ).
Energy balance, control model.

24. "Effect of the Use of Oxygen on the Heat Balance of an Open-Hearth Furnace", E. Amelung, Stahl Eisen 1970, 90(9), 458-60.
Plant data analyzed.

25. The Making, Shaping and Treating of Steel, 9th Ed., U.S. Steel, Pittsburgh, Pa., 1971, p. 534.
General open-hearth heat balance.

26. "Thermal Operation of a 600-ton Open-Hearth Furnace", V. Grankovskii, V. Naidek, V. Pereloma, and L. Pyradkin, Izv. Vyssh. Ucheb. Zaved., Chern. Met. 1971, 14(3), 150-1 (Russ).
Energy and material balances, operating data.

27. "Operation of an Open-Hearth Furnace During Intensive Oxygen Blowing Through the Bath", V. Grankovskii, B. Yupko, P. Shchastnyi, E. Shvets, Metallurg (Moscow) 1971, 16(1), 18-21 (Russ).
Energy and material balances, plant data.

28. "Accumulation of Heat by the Brickwork of an Open-Hearth Furnace Chamber During Melting", I. Zavarova, L. Lomakin, M. Gordon, Izv. Vyssh. Ucheb. Zaved., Chern. Met. 1971, 14(8), 157-62 (Russ).
Analog simulation.

29. "Investigations by Radioisotope Tracers of the Behavior of Chromium During the Melting of 08ZX Steel in an Open-Hearth Furnace", Z. Bazaniak and J. Michalik, Hutnik 1971, 38(5), 240-5 (Pol).
Material balance on Cr. Plant scale study.

30. "Desulfurization of O-H Steel", G. Kamyshev, Stal' 1972, (9), 806-7.
Sulfur balance from plant data.

31. "Scrap-Oxygen Process, a New Technology of Open Hearth Steelmaking", J. Miko, Banyasz. Kohasz. Lapok, Kohasz. 1973, 106(1), 29-34 (Hung).
Heat and O balances are given for experiments in 250 and 500 ton furnaces.

K. *Tandem Process:*

1. "Importance of the Tandem Process", Z. Boehn, Met. ABM (Ass. Brasil. Metais) 1972, 28(172), 171-80 (Port).
Material balances, plant data.

2. "Heat and Temperature Regimes of a Two-Bath Steelmaking Furnace", V. Antipin, N. Ivanov, V. Lorman, A. Blokhin, N. Bazhenov, A. Sergeev, Yu. Snegirev, G. Zakharov, K. Nosov,

et al, Izv. Vyssh. Ucheb. Zaved., Chern. Met. 1973, (7), 41-5 (Russ).
Heat balances for each chamber of a 2-bath steel-melting furnace. Material balances were used for calculating the heat balances.

C.3.21 TIN

1. "Electrolytic Refining of Crude Tin Containing Heavy Metals", N. Golikov, Tsvetn. Metal. 35, No. 2, 40-3 (1962).
Material balances, plant data.

C.3.22 TITANIUM

1. "Heat Balance of an Electric Shaft Furnace", N. Galitskii, Titan i ego Splavy, Akad. Nauk SSSR, Inst. Met. 1961, No. 5, 254-66.
Energy balances, plant data.

2. "Thermal Balance of Electric Shaft Furnace and Condensing System Used in Chlorination of Titanium-Bearing Slags", N. Galitskii, Tr., Vses. Nauchn.-Issled. Alyumin.-Magnievyi Inst. 1962 (48), 132-9.
Energy balance.

3. "Material and Heat Balances for Metallothermic Melts", V. Bogolyubov, Fiz.-Khim. Osnovy Met. Protsessov, Komis, po Fiz.-Khim. Osnovam Proizv. Stali, Sb. Statei 1964, 72-6.
Energy and material balances.

4. "Electrolytic Refining of Industrial Titanium Alloy Wastes", Yu. Olesov, V. Ustinov, A. Rubtsov, and A. Suchkov, Tsvetn. Metal. 39(5), 69-72 (1966) (Russ).
A material balance of electrolysis is given.

5. "Material and Heat Balances in the Operation of a Furnace for Melting Titanium Slags", L. Lekalova, Tsvet. Metal. 40(5), 69-72 (1967).
Energy and material balances, plant data.

C.3.23 VANADIUM

1. "Material and Heat Balances for Metallothermic Melts", V. Bogolyubov, Fiz.-Khim. Osnovy Met. Protsessov, Komis. po Fiz.-Khim. Osnovam Proisz. Stali, Sb. Statei 1964, 72-6.
Energy and material balances.

C.3.24 ZINC

1. Hydrometallurgy of Base Metals, G. Van Arsdale, ed., McGraw-Hill Book Co., Inc., 1953, pp. 167-262.
Material balances for several plants. Discusses copper and zinc industries utilizing various processes.

2. "Cyclone Smelting of Copper-Zinc Concentrates", P. Myasnikov, A. Okunev, and D. Lutokhin, Trudy Inst. Energet., Akad. Nauk

Kazakh. S.S.R. 2, 274-84 (1960).
Energy and material balances, review.

3. "Electrothermal Method for Making Zinc Dust", L. Rabicheva, B. Slonimskii, V. Lazarev, E. Alyushin, and G. Poletaev, Sb. Nauchn. Tr. Gos. Nauchn.-Issled. Inst. Tsvetn. Metal. No. 18, 165-74 (1961).
Energy balance, plant data.

4. "Testing on Commercial Scale of Processing Altai Polymetallic Ores Following the Flow Sheet of Magnitogorsk Combine", I. Tsygoda, V. Kazakov, N. Kolesnikov, N. Bryukhanov, A. Burba, V. Sadykov, and A. Pigarev, Tsvetn. Metal. 36(12), 12-15 (1963).
Material balances, plant data.

5. "Design of Industrial Cyclone Reactor by Using Data Obtained in Pilot-Plant Installation", V. Holy, Hutnicke Listy 4(19), 262-8 (1964).
Energy balances, pilot plant.

6. "Mathematical Model of Metallurgical Conversion Processes in a Copper Smelter", I. Rabinovich, Tsvet. Metal. 40(3), 46-51 (1967).
Material balances, theoretical model.

7. "Material and Heat Balances of Fluidized-Bed Reactor for Zinc Electrolysis in Kosovska Mitrovica", Z. Popovic and V. Sekulovic, Metalurgija (Sisak, Yugoslavia) 1971, 10(1), 19-27 (Croat).
The main features of the Kosovska Mitrovica electrolysis plant, including the material and heat balances, are presented.

8. "Heat Balance of Zinc Furnaces and Effect of Blast Oxygen-Enrichment on the Excess Heat of the Bed", V. Zorkov, L. Pakhomov, I. Tsarev, T. Tserikov, Sb. Nauch. Tr. Gos. Nauch-Issled. Inst. Tsvet. Metal. 1971, No. 32, 109-17 (Russ).
Energy balance.

9. "Simultaneous Reduction of Metal Oxides by Gases", W. Ptak and M. Sukiennik, Zesz. Nauk. Akad. Gorn.-Hutn., Cracow, Met. Odlew. 1973, No. 50, 17-30 (Pol).
Energy and mass balances, theoretical model.

10. "Heat Conservation in Zinc and Lead Extraction and Refining", S. Woods, Metals & Materials 1974, 8(3), 187-9.
Suggestions for conserving energy.

C.3.25 ZIRCONIUM

1. "Material and Heat Balances for Metallothermic Melts", V. Bogolyubov, Fiz.-Khim. Osnovy Met. Protsessov, Komis, po Fiz.-Khim. Osnovam Proisz. Stali, Sb. Statei 1964, 72-6.
Energy and material balances.

C.4 NONMETALLIC MATERIALS PRODUCTION

C.4.1 CEMENT

1. The Technology of Cement and Concrete, Volume I, Concrete Materials, R. Blanks and H. Kennedy, Wiley Publishers (1955) pp. 129-137, 138-147.
   Material balance for four-component Portland Cement. Heat balance for burning of Portland cement.

2. "Calculation of the Heat Consumption of Coke-Fired Cement Shaft Kilns", F. Linhoff, Zement-Kalk-Gips 12, 151-4 (1959).
   Energy balance.

3. "Thermal Efficiency of Rotary Cement Kilns and the Composition of the Escape Gases", A. Bloda, Cemento y Hormigón (Barcelona) 27, 4-17, 56-70, 237 (1959).
   Energy and material balances, plant data.

4. "Principles of the Thermophysics of the Cement Shaft Kiln", H. Eigen, Zement-Kalk-Gips 13, 458-66 (1960).
   Energy balance.

5. "Progress in the Utilization of Waste Heat from Dry-Process Kilns", G. Ruppert, Zement-Kalk-Gips 13, 366-75 (1960).
   Energy consumption, plant data.

6. "Scientific and Process Engineering Problems in the Calcination of Portland Cement Clinker from Quicklime", J. Wuhrer, Zement-Kalk-Gips 13, 181-92 (1960).
   Energy balance, plant data.

7. "The Influence of Clinker Precooling in the Rotary Kiln on the Heat Consumption", H. Eigen, Zement-Kalk-Gips 13, 226-9 (1960).
   Energy balance.

8. "Double or Single Air Percolation in the Grate Cooler of the Wet Cement Kiln with 32% H_2O in the Sludge", H. Eigen, Tonind.-Ztg. u. keram. Rundschau 85, 230-1 (1961).
   Energy balance.

9. "Wrong and Right Applications of Heat-Consumption Formulas for the Cement Shaft and Rotary Kilns", H. Eigen, Tonind.-Ztg. u. Keram. Rundshau 85, 334-8 (1961).
   Energy balance.

10. "A Horizontal Reciprocating Grate (Clinker) Cooler", I. Cherep, Tsement 27, No. 1, 14-18 (1961).
    Energy balance, industrial clinker cooler.

11. "The Heat Consumption of Shaft and Rotary Kilns", A. Pluss, Zement-Kalk-Gips 14, 297-305 (1961).
    Heat Consumption.

12. "Thermophysics of the Wet Portland Cement Rotary Kiln", H. Eigen, Radex Rundschau 1961, 529-35.
    Heat consumption.

13. "Differential Thermal Analysis as Applied to Study of Thermal Efficiency of Kilns", V. Ramachandran and N. Majumdar, J. Am. Ceram. Soc. 42, 96 (1961).

14. "The Functional Relations Between the Thermodynamic Factors Affecting the Burning of Cement", H. zur Strassen, Zement-Kalk-Gips 15, 365-77, 379-82 (1962). "Discussion", H. Eigen, Ibid. 378-9
Mathematical treatment.

15. "The Heat Economy of Cement Burning", H. Eigen, Tonind.-Ztg. u. Keram. Rundschau 86, 37-43 (1962).
Heat consumption.

16. "A Mathematical Model of a Rotary Cement Kiln", L. Weeks. Quart. Colo. School Mines 59(4), 493-503 (1964).
Energy and material balances, mathematical model for control.

17. "New Approach to Some Values Appearing in the Heat Balance of the Hoffmann Kiln", M. Kakol, W. Nowak, and R. Sobanski, Szklo Ceram. 17(1), 18-21 (1966) (Pol).
Energy balance data.

18. "The Utilization of Waste Heat in Rotary Cement Kilns", H. Huckauf, Silikattechnik 17(3), 69-75 (1966) (Ger).
In the rotary cement kiln discussed as an example of the dry process, it is estimated that it takes 1328 kcal/kg. of clinker if no waste heat recovery is attempted. The possibility of using the waste heat again at some place in the cement-burning process is studied.

19. "Heat Losses Through the Walls of Short Dry-Process Rotary Kilns with Preheaters", D. Optiz, Zem.-Kalk-Gips 20(4), 177-85 (1967) (Ger).
Heat losses.

20. "Thermotechnical Means in Development of a Highly Efficient Method for Production of Fused Cement Clinker", A. Klyuchnikov, Yu. Kazanskii, V. Khokhlov, V. Shelud'ko, and Z. Entin, Izv. Vyssh. Ucheb. Zaved., Energ. 10(4), 76-81 (1967) (Russ).
Theoretical calculations.

21. "Heat Balance System for Rotary Kilns in the Cement Industry", G. Bornschein, Silikattechnik 18(3), 69-71 (1967) (Ger).
Energy balance.

22. "Mechanical and Thermophysical Properties of (Cement) Raw Materials", A. Malyshev, Tsement 1968, 34(2), 5-7 (Russ).
Clinker thermal and mechanical data.

23. "Application of Digital Computers for the Evaluation of Industrial Thermal Measurements", J. Novak, M. Vrestalova, V. Stefkova, Stavivo 1970, 48(1), 9-10 (Czech).
Energy and material balances, theoretical model.

24. "Combustion and Heat Transfer in Rotary Cement Kilns", P. Sunavala, J. Mines, Metals Fuels 1971, 19(2), 48-52.
Energy and material balances, plant data.

25. "Comparative Technical and Economic Analysis of Dry Methods for Clinker Production in Kilns with External Heat Exchangers Operating on the Principles of Heat Transfer Between Gases and Raw Material Suspended in Exhaust Gases", K. Cichon and S. Scieranski, Biul. Inform., Osrodek Inform. Tech. Ekon., Inst. Przem. Wiazacych Mater. Budowlanych, Krakow 1971, No. 6, 131 pp. (Pol).
Heat consumption and losses, review.

26. "Thermal Balance for Industrial Kilns. II. Kilns. Combustion and Firing. Crude Baking. Installations. Heat and Material Balances. Conclusions.", J. de Assumpcao Santos, Cem.-Hormigon 1971, 42(443), 119-33 (Span).
Energy and material balances, plant data.

27. "Pyzel Process for Cement Clinker Burning", G. Schroth, Zem.-Kalk-Gips 1971, 24(12), 571-3 (Ger).
Energy and material balances, plant data.

28. "Monitoring the Rotary Furnace Operation Using Some Gas-Phase Characteristics", G. Val'berg and A. Glozman, Tsement 1974, (8), 17-18 (Russ).
Energy balance.

C.4.2 GLASS AND CERAMICS

1. Ceramics, Industrial Processing and Testing, J. Jones and M. Berard, The Iowa State University Press, Ames, Iowa (1972) pp. 120-122.
Material balance for determing batch composition of a typical ceramic.

2. The Handbook of Glass Manufacture, Volume A, F. Tooley, Books for the Industry, Inc., New York (1974).
General equations for calculating heat losses from furnaces. pp. 240-242. Heat balance on a continuous regenerative furnace. pp. 242-246. Total energy balance applied to fluids. pp. 192-210.

C.4.3 LIME

1. "New Knowledge with Regard to the Coke-Fired Lime Shaft Kiln", H. Eigen, Zement-Kalk-Gips 12, 509-15 (1959).
Heat consumption.

2. "Heat Losses with the Volatile Substances in Lime-Burning Shaft Kilns", N. Tabunshchikov, Khim. Prom. 1960, 425-6.
Heat losses.

3. "Limestone Burning in a Vertical Lime Kiln with Addition of Auxiliary Gas", H. Eigen, Tonind.-Ztg. Keram. Rundscahu 86, 333-9 (1962).
Energy balance, plant data.

4. "Hot Cyclone Development Improves Yield of Chemical Lime", M. Shafer and N. Brandt, Zement-Kalk-Gips 53(11), 515-19 (1964) (Ger).
Energy balance.

5. "The Uniflow Regenerative Kiln for Burning Limestone", H. Hofer, F. Bartu, and L. Hummler, Zement-Kalk-Gips 54(8), 395-403 (1965) (Ger).
Energy balance.

6. "Possibilities of Increasing the Utilization of Heat in Lime Kilns", O. Gabriel, Stavivo 44(9), 325-7 (1966).
Energy balances.

7. "Modifications of the Combustion Zone of Furnaces for Calcined Soda at Factory 'K. Marx.'", Vl. Khadzov, Tsv. Torbov, D. Peev, and V. Mitov., Khim. Ind. (Sofia) 38(1), 33-7 (1966) (Bulg).
Heat and material balances, plant data.

8. "Waste-Heat Boiler Behind a Lime Kiln", V. Kuyanskii, Stal' 1969, 29(9), 857-9.
Installation involving two rotary kilns and three waste-heat boilers and its operations are briefly described, indicating advantages of using boilers in connection with kilns.

9. "Heat Balance of Carbonate-Burning Furnaces (with Special Regard of the Recuperation of the Heat of the Product and its Effect on Fuel Saving and Output Raising)", S. Cernoch, Zb. Ved. Pr. Vys. Sk. Tech. Kosiciach 1972, 79-94 (Slo).
Energy balance, plant data.

10. "Burning of Crushed Limestone in a Suspension", G. Zakharov, Stroit. Mater. 1973, (4), 11-12 (Russ).
Experimental heat balance.

11. "Use of Heat in Lime-Burning Kilns of Various Types", N. Tabunshchikov, Stroit. Mater. 1974, (11), 24-6 (Russ).
Heat losses.

Index

G

H

U

V

W

Y